Principles of Marine Bioacoustics

For other titles published in this series, go to
www.springer.com/series/3754

Modern Acoustics and Signal Processing

Principles of Marine Bioacoustics

Whitlow W. L. Au
Hawaii Institute of Marine Biology, Kaneohe, HI, USA

Mardi C. Hastings
Pennsylvania State University, State College, PA, USA

Springer

Whitlow W. L. Au
Hawaii Institute of Marine Biology
Kaneohe, HI
USA
wau@hawaii.edu

Mardi C. Hastings
Pennsylvania State
University, State College, PA
USA

ISBN: 978-0-387-78364-2 e-ISBN: 978-0-387-78365-9
DOI: 10.1007/978-0-387-78365-9

Library of Congress Control Number: 2008931428

Printed on acid-free paper

springer.com

We dedicate this book to the memory of our parents, Tai-Hing and Violet Au, and Forrest and Hilda Cox, for ensuring that we received the education that they were never privileged to obtain.

And God said, "Let the water teem with living creatures, and let birds fly above the earth across the expanse of the sky." So God created the great creatures of the sea and every living and moving thing with which the water teem, according to their kinds... And God saw that it was good.

Genesis 1:20–21

Preface

Humans have always been fascinated by marine life, from extremely small diatoms to the largest mammal that inhabits our planet, the blue whale. However, studying marine life in the ocean is an extremely difficult proposition because an ocean environment is not only vast but also opaque to most instruments and can be a hostile environment in which to perform experiments and research. The use of acoustics is one way to effectively study animal life in the ocean. Acoustic energy propagates in water more efficiently than almost any form of energy and can be utilized by animals for a variety of purposes and also by scientists interested in studying their behavior and natural history. However, underwater acoustics have traditionally been in the domain of physicists, engineers and mathematicians. Studying the natural history of animals is in the domain of biologists and physiologists. Understanding behavior of animals has traditionally involved psychologists and zoologists. In short, marine bioacoustics is and will continue to be a diverse discipline involving investigators from a variety of backgrounds, with very different knowledge and skill sets. The inherent inter-disciplinary nature of marine bioacoustics presents a large challenge in writing a single text that would be meaningful to various investigators and students interested in this field. Yet we have embarked on this challenge to produce a volume that would be helpful to not only beginning investigators but to seasoned researchers. Most of the material comes from the vast number of combined years of research in this field.

Investigators in the field of marine bioacoustics should have some knowledge of the mathematical concepts of acoustics starting with the wave equation and how acoustic waves and signals propagate in an oceanic medium. Knowledge of how sounds are produced by the conversion of electrical energy into mechanical energy by transducers and how sounds are detected by hydrophones is necessary. Knowledge of some elementary electronics, especially on how acoustic signals are converted from the analog into digital domain so that meaningful and useful information can be stored and analyzed by computers is needed as well. It is also important to understand how digital signals from a computer can be transformed into analog signals that can drive a sound projector. Knowledge of basic signal processing is also important in order to gain understanding of the properties and characteristics of sounds used by marine animals. Finally, the hearing and sound production capabilities of marine animals are important areas in marine bioacoustics. Unfortunately, whole volumes have been devoted to the various areas of knowledge that a good bioacoustician should master. We have attempted to "cut through the chase" in writing this text and hopefully we have been successful and this book will be a good contribution to this marvelous field.

W.W.L. Au
M.C. Hastings

Contents

Part I
Principles and Methodology

1

Introduction

1.1 What Is Marine Bioacoustics?

Simply stated, marine bioacoustics can be defined as the use of acoustics to study any facets of marine animals such as auditory capabilities, sound production, communications, foraging behavior, in short, anything having to do with their natural history. Acoustic energy propagates in water more efficiently than almost any form of energy and can be utilized by animals for a variety of purposes and also by scientists interested in studying animal life in the ocean. Electromagnetic, thermal, light and other forms of energy are severely attenuated in water. Therefore, the most effective method for an animal to perform various functions associated with its life cycle such as communications, navigation, obstacle avoidance, prey avoidance and detection is by the use of acoustics in either an active echolocation or passive (listening) manner. Similarly, the location, movement, social interaction, density and other facets of animal life in the sea can be studied by the use of acoustical techniques.

Bioacousticians are interested in a wide variety of questions concerning the reception and use of sounds by different animals. Water being a much denser medium than air presents some interesting problems for the reception and transduction of sounds into the auditory organs of animals. Therefore, the anatomical structure of the hearing system must have certain features that are very different than the structure of the hearing organ in terrestrial animals. The physics of sound conduction starting with the transduction of sound from the water into the head of an animal continuing into the hearing organ and the mechanisms involved within the hearing organ that send nerve impulses to the brain are of interest to marine bioacousticians.

One of the most fundamental properties in the use of sound by animals deals with how well and over what frequency range an animal can hear. Other properties of the auditory system, such as the capabilities for frequency discrimination, intensity discrimination, directional hearing, discrimination of complex sounds such as pulse sounds and discrimination of both frequency- and amplitude-modulated sounds, are also of interest. Knowledge of these different auditory properties is important and essential in order to

W.W.L. Au, M.C. Hastings, *Principles of Marine Bioacoustics*,
DOI: 10.1007/978-0-387-78365-9_1, © Springer Science+Business Media, LLC 2008

understand how animals utilize sounds and the meaning they attach to different sounds, whether from con-specifics or other animals, including humans. The characteristics of an organism hearing system are most often determined by psychophysical experiments in which the techniques developed in psychology are used in conjunction with different acoustic stimuli.

All sounds, whether they are produced by a non-human biological source, a natural source (such as wind, rain, earthquakes, etc.) or humans, convey information and may be used by animals in their struggle for survival. Some sounds, especially those produced by con-specifics and predators, have very special meaning to marine animals. Sounds from con-specifics can and do carry a considerable amount of information varying from the emotional state of the producer to warning of danger and immediate threats within the environment. However, unlike our understanding of the role of sounds in certain terrestrial animals, marine bioacousticians have a very poor understanding of the social context of sounds used by animals. Bioacousticians are interested in the meaning of sounds used by both con-specifics and other species and how these sounds affect the behavior of animals.

Electrophysiological techniques have increasing played a wider role in the study of hearing with marine animals (Supin et al., 2001). Measuring the response of auditory nerve fibers to different acoustic stimuli can provide the researcher with important insights into the hearing mechanisms of certain species. Measuring the auditory brain stem-evoked potentials produced by a synchronized auditory stimulus may be one of the most practical ways of evaluating the hearing sensitivity of large whales. It is certainly within the realm of possibilities to conduct hearing experiments with beached or entrapped whales.

Sonar operators listening and monitoring sounds from the bowels of their sleek, silent-running submarines are well aware of the wide variety of different sounds that marine animals produce. The term "biologics" have been traditionally attached to these biologically generated sounds. Since World War II until the mid-1970 s, the U.S. Navy has expended considerable resources in characterizing the properties of different biological sounds. Marine bioacousticians are also interested in the different kinds of sounds produced by marine animals but for different reasons than the Navy. Not only the characteristics of the sounds but also the possible kinds of meanings different sounds can convey, conditions under which these sounds are produced and the functional use of sounds are of interest.

The mechanisms involved in the generation of sounds are of considerable interest to bioacousticians. There is still a poor understanding of how sounds are produced by dolphins and whales. Many theories exist but data to support the different theories are essentially non-existent in many cases. Understanding the anatomy of possible sound production organs is essential in trying to understand the physics of sound production, not only in cetacean but in all marine animals. What specific tissues are vibrating to produce sounds, what are the sources of energy for sound production, how are sounds modulated and how is the acoustic energy coupled into the water are among the many unanswered questions involving sound production by marine animals.

Underwater acoustics principles, techniques and technology can also be used to detect, localize, track and obtain statistical data on marine animals. As fish resources become progressively depleted, active sonars have been increasingly used to detect and localize fish schools and to also determine population density and size distribution. Active sonar can be used in similar ways to study marine mammals belonging to the top of the ocean food chain to zooplankton at the bottom of the chain. Passive sonar arrays can be used to detect and track the movement of different marine mammals as well as certain sound-producing fish.

An area of bioacoustics that has recently gained considerable interest in both the scientific and non-scientific communities is the effect of noise and other sounds on both marine and terrestrial animals. As humans increase the amount of activity in the sea, whether for pleasure, research or economic interest, the amount of sounds that are produced continues to increase. How these sounds impact the everyday life of marine animals is being studied at a greater level than in the past. We will not spend much time on this issue since there is already a whole book devoted to this subject. Interested readers are encouraged to delve into the book written by Richardson et al. (1995).

From this brief discussion about marine bioacoustics, it is obvious that this is a multidiscipline field involving not only scientists from many technical disciplines, but individuals who are "jacks of all trade," and master of many. Knowledge in certain aspects of biology, psychology, physiology, physics, veterinary medicine, mathematics and statistics and electrical engineering is important in performing research in marine bioacoustics. Because of the wide diversity of disciplines involved with marine bioacoustics, it is difficult to find a single text that contains information about the many disciplines involved. In this text, we will attempt to cover a broad range of topics that we think is fundamental to the field of marine bioacoustics. Part I of this book concentrates mainly on principles, techniques and methodologies from various disciplines that are useful in bioacoustics. These chapters are not meant to be an exhaustive and comprehensive treatment of the subject at hand, but are meant to introduce and equip the reader with some important tools for work in bioacoustics. A wide variety of subjects will be addressed in various chapters; however, fish hearing and sound production will not be considered in detail. Those interested in fish acoustics should refer to a recent book edited by Webb et al. (2008). Part II will concentrate on the acoustics of marine animals discussing the properties of hearing and sound production of various marine animals.

1.2 Introduction to Underwater Acoustics

Acoustic energy in water consists of molecular vibrations that travel at the speed of sound. The vibrations in water and other fluids are in the direction of propagation and are therefore referred to as longitudinal waves. In inelastic

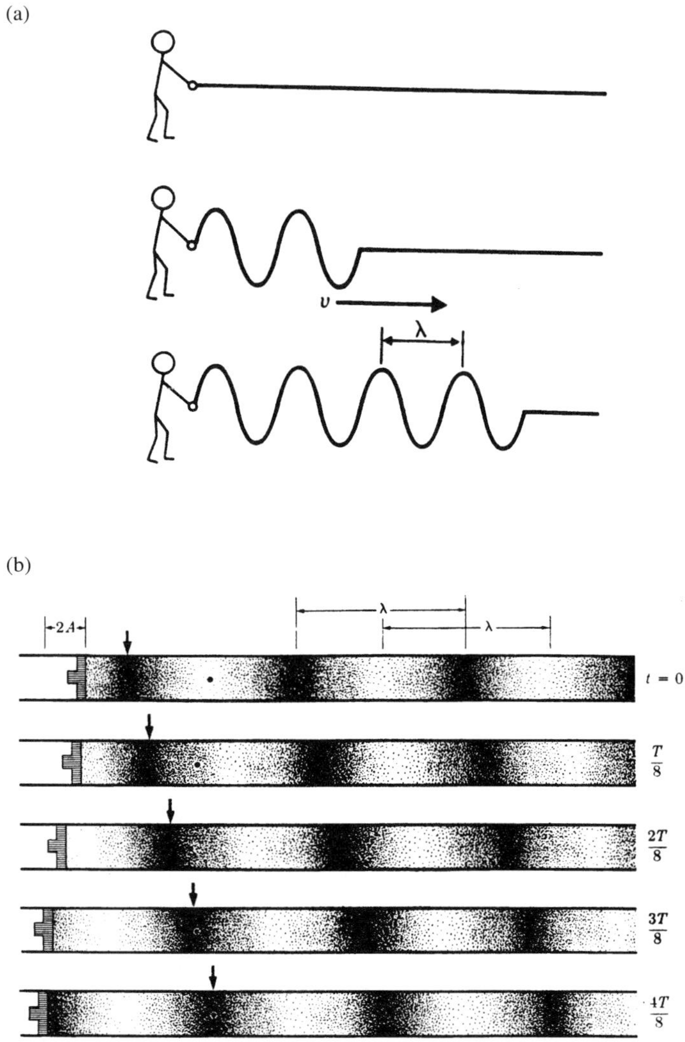

FIGURE 1.1. (a) Example of a transverse wave traveling down a stretched piece of rope. (b) Example of compression and refraction caused by a longitudinal wave traveling through the medium (Adapted from Rossing, 1990).

materials, acoustic vibrations can also occur in a direction perpendicular to the direction of propagation; such vibrations are referred to as transverse waves. A simple visual example of a transverse wave can be seen in Fig. 1.1a. One end of a rope is tied to a rigid support and the other end is held in a way to keep the rope taut. A transverse wave can be generated with a flick of the wrist causing a disturbance to travel down the rope. The motion of the rope will be perpendicular to the direction in which the disturbance travels. Longitudinal waves consist of fluid particles vibrating along the direction of energy propagation and are more difficult to visualize than transverse waves. In Fig. 1.1b, the amplitude variation of a longitudinal wave as a function of distance at an instant of time is shown along with the compression (increase in particle density) and rarefaction (decrease in particle density) of the fluid particles as the wave propagates. The positive peaks of the pressure wave correspond to the locations where the particles are bunched together and the negative peaks to the location where the particles are spread out. This series of compression and rarefaction will propagate as a longitudinal wave away from the source, transferring energy without transporting any mass.

1.2.1 Derivation of the Wave Equation

The wave equation is the most fundamental differential equation describing the propagation of sound through a medium. We shall derive this basic equation to gain an appreciation of some of the physical principles governing the propagation of acoustic energy. There are several different ways to derive the wave equation; we will take the approach presented in the book *Physics of Sound in the Sea* written at the end of World War II by some of the eminent acousticians and in the book of Medwin and Clay (1998). In the derivation, we shall assume a plane wave, the simplest type of wave motion in which the properties of the wave, such as pressure, particle displacement, density changes, etc., have common phase and amplitudes at all points on any given plane perpendicular to the direction of wave propagation. The wave equation will be derived by considering the *equation of continuity* which is associated with the law of conservation of mass, *Newton's second law* applied to the small particles of a disturbed fluid, the *equation of force* which relates the fluid pressure inside a small volume of the fluid to the external forces acting on the volume and the *equation of state* which relates pressure changes inside a fluid to density changes.

Equation of Continuity: *Acoustic* energy propagating in water causes a disturbance in the pressure and density of the medium. The law of the conservation of mass states that no disturbances of a fluid can cause mass to be either created or destroyed. Consider a small cubic volume fixed in space, and acoustic energy causes mass to flow into one face and out the other, as shown in Fig. 1.2. The mass of fluid entering into face A per unit time moving in the x direction is equal to $\rho_T u_x A$ and the mass of fluid per unit time leaving the volume through the opposite face from A in the x direction is

$$\left[\rho_{\mathrm{T}} u_x + \frac{\partial}{\partial x}(\rho_{\mathrm{T}} u_x)\Delta x\right] A \tag{1.1}$$

The net increase in the mass inside the volume is the difference between the mass flowing into the volume through face A and that flowing out of the opposite face to A and can be expressed as

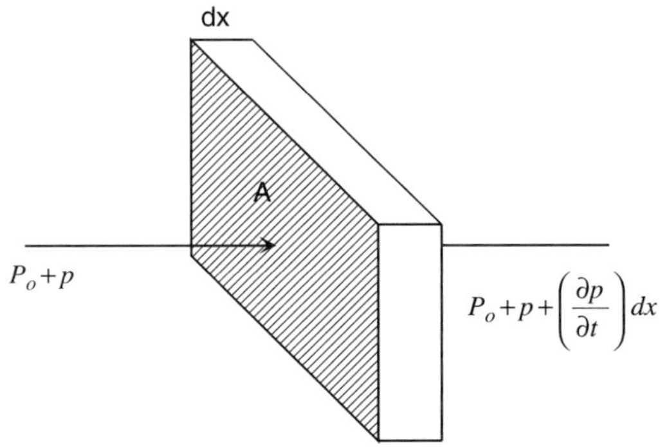

FIGURE 1.2. A hypothetical volume of a fluid with an acoustic wave traveling through it.

$$-\frac{\partial}{\partial x}(\rho_T u_x)\, \Delta x\, A \tag{1.2}$$

The total density, ρ_T, is equal to the ambient density plus the acoustic density or

$$\rho_T = \rho_0 + \rho \tag{1.3}$$

where ρ_0 is the ambient density and ρ is the acoustic density, and $\rho_0 >> \rho$. Water is compressible so that more mass may flow in than out, and the density within the volume may increase. This rate of deposit of mass must result in a corresponding change in the average density ρ_T inside the volume so that

$$\frac{\partial}{\partial t}(\rho_T V) = -\frac{\partial}{\partial x}(\rho_T u_x)\, V \tag{1.4}$$

Canceling V on both sides of the equation, we arrive at

$$\frac{\partial \rho_T}{\partial t} = -\frac{\partial}{\partial x}(\rho_T u_x) \tag{1.5}$$

Inserting Eq. (1.3) into Eq. (1.5) and recognizing that $\rho_0 \partial u/\partial x >> \partial(\rho u)/\partial x$ and $\partial \rho_0/\partial t = 0$ because ρ_0 is essentially constant, we obtain

$$\frac{\partial \rho}{\partial t} = -\rho_0 \frac{\partial u_x}{\partial x} \tag{1.6}$$

Equation (1.6) is the acoustics equivalent of the conservation of mass equation.

Equation of Motion: Newton's second law states that the product of the mass of a particle by its acceleration is equal to the force acting on the particle. For a plane wave traveling in the x direction and with the assumptions that second-order products of displacements, particle velocities and pressure changes are negligible, the acceleration of a particle of water can be approximated by $\partial u_x/\partial t$ so that the force per unit volume inthe x direction can be expressed as

$$f_x = \rho_0 \frac{\partial u_x}{\partial t} \tag{1.7}$$

Equation of Force: *Non*-viscous fluids, such as water, cannot support shear stress and an acoustic waves propagating in the fluid will subject fluid particles to compression and expansion forces. For an acoustic wave propagating in the x direction, the force per unit volume is related to the pressure acting on a unit volume according to the equation

$$f_x = -\frac{\partial p_T}{\partial x} \tag{1.8}$$

where $p_T = p_0 + p$, and p_0 is the ambient pressure which does not change over a small volume, p is the acoustic pressure and $p_0 >> p$. Equating Eqs. (1.7) and (1.8), we obtain

$$\frac{\partial p}{\partial x} = -\rho_0 \frac{\partial u_x}{\partial t} \tag{1.9}$$

Equation of State: The equation of state for a fluid describes the pressure of the fluid as a function of its density and temperature. We assume that the acoustic wave propagating in a fluid causes the fluid to deviate slightly from its state of equilibrium and a small change in pressure is proportional to the fractional change in density. We further assume that the changes caused by the acoustic wave take place so rapidly that there is no conduction of heat. Hooke's law, which states that stress is proportional to the strain, can be applied, where stress (force per unit area) is the acoustic pressure, p, and the strain (relative change of dimension) is the relative change of density, ρ/ρ_0. The proportionality constant is the bulk modulus of elasticity, E, for small-amplitude acoustic waves. This relation between acoustic pressure and acoustic densi4y is referred to as the acoustical equation of state:

$$p = \left(\frac{K}{\rho_0}\right)\rho \tag{1.10}$$

where K is the bulk modulus of the fluid. For a homogeneous medium, the speed of sound is given by $c = \sqrt{[\partial p/\partial \rho]_s}$ where the subscript s indicates that the local compression and rarefaction caused by the acoustic wave occurs at constant entropy (i.e., an ideal process). Applying this definition of the speed of sound to Eq. (1.10), we get the following expression for the speed of sound in a fluid:

$$c = \sqrt{\frac{K}{\rho_0}} \tag{1.11}$$

We now incorporate the results from the four laws to derive the one-dimensional wave equation by taking the $\partial/\partial t$ of Eq. (1.6) and the $\partial/\partial x$ of Eq. (1.9) and eliminating the common second derivative $\partial^2/\partial x\partial t$ term to get

$$\frac{\partial^2 p}{\partial x^2} = \frac{\rho_0}{K} \frac{\partial^2 p}{\partial t^2} \tag{1.12}$$

The speed of sound in water, c, is not dependent on the direction of propagation and is equal to $\sqrt{K/\rho_0}$, so that the one-dimensional form of the wave equation can be rewritten as

$$\frac{\partial^2 p}{\partial x^2} = \frac{1}{c^2} \frac{\partial^2 p}{\partial t^2} \tag{1.13}$$

The three-dimensional form of the wave equation can also be derived in a similar manner as in the one-dimensional case by considering a wave traveling in the y and z directions. The wave equation in three-dimensional space can be expressed for a rectangular coordinate system, as

$$\left(\frac{\partial^2 p}{\partial x^2} + \frac{\partial^2 p}{\partial y^2} + \frac{\partial^2 p}{\partial z^2}\right) = \frac{1}{c^2} \frac{\partial^2 p}{\partial t^2} \tag{1.14}$$

Equation (1.14) is a second-order partial differential equation describing how the acoustic pressure will vary as a function of space and time. The wave equation can be derived in terms of a number of variables such as particle displacement ξ, particle velocity u and incremental temperature ΔT. However, the wave equation is often expressed in terms of acoustic pressure since most hydrophones respond directly to acoustic pressure rather than to other variables.

The wave equation is often expressed in terms of a divergence operator ∇ in vector analysis and is defined as

$$\nabla = \mathbf{i} \frac{\partial}{\partial x} + \mathbf{j} \frac{\partial}{\partial y} + \mathbf{k} \frac{\partial}{\partial z} \tag{1.15}$$

$$\nabla^2 = \nabla \cdot \nabla = \frac{\partial^2}{\partial x^2} + \frac{\partial^2}{\partial y^2} + \frac{\partial^2}{\partial z^2} \tag{1.16}$$

where **i**, **j** and **k** are unit vectors along the x, y and z axes, respectively (see Appendix for a description of the vector notation). Using ∇^2 in Eq. (1.14), the wave equation in three-dimensional volume of fluid can be expressed as

$$\nabla^2 p = \frac{1}{c^2} \frac{\partial p}{\partial t^2} \tag{1.17}$$

A considerable amount of research effort since and during World War II has been devoted to various ways of solving the wave equation for different ocean conditions and in different coordinate systems. Interested readers should consider the AIP press book by Jensen et al. (1994) and the edited volume of Robinson and Lee (1994). Other classical volumes dealing with ocean propagation include the National Research Council volume 8 (1946) and Officer (1958). We will not examine the different methods by which the wave equation can be solved; it would be beyond the scope of this book to do so.

1.2.2 A Simple Harmonic Solution to the One-Dimensional Wave Equation

We shall now consider the one-dimensional form of the wave equation described by Eq. (1.13) and merely give a particular solution without actually solving the equation. The most general solution of Eq. (1.13) is in the form

$$p = f_1\left(t - \frac{x}{c}\right) + f_2\left(t + \frac{x}{c}\right) \tag{1.18}$$

where f_1 and f_2 are arbitrary but suitable functions having continuous derivatives. The first function $f_1(t - x/c)$ represents a wave traveling in the positive direction and the second function represents a wave traveling in the negative direction. Consider the solution $f_1(t - x/c)$. The function will have a constant value for all values of x and t which satisfy the relationship $t = x/c +$ constant or $x = ct +$ constant. Therefore, as t increases, x becomes progressively larger indicating propagation in the positive x direction. Similarly, for the solution $f_2(t + x/c)$, as t increases, $x = -ct +$ constant becomes progressively more negative, indicating propagation in the negative x direction.

Practical functions represented by f_1 and f_2 in Eq. (1.18) for continuous wave (CW) are pure tone cosine and sine waves of cycle $2\pi f(t - x/c)$ or $\omega(t - x/c)$, where $\omega = 2\pi f$ is the radian frequency. From Euler's formula in Complex Variable Theory (see Appendix for some basics in complex arithmetic)

$$e^{j\omega(t-x/c)} = \cos \omega(t - x/c) + j \sin \omega(t - x/c) \tag{1.19}$$

where $j = \sqrt{-1}$. The general solution of Eq. (1.18) can be expressed as

$$p = A\, e^{\,j\omega(t-x/c)} + B\, e^{\,j\omega(t+x/c)} \tag{1.20}$$

where the first term represents a wave traveling in the $+x$ direction and the second term represents a wave traveling in the $-x$ direction. When a wave is expressed as a complex exponential, it is to be understood that the actual physical quantity is represented by the real part or the imaginary part of the complex number. For example, if the pressure is given as

$$p = Ae^{j\omega\left(t-\frac{x}{c}\right)} \tag{1.21}$$

where A is a real number representing the peak amplitude of the sinusoidal signal, then the actual physical pressure is

$$p = A \cos[\omega(t - x/c)] \tag{1.22}$$

If we were to observe a sinusoidal wave as expressed in Eq. (1.22) from a fixed location in space the instantaneous pressure as a function of time would oscillate sinusoidally between A and $-A$, as depicted in Fig. 1.1. Conversely, if the wave is frozen at any instant of time, we would observe a sinusoidal variation of the instantaneous pressure with x, as indicated in Fig. 1.3. The distance of one cycle is referred to as a wavelength, λ. Note in Fig. 1.3 that

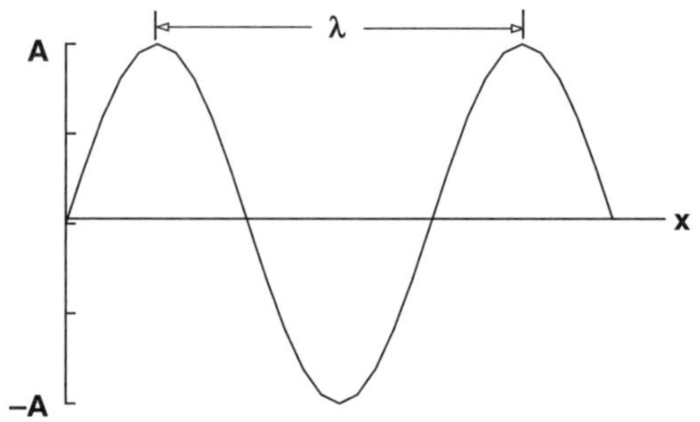

FIGURE 1.3. A cw acoustic wave.

when x changes by an amount λ, the phase angle in Eq. (1.22) changes by 2λ radians. Therefore, we can write

$$2\pi f\left(t - \frac{x}{c}\right) - 2\pi f\left(t - \frac{x + \lambda}{c}\right) = 2\pi \tag{1.23}$$

and after eliminating the appropriate terms we arrive at the important relationship between frequency, wavelength and the speed of sound of an acoustic signal

$$f\lambda = c \tag{1.24}$$

Equation (1.24) is often written in a slightly different way by expressing the frequency in terms of radian frequency ω. Let

$$k = \omega/c = 2\pi/\lambda \tag{1.25}$$

so that Eq. (1.22) can be expressed as

$$p = A\cos(\omega t - kx) \tag{1.26}$$

where k is defined as the wavenumber.

1.2.3 Particle Displacement and Velocity

Particle displacement ξ is defined as the displacement of the particles of the fluid from the equilibrium due to the acoustic wave. It is governed by the same wave equation as the acoustic pressure and can be derived in a similar manner. In one dimension the particle displacement wave equation can be expressed as

$$\frac{\partial^2 \xi}{\partial x^2} = \frac{1}{c^2}\frac{\partial^2 \xi}{\partial t^2} \tag{1.27}$$

Its direction is parallel to the direction of energy propagation. Let u be the particle velocity which is the change of ξ with time so that

$$u = \frac{\partial \xi}{\partial t} \tag{1.28}$$

The solution for ξ in Eq. (1.27) will have the form

$$\xi = B\,\mathrm{e}^{\mathrm{j}\omega\left(t - \frac{x}{c}\right)} \tag{1.29}$$

The constant B can be determined by differentiating Eq. (1.28) with respect to t using Eq. (1.29) and inserting the results into Eq. (1.9) which is the relationship between acoustic pressure and particle velocity, so that

$$-\frac{j\omega\, B}{c}\,\mathrm{e}^{\mathrm{j}\omega\left(t - \frac{x}{c}\right)} = -\frac{p}{\rho_0\, c^2} = -\frac{A}{\rho_0\, c^2}\,\mathrm{e}^{\mathrm{j}\omega\left(t - \frac{x}{c}\right)} \tag{1.30}$$

Canceling out the appropriate terms in Eq. (1.30), B can be expressed as

$$B = -\mathrm{j}\,\frac{A}{\omega\,\rho_0\, c} \tag{1.31}$$

Equation (1.25) can now be written as

$$\xi = -\mathrm{j}\,\frac{A}{\omega\,\rho_0\, c}\,\mathrm{e}^{\mathrm{j}\omega\left(t - \frac{x}{c}\right)} \tag{1.32}$$

Since $\mathrm{e}^{-\mathrm{j}\pi/2 = -\mathrm{j}}$ (see Eq. (1.67) in Appendix), we can express the particle displacement as

$$\xi = -\mathrm{j}\,\frac{A}{\omega\,\rho_0\, c}\,\mathrm{e}^{\mathrm{j}\omega\left[\left(t - \frac{x}{c}\right) - \frac{\pi}{2}\right]} \tag{1.33}$$

Equation (1.33) indicates that the particle displacement lags the acoustic pressure by a phase angle of 90°. The real displacement is

$$\xi = \frac{A}{\omega\,\rho_0\, c}\,\sin\left[\omega\left(t - \frac{x}{c}\right)\right] \tag{1.34}$$

In order to derive the particle velocity, we merely need to take the derivative of Eq. (1.33) with respect to t to get

$$u = \frac{A}{\rho_0\, c}\,\mathrm{e}^{\mathrm{j}\omega\left(t - \frac{x}{c}\right)} \tag{1.35}$$

Substituting Eq. (1.21) into Eq. (1.35), we obtain the relationship between acoustic pressure and particle velocity for a plane wave,

$$p = \rho_0 c u \tag{1.36}$$

Equation (1.33) indicates that the acoustic pressure and the particle velocity are in phase so that at the peak of the sinusoidal acoustic wave of Fig. 1.3 the particle velocity will be maximum in the positive x direction, and at the minimum of the acoustic wave, the particle velocity will be maximum in the negative x direction.

1.2.4 Acoustic Intensity and Acoustic Impedance

Acoustic intensity is defined as the rate of flow of energy through a unit area normal to the direction of the wave propagation. Let S be the area (normal to the direction of wave propagation) of a small volume of fluid, that is, in which the intensity is defined and $\mathrm{d}E/\mathrm{d}t$ be the rate of flow of acoustic energy across area S; the intensity of the acoustic wave is defined as

$$I = \lim_{S \to 0} \frac{\mathrm{d}E/\mathrm{d}t}{S} \tag{1.37}$$

The energy E of the acoustic wave can be thought of in terms of work on a sheet of fluid particles by the acoustic pressure. Since work is equal to *Force* × *Distance*, the flow of energy or work per unit time caused by the acoustic wave is

$$\frac{\mathrm{d}E}{\mathrm{d}t} = F\frac{\mathrm{d}x}{\mathrm{d}t} \tag{1.38}$$

Since the acoustic pressure is related to the force on the fluid ($p = F/S$) and $u = \mathrm{d}x/\mathrm{d}t$, the intensity of an acoustic wave can be expressed in terms of the acoustic pressure as

$$I = pu = \frac{p^2}{\rho_0 c} \tag{1.39}$$

The concept of impedance has been used for many years in alternating current (ac) electrical circuit analysis, and constructing an analogy between an ac electrical circuit and a plane acoustic wave can be very useful. Therefore, the following analog will be used, where i = electrical current and R = resistance.

Acoustics	Electrical
pressure ⇒ $p = \rho c u$	voltage ⇒ $\nu = R\,i$
intensity ⇒ $I = pu$	power ⇒ $P = \nu i$
$= \rho c u^2$	$= Ri^2$
$= p^2/\rho c$	$= v^2/\mathrm{R}$
velocity ⇒ u	current ⇒ i

The ratio between the acoustic pressure and fluid particle velocity can be defined as the specific acoustic impedance

$$z = \rho c \tag{1.40}$$

and is analogous to electrical resistance. For the plane wave situation considered here, the specific acoustic impedance is resistive since the acoustic pressure and particle velocity are in phase. However, for other types of waves, such as spherical waves, the pressure and particle velocity are not in phase and therefore Eq. (1.40) will have both resistance and reactance components.

1.2.5 The Decibel and Sound Pressure Level

Many of the conventions used in specifying acoustic quantities have their root in electrical ac network analysis. Since ac power is defined as v^2/R, the average power taken over a period of time T delivered to a resistor from an ac electrical signal generator is

$$P_{\text{ave}} = \frac{[v^2]_{\text{ave}}}{R} = \frac{1}{\mathrm{T}} \int_0^{\mathrm{T}} \frac{v^2(t)}{R} \mathrm{d}t \tag{1.41}$$

The effective or root mean square (rms) voltage that can be used directly to compute the average power delivered to a load is then defined as

$$v_{\text{rms}} = \sqrt{\frac{1}{\mathrm{T}} \int_0^{\mathrm{T}} v^2(t) \mathrm{d}t} \tag{1.42}$$

By convention, the subscripts (ave) and (rms) in Eq. (1.39) are usually not included and it is understood that P is the average power and v is the rms voltage, unless specified otherwise. Similarly, in acoustics, intensity is usually defined in terms of its averaged values so that

$$I_{\text{ave}} = \frac{[p^2]_{\text{ave}}}{\rho c} = \frac{1}{\mathrm{T}} \int_0^{\mathrm{T}} \frac{p^2(t)}{\rho c} \mathrm{d}t \tag{1.43}$$

The effective or rms pressure used to calculate the average intensity is

$$p_{\text{rms}} = \sqrt{\frac{1}{\mathrm{T}} \int_0^{\mathrm{T}} p^2(t) \mathrm{d}t} \tag{1.44}$$

Once again, the subscripts (ave) and (rms) are not usually included and by convention, intensity I is understood to mean the average intensity and pressure p is understood to mean the rms pressure unless specifically stated otherwise. If the acoustic signal is a sinusoidal wave in the form of Eq. (1.26) having a peak amplitude of A, a frequency of f and a period of T, the rms pressure over a cycle will be

$$p = \sqrt{\frac{A^2}{\mathrm{T}} \int_0^{\mathrm{T}} \cos^2(2\pi f t)\, dt} = \sqrt{\frac{A^2}{\mathrm{T}} \int_0^{\mathrm{T}} \frac{1 + \cos(4\pi f t)}{2}\, dt} \tag{1.45}$$

$$= \sqrt{\frac{A^2}{\mathrm{T}} \left[\frac{\mathrm{T}}{2} + \frac{\sin 4\pi}{8\pi f}\right]} = \frac{A}{\sqrt{2}} = 0.707\, A \tag{1.46}$$

Therefore, the rms value for a continuous sinusoidal wave or a sine pulse is simply 0.707 times the peak amplitude. Most ac meters and other electronic measuring instruments are calibrated to measure rms voltage or current.

Pressure is a scalar quantity which describes the force perpendicular or normal to a surface per unit area, or the normal force per unit area. Therefore, pressure has the unit N/m^2 which is equivalent to kg m/s^2, where N refers to

Newton, the unit of force named after the great physicist, Sir Isaac Newton (1642–1727). Another unit of pressure is the Pascal (Pa) named after the French scientist Blaise Pascal (1623–1662). One atmosphere of pressure is the average pressure that the earth's atmosphere exerts at sea level and is equal to 101.3 kPa, which is equivalent to $14.7\,\mathrm{lb/in}^2$. Acoustic pressure is usually very small in comparison to the ambient pressure and therefore, the unit μPa, which is equal to 10^{-6} Pa, is used to describe its magnitude. The μPa unit was adopted as the American National Standard in 1968 as the unit for acoustic pressure. Previously, the common unit was 1 μbar (1 dyne/cm^2) for underwater acoustics and 0.000204 μbar for airborne acoustics.

The decibel notation, dB, has traditionally been used to describe the intensity and pressure of acoustic waves. The original and traditional use of decibels involved power ratios in electrical AC circuitry as given by the equation

$$\text{power ratio in dB} = 10\,\log(P_1/P_2) \tag{1.47}$$

where P_1/P_2 is the ratio of two powers. Throughout this book *log* refers to common logarithm and *ln* to natural logarithms. The logarithmic nature of the decibel notation provides a convenient way of expressing large changes in pressure. They also permit quantities that would otherwise be multiplied or divided to be simply added or subtracted, respectively. Finally, decibels are a convenient measure of ratios: in underwater acoustics, the primary interest is often in ratios rather than in absolute quantities.

Intensity ratios are defined similarly as power ratios:

$$\text{intensity ratio in dB} = 10\,\log(I_2/I_2) \tag{1.48}$$

The unit for intensity in Eq. (1.48) is dB re 1 W/m^2. Taking the definition of intensity found in Eq. (1.39), a pressure ratio in decibels can be expressed as

$$\text{pressure ratio in dB} = 20\,\log(p_2/p_2) \tag{1.49}$$

The most often measured quantity in acoustics is the acoustic pressure or the sound pressure level (SPL) which can be expressed as

$$\mathrm{SPL} = 20\,\log(p/p_0) \tag{1.50}$$

where p_0 is a reference pressure of 1 μPa. Since 1 atm $\approx 1\times10^5$ Pa $= 1\times10^{11}$ μPa, it is equal to 220 dB re 1 μPa. In order to convert SPL expressed in dB re 1 μbar into SPL in dB re 1 μPa (re is a short-hand notation for referenced), add 100 dB to the SPL expressed in dB re 1 μbar. A pictorial representation of the relationships between μPa, μbar and 20 μPa for in air reference is shown in Fig. 1.4.

A voltage level can be expressed in dB by letting the reference be 1 V, so that the voltage is expressed as dBV, which reads as "dB re 1 V." For a sinusoidal signal we see from Eq. (1.41) that the rms value will be 3 dB less than the peak value, and since the peak-to-peak amplitude is twice the peak amplitude, the rms value in dB is approximately 9 dB less than the peak-to-peak value expressed in dB.

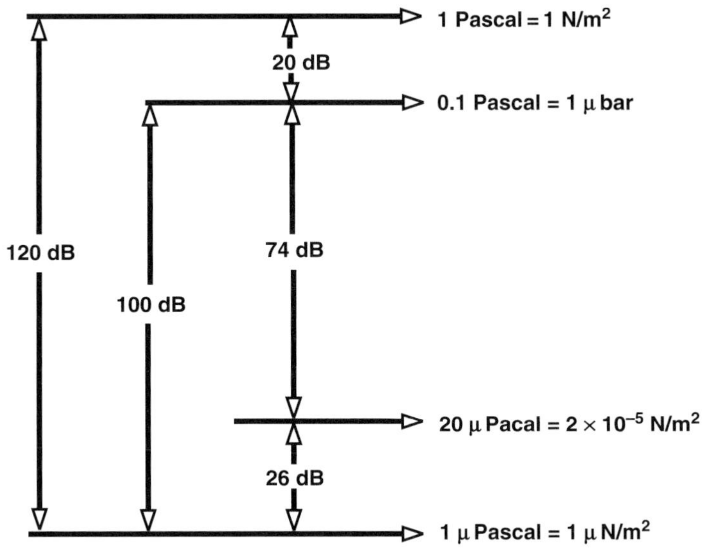

FIGURE 1.4. A pictorial description of the relationship between μPa, μbar and 20 μPa in dB.

There will be occasions when one would like to compare the effects of underwater acoustics pressure with airborne acoustic pressure. For example, one may want to compare the difference in the hearing sensitivity of a marine mammal to a terrestrial animal or to compare the underwater hearing sensitivity to the in air sensitivity for an amphibian. Two factors must be taken into account, the first is related to the reference level of 20 μPa used in airborne acoustics to determine dB levels in contrast to the reference of 1 μPa used in underwater acoustics. Therefore, 26 dB (20 log 20) must be added to the dB level measured in air in order to have the same reference level (see also Fig. 1.4). Secondly, water is much denser than air, so the effects of a specific acoustic pressure will be different in both media. The appropriate parameter to compare is the acoustic intensity in both media. The acoustic intensity is given in Eq. 1.36 and the ratio of intensity in air and under water can be written as

$$\frac{I_{\text{air}}}{I_{\text{water}}} = \frac{\left(\frac{p^2}{\rho c}\right)_{\text{air}}}{\left(\frac{p^2}{\rho c}\right)_{\text{water}}} \tag{1.51}$$

At 20°C the density of air at sea level is 1.21 kg/m^3 and the speed of sound is 344 m/s, so that the characteristic impedance of air is

$$(\rho c)_{\text{air}} = 416 \text{ Pa s/m} \tag{1.52}$$

The density of seawater at 20°C is 1,026 kg/m^3 and the speed of sound is 1,500 m/s, so that the characteristic impedance is

$$(\rho c)_{\text{water}} = 1.5 \times 10^6 \text{ Pa s/m} \tag{1.53}$$

Therefore the ratio of characteristic impedances is

$$\frac{(\rho\, c)_{\text{water}}}{(\rho\, c)_{\text{air}}} = \frac{1,500,000}{416} = 3,606 \tag{1.54}$$

This means that if the acoustic pressure in both medium is the same, the intensity in air is 3,606 times or 35.6 dB greater than the corresponding intensity in water. If the intensity is the same in both media then the ratio of acoustic pressures will be

$$\frac{p_{\text{water}}}{p_{\text{air}}} = \sqrt{3,606} = 60 \tag{1.55}$$

which means that the sound pressure level in water is 60 times or 35.6 dB (exactly the same dB difference as the difference in intensities) greater than the corresponding value in air.

1.2.6 Spherical Spreading Transmission Loss

Transmission loss refers to the decay of acoustic intensity as a signal travels from a source. If the intensity of the acoustic wave at a point 1 m from the source is I_0, and I_1 is the intensity at a distant point, then the transmission loss (TL) in dB is

$$\text{TL} = 10 \log_{10} I_0/I_1 = 20 \log_{10} p_0/p_1 \tag{1.56}$$

where p_0 and p_1 are the corresponding sound pressure levels at 1 m and at the distant point. Transmission loss may be considered to be the sum of a loss due to *spreading* and a loss due to *absorption* or *attenuation*.

The simplest kind of spreading loss is spherical spreading loss. We can examine spherical spreading loss assuming a source emitting signals of equal intensity in all directions so that the wave equation (1.13) will be only a function of the radial coordinate. The wave equation (see Appendix) in a spherical coordinate system reduces to

$$\frac{\partial^2 p}{\partial r^2} + \frac{2}{r}\frac{\partial p}{\partial r} = \frac{1}{c^2}\frac{\partial^2 p}{\partial t^2} \tag{1.57}$$

The solution to the wave equation is

$$p = \frac{A}{r}\, e^{j(\omega t - kr)} \tag{1.58}$$

where A is a constant representing the peak amplitude of the acoustic wave. One can verify that Eq. (1.58) is a solution to the wave equation by using it and performing the operations on it as dictated by Eq. (1.57). The transmission loss can be derived by placing the source of the spherical wave at the center of an imaginary sphere of radius 1 m and another larger sphere of radius R. If for the moment we assume that there are no absorption losses, the power crossing both spheres must be the same. Since power equals intensity times area, then

$$4\pi R_0^2 I_0 = 4\pi \mathrm{R}_1^2 \mathrm{I}_1 \tag{1.59}$$

Since R_0 is equal to 1 m, the transmission loss (TL) due to spreading, from Eqs. (1.48) and (1.50), can be written as

$$\mathrm{TL} = 10\ \log_{10} R_1^2 = 20\ \log_{10} R_1 \tag{1.60}$$

The spherical spreading loss is also referred to as *inverse square* loss since the acoustic intensity decreases as the square of range so that transmission loss increases as the square of range.

1.3 Appendix: Some Mathematics

1.3.1 Introduction to Complex Variables

Complex variable theory in mathematics has been around for over two centuries but the fundamental concepts are often confusing and difficult to understand to the uninitiated. Students in physics, engineering and mathematics are introduced to complex variable theory early in their undergraduate training whereas those in the life science may never encounter this field of mathematics until they enter some interdisciplinary field such as bioacoustics. This appendix is written mainly for the "biological type" and will take a somewhat unorthodox approach to this subject.

An analogy to a complex variable is the *vector* quantity. A vector has a magnitude and direction as is depicted in Fig. 1.5a. The vector magnitude is r and its direction is θ. In a rectangular coordinate system the vector has an x-component equal to r cos θ and a y-component equal to r sin θ. Like a vector, a complex variable also has two quantities to describe it, a magnitude and a phase. The term "complex" can be misleading since it refers to a variable having two quantities ascribed to it rather than being complicated or

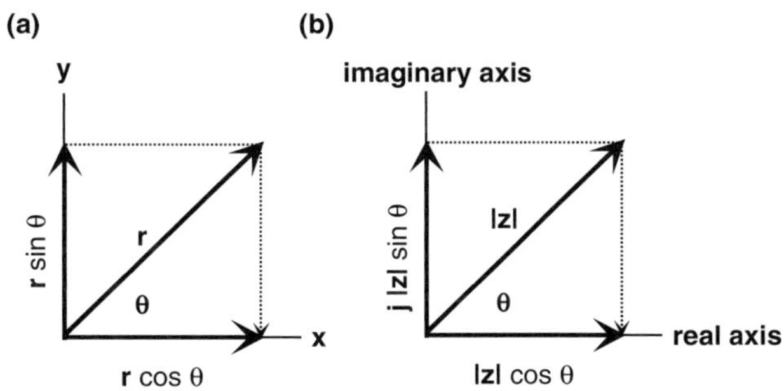

FIGURE 1.5. **(a)** Graphical representation of a vector. **(b)** Graphical representation of a complex variable.

intricate. Like a vector, a complex variable can be described in a rectangular coordinate system where the horizontal axis is the *real* axis and the vertical axis is the *imaginary* or **j** axis as shown in Fig. 1.5b. The imaginary descriptor j is equal to $\sqrt{-1}$. Mathematicians and physicist usually use the alphabet i as the imaginary descriptor whereas engineers use j. The term "imaginary" can also be confusing since it does not refer to an imagined or fictitious quantity, but is an actual quantity but one in the imaginary plane.

A complex variable is typically represented in the form

$$z = x + \mathrm{j}y \tag{1.61}$$

where x is the real part of z and is often denoted as $x = \mathrm{Re}\ z$, and y is the imaginary part of z denoted as $y = \mathrm{Im}\ z$. A complex variable can also be represented in a polar form, having a radius and an exponential phase angle

$$z = \sqrt{x^2 + y^2}\, \mathrm{e}^{\mathrm{j}\,\phi} \tag{1.62}$$

where

$$|z| = \sqrt{x^2 + y^2} \tag{1.63}$$

$$\phi = \tan^{-1} \frac{y}{x} \tag{1.64}$$

A useful expression relating the exponential in Eq. (1.62) to sine and cosine functions is Euler's formula which can be written as

$$\mathrm{e}^{\pm \mathrm{j}\phi} = \cos\phi \pm \mathrm{j}\ \sin\phi \tag{1.65}$$

From here on, the analogy between a vector and a complex variable will depart substantially and we will not refer to this analogy again.

The notion of a complex variable being defined as a variable with a magnitude and phase is convenient to apply in many areas of physics, engineering and signal processing, especially in Fourier analysis. Consider the simple series RLC circuit in Fig. 1.6. The input impedance of the RLC can be expressed as a complex variable

$$Z = R + \mathrm{j}X \tag{1.66}$$

where $X = X_\mathrm{L} - X_\mathrm{C}$ and

$$\begin{aligned} X_L &= \omega\, L \\ X_C &= \frac{1}{\omega\, C} \end{aligned} \tag{1.67}$$

The real part of the complex input impedance is the resistor R and the imaginary part is the reactance X. Only the resistor R contributes to any energy losses, and the reactance contributes to the complex or phase portion

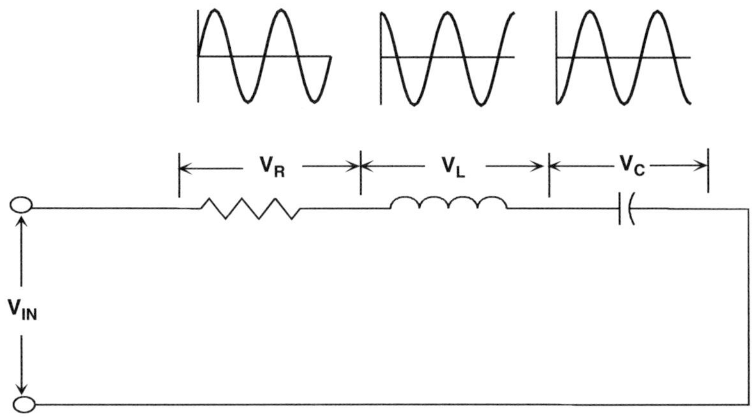

FIGURE 1.6. A series R, L, C electrical network with V_{IN} being a sinusoidal signal. The signal across each element is shown above the elements.

of the impedance. The voltage waveforms across the different components are also shown in Fig. 1.6. Also shown in the figure is the signal across the various elements. The voltage waveform across the inductor has a plus 90° phase shift and the waveform across the capacitor has a minus 90° phase shift compared to the waveform across the resistor.

Therefore, the imaginary part of Eq. (1.66) represents signals that are shifted by 90° from the signal across the real part of z, namely the resistor R.

1.3.2 *Arithmetic Operations of Complex Numbers*

The ordinary rules of algebra hold for complex variables. if

$$z_1 = x_1 + \mathrm{j}y_1 \tag{1.68}$$

$$z_2 = x_2 + \mathrm{j}y_2 \tag{1.69}$$

then the sum and difference between z_1 and z_2 can be expressed as

$$z_1 \pm z_2 = (x_1 \pm x_2) + \mathrm{j}(y_1 \pm y_2) \tag{1.70}$$

The real parts add or subtract from each other separately and then the imaginary parts also add or subtract from each other. When multiplying complex variables, we need to realize that

$$\mathrm{j}^2 = -1 \quad \text{and} - \mathrm{j} \cdot \mathrm{j} = 1 \tag{1.71}$$

so that

$$z_1 z_2 = (x_1 x_2 - y_1 y_2) + \mathrm{j}(x_1 y_2 + x_2 y_1) \tag{1.72}$$

In the polar format we have

$$z_1 z_2 = |z_1| \cdot |z_2| \mathrm{e}^{\mathrm{j}(\phi_1 + \phi_2)} \tag{1.73}$$

The complex conjugation of a complex variable is defined as the complex variable with the sign of its imaginary comments multiplied by –1. Therefore, the complex conjugate of Eq. (1.61) is merely

$$z_1^* = x_1 - \mathrm{j}y_1 \tag{1.74}$$

where the * denotes that the complex conjugation of the variable has been performed. The complex conjugate comes in handy since the product of a complex variable and its conjugate is the magnitude of the complex number squared, or

$$z_1 z_1^* = x_1^2 + y_1^2 = |z_1 z_1| = |z_1^2| \tag{1.75}$$

The division of a complex variable by another complex variable can be achieved by multiplying the numerator and denominators by the complex conjugate of the denominator, or

$$\begin{aligned} \frac{z_2}{z_1} &= \frac{x_2 + \mathrm{j}y_2}{x_1 + \mathrm{j}y_1} = \frac{x_2 + \mathrm{j}y_2}{x_1 + \mathrm{j}y_1} \cdot \frac{x_1 - \mathrm{j}y_1}{x_1 - \mathrm{j}y_1} \\ &= \frac{(x_1 x_2 + y_1 y_2) + \mathrm{j}x_1 y_2 - x_2 y_1}{x_1^2 + y_1^2} \end{aligned} \tag{1.76}$$

In the polar representation, a division of two complex numbers can be expressed as

$$\frac{z_2}{z_1} = \frac{|z_2|}{|z_1|} e^{j(\phi_2 - \phi_1)} \tag{1.77}$$

1.3.3 Wave Equation in Different Coordinate Systems

If **A** is a vector of magnitude A, its unit vector is a vector having a magnitude of 1 but pointing in the same direction as the vector **A** and will be $\mathbf{A}/A$. In a two-dimensional rectangular coordinate system as shown in Fig. 1.5, an important set of unit vectors are those having the same directions as the positive x and y and are denoted as **i** and **j**, respectively. For a three-dimensional rectangular coordinate system as shown in Fig. 1.7,a third axis, the z axis, is added and it has a unit vector **k**. The dot product of two vectors **A** and **B** is denoted as $\mathbf{A}\cdot\mathbf{B}$ defined as the product of the magnitude of **A** and **B** and the cosine of the angle between them. Therefore, since the unit vectors **i**, **j** and **k** are orthogonal to each other the dot products between them can be expressed as

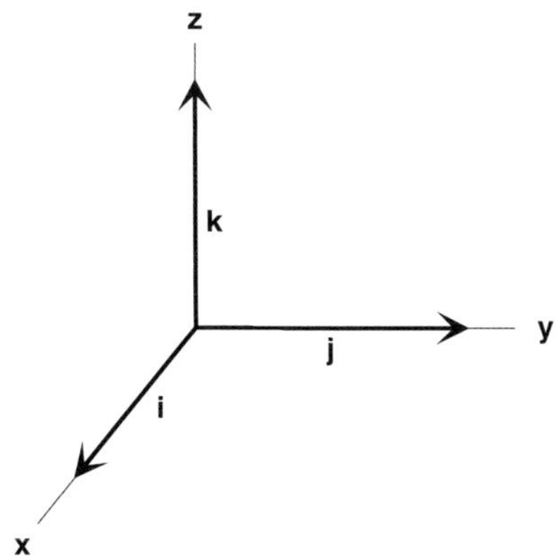

FIGURE 1.7. Rectangular coordinate system showing the unit vectors **i**, **j** and **k**. Unit vector of **k**.

$$\begin{aligned} &\mathbf{i}\cdot\mathbf{i}=1 \quad \mathbf{i}\cdot\mathbf{j}=0 \quad \mathbf{i}\cdot\mathbf{k}=0 \\ &\mathbf{j}\cdot\mathbf{i}=0 \quad \mathbf{j}\cdot\mathbf{j}=1 \quad \mathbf{j}\cdot\mathbf{k}=0 \\ &\mathbf{k}\cdot\mathbf{i}=0 \quad \mathbf{k}\cdot\mathbf{j}=0 \quad \mathbf{k}\cdot\mathbf{k}=1 \end{aligned} \tag{1.78}$$

The dot product of the unit vector with itself is equal to 1 since $\cos 0^\circ = 1$, and the dot product of a unit vector with any other unit vector except itself is equal to 0 since $\cos 90^\circ = 0$.

The divergence operator ∇ (del) in the rectangular coordinate system is written as

$$\nabla = \mathbf{i}\frac{\partial}{\partial x} + \mathbf{j}\frac{\partial}{\partial y} + \mathbf{k}\frac{\partial}{\partial z} \tag{1.79}$$

so the Laplacian operator, which is $\nabla^2 = \nabla \cdot \nabla$ can be written as

$$\nabla^2 = \nabla \cdot \nabla = \frac{\partial^2}{\partial x^2} + \frac{\partial^2}{\partial y^2} + \frac{\partial^2}{\partial z^2} \tag{1.80}$$

Using ∇^2, the wave equation in three-dimensional volume of fluid can be expressed as

$$\nabla^2 p = \frac{1}{c^2}\frac{\partial p}{\partial t^2} \tag{1.81}$$

The acoustic pressure should have been expressed as $p(x,y,z,t)$ showing it to be a function of x, y, z and t, and for simplicity sake the parenthesis was excluded. If a rectangular coordinate system is used, then the Laplacian operator of Eq. (1.80) should be used in Eq. (1.81).

The wave equation can be written for different coordinate systems, two of the most common besides the rectangular coordinate system are the cylindrical and the spherical coordinate systems. In the cylindrical coordinate

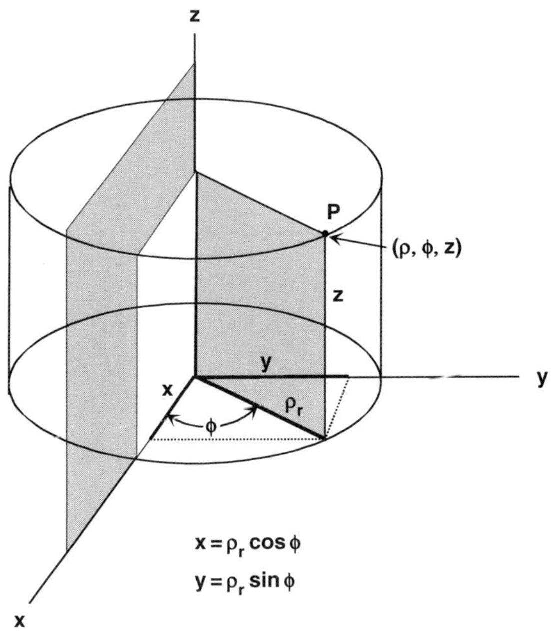

FIGURE 1.8. Cylindrical coordinate system.

system shown in Fig. 1.8, the location of a point is defined by the coordinates ρ_r, θ, and z, where

$$\rho_r = \sqrt{x^2 + y^2} \qquad \phi = \tan^{-1}\left(\frac{y}{x}\right) \tag{1.82}$$

The wave equation in the cylindrical coordinate system is

$$\frac{\partial^2 p}{\partial \rho_r^2} + \frac{1}{\rho_r}\frac{\partial p}{\partial \rho_r} + \frac{1}{\rho_r^2}\frac{\partial^2 p}{\partial \theta^2}\frac{\partial^2 p}{\partial z^2} = \frac{1}{c^2}\frac{\partial^2 p}{\partial t^2} \tag{1.83}$$

Let us now consider the spherical coordinate system shown in Fig. 1.9. A point in the spherical coordinate system will have coordinates described by r, θ, ϕ, where

$$x = r\cos\theta\, s\,in\phi \qquad y = r\sin\theta\,\sin\phi \qquad z = r\cos\phi \tag{1.84}$$

Without going into the details, the wave equation can be expressed in the spherical coordinate system as

$$\frac{\partial^2 p}{\partial r^2} + \frac{2}{r}\frac{\partial p}{\partial r} + \frac{1}{r^2}\frac{\partial^2 p}{\partial \phi^2} + \frac{\cot\phi}{r^2}\frac{\partial p}{\partial \phi} + \frac{1}{\sin^2\phi}\frac{\partial^2 p}{\partial \theta^2} = \frac{1}{c^2}\frac{\partial^2 p}{\partial t^2} \tag{1.85}$$

Equation 1.85 can be rewritten as

$$\frac{1}{r^2}\left[\frac{\partial}{\partial r}\left(r^2\frac{\partial p}{\partial r}\right) + \frac{1}{\sin\phi}\frac{\partial}{\partial \phi}\left(\sin\phi\frac{\partial p}{\partial \phi}\right) + \frac{1}{\sin^2\phi}\frac{\partial^2 p}{\partial \theta^2}\right] = \frac{1}{c^2}\frac{\partial^2 p}{\partial t^2} \tag{1.86}$$

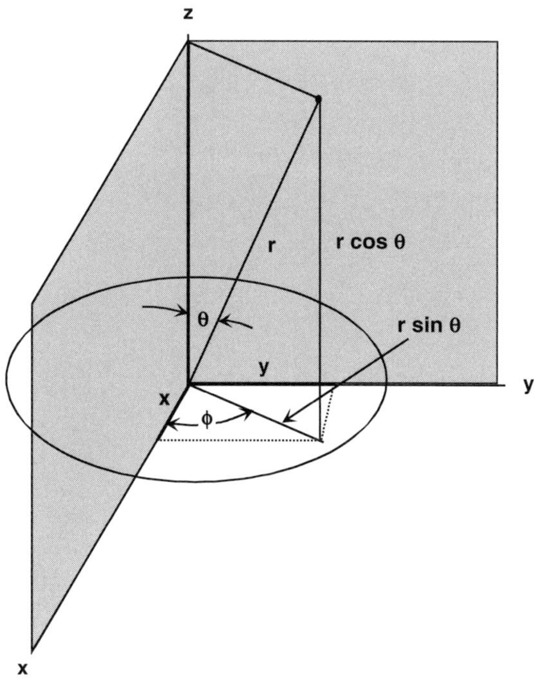

FIGURE 1.9. Spherical coordinate system.

References

Jensen, F. B., Kuperman, W. A., Porter, M. B., and Schmidt, H. (1994). *Computational Ocean Acoustics*, Springer-Verlag, N.Y.

Medwin, H. and Clay, C. S. (1998). *Fundamentals of Acoustical Oceanography* (Academic Press, San Diego).

National Defense Research Committee Volume 8. (1946). *Physics of Sound in the Sea*, Summary Technical Report of Division 6, Navy Public. & Forms Center.

Officer, C. B. (1958). *Introduction to the Theory of Sound Transmission with Application to the Ocean* (McGraw Hill, New York).

Richardson, W. J., Greene Jr., C.R., Malme, C.I., and Thomson, D. H. (1995). *Effects of Noise on Marine Mammals* (Academic Press, New York).

Robinson A. R. and D. Lee (Editors). (1994). *Oceanography and Acoustics* (American Institute of Physics, Woodbury, New York).

Rossing, T. D. (1990). *The Science of Sound (2nd Ed)* (Addison-Wesley, Reading, MA).

Supin, A. Ya, Popov, V. V., and Mass, A. M. (2001). *The sensory Physiology of Aquatic Mammals* (Kluwer Academic Publishing, Boston).

Webb, J. F., Fay, R. R., and Popper, A. N. (Eds). (2008). *Fish Bioacoustics* (Springer-Verlag, NY).

2

Measurement and Generation of Underwater Sounds

2.1 Electroacoustic Transducers

Acoustic signals are detected and generated with devices called electroacoustic transducers. These devices convert electrical energy into acoustic energy or acoustic energy into electrical energy. Transducer that converts acoustic energy into electrical energy has traditionally been called hydrophones or receiving transducers. Likewise, devices that convert electrical energy into acoustic energy are referred to as projectors or transmitting transducers. Many electroacoustic devices are used for both receiving and transmitting acoustic signals and are referred to as transducers.

In many instances, a bioacoustician has to construct his/her own transducer because of a special application requiring an unusual design or size, or because of budgetary constraints. The goal of this chapter is to provide sufficient information so you could, if necessary, construct transducers with a reasonable certainty of being successful, but without the need to become a transducer "expert." It is not our goal to provide an in-depth analysis of transducer material and design, which would be beyond the scope of this book. Those interested in a deeper discussion of underwater acoustic transducer material and design should refer to books by Wilson (1988), Bobber (1970), Huerter and Bolt (1955) and Cady (1964).

Most transducers convert energy from one form into another by using (a) materials with *electrostriction* on *piezoelectric* properties, (b) materials with *magnetostriction* properties, and (c) electrodynamics principles. Piezoelectric properties are found in some crystalline substances such as quartz, ammonium dihydrogen phosphate (ADP), and Rocelle salt. These crystals acquire an electrical charge between certain crystal surfaces when placed under pressure, or they acquire a stress when a voltage is placed across them. Electrostrictive materials exhibit the same effect as piezoelectric crystals, but are polycrystalline ceramics that have to be properly polarized by subjecting them to a high electrostatic field while in a melted form under high temperature and cured into a solid with the electrostatic field still being impressed. Popular electrostrictive materials are barium titanate and lead zirconate titanate, and they are commonly referred to as piezoelectric

W.W.L. Au, M.C. Hastings, *Principles of Marine Bioacoustics*,

DOI: 10.1007/978-0-387-78365-9_2, © Springer Science+Business Media, LLC 2008

ceramics. Magnetostrictive materials change dimensions, producing stresses when placed in a magnetic field; and conversely, elastic strain changes the flux density and induces an electromotive force in a conductor that surrounds the material. Electrodynamic transducers operate like in-air speakers, but are adapters for operation under water. They are particularly effective for the generation of relatively low-frequency sounds.

Although many different materials have been the subject of research and development for use in underwater transducers, most of the modern devices are constructed out of piezoelectric ceramics, either barium titanate (BaTi) or lead zirconate titanate (PZT). Over the past two decades, the use of lead zirconate titanate has far outstripped that of barium titanate. One distinct advantage of piezoelectric ceramics over piezoelectric crystals and magnetostrictive materials is the relative ease at which they can be molded into desirable shapes such as hemispheres, cylinders, and plates. Still easier to shape are the piezoelectric plastics and rubbers, which will be discussed later in this section. These materials are becoming increasingly popular as individuals become accustomed to making and using them for the reception of acoustic signals.

2.2 Sensitivity and Frequency Response of Piezoelectric Elements

Underwater acoustics can be used for many different purposes such as navigation, communication, target location, intruder detection, and even the monitoring of marine life, requiring the availability of many different types of acoustic sensors and sound generator devices. The major differences between a general measurement transducer and other special purpose transducers are the sensitivity and frequency-response characteristics. High sensitivity over a very wide band of frequency is desired in general purpose measurement transducers, whereas special purpose transducers often operate over a very narrow band of frequency. Some general purpose hydrophones have good sensitivity over a four-decade range of frequency.

Although most piezoelectric ceramic are isotropic and not piezoelectric prior to poling (polarization), after poling they become anisotropic; their electromechanical properties differ for electrical or mechanical excitation along different axes of a crystallographic coordinate system. There are also coupling or interaction between properties along the different axes so that a system of nine linear partial differential equations is necessary to describe the piezoelectric action in the presence of a sound field. We will instead take a simplified approach to the discussion of piezoelectric ceramics, but still attempt to include the most important issues involved in the design, construction, and use of piezoelectric transducers.

The sound pressure level (SPL) measured by a hydrophone can be determined from a knowledge of the hydrophone free field sensitivity, the gain in the measurement system and the amount of voltage measured. Let M_h be the free field voltage sensitivity of a hydrophone in dB re 1 V/μPa, and assume that the hydrophone is connected to an amplifier of gain G in dB, then the SPL in dB re 1 μPa is given by the equation

$$\mathrm{SPL} = |M_h| - \mathrm{G} + 20 \log V, \tag{2.1}$$

where V is the voltage read off an oscilloscope or voltmeter. It is conventional to measure the rms voltage so that SPL will be the rms SPL. However, there are times when it is considerably more convenient to measure either peak or peak-to-peak voltages. In these cases, SPL will be either the peak or peak-to-peak SPL in dB re 1 μPa. The input impedance of an amplifier should be much higher than the impedance of the hydrophone so that "loading" effects will not affect the measurements. In most broadband hydrophones, the impedance decreases with frequency. If acoustic measurement of a sound source was conducted with a hydrophone located at a range of R from the source and the source level (SL), defined as the SPL at a reference range of 1 m, is desired, the transmission loss must be taken into account. If the hydrophone is not very far from the source (less than several tens of meters at most), we can assume spherical spreading loss so that using Eq. (1.51) the source level can be written as

$$\mathrm{SL} = \mathrm{SPL} + 20 \log R + \alpha R, \tag{2.2}$$

where α is the sound absorption coefficient in dB/m, and will be discussed later in Chapter 3.

Sounds can be produced in water by driving a transducer with a signal generator and a power amplifier. The transmit sensitivity of a transducer is generally given in terms of its source level per volt or per ampere of drive. Let S_v be the voltage transmit sensitivity in dB re 1 μPa/volt and S_i be the current transmit sensitivity in dB re 1 μPa/amp; the source level in dB re 1 μPa will be

$$\mathrm{SL} = S_v + 20 \log V_{in}, \tag{2.3a}$$

$$\mathrm{SL} = S_i + 20 \log I_{in}, \tag{2.3b}$$

where V_{in} is the voltage and I_{in} is the current at the input of the transducer cable. The SPL at any distance from the transducer can be calculated by subtracting the transmission loss from Eq. (2.3). The use of the nomenclature M for the hydrophone (receive) sensitivity and S for the projector sensitivity is a holdover from airborne acoustics where M stands for microphone and S for speaker. Throughout this chapter for convenience sake, M and S will be used interchangeably on a dB and linear scale.

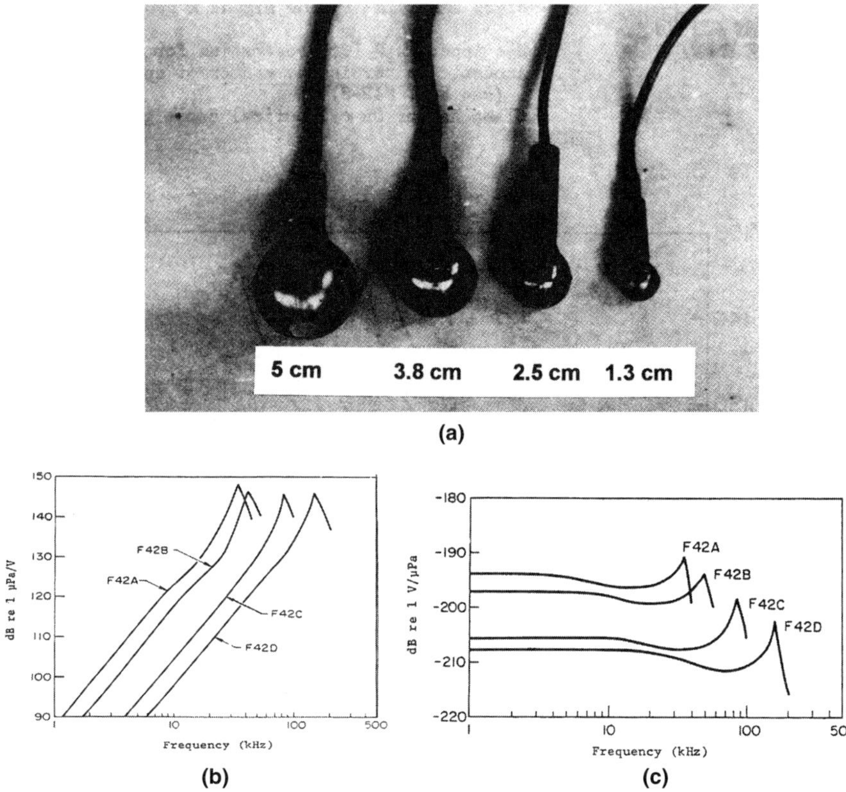

FIGURE 2.1. **(a)** Photograph of the USRD family of F42 spherical hydrophones; **(b)** the transmit response curves; **(c)** the receive response curves.

A family of F-42 spherical hydrophones of different diameters (standard hydrophones from the U. S. Navy's Underwater Sound Reference Detachment [USRD]) is shown in Fig. 2.1a, with the transmit and receive sensitivity curves as a function of frequency shown in Fig. 2.1b,c, respectively. The diameter of the elements are 5.00, 3.81, 2.54, and 1.27 cm for the model A, B, C, and D, respectively. The sensitivity curves indicate that the larger the diameter the more sensitive the response, but with a smaller frequency range. The positive peak in each response curve occurs at the resonance frequency of the different elements. Resonance has to do with the vibrational characteristic of a structure. Any structure will have certain frequencies at which larger vibrations can be induced with a small driving function having minimal energy. These characteristic frequencies are called resonance frequencies, and their exact values will depend on the shape, size, and material composition of the structure. Resonance in transducer elements will be treated in more detail in a later section of this chapter. The receiving response of a spherical hydrophone is relatively flat for frequencies below the resonance frequency, peaks at the resonance frequency and then drops off rapidly for frequencies above the resonance frequency.

2.2.1 Equivalent Circuit and Resonance

The shape of the receiving sensitivity curves shown in Fig. 2.1 can be explained by considering a simple equivalent circuit for a piezoelectric transducer. In Section 1.2.4, we used an analogy between an electrical circuit and a plane acoustic wave propagating in a fluid. Acoustic pressure was said to be analogous to voltage, and the velocity of a particle under the influence of an acoustic pressure wave was analogous to current, and in this manner derived the specific acoustic impedance expressed by Eq. (1.40). The impedance of an electrical circuit is the ratio of voltage and current and can be characterize by its resistance, inductance, and capacitance. In an analogous manner, we can also characterize a piezoelectric element by its mechanical impedance, defining this impedance as the ratio of the force on and the velocity of the element. We can define a mechanical impedance for the piezoelectric element as

$$Z_{\mathrm{m}} = R_{\mathrm{m}} + j\left[\omega L_{\mathrm{m}} - \frac{1}{\omega C_{\mathrm{m}}}\right] = R_{\mathrm{m}} + j\left[\omega m - \frac{s}{\omega}\right], \tag{2.4}$$

where R_{m} is the mechanical resistance associated with mechanical losses in the element, $L_{\mathrm{m}} = m$ is the effective mass of the vibrating element and is analogous to electrical inductance, $C_{\mathrm{m}} = 1/s$ is the mechanical compliance or stiffness of the material and is analogous to the electrical capacitance, and ω is the radian frequency defined as $\omega = 2\pi f$. The mechanical impedance is a complex quantity with a real and imaginary part, and is referred to as the mechanical resistance and reactance, respectively. Its unit is not ohms, since the quantity is a ratio between force and velocity rather than voltage and current. If a sinusoidal force (either electrical or acoustics) is applied to the element, the element will vibrate at a velocity of

$$u = \frac{F e^{j\omega t}}{R_{\mathrm{m}} + j\left[\omega L_{\mathrm{m}} - \frac{1}{\omega C_{\mathrm{m}}}\right]} = \frac{F e^{j\omega t}}{R_{\mathrm{m}} + j\left[\omega m - \frac{s}{\omega}\right]}, \tag{2.5}$$

where F is the magnitude of the sinusoidal force. There will be a frequency, called the resonance frequency, at which $\omega L_{\mathrm{m}} = 1/\omega C_{\mathrm{m}}$ ($\omega m = s/\omega$). At this frequency, the impedance is real and has its minimum value so that the velocity (Eq. (2.5)) is at its maximum. The relative velocity amplitude as a function of frequency is shown in Fig. 2.2, where f_0 is the resonance frequency and Δf is the 3-dB bandwidth. Another important parameter used to describe resonance is the Q of the element, which is defined as

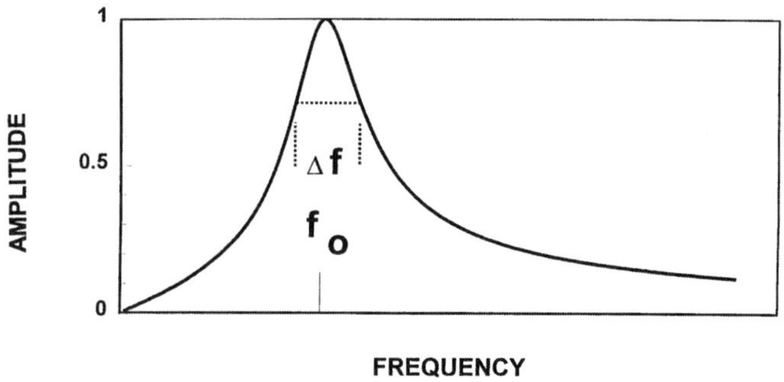

FIGURE 2.2. An example of a resonance response curve.

$$Q = \frac{f_0}{\Delta f}. \tag{2.6}$$

The parameter Q is also used to describe the relative bandwidth of resonance in electrical circuits. The bandwidth of the resonance curve is inversely proportional to the mechanical resistance of the element. If R_m is small, the curve falls off very rapidly and the resonance is sharp. If R_m is large, the curve falls off more slowly and the resonance is broad.

A piezoelectric transducer in an acoustic field can be thought of as having an *electrical component* and a *mechanical component*. An acoustic wave will apply a stress on the piezoelectric element, which in turn will cause the element to be strained. The strain on the element will cause an electrical voltage to appear across the electrodes of the transducer. If we let ϕ be the electromechanical voltage to force conversion ratio (a given force on the hydrophone element will generate an electrical voltage across the electrodes), we can depict a piezoelectric transducer by the simple equivalent circuit shown in Fig. 2.3a, where C_e is the capacitance across the terminal of the transducer. The sensitivity of a receiving transducer (M_h) is merely the ratio of the acoustic pressure and the open-circuit voltage generated across the electrodes ($M_h = V_{0c}/p$). For the equivalent circuit of Fig. 2.3, the receiving sensitivity can be expressed as

$$M_h = \frac{V_{0c}}{p} = \frac{\varphi A}{j\omega C_e \varphi^2 \left(R_m + j\left[\omega m - \frac{s}{\omega}\right]\right) + 1}. \tag{2.7}$$

At all frequencies except perhaps near resonance, $j\omega C_e \phi^2 Z_m >> 1$ (Bobber, 1970), so that the receiving sensitivity is nearly inversely proportional to ωZ_m. At very low frequencies, $s/j\omega >> [R_m + j\omega m]$ so that

$$M_h \approx \frac{A}{\phi C_e s} = \frac{A}{\phi C_e / C_m}. \tag{2.8a}$$

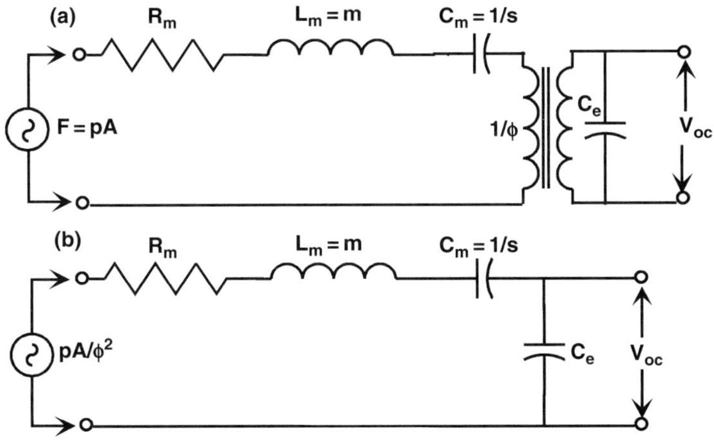

FIGURE 2.3. (**a**) A simple equivalent circuit for a piezoelectric element used as a hydrophone; (**b**) same equivalent circuit with the transformation constant reflected back to the source.

A generalized receiving response piezoelectric element as a function of frequency is shown in Fig. 2.4, relating the various areas of the response curve to the relative values of the components making up the mechanical impedance. Above the resonance frequency, $j\omega m >> (R_m + 1/j\omega C_m)$ (Bobber, 1970), so that the sensitivity falls off rapidly with frequency with a slope that is inversely proportional to the square of the frequency as can be seen in Eq. (2.8b). The generalized

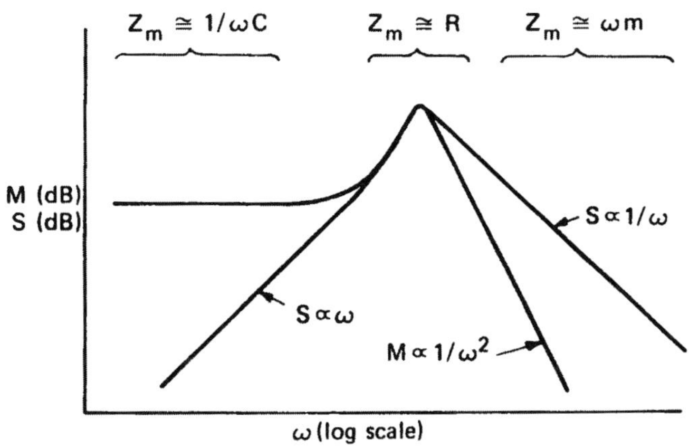

FIGURE 2.4. Generalized response curves for a piezoelectric transducer (adapted from Bobber, 1970).

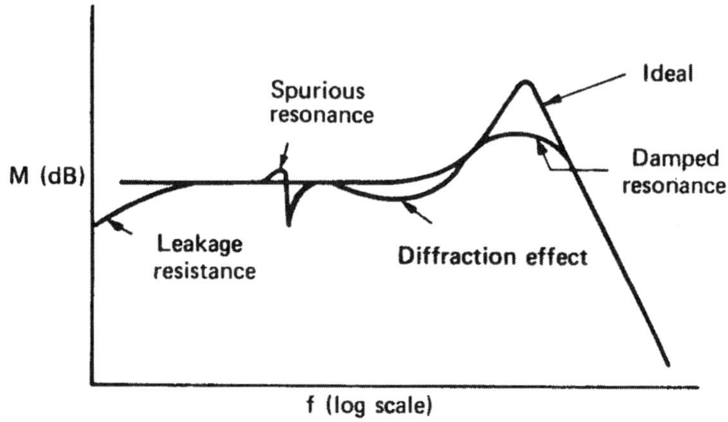

FIGURE 2.5. Typical effects of resonant damping, diffraction, spurious resonances and leakage.

$$M_\mathrm{h} \approx \frac{-\mathrm{A}}{\omega^2 C_\mathrm{e} \varphi m} = \frac{-\mathrm{A}}{\omega^2 C_\mathrm{e} \varphi L_\mathrm{m}} \tag{2.8b}$$

transducer response curves of Fig. 2.4 pertain to an ideal situation. Real calibration curves seldom look like the idealized curves, as can be seen by comparing the curves of Figs. 2.1 and 2.4. The resonant peak is damped and spurious resonances may appear as shown in Fig. 2.5. At very low frequency, the capacitive reactance of the element will become larger than the resistivity of, or the leakage resistance around, the element causing the response to roll off at a 6-dB per octave rate as the frequency decreases. The diffraction effect has to do with the presence of the transducer disturbing the acoustic field, causing perturbation of the field as the wave propagates around the transducer.

Piezoelectric elements are reciprocal devices, which mean that they can be used for both receiving and transmitting acoustic signals. Sounds are produced by driving the element with a signal generator connected across the electrodes of the element. The electrical voltage will cause the element to vibrate and couple acoustic energy into the water.

We will now consider the transmit sensitivity response of a piezoelectric element. The transmit sensitivity curves shown for the F-42 family of transducers in Fig. 2.1b are typical of the piezoelectric elements. There is a parameter called the reciprocity parameter and is usually designated as J, which relates the receive sensitivity to the current transmit sensitivity (S_i) of a piezoelectric element, and is defined as

$$J = \frac{M_\mathrm{h}}{S_\mathrm{i}}. \tag{2.9}$$

For this discussion, J can be expressed as (Bobber, 1970)

$$J = \frac{b}{f}, \tag{2.10}$$

where f is the frequency and b is a constant. Inserting Eqs. (2.9) and (2.10) into Eq. (2.7), we get

$$S_{\mathrm{i}} = \frac{\varphi A\omega/b'}{j\omega C_{\mathrm{e}}\varphi^2(R_{\mathrm{m}} + j[\omega m - s/\omega]) + 1}, \tag{2.11}$$

where $b' = 2\pi b$. At all frequencies expect perhaps near resonance, $j\omega C_{\mathrm{e}}\phi^2 Z_{\mathrm{m}} >> 1$ (Bobber, 1970), and at very low frequencies, $s/j\omega >> [R_{\mathrm{m}} + j\omega m]$. Therefore, the mechanical impedance is dominated by the stiffness of the material so that the Eq. (2.11) can be reduced to

$$S_{\mathrm{i}} \approx \frac{A\omega}{b' C_{\mathrm{e}}\varphi s}. \tag{2.12}$$

The generalized transmission response curve for a piezoelectric transducer is depicted by the curve denoted as S in Fig. 2.4. At all frequencies except perhaps near resonance, $j\omega C_{\mathrm{e}}\phi^2 Z_{\mathrm{m}} >> 1$ (Bobber, 1970), and at very low frequencies, $1/j\omega C_{\mathrm{m}} >> [R_{\mathrm{m}} + j\omega m]$. Therefore, below resonance, the current transmit sensitivity varies linearly with frequency until the resonance frequency is approached. Above the resonance frequency, $j\omega m >> (R_{\mathrm{m}} + 1/j\omega C_{\mathrm{m}})$ (Bobber, 1970), so that the sensitivity falls off rapidly with frequency with a slope that is inversely proportional to the frequency. From Eqs. (2.3a,b), the voltage transmit sensitivity, $S_{\mathrm{v}} = I_{\mathrm{in}}/V_{\mathrm{in}}\ S_{\mathrm{i}} = S_{\mathrm{i}}/Z_{\mathrm{in}}$. At low frequencies below resonance, Z_{in} decreases linearly with frequency so that S_{v} will be a function of ω^2 or f^2, as can be seen in the voltage transmit sensitivity curves of the spherical transducers of Fig. 2.1.

2.2.2 Resonance of Different Shaped Elements

It is important to have some understanding of the resonance of transducer elements because of the large role resonance plays in the response characteristics of transducer elements. The specific resonant frequency of a transducer element will depend on its shape, size, and material composition. If we are interested in projecting high-amplitude signals into the water, it would be best to operate close to the resonance frequency of the element. Conversely, if we are interested in receiving signals over a wide frequency band with the response of the hydrophone being relatively flat over the frequency band, then we would want to operate as far below the resonant frequency of the element.

A free-standing element of a given shape will vibrate in many different modes, and a resonant frequency will be reached whenever the wavelength of the signal in the element is a integer number of half a wavelength. Therefore, the size and shape of the element along with the speed of sound of the material associated with different modes are important in determining its resonant frequencies. Figure 2.6 shows the modes of some simple element shapes along with the fundamental resonant frequency for the specific mode of vibration. The parameter N_{n} is related to the sound velocity of the material and is referred to as the frequency constant. This parameter is usually provided by the piezoelectric ceramic manufacturer and different for different material compositions and modes of vibration. Some typical values for the barium titanate and lead titanate zirconate compositions are given in Table 2.1. Some coupling is always present between the various modes of vibration depicted in Fig. 2.6, and the upper frequency limit of the relatively flat response portion of a receiving hydrophone

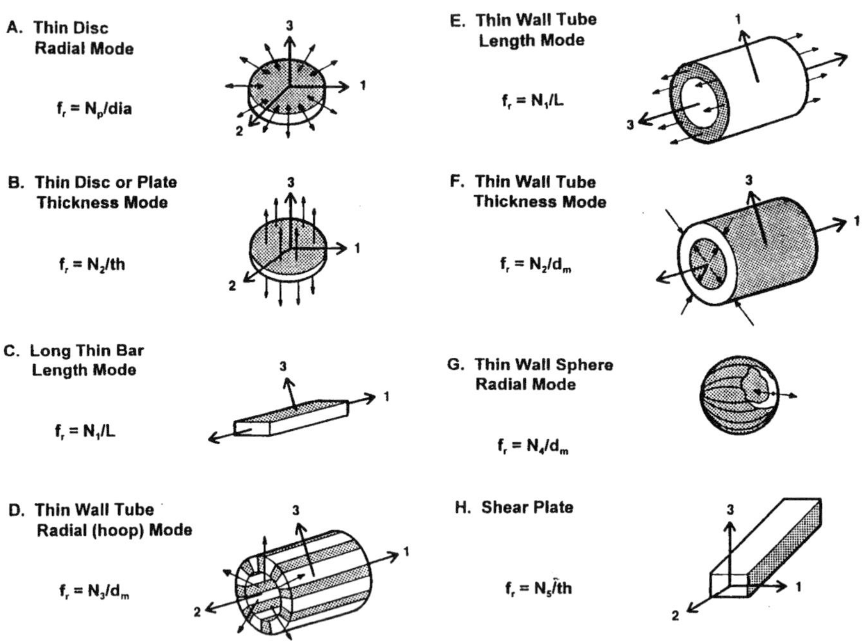

FIGURE 2.6. Vibration modes of piezoelectric elements. The direction of polarization is along the **3**-axis usually coinciding with the *z*-axis and the **1** and **2**-axes coincide with the *x* and *y*-axes of a Cartesian coordinate system, respectively (with permission of Edo-Western Corp.).

TABLE 2.1. Typical Values of Frequency Constant for Barium Titanate and Lead Zirconate Titanate

Parameter	Vibrational mode	Freq. const. (Hz/m) barium titanate	Freq, const. (Hz/m) lead zirconate titanate
N_1	Length mode	1321–2324	1359–2337
N_2	Thickness mode	1727–2845	1765–2858
N_3	Circumferential mode	851–1461	876–1473
N_4	Radial mode	1435–2438	1473–2464
N_5	Shear mode	1067–1575	1092–1600
N_p	Planar mode	1943–3150	1961–3188

sensitivity curve is somewhat below the frequency of the lowest natural mode of vibration of the element.

2.2.3 Measurement of Resonant Frequency

The lowest resonant mode of a piezoelectric element can be easily measured using the circuit shown in Fig. 2.7 and performing the measurement in air. The oscillator should generate a constant voltage, and in the ideal case, $R << Z_{res}$, where Z_{res} is the impedance of the element at resonant. The variation of impedance as a function of frequency showing the effects of resonance and anti-resonance is shown in Fig. 2.7b. The piezoelectric material behaves capacitively below f_r and above f_a, and between the two frequencies it behaves inductively. The phase angle of the element also undergoes a sign change at both the resonance frequencies, and can be used to measure resonant frequencies. While the ideal case is easy to realize in measuring either resonant or anti-resonant conditions, it is more difficult to accurately measure both with a single R. The high Q of the element in air makes it difficult to find a value of R that results in good signal-to-noise ratio anti-resonance and yet meet the condition $R << Z_{res}$ at resonance.

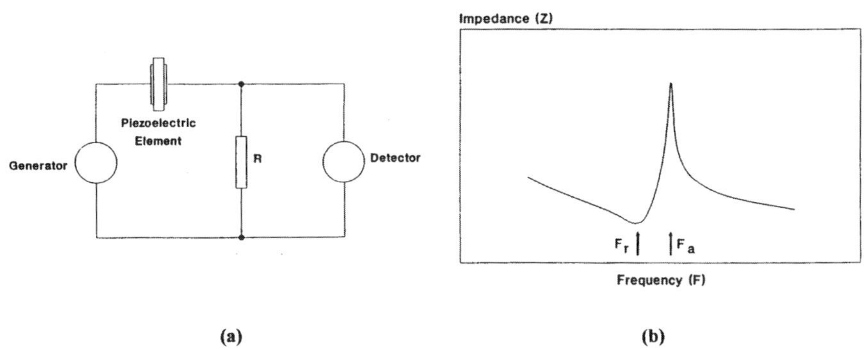

FIGURE 2.7. **(a)** Constant voltage generator circuit for measuring resonant frequency; **(b)** variation of impedance with frequency.

2.3 Hydrophone Sensitivity Using Piezoelectric Parameters

A piezoelectric ceramic is a complex electromechanical device that requires many parameters, such as the dielectric constant, loss tangent, piezoelectric charge coefficients, piezoelectric coupling coefficients, elastic constants to adequately describe its electromechanical properties. Since a piezoelectric ceramic is anisotropic, some of the parameters will have different values along the different axes of the crystallographic coordinate system. It would be far beyond the scope of this book to attempt an in-depth examination of the properties of piezoelectric material. Fortunately, the receiving sensitivity of some simple hydrophone geometries can be calculated from the piezoelectric voltage coefficient usually supplied by ceramic manufacturers. This coefficient is normally denoted as $\mathbf{g_{ij}}$, where the **i** subscript refers to the direction in which the electrode is perpendicular to and **j** is the direction of the applied stress caused by the acoustic wave. The **i** and **j** subscripts can have values of 1, 2 or 3 corresponding to the x, y, and z-axis, respectively (see Fig. 2.6).

We will consider some simple cases, assuming that the piezoelectric element has dimensions that are much smaller than the wavelength of the sound in a fluid, so that the acoustic pressure is essentially uniform over the dimension of the element. Usually, the frequency of operation of a hydrophone receiver is well below the frequency of mechanical resonance, so that wave propagation in the ceramic can be neglected. For all practical purposes, the receiving sensitivity we are examining can be considered as a low-frequency estimate. Consider a very small rectangular block of piezoelectric ceramic of dimensions w, l, and t, shown in Fig. 2.8a, polled along the t-direction. Since the element is small compared to a wavelength, the acoustic pressure will have the same amplitude and phase on all sides. For such a situation, the receiving sensitivity can be determined using the piezoelectric voltage coefficients, g_{31} to account for the acoustic pressure on the side faces and g_{33} to account for the pressure on the electrode face. The receiving sensitivity will be (Wilson, 1988)

$$M_0 = \frac{V_0}{p} = (2g_{31} + g_{33})t. \qquad (2.13)$$

Now let us consider the case in which all the non-electrode faces of the piezoelectric block are isolated by covering these faces with a acoustically reflective material. In this case, the acoustic pressure acts only on the electrode faces (see Fig. 2.8b), and the receiving sensitivity will be

$$M_0 = g_{33}t. \qquad (2.14)$$

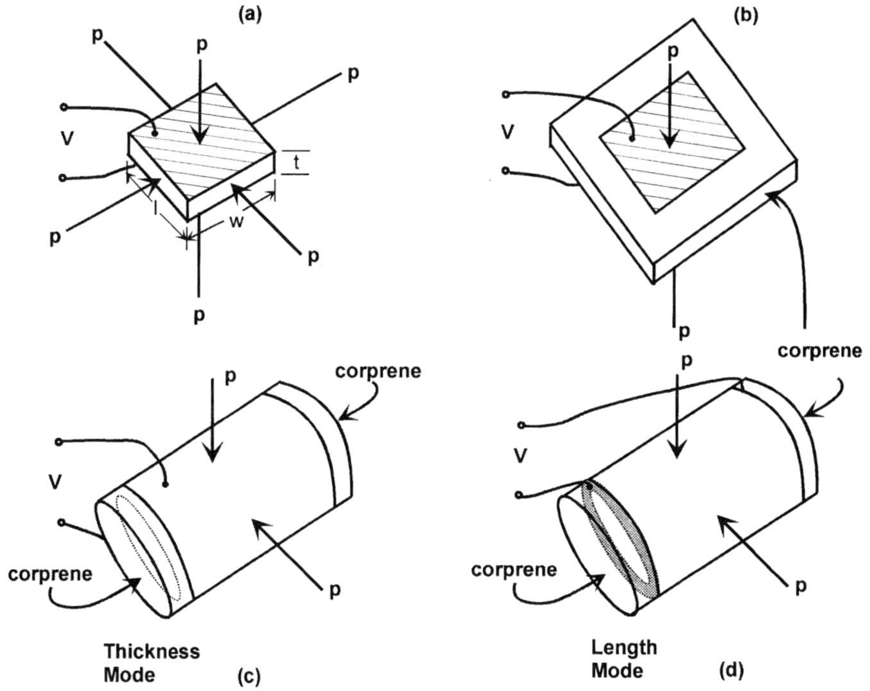

FIGURE 2.8. Simple geometry of piezoelectric elements for the calculation of hydrophone receiving sensitivity.

Equations (2.13) and (2.14) will also apply to a circular disk element, similar to that shown in Figs. 2.6a and 2.6b. The hydrophone depicted in Fig. 2.8b will be more sensitive than the one depicted in Fig. 2.8a, because g_{31} will have a negative value. For example, typical values of the piezoelectric voltage coefficients for a Navy Type 1 PZT material are $g_{31} = -10.5 \times 10^{-3}$ (volt-m/Newton) and $g_{33} = 24.5 \times 10^{-3}$ (volt-m/Newton). Therefore, for the element in Fig. 2.8a, $M_0 = 3.5 \times 10^{-3} t$ (volt-m/Newton) and for Fig. 2.8b, $M_0 = 24.5 \times 10^{-3}\ t$ (volt-m/Newton).

We will now consider hollow cylindrical piezoelectric elements with an inner radius of a and an outer radius of b. For a cylinder that is radially polarized as shown in Fig. 2.8c, and with the ends and interior surfaces acoustically isolated, the receiving sensitivity can be estimated with the equation (Wilson, 1988)

$$M_0 = b\left(g_{31} + g_{33}\left[\frac{1-\rho}{1+\rho}\right]\right), \tag{2.15}$$

where $\rho = a/b$. Since the piezoelectric voltage coefficients g_{31} and g_{33} have opposite signs, the sensitivity will depend on the ratio of both a/b and the piezoelectric voltage coefficients. The opposite signs of the g coefficients will cause the sensitivity to be equal to zero for certain values of ρ. Setting the expression within the parenthesis in Eq. (2.14) to zero, we get

$$\frac{1+\rho}{1-\rho} = -\frac{g_{33}}{g_{31}}. \tag{2.16}$$

Solving for ρ, we obtain values of ρ that will make the sensitivity equal to zero,

$$\rho = -\frac{1+(g_{33}/g_{31})}{1-(g_{33}/g_{31})}. \tag{2.17}$$

Using the g coefficients for the Navy Type I PZT material in the previous example, $g_{33}/g_{31} = -2.3$, so that the cylindrical element will have zero sensitivity when $a/b = 0.399$.

For a longitudinally polarized hollow cylinder in which the ends and interior are acoustically shielded (Fig. 2.8d), the receiving sensitivity can be estimated with the expression

$$M_0 = \frac{2\,Lg_{31}}{1-\rho^2} \tag{2.18}$$

(Wilson, 1988). This expression indicates that the sensitivity is proportional to the length L and inversely proportional to $(a/b)^2$. Therefore, the thinner the wall the greater the sensitivity.

The final case we will consider is that of a hollow spherical piezoelectric element. Again, with $p = a/b$, the receiving sensitivity can be estimated by the expression

$$M_0 = \frac{b}{2(1+\rho+\rho^2)}[g_{33}(\rho^2+\rho-2) - g_{31}(\rho^2+\rho+4)]. \tag{2.19}$$

Like the radially polarized cylinder, the sensitivity of the spherical element will depend on both the piezoelectric voltage coefficients and the ratio of inner and outer radii. The sensitivity can also be zero for a certain ratio of a/b. However, this ratio cannot be easily derived with an analytical expression, but must be solved numerically using an iterative process.

2.4 Piezoelectric Polymer Material

Thus far we have considered only ceramics and crystalline materials as electroacoustic elements. Recently, a new technology has been developed, which is slowly gaining popularity as an underwater sensing element. Kawai

(1969) discovered that there were large useful piezoelectric pyroelectric effects in the polymer polyvinylidene fluoride (PVDF). This material comes in the form of a thin film, like a sheet of plastic, which is stretched and polarized with a thin layer of electrode electroplated on each side of the sheet. After polarization, the material exhibits a piezoelectric property that is stable over time, and can be used to project and receive sounds. PVDF elements have been applied to loudspeakers, headphones, and microphones, high-frequency broadband ultrasonic transducers, thermal detector for optical imaging systems, and a variety of force, pressure, acceleration, and strain transducer.

PVDF material has some distinct advantages over piezoelectric ceramics in underwater applications. Besides being relatively inexpensive, it is tough, flexible, lightweight and available in large sheets up to several square meters. Its specific acoustic impedance of 1.5×10^6 rayls is very close to water so that it will provide only a minimum distortion to a sound field. Its thin, flexible structure makes it easy to literally take a scissor and cut out any desirable aperture shape for a transducer element. The material can also be easily laid over and made to conform to structures of different shapes. Finally, PVDF films are inherently broadband devices that can operate in the megahertz frequency range. A PVDF element, because of its low compliance and flexible nature, is not subject to the various resonance modes depicted in Fig. 2.6, and is affected mainly by the thickness mode of resonance. However, PVDF materials also have some distinct disadvantages compared to piezoelectric ceramics. The piezoelectric constants of PVDF are significantly smaller than those of PZT ceramic, and the material cannot be used at temperatures much above 80°C because of the poor thermal stability of some of its properties (Wilson, 1988).

PVDF materials are used mainly as hydrophone elements because its transmission properties are relatively poor for generating underwater sounds. For a planar element of thickness t of almost any shape, the low-frequency open-circuit sensitivity is (Moffett et al., 1986)

$$M_0 = 20 \log |\mathrm{m}_0| - 120, \tag{2.20}$$

$$m_0 = g_\mathrm{h}\boldsymbol{t}, \tag{2.21}$$

where $g_\mathrm{h} = g_{31} + g_{32} + g_{33}$ is the volume expansion piezoelectric constant. The constant of 120 in Eq. (2.20) is used to convert the reference from Pa to μPa. The unit for M_0 is dB re 1 V/μPa.

Besides being available as plain sheets, PVDF are often supplied as planar elements attached to stiffening electrodes, as shown in Fig. 2.9a,b, and as cylinders, as shown in Fig. 2.9c. The stiffening electrodes are used to minimize the in-plane or transverse (g_{31} and g_{31}) sensitivity and to maximize g_h. The elements depicted in Fig. 2.9b are connected in parallel to increase the overall

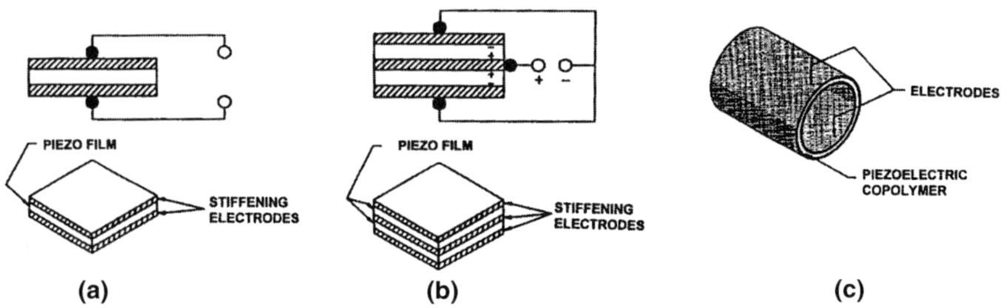

FIGURE 2.9. **(a)** PVDF material with stiffening electrodes; **(b)** PVDF material in parallel with stiffening electrodes; **(c)** PVDF material in a cylindrical form.

capacitance of the element, which in turn will give the element a greater capacity to drive cables of longer lengths. The capacitance of PVDF material is usually provided by the manufacturer in term of Farad per unit area. Its sensitivity will be the same as the single element. The sensitivity of the cylindrical element having a wall thickness of t can be calculated from Eqs. (2.20) and (2.21). Its capacitance is given by the expression

$$C = 2\varepsilon L/\log(b/a), \tag{2.22}$$

where L is the length of the cylinder, ε is the permittivity of the PVDF material, a is the inner radius and b is the other radius.

There are some synthetic rubbers that have been demonstrated to have piezoelectric properties after proper polarization. However, this technology is still in its infancy, and considerable work needs to be done before it can be used as a hydrophone. Therefore, we will not discuss this topic any further.

2.5 Transducer Configuration

In the preceding sections, we discussed transducers from a single-element point of view. However, many, if not most, transducers are configured with an array of elements, usually connected in parallel. The use of multiple elements serves three purposes: first, as a means to construct electroacoustic devices with large apertures that will have a desired directivity property (directivity will be covered in the next chapter) but still have a desired bandwidth; second, to lower the output impedance of the parallel connected element so that the array will be able to drive a longer cable than a single element; and third, to lower the output impedance of parallel connected elements so that the input impedance of a preamplifier can be kept low and therefore reduce the noise limit of a hydrophone–amplifier pair. From the equivalent circuit Fig. 2.3 in which we neglect the leakage resistance of an element, each element in a parallel-connected configuration acts basically as a charged capacitor. Therefore, a number of

FIGURE 2.10. Piezoelectric ceramic element array, USRD type F33 transducer (adapted from Bobber, 1970).

capacitors connected in parallel will cause the total capacitance of the array to be larger, so that the total capacitance will be the sum of the individual capacitance of each element. Since the electrical impedance is inversely proportional to capacitance, an increase in the output capacitance will mean a lower output impedance and the ability to drive a longer cable. An array of parallel elements will also circumvent resonance problems that would be inherent in a single, large ceramic element. The resonance frequency of an array of elements will be determined by the resonance frequency of each individual element. Therefore, an array of elements will have a higher resonance frequency (leading to greater bandwidth) than a single, large element of comparable size as the array of elements. Finally, using an array of parallel elements is often more cost-effective than trying to produce a large, single piezoelectric ceramic element. The discussion so far does not apply to PVDF since PVDF is usually made in relatively large sheets and does not tend to resonate in other than the thickness mode. However, as stated in Section 2.4, PVDF is not usually used for the projection of sounds, but is used mainly in hydrophones.

Piezoelectric elements with circular and rectangular cross-sectional area, as well as cylinders (and rings), are usually used in multi-element transducers.

An example of a circular-shaped transducer composed of both rectangular and circular piezoelectric elements is the USRD F-33 transducer shown in Fig. 2.10. The inner array is composed of nine circular PZT disks, and the outer array is constructed of 64 BaTi rectangular disks. The individual arrays can be used separately or joined together. Each of the ceramic plates are cemented to a steel backing embedded in butyl rubber. Pressure release corprene is used between the individual ceramic elements. Both sections are sealed in transparent polyurethane with castor oil, providing the coupling between the polyurethane potting material and the butyl rubber acoustic window. An example of PZT ceramic rings configured in an array is the Atlantic Research LC-32 hydrophone shown schematically in Fig. 2.11. The rings are separated by a narrow gap with pressure-release corprene occupying the inner portion of the rings. The array is encapsulated in a degassed polyurethane compound. These two examples are just part of an imaginable number of a different transducer configuration that is in existence.

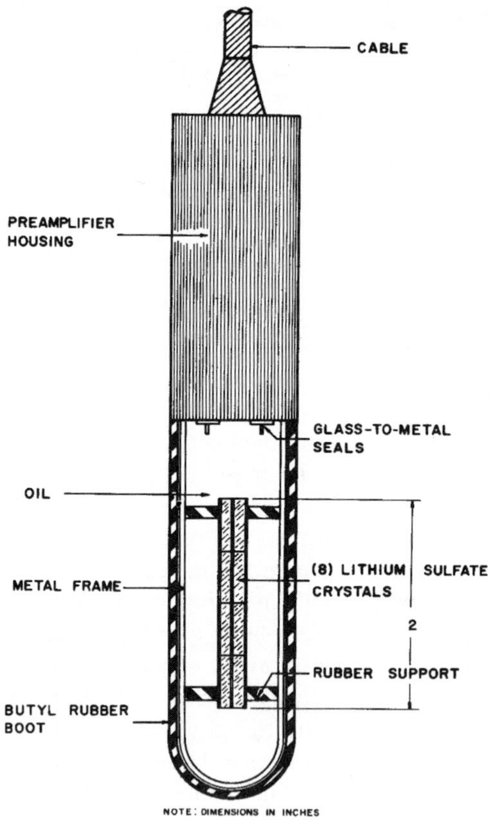

FIGURE 2.11. Schematic of an Atlantic Research Corporation type LC32 hydrophone (adapted from Bobber, 1970).

2.6 Projection of Low-Frequency Sound

Our discussion so far has been on piezoelectric transducers, and have not considered transducers operating on electrodynamic principles. Moving-coil transducers that operate like airborne speakers can be very effective in producing low-frequency acoustic signals from tens of Hertz to approximately 10–20 kHz. There is often a need to produce low-frequency sound to measure the hearing sensitivity of marine animals and to perform playback experiments using the natural sounds of whales, dolphins, and other marine mammals. Moving-coil transducers are more effective in producing low-frequency underwater sound than piezoelectric ceramics, because of the inherently high-resonance frequencies of piezoelectric elements that would require a transducer to be very large in order to achieve resonance at low frequencies.

A well-known and well-used moving-coil transducer is the J-9, developed at the U.S. Navy's Underwater Sound Reference Detachment in Orlando by

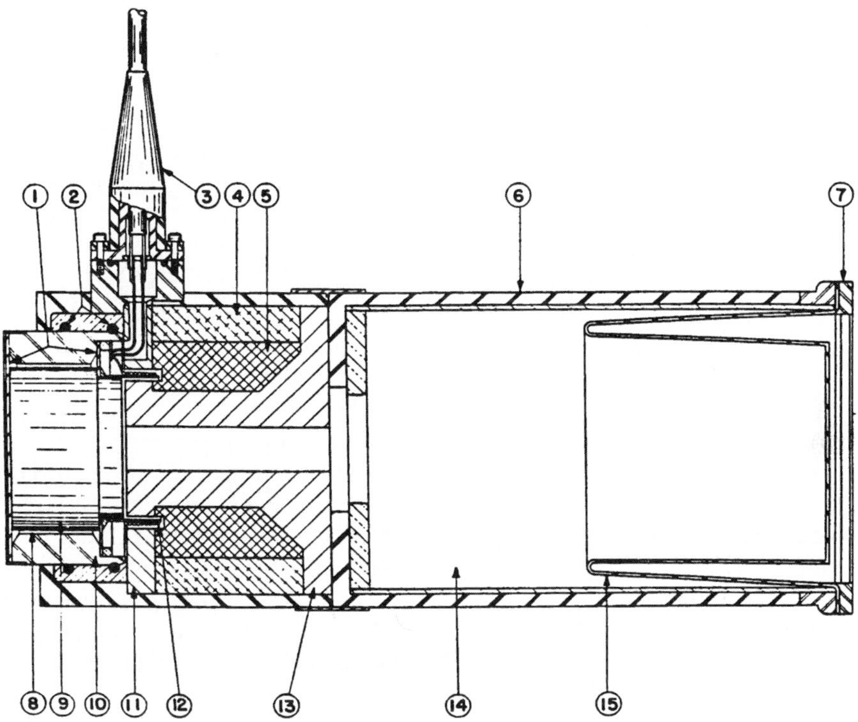

FIGURE 2.12. Assembly drawing of the USRD J9 transducer; 1, rubber seals; 2, rubber O-rings; 3, cable gland; 4, magnet; 5, lead; 6, rubber jacket; 7, grille; 8, slit filled with silicone oil response 9, magnesium diaphragm; 10, diaphragm housing; 11, front pole pieces; 12, coil; 13, back pole piece; 14, compensating air chamber; 15, rubber compensating bag. Overall length, 11 in.; diameter, 4-1/2 in. (adapted from Bobber, 1970).

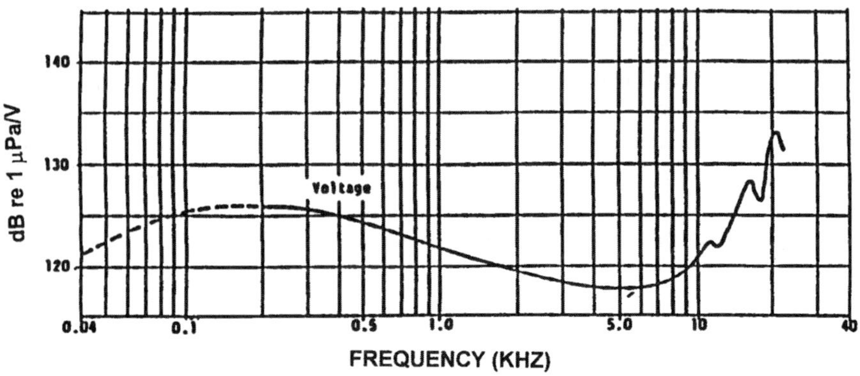

FIGURE 2.13. Typical transmitting voltage response of the J9 transducer.

C. C. Sims (1959). This transducer was designed to be operated in the infrasonic and audio frequency range from 40 to 20 kHz. A cut-away diagram of the J-9 is shown in Fig. 2.12, and the voltage transmit sensitivity is shown in Fig. 2.13. Being an electrodynamic or moving-coil transducer, it can provide the large volume displacement of water needed to produce low-frequency sound pressures without resorting to a very large radiating area that would be needed for piezoelectric ceramic transducers. When the transducer is submerged, water enters the rear compensation chamber and compresses the butyl rubber bag until the internal air pressure is equal to the external water pressure up to a depth of about 24 m. At deeper depths, a scuba bottle must be used. The compliance of the air inside the transducer is important for the highly damped resonance below 200 Hz. Changes in this compliance cause the response below 200 Hz to be a function of depth. The peak in the response at 20 kHz is due to a flexure resonance in the diaphragm. At still higher frequencies, other resonance causes wild fluctuations in the response curve. Although the J-9 is reversible, its performance as a hydrophone is relatively poor so that it is used almost exclusively as a projector.

The maximum input power to the J-9 is 24 W at frequencies above 200 Hz. If higher power is required, several J-9 can be clustered together or a larger projector can be used. The J-9 has several larger "brothers," the J-11 and J-13. Both of these transducers operate on the same principle, but are much larger and heavier than the J-9. For example, the in-air weight of the J-9 is 20 lbs while the J-11 weighs 125 lbs and the J-13, 121 lbs. The maximum input power to the J-11 and J-13 are approximately 80 and 150 watts, respectively.

2.7 Calibration of Transducers

There are special facilities that have anechoic properties or are large enough so that reflections from boundaries will not seriously affect any transducer calibration. However, many of us will not have the access to a large and deep testing facility but will be restricted to relatively small tanks in which to calibrate transducers. Therefore, in order to do meaningful and accurate calibration of transducers, we must use pulsed pure tone signals so that the direct signal from the projector can be separated out from the reflected signal. However, when using pulsed tone signals, special consideration of the Q of the projector needs to be taken into account. The duration of the signal must be long enough for a steady-state condition to be reached in the projector. A typical example of a received hydrophone signal from a projector driven by a pulsed tone signal is shown in Fig. 2.14a. At the beginning of the pulse there is a transient effect present before steady state is reached. At the end of the pulse, there is another transient oscillation, which in this case is smaller than the steady-state response, but in some higher Q transducer this oscillation may be larger than the steady-state response. The transducer should be calibrated using the steady-state portion of the signal. The transient period will generally consist of Q cycles (Bobber, 1970). The signal of Fig. 2.14a was turned on and off at a zero crossing, which would minimize transient effects. Some signal generation system will gate a continuous wave (cw) signal to produce a pulsed tone signal without regard to the zero crossing of the cw signal. Such a signal may produce considerable more transient effects, since the turn-on time from zero volt to a specific level and the turn-off time from a specific level to zero volt are instantaneous. The presence of a turn-on and turn-off transient can be extremely important when measuring the hearing sensitivity of a marine animal, since it could cue the animal to the onset of the signal. One method to reduce or even eliminate the transient effects is to turn the signal on and off gradually. The hydrophone signal in Fig. 2.14b is for the similar situation as in the trace above it, except the signal was turned on gradually using a trapezoidal amplitude-modulated signal.

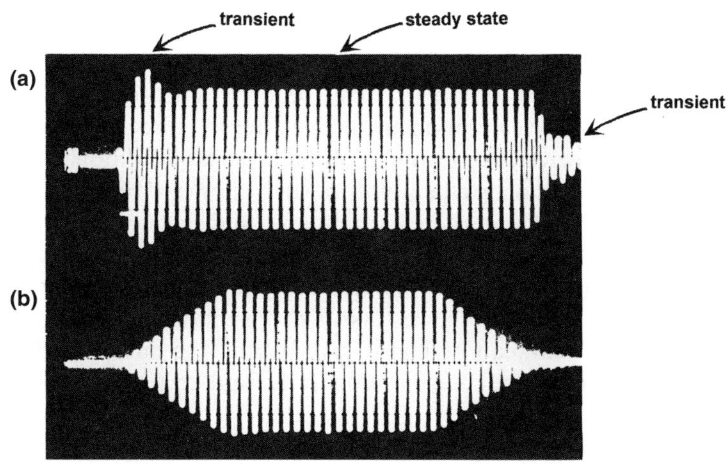

FIGURE 2.14. (**a**) Received hydrophone signal from a projector driven by a pulsed tone signal that is turned on and off abruptly; (**b**) received hydrophone signal form projector driven by a trapezoidal modulated signal.

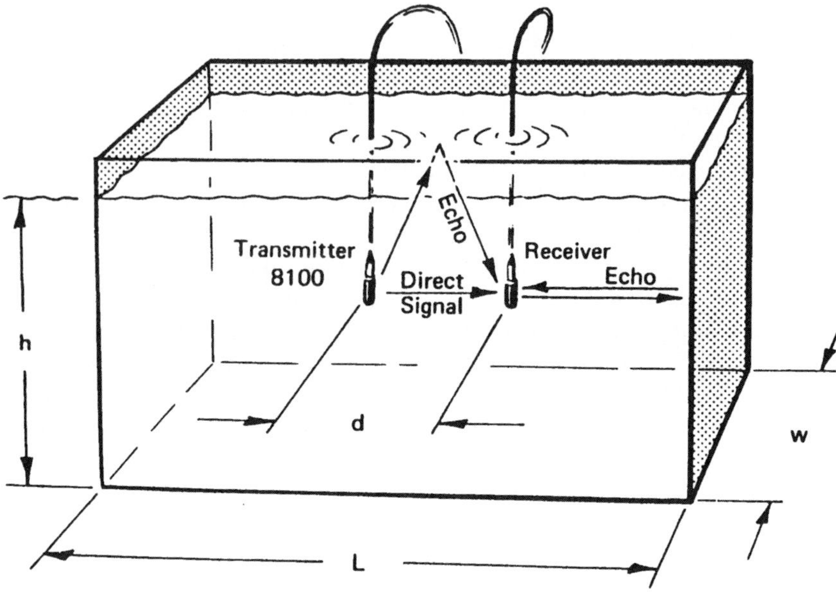

FIGURE 2.15. Sketch of acoustic propagation in a small tank showing some of the paths of propagation from a projector to a hydrophone.

A typical tank situation for calibration is depicted in Fig. 2.15 showing some of the paths that one must be concerned about. One important constraint on the minimum size of a tank is the necessity of having the hydrophone located in the far field of the projector. The far field of a transducer will be discuss in the next chapter; it suffices here to state that the far field of a transducer exists at distance greater than a^2/λ, where a is the largest active dimension of the transducer, and λ is the wavelength in water. The size of a tank could impact on the minimum pulse duration that can be used, since it would be most desirable to have reflective components arrive after the direct signal has completely passed the receiver. An example of the acoustic condition in a 2.44-m diameter, circular tank having a depth of 1.83 m is shown in Fig. 2.16. Both the projector and the hydrophone were at mid-depth and separated by 1.2 m. The top trace shows the signal fed to the projector and the bottom signal the received hydrophone signal. The first received pulse is from the direct path, and the following pulses are the results of propagation along the reflective paths. Since a small tank can be highly reverberant, the pulse repetition rate must be slow enough for the reverberation in the tank to totally decay.

2.7.1 Calibration with a Calibrated Hydrophone

There are several methods by which a transducer can be calibrated; the simplest method uses a calibrated hydrophone. In order to measure the

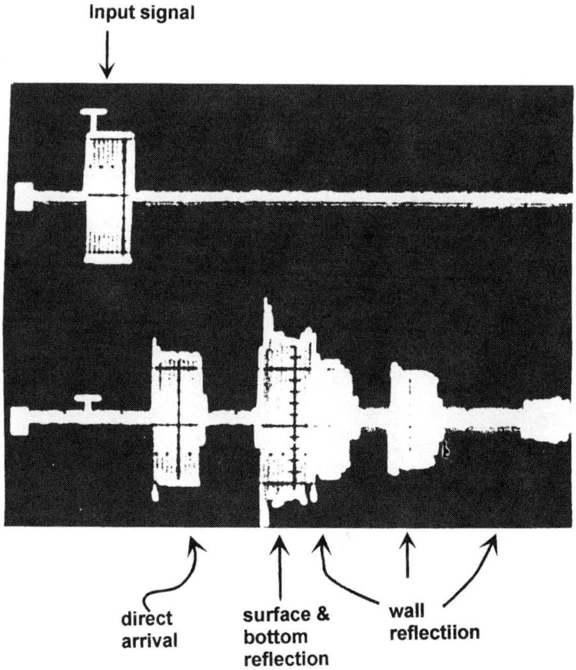

FIGURE 2.16. *Top* trace shows the signal sent to a projector and the *bottom* trace shows the received signal from a hydrophone that is 1.2 m from the projector. The first pulse is the direct signal from the projector, the next pulse is the surface reflected signal, and the remaining pulses are echoes from different parts of the circular tank boundary.

receive response of an uncalibrated hydrophone, the calibrated and uncalibrated hydrophone can be situated side by side with a projector emitting an acoustic signal of a specific frequency at both of them. Both hydrophones will receive the same acoustic signal so that, from Eq. (2.1), the received sound pressure level will be

$$\mathrm{SPL} = |M_1| - G_1 + 20\log V_1 = |M_{\mathrm{unk}}| - G_2 + 20\log V_2, \tag{2.23}$$

where the subscript 1 is used for the calibrated hydrophone and subscript 2 for the uncalibrated hydrophone, and G is the gain of the receiving system in dB. Solving for M_{unk}, we get

$$|M_{\mathrm{unk}}| = |M_1| - (G_1 - G_2) + 20\log\left(\frac{V_1}{V_2}\right). \tag{2.24}$$

The receive sensitivity will be a negative quantity, and absolute values are used to avoid confusion in using Eqs. (2.23) and (2.24). Typically, the response characteristics is given in terms of rms voltage; however, since the ratio of V_1 and V_2 is measured, the peak-to-peak values can be read off an oscilloscope.

The voltage projection or transmit sensitivity of a transducer can easily be measured using a calibrated hydrophone. Suppose a signal V_{in} directed to a projector to be calibrated and the calibrated hydrophone is at a distance r from the project, we have from Eqs. (2.1) and (2.3),

$$\mathrm{SPL} = |M_1| - G_i + 20\log V_1 = S_{unk} + 20\log V_{in} - 20\log r, \tag{2.25}$$

where again, the subscript 1 refers to the calibrated transducer. Rearranging Eq. (2.25), we have for the transmit sensitivity

$$S_{vunk} = |M_1| - G_1 + 20\log\left(\frac{V_1}{V_{in}}\right) + 20\log r. \tag{2.26}$$

The transmit sensitivity is usually a positive quantity. In a similar manner, the current transmit sensitivity can be measured, using Eqs. 2-1 and 2-3 to get

$$\mathrm{SPL} = |M_1| - G_1 + 20\log V_1 = S_{iunk} + 20\log I_{in} - 20\log r. \tag{2.27}$$

The input current can be measured by measuring the voltage across a resistor R placed in series with the input so that

$$S_{iunk} = |M_1| - G_1 + 20\log\left(\frac{V_1}{V_r}\right) + 20\log R + 20\log r. \tag{2.28}$$

A hydrophone can also be easily calibrated with a calibrated projector. From Eqs. (2.1), (2.2) and (2.3), we have in this case

$$\mathrm{SPL} = S_1 + 20\log V_{in} - 20\log r = |M_{unk}| - G_2 + 20\log V_2. \tag{2.29}$$

Solving for the unknown hydrophone response, we get

$$|M_{unk}| = S_1 + G_2 + 20\log\left(\frac{V_{in}}{V_2}\right) - 20\log r, \tag{2.30}$$

where again, the subscript 1 refers to the calibrated transducer and subscript 2 to the uncalibrated hydrophone, and r is the separation distance between the projector and the hydrophone. Although we can calibrate a hydrophone with a calibrated projector, the same projector cannot be used to calibrate another projector, only a calibrated hydrophone must be used.

2.7.2 Spherical Reciprocity Parameter

Transducers can be calibrated in the absence of a calibrated transducer by making use of the reciprocity property of most electroacoustic devices. To be a reciprocal transducer, the element must be linear, passive, and reversible. These conditions are satisfied by piezoelectric ceramics. Let us use a spherical transducer as a projector driving it with a harmonic input current I_{in}, which causes the surface of the spherical element to vibrate with a velocity u and producing a spherical acoustic signal p at a distance r from the element as shown in Fig. 2.17a. Keep in mind that the input is harmonic or sinusoidal, so that u and p will also be sinusoidal. Now let us use this spherical element as a receiver so that an acoustic force F produces a voltage V_{0c} at the terminal as depicted in Fig. 2.17b. The electroacoustic reciprocity theorem states that

$$\frac{I_{in}}{u} = \frac{F}{V_{0c}}. \tag{2.31}$$

It will be more convenient to invert Eq. (2.31) so that

$$\frac{u}{I_{in}} = \frac{V_{0c}}{F}. \tag{2.32}$$

Now we need to snatch an expression for the relationship of a vibrating sphere and the resulting acoustic pressure. From Kinsler et al. (1982) and almost any physical acoustics book, we find that the transmitted pressure at a distance r produced by a sphere of surface area A vibrating at a frequency f is given by the expression

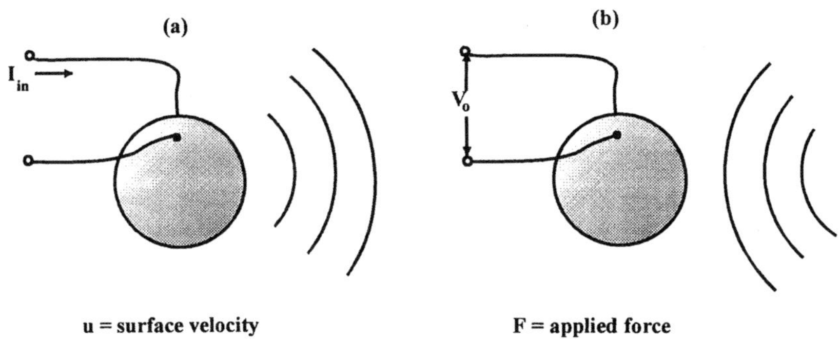

FIGURE 2.17. (**a**) A spherical transducer as a projector driven by a harmonic current I_{in} causing the surface of the sphere to vibrate with a velocity u and producing an acoustic pressure p_t; (**b**) an acoustic force F causing the voltage V_0 at the hydrophone terminals.

$$p = \frac{\rho c}{2r\lambda} Au, \tag{2.33}$$

where ρ is the density and c is the sound velocity in water. Since the acoustic force in Eq. (2.31) is nothing more than the incident acoustic pressure p times the area of the sphere ($F = pA$), Eq. (2.32) can be rewritten, using the relationship of Eq. (2.33) as

$$\frac{p}{I_{in}}\left(\frac{2r\lambda}{\rho c A}\right) = \frac{V_{0c}}{F} = \frac{V_{0c}}{pA}. \tag{2.34}$$

From the definition of voltage receive sensitivity ($M_h = V_{0c}/p$) and current transmit sensitivity ($p = S_i I_{in}$), Eq. (2.34) can be rewritten as

$$S_i\left(\frac{2r\lambda}{\rho c}\right) = M_h. \tag{2.35}$$

The spherical reciprocity parameter was defined earlier as the ratio of M_h and S_i in Eq. (2.9), so that from Eq. (2.35) we have

$$J = \frac{M_h}{S_i} = \frac{2r\lambda}{\rho c} = \frac{2r}{\rho f}. \tag{2.36}$$

The reciprocal parameter J can now be used in the calibration of transducers without the necessity of having a calibrated transducer present.

The reciprocal parameter is slightly different for cylindrical and planar elements and can be found in Urick (1983). However, if the calibration is done in the far field, i.e., $r >> L^2/\lambda$ for cylindrical elements and $r >> A/\lambda$ for planar elements, then the spherical reciprocal parameter could still be used with little error. This means that both the frequency of calibration and the distance between elements must be seriously considered. The reciprocity parameter can now be used in the calibration of transducers without the necessity of having a calibrated transducer present.

2.7.3 Three-Transducer Spherical Wave Reciprocity Calibration

Three transducers are needed in this reciprocity calibration technique to determine the free-field sensitivity of a hydrophone; the hydrophone H, a projector P and a reciprocal transducer T. Three measurements are also required as depicted in Fig. 2.18. In the first and second measurements, a current I_P is fed into projector P to produce an acoustic pressure of p_P at

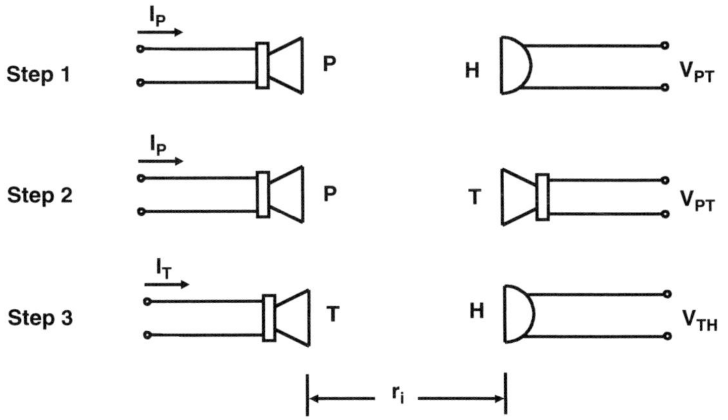

FIGURE 2.18. The three steps used in the three-transducer reciprocity calibration method.

a distance r_1, producing the following voltages at the terminals of hydrophone H and transducer T. From these measurements we have

$$V_{\mathrm{PH}} = M_{\mathrm{H}}\, p_{\mathrm{P}} = \frac{M_{\mathrm{H}}\, I_{\mathrm{P}}\, S_{\mathrm{P}}}{r_1}, \tag{2.37}$$

$$V_{\mathrm{PT}} = M_{\mathrm{T}}\, p_{\mathrm{P}} = \frac{M_{\mathrm{T}}\, I_{\mathrm{P}}\, S_{\mathrm{P}}}{r_1}, \tag{2.38}$$

where M_{H} and M_{T} are the received responses of transducers H and T, respectively, and S_{P} is the transmit sensitivity of transducer P. Dividing Eq. (2.37) by Eq. (2.38), we get

$$\frac{V_{\mathrm{PH}}}{V_{\mathrm{PT}}} = \frac{M_{\mathrm{H}}}{M_{\mathrm{T}}}. \tag{2.39}$$

For the reciprocal transducer T, we have from Eq. (2.36)

$$J = \frac{M_T}{S_T} = \frac{2 r_1 \lambda}{\rho c}. \tag{2.40}$$

Substituting Eq. (2.40) into Eq. (2.39), we get

$$M_{\mathrm{H}} = J S_{\mathrm{T}} \frac{V_{\mathrm{PH}}}{V_{\mathrm{PT}}}. \tag{2.41}$$

For the third measurement, the reciprocal transducer T is driven by current i_T to produce an acoustic pressure p_T, which will be received by the hydrophone H to produce open-circuit voltage

$$V_{TH} = M_H p_T = \frac{M_H i_T S_T}{r_1}. \tag{2.42}$$

Solving for M_H from Eqs. (2.41) and (2.42), we arrive at the following expression

$$M_H = \left(\frac{V_{TH} V_{PH}}{V_{PT} i_T} J r_1 \right)^{1/2}$$

The current I_T can be measured as a voltage drop across a series resister R so that

$$M_H = \left(\frac{V_{TH} V_{PH}}{V_{PT} i_T} R J r_1 \right)^{1/2}. \tag{2.44}$$

Since the four voltages appear as a dimensionless ratio, they can be measured by an uncalibrated voltmeter that is linear.

2.7.4 Two-Transducer Reciprocity and Self-Reciprocity

If two transducers are identical, one can be used as a hydrophone and the other as a reciprocal transducer so that only one measurement is needed, which is depicted in step three of Fig. 2.18. With transducer T driven by current I_T an open-circuit V_0 will be produced across the terminals of hydrophone H, so that from Eq.(2.43) along the assumption that $M_H = M_T$, we have

$$M_H = \left(\frac{V_{TH}}{I_T} r J \right)^{1/2}. \tag{2.45}$$

Once again, I_T can be measured by a voltage drop V_T across a series resister R so that

$$M_H = \left(\frac{V_{TH}}{V_T} r R J \right)^{1/2}. \tag{2.46}$$

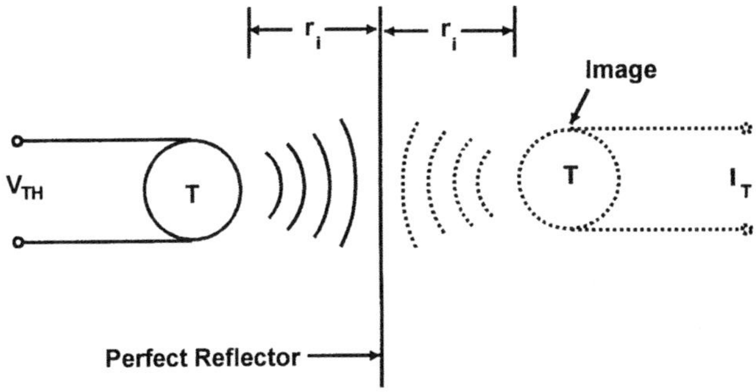

FIGURE 2.19. Schematic for a self-reciprocity calibration of a transducer.

This calibration should be considered as estimate unless one can show that the two transducers are indeed identical. This can be done by using a third transducer to project an acoustic signal that the first two transducers can measure. The results should be identical if the first two transducers are identical.

The two-transducer reciprocal method can be a true primary method if the same transducer is used as both a hydrophone and a projector. This can be done by reflecting a transmitted signal back to the transducer so that it receives its own signal. The self-reciprocity calibration is shown schematically in Fig. 2.19 where the mirror image can be considered as a second transducer at the same distance from the reflecting surface. The reflection must be perfect so that the transmitting current response of the image is identical to that of the real transducer. Pulsed sounds must be used in this technique. The receiving sensitivity is given by Eq. (2.46), where $r = 2r_1$. Patterson (1967) used this method to calibrate a transducer at sea, transmitting the signal toward the surface and receiving the reflection from the surface. He obtained an accuracy of 2 dB. However, using the sea surface requires the presence of a calm sea with a flat surface and the absence of a prominent supporting platform over the transducer.

References

Bobber, R. J. (1970). *Underwater Electroacoustic Measurements* (Naval Research Lab, Washington, DC).

Cady, W. G. (1964). *Piezoelectricity* (Dover, New York).

Hueter, T. F. and Bolt, R. H. (1955). *Sonics: Techniques for the use of Sound and Ultrasound in Engineering and Science* (John Wiley & Sons, Inc., New York).

Kawai, H. (1969). "The piezoelectricity of poly (vinylidene fluoride)," Jpn. J. Appl. Phys. **8**, 975.

Kinsler, L. E., Frey, A. R., Coppens, A. B., and Sanders, J. V. (1982). *Fundamentals of Acoustics* (John Wiley and Sons, New York).

Moffett, M. B., Powers, J. M., and McGrath, J. C. (1986). "A Δc hydrophone," J. Acoust. Soc. Am. **80**, 375–381.

Patterson, R. B. (1967). "Using the ocean surface on a reflector for a self-reciprocity calibration of a transducer," J. Acoust. Soc. Am. **42**, 653.

Sims, C. C. (1959). "High-fidelity underwater sound transducers," Proc. IRE **47**, 866.

Urick, R. J. (1983). *Principles of Underwater Sound*, 3rd ed. (McGraw-Hill, New York).

Wilson, O. B. (1988). *Introduction to Theory and Design of Sonar Transducers* (Peninsula, Los Altos, CA).

3

Transducer Properties and Utilization

3.1 Near and Far Acoustic Fields

In Chapter 2, we concentrated on the properties and behavior of the electro-acoustic elements that are used to construct a transducer. In this chapter, we will consider the transducer as a whole, whether it is constructed with an array of elements or is made up of a single element. We will begin this chapter with a discussion of the acoustic near and far fields of a transducer. The concept of a *near field* is often difficult to grasp since many explanations are embroiled in mathematics that seem to confuse rather than to clarify the issue. In this section, a simple model will be used to discuss the notion of the near and far portion of an acoustic field generated by an acoustic radiator, in the hopes of giving the reader a physically intuitive understanding of the topic. This whole analysis is also applicable to a transducer used as a receiver.

Consider a circular disc of radius a shown in Fig. 3.1, and assume that there are two point sources of acoustic energy spaced at distance a apart, one at the origin and the other at the edge of the disc. This model will be used to describe the acoustic near field generated by the two point sources and by the disc transducer. Both point sources are being driven by a harmonic or sinusoidal signal generator and each source produces an acoustic pressure of $Ae^{j(\omega t - kr)}$ (see Eq. (1.58)). The acoustic pressure at an observation point on the y-axis will be the sum of the pressure from both point sources and can be expressed as

$$p = Ae^{j\omega t}\left(\frac{e^{-jkr}}{r} + \frac{e^{-jkr_s}}{r_s}\right), \tag{3.1}$$

where r_s is the slant range from point source #2 to the observation point and is equal to

$$r_s = \sqrt{a^2 + r^2}. \tag{3.2}$$

The amplitude of the pressure is the absolute value of the expression in Eq. (3.1). Using Euler's formula, $e^{jx} = \cos x + j \sin x$, the absolute value of the pressure amplitude can be expressed as

W.W.L. Au, M.C. Hastings, *Principles of Marine Bioacoustics*,
DOI: 10.1007/978-0-387-78365-9_3,

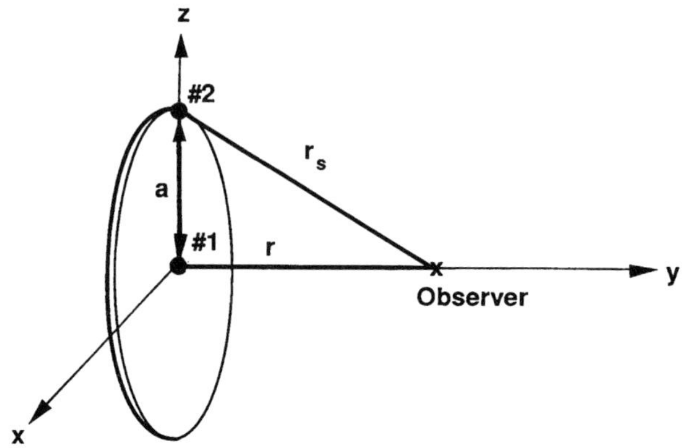

FIGURE 3.1. Simple model of a disc of radius a with two point sources of acoustic energy separated by the radius of the disc. Each point source is driven by the same sinusoidal signal generator.

$$|p(r)| = A\sqrt{\left(\frac{\cos kr}{r} + \frac{\cos kr_s}{r_s}\right)^2 + \left(\frac{\sin kr}{r} + \frac{\sin kr_s}{r_s}\right)^2}. \tag{3.3}$$

The acoustic pressure for a cw signal as expressed in Eq. (3.3) is plotted as a function of r/a in Fig. 3.2. In the near field, the acoustic pressure oscillates with range, exhibiting a number of peaks and nulls. These oscillations are present

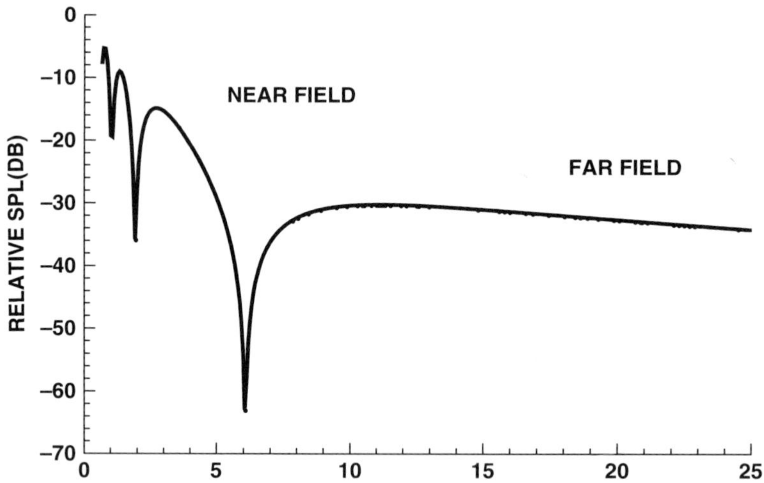

FIGURE 3.2. Acoustic pressure versus range for the two-point source model shown in Fig. 3.1.

because the difference in the distances of the two point sources to the observation point becomes larger as the observer moves toward the disc and the sinusoidal waves arriving at the observer are out of phase with each other, causing either constructive or destructive interference. The pressure amplitude drops off smoothly with range in the *far field*, because the difference in the distance from the two point sources will approach zero as the observer moves further away from the origin. Both lines from the observation point to each point source will be nearly parallel. The phase difference is related to the range difference ($\Delta r = r_s - r$) and the wavelength (λ) of the signal and can be expressed as

$$\Delta\phi = \Delta r \frac{2\pi}{\lambda}. \tag{3.4}$$

When Δr equals an integer multiple of $\lambda/2$, the two waves will be 180° out of phase and will experience the maximum amount of destructive interference; and when Δr equals zero or an integer multiple of λ, the two waves will be in phase and will experience the maximum amount of constructive interference. For short ranges, Δr and $\Delta\varphi$vary considerably with small changes in range whereas at long ranges they hardly vary with range so that the waves from the point sources arrive nearly in phase. In the situation where $r>>a$ we have

$$r_s = \sqrt{a^2 + r^2} \approx r, \tag{3.5}$$

and the pressure of Eq. (3.3) can now be expressed as

$$|p(r)| = 2\frac{A}{r}. \tag{3.6}$$

The far field is usually defined as starting at the range at which the acoustic pressure begins to decay as a function of $1/r$. This range cannot be defined precisely since the acoustic field makes a gradual transition from the near field to the far field. The near field is also referred to as the *Fresnel zone* and the far field as the *Fraunhofer zone* in the wave propagation studies associated with optics, acoustics, and electromagnetics.

We will now continue our discussion of the near and far field by considering a circular disk mounted flush with the surface of an infinite baffle and vibrating with simple harmonic motion. Consider the geometry of Fig. 3.3 depicting a disk of radius a and an observation point along the y-axis. To compute the pressure at the observation point p, we divide the radiating face into infinitesimal elements dS having an acoustic strength of AdS, where A is now a measure of the pressure produced per unit area, and d$S = \rho$dρdφ is the area of each infinitesima. The pressure at point p can be expressed as

$$p(r,t) = A\mathrm{e}^{j\omega t}\int_0^a\int_0^{2\pi}\frac{\mathrm{e}^{-jk\sqrt{r^2+\rho^2}}}{\sqrt{r^2+\rho^2}}\rho\,\mathrm{d}\rho\,\mathrm{d}\varphi. \tag{3.7}$$

The intergrand over ρ is a perfect differential since

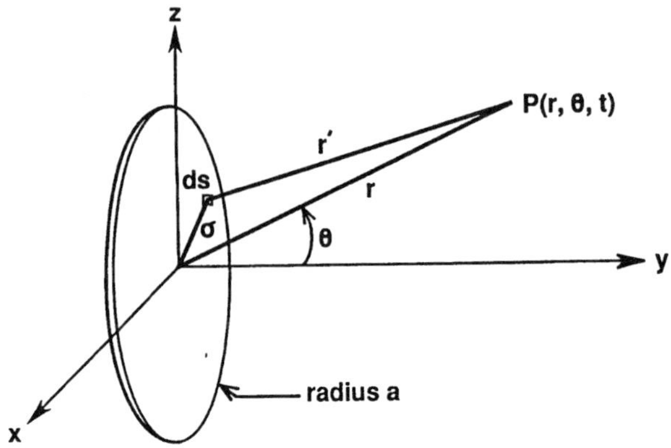

FIGURE 3.3. Geometry used in deriving the radiation characteristics of a circular piston.

$$\frac{\rho}{\sqrt{r^2+\rho^2}}\mathrm{e}^{-jk\sqrt{r^2+\rho^2}} = -\frac{\mathrm{d}}{\mathrm{d}\rho}\left(\frac{\mathrm{e}^{-jk\sqrt{r^2+\rho^2}}}{jk}\right), \tag{3.8}$$

and the pressure at point p can now be expressed as

$$p(r,t) = \frac{j2\pi A}{k}\mathrm{e}^{j\omega t}\left[\mathrm{e}^{-kkr} - \mathrm{e}^{-jk\sqrt{r^2+a^2}}\right]. \tag{3.9}$$

The pressure amplitude on the axis of the piston is the magnitude of Eq. (3.9)

$$|p(r)| = \left|\frac{j2\pi A}{k}\left[\cos(kr) - \cos(k\sqrt{r^2+a^2}) - j\left(\sin(kr) - \sin(k\sqrt{r^2+a^2})\right)\right]\right|. \tag{3.10}$$

The magnitude can be evaluated by taking the square root of the sum of the square of the real and imaginary parts so that

$$|p(r)| = \frac{2\pi A}{k}\left(2 - 2\left[\cos(kr)\cos\left(k\sqrt{r^2+a^2}\right) + \sin(kr)\sin\left(k\sqrt{r^2+a^2}\right)\right]\right)^{1/2}. \tag{3.11}$$

We can now use the following two trigonometric identities, $\cos(x-y) = \cos x \cos y + \sin x \sin y$, and $\sin(x/2) = \sqrt{[1-\cos(x)]/2}$ in Eq. (3.11) to arrive at the expression for the acoustic pressure at point p

$$|p(r)| = \frac{4\pi A}{k}\left|\sin\left[\frac{kr}{2}\left(\sqrt{1+\left[\frac{a}{r}\right]^2} - 1\right)\right]\right|. \tag{3.12}$$

Equation (3.12) is the general expression for the acoustic pressure along the acoustic axis (y-axis). For $r >> a$, the square root can be simplified by writing it in terms of its binomial expansion and ignoring the higher order terms so that

$$\sqrt{1+\left(\frac{a}{r}\right)^2} \approx 1 + \frac{1}{2}\left(\frac{a}{r}\right)^2. \tag{3.13}$$

If $r/a >> ka$ so that the range of the observer is much greater than both the radius of the piston and the wavelength of the signal, then $\sin[ka(a/r)] \approx ka(a/r)$ and the pressure along the acoustic axis can be expressed as

$$p(r) = \frac{Aa^2}{2r}. \tag{3.14}$$

The variations in the acoustic pressure as a function of range along the y-axis, calculated with both Eqs. (3.12) and (3.14), are depicted in Fig. 3.4. The pressure varies drastically in the near field, exhibiting a number of oscillations with deep nulls and decays gradually with range into the far field. In the far field, the pressure decays with range in a $1/r$ manner. From Fig. 3.4, we can see that the acoustic pressure makes a gradual transition into the far field, and the range at which the far field starts is proportional to the size of the piston and inversely proportional to the signal frequency. At a range of $r = 2a^2/\lambda$, the acoustic pressure approaches the $1/r$ spherical spreading loss curve, making a gradual transition into the far field. At a range of $r = \pi a^2/\lambda$ (area/λ), the observation point is definitely in the far field of the circular piston.

The American National Standards Institute has decided that the critical range defining the beginning of the far field for a circular piston is

$$R_c = \frac{\pi a^2}{\lambda}, \tag{3.15}$$

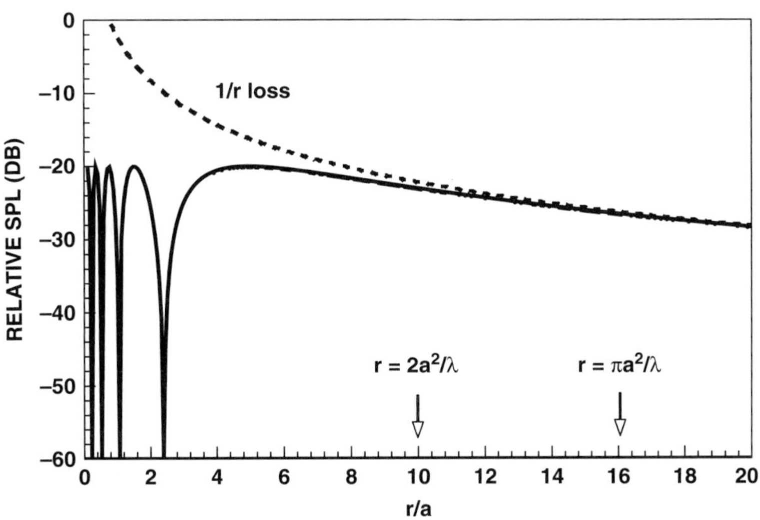

FIGURE 3.4. Variations in the on-axis acoustic pressure for a circular piston with $a/\lambda = 5$. Also included is the $1/r$ spherical spreading loss calculated with Eq. (3.14).

the same as area over wavelength. The criterion of Eq. (3.15) can be generalized to a piston of area A in any shape so that

$$R_c = \frac{\text{Area}}{\lambda}. \tag{3.16}$$

For a line or thin cylindrical transducer, the critical range is

$$R_c = \frac{l^2}{\lambda} \tag{3.17}$$

It is important to understand that the critical ranges for the beginning of the far field of a transducer given by Eqs. (3.15)–(3.17) apply to the size of the acoustic element and not to the ultimate exterior dimensions of the transducer. It is not unusual, for example, that a cylindrical shaped transducer to have an exterior length that is much longer than the cylindrical element. These criteria are relatively conservative and other criteria have been used. The reader should refer to Bobber (1970) for a more detailed discussion on the criteria used for determining the beginning of the far field.

3.2 Directivity Pattern of Simple Transducers

A transducer can be constructed with elements having different shapes as depicted in Fig. 2.6. Arrays of elements can also be assembled to have an overall shape that is independent of the shapes of the individual elements. The simplest shaped element from a directivity standpoint is the sphere, which has no directivity at all. Signals arriving from any direction are received equal with a spherical hydrophone. Conversely, driving a spherical transducer with a time-varying signal will cause the transducer to radiate acoustic energy equally in all directions. Such a transducer is called an omnidirectional or non-directional transducer. Its beam pattern for transmission and reception in any plane passing through the origin of the element is a circle having a common origin as the element. In the following sections, we shall consider the beam pattern of a cylindrical transducer and of the planar transducers shaped as circular and rectangular pistons. The beam patterns will apply equally to the transducers used as projectors or hydrophones.

3.2.1 Beam Pattern of a Thin Cylindrical Transducer

The first directional beam pattern we shall consider is that of a thin cylinder depicted in Fig. 3.5. We divide the cylinder into infinitesimal elements of length $d\delta$ and treat each of these elements as omnidirectional sources of strength $2\pi a A d\delta$, where A is a measure of the pressure produced per unit length and a is the radius of the cylinder. In the far field with $r >> L$,

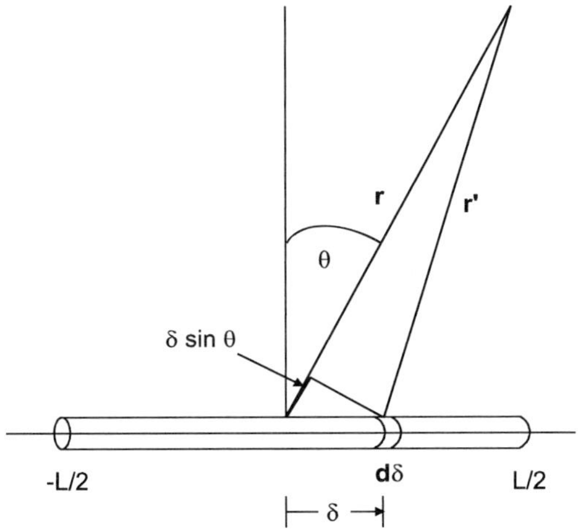

FIGURE 3.5. Geometry used in deriving the radiation beam pattern of a cylindrical element.

$$\mathrm{r}' \approx \mathrm{r} - \delta \sin\theta \epsilon \tag{3.18}$$

so that at the field point p the acoustic pressure generated by the element can be expressed as

$$\mathrm{d}p(\theta) = \frac{2\pi a \mathrm{A}\,\mathrm{d}\delta}{r'} \mathrm{e}^{j(\omega t - kr + k\delta \sin\theta)}. \tag{3.19}$$

The total pressure is found by integrating dp over the length of the element

$$p(\theta) = \frac{2\pi a A}{r'} \mathrm{e}^{j(\omega t - kr)} \int_{-L/2}^{L/2} \mathrm{e}^{jk\delta \sin\theta} \mathrm{d}\delta. \tag{3.20}$$

Under the assumption that $r >> L$, r' in the denominator can be replaced by its approximate value r since this amounts to a very small error in the amplitude of the acoustic fields generated by each element. However, in the exponential, this simplification cannot be made because the relative phase of the elements is a very strong function of angle. Equation 3-20 integrates to

$$p(\theta) = \frac{2\pi a A L}{r} \left[\frac{\sin\left(\frac{KL}{2}\sin\theta\right)}{\frac{kL}{2}\sin\theta} \right] \mathrm{e}^{j(\omega t - kr)}. \tag{3.21}$$

At $\theta = 0°$, the term in the bracket takes on its maximum value of one and the pressure is maximum, so that

$$p_0 = \frac{2\pi a A L}{r} \mathrm{e}^{j(\omega t - kr)}. \tag{3.22}$$

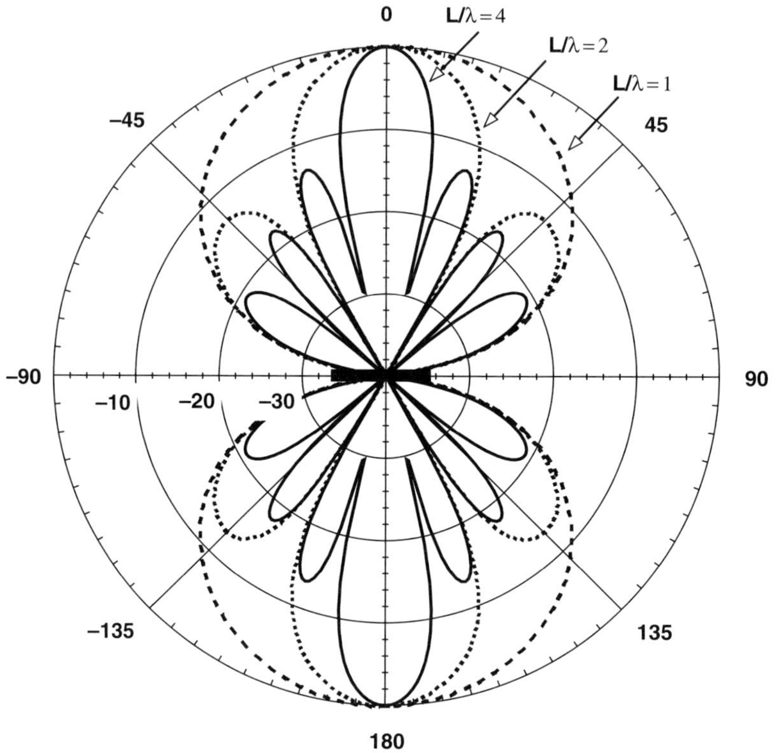

FIGURE 3.6. Beam patterns for a thin cylindrical transducer for different values of L/λ. The dark line through the origin represents the cylindrical transducer.

The relative intensity response or beam pattern of the cylindrical transducer is

$$bp(\theta) = \left(\frac{p(\theta)}{p_0}\right)^2 = \left[\frac{\sin\left(\frac{KL}{2}\sin\theta\right)}{\frac{kL}{2}\sin\theta}\right]^2. \tag{3.23}$$

Examples of three beam patterns are shown in Fig. 3.6 for L/λ of 4, 2, and 1. From this figure and Eq. (3.23), we can see that the shape of the beam pattern is a function of the L/λ ratio, becoming narrower as the ratio increases in value. Therefore, for a cylinder of fixed length, the beam will be narrower as the signal frequency increases, and conversely, for a fixed frequency, the beam will be narrower as the length of the cylinder increases. The 3-dB beamwidth can be determined by letting

$$\left[\frac{\sin\left(\frac{kL}{2}\sin\theta\right)}{\frac{kL}{2}\sin\theta}\right]^2 = 0.5. \tag{3.24}$$

From a $\sin(x)/x$ table, we see that when $x = \pi/2.26$, $\sin(x)/x = 0.707$ so that at the 3-dB point,

$$\theta = \sin^{-1}\left(\frac{0.443}{L/\lambda}\right). \tag{3.25}$$

The beamwidth is equal to 2θ.

3.2.2 Beam Pattern of a Circular Piston

The next transducer we will consider is the circular piston mounted flush with the surface of the infinite baffle and vibrating with simple harmonic motion. The geometry of a circular piston is depicted in Fig. 3.7, with the piston in the x–z plane and the observation point P in the y–z plane. To compute the pressure $p(\theta)$ at point P, we divide the radiating surface into infinitesimal elements dS and consider each element to be an omnidirectional source of strength $A\mathrm{d}S$, where A is the pressure amplitude per unit area. The observation point is in the far field at a range r and the amplitude of the pressure at P due to each element is $A\mathrm{d}S/r$, where $r >> a$. The infinitesimal element is at point Q on the surface of the piston, and the observation point is far enough away that the line from point Q to point P is essentially parallel to the line from the origin of the piston to point P. The area of the infinitesimal element is

$$\mathrm{d}S = \rho \mathrm{d}\rho \mathrm{d}\varphi. \tag{3.26}$$

From Fig. 3.8b, the length of the line QP is

$$\mathrm{QP} \approx \mathrm{r} - \rho \cos\varphi \sin\theta \tag{3.27}$$

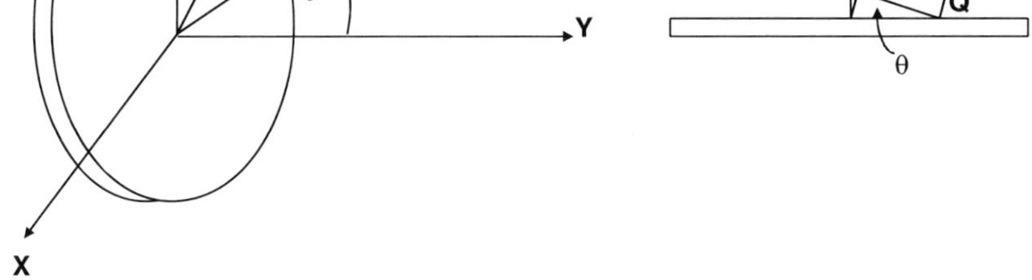

FIGURE 3.7. Geometry for deriving the beam pattern of a circular piston.

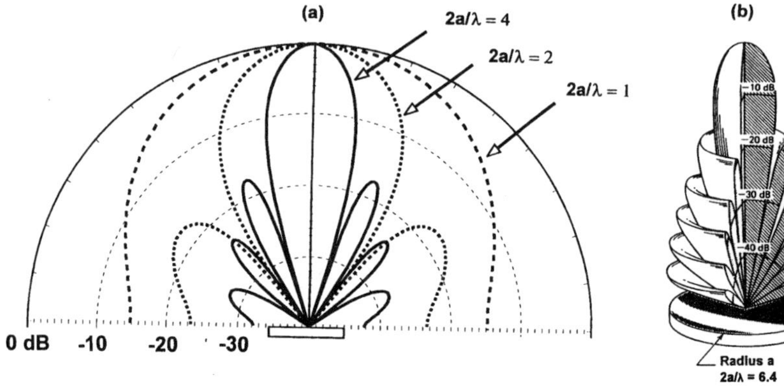

FIGURE 3.8. (**a**) Polar plots of the beam pattern of a circular disk for $2a/\lambda = 1, 2$, and 4. (**b**) Three-dimensional plot of the beam pattern of a circular piston for $2a/\lambda = 6.4$ (adapted from Eckart, 1968).

The incremental pressure at P can be expressed as

$$\mathrm{d}p(\theta) = \frac{A}{r}\, e^{j(\omega t - kr + k\rho\cos\phi\sin\theta)}\, \rho \mathrm{d}\rho \mathrm{d}\phi. \tag{3.28}$$

The resultant pressure is obtained by integrating $\mathrm{d}p(\theta)$ over the area of the piston

$$p(\theta) = \frac{A}{r} e^{j(\omega t - kr)} \int_0^a \rho \mathrm{d}\rho \int_0^{2\pi} e^{jk\rho\sin\theta\cos\phi} \mathrm{d}\phi. \tag{3.29}$$

The second integral can be evaluated by using a special integral relationship

$$\frac{1}{2\pi}\int_0^{2\pi} e^{jx\cos\psi}\mathrm{d}\psi = J_0(x), \tag{3.30}$$

where $J_0(x)$ is a Bessel function of order *0*. Equation (3.29) can now be expressed as

$$p(\theta) = \frac{2\pi A}{r} e^{j(\omega t - kr)} \int_0^a J_0\left(k\rho\,\sin\theta\right) \rho d\rho \tag{3.31}$$

The integral of Eq. (3.31) can be evaluated by using the relationship

$$\int xJ_0(x)\mathrm{d}x = xJ_1(x). \tag{3.32}$$

Letting $x = k\rho\sin\theta$, Eq. (3.31) can be expressed as

$$p(\theta) = \frac{2\pi A}{r} e^{j(\omega t - kr)} \left[\frac{J_1(ka\sin\theta)}{ka\sin\theta}\right]. \tag{3.33}$$

The acoustic pressure on the acoustic axis can be evaluated by setting $\theta = 0$, realizing that $J_1(x)/x = 1/2$ for $x = 0$ so that

$$p_0 = \frac{\pi A a^2}{r} e^{j(\omega t - kr)}. \tag{3.34}$$

The beam pattern can finally be expressed as

$$bp(\theta) = \left(\frac{p(\theta)}{p_0}\right)^2 = \left(\frac{2J_1(ka \sin\theta)}{ka \sin\theta}\right)^2. \tag{3.35}$$

This equation is similar in form as Eq. (3.23) for the cylindrical transducer, except in this case we are dealing with a $J_1(x)/x$ variation rather than a $\sin(x)/x$. Polar plots of beam patterns for $2a/\lambda = 1$, 2, and 4 are shown in Fig. 3.8a, and a three-dimensional beam pattern for $2a/\lambda = 6.4$ is shown in Fig. 3.8b. Note that the beams are symmetrical about the angle θ depicted in Fig. 3.7. From Fig. 3.8a we can see that as the frequency increases ($2a/\lambda$ increases), the major lobe becomes narrower; however, more side lobes appear in the beam pattern. Side lobes must be taken into account when interpreting both passive and active sonar information. For example, if a circular piston receiver is rotating or scanning in azimuth, a signal detected in a side lobe may be mistaken for being detected by the major lobe. As the transducer continues to scan, the signal source will eventually be detected by the main lobe, and one could easily reach an erroneous conclusion that two signals sources are present, separated in azimuth by the angular position of the side lobe, with one of the source being more intense than the other.

The 3-dB beamwidth of the main lobe can be obtained by setting Eq. (3.35) equal to 0.5 and determining the value of θ that will satisfy the relationship. Using a published table of values for $J_1(x)$ from a handbook, we find that $2\,J_1(x)/x = 0.707$ when x or $ka\sin\theta = 1.6$, so that

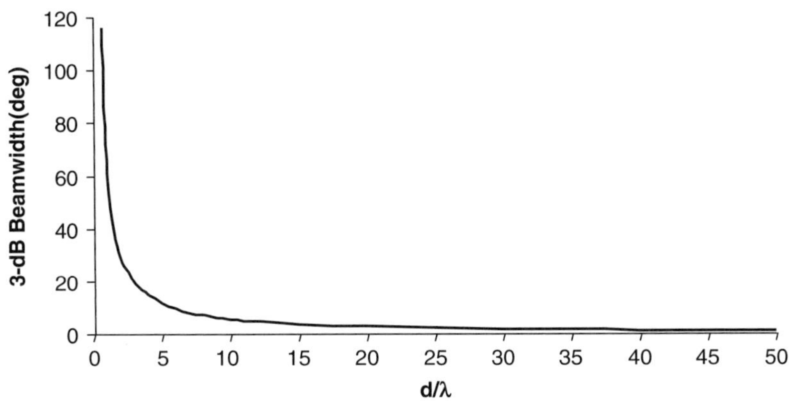

FIGURE 3.9. 3-dB beamwidth of a circular disk transducer as a function of d/λ.

$$\theta = \sin^{-1}\left(\frac{0.509}{d/\lambda}\right) \tag{3.36}$$

The beamwidth is equal to 2θ. The 3-dB beamwidth as a function of d/λ is shown in Fig. 3.9. This indicates that the beamwidth changes slowly for $d/\lambda > 10$ and changes rapidly as d/λ decreases below approximately 3.

3.2.3 Beam Pattern of a Rectangular Piston

We will now consider a rectangular piston mounted on an infinite baffle vibrating with simple harmonic motion as depicted in Fig. 3.10. The radiation pattern can be derived as the product of the beam pattern of two thin cylindrical transducers, one aligned along the z-axis and the other along the x-axis. We assume that the entire surface of the piston moves in phase and with the same amplitude so that the radiation pattern can be expressed as a product of the beam pattern (Eq. (3.24)) of two cylindrical elements

$$bp(\theta, \phi) = \left(\frac{\sin\left(\frac{kH}{2}\sin\theta\right)}{\frac{kH}{2}\sin\theta}\,\frac{\sin\left(\frac{kW}{2}\sin\phi\right)}{\frac{kW}{2}\sin\phi}\right)^2. \tag{3.37}$$

The 3-dB beamwidth in the vertical and horizontal planes can be determined by using Eq. (3.25) and the appropriate linear dimension of the rectangular piston shown in Fig. 3.9.

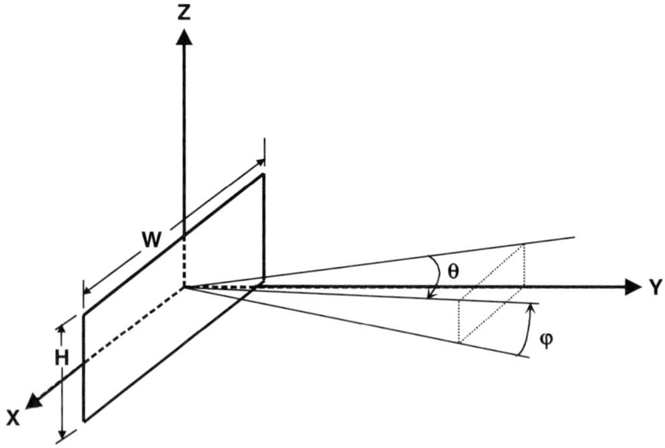

FIGURE 3.10. Geometry for deriving the beam pattern of a rectangular transducer.

3.3 Linear Transducer Arrays

3.3.1 Beam Pattern of a Dipole Array

Consider the two omnidirectional point sources of acoustic energy depicted in Fig. 3.10a, each point source being of strength A. The pressure at an observation point P will be the sum of the pressure produced by each point source,

$$p(\theta) = \frac{A}{r_1} e^{j(\omega t - kr_1)} + \frac{A}{r_2} e^{j(\omega t - kr_2)}. \tag{3.38}$$

We are interested in the situation where the observation point P is a long distance from the dipole so that the lines connecting p to the point sources and the origin of the coordinate system are essentially parallel, as depicted in Fig. 3.11b. From Fig. 3.11b, we have

$$\begin{aligned} r_1 &\approx r + \frac{d}{2} \sin\theta \\ r_2 &\approx r - \frac{d}{2} \sin\theta. \end{aligned} \tag{3.39}$$

Substituting for r_1 and r_2 in Eq. (3.38), we have

$$p(\theta) = \frac{A}{r} \left(e^{j\frac{kd}{2}\sin\theta} + e^{-j\frac{kd}{2}\sin\theta} \right) e^{j(\omega t - kr)}. \tag{3.40}$$

From the relationship $\cos x = (e^{jx} + e^{-jx})/2$, Eq. (3.40) can be expressed as

$$p(\theta) = \frac{2A}{r} e^{j(\omega t - kr)} \cos\left(\frac{kd}{2} \sin\theta \right). \tag{3.41}$$

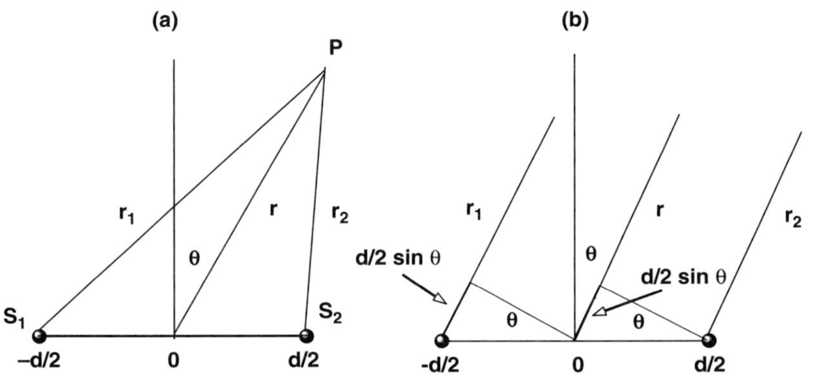

FIGURE 3.11. **(a)** Geometry of a dipole source; **(b)** far field approximation of the dipole geometry in which the lines r_1, r_2, and r are essentially parallel.

The pressure is maximum at $\theta = 0$, so that

$$p(\theta) = \frac{2A}{r} e^{j(\omega t - kr)}. \tag{3.42}$$

To simplify the notation, let

$$\psi = \frac{kd}{2} \sin \theta. \tag{3.43}$$

The beam pattern can now be expressed as

$$bp(\theta) = \left(\frac{p(\theta)}{p_0}\right)^2 = \cos^2 \psi = \left(\frac{\sin 2\psi}{2 \sin \psi}\right)^2 \chi. \tag{3.44}$$

In this simple dipole model, we assumed that the two point sources are driven in phase. We will now consider the case in which the point sources are 90° out of phase (quadrature phase case), where the phase of S_1 is advanced by $\pi/4$ and the phase of S_2 is retarded by $\pi/4$, so that Eq. (3.40) can be expressed as

$$p(\theta) = \frac{A}{r} \left(e^{j\left[\psi + \frac{\pi}{4}\right]} + e^{-j\left[\psi + \frac{\pi}{4}\right]}\right) e^{j(\omega t - kr)}. \tag{3.45}$$

The beam pattern can now be expressed as

$$bp(\theta) = \cos^2\left(\psi + \frac{\pi}{4}\right) \qquad 90^{\circ}\text{phase shift} \quad . \tag{3.46}$$

Another case of interest is the one in which the two point sources are 180° out of phase, so that the beam pattern can be expressed as

$$bp(\theta) = \sin^2 \psi \qquad 180^{\circ}\text{phase shift} \quad . \tag{3.47}$$

Examples of beam patterns for dipole sources having different phase shifts between the two point sources are shown in Fig. 3.12. The elements of the dipole in each case are separated by $\lambda/2$. When the two point sources are driven in phase, the main lobes are directed perpendicular to the line joining the sources. When the two sources are 180° out of phase, the main lobes are directed along the line joining the sources. The latter case is often referred to as an "end fire" array. These examples indicate that the radiation pattern can change considerably with how the two elements are driven. Although we derived the beam pattern for a dipole as a projector, the results also hold for a dipole receiver. Phase differences in the receiving case can be achieved by attaching the appropriate circuitry to the output of each dipole element.

3.3.2 Beam Pattern of an N-element Array

We will now consider a linear array of n equally spaced ominidirectional point sources. All the elements are assumed to be identical with a spacing

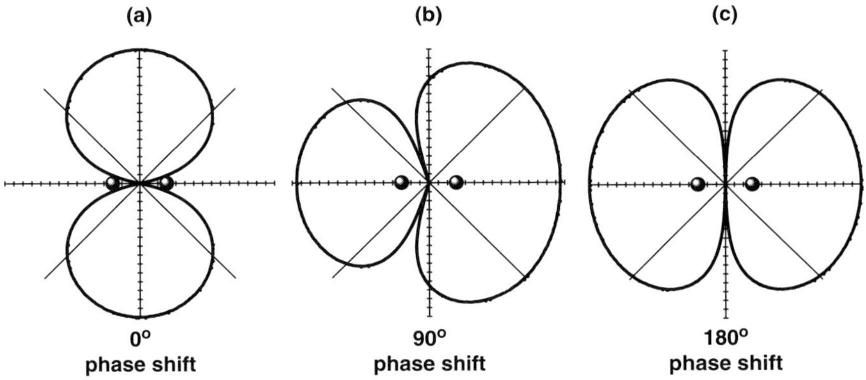

FIGURE 3.12. Radiation from a dipole where the elements are separated by a distance of $\lambda/2$. (**a**) Source driven in phase. (**b**) Sources have a 90° phase shift between them. (**c**) Sources are 180° out of phase.

of d between adjacent elements. We will further assume that n is an even number although the results will apply for both even and odd number of elements. The length of the array for n being even is simply $(n - 1)d$. The origin of the coordinate system is at the midpoint of the array so that the distance of the two furthest elements from the origin is $(n - 1)d/2$. The distance from the origin of the next two furthest element is $(n - 3)d/2$, and so on until we reach the two closest elements having a distance of $d/2$ from the origin. The geometry of the n-element array is shown in Fig. 3.13, where the observation point P is assumed far enough away that the lines from

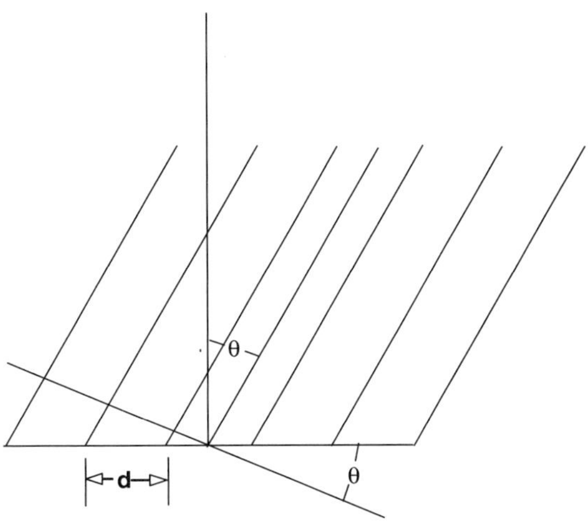

FIGURE 3.13. Geometry of an n-element array of point sources. The observation P is far enough away that the lines from each source to P are essentially parallel.

each element to point P are essentially parallel. In order to explain the derivation of the beam pattern for this geometry, it will be helpful to briefly consider an array composed of a specific number of elements. In the case where $n = 6$ and ψ defined by Eq. (3.43), the resultant pressure at point P starting with the upper most element in Fig. 3.13 is merely an extension of Eq. (3.40),

$$p(\theta) = \frac{A}{r}\,\mathrm{e}^{j(\omega t - kr)}\left(\mathrm{e}^{j5\psi} + \mathrm{e}^{j3\psi} + \mathrm{e}^{j\psi} + \mathrm{e}^{-j\psi} + \mathrm{e}^{-j3\psi} + \mathrm{e}^{-j5\psi}\right). \tag{3.48}$$

We can reduce Eq. (3.48) into two alternative forms. First, from the formula $\cos x = (\mathrm{e}^{jx} + \mathrm{e}^{-jx})/2$,

$$p(\theta) = \frac{2A}{r}\,e^{j(\omega t - kr)}\,[\cos\psi + \cos 3\psi + \cos 5\psi]. \tag{3.49}$$

To obtain the alternate expression, factor out $\mathrm{e}^{j5\psi}$ from Eq. (3.48) to obtain

$$p(\theta) = \frac{A}{r}\,\mathrm{e}^{j(\omega t - kr)}\mathrm{e}^{j5\psi}\left[1 + \mathrm{e}^{-j2\psi} + \mathrm{e}^{-j4\psi} + \mathrm{e}^{-j6\psi} + \mathrm{e}^{-j8\psi} + \mathrm{e}^{-j10\psi}\right]. \tag{3.50}$$

We can now generalize for n-elements and use the form of Eq. (3.50) to write

$$p(\theta) = \frac{A}{r}\,\mathrm{e}^{j(\omega t - kr)}\mathrm{e}^{j(n-1)\psi}\left[1 + \mathrm{e}^{-j2\psi} + \mathrm{e}^{-j4\psi} + \cdots + \mathrm{e}^{-j2(n-1)\psi}\right]. \tag{3.51}$$

The maximum for $p(\theta\Delta)$will occur at $\theta = 0$ and the sum of the n terms in the bracket, each with a value of one, will be n, so

$$p_0 = n\frac{A}{r}\,\mathrm{e}^{j(\omega t - kr)}. \tag{3.52}$$

Dividing $p(\theta)$ by p_0 we have

$$\frac{p(\theta)}{p_0} = \frac{\mathrm{e}^{j(n-1)\psi}}{n}\left[1 + \mathrm{e}^{-j2\psi} + \mathrm{e}^{-j4\psi} + \cdots + \mathrm{e}^{-j2(n-1)\psi}\right]. \tag{3.53}$$

If we let $x = \mathrm{e}^{-j2\psi}$, then the series in the bracket has the same form as

$$1 + x + x^2 + x^3 \cdots + x^{n-1} = \frac{1 - x^n}{1 - x}. \tag{3.54}$$

Therefore, we can express Eq. (3.53) as

$$\frac{p(\theta)}{p_0} = \frac{\mathrm{e}^{j(n-1)\psi}}{n}\,\frac{[1 - \mathrm{e}^{-2jn\psi}]}{[1 - \mathrm{e}^{-2j\psi}]} = \frac{\mathrm{e}^{j(n-1)\psi}}{n}\,\frac{\mathrm{e}^{-n\psi}[\mathrm{e}^{jn\psi} - \mathrm{e}^{-jn\psi}]}{[\mathrm{e}^{j\psi} - \mathrm{e}^{-j\psi}]}. \tag{3.55}$$

From the trigonometric identity

$$\sin\psi = \frac{\mathrm{e}^{j\psi} - \mathrm{e}^{-j\psi}}{2j}, \tag{3.56}$$

we can finally express the beam pattern of an n-element linear array as

$$bp(\theta) = \left(\frac{p(\theta)}{p_0}\right)^2 = \left(\frac{\sin n\psi)}{n\ \sin\psi}\right)^2. \tag{3.57}$$

In order to gain some insights as to the functioning of multi-element arrays, we will consider an array of six elements. Beam patterns for an array having a fixed spacing between adjacent elements but for different signal frequency are shown in Fig. 3.14. The most optimal spacing for a linear array of ominidirectional element is $\lambda/2$ (Kraus, 1950). The beam pattern in the upper left-hand corner of Fig. 3.14 is for the case in which the spacing between elements is exactly $\lambda/2$. The next two beam patterns on the upper part of the figure covers a frequency range of one octave with element spacing of 2/3 λ and 1/3 λ, respectively. The beam pattern in the lower left-hand corner is for an element spacing of λ followed by a beam pattern for an element spacing of $\lambda/4$. The beam patterns of Fig. 3.14 indicate that a multi-element linear array is not a very broadband device. It can cover an octave frequency range without the pattern changing much. However, as the spacing approaches a wavelength, the side lobes at ±90° can become extremely large and even dominate the major lobe. As the spacing becomes small, approaching of $\lambda/4$, the side lobes at ±90° can become extremely wide, affecting the overall shape of the beam.

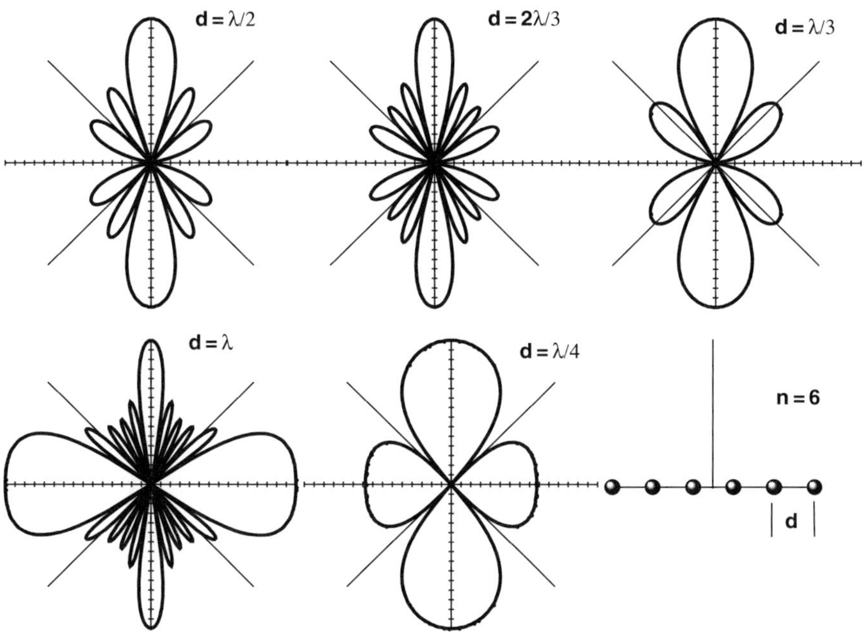

FIGURE 3.14. Beam patterns for a six-element array with different spacing between elements. The vertical axis has a range of 30 dB with 2 dB per tic mark.

Therefore, the results depicted in the figure suggest that a linear array should be most effective for frequencies at which the spacing is between $\lambda/3$ and $2\lambda/3$, with the optimal spacing being $\lambda/2$.

We will now consider how the number of elements affects the beam pattern. Using Eq. (3.57), the beam pattern for a linear array having elements spaced $\lambda/2$ apart are shown in Fig. 3.15. for different number of elements. Only the results for the first quadrant are shown in the figure since the beam patterns are symmetrical as can be seen in Fig. 3.14. The results indicate that as the number of elements increase, the main lobe and the side lobes become narrower and the number of side lobes also increase.

The presence of side lobes is usually undesirable and these side lobes can be reduced by a technique known as *amplitude shading*. Different signal amplitudes are applied to each element, with the maximum amplitudes usually applied to the center elements and a tapering off of amplitude towards the end elements. To understand the effects of amplitude shading on the beam pattern, let us consider a six-element array each being driven at a different amplitude and use Eq. (3.49) to arrive at

$$bp(\theta) = \left(\frac{A_1 \cos\psi + A_2 \cos 2\psi + A_3 \cos 3\psi}{A_1 + A_2 + A_3}\right). \tag{3.58}$$

An interesting shading is the binomial shading, with the amplitudes being proportional to the coefficient of a binomial expansion. For this shading, the relative amplitudes of the elements starting with the outer most element and

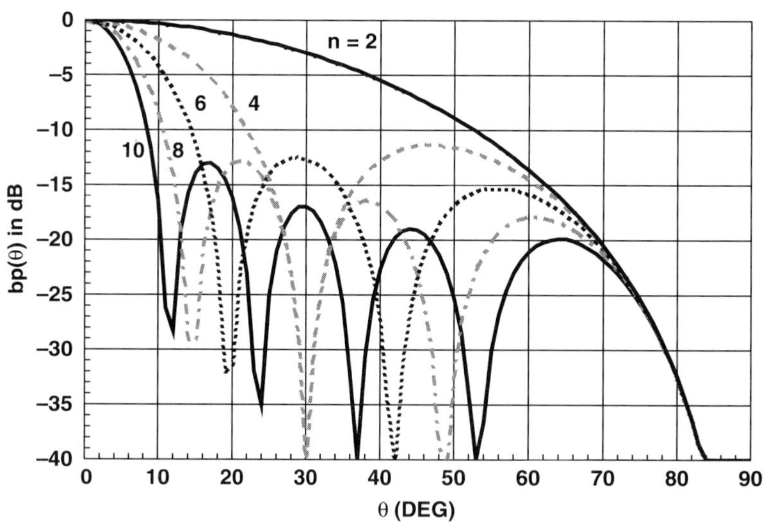

FIGURE 3.15. Beam pattern of a linear array for different number of elements, each element spaced at a distance of $\lambda/2$ from adjacent elements. Only the first quadrant is shown for each beam.

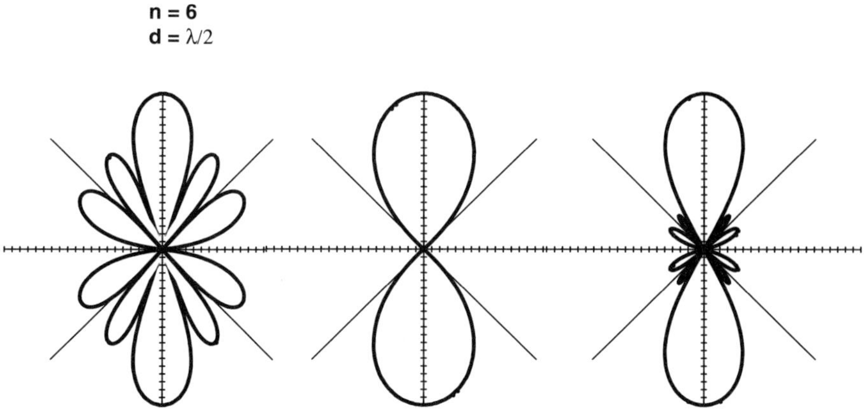

FIGURE 3.16. Beam patters of a six-element line array of half wavelength spacing between elements for different amplitude shading.

progressing toward the other end will be 0.1, 0.5, 1, 1, 0.5, 0.1 so that $A_1 = 1$, $A_2 = 0.5$, and $A_3 = 0.1$ (Urick, 1983). The beam pattern for a binomial shaded six-element array is shown in Fig. 3.16, along with beam pattern for an unshaded and a Dolph–Chebyshev shaded array. The relative amplitude for the Dolph–Chebyshev shading is 0.30, 0.69, 1, 1, 0.69, 0.30, so that $A_1 = 1$, $A_2 = 0.69$, and $A_3 = 0.30$ (Urick, 1983). The binomial shaded array has no side lobe; however, its major lobe is much wider than for either the unshaded or Dolph–Chebyshev shaded array. The 3-dB beamwidths for the beam patterns shown in Fig. 3.16 are 17°, 27°, and 22°, for the unshaded, bionomial shaded, and Dolph–Chebyshev shaded array, respectively. The Dolph–Chebyshev shading is compromised between the unshaded and the bionomial shaded array, having a slightly broader beamwidth than the unshaded array but with a more suppressed side lobe structure than the unshaded array. Those interested in learning more of the Dolph–Chebyshev shading should refer to the classic antenna book by Kraus (1950).

3.3.3 Product Theorem

In the previous section, we considered a linear array composed of simple omnidirectional point sources. When an array is constructed of identical directional transducers, the beam pattern is given by the product theorem which states that the beam pattern of an array of identical, equally spaced directional hydrophones is the product of the beam pattern of one of the directional transducers and the beam pattern of an array of omnidirectional transducers having identical geometry as the array of directional transducers. Therefore, if $bp_D(\theta,\varphi)$ is the beam pattern of the directional transducer and $bp_A(\theta)$ is the beam pattern of an array of omnidirectional transducers having an identical geometry, then the beam pattern of the array of directional transducer will be

$$\mathrm{bp_T}(\theta,\varphi) = \mathrm{bp_D}(\theta,\varphi)\mathrm{bp_A}(\theta). \tag{3.59}$$

As an example, let us consider an array of n rectangular pistons, each piston having a height H and a width W, as depicted in Fig. 3.10, so that from Eqs. (3.37) and (3.57) we have

$$bp(\theta) = \left(\frac{\sin\psi_H}{\psi_H}\frac{\sin\psi_W}{\psi_W}\right)^2\left(\frac{\sin n\psi_A}{n\,\sin\psi_A}\right)^2, \tag{3.60}$$

where $\psi_H = (kH/2)\sin\theta$, $\psi_w = (kw/2)\sin\varphi$ and $\psi_A = (kd/2)\sin\theta$.

3.3.4 Electrically Steered Beam

The main lobe of a linear array of n-identical omnidirectional elements spaced a distance d apart can be electrically steered to any angle θ without the need of mechanically rotating the whole array. Consider the linear array of elements A, B, C, n depicted in Fig. 3.17a as a linear array of projectors. Suppose the signal to element B is delayed by a time τ relative to element A and the signal to element C is delayed relative to B and the signal to D is delayed relative to C, etc., by the same time τ. From Fig. 3.17a, we see that the wave front will be along the line RST and the direction of the signal is toward AR, making an angle θ_0 with the normal to the array. The angle θ_0 is equal to

$$\sin\theta_0 = \frac{c\tau}{d}. \tag{3.61}$$

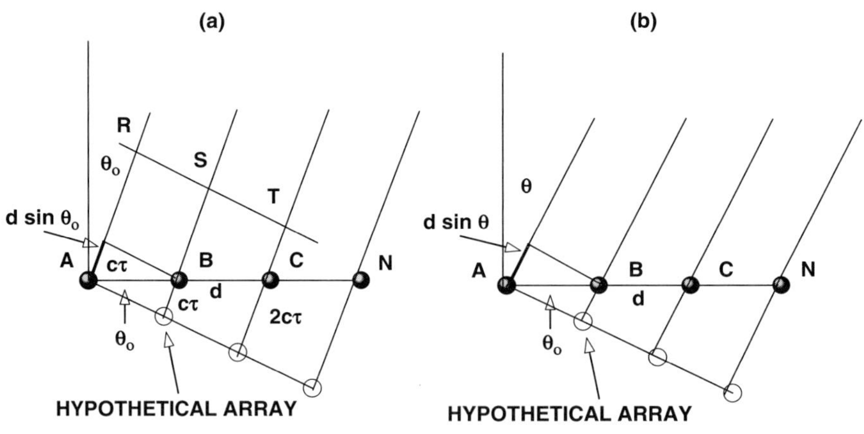

FIGURE 3.17. (**a**) Geometry of an n-element array with the signal directed to each element delayed by a time τ relative to the adjacent element. (**b**) Geometry to compute the beam pattern of an electrically steered array of n-elements.

From Fig. 3.17a, the difference in the distance from an element to the wave front RST and the similar distance of the adjacent element to the wave front RST can be expressed as

$$\begin{aligned} &\mathrm{AR} - \mathrm{BS} = c\tau = \mathrm{d}\ \sin\ \theta_0, \\ &\mathrm{BS} - \mathrm{CT} = c\tau = \mathrm{d}\ \sin\ \theta_0, \\ &\text{etc.} \end{aligned} \tag{3.62}$$

The effects of the delay τ is to steer the beam axis through the angle θ_0 given by Eq. (3.61). By reversing the direction of delay, the beam axis can be steered in the opposite direction. Since a delay is equivalent to a phase shift in the signal, the phase shift due to the delay τ can be expressed as

$$\psi_0 = \frac{2\pi}{\lambda}\, c\tau = \frac{2\pi\, d}{\lambda} \sin\ \theta_0. \tag{3.63}$$

The successive delay between each element is equivalent to taking the original array and creating a hypothetical array that is rotated by an angle θ_0 as indicated in Fig. 3.17b. The elements of the hypothetical array are all driven in phase without any delay between them. In order to determine the beam pattern of this electrically steered array, compare the phase of the wave radiated from element A with that of B. Let the signal from element A propagating at an angle θ from the normal to the array, travel a distance AL given by the equation

$$\mathrm{AL} = \mathrm{d}\ \sin\ \theta_0. \tag{3.64}$$

Since the signal from element B is delayed by a time τ relative to the signal from element A, the signal from element B would have traveled a distance of (AL–$c\tau$) so that the difference in the effective distance traveled is

$$\Delta r = d(\sin\ \theta_0 - \sin\ \theta), \tag{3.65}$$

and the phase difference is

$$\Delta\psi = \frac{2\pi\, d}{\lambda}\ (\sin\ \theta - \sin\theta_0). \tag{3.66}$$

The same phase difference occurs between the signal from elements B and C and between the signal from elements C and D, etc. However, if there was no delay between the elements, then the phase difference between the signals from elements A and B would merely be

$$\Delta\psi = \frac{2\pi\, d}{\lambda} \sin\ \theta. \tag{3.67}$$

The effect of the time delay with its associated phase delay is to change sin θ to (sin θ–sin θ_0). Therefore, the beam pattern of an n-element array given by Eq. (3.57) can be applied to the same n-element array electrically steered by an angle θ_0, so that

$$bp(\theta) = \left(\frac{\sin n\psi}{n\ \sin\psi}\right)^2, \tag{3.68}$$

where $\psi = (2\pi d/\lambda)$ (sin θ–sin θ_0) and sin $\theta_0 = c\tau/d$. Equation 3-68 also applies to an array of omnidirectional receiving elements. For a receiving array of hydrophones, the output of each hydrophone can be delayed relative to its neighbor by a fixed amount to steer the major lobe of the receiving beam. Let us now consider an example using an array of six-elements omnidirectional hydrophones, each element spaced at a distance of $\lambda/2$ apart. The beam pattern for different values of θ_0 between 15° and 90° are displayed in Fig. 3.18. The beam pattern for all the elements driven simultaneously has already been depicted in Fig. 3.14. Several features of the steered beams are obvious in Fig. 3.18. First, the greater the angle through which the beam is steered (θ_0), the wider the beam becomes. Secondly, for steering angles greater than about 45°, the main lobes in the first and fourth quadrant begin to merge to form a wide beam aligned along the axis of the array and a lobe directed in the opposite direction along the axis of the array begins to develop. For steering angles between 0° and 45°, there is a front–back ambiquity making

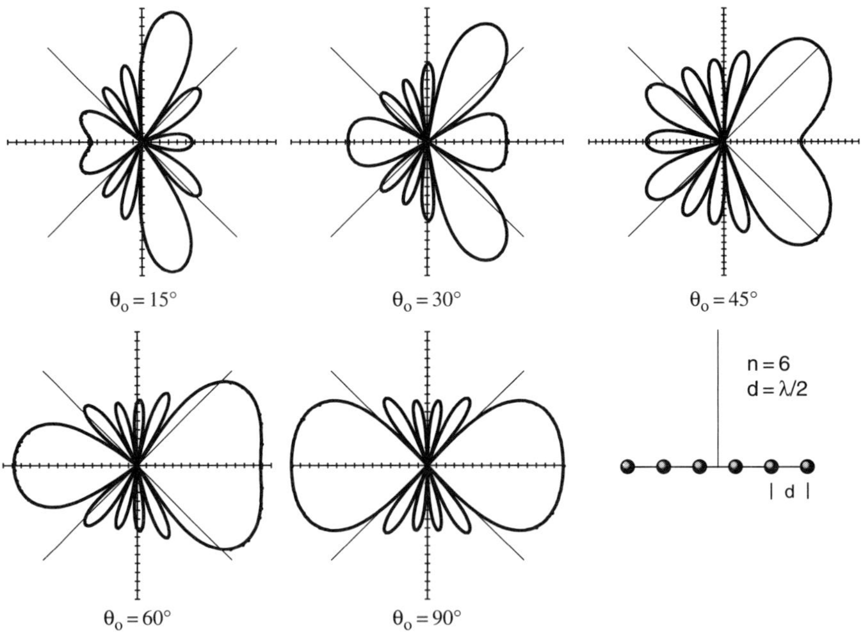

FIGURE 3.18. Beam patterns of a six-hydrophones array for different steering angles. The spacing between hydrophones is one half wavelength.

it impossible to tell whether a signal is being received by the front or back lobe, and for steering angles greater than 45°, the beam becomes so broad that it would be difficult to localize the azimuth of a sound source. Finally, the major lobes of steered beams are no longer symmetrical about the line joining the elements, but form a conical structure about this line, as can be seen by the beam patterns between 15° and 45°.

Equation (3.68) is convenient for determining the beam pattern of an electrically steered line array. However, if we are interested in determining the acoustic pressure of a signal arriving at the array at an angle θ, we can approach the problem in a slightly different manner. Figure 3.19 is a schematic of the conventional delay and sum technique for array beamforming in which the axis of the four-element line array is pointing in the direction θ. The delay process essentially ensures that the signal from the wave front arrives simultaneously at the input of the summer. The delay in the time of arrival of the signal for adjacent hydrophones is

$$\tau = \frac{d}{c}\sin\theta. \tag{3.69}$$

Therefore, the output of each successive hydrophone, starting with hydrophone $m = 0$ must be delayed by an amount

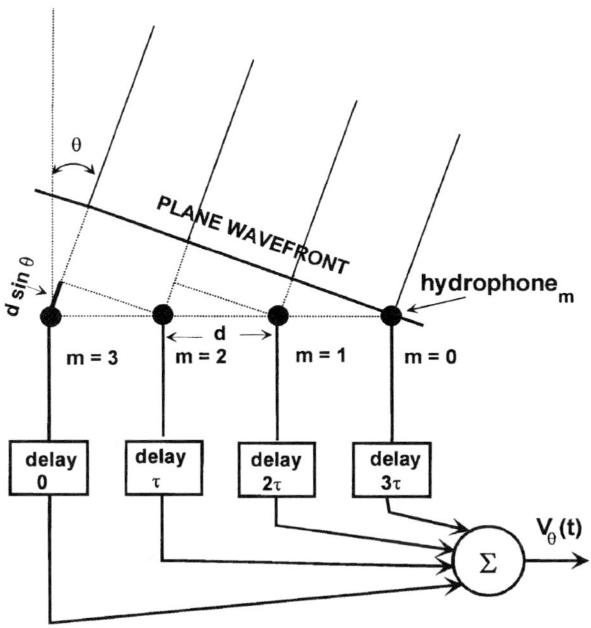

FIGURE 3.19. Schematic of a conventional delay and sum technique for beamforming (adapted from Brigham (1988).

$$m\tau = \frac{md}{c}\sin\theta. \tag{3.70}$$

Signal recombination or spatial array beamforming is then achieved by coherently summing the delayed hydrophone outputs, so that

$$V_\theta(t) = \sum_{m=0}^{N} v_m(t - m\tau). \tag{3.71}$$

In Eq. (3.71) we use the nomenclature $v_m(t–m\tau)$ to indicate a delay of $m\tau$ with the understanding that $v_m(t–m\tau)$ becomes valid only for time $t \geq m\tau$. In Chapter 6, where Fourier analysis will be discussed, we will show how from Fourier analysis a simple technique can be applied to beam steering.

The process of amplitude shading and steering of beam in an array is often referred to as beamforming. Arrays that are steered electronically are often referred to as phased array, since the processes of delaying either the transmitted or received signal of an element relative to another element will introduce a phase shift when working with single-frequency signals. In this chapter, we have considered only arrays that have elements aligned along a line, typically referred to as linear arrays. However, elements can be arranged in a two-dimensional pattern in a plane, a three-dimensional pattern within a volume, or around the perimeter of a ring or cylinder, etc. The analysis of these type of more complex arrays are beyond the scope of this book. Readers interested in more complex array geometries can consult a text by Ziomek (1985).

3.3.5 Multibeam Sonar

The principles involved with electronically steering an acoustic beam can be applied to an active sonar array to produce a "multibeam" system. A multibeam sonar is one in which azimuth and range of a target can be obtained with the echoes produced from a single ping. Therefore, with such a sonar, mechanical scanning of the sonar head is not necessary in order to obtain azimuthal information. As an illustration of the multibeam process, consider the six-element line array of Fig. 3.14 in which the elements are attached to a corprene backplane so that the back lobes are suppressed. With the elements spaced at a distance of $\lambda/2$, the effective beams from such an array that is designed as a multibeam receiver is shown in Fig. 3.20. The beam patterns were calculated using Eqs. 3.66 and 3.68. The side lobes can be suppressed by appropriate amplitude shading or weighting as discussed in Section 3.3.2. There are nine beams in Fig. 3.20, with each beam having a width that is the same as the basic six-element array of Fig. 3.14. The width of each beam is determined by the effective length of the array, which will be directly proportional to the number of elements spaced $\lambda/2$ apart.

In the age of digital electronics, obtaining a multibeam function with an array of hydrophones is fairly straightforward. We will use the process that is shown schematically in Fig. 3.19.

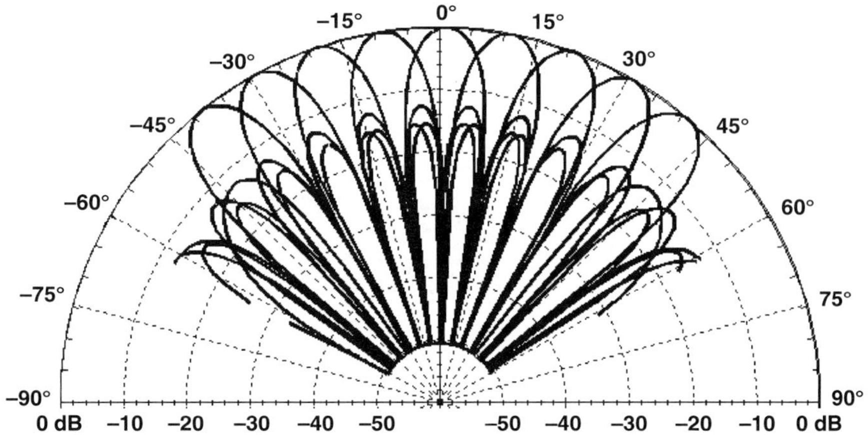

FIGURE 3.20. An example of the effective beam pattern of a six-element array with the elements spaced $\lambda/2$ apart and the output of the elements processed to form a multi-beam pattern.

Assume that the signal received by each element of a narrow-band sonar is demodulated in a process called quadrature sampling (other names include complex demodulations, quadrature heterodyning) so that the carrier frequency is removed. The demodulated signal will be much lower in frequency than carrier so that the A/D sampling rate can be reduced considerably than if signals at the carrier frequency were being digitized. The quadrature sampling process is beyond the scope of the book; interested readers should refer to several excellent books written by Brigham (1988), Lyons (1997), and Proakis and Manolakis (2006).

The digitized data from the M hydrophones are stored as a $\mathbf{N} \times \mathbf{M}$ matrix (see Fig. 3.21a), where each row of the matrix represents the value of the demodulated echo at some time $k\Delta t$ where $k = 1, 2, \ldots M$, and Δt is the sampling interval. Multiple beams can be formed by selecting the azimuth limits over which θ in Fig. 3.19 will vary and the step-size or $\Delta\theta$ (usually the beamwidth) over which the whole range of azimuthal angle will be covered. If we designate the angle of the i-th beam as θ_i, the appropriate delays (from Eq. 3.69) will be

$$\tau_i = \frac{d}{c}\sin\,\theta_i, \tag{3.72}$$

and the output from each hydrophone should be multiplied by $m\tau_i$ as described by Eq. (3.70). In order to continue our discussion we need to either jump ahead and borrow an expression and idea from Chapter 6, or wait until we reach Chapter 6 to continue our discussion. Since we are on the topic, we believe it would be better to jump ahead and consider a theorem from

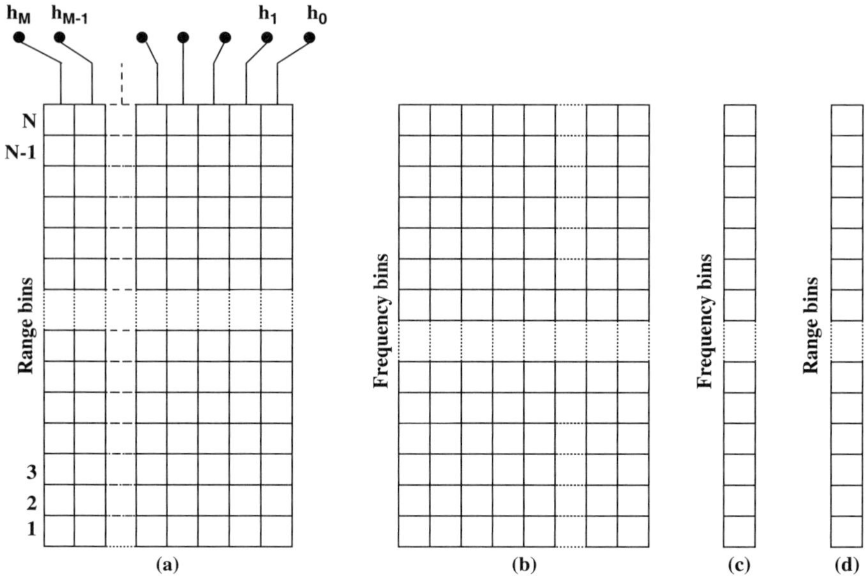

FIGURE 3.21. (**a**) Schematic of the digitized output of N hydrophones stored as an N × M matrix, where the each row represent the digitized value for at a specific time or range for each hydrophone. Therefore, each column represents the base band signals from a specific hydrophone; (**b**) the N × M matrix after each column is Fourier-transformed; (**c**) column vector after each row is summed and (**d**) amplitude versus range or time of the signal arriving at an angle θ_i.

Fourier analysis, which states that the Fourier transform of a time-shifted signal $s(t{-}t_0)$ will be (From Eq. (6.48))

$$\Im[s(t - \tau_i)] = e^{-j2\pi f \tau_i} S(f), \tag{3.73}$$

where $\Im$ is the Fourier transform symbol operating on the variable in the brackets and $S(f)$ is the resultant Fourier transformed variable of $s(t)$ and is a complex variable. Recall that the columns of our **M** × **N** data matrix are the digitized values from each hydrophone. Therefore, if the Fourier transform is applied to each column of our **N** × **M** data matrix and each row of each column is multiplied by $e^{-j2\pi f m \tau_i}$, then the **N** × **M** data matrix represent the frequency representation of a signal arriving at an angle of $\theta_{i.}$. The summation of the values in a specific row is equivalent to the summation in Eq. (3.71), representing what the signal received by the array arriving from an azimuth of θ_i at a specific time. Therefore, we can sum across each row to form the column vector shown in Fig. 3.21b, which now represents the Fourier transform of the signal. By progressively multiplying each Fourier transformed columns by $e^{-j2\pi f m \tau_i}$ and compute the inverse Fourier transform on the column vector, we arrive at the signal arriving at an azimuth of θ_i in different range bin. By stepping through the azimuthal angle in increments of $\Delta\theta$, the echoes as a function of the angle θ and range can be determined for a single transmission. For a more detail discourse on multi-beam processing, the reader is referred to Ziomek (1985).

3.3.6 Directivity Index

The degree of directionality of either a receiving or transmitting transducer can be described by its directivity index. For a receiving hydrophone, it is the ratio of the power received by an omnidirectional receiver to that received by a directional receiver in the same isotropic noise field, as described by the equation

$$\mathrm{DI} = 10\ \log D = 10\ \log\left(\frac{P_0}{P_D}\right), \tag{3.74}$$

where D is the directivity of the hydrophone, P_0 is the total acoustic power received by an omnidirectional hydrophone, and P_D is the total acoustic power received by a directional hydrophone. Since a directional hydrophone will receive less noise than a nondirectional hydrophone of the same sensitivity, the directivity index will be a positive quantity. The larger the value of DI, the more directional the hydrophone. Assume that an omnidirectional and a directional hydrophone are located at the origin of the coordinate system shown in Fig. 3.22, where the acoustic axis of the directional transducer is pointed in the z-direction. The amount of acoustic power received by the omnidirectional hydrophone in an isotropic noise field of intensity I_0 is I_0A, where A is the area of the sphere about the origin. Therefore, the power received by the omnidirectional hydrophone will be

$$\mathrm{P_0} = 4\pi \mathrm{r}^2 \mathrm{I_0}. \tag{3.75}$$

The power received by the directional transducer will be

$$P_D = \int_0^{4\pi} \left(\frac{p(\theta,\ \phi)}{p_0}\right)^2 I_0 \mathrm{d}S, \tag{3.76}$$

where $[p(\theta,\varphi)/p_0]^2$ is the beam pattern of the directional transducer and dS is the incremental area of the sphere shown in Fig. 3.22. The receiving intensity of the directional hydrophone is merely the beam pattern multiplied by the isotropic intensity I_0 in Eq. (3.74). From Fig. 3.19, the incremental area on the surface of the sphere can be expressed as

$$\mathrm{dS} = \mathrm{r}^2 \sin\theta\ \mathrm{d}\theta\ \mathrm{d}\varphi. \tag{3.77}$$

Therefore, the directivity can be expressed as

$$D = \frac{4\pi}{\int_0^{2\pi}\int_0^{\pi}\left(\frac{p(\theta,\phi)}{p_0}\right)^2 \sin\theta\, \mathrm{d}\theta\, \mathrm{d}\phi}. \tag{3.78}$$

Although the directivity index equation is relatively simple, using it to obtain numerical values can be a formidable task unless transducers of relatively simple shapes such as cylinders, lines, and circular pistons, having symmetry about one axis, are involved. Otherwise, the beam pattern would need to be measured as a function of both θ and φ. This is done by choosing various discrete values of θ and measuring the beam pattern as a function of φ, a rather tedious process.

We will consider a simple example involving the circular piston of radius a with the piston located at the origin of the coordinate system of Fig. 3.22 and facing upward. We shall assume an ideal infinite battle that confines the sound field to the forward hemisphere. Under this assumption, we can insert Eq. (3.35) into Eq. (3.76), we get for the directivity

$$D = \frac{4\pi}{\int_0^{2\pi} d\phi \int_0^{\pi/2} \left(\frac{2J_1(ka\sin\theta)}{ka\sin\theta}\right)^2 \sin\theta\, d\theta}. \tag{3.79}$$

Performing the integration, we arrive at (Kinsler et al., 1982)

$$D = \frac{(ka)^2}{1 - \frac{J_1(2ka)}{ka}} \tag{3.80}$$

For the case in which the diameter of the piston is much greater than the wavelength ($ka >> 1$), the Bessel function becomes very small and we have

$$D = (ka)^2 = \frac{4\pi A}{\lambda^2}. \tag{3.81}$$

where A is the area of the circular piston.

The expression for the directivity of a circular piston indicates that for a fixed frequency the larger the piston the more directional the beam, or for a fixed dimension, the higher the frequency the narrower the beam. These relationships between size and directivity and frequency and directivity are valid for any directional transducer. Expressions for the beam pattern and directivity index of different types of transducers are included in Table 3.1. The directivity index for the n-element array is valid for both a shaded and an

TABLE 3.1. Beam Pattern and Directivity Index for Different Transducer Shapes

Type	$bp(\theta)$	$\mathrm{DI} = 10 \log bp$
Continuous line of length L $L>>\lambda$ and $\psi = kL/2 \sin\theta$	$[\sin\psi/\psi]^2$	$2L/\lambda$
Circular piston of radius a $a>>\lambda$ and $\psi = a\sin\theta$	$[2J_1(\psi)/\psi]^2$	$4\pi A/\lambda^2$
Rectangular piston of dimensions H and W $\psi_H = kH/2 \sin\theta$ $\psi_W = kW/2 \sin\varphi$	$\left(\frac{\sin\psi_H \sin\psi_W}{\psi_H \psi_W}\right)^2$	$\frac{4\pi A}{\lambda^2}$
Line array of n elements of equal spacing d and $\psi = kd/2 \sin\theta$		

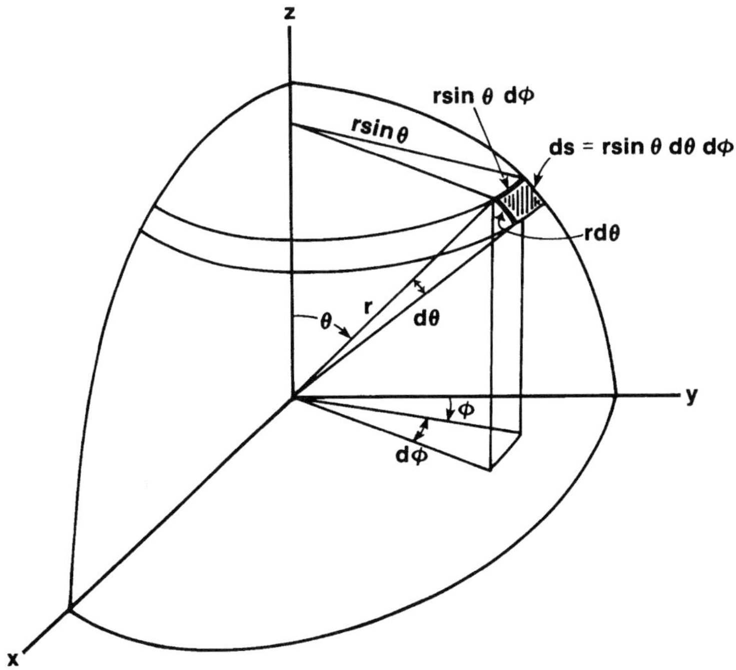

FIGURE 3.22. Coordinate system used to derive the directivity index of a transducer. The incremental area dS is located on the surface of the sphere.

unshaded array. Although amplitude shading reduces the side lobes, the main lobe becomes broader and this effect tends to keep the directivity index from changing significantly for a shaded array.

References

Bobber, R. J. (1970). *Underwater Electroacoustic Measurements* (Naval Research Lab, Washington, D.C.)

Brigham, E. O. (1988). *The Fast Fourier Transform and Its Applications* (Prentice Hall, Englewood Cliffs, New Jersey).

Eckart, C., (Ed.). (1968). *Principles and Applications of Underwater Sound* (Department of the Navy, NAVMAT P-9674, Government Printing Office, Washington, DC).

Kinsler, L. E., Frey, A. R., Coppens, A. B., and Sanders, J. V. (1982). *Fundamentals of Acoustics*, 3rd ed. (John Wiley & Son, New York).

Kraus, J. D. (1950). *Antennas* (McGraw Hill, New York).

Lyons, R. G. (1997). *Understanding Digital Signal Processing* (Addison-Wesley, Reading, Massachucetts).

Proakis, J. G. and Manolakis, D. G. (2006). *Digital Signal Processing: Principles, Algorithms, and Applications*, 4th ed. Prentice Hall, (New York).

Urick, R. J. (1983). *Principles of Underwater Sound* (McGraw Hill, New York).

Ziomek, L. J. (1985). *Underwater Acoustics: A Linear Systems Theory Approach* (Academic Press, New York).

4

Acoustic Propagation

4.1 Basic Principles

We discussed in Chapter 1 the fact that sound consists of mechanical vibrations that travel as a wave. It will not propagate in a vacuum, but requires some material to act on for transmission. The medium for propagation can be either solid or fluid. In fluids such as water and air the propagation is only longitudinal because fluids cannot support shear wave and so the fluid particles vibrate along the direction of energy propagation. Sound is usually considered to be just a pressure wave, primarily because pressure is an easily measured quantity; however, sound has various other descriptors. If pressure changes locally in a fluid, the temperature and density will also change. Thus temperature and density are also acoustic variables. In addition, we showed in Chapter 1 that propagation of sound has a vector description, meaning that both magnitude and direction are associated with the acoustic particle velocity, displacement and acceleration. All acoustic quantities, vectors (e.g., particle velocity, displacement and acceleration) and scalars (e.g., pressure, density and temperature), propagate through a fluid at the sound speed, *c*. The time and spatial variations of these acoustic quantities are all defined by the wave equation for the specific quantities.

4.1.1 Plane Waves and Acoustic Impedance

Plane waves are planes of constant pressure that propagate longitudinally as illustrated in Fig. 4.1. Longitudinal propagation means that the direction of propagation is perpendicular to the plane of constant pressure. A plane wave does not have to be a harmonic function; it can be any function of time as long as it repeats itself in space. In Section 1.2, we treated particle velocity from a simple perspective, assuming a plane wave in the far field of a source. In this chapter we will take a more general approach, describing the spatial variable as a vector, $\vec{r}$, so that the pressure and particle velocity for a forward traveling plane wave are

$$p\left(\vec{r},t\right) = f\left(t - \frac{\vec{n}\cdot\vec{r}}{c}\right) \tag{4.1}$$

W.W.L. Au, M.C. Hastings, *Principles of Marine Bioacoustics*,
DOI: 10.1007/978-0-387-78365-9_4,

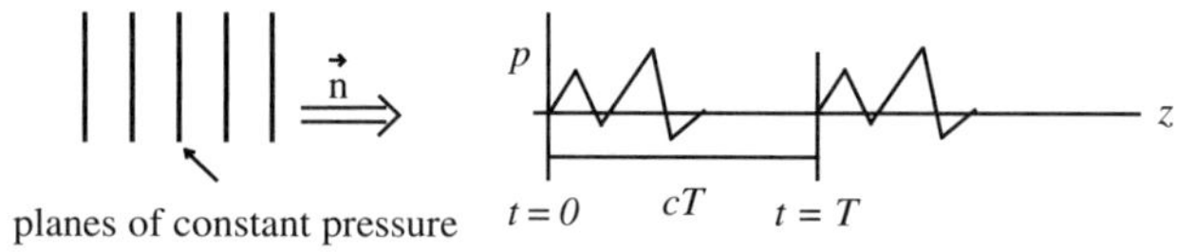

FIGURE 4.1. Plane waves propagate longitudinally and can have any shape as long as they repeat themselves in time and space.

$$\vec{u}\left(\vec{r},t\right) = \frac{\vec{n}}{\rho_0 c} p\left(\vec{r},t\right) \tag{4.2}$$

where f denotes any function and is the forward traveling wave in Eq. (1.18), $\vec{n}$ is the unit vector perpendicular to the planes of constant pressure and pointing in the longitudinal direction and $\vec{n} \cdot \vec{r}$ is the direction of propagation. For a forward traveling plane wave, the acoustic pressure and particle velocity are in phase; but, for a wave traveling in the opposite direction the pressure and particle velocity are 180° out of phase. The characteristic impedance, $\rho_0 c$, is analogous to electrical resistance in relating acoustic pressure ("voltage") to particle velocity ("current").

Spherically symmetric waves propagate radially outward such that the pressure is a function of radius (r) from the center of the source. They form spherical surfaces of constant pressure. The wave equation in the spherical coordinate system is given by Eq. (1.86) in the Appendix to Chapter 1. For spherically symmetric waves with no θ and ϕ dependency the wave equation reduces to

$$\frac{\partial^2 p}{\partial r^2} + \frac{2}{r}\frac{\partial p}{\partial r} = \frac{1}{c^2}\frac{\partial^2 p}{\partial t^2} \tag{4.3}$$

Because energy is conserved and the wave is spread out over larger and larger surface areas as it propagates, the pressure decreases as the distance (radius r) from the source increases. This relationship is given by the solution to Eq. (4.3) which is similar to Eq. (1.18):

$$p(r,t) = \frac{f(t - r/c)}{r} \tag{4.4}$$

The decrease in pressure and acoustic intensity with increasing radius is called spherical spreading. Note that the vector notation has been dropped for convenience.

The acoustic particle velocity generated by a spherical source can be determined from the equation for conservation of momentum in an ideal fluid (Euler's equation) as was done in arriving at Eq. (1.9) but expressed in a more generalized form:

$$\rho_0 \frac{\partial v_r}{\partial t} = \frac{-\partial p}{\partial r} \tag{4.5}$$

Using Eq. (4.4) to evaluate the right-hand side of Eq. (4.5) gives

$$\rho_0 \frac{\partial u_r}{\partial t} = \frac{f'(t - {}^r/c)}{cr} + \frac{f(t - {}^r/c)}{r^2} \tag{4.6}$$

Then integration of Eq. (4.6) yields

$$\vec{u}(r,t) = \left[\frac{p(r,t)}{\rho_0 c} + \frac{\int\limits_{-\infty}^{t} p(r,\tau)\mathrm{d}\tau}{\rho_0 r} \right] \vec{e}_r \tag{4.7}$$

where $\vec{e}_r$ is the unit vector in the radial direction (i.e., the direction of wave propagation). Thus the particle velocity of the spherically symmetric wave has two parts, one that propagates because of the compressibility of the fluid and the other that is a bulk flow near the source. Far away from the source ($r >> \lambda$ for harmonic waves) in the "far field," the relationship between pressure and particle velocity is the same as that for a plane wave, $u_r = p/\rho_0 c$, which is a generalized form of Eq. (1.36). Closer to the source in what is called the "near field," the second term (hydrodynamic flow) in Eq. (4.7) is important. For example, at 50 Hz, $\lambda = 30$ m in water and usually measurements are made only 1 m away from the source to characterize it. At a frequency of 50 Hz, 1 m is in the near field of the source. The impedance for a harmonic spherical wave is

$$Z = \frac{p}{u_r} = \rho_0 c \frac{1}{1 - {}^\mathrm{j}/kr} = \frac{\rho_0 c(1 + {}^\mathrm{j}/kr)}{1 + ({}^\mathrm{j}/kr)^2} \tag{4.8}$$

Note that the impedance is a complex quantity.

4.1.2 *Surface and Bottom Reflections*

Acoustic plane waves are often depicted by a ray perpendicular to the plane of constant pressure and pointing in the direction of propagation. Figure 4.2 uses rays to show incident (Subscript i), reflected (Subscript r)and transmitted (Subscript t) waves traveling from fluid medium 1 to fluid medium 2 going across a fluid interface. At the boundary between two fluids, such as air and water, which has two different sound speeds (c_1 and c_2) and densities (ρ_1 and ρ_2), Snell's law applies:

$$\frac{\cos\theta_\mathrm{i}}{\cos\theta_\mathrm{t}} = \frac{c_1}{c_2} = \frac{k_2}{k_1} \tag{4.9}$$

where the wave number is $k = \frac{\omega}{c} = \frac{2\pi}{\lambda}$. From the figure, we see that waves are refracted by passing through the interface so that the transmission angle, θ_t, will be different than the incident angle, θ_i, whereas the reflection angle will be the same as the incident angle. Figure 4.2 also depicts the wavefront as

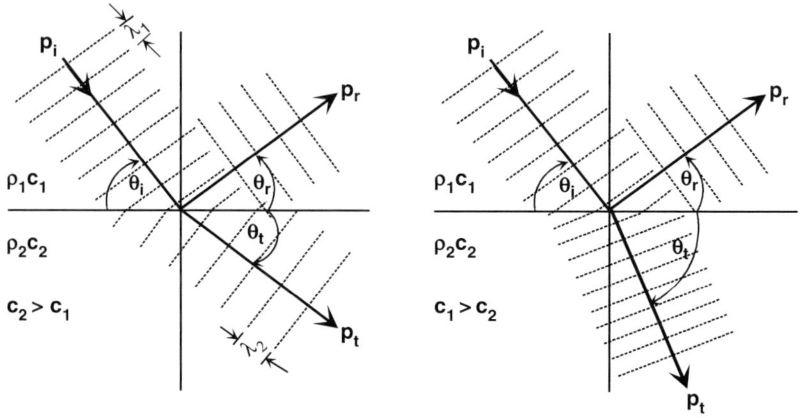

FIGURE 4.2. Reflection and transmission of an incident plane wave at a fluid–fluid interface.

parallel lines that are perpendicular to the direction of propagation. Figure 4.2a depicts the acoustic signal traveling from a slower medium to a faster medium with the transmitted ray refracting or bending toward the medium with the lower sound velocity. This figure also provides a graphical illustration of the refraction process. The portion of the wavefront in the medium with the higher sound velocity begins to move faster than the portion in the slower medium. The opposite situation is shown in Fig. 4.2b with the acoustic signal propagating from a medium of higher sound velocity into a medium of lower sound velocity. The refraction process can also be seen in the cases in which portions of a wavefront are in both media at the same time. Again, the transmitted signal bends toward the medium of lower sound velocity. Although the wavelength of the signal is shorter in the faster medium, the frequency of the incident, reflected and transmitted signals are the same since the difference in wavelength is exactly offset by the difference in sound velocity.

At the boundary between the two media, there are two boundary conditions that must be satisfied: (1) pressure must be continuous at $y = 0$ and (2) normal surface velocity must be continuous at $y = 0$. The continuity condition is merely a statement that a variable at an infinitesimally small distance from the boundary in one medium (at the boundary) will have the same value at an infinitesimally small distance into the second medium. If we let p_i be the pressure of the incident acoustic wave, p_r be the pressure of the reflected wave and p_t be the pressure of the transmitted wave, the first boundary condition states that $p_i + p_r = p_t$ which results in

$$1 + R = T \tag{4.10}$$

where $R = p_r/p_i$ is the reflection coefficient at the boundary and $T = p_t/p_i$ is the transmission coefficient. From Snell's law (Eq. (4.9)), $k_1 \cos\theta_i = k_2 \cos\theta_t$. Using the trigonometric identity, $\cos^2\theta + \sin^2\theta = 1$, we have for the transmission angle

$$\sin\theta_t = \sqrt{1 - \left(\frac{c_2}{c_1}\right)^2 \cos^2\theta_i} \tag{4.11}$$

This equation has three implications. (1) If $c_1 > c_2$ the angle of transmission θ_t is real and less than the angle of incidence. Therefore, the transmitted wave is bent away from the interface. (2) If $c_1 < c_2$ and $\theta_i < \theta_c$ where θ_c is the *critical angle* defined by

$$\cos\theta_c = \frac{c_1}{c_2} \tag{4.12}$$

the angle of transmission is again real but greater than the angle of incidence so that the transmitted wave is bent toward the interface for all incident angles less than the critical angle. (3) If $c_1 < c_2$ and $\theta_i > \theta_c$, then Eq. (4.11) becomes purely imaginary and the incident wave is totally reflected.

The second boundary condition results in

$$u_{n,t} = u_{n,i} + u_{n,r} \tag{4.13}$$

From Eq. (4.2), the particle velocities normal to the boundary can be expressed as

$$u_{n,i} = \frac{-p_i \sin\theta_i}{\rho_1 c_1}\, u_{n,t} = \frac{-p_t \sin\theta_t}{\rho_2 c_2} \textit{ and } u_{n,r} = \frac{p_r \sin\theta_i}{\rho_1 c_1} \tag{4.14}$$

Substituting these latter expressions into Eq. (4.13) and using the definition of reflection and transmission coefficients and the expression for $\cos\theta_t$ given by Eq. (4.11), we obtain

$$\frac{\sin\theta_i}{\rho_1 c_1}(1 - R) = \frac{T\sqrt{1 - \left(\frac{c_2}{c_1}\right)^2 \sin^2\theta_i}}{\rho_2 c_2} \tag{4.15}$$

An expression for the pressure reflection coefficient is found by simultaneously solving Eqs. (4.10) and (4.15) for $\theta_i < \theta_c$:

$$R = \frac{\rho_2 c_2 \sin\theta_i - \rho_1 c_1 \sin\theta_t}{\rho_2 c_2 \sin\theta_i + \rho_1 c_1 \sin\theta_t} \tag{4.16}$$

Similarly, the transmission coefficient can be expressed as

$$T = \frac{2\rho_2 c_2 \sin\theta_i}{\rho_2 c_2 \sin\theta_i + \rho_1 c_1 \sin\theta_t} \tag{4.17}$$

The reflection and transmission coefficients can be written in terms of the specific acoustic impedance of both medium as

$$R = \frac{z_2 \sin\theta_i - z_1 \sin\theta_t}{z_2 \sin\theta_i + z_1 \sin\theta_t} \qquad T = \frac{2z_2 \sin\theta_i}{z_2 \sin\theta_i + z_1 \sin\theta_t} \tag{4.18}$$

For the case in which $\theta_i = 90°$ (the incident signal is perpendicular to the interface), we have

$$R = \frac{z_2 - z_1}{z_2 + z_1} \qquad T = \frac{2z_2}{z_2 + z_1} \tag{4.19}$$

If a plane wave is traveling from air into water then $\rho_2 c_2 >> \rho_1 c_1$ ($Z_2 >> Z_1$) and from Eq. (4.19) $R \approx 1$; likewise if the wave is traveling from water into air then $\rho_1 c_1 >> \rho_2 c_2$ ($Z_1 >> Z_2$), $R \approx -1$. The reflection coefficient of $R \approx -1$ dictates that the pressure at the air–water interface is essentially zero. Thus it is very difficult to make measurements near the water surface. Although the water surface is an excellent reflector of sound, the reflected wave is 180° out of phase at the surface because the acoustic pressure is extremely small since the surface-reflected sound tends to cancel the direct sound. The bottoms of bodies of water are usually poor reflectors ($R < 0.5$).

4.1.3 *Absorption and Refraction*

Sound waves traveling in the ocean can suffer an absorption loss in which acoustic energy is converted into heat. Absorption losses in seawater are caused by the effect of shear and volume viscosity, ionic relaxation of magnesium sulfate ($MgSO_4$), molecules and a complicated boric acid ionization process. In the frequency range between 10 and 200 kHz, the dominant absorption is due to ionic relaxation of $MgSO_4$. The attenuation rate in this frequency range was derived by Fisher and Simmons (1977) as

$$\alpha = \frac{A_2 f_2 f^2}{f_2^2 + f^2} \quad \text{dB/m} \tag{4.20}$$

where

$A_2 = (48.8 \times 10^{-8} + 65.4 \times 10^{-10}\, T)$ s/m
$f_2 = 1.6 \times 10^7 (T + 273)\exp[-3{,}052/(T + 273)]$ Hz
f = frequency in Hz
T = temperature in °C

Examples of typical acoustic absorption coefficients for shallow depths (<300 m) are plotted in Fig. 4.3 as a function of frequency for several water temperatures. The effect of depth on absorption is relatively small: absorption decreases about 2% for every increase of 300 m in depth. Transmission loss of an acoustic signal at a particular frequency that has traveled a distant

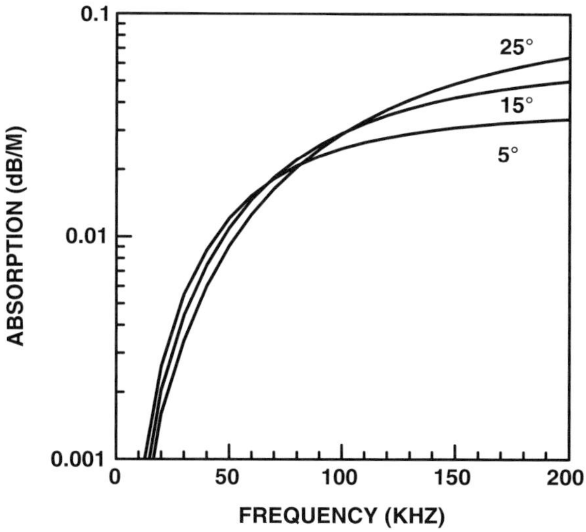

FIGURE 4.3. Absorption coefficient as a function of frequency for several different temperatures.

r will consist of a spreading loss term (20 log r for spherical spreading) plus an absorption term

$$\text{TL} = \text{spreading loss} + \alpha \text{r} \tag{4.21}$$

Spreading loss is no more than the spreading of acoustic energy over a larger area so that the acoustic intensity decreases as the wave propagates away from a source. The effects of spherical spreading was discussed in Chapter 1.

Refraction of sound is bending of the wave caused by a change in sound speed at an interface or a sound speed gradient, usually created by gradual changes in fluid temperature. The sound velocity as a function of depth in the ocean is a function of temperature, salinity and pressure (depth) and it will not be constant. A typical deep-sea sound velocity profile is shown in Fig. 4.4. The sound velocity profile can be divided into a *surface layer* in which the velocity of sound is susceptible to daily and local changes of heating, cooling and wind action, the *seasonal thermocline* (a layer in which temperature changes with depth) characterized by a negative temperature gradient and affected by seasons, a *main thermocline*under the seasonal one and finally the *deep isothermal layer* of constant temperature of about 39°F which extends to the bottom. The velocity of sound in the deep isothermal layer increases with depth because of the effect of pressure on sound. The change in sound velocity with depth is not very large; there is only a 25 m/s difference between the sound velocity at the surface and at a depth of 1,000 m, where the minimum in sound velocity is located in Fig. 4.4.

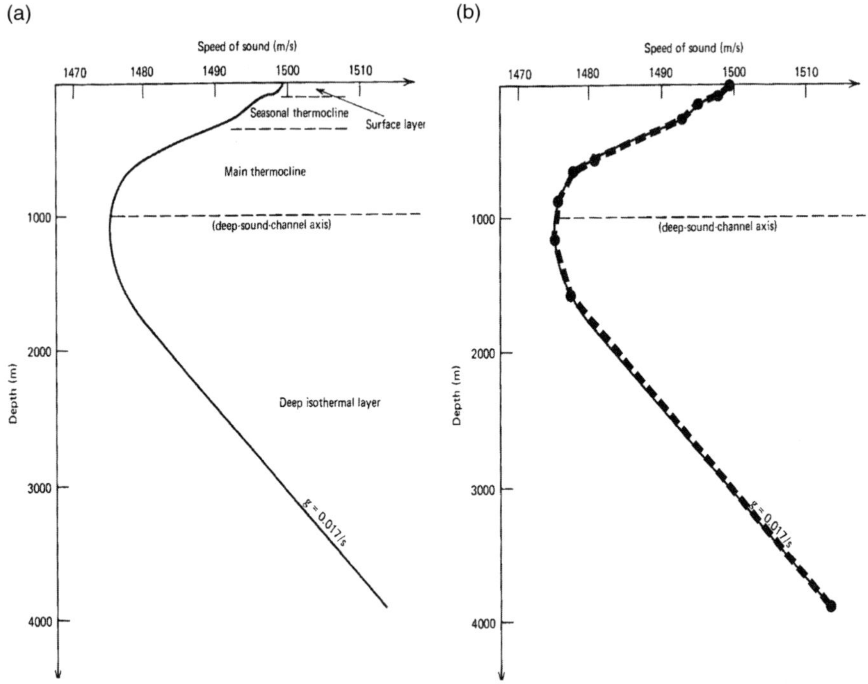

FIGURE 4.4. (a) Representative deep ocean sound velocity profile (from Urick, 1983), (b) a linearized version.

There are various expressions for sound velocity as a function of temperature, salinity and depth, depending on their range. A simple expression for sound velocity is given by Medwin (1975) for temperatures greater than 0°C but less than 35°C, salinity less than 45 parts per thousand and depth less than 1,000 m:

$$c(z) = 1,449.2 + 4.6T - 0.055T^2 + 0.00029T^3 + (1.34 - 0.01T)(S - 35) + 0.016z \tag{4.22}$$

where T is the temperature in °C, S is the salinity in parts per thousand and z is the depth in m.

4.2 Propagation of Sound in the Ocean

A simple model of sound velocity in the ocean is one in which the sound velocity varies only with depth and not horizontally so that the ocean can be stratified in depth and modeled by horizontal layers of different sound velocities. From such a simple model and Snell's law, one can easily come to the conclusion that sounds emanating from a source will undergo refraction as it propagates outward in a horizontally stratified ocean.

The surface and bottom will also affect the propagation of sound. Therefore, in order to determine how sound will propagate it is necessary to solve the wave equation given by

$$\nabla^2 p = \frac{1}{c(z)^2} \frac{\partial^2 p}{\partial t^2} \tag{4.23}$$

Determining simple ways of solving the wave equation for realistic ocean models has been the source of intense research under the sponsorship of the U.S. Navy for several decades after World War II. There are presently three fundamental approaches to predicting acoustic propagation in the ocean. The first involves *ray theory* in which the sound field is described by *rays*in an analogous manner as in optics. The second involves a waveguide approach called the *normal mode* approach and the third is the *parabolic equation* approximation to the wave equation which is then solved numerically using an iterative approach. The parabolic equation approach is often referred to as the PE model (Medwin and Clay, 1998). Most of the present-day ocean propagation software programs utilize the PE model. It would be beyond the scope of this book to dwell on these approaches; interested readers should refer to two excellent books on computational acoustics, edited by Lee and Schutz (1995) and Jensen et al. (1994), respectively. We will consider ray theory since it is the simplest approach for the neophyte to grasp and provides an intuitive understanding of the underlining physical principles involved with acoustic propagation in the ocean.

4.2.1 Ray Theory

Ray theory involves the application of Snell's law of refraction to a horizontally stratified ocean, each strata having a slightly different sound velocity profile. A convenient model is one in which each layer is described by a linearly varying sound velocity profile. It can produce a fairly accurate approximation to a sound velocity profile, as can be seen in Fig. 4.4b. We will now show that the path of sound traveling through a layer in which the sound velocity varies linearly with depth will be an arc of a circle. Considering the geometry in Fig. 4.5a, assume that the ray is horizontal at c_0 so that Snell's law in Eq. (4.9) can be written as

$$\cos\theta_1 = \frac{c_1}{c_0} \qquad \cos\theta_2 = \frac{c_2}{c_0} \tag{4.24}$$

The equation for the sound velocity is given by

$$c(z) = c_0 + gz \tag{4.25}$$

where g is the constant gradient in the sound velocity given by

$$g = \frac{c_1 - c_0}{d_1} = \frac{c_2 - c_0}{d_2} \tag{4.26}$$

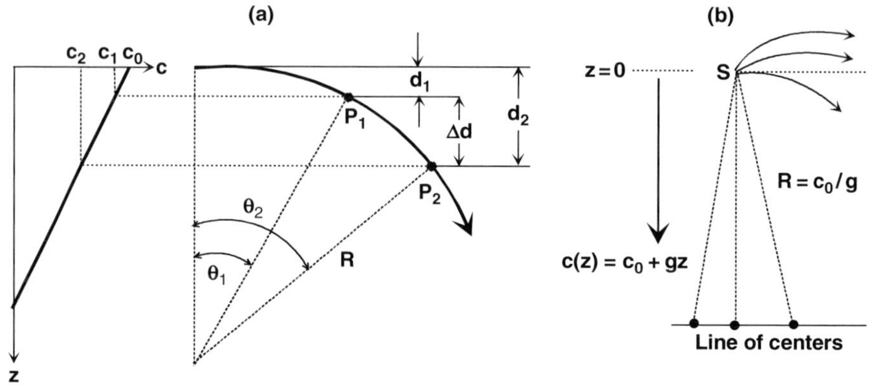

FIGURE 4.5. Diagram used in describing a ray path in a layer in which the sound velocity varies linearly with depth.

so that

$$c_1 = c_0 + gd_1 \qquad c_2 = c_0 + gd_2 \tag{4.27}$$

From the ray path diagram of Fig. 4.5b, we have

$$d_2 = R[1 - \cos\theta_2] \tag{4.28}$$

$$d_1 = R[1 - \cos\theta_1] \tag{4.29}$$

Therefore,

$$\Delta d = R[\cos\theta_2 - \cos\theta_1] \tag{4.30}$$

From Eq. (4.27) we have

$$d_2 - d_1 = \frac{c_2 - c_1}{g} \tag{4.31}$$

Eliminating $d_2 - d_1$ from Eqs. 4.31 and 4.30 and using Snell's law given by Eq. (4.24), we arrive at

$$R = \frac{c_0}{g} = \frac{c}{g\cos\theta} \tag{4.32}$$

R is a constant, therefore the ray path is an arc of a circle. The center of curvature of the rays leaving a source S where the velocity is c_0 is shown in Fig.4.5b to be lying along a horizontal line at a distance c_0/g below the source. If the gradient was reversed, then the line of centers would be above the source.

Consider now a ray of sound leaving its source at an angle of θ_0, where the sound velocity is c_0; the sound path will be as shown in Fig. 4.6. The angle of

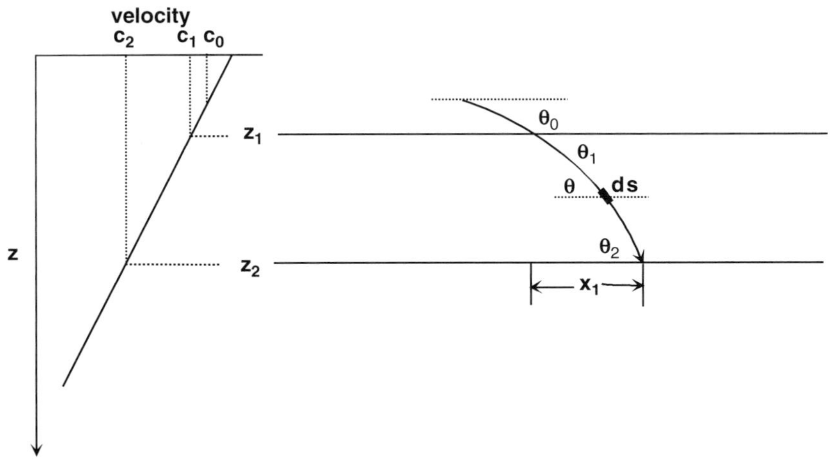

FIGURE 4.6. The ray path in a layer of linearly decreasing sound velocity versus depth.

incident and the angle of transmission from the layer can be determined from Snell's law to be

$$\theta_1 = \cos^{-1}\left[\frac{c(z_1)\cos\theta_0}{c_0}\right] \tag{4.33}$$

$$\theta_2 = \cos^{-1}\left[\frac{c(z_2)\cos\theta_0}{c_0}\right] \tag{4.34}$$

where $c(z_1)$ and $c(z_2)$ are calculated from Eq. (4.25). The ray in the layer will have a radius given by Eq. (4.32), $R = -c/(g\cos\theta)$. The negative sign comes from the convention of a downward-bending ray having a negative value and an upward-bending ray having a positive value. We will now determine the horizontal range, x_1, the ray traveled in the layer. Consider the small distance ds shown in Fig. 4.6, the ray travels horizontally a distance $dx_1 = ds^\circ \text{Cos}\,\theta$, and the distance along the arc is equal to $ds = R\,d\theta$. Therefore, we have for the horizontal distance traveled by the ray

$$\begin{aligned} x_1 &= \int_{\vartheta_1}^{\vartheta_2} \frac{-c_1}{g\cos\theta_1}\cos\theta\, d\theta \\ &= \frac{c_0}{\cos\theta_0}\,\frac{\sin\theta_1 - \sin\theta_2}{g} \end{aligned} \tag{4.35}$$

where $c_1/\cos\theta_1 = c_0/\cos\theta_0$ from Snell's law. We can generalize Eq. (4.35) for a multilayer situation in which the horizontal distance traveled by a ray within the ith layer can be expressed as

$$x_i = \frac{c_0}{\cos \theta_0} \frac{\sin \theta_i - \sin \theta_{i+1}}{g_i} \tag{4.36}$$

where $i = 0, 1, 2, \ldots, N$ represents the ith layer of N layers. Therefore, a ray propagating through several layers will have to travel a horizontal distance equal to

$$x = \sum_{i=0}^{N} x_i = \frac{c_0}{\cos \theta_0} \sum_{i=0}^{N} \frac{\sin \theta_i - \sin \theta_{i+1}}{g_i} \tag{4.37}$$

Let us consider a simple example illustrated by Fig. 4.7 showing N horizontal layers of linear velocity profiles. In order to determine the ray path for a single ray, we first choose a value for θ_0 and determine the various angles of the ray at the different interfaces using Eq. (4.33). Once all the angles are determined, Eq. (4.36) can be used to determine the horizontal distance the ray will travel in each layer. Another value for θ_0 can now be chosen and the values of θ_i computed until all the desired values of θ_0 have been used or the appropriate values of θ_0 have been chosen so that the rays surround a point of interest, such as the hydrophone in Fig. 4.7. Caution must be taken for rays in the first layer that has an upward direction in reference to the horizon because these rays may intersect and be reflected at the surface. The point at which the intersection of the surface must be determined along with the angle of incident to the surface at some angle θ will be reflected at the

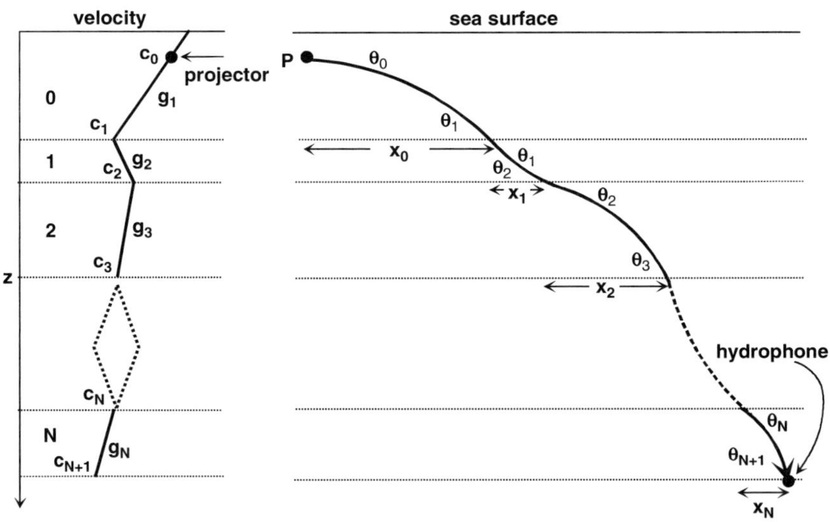

FIGURE 4.7. An example of a linearized sound velocity profile on the left and the corresponding ray path on the right for a single ray leaving the projector as a specific value of θ_0.

same angle but will experience a 180° phase shift. A ray can also intersect the bottom and be reflected back into the *N*th layer. The reflection coefficient and the amount of phase shift will depend on the specific type of bottoms.

In practice the horizontal range is much greater than the depth so that one would not plot a number of circular arcs as a ray goes through various layers. Rather, the angle of entry and exit of a ray through each layer and the horizontal range in the layer are the variables of interest. Since the depth of each layer is known and the horizontal range of a ray through each layer can be calculated, the trajectory of a ray can be plotted by connecting straight lines between each (x_i, z_i) pairs, where x_i is the horizontal distance a specific ray travels through the ith layer and z_i is the depth on the lower vertical boundary of the ith layer for a downward-bending ray or the upper vertical boundary for an upward-bending ray.

The geometric spreading loss of a signal traveling through horizontal stratified layers of constant gradient sound velocity profiles can be determined in a similar manner as in Section 1.2.6 for spherical waves in three-dimensional space. First we determine the appropriate angles θ_{0a} and θ_{0b} at which rays will pass on either side of the desired range and depth of a hydrophone as shown in Fig. 4.8. Next we determine the intensity of the sounds between rays launched from the source at these two angles. The intensity of the acoustic field between the two rays at the surface of the sphere is equal to

$$I_1 = \frac{\Delta P}{\Delta A_1} \tag{4.38}$$

where $\Delta A_1$is the area on the sphere subtended by the pair of rays and ΔP is the acoustic power between the two rays. At H, the intensity is similarly

$$I_2 = \frac{\Delta P}{\Delta A_2} \tag{4.39}$$

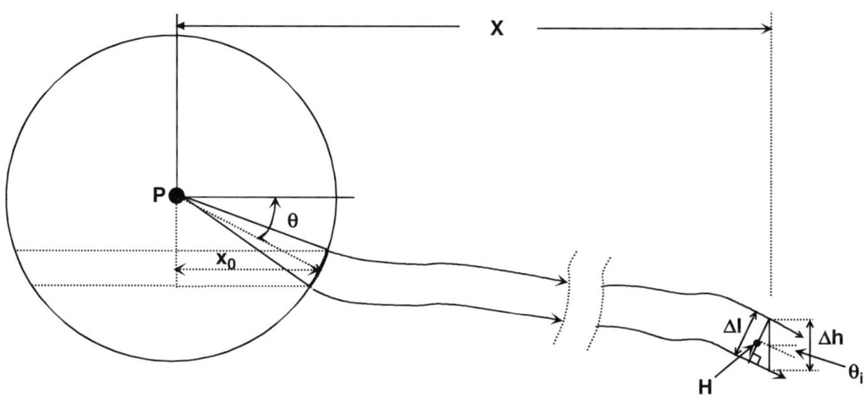

FIGURE 4.8. Geometry for calculating spreading loss from ray diagrams showing two rays from the propagating source passing an imaginary sphere of radius *r*.

where $\Delta A_2$is the area included by the ray pair in the vicinity of H, taken normal to the rays. Therefore, the transmission loss at H resulting from the spreading of the rays is

$$\mathrm{TL} = 10\log\frac{I_1}{I_2} = 10\log\frac{\Delta A_2}{\Delta A_1} \tag{4.40}$$

From Fig. 4.8 $x_0 = r\cos\theta$ and the arc length between the two rays is equal to $r\Delta\theta$. Therefore, the area subtended by the arc length intersecting the sphere is equal to

$$\Delta A_1 = 2\pi r\cos\theta\,\Delta\theta \tag{4.41}$$

By convention, the radius of the sphere is equal to 1 m so that the r term is replaced by unity. In a similar manner

$$\Delta A_2 = 2\pi x\,\Delta l = 2\pi x\,\Delta h\cos\theta_2 \tag{4.42}$$

where x is the horizontal distance from P to H and Δl is the perpendicular distance between the two rays at H. The transmission loss can now be expressed as

$$\mathrm{TL} = 10\log\frac{x\Delta h}{\Delta\theta}\frac{\cos\theta_2}{\cos\theta_1} \tag{4.43}$$

If c_1 is the sound velocity at P and c_2 is the sound velocity at H, then from Snell's law, we have $\cos\theta_2/\cos\theta_1 = c_2/c_1$, so that the transmission loss can be expressed as

$$\mathrm{TL} = 10\log\frac{x\Delta h}{\Delta\theta}\frac{c_2}{c_1} \tag{4.44}$$

Note that x and Δh are in meters and $\Delta\theta$ is in radians. In this derivation of transmission loss due to spreading of the rays, we assumed that there is no crossing of acoustic energy between rays so that all the energy emitted by the source stays within a pair of rays remains within the rays.

4.2.2 Lloyd Mirror Effect

The sea surface can have a profound effect on the acoustic field of a near-surface source because of interference caused by the summation of a surface-reflected signal and a direct signal. We will consider a special case in which we have isovelocity for depths close to the surface and a geometry shown in Fig. 4.9. An omnidirectional source at depth d_1 produces a signal that is received by a hydrophone at range x and at depth d_2. The hydrophone will detect a direct signal and a surface-reflected signal. The angle of reflection of the downward ray at the surface will be the same as the angle of incident of the upward ray. In such a situation, the surface reflected can be thought of as being produced by an image source at a distance d_1 above the surface.

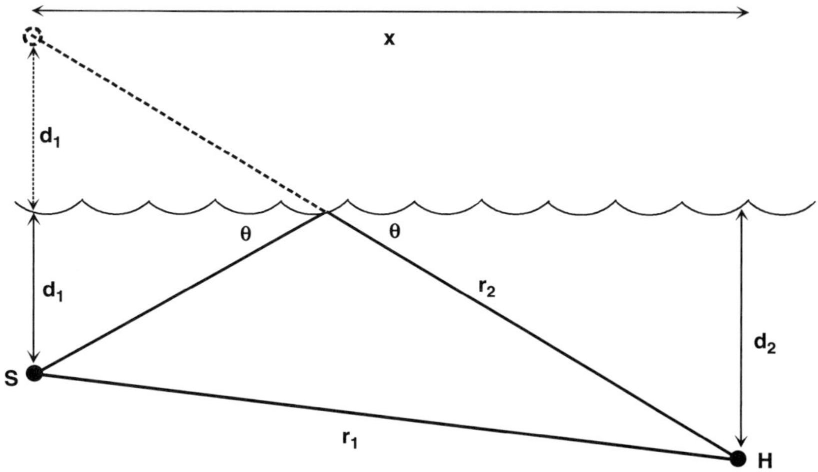

FIGURE 4.9. Geometry for propagation with a surface-reflected component.

Let us assume that the point source emits a cw sinusoidal acoustic signal $p = A \sin 2\pi ft$ so that the acoustic pressures can be expressed as

$$p_1 = \frac{A}{r_1} \sin 2\pi f(t + \tau_1) \tag{4.45}$$

$$p_2 = \frac{\mu A}{r_2} \sin 2\pi f(t + \tau_2) \tag{4.46}$$

where

$$r_1 = \sqrt{(d_1 - d_2)^2 + x^2}, \; r_2 = \sqrt{(d_1 + d_2)^2 + x^2} \tag{4.47}$$

and μ is the reflection coefficient at the sea surface. The time delays τ_1 and τ_2 are associated with the time for the direct and surface-reflected signals to reach the hydrophone. The total pressure at the hydrophone can now be expressed as

$$p_T = p_1 + p_2 = \frac{A}{r_1} \sin \omega t + \frac{\mu A}{r_2} \sin \omega (t + \tau) \tag{4.48}$$

where $\omega = 2\pi f$ is the radian frequency. The delay time τ_1 was set to zero and τ_2 was designated as τ, the time delay between the reception of the direct and surface-reflected signals. The intensity of the received signal is then equal to

$$I = \frac{A}{\rho c} \frac{1}{T} \int_0^T \left(\frac{1}{r_1} \sin \omega t + \mu \frac{1}{r_2} \sin \omega (t + \tau) \right)^2 \mathrm{d}t \tag{4.49}$$

where the integration is taken over a cycle of time T.

Before proceeding any further, it would be appropriate to divide the acoustic field into three parts: (1) the *near field*, where the surface reflection is negligible and spherical spreading exists; (2) the *interference field*, where constructive and destructive inferences between the direct and surface-reflected signals occur; and (3) the *far field*, where the field decays in a smooth manner. For convenience, we shall assume that the surface reflection coefficient $\mu = -1$.

In the near field where $r_2 >> r_1$, the second term of Eq. (4.49) is negligible and the intensity decays in a $1/r^2$ fashion. As the horizontal range increases the second term will begin to be a factor. We can define the limit of the near field to be the range at which the first term is twice as large as the second term, which is equivalent to stating that the second term is ½ or 3 dB below the amplitude of the first term. This limit between the near field and the interference field is defined by

$$\left(\frac{r_2}{r_1}\right) = 2 \tag{4.50}$$

From the definition of r_1 and r_2 in Eqs. (4.47) and (4.50), the outer limit of the near field before the interference field is reached can be expressed as

$$r_1 = 2\sqrt{d_1 d_2} \tag{4.51}$$

In the interference field, the integral of Eq. (4.58) needs to be evaluated. First let us simplify the expression by assuming that $r_1 \approx r_2$ and perform the square operation in the integral to get

$$I = \frac{A}{\rho c r^2}\frac{1}{T}\int_0^T \left[\sin^2 \omega t + \sin^2 \omega(t+\tau) - 2\sin \omega t \sin \omega(t+\tau)\right] \mathrm{d}t \tag{4.52}$$

Integrating the $\sin^2$ terms over a period will be equivalent to

$$\int_0^T \sin^2 \omega t \, dt = \int_0^T \left(\frac{1-\cos 2\omega t}{2}\right) dt = \frac{T}{2} \tag{4.53}$$

so that the intensity can now be written as

$$I = \frac{A}{\rho c r^2}\left(1 - \frac{2}{T}\int_0^T \sin \omega t \sin \omega(t+\tau) \mathrm{d}t\right) \tag{4.54}$$

Now we use the following trigonometric identities:

$$\begin{aligned} \sin\omega(t+\tau) &= \sin\omega\tau \sin\omega t + \sin\omega\tau \cos\omega t \\ \sin\omega t \sin\omega(t+\tau) &= \sin\omega t \sin^2\omega\tau + \sin\omega\tau \sin\omega\tau \cos\omega\tau \\ \sin\omega t \cos\omega t &= \frac{1}{2}\sin 2\omega t \end{aligned} \tag{4.55}$$

Finally, the intensity can be expressed as

$$I = \frac{I_0}{r^2} 2(1 - \cos 2\pi f \tau) \tag{4.56}$$

where $I_0 = A/\rho c$ and is the intensity at 1 m from the source. We can simplify Eq. (4.56) by noting that $\tau = (r_2 - r_1)/c = \Delta r/c$ and that $r_2^2 - r_1^2 = (r_2 - r_1)(r_2 + r_1) = \Delta r(r_1 + r_2)$. Now from our definition of r_1 and r_2 we have

$$\Delta r = \frac{r_2^2 - r_1^2}{r_2 + r_1} = (d_2 + d_1)^2 + x^2 - (d_2 - d_1)^2 - x^2 = \frac{4d_1 d_2}{r_2 + r_1} \tag{4.57}$$

Inserting Eq. (4.57) into Eq. (4.56), we have

$$I = \frac{I_0}{r^2} 2\left(1 - \cos \frac{8\pi f d_1 d_2}{c(r_2 + r_1)}\right) = \frac{I_0}{r^2} 2\left(1 - \cos \frac{\pi r_0}{r}\right) \tag{4.58}$$

where $r_0 = 4fd_1d_2/c$ is a reference range and $2r \approx (r_2 + r_1)$. The value in parenthesis will equal zero whenever $r_0/r = 2, 4, \ldots$ (even integers) and will equal 2 whenever $r_0/r = 3, 5, \ldots$ (odd integers). Therefore, the interference field will consist of a series of maxima and minima as shown in Fig. 4.10. The term in the parenthesis is typically called the *transmission anomaly*, a term from World War II referring to unexplained deviation of an acoustic field from spherical spreading. The interference field continues until $r_0/r = 1$, after which we make a transition into the far field. Within the interference field, the sound pressure level can dip in magnitude by over 30 dB.

4.2.3 Propagation of Sound in the Deep Ocean

The propagation of sound in the ocean is an extremely complex phenomenon and cannot be easily generalized because conditions in the ocean depend on so

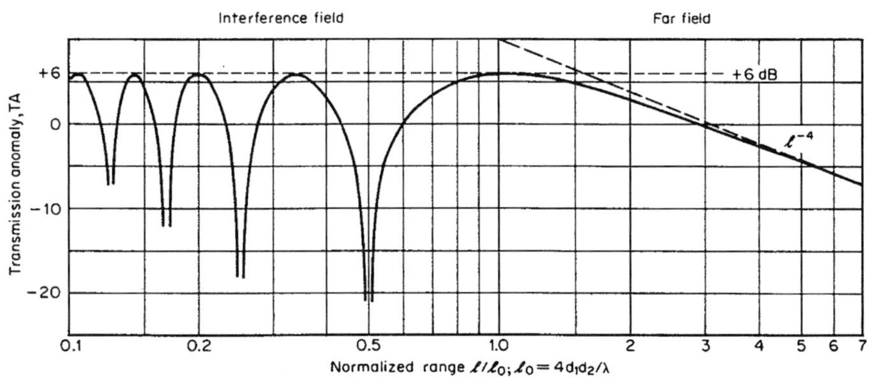

FIGURE 4.10. Theoretical Lloyd mirror propagation showing the interference and far fields (from Urick, 1983).

many factors such as latitude, time of day, seasons, wind and wave conditions, weather conditions and so on. All these factors affect the velocity profile at any given time and location. It would be considerably beyond the scope of this book to consider an in-depth discussion of the many different types of propagation conditions that can occur in the ocean. Interested readers should refer to such outstanding books as Urick (1983), Jensen et al. (1994) and Lee and Schultz (1995). In this section we will only consider some very general ocean propagation conditions.

4.2.3.1 Mixed Surface Layer

It is not unusual to find weather and oceanographic conditions such that a layer of water close to the surface is isothermal in which the temperature does not vary with depth. This layer is created and maintained by turbulent winds mixing the near-surface water of the ocean and is simply referred to as the *mixed layer*. In this layer, the velocity of sound increases with depth because of the effect of pressure on sound velocity as can be seen in Eq. (4.22). From our discussion of Snell's law we know that downward-directed rays in this layer will be refracted upward toward the surface since sound waves bend toward lower velocities. Therefore, a surface duct of sound propagation is formed. Once this duct is formed it will remain in place until heating from the sun begins to affect the duct decreasing the gradient. This usually occurs in the afternoon and has become known as the "afternoon effect." During the night, surface cooling and wave mixing once again allow the mixed layer to form. The gradient in the mixed layer usually does not exceed 0.016 s^{-1}.

We can model the sound velocity profile with two layers, one having a positive constant gradient (the surface layer) and the other having a negative constant gradient of velocity change. Such a model and its effects on sound propagation is shown in Fig. 4.11. An interesting feature of sound propagation in the mixed layer is the fact that the angle θ does not change in this layer. There is a critical angle after which the sounds will not be refracted upward but will escape the layer and refract toward deeper waters. The ray associated with the critical angle is referred to as the *critical ray*. There is also a *shadow zone* in which no rays will be found. However, this simple model does not truly describe the shadow zone since scattering from bubbles and the rough sea surface and the presence of a phenomenon called *internal waves* will cause the critical depth D to change horizontally so that some acoustic energy diffracts into the shadow zone.

At very low frequencies, sound ceases to be trapped in the mixed layer. As frequency decreases the wavelength increases and when the wavelength approaches the thickness of the layer, propagation can cease. A handy rule of thumb obtained from work with radio wave propagation in an atmospheric duct, specifying the maximum wavelength that the mixed layer will support, is given in Urick (1983), which is

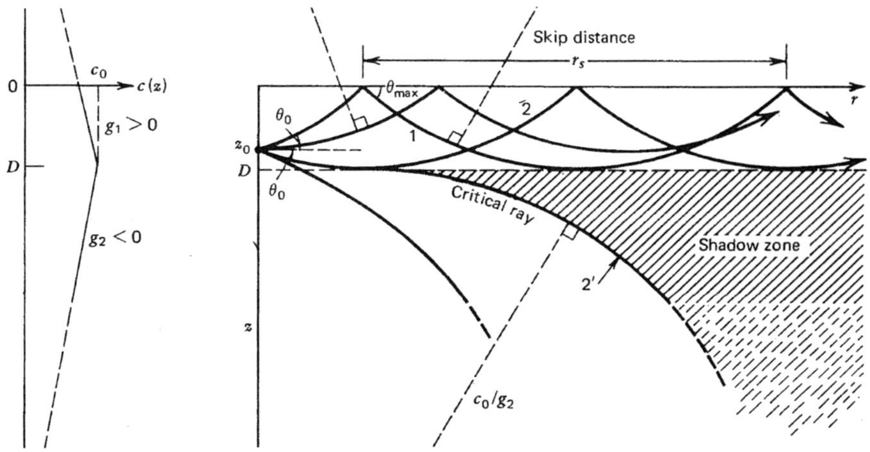

FIGURE 4.11. Sound propagation in a mixed layer (adapted from Kinsler et al., 1982).

$$\lambda_{max} = 4.7 \times 10^{-3} D^{3/2} \tag{4.59}$$

Therefore, the minimum frequency of an acoustic signal that a mixed layer will trap is equal to

$$f_{min} = 213 \frac{c}{D^{3/2}} \tag{4.60}$$

where c is the sound velocity and D is the critical depth of the mixed layer.

An expression for the transmission loss for a mixed layer propagation can be derived from simple geometric considerations. In Fig. 4.12 we have an omnidirectional source of sound in a mixed layer with an imaginary sphere of 1 m radius around it and two limiting rays at angles θ intersecting the surface of the sphere. If we assume no absorption and leakage losses, the same amount of power will be contained within the area of the 1 m sphere and the area at some range r.

Let A_1 be the area on the reference sphere and A_2 be the area associated with the surface duct at range r so that the transmission loss due to spreading can be expressed as

$$\mathrm{TL} = 10 \log \frac{I_1}{I_2} = 10 \log \frac{A_2}{A_1} \tag{4.61}$$

At the reference sphere A_1 is given by

$$A_1 = 2\pi r_1 \int_{-\vartheta}^{\vartheta} \cos\theta \, d\theta = 4\pi r_1 \sin\theta \tag{4.62}$$

At the horizontal range r, the area associated with the layer of thickness H is simply

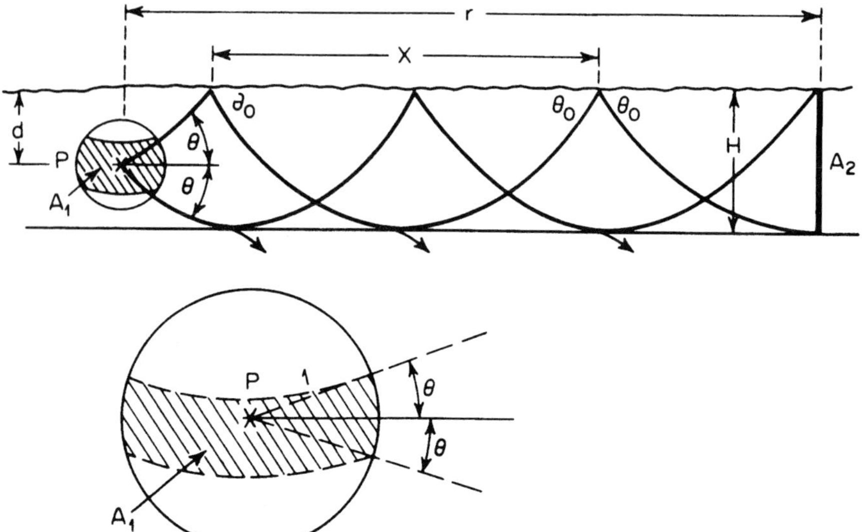

FIGURE 4.12. Geometric spreading propagation loss in a mixed layer (adapted from Urick, 1983).

$$A_2 = 2\pi rH \tag{4.63}$$

Therefore, the transmission loss due to geometric spreading (letting $r_1 = 1$ m) can be expressed as

$$\mathrm{TL_{gs}} = 10\log\frac{rH}{2\sin\theta} = 10\log rr_0 = 10\log r + 10\log r_0 \tag{4.64}$$

where $r_0 = H/2\sin\theta$ is the *transition range* at which the geometric spreading loss makes a transition from being spherical to cylindrical. The total transmission loss will be

$$\mathrm{TL} = 10\log r_0 + 10\log r + (\alpha + \alpha_L)r \tag{4.65}$$

where α_L is the leakage loss and is the per meter rate at which acoustic energy leaks out of the duct.

There are several variables that are important in mixed layer propagation. From Fig. 4.12 the skipped distance is the distance from consecutive surface reflection of a ray. The maximum angle of the limiting ray θ_{max} is at the source depth. In most situations the horizontal range r will be much greater than the layer depth H and θ_{max} is small ($\sin\theta_{max} << 1$) so that certain approximations can be made ($r >> H$), for the skipped distance, the angle of the limiting ray and the transition range. From our previous discussion we know that the acoustic wave traveling in a layer of a constant sound velocity gradient will travel in an arc of radius $R = c_0/g$. From Urick (1983) we obtain the following expressions:

$$x = \sqrt{8RH} = \text{skip distance of the limiting ray}$$

$$\theta_{max} = \sqrt{\frac{2(H-d)}{R}} = \text{angle of limiting ray in radians} \quad (4.66)$$

$$r_0 = \sqrt{\frac{RH}{2}}\sqrt{\frac{H}{H-d}} = \text{transition range}$$

As an example, assume that we have a source at the surface ($d = 0$) and the layer is 90 m in height, the sound velocity at the surface is 1,500 m/s and $g = 0.016\,\text{m}^{-1}$, then $R = 1{,}500/0.016 = 93{,}740$ m, $x = 2{,}598$ m, $\theta_{max} = 2.5°$ and $r_0 = 2{,}054$ m.

4.2.3.2 The Deep Sound Channel

The sound velocity profile shown in Fig. 4.4 is typical of the deep ocean at mid-latitudes. There is a sound velocity minimum at a fairly deep depth, approximately 1,000 m in Fig. 4.4. If a sound source is placed at the depth of the minimum sound velocity then a large number of rays leaving the source in both upward and downward angles will be refracted back toward the depth of the source since we found that sound waves bend toward a lower sound velocity region as they propagate. Therefore, a duct or channel is formed in the region of the minimum sound velocity. These rays that are bent back toward the region of minimum sound velocity are essentially trapped in a duct and will experience low spreading losses. This deep sound channel is often referred to as the sofar channel, and the depth at which the sound velocity is minimum is called the sofar axis. This axis occurs at shallower depths at higher latitudes. In the arctic the axis is at or near the ice-covered surface. An example of how sounds travel in the sofar channel is shown in Fig. 4.13, with the sound velocity profile shown in the right-hand side of the figure. In this example, the source is on the sofar axis and rays leaving the source at angles of $\pm 11°$ and less are trapped in the sound channel. The ray paths shown in Fig. 4.13 clearly indicate that although rays are trapped in the channel, the rays at the higher angles will travel a longer distance than the lower angle rays for a specific horizontal range. However, rays with the greatest excursion from the axis have the shortest travel time while the shortest path straight down the sofar axis has the longest travel time since it travels entirely at the depth of minimum sound velocity. Rays that leave the source at absolute angles greater than 11° will either reflect off the ocean surface or eventually be refracted upward after traveling to very deep depths.

The transmission loss for sound propagation in the deep sound channel can be calculated in much the same way as it was done with the mixed layer case. The propagation in the deep sound channel can be envisioned as starting with spherical spreading out to a transition range r_0, after which cylindrical

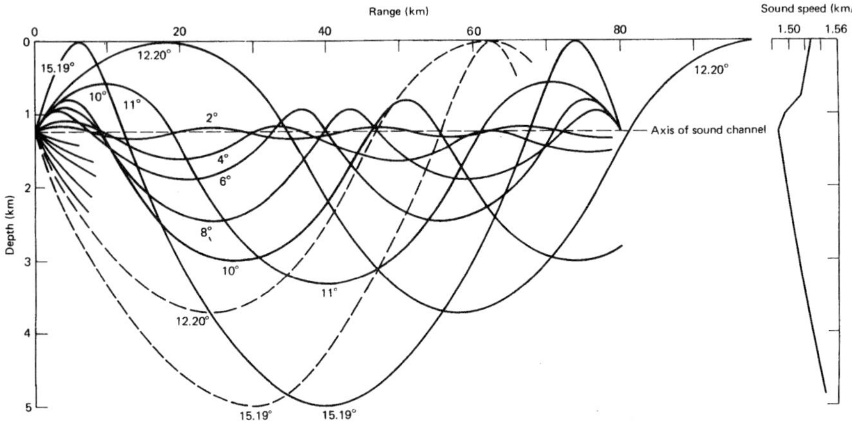

FIGURE 4.13. Ray diagram of transmission in the deep sound channel for a source on the axis (from Ewing and Worzel, 1948).

spreading occurs. The rays emitted by the source at different angles will arrive at a destination at different times, the energy will be spread out in time and the signal will suffer distortion. Therefore, the geometric spreading loss is more properly described in terms of energy. The acoustic energy flux density is often defined as either

$$E = \frac{1}{\rho c}\int_0^T p^2(t)\mathrm{d}t \text{ in p J/m}^2 \tag{4.67}$$

or

$$E = \int_0^T p^2(t)\mathrm{d}t \text{ in } \mu\text{Pa}^2\text{s} \tag{4.68}$$

where T is the duration of the signal. From Eq. (4.64) the transmission loss for the deep sound channel can be expressed as

$$\mathrm{TL} = 10\log\frac{E_1}{E_r} = 10\log r_0 + 10\log r + \alpha r \tag{4.69}$$

where E_1 is the source energy flux density 1 m from the source and E_r is the energy flux density at the horizontal range r. The remarkable properties of the deep sound channel have been utilized in many different ways. During WWII it was used to localize downed Navy aviators. A small explosive charge was dropped at sea by a downed aviator and programmed to detonate close to the deep sound channel axis. The explosion could then be detected by shore stations thousands of miles away and by taking the time difference of arrival of the explosive sound at two or more stations. The term *sofar* came from the technique of *so*und *fi*xing and *r*anging. Explosive sources denoting close to the sofar axis have been used by oceanographers for geodetic distance determination (Bryan et al., 1963) and for location of missile impact (Baker, 1975). Recently, a worldwide sound propagation experiment was conducted in which a sound source consisting of an array of 5 transducers was lowered from the center well of a ship off Heard Island, an island in the southern Indian Ocean relatively close to Antarctica (Munk et al., 1994). The array was lowered into the sofar axis at a depth of 175 m. Three different signals were used, a pure tone 57 Hz signal and two specific types of phase-modulated 57 Hz signal. The signals projected from each transducer had a source level of 205 dB re 1 μPa providing an overall source level of approximately 220 dB for the 5-transducer array. The transmitted signals were received at various locations around the world as shown in Fig. 4.14. Single dots indicate sites with single receivers. Dots connected by horizontal lines designate horizontal bottom-mounted arrays, vertical line designated vertical arrays and slanted line designated arrays towed in the direction of the arrow (Munk et al., 1994). The Heard Island test was only a short-term feasibility test lasting 6 days and 35 transmissions, each lasting 20 min. This experiment clearly demonstrated how low-frequency sounds can travel to great distances in the sofar channel.

The Heard Island feasibility test was followed by the ATOC (Acoustic Thermometry of Ocean Climate) feasibility experiment in which a single transducer was placed in the sofar axis by mounting it on the Pioneer Seamount off Point Sur in California at a depth of approximately 900 m and another transducer of the same type about 7 miles off the island of Kauai at a depth of 850 m. A phase-modulated signal with a carrier frequency of 75 Hz was projected at a source level of 195 dB re 1 μPa. The signals were

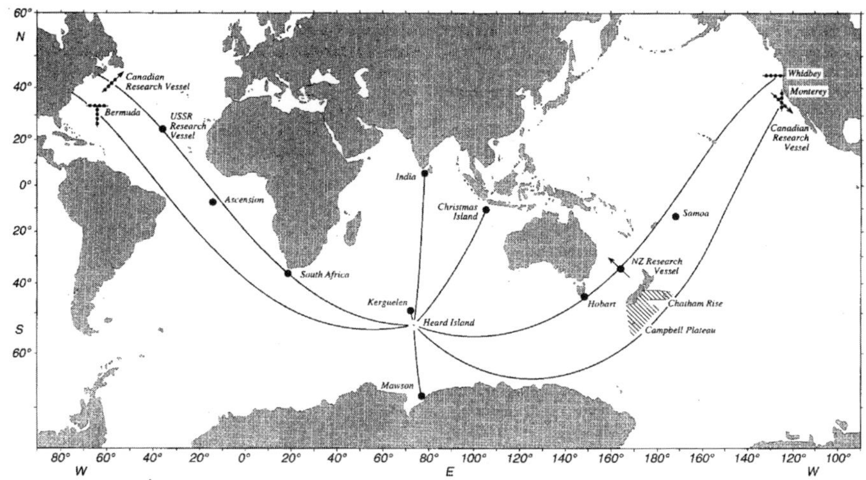

FIGURE 4.14. The source location and various receiver sites for the Heard Island test (from Munk et al., 1994).

received at various locations in the northern hemisphere at ranges of up to several thousand miles.

4.2.3.3 The Arctic Sound Channel

The Arctic is a region of the world that is extremely rich in sea life and the home of many species of marine mammals. The Arctic is roughly a circular sea with depth down to 3.6 km. Sea-ice occurs everywhere in the winter and over much of it in the summer. The environmental conditions in the Arctic are very unique and the typical sound velocity profile has some very interesting characteristics. Characteristics of both a mixed layer and a deep sound channel are present in these waters. A positive velocity gradient extends up to shallow depths of a few tens of meter during the summer and up to the ice surface during the winter. Sounds can propagate to long ranges by repeated reflection at the ice under-surface which in turn is refracted upward at a shallow depth.

An example of a typical sound velocity profile and the corresponding ray diagram are shown in Fig. 4.15. The velocity profile shown in the figure is relatively free from diurnal and seasonal effects below a shallow depth. Close to the surface the rays resemble a mixed layer propagation, yet signals are also refracted from deeper waters. This type of propagation will cause a pulse signal to be extended considerably as it propagates to long distances. Marsh and Mellen (1963) found that the short transient of an explosive source stretched to about 15 s after traveling approximately 1,000 miles. Another peculiarity with Arctic propagation is the attenuation of both high- and low-frequency signals. Reflection losses at the ice-cover affect high-frequency signals while low-frequency signals are not effectively

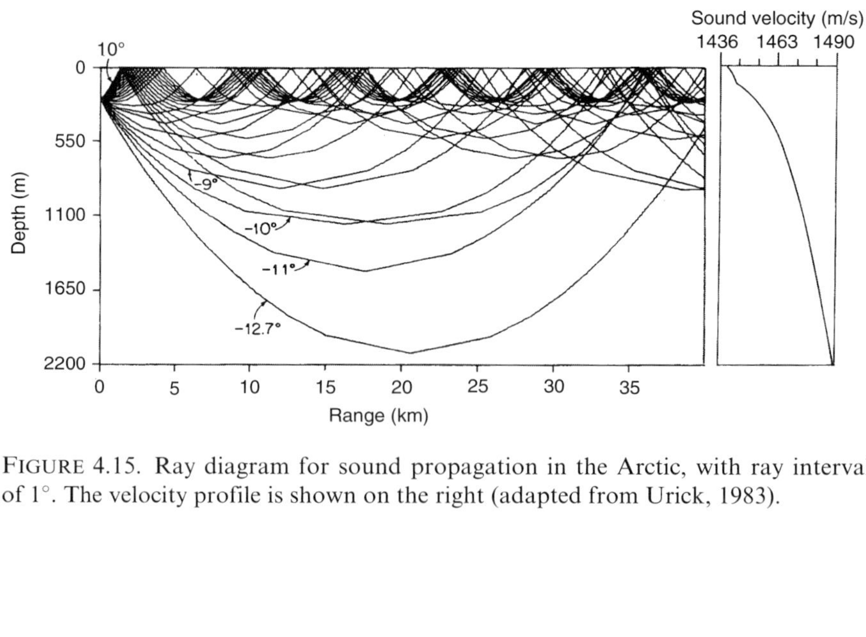

FIGURE 4.15. Ray diagram for sound propagation in the Arctic, with ray interval of 1°. The velocity profile is shown on the right (adapted from Urick, 1983).

trapped in the channel. The best frequency of propagation has been found to occur in the band between 15 and 30 Hz. Therefore, the Arctic Ocean behaves like a bandpass filter. However, even in this bandpass, frequency and time dispersions occur. The explosive signals of different frequencies have different group velocities, with the low frequencies traveling faster than the high frequencies. Most of the information and understanding of propagation in the Arctic have come from experiments conducted by Marsh and Mellon (1963) and Buck and Greene (1964) and others using explosive charges. The results of these and other experiments are nicely summarized by Buck (1968) with the results shown in Fig. 4.16 for different frequencies. These curves show the effects of frequency on the propagation of sounds, being best at 20 Hz and progressively deteriorating for higher frequencies. The propagation at any frequency starts off with lower transmission loss as a function of range than spherical spreading loss but eventually a range is reached in which the loss becomes greater than the spherical spreading loss. The higher the frequency the shorter this range. For a 20 Hz signal the range at which the transmission loss equals the spherical spreading loss is about 3,000 km, whereas at 3.2 kHz, the range is equal to about 50 km. Therefore, at short and moderate ranges, ducting improves the propagation of sounds, but at long ranges the repeated reflection off the under-ice surface begins to degrade transmission of sounds.

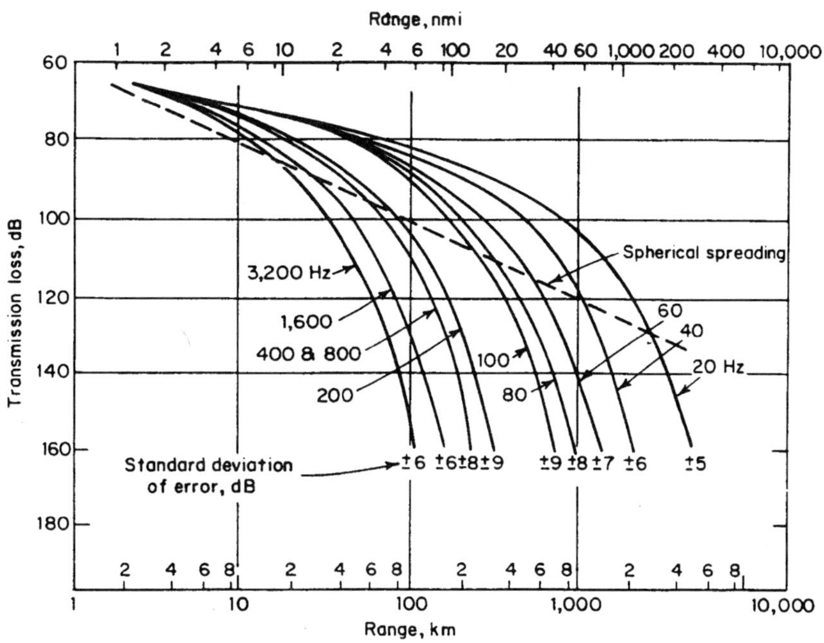

FIGURE 4.16. Average transmission loss in the Arctic (from Buck, 1968).

4.3 Propagation in Shallow Water

Acoustic propagation in shallow water consists of repeated reflections off the water surface and the bottom as the wave propagates outward. The transmission loss of sound propagation in shallow water depends upon the property of the bottom, the roughness of the sea surface and the depth of the water. Shallow water propagation is a very complex topic mainly because the ocean bottom is so highly variable and complex. In addition to specular reflection, the surface and bottom (and other objects) may "scatter" sound in other directions owing to surface roughness. The sea bottom is highly variable in its composition, varying from hard rock to soft mud. The smoothness of the sea bottom is also extremely varied from being relatively smooth to being very rough with large boulders, outcrops and other geological formations. The slope of the bottom can vary tremendously, even in a relatively enclosed location. Furthermore, the ocean bottom is often layered with different densities as a function of depth into the bottom. Therefore, the complexity of the bottom makes general modeling of propagation extremely difficult.

The approach taken by Marsh and Schulkin (1962) consist of making acoustic propagation measurements in a wide variety of locations and using the results to empirically derive expression for transmission loss. They made about 100,000 measurements in shallow waters in the frequency range of 0.1–10 kHz and derived three equations from their results. Letting $H = [1/8(D + L)]^{1/2}$ where D is the water depth in feet, L is the layer depth in feet and H is in kiloyards, they found that at short ranges where r is less than H, the transmission loss is approximately

$$\mathrm{TL} = 20\log r + \alpha r + 60 - k_{\mathrm{L}} \tag{4.70}$$

Where r is in kiloyards, α is the absorption coefficient is dB/kiloyard and k_{L} is a "near-field anomaly" that is dependent on the sea state and bottom type. This near-field anomaly is plotted in Fig. 4.17 as a function of the signal frequency in kHz for sea states of 1, 3 and 5 and a sand and mud bottom. From Fig. 4.17, we can see that k_{L} is larger for lower sea states, lower for mud compared to sand and decreases with frequency.

At intermediate ranges where $H \leq r \leq 8H$, the transmission loss is

$$\mathrm{TL} = 15\log r + \alpha r + \alpha_{\mathrm{T}}(r/H - 1) + 5\log H + 60 - k_{\mathrm{L}} \tag{4.71}$$

where r and H are in kiloyards and α_{T} is a shallow water attenuation coefficient that is shown in Fig. 4.17 as a function of frequency. The attenuation coefficient increases with frequency and does not appear to level off at 10 kHz.

For long ranges, $r > 8H$, the transmission loss is

$$\mathrm{TL} = 10\log r + \alpha r + \alpha_{\mathrm{T}}(r/H - 10) + 10\log H + 64.5 - k_{\mathrm{L}} \tag{4.72}$$

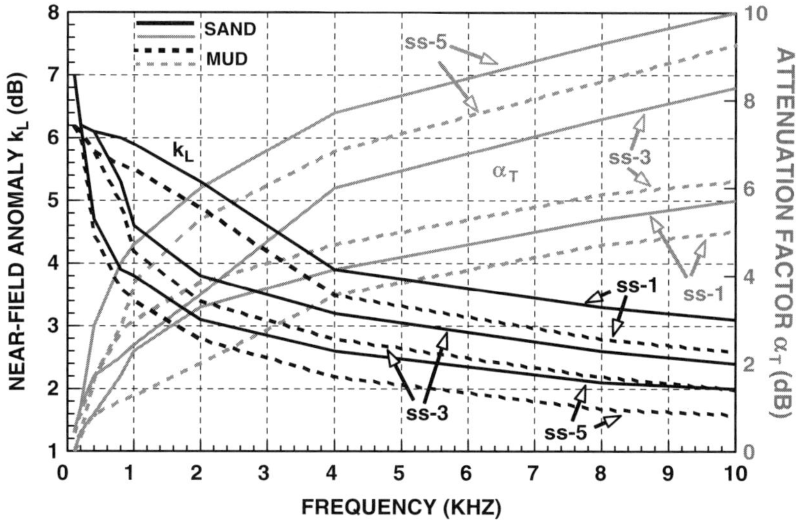

FIGURE 4.17. The near-field anomaly and shallow water attenuation factor (taken from data presented in Urick, 1983).

The long-range transmission loss has the quality of cylindrical spreading while the short-range expression in Eq. (4.68) is related to spherical spreading with the addition terms accounting for the repeated surface and bottom reflections.

The propagational characteristic of sound waves in shallow water can best be approached from a normal mode perspective where the sound channel is essentially treated like an acoustic wave guide. The wave equation must be solved as a boundary value problem. We will not discuss the normal mode theory in any detail except to simply state that the solution to the wave equation consists of a finite sum of normal modes. Each mode is a standing wave in the vertical direction that propagates in the horizontal direction with its own frequency-dependent speed. Each mode has a cutoff frequency below which it cannot propagate. No sound can propagate at frequencies below the cutoff frequency (f_c) for the first mode:

$$f_c = \frac{c_w/4D}{\sqrt{1 - c_w^2/c_s^2}} \tag{4.73}$$

where c_w is the sound speed in water, D the water depth and c_s the sound speed of the bottom. Figure 4.18 shows the cutoff frequency for different types of bottom materials. For example, with a fine sand bottom the cutoff frequency is about 1,000 Hz for a water depth of 1 m. This means sound at frequencies below 1,000 Hz will not propagate and can be detected only at ranges close to the source.

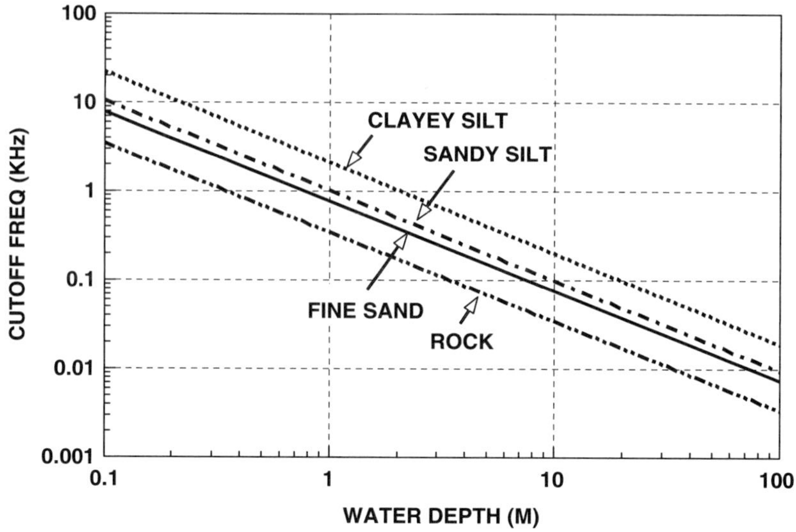

FIGURE 4.18. Cutoff frequencies for propagation of sound for different bottom materials; at frequencies below cutoff, sound will not propagate in shallow water.

4.3.1 Ocean Acoustics Software

In this section on acoustic propagation in the ocean, we have merely introduced some concepts and principles that govern the propagation of sounds in the ocean. As previously mentioned detailed discussion on various techniques and models for computation can be found in two books, one edited by Jensen et al. (1994) and the other by Lee and Schultz (1995). Interested readers are also encouraged to refer to the Ocean Acoustics Library at "www.**oalib**.hlsresearch.com." The Ocean Acoustics library contains acoustic modeling software and data. It is supported by the Office of Naval Research (*Ocean Acoustics Program*) as a means of publishing software of general use to the international ocean acoustics community. Various ray tracing, normal mode, parabolic equation and wavenumber integration software can be found. Since this website is supported by the ONR, its long-term existence should be secured. Other types of computational models can be found on this website.

4.4 Sound Fields in Tanks

Many underwater laboratory experiments take place in a tank, waveguide or small pool. Because of the relatively long acoustic wavelengths in water and the presence of many boundaries, acoustic fields within tanks may consist of a combination of many standing waves. The consequence is that acoustic particle velocity which is directly proportional to the pressure gradient may be extremely large or nearly zero from point to point inside the tank. Therefore special precautions must be taken in order to minimize the effects of the tank wall. Since tanks come in a myriad of different shapes, sizes and depths, it is difficult to describe the acoustic field in a general manner. However, some general guidelines can be given in terms of performing experiments in a pool or tank. These guidelines are as follows:

1. Use as large and as deep a tank as possible.
2. Use a directional transducer or sound projector when possible. Difficulties arise when hearing studies are conducted at low frequencies because it is difficult if not impossible to have a directional transducer that would fit in a pool for frequencies less than about 1 kHz. Directional transducers are readily available and small enough to conveniently handle for frequencies of about 10 kHz and greater. However, caution must be used to ensure that at the highest frequency of interest the beamwidth of the transducer will cover the area of the subject.
3. The directional transducer should be located close to or against a wall of the tank and the subject located directly in front of the transducer about 2–3 m away. If the subject is too close to the transducer, small changes in distance from the transducer can introduce large difference in the sound pressure level from trial to trial. For example, the difference between the sound pressure level at 1 and 1.5 m is about 3.5 dB whereas the difference between 2 and 2.5 m is only 1.9 dB.
4. The subject should be far away from the wall directly behind it (assuming a rectangular tank) to take advantage of spreading loss. For example, assume that we have a subject 2 m from the directional sound source and the wall behind the subject is at a distance of 7 m from the projector. The SPL of the reflected signal off the back wall will be about 21 dB lower than the direct signal impinging on the subject, and its effect will be small.
5. Take care to eliminate or reduce the surface-reflected component of the signal reaching the subject by the use of appropriate baffles. In most psychoacoustic experiments with marine mammals, the transducer depth and the animal stationing apparatus are both generally within 1–2 m from the water surface. Therefore, the surface-reflected component of the projected signal can severely affect what the subject will receive. However, the effects of the surface-reflected component can be minimized by having a baffle midway between the subject and the sound source as shown in Fig. 4.19. Also shown in Fig. 4.19 are techniques that can be used to reduce the effect of bottom reflections. The effects of reflections from the back wall can also be reduced by placing a reflecting "V" screen or a flat screen that would reflect the signal to another portion of a pool. However, these techniques with baffles and reflectors only work if the frequency is high enough. Low-frequency signals having large wavelength can and will diffract around a screen.

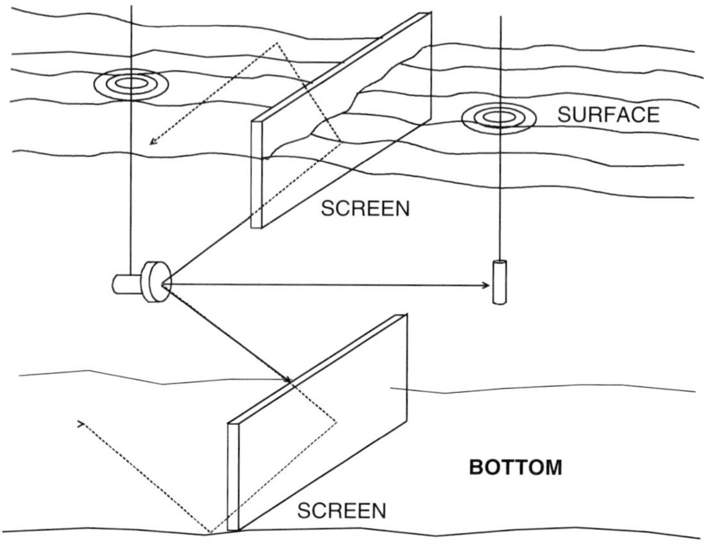

FIGURE 4.19. Use of screens to block surface and bottom reflections from a directional transducer.

6. Use short pulse signals to reduce the possibility of creating standing waves. Short tone burst with directional transducers and reflecting screens can be very effective in presenting a subject with a stable sound field.

A general rule of thumb defining low frequency is the ratio of the wavelength and the smallest dimension of the pool. A conservative estimate of when a low-frequency problem will be present is when the smallest dimension of a pool is equal to a wavelength of the acoustic signal. When such a signal is used, standing waves will be generated in a pool, even if sine pulses are used. In such a case we do n0ot have a free field or plane wave situation and the pressure in the pool will have steep gradients. The water particles will have velocities that vary in magnitude and direction in almost a random manner so that velocity is no longer in phase with the pressure. The issue of determining the intensity of an acoustic field at any specific point in a pool becomes more complex than in a free field situation in which plane wave propagation prevails. Sound intensity is the time-averaged product of the pressure and particle velocity, which we express as

$$I = \overline{p \cdot u} \tag{4.74}$$

where the bar indicates a time average process and u is the particle velocity. The particle velocity from Eq. (1.9) can be expressed as

$$u = \frac{1}{\rho}\int \frac{\partial p}{\partial r}\mathrm{dt} = \frac{1}{\rho}\int \frac{\Delta p}{\Delta r}\mathrm{dt} \tag{4.75}$$

where the partial derivative of pressure with respect to distance is approximated by the pressure gradient or change in pressure over a small distance. The pressure gradient can be measured with two identical hydrophones spaced a distance Δr apart in a specific direction. Let p_1 be the pressure measured by the first hydrophone and p_2 be the pressure measured by the second hydrophone, and p is the average of the two pressure measurements so that $p = (p_1 + p_2)/2$, then the intensity at that location can be expressed as

$$I = \frac{p_1 + p_2}{2\rho\Delta r} \int (p_2 - p_1)\,\mathrm{d}t \tag{4.76}$$

4.5 Sound Fields in Small Tanks

What constitutes a large versus small tank depends mainly on the wavelength of the signal being used and the dimensions of the tank. If one is dealing with fish acoustics small aquaria used by hobbyists may be involved. Since fish acoustics usually involves frequencies of several kHz and less, the wavelengths are often larger than the tanks. The techniques or guidelines of the previous section would not be meaningful for small aquaria where the largest dimension maybe about 1–2 m. Many fishes produce sounds in short pulses and these sounds often will strike one of the boundaries even before the entire sound is emitted. The specific characteristics and properties of fish sounds will be discussed in a latter chapter.

Reverberation in a small tank will be the major problem. An acoustic signal (either from a fish or from a transducer) will undergo multiple reflections that can create standing waves and the tank will be in a resonance mode where several resonance frequencies can be established (Kinsler et al., 2000). The derivation of the resonance frequency is not difficult but would add unnecessary details to the primary message of this section (see Kinsler et al., 2000). It suffices to state that the three-dimensional wave equation must be solved and the appropriate boundary conditions inserted. For a rectangular tank of length L, width W and water height of D, the resonance frequencies will be given by the equation

$$f_{lmn} = \frac{c}{2}\sqrt{\left(\frac{l}{L}\right)^2 + \left(\frac{m}{W}\right)^2 + \left(\frac{n}{D}\right)^2} \tag{4.77}$$

where l, m and n are integers that specify the mode of interest. The lowest resonance frequency occurs when l, m and n are equal to 1.

Akamatsu et al. (2002) discussed some of the issues involved with small tank acoustics and some of their results will be shown here. Their 170 l tank was rectangular with a length of 88 cm, width of 43 cm and a depth of 45 cm. In one case, a projector and hydrophone were separated by 44 cm along the length and spaced 22 cm from the side walls. The water depth was varied and

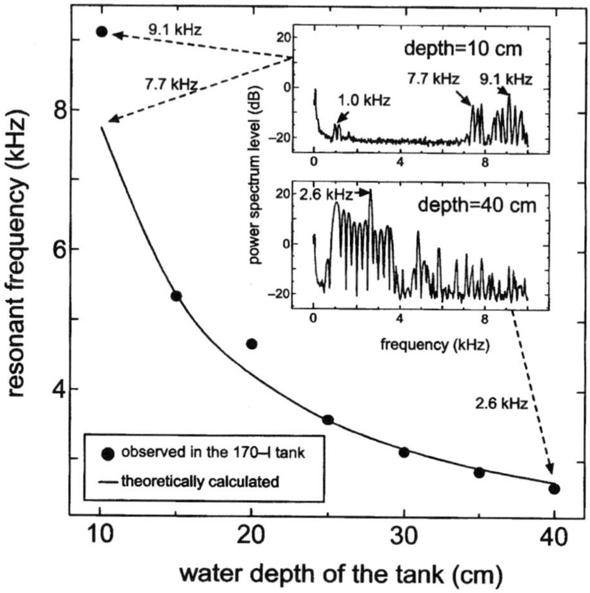

FIGURE 4.20. Resonance frequency as a function of water depth of a 88 × 43 × 45 cm rectangular tank (Akamatsu et al., 2002).

the resonance frequency was determined by considering the spectrum of the received hydrophone signal. The results of this simple experiment are shown in Fig. 4.20 where the resonance frequency is plotted against the water depth. The solid line is the calculated minimum resonance obtained by setting $l = m = n = 1$ in Eq. (4.77). There was good agreement between the theoretical resonance frequency and what Akamatsu et al. (2002) measured. The two inserts are spectral plots showing the resonance frequencies for two different water depths. The top insert indicates resonance frequencies of 7.7 and 9.1 kHz. The bottom insert is for a water depth of 40 cm and indicates a resonance frequency of 2.6 kHz.

The next set of measurements by Akamatsu et al. (2002) involved the measurement of the spectrum as a function of distance between the projector and hydrophone. The projector was placed at the center of the tank and the water depth was 40 cm. The measuring hydrophone was progressively moved away from the projector and the spectrum of the sound measured. The results of the measurements are shown in Fig. 4.21 showing the power spectrum as a function of the separation distance between the hydrophone and projector. From Fig. 4.21, the resonance frequency for a water depth of 40 cm was 2.6 kHz. The signal received by the hydrophone at 5 cm started to become apparent. By the time the signal propagated to 50 cm, the resonance component is the more dominant signal. The graph in part (e) shows that the 1 kHz component decreased monotonically with distance from the projector but at a

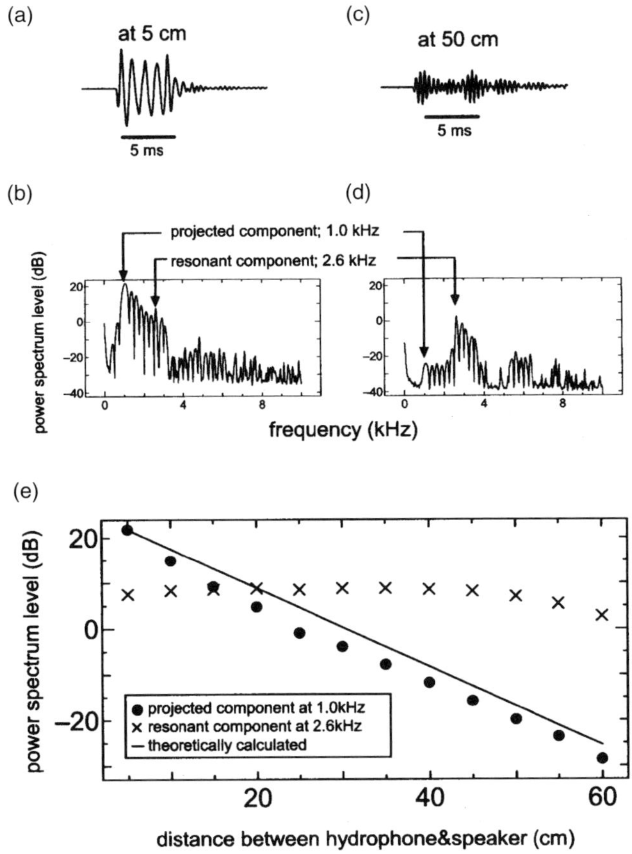

FIGURE 4.21. Waveform and power spectra of the received hydrophone signal, (a, c) 5 cm from projector, (b, d) 50 cm from the projector. (e) The power spectrum as a function of the hydrophone distance, solid circle for the 1 kHz signal, X are for the 2.6 kHz component (from Akamatsu et al., 2002).

much faster rate than for spherical spreading. The solid line is a linear regression line of the data described by the equation

$$S = 20X/D \tag{4.78}$$

where S is the spectral level, X is the distance from the projector and D is defined as the attenuation distance and is equal to 23 for a frequency of 1 kHz. The attenuation distance was defined as the distance at which the spectral level decreased by 20 dB for every integer multiple of D. The 2.6 kHz component is shown not to decrease with distance from the projector. This is not surprising because the 2.6 kHz component was produced because of resonance and is a standing wave.

The overall conclusions that one can draw from the measurements of Akamatsu et al. (2002) are that the acoustics of small tanks can be very complex and that a tank needs to be calibrated before being used in any acoustic experiments. Akamatsu et al. (2002) also included a number of recommendations for use of small tanks.

References

Akamatsu, T., Okumura, T., Novarini, N., and Hong, Y. Y (2002). "Empirical refinements applicable to the recording of fish sounds in small tanks." J. Acoust. Soc. Am. **112**, 3073–3082.

Baker, W. F. (1975). "New formula for calculating acoustic propagation loss in a surface duct," J. Acoustic. Soc. Am. **57**, 1198–1200.

Buck, B. M. and Greene, C. R. (1964). "Arctic deep-water propagation measurements," J. Acoust. Soc. Am. **36**, 1526–1533.

Buck, B. M. (1968). "Arctic acoustic transmission loss and ambient noise," in *Arctic Drifting Stations*, J. E. Sater, ed. (Arctic Institute of North America, Arlington, VA).

Byran, G. M., Truchan,M., and Ewing, J. L. (1963). "Long-Range SOFAR studies in the south Atlantic Ocean." J. Acoust. Soc. Am. **35**, 273–278.

Ewing, M. and Worzel, J. L. (1948). "Long-range sound transmission," Geol. Soc. Am. Mem. **27**, 1–35.

Fisher, F. H. and Simons, V. P. (1977). "Sound absorption in sea water." J. Acoust. Soc. Am. **62**, 558–564.

Jensen, F. B., Kuperman, W. A., Porter, M. B., and Schmidth, H. (1994). *Computational Ocean Acoustics*. Springer-Verlag, N.Y.

Kinsler, L. E., Frey, A. R., Coppens, A. B., and Sanders, J. V. (2000). *Fundamentals of Acoustics, 3rd Edition* (John Wiley & Sons, New York).

Lee, D. and Schutz, H. (1995). *Numerical Ocean Acoustics Propagation in Three Dimensions*. World Scientific., Hackensack, NJ.

Marsh, H. W. and Schulkin, M. (1962). "Shallow water transmission," J. Acoust. Soc. Am. **34**, 863–887.

Marsh, H. W. and Mellen, R. H. (1963). "Underwater sound propagation in the Arctic Ocean," J. Acoust. Soc. Am. **35**, 552–556.

Medwin, H. (1975). "Speed of sound in water: A simple equation for realistic parameters." J. Acoust. Soc. Am. **58**, 1318–1319.

Medwin, H. and Clay, C. S. (1998). *Fundamentals of Acoustical Oceanography* (Applications of Modern Acoustics). Academic Press, San Diego.

Munk, W. H., Spindel, R. C., Baggeroer, A., and Birdsall, T. G. (1994). "The heard Island feasibility test," J. Acoust. Soc. Am. **96**, 2330–2342.

Urick, R. J. (1983). *Principles of Underwater Sound* (McGraw-Hill, New York).

5

Signal Recording and Data Acquisition

5.1 Measurement of Underwater Sounds

A simple fairly generic system for measuring and storing underwater acoustic signals is depicted in Fig. 5.1. The first componnt in the signal acquisition chain consists of a sensing element, usually a hydrophone that has integrated within its housing a pre-amplifier and line driver or a hydrophone connected to a relatively short cable (up to 10–20 m depending on the capacity of the sensor to drive the cable). The next component is usually a filter, a high pass, low pass, or bandpass filter followed by an amplifier. After the amplifier, there is usually a monitor of some sort in parallel with some type of recording device. Typical monitors include oscilloscopes, voltmeters, and spectrum analyzers. The storage device can be a strip chart recorder, an analog tape recorder, a digital audio tape (DAT) recorder, a digital oscilloscope with memory and disc storage capabilities, a digital spectrum analyzer with memory and disc storage capabilities, or an analog-to-digital (A/D) converter with disc storage or magnetic tape storage capabilities, or any combination of the above devices. This simple system will be able to accurately measure and faithfully record underwater acoustics signals only if its characteristics are consistent with the properties of the signals under observation. Some of the critical characteristics a measurement system must have are listed below:

1. Bandwidth – usually measured from the lower frequency cutoff (lower 3 dB roll off frequency) to the upper frequency cutoff (upper 3 dB roll-off frequency). The measurement system's bandwidth should be wider than the bandwidth of the observed signal or the system will not accurately measure the acoustic signal. The device in the signal acquisition chain with the highest low-frequency roll off will determine the low-frequency limit of the system. Conversely, the device with the lowest high-frequency roll off will determine the highest frequency limit of the system. Often the frequency range of the signal is not known exactly so that a system with a broad bandwidth is required.
2. Noise floor – the noise produced by the various devices in the signal acquisition chain will usually determine the smallest signal that the measurement

W.W.L. Au, M.C. Hastings, *Principles of Marine Bioacoustics*,
DOI: 10.1007/978-0-387-78365-9_5, © Springer Science+Business Media, LLC 2008

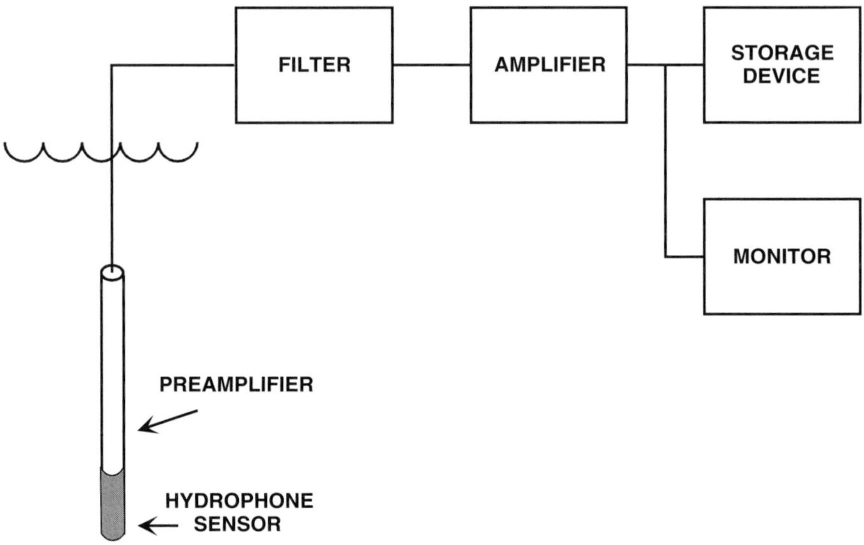

FIGURE 5.1. Components of a fairly simple, generic underwater acoustic measurement system.

system will be able to detect. The device with the highest noise floor will determine the noise floor of the system. However, there will be instances in which interfering noise, either from the ambient background or from man-made sources will actually dictate the minimum amplitude of any acoustic signal that can be detected.

3. Signal ceiling – the largest signal that the measurement system can handle without any distortions. The device with the lowest ceiling will determine the ceiling of the system. The ratio in dB of the largest signal that can be measured without any distortion and the minimum signal that can be detected and separated from the noise of the system is the dynamic range of the system. In order to obtain the maximum capability of a measuring system, the dynamic range of the system should be greater than the dynamic range of the signals to be measured. Often the dynamic range of the recording device may dictate the overall dynamic range possible with a specific measurement system.
4. Harmonic distortion level – The devices in the measurement system must not introduce any distortions to the signals in order to observe the signals with good fidelity. Distortion can be introduced when the gain of either the pre-amplifier or any amplifier is so high that the amplifier or any device following the amplifier cannot handle the high amplitude of the signal. Distortion can also be introduced when the desired signal has a bandwidth that is greater than the bandwidth of the system.
5. Sensitivity – in many situations the desired signals are so minute that the measurement system cannot detect them. Usually, the sensitivity and noise

level of the hydrophone pre-amplifier are the limiting factors in these cases. Unless the hydrophone sensitivity is high enough so that an acoustic signal in the band of interest will cause the hydrophone voltage to be higher than the noise of the pre-amplifier/line driver, the signal will usually not be detected. There are techniques with repeatable signals that can be used to "pull" signals out of noise such as averaging of the signal and noise, or having a long integration window or narrow frequency filter to extract narrow band signals from noise. These techniques will be discussed further in this chapter.

The simple system depicted in Fig. 5.1 can quickly become more complex by having an array of several hydrophones, each having its own pre-amplifier, filter, and amplifier, all connected simultaneously to a multi-channel storage device.

5.2 Underwater Acoustic Noise

Acoustic noise can be considered as any sound other than the desired sounds. The ocean and many other bodies of water are very noisy with sounds being generated by a variety of different sources. The acoustic noise in the ocean is often referred to as ambient or background noise and is normally measured in terms of intensity in dB detected by an omnidirectional hydrophone. Noise intensity may be measured in different frequency bands; however, ambient levels are usually reduced to a 1 Hz frequency band and referred to as spectrum levels. Suppose we use an omnidirection hydrophone to measure the ambient noise of a body of water with a system having a bandwidth of Δf Hz. If N is the intensity of the acoustic noise in dB measured by our system, the noise spectrum level will be

$$N_{\mathrm{J}} = N - 10 \log \Delta f, \tag{5.1}$$

and the units will be dB re 1 μPa^2/Hz, dB re 1 μPa/ Hz$^{1/2}$, or simply spectrum level in dB re 1 μPa, with the understanding that the noise level has been normalized into a 1 Hz band. The second set of units is a hold over from the concept of spectral levels of electrical noise voltage, which is expressed in V/ Hz$^{1/2}$. The second set of units also appeals to those that consider the noise spectrum level from a noise pressure level instead of an intensity point of view. Nevertheless, normalizing the measured noise to a 1-Hz band makes it possible to relate to the noise measurements made by systems having different bandwidths since the bandwidth of the measuring system is normalized out as seen in Eq. (5.1). The equation also indicates that the wider the bandwidth of the measurement system, the greater will be the total noise intensity measured.

Besides spectrum levels, noise levels are also commonly reported in terms of a sound pressure level within a specific frequency band. Some common sets of frequency bands are octave bands, 1/2-octave bands, and 1/3-octave bands. If f_{l} is the lower frequency of a particular frequency band and f_{u} is the upper frequency of the band, then the relationship between the upper and lower cutoff frequencies for an octave-related band can be expressed as

$$\frac{f_u}{f_l} = 2^n, \tag{5.2}$$

where $n = 1/3$, ½, or 1 for a 1/3-octave, 1/2-octave, or 1-octave band, respectively. The center frequency of the band can be expressed arithmetically $\sqrt{f_l f_u}$. The preferred American National Standards center frequencies of octave bands are 31.5, 63, 125, 250 Hz, and so on until 16 kHz. Then the preferred center frequencies become 31.5, 63, 125 kHz, and so on until 160 kHz. The preferred center frequencies for 1/3-octave bands are 100, 125, 160, 200, 250, 315, 400, 500, 630, and 800 Hz. The next 10-to-1 set starts with 1 kHz and continues by multiplying the first set by 10, 100, and so on. Measurement of ambient noise with a digital acquisition system will be discussed in Section 6.6.2 of Chapter 6.

5.2.1 Ambient Noise in the Deep Ocean

Many sources contribute to the ambient background noise in the ocean. A conceptual diagram showing some of the sources of ambient noise is depicted in Fig. 5.2. We will only briefly discuss the various sources and their effects. Readers interested in a detailed treatment of ambient noise in the ocean should refer to the excellent monograph by Urick (1984). We will begin our discussion with an average representative ambient-noise spectra depicted in Fig. 5.3 for different conditions in deep-waters (beyond the continental shelf). The noise in the infrasonic frequency range (below human audible range) from 1 Hz to slightly above 20 Hz is caused by seismic activity. A single line is drawn since this noise does not depend strongly on wind speed. Ship noise also occupies the lower frequencies and extends to about 1 kHz. The heavy shipping curve

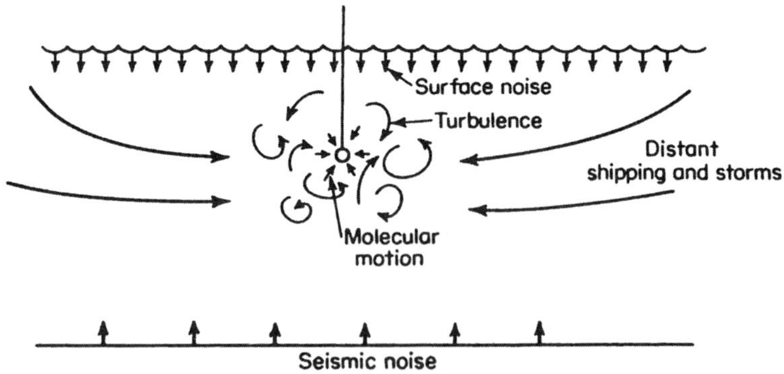

FIGURE 5.2. Conceptual diagram of the sources of deep-water ambient noise (adapted from Urick, 1983).

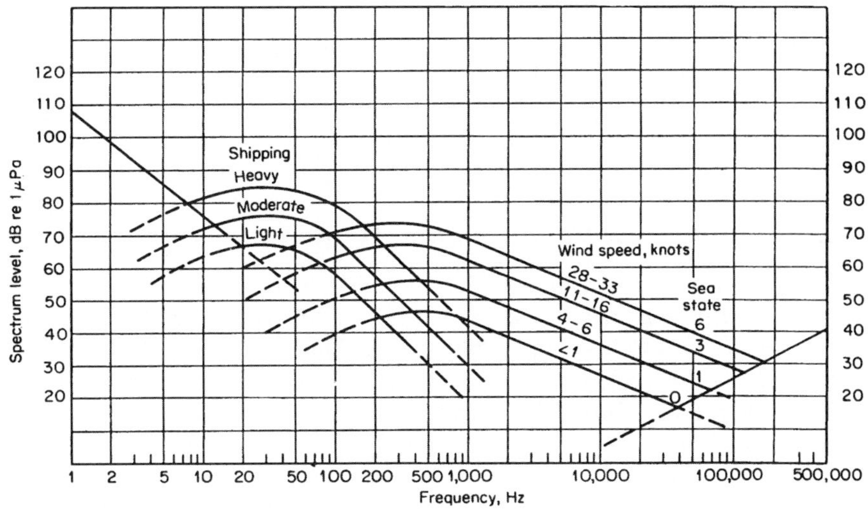

FIGURE 5.3. Average deep-water ambient noise spectra (adapted from Urick, 1983).

applies to various shipping lanes of high activity in the North Atlantic and the North Pacific. At the higher frequencies, most of the ambient noise is caused by the roughness of the sea surface, which is connected to sea state or wind force. Although the sea surface conditions correlate well with acoustic noise levels, the process by which the noise is generated is still uncertain (Urick, 1983). Processes that contribute to the noise in this frequency range include breaking whitecaps, flow noise produced by wind in blowing over a rough sea surface, cavitation or collapse of air bubbles formed by turbulent wave action, and wave generation action of the wind on the sea surface (Urick, 1983). The straight line in the far right side of Fig. 5.3 is the thermal noise limit caused by molecular motion of the water. Mellen (1952) showed theoretically that for a nondirectional hydrophone with perfect efficiency in converting acoustic energy into electrical energy, the equivalent thermal noise spectrum level can be expressed as

$$\mathrm{NL} = -15 + 20\log f(kHz), \tag{5.3}$$

where f (kHz) is the frequency in kHz. NL is expressed in terms of dB re 1 μ Pa^2/Hz. The thermal noise that rises at a rate of 6 dB per octave places a threshold on the minimum observable sound pressure levels at high frequencies.

At higher frequencies where most of the noise is generated by wind-related surface roughness, it would not be surprising to find that the ambient noise decreases with hydrophone depth. Measurements by Urick et al. (1972) with a vertical string of six hydrophones placed 610 m apart off St. Croix, Virgin

Islands, showed that the noise level decreased with depth at all frequencies, with the greatest decrease in the first 610 m. Measurements by Morris (1978) off San Diego showed very similar decrease in the ambient noise with depth. The decrease in the noise with depth should also be stronger at higher frequencies since acoustic absorption losses are dependent on frequency.

5.2.2 Ambient Noise in Shallow Waters

The term shallow waters usually refer to coastal waters within the continental shelf. This definition includes bays, harbors, and inlets. Ambient noise in shallow water tends to be highly variable, both in time and location. As in the deep ocean, wind speed seems to be the major factor in determining the levels of the ambient noise above a frequency of about 100 Hz. The noise spectral levels from several studies of ambient noise in shallow waters are shown in Fig. 5.4. Piggott (1965) performed a 1-year study on the Scotian shelf in waters about 46 m in depth. The measurements of Knudsen et al. (1948) off Fort Lauderdale, Florida, in 183 m water depth are also included in Fig. 5.4 along with the data from Wenz (1962) for several Pacific locations.

The data in Fig. 5.4 indicate that the noise is dependent on wind speed at all frequencies between 10 and 100 kHz. Comparison with the deep water spectra in Fig. 5.3 indicates that ambient noise in shallow water is about 5–10 dB higher than in deep water for frequencies greater than 500 Hz. In the absence of shipping and biological noise sources, shallow water ambient noise levels are less variable and very similar with noise levels in deep waters (Urick, 1983). In shallow waters, the measured noise levels are independent of hydrophone

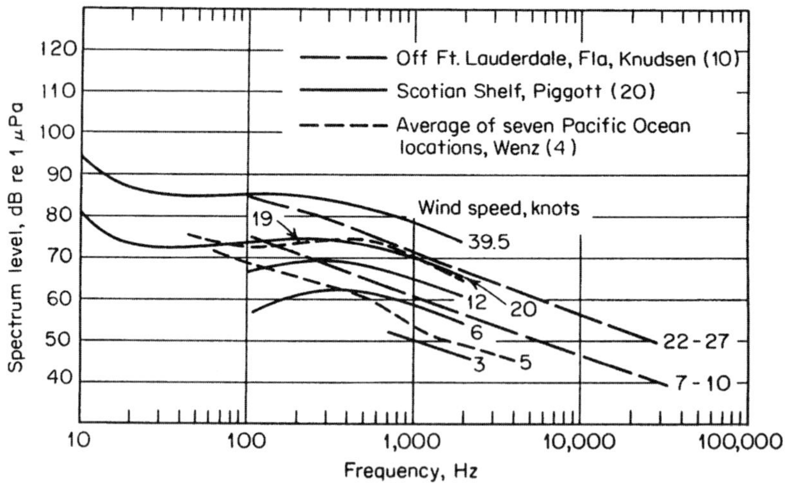

FIGURE 5.4. Noise spectra of coastal locations, with wind speed as a parameter (adapted from Urick,1983).

depth, water depth, and depend only on wind speed. The noise intensity increases as the square of the wind speed above a wind speed of about 5 knots.

5.2.3 The Effects of Rain on the Ambient Noise

Franz (1959) found that an object like a rain droplet impacting an air–water interface creates noise by two processes: by the impact itself and by the oscillations of the entrained air carried by the droplet below the surface. The noise created by rain will depend on the size and velocity of the water droplets, which in turn depends on the rate of rainfall. An increase of almost 20 dB is shown in Fig. 5.5 along with the rain-free spectrum for wind speed of 20–40 knots. Note that the dashed curve is a spectrum in the absence of rain. The ambient noise spectral density data of Heindsman et al. (1955) taken in different rain conditions are for the heavy rain condition; the noise spectrum is relatively flat over a wide frequency range and can be considered as white noise. Rain noise at low frequencies (1 kHz and lower) also seems to be relatively independent on the rate of rainfall since the noise levels for different conditions seem to converge at these frequencies. However, at higher frequencies, the increase in the ambient noise caused by rain is highly dependent on the rate of rainfall.

5.2.4 Noise Caused by Ice

Many different species of marine mammals inhabit the nutrient-rich waters of the near-polar latitudes where the waters are usually or often covered with different types of ice. The underwater noise in ice-covered waters can vary substantially depending on the type of ice and weather conditions. In continuous shore-fast ice with little wind, the noise levels can be extremely low,

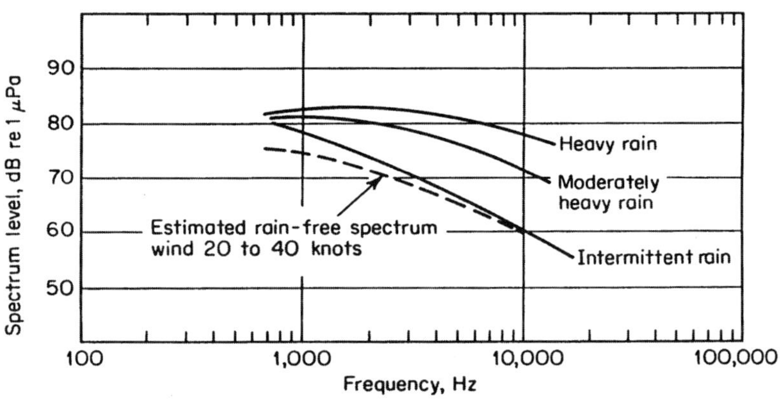

FIGURE 5.5. Ambient-noise spectra in rain observed in Long Island Sound (adapted from Heindsman et al., 1955).

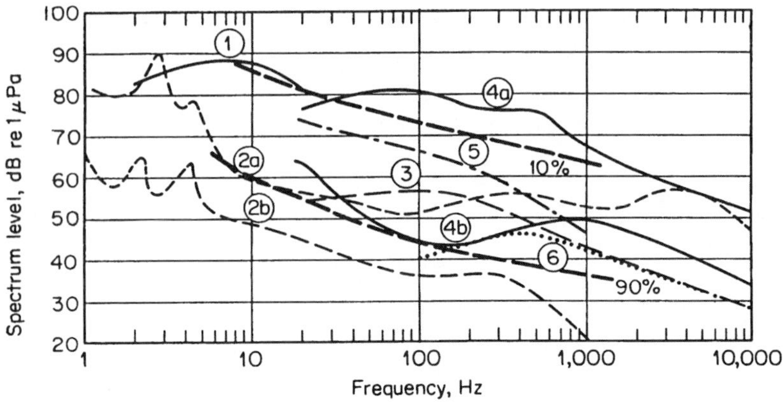

FIGURE 5.6. Spectra of ambient noise observed under ice. (1) 80–90 in. ice, no cracking noise, Barrow Strait (Milne, 1960). (2) Old polar ice with frozen leads and pressure ridges, (2a) noisy period, (2b) quiet periods (Milne and Ganton, 1965). (3) 70% 1-year ice 3–5 ft thick, 30% polar floes (Milne and Ganton, 1965). (4) Shore-fast ice like that of (2), cracking noises prevalent (4a), absent (4b) (Milne and Ganton, 1965). (5) average over a 2-week period in Beaufort Basin (Greene and Buck, 1964). (6) Ice-free deep water, sea state 0. Curves labeled 10% and 90% show the estimated percentage of time that a given level will be exceeded, based on all available Arctic data (adapted from Urick, 1983).

lower than typical ice-free water at sea state 0. However, in ice packs, noise levels can be 5–10 dB higher than those measured in comparable sea states for ice-free waters. When the air temperature decreases and cracks in the ice are formed, the noise is often spiky or impulsive. Urick (1983, 1984) presented an interesting graphical summary of noise levels for different ice conditions and his graph is reproduced in Fig. 5.6. From this figure, one can see how variable the noise levels can be, depending on the ice and weather conditions. Some of the curves (2a, 2b, and 3) are close to or below the Knudsen sea state 0 noise curve for frequencies greater than several hundred Hz, and some (1 and 4a) are close to or above the sea state 6 noise curve.

5.2.5 *Other Sources of Acoustic Noise*

A common source of acoustic noise is often referred to as "self" noise, and it is the noise usually caused by the method in which sounds are being measured. Hydrophones that are extended overboard a moving boat will often detect the noise caused by water flow around the vessel, noise produced by the engine and airborne sounds produced on the vessel that is coupled into the water. Dragging a hydrophone in water can also produce the flow noise caused by turbulence produced by the hydrophone itself as it moves in the water. In such situations, careful design considerations must be taken to ensure that the hydrophone or an underwater vessel containing a hydrophone does not produce turbulent flow as it moves in the water. Noise produced by the hydrophone cable "strumming" as the hydrophone is dragged in the water, or is "stationary" in waters with high fluctuating currents, is another common source of noise. These self-noises can usually be reduced by careful design of the structures associated with the hydrophone and its deployment.

Another common source of acoustic noise has been labeled as "biologics" by underwater acousticians and sonar operators. These are the sounds produced by various marine organisms that tend to interfere with the measurement of certain natural or man-made sounds by experimenters. However, what is noise to a group of acousticians may be signals for bioacousticians. "Noise" caused by fish, crustaceans, and marine mammals such as cetaceans and pinnipeds are signals of interest for bioacousticians. These sounds are many and extremely varied, from the sounds of dolphins and porpoises to the broadband impulses of snapping shrimps. The acoustic emissions of these marine animals will be discussed in detail in a later chapter and are only briefly mentioned here.

Finally, the source of underwater acoustic noise we will consider has to do with seismic events and volcanic eruptions. These noises tend to be very low in frequencies, in the infrasonic range (1–50 Hz). However, the magnitudes of these sounds can be relatively high and can interfere with the measurements of some very low-frequency sounds produced by baleen whales. McGrath (1976) observed over 500 seaquakes over a 7-day period and found the broadband noise level in the band between 5 and 32 Hz to raise at least 20 dB in an area of extreme seismic activity. Northrup (1974) monitored the activity of a submarine volcano in the western Pacific and found most of the acoustic energy in an octave band centered near 10 Hz. The sound pressure level would certainly depend on the distance of the observer from the volcanic activity.

5.3 Electronic Noise

Aside from the environmental noise considered in the previous section, the noise limitation of an acoustic measurement system will be determined by the noise contributed by the various components of the system. The component of the measurement system that will probably have the most influence on the minimum noise floor of the system is the hydrophone–pre-amplifier combination. A simplified equivalent circuit of a piezoelectric sensor connected to a pre-amplifier is depicted in Fig. 5.7, with the impedance of the piezoelectric sensor assumed to be purely capacitance (not a bad assumption for low frequencies). The voltage generator E_s depicts the conversion of acoustic energy into electrical energy by the piezoelectric sensor. C_s is the capacitance of the sensor, and R_{in} is the input shunting resistor to the pre-amplifier. Electrons in every conductor are in random motion and, according to the well-known Johnson–Nyquist equation, this motion will generate a mean-square thermal-noise voltage of

$$E_t^2 = 4kTR_{in}\Delta f, \qquad (5.4)$$

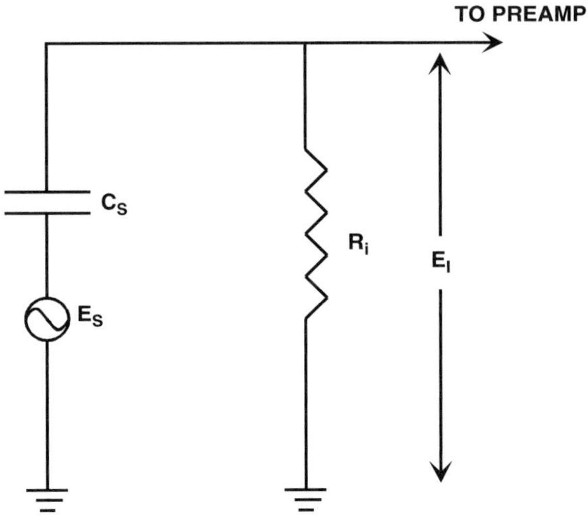

FIGURE 5.7. A simplified equivalent circuit of a piezoelectric sensor connected to a pre-amplifier.

where

k = Boltzmann's constant, 1.38×10^{-23} Joules/°K
T = temperature of the conductor in degrees K
R = resistance
Δf = noise bandwidth in Hz.

Therefore, the input resistance will play a major role in terms of the noise voltage of the pre-amplifier, the larger the resistance the larger the thermal noise level. Whatever noise present at the input of the pre-amplifier will be increased by the gain of the pre-amplifier. The voltage input to the pre-amplifier can be expressed by the equation

$$e_{\mathrm{i}} = \frac{R_{\mathrm{i}}}{R_{\mathrm{i}} + \frac{1}{j\omega C_{\mathrm{s}}}} e_{\mathrm{s}} = \frac{j\omega R_{\mathrm{i}} C_{\mathrm{s}}}{1 + j\omega R_{\mathrm{i}} C_{\mathrm{s}}} e_{\mathrm{s}}. \tag{5.5}$$

The absolute value of the voltage ratio $e_{\mathrm{i}}/e_{\mathrm{s}}$ can be expressed as

$$\left|\frac{e_{\mathrm{i}}}{e_{\mathrm{s}}}\right| = \frac{\omega R_{\mathrm{i}} C_{\mathrm{s}}}{\sqrt{1 - \omega^2 R_{\mathrm{i}}^2 C_{\mathrm{s}}}}. \tag{5.6}$$

When the frequency is such that $\omega = 1/(R_{\mathrm{i}} C_{\mathrm{s}})$ or

$$f = \frac{1}{2\pi R_{\mathrm{i}} C_{\mathrm{s}}}, \tag{5.7}$$

then Eq. (5.6) is reduced to

$$\left|\frac{e_{\mathrm{i}}}{e_{\mathrm{s}}}\right| = \frac{1}{\sqrt{2}} = 0.707. \tag{5.8}$$

Therefore, Eq. (5.7) specifies the lower 3-dB roll off frequency of the piezoelectric–pre-amplifier assembly. Note from Eq. (5.7) that the higher the capacitance of the sensor the lower the resistance has to be to achieved for a particular low-frequency roll off. The lower the input resistor, the lower will be the noise of the pre-amplifier. In reality, the sensor must also have a high sensitivity and high capacitance in order to achieve the lowest noise levels possible from the combination of the sensor and pre-amplifier.

We will now consider an example of typical trade-off between frequency bandwidth and the sensitivity of the piezoelectric sensor in a hydrophone. The noise pressure (equivalent sound pressure of an acoustic noise in a 1 Hz band, which is equal to the noise voltage) along with the receiving sensitivity of two standard hydrophones developed at the Naval Underwater Sound Reference Detachment in Orlando, Florida, are shown in Fig. 5.8 for the H-56 and Fig. 5.9 for the H-52. The pre-amplifier gain for both sensitivity curves were approximately the same, which indicates that the piezoelectric element of the H-56 is about 18–19 dB more sensitive than the element for the H-52. The noise pressure for the H-56 is also about 18–19 dB lower than for the noise pressure for the H-52. Since the piezoelectric elements

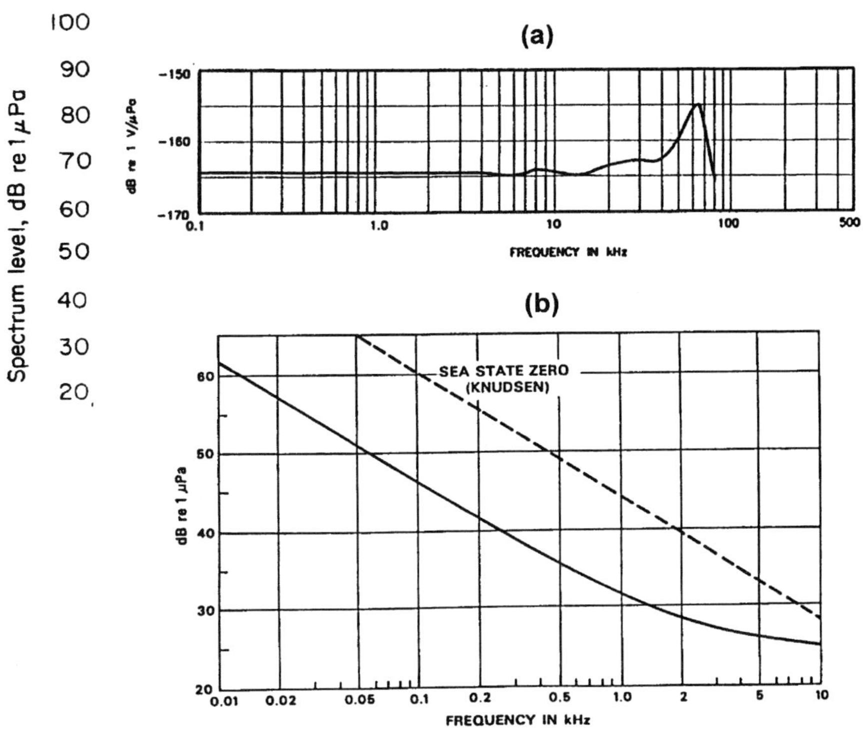

FIGURE 5.8. (a) Receiving sensitivity and (b) noise pressure for an H-56 low-noise hydrophone.

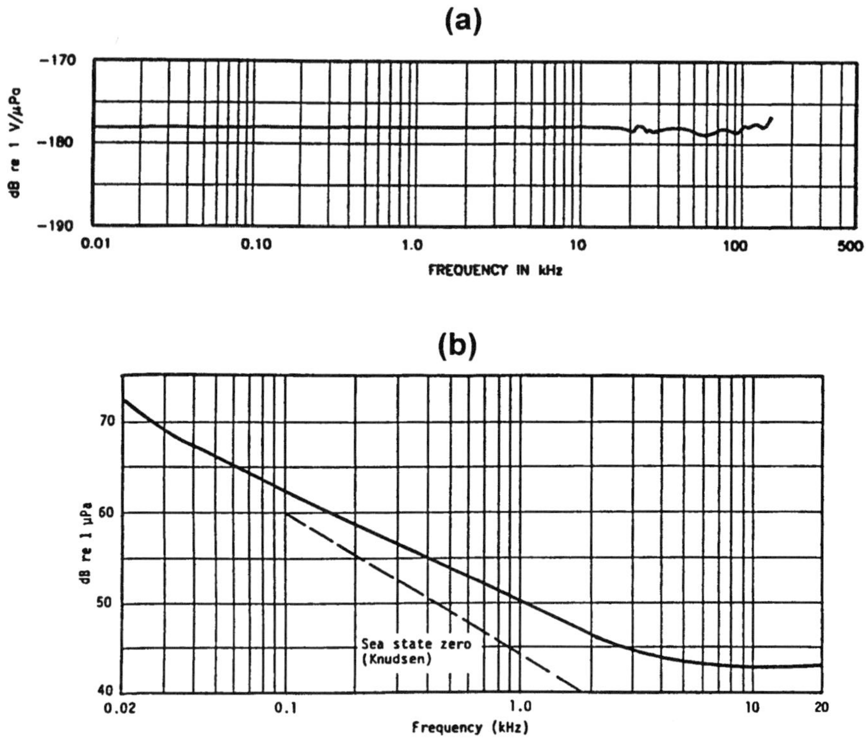

FIGURE 5.9. (a) Receiving sensitivity and (b) noise pressure for an H-52 broadband hydrophone.

do not introduce any significant amount of noise, the higher sensitivity of the H-56 element resulted in much lower noise pressure for it than for the H-52. However, the H-52 is a much broader hydrophone than the H-56, with its sensitivity being relatively flat out to about 200 kHz, whereas the H-56 rolls off at about 80 kHz with a resonance peak of about 10 dB at 65 kHz. A broader frequency response usually requires smaller elements that translate to lower sensitivity. Therefore, there are definite trade-offs between sensitivity, frequency response, and noise pressure, which must be made in choosing the most appropriate hydrophone for a given task.

5.4 The Sonar Equation

When performing any underwater acoustic experiment or making underwater acoustic measurements, we will eventually need to use the sonar equation in order to obtain an estimate of how far a sonar may detect a marine animal, how far we will be able to hear and detect a sound producing animal, or to study the effectiveness of the sonar system of odontocetes. The noise-limited form of the sonar equation can be considered like an accountant's ledger. In one column we can compute the expected signal level, and in the other column we can estimate the received noise from the ambient conditions. There are two forms of the sonar equation: the active form in which we are interested in pinging and receiving echoes from targets and the passive form in which we are interested in listening to sounds.

5.4.1 Active Form of the Sonar Equation

The noise-limited form of the sonar equation is used when the detection range of a sonar is limited by environmental noise and it can be expressed in units of dB as (Urick, 1983)

$$\mathrm{DT} = \mathrm{EL} - \mathrm{NR}, \tag{5.9}$$

where

DT = the detection thre.shold and will depend on the specific sonar system,
EL = the intensity of the echo
NR = the received noise spectral density.

The intensity of the echo from a target is equal to the source level, SL (SPL at 1 m from the source), of the transmitted signal minus the transmission loss encountered by the signal traveling to a target and back, plus the target strength, which is related to the fraction of energy reflected by the target back to the source (usually a negative quantity). Therefore, the echo level can be written as

$$\mathrm{EL} = \mathrm{SL} - 2\,\mathrm{TL} + \mathrm{TS}, \tag{5.10}$$

where

SL = source level or intensity at 1 m from the source
TL = one-way transmission loss
TS = target strength.

Target strength is defined as the ratio in dB of the echo intensity measured 1 m from the target to the intensity of the incident signal at the location of the target, or

$$\mathrm{TS} = 10 \log \frac{\text{echo intensity 1 m from target}}{\text{incident intensity}}. \tag{5.11}$$

The background noise is assumed to be isotropic, which means that the noise arriving at the receiver is the same from all directions in three-dimensional space. The ambient noise term is associated with the noise measured at the terminals of a hydrophone and does not include the effects of any filtering that may reduce the noise. This procedure is taken deliberately to make the sonar equation applicable in a general situation and not be associated with any particular sonar system. If the receiving transducer has a directional receiving beam, then only noise that is within the beam will interfere with the reception of the signal, so that the received noise is equal to the isotropic noise minus the directivity index of the receiving hydrophone, or

$$\mathrm{NR} = \mathrm{NL} - \mathrm{DI}. \tag{5.12}$$

Combining Eqs. (5.10) and (5.11) into Eq. (5.9), the noise-limited form of the sonar equation can be expressed as

$$\mathrm{DT} = \mathrm{SL} - 2\,\mathrm{TL} + \mathrm{TS} - (\mathrm{NL} - \mathrm{DI}). \tag{5.13}$$

If we know the detection threshold for a particular sonar, we can calculate how far we should be able to detect a target having a particular target strength in a particular noise environment. Conversely, we can also determine how much noise our system could tolerate before we will not be able to detect a target at a specific range, or we can calculate how large a target needs to be before we can detect at a desired range.

If reverberation is the limiting factor in terms of detecting a target, then we would need to use the reverberation form of the sonar equation. In this case, the terms NL – DI can be replaced by an equivalent plane-wave reverberation level RL. The reverberation-limited form of the sonar equation can now be expressed as (Urick, 1983)

$$\mathrm{DT_R} = \mathrm{SL} - 2\,\mathrm{TL} + \mathrm{TS} - \mathrm{RL}. \tag{5.14}$$

This equation is actually much more complicated than the noise-limited form of the sonar equation. An example of how this equation can be used will be shown in Chapter 13, when we discuss echolocation target detection by dolphins in a reverberation-limited situation.

5.4.2 Passive Form of the Sonar Equation

In the passive case when we are only listening to the sounds produced by animals, we are interested in the one-way transmission loss only. The received signal will be equal to the source level of the animal minus the transmission loss, or

$$\mathrm{EL} = \mathrm{SL} - \mathrm{TL}. \tag{5.15}$$

Our ability to detect the signal will be limited by the received noise of our measurement system. Therefore, the passive form of the sonar equation can be expressed as

$$DT = SL - TL - (NL - DI). \tag{5.16}$$

The ambient noise level, NL, is a density expressed in term of its per Hertz value. If we had a specific measurement system with a limited but specific bandwidth, the effect of the filter can be added to Eq. (5.16) so that we can rewrite the sonar equation applicable to our system as

$$DT_A = SL - TL - (NL - DI + \Delta f), \tag{5.17}$$

where Δf is the filter bandwidth of the system and DT_A is the new detection threshold. If we let $DT_A = 0$, then the received signal and receive noise will have the same intensity and a mammalian auditory system should be able to detect the signal about 50% of the time. This is similar to the definition of critical ratio for an auditory system, which will be discussed in Chapter 10. If we choose a DT_A of 3 dB, then the signal intensity will be twice as large as the noise intensity, and the signal should be easy to detect. By letting $DT_A = 3$ dB, we can now estimate how our system can tolerate the transmission loss in order to detect the signal.

If our interest is in estimating how far a marine mammal can detect a sound, we can use Eq. (5.17) and use the critical ratio for the system bandwidth. We can choose a conservative estimate of a mammalian auditory system by using a bank of 1/3-octave wide filters. A 1/3-octave estimate would be appropriate for the critical bandwidth of a mammalian auditory system, but would be wider than the critical ratio. If we use a 1/3-octave band estimate of the marine mammal auditory filter, then we can let $DT_A = 0$ to get a conservative estimate of how far a marine mammal will detect a sound. The bandwidth of a constant-Q filter, which is a fraction of a octave, can be derived from Eq. (5.2) and the relationship $f_c = \sqrt{f_u f_l}$ resulting in

$$\Delta f = \frac{2^n - 1}{2^{n/2}} f_c, \tag{5.18}$$

where n is the filter bandwidth expressed in terms of an octave, i.e., $n = 1/3$ for a 1/3-octave wide filter. The transmission loss that a marine mammal will be able to tolerate and still hear a signal can be expressed as

$$TL = SL - (NL - DI + 10 \log \Delta f). \tag{5.19}$$

5.5 Recordings of Sounds on Magnetic Tape

5.5.1 Linear Analog Magnetic Tape Recorders

Analog tape recording technology has been a part of our society for decades. Over the years, a wide variety of different types of tape recorders has been developed, from the very expensive multi-channel instrumentation tape recorders to the very inexpensive cassette tape recorders. It is not our intension to discuss the various types of recorders but to highlight the important characteristics a tape recorder needs to possess in order to be useful in making the recordings of bioacoustics sounds.

The first type of recorders that will be considered is the type known as instrumentation tape recorders. These recorders also come in many shapes and forms. Generally, 16-channel recorders use 1-in wide magnetic tapes, 8-channel recorder uses 1/2-wide tapes, and 4-channel recorders use 1/4-in wide tapes. Two different recording techniques are generally used and are referred to as direct record and FM record. In the direct record technique, the intensity of the magnetization on the tape is directly proportional to the amplitude of the signal. The FM record technique frequency modulates the input signal about a specific carrier frequency, and the amplitude of the input signal is encoded as a frequency deviation about the center frequency. Therefore, a constant amplitude signal is recorded on tape with the frequency of the signal encoding the amplitude of the input signal. The direct record technique has the advantage of having a wider bandwidth for a given tape speed. Its disadvantages include a lower frequency cutoff and lower dynamic range. The FM record technique has the advantage of being able to record dc signals (signals with a frequency of zero) and has a wide dynamic range. Its disadvantage is its low-frequency response for a given tape speed. The specific mode of operation of each channel of a multi-channel instrumentation tape recorder can be controlled by using the appropriate record and reproduce plug-in cards. Therefore, it is possible to have a mixed mode recording in which some channels are direct record and some are FM record channels. Some common characteristics of instrumentation tape recorders are given in Table 5.1. From the table, we can see that bandwidth and dynamic range are both affected by recording speed. Bandwidth and speed are directly proportional; the higher the speed the wider the bandwidth, for both direct and FM record modes. This principle will hold for any type of magnetic tape recorder,

TABLE 5.1. Some Characteristics of Instrumentation Tape Recorders

	Direct Record		Interm-Band FM		Wide-Band FM	
Tape speed) (ips)	Bandwidth	Dynamic range (dB)	Bandwidth	Dynamic range (dB)	Bandwidth	Dynamic range (dB)
120	400 Hz–600 kHz	40	DC–40 kHz	52	DC–80 kHz	50
60	300 Hz–300 kHz	40	DC–20 kHz	52	DC–40 kHz	50
30	200 Hz–150 kHz	40	DC–10 kHz	52	DC–20 kHz	50
15	100 Hz–75 kHz	40	DC–5 kHz	50	DC–10 kHz	50
7–1/2	100 Hz–38 kHz	38	DC–2.5 kHz	50	DC–5 kHz	48
3–3/4	100 Hz–19 kHz	38	DC–1.3 kHz	50	DC–2.5 kHz	48
1–7/8	100 Hz–9.5 kHz	38	DC–625 Hz	46	DC–1.3 kHz	46
15/16	100 Hz–4.8 kHz	36	DC–312 Hz	45	DC–625 Hz	44

whether it be an instrumentation, professional, or consumer-type tape recorder. The bandwidth for a given tape speed listed in Table 5.1 represents the state-of-the art values, so that it would be extremely uncommon to have a tape recorder with wider bandwidth for a given tape speed than what is shown in the table. The dynamic range values are usually given for the recording and playback of a tonal signal with a bandpass filter connected to the playback channel. If broadband signals are recorded, then the filter in the playback end must be broader than the bandwidth of the signal, so the playback noise level will be higher and therefore reduces the dynamic range values from those shown in the table by about 10–15 dB, depending on the specific filter bandwidth.

Having good constant speed control is essential for producing high-fidelity recordings and for accurate reproduction or playback of the recorded signals. If the playback and record speed are not the same, then the frequency of any recorded signal will not be the same when played back. One easy and common method to ensure that the record and reproduce speed are the same when using a multi-channel recorder is to record a reference signal of known frequency and amplitude and to examine that channel during playback to determine if the reproduce signal has the same frequency as the record signal. If the two speeds are not exactly the same, then the frequency of the reproduce signal will not match the frequency of the record signal and the deviation in speed can be determined by considering the difference in the frequencies of both signals. Wow and flutter are also two terms commonly used to describe changes in the signal output frequency caused by tape speed variations that are not smooth and constant but exhibit fluctuations. Speed variations up to 10 Hz are called "wow" and speed variation above 10 Hz are considered as flutter.

Instrumentation tape recorders have potentiometers that can be used to adjust the record and reproduce gains so that the record/reproduce system can be optimized for the type of signals being recorded. However, having such a capability can also cause problems if the gain settings drift or are readjusted by others for recorders that are shared by several users. A good habit to develop in using tape recorders to record acoustical and electrical signals is to record on the beginning portion of a tape a reference sine signal of known amplitude and frequency. Then any adjustments made to the tape recorder after the recording is done can be accounted for by playing back the reference signal. This technique is also helpful when recording on one tape recorder and playing back on another recorder.

There are many different types of consumer and professional tape recorders that use 1/4-in wide tapes. These recorders can have up to four channels and also have multiple speed control. However, the bandwidth and dynamic range of these recorders will depend on the record and reproduce speed, and will not be very much different than the values shown in Table 5.1 for the corresponding speed and method of recording. Cassette tape recorders also come in many different models. Generally, these recorders are restricted to a single speed and have limited bandwidth, with upper frequency cutoff of about 15 kHz. The speed control of these recorders also tend not to be of high precision. Many cassette tape recorders have automatic gain control so that the gain of the recorder varies inversely with the amplitude of the incoming signal. Such a recorder would not be appropriate if the intensity of the acoustic signal is desired.

5.5.2 Helical Scan Tape Recorders

The audio channels of video tape recorders can also be used to record analog signals with the data stored on VHS cassette tape and 8-mm cassette tape. The audio channels on most advanced consumer VHS recorders and camcorders with four heads usually employ an FM record technique in conjunction with a helical scan so that high-fidelity audio recording can be made with many hours of recording stored on a single cassette. The tape path for helical scanning is depicted in Fig. 5.10. The tape takes a slanted or helical path around a rotating-record and reproduce head, which increases the overall head-to-tape contact speed that in turn allows the longitudinal tape speed to be as low as 0.44 ips. With four heads mounted on the head wheel, one set is used for recording and playback of FM sounds and the other set is used for recording video images. A drawing of the position of the video and

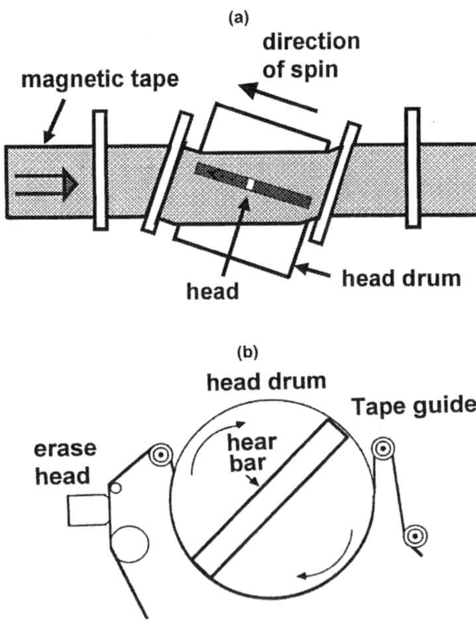

FIGURE 5.10. Tape path for helical scanning. (a) portrayal of a slanted track, (b) schematic of tape path around a rotating drum.

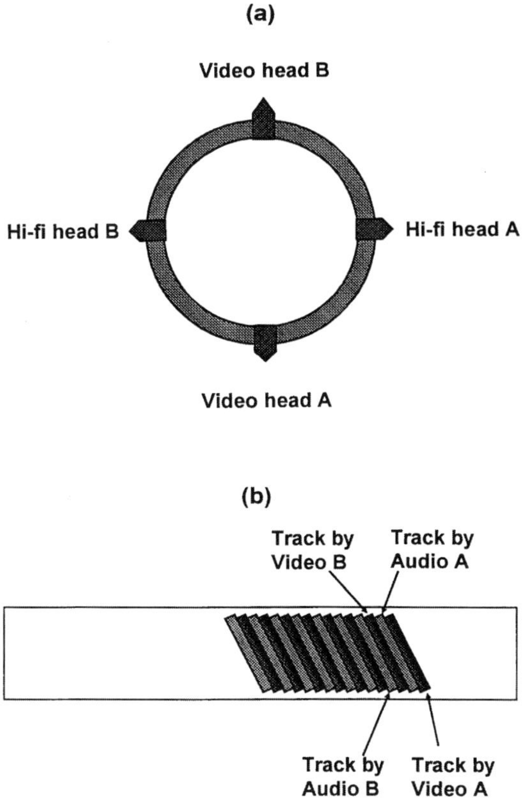

FIGURE 5.11. (a) Diagram of the video and audio head of a four head VC and (b) an example of the slanted recorded tracks on the VCR tape.

audio heads on the rotating drum is shown in Fig. 5.11. The FM record mode allows for large dynamic range. Bandwidths from DC to 20 kHz are typical and, in some instances, the internal low-pass filter of some machines has not been adjusted to 20 kHz so that the upper frequency limit for recording analog signals can be as high as 35 kHz. The rotating head wheel spins at about 1,800 revolutions per minute, effectively covering about 230 ips.

Some manufacturers have produced multi-channel instrumentation analog tape recorders using the slant head technology and either VHS or 8-mm cassette tapes. The capability for different slanted recorded tracks on the VCR tape makes this possible. Tape speed is usually included in these machines. Analog bandwidths from DC to 150 kHz are also available. However, in order to achieve such wide bandwidth, the tape speed must be relatively high so that record time is limited. Models with 16, 21, and 28 channels have also been developed.

5.5.3 Digital Audio Tape (DAT)

A relatively new type of recorder that came upon the scene in about 1986 is the digital audio tape (DAT) recorder. This type of recorder combines the helical scan technique of videocassette recorders with digital data acquisition, which centers around analog-to-digital (A/D) conversion principles that will be discussed further in the next section. The A/D and digital-to-analog (D/A) converters on DAT recorders operate at three standard sampling rates: 32, 44.1, and 48 kHz. It can digitize and record analog signals at 32 and 48 kHz sample rates, but can reproduce digitally recorded tape at all three standard sampling frequencies. The 44.1 kHz sampling rate is reserved for prerecorded DAT tapes in order to discourage unlawful copying of prerecorded programs (Huber and Runstein, 1989). Most DAT recorders operate with a 16-bit A/D converter so that the dynamic range of the recorder is very good, more than 90 dB. Having the analog signals stored in digital format also reduces wow and flutter problems since speed control is not at issue here, with the digitization and reproduction being controlled by relatively stable clocks that keep the sampling rates nearly constant. Some DAT manufacturers along with third-party manufacturers have PC (personal computer) interface cards that can be used to extract the digital data directly from the recorder and read into a PC. This capability eliminates the need to redigitize any recorder signals.

The helical scan technique adopted from videocassette tape recording technique makes it possible to have up to 2 hours of recording time on a cassette tape that is as small as 7.3 × 5.4 × 1 cm. The slanted or helical path around a rotating-record and reproduce head increases the overall head-to-tape contact speed, which in turn allows the longitudinal tape speed to be as low as 1/4 ips. The actual track width of a DAT's helical scan is about 1/10 the thickness of a human hair, allowing a density of 114 million bits per square inch (Huber and Runstein, 1989). A sophisticated tracking correction scheme is used to keep the heads directly over the tape's scan path. Generally, two recording and reproduce heads are mounted on the rotating head and a wrap angle of 90° is used (see Fig. 5.12). A wrap angle of 90° means that there will be a portion of time in which a head is not in contact with the tape. A special time compression technique is used in which data go into a buffer memory to temporarily store the digital information during the portion of time that a head is not in contact with the tape, and that buffer is downloaded in a burst when a head is again in contact with the tape. In addition to the digital data, other information can be written on the track, which can be used for identification and search functions. During playback, a similar buffer memory stretches the data back into a continuous digital bit stream.

A strong advantage of the DAT technology is the small size of the recorders. DAT recorders tend to be smaller than their analog counter part by a factor of three to four or more. Their weights also tend to be less by a factor of five or more. One of the reason for the smaller size and lighter weight has to do the helical scan technique that requires much less tape so that tape

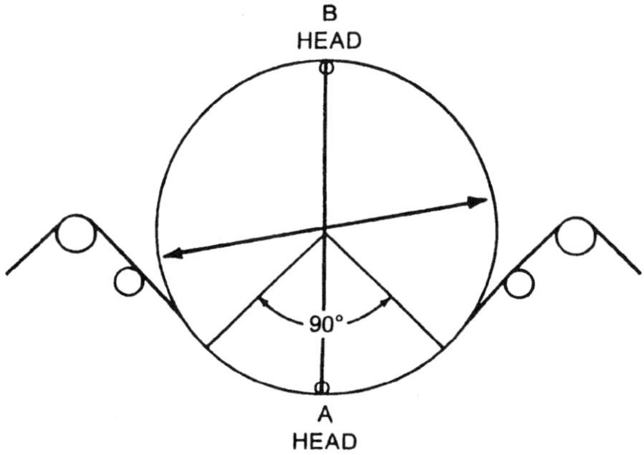

FIGURE 5.12. A schematic of the 90° wrap angle used in DAT recorders along with a schematic of the recorder tracks.

cassette is much smaller than a typical reel of an analog multi-channel instrumentation tape recorder. The electronics used in DAT recorders are also about two generations more advanced than typical analog instrumentation tape recorders, and therefore are much smaller and compact in design.

The major disadvantage of DAT recorders is the relatively low analog bandwidth of about 22 kHz. However, this technology is rapidly changing as manufacturers discovered new ways to increase bandwidth. For example, some manufacturers have produced advanced professional models with twice the standard 44.1 kHz sampling rate so that the analog upper frequency limit for these recorders is about 44 kHz. Still other manufacturers are experimenting even faster sampling rates, which would increase the upper frequency limit but shorten the running time of a tape. The DAT technology is also being used by electronic instrumentation manufacturers who produce multichannel recorders that can have up to 64 channels of recording, with a bandwidth from dc to 100 kHz for a single channel. It would be futile to specify any specific manufacturer and model number, since the information would soon be obsolete as the technology surges ahead.

5.5.4 *Digital Audio Recorders*

A recent consumer electronic item for music lovers that has gained wide spread popularity is the digital audio recorder (DAR) often referred as MP3 players. These DARs usually have an imbedded microcontroller that control the digitization (16 bits and higher) of stereo analog signals and the storage of the signals on a hard disc or compact flash cards. Data storage of these recorders are often based on the MP3 data compression technique that uses perceptual audio coding to compress CD-quality sound by a factor of about 10, while providing almost the same high fidelity. However, many machines also store data in a WAV format. Previously, digitized signals can also be downloaded from the Internet or other digital devices. The presence of an imbedded microcontroller allows a DAR to have considerable flexibility that seems almost endless from a music entertainment perspective. One important aspect of a DAR is the ability to control the sampling rate of the data acquisition system. Typical sampling rates are $44.1/2^n$ kHz, where $n = 0$, 1, 2, ... N, and N for stereo recordings depends on the specific recorder. The fastest recorder can sample at a rate of 96 kHz. Therefore, a sampling rate of 96 or 44.1 kHz can be used to monitor social sounds of odontocetes, and a lower sampling rate can be used for monitoring baleen whale sounds and even lower sampling rates can be used to monitor fish sounds. Another useful feature of these recorders is the capability to download recorded data to a computer using either a USB-2, a firewire port or reading a compact flash directly into a computer, depending on the specific recorder.

Several cautions should be taken in using the commercial consumer product type DAR. The first caution is to ensure that the recorder does not have automatic gain control or that this feature can be turned off. With automatic gain control, it is extremely difficult to know the gain of the recording system at any specific time so that absolute measurements cannot be done. The second caution is to ensure that the data can be stored in a different format than the MP3 compression format. Individuals have reported that there are some distortions involved with playing back MP3 formatted data. In order words, the data compression format involved with the MP3 data is not reliable for the storage of scientific data.

There are some companies that have developed units that can control specific brands of DAR in such a manner that controls the recorder and can specify a duty cycle of recording periods in between a low-power mode and sleep state. Such timing control will allow for the conservation of both power and storage space. The total recording time will depend on the sample rate and the available disc space per channel. If we let M = the total disc space capacity, SR = A/D sampling rate, and allow for 2 bytes per channel for each analog-to-digital conversion, the maximum amount of time in hours that a recorder can be operated continuously can be expressed as

$$\text{hours of recording} = \frac{M/2}{2x\text{SR}x\ 3600}. \tag{5.20}$$

5.6 Principles of Digital Data Acquisition: A/D Conversion

We live in a world in which electronics is progressively becoming more digital at a rapid rate. The appearance of personal computers (PC) slightly over a decade ago has completely changed the manner in which our society functions, including our science. In less than 40 short years since the development of the first general purpose digital computer, the ENIAC, computer power, speed, and availability have literally exploded to the point that practically every scientist personally has access to several PCs. Marching along with the computer explosion has been the continual development and improvement of digital electronics such as analog-to-digital converters and digital-to-analog converters, which make it possible to interface the analog world of acoustic sensors and other electronic instrumentation to the digital world of computers, and vice versa. Therefore, it becomes imperative for scientists and engineers to understand the principles of digital data acquisition, namely analog-to-digital and digital-to-analog conversion. In this section, we will broadly and briefly discuss some of the important principles of digital data acquisition. Those who would like to obtain more in-depth information should refer to some of the many excellent books on this subject.

5.6.1 *Sampling*

The heart of a digital data acquisition system is the clock that controls the *sample rate*, also referred to as the *sample frequency*, or the *sample time*, which is the inverse of the sample rate. At the beginning of each clock pulse, the analog signal is sampled and its amplitude determined. In the example of Fig. 5.13,

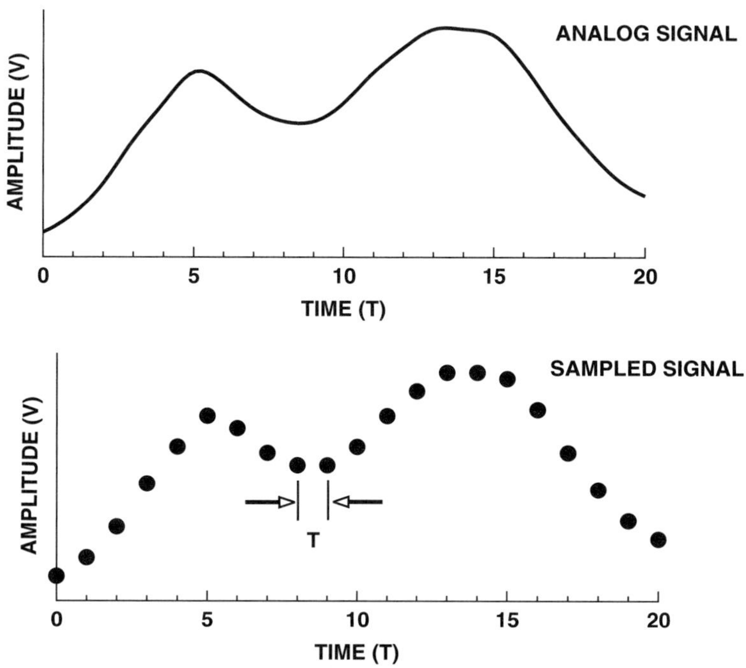

FIGURE 5.13. (a) An analog signal with amplitude plotted against time in T seconds. (b) A sampled version of the analog signal with the sampling interval equal to T seconds.

the top curve is an analog signal with the amplitude plotting as a function of time. The bottom curve is the sampled version of the analog signal with the ting sampling taking place at every Δt seconds. Therefore, the continuous signal $x(t)$ is now represented by its sampled version $x(n\Delta t)$, where $n = 0, 1, 2, \ldots, N-1$. From the figure, it seems that the smaller the sampling period, the more accurately the sampled version of the signal would represent the actual signal. However, if the analog signal is band-limited (its amplitude equal to zero for frequencies greater than a specific maximum frequency, $f_{\max}$) and the sampling frequency is greater than twice the maximum frequency ($f \geq 2f_{\max}$), then the sampling theorem of Shannon—Nyquist states that the original signal can be fully recovered from the sampled signal by the expression

$$x(t) = \Delta t \sum_{n=0}^{N-1} x(n\Delta t) \frac{\sin 2\pi f_c(t - n\Delta t)}{\pi(t - n\Delta t)}, \tag{5.21}$$

where Δt is the sampling interval and $1/\Delta t$ is the minimum sampling rate in order to faithfully reproduce the original signal, which is called the *Nyquist* frequency, which is 1/2 of the maximum frequency in the signal. In other words, if a sample frequency f_s is used in a data acquisition system, a proper *anti-aliasing* low-pass filter must be used to limit any frequency components of the signals beyond the Nyquist frequency, which is $f_s/2$. Any frequency components of the signal being sampled that is greater than the Nyquist frequency will cause *aliasing* to occur. An excellent example of aliasing is given in the application notes for the Tucker Davis' TDT system II, and is presented in Fig. 5.14. Assume that our data acquisition system has a sampling frequency $f_s = 1/\Delta t$ so that any input signal is sampled at intervals of Δt seconds. If a signal of frequency 7/8 f_s, a frequency 7/4 time greater than the Nyquist frequency of $f_s/2$, is fed into the data acquisition system, the results shown in Fig. 5.14 will occur. The points of the input signal is shown along with the aliased signal. Aliasing has caused the input signal to "fold-over" or be reflected about the Nyquist frequency so that the resultant signal will be of a lower frequency but still maintain the same distance from the Nyquist frequency as the original signal. For example, if we assume that a signal of frequency f is greater than the Nyquist frequency ($f_s/2$), then the signal when sampled will have an aliased frequency that is folded about the Nyquist frequency so that

$$\mathrm{f_a} = \mathrm{f_s}/2 - f, \tag{5.22}$$

where f_a is the frequency of the aliased signal. In the example of Fig. 5.14, $f = 7/8\, f_s$ so that from Eq. (5.22), $f_a = f_s/2 - 7/8\, f_s = 1/8\, f_s$. From this contrived example, one can see that without a proper anti-aliasing filter or for too low a sampling rate, the sampled data can be very different from the signal being sampled. The anti-aliasing filter should ideally have a steep roll off since any signal with frequency greater than the Nyquist frequency that is not sufficiently attenuated by the filter will be aliased. An example of this phenomenon is illustrated in Fig. 5.15, which shows the transfer function of

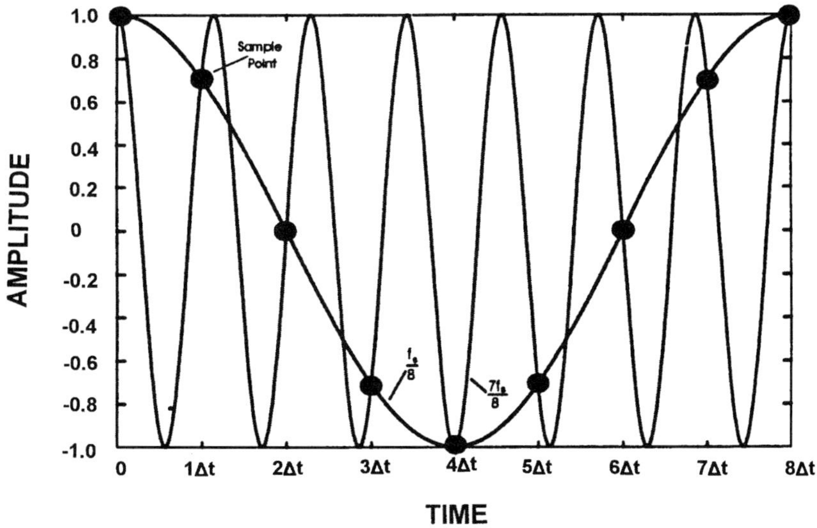

FIGURE 5.14. An example of aliasing with a sampling frequency of $f_s = 1/t$, an input signal of frequency 7/8 f_s, the sampled points (*solid circles*) and the resultant aliased signal (adapted from Varosi and Tucker, 1994).

a low-pass filter having only a 24 dB/octave roll off. All component of a signal with frequencies greater than the Nyquist frequency will be folded over. Many hydrophones also have very steep high-frequency roll off and these hydrophones themselves can be used both as a sensor and an

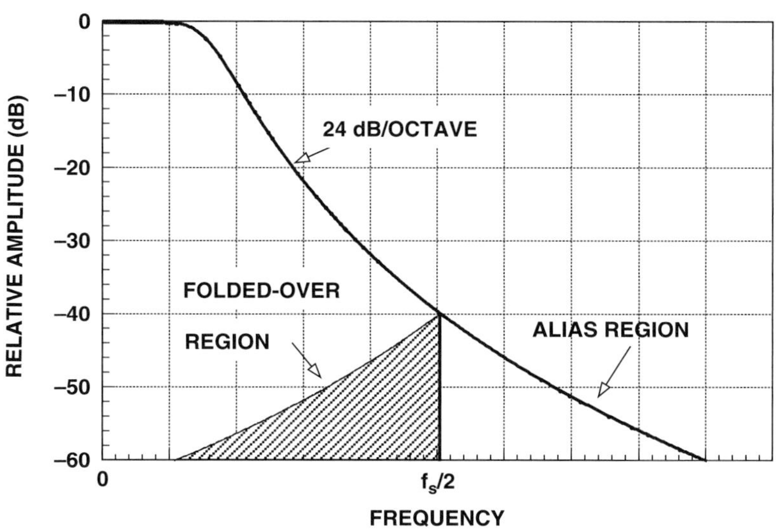

FIGURE 5.15. Example of aliasing of a portion of a white noise signal is not sufficiently attenuated by an anti-aliasing filter.

anti-aliasing filter, providing no additional signals are added into the system after the output of the hydrophone.

5.6.2 *Binary Representation*

The end result of using a digital data acquisition system is the conversion of analog signals into a digital format. Digital data are usually represented in a binary form, so it would be useful at to have a brief review of the binary numbering system. A binary number consists of a horizontal array of bits, each bit having a value of either 0 or 1. The number of bits used to represent an integer will determine the maximum value that a binary number can represent. A *nibble* is commonly used to refer to a 4-bit, a *byte* to an 8-bit, and a *word* to a 16-bit binary number. Each bit represents a specific integer value, depending on its position in the array, as shown below. Bit b_i is the ith bit, LSB stands for the least significant bit, MSB stands for the most significant bit, and I_u is the integer value of the unsigned binary number. If n is the number of bits in a binary number, then I_u can be determined as shown below and by Eq. (5.23)

		MSB							LSB	
↓									↓	
b_{n-1}	.....	b_7	b_6	b_5	b_4	b_3	b_2	b_1	b_0	← bits
|		|	|	|	|	|	|	|		
2^{n-1}		2^7	2^6	2^5	2^4	2^3	2^2	2^1	2^0	← Integer values

$$I_u = \sum_{i=0}^{n-1} b_{n-1-i}\, 2^{n-1-i} = \sum_{i=0}^{n-1} b_i 2^i. \tag{5.23}$$

The first summation in Eq. (5.23) counts downward starting with b_n TM12^{n-1} while the second summation counts upward starting with $b_0 2^0$. As an example, the integer value of 11101101 can be evaluated as: $I_u = 1\times2^7 + 1\times2^6 + 1\times2^5 + 0\times2^4 + 1\times2^3 + 1\times2^2 + 0\times2^1 + 1\times2^0 = 237$.

The binary representation discussed above is called the *unsigned binary* representation. The unsigned binary form is ideal for representing the digital values of unipolar signals that have only one polarity so that possible values are either all positive or all zeroes. However, the unsigned binary format can also be used to represent bipolar signals that can have positive and negative polarities. In this case, the binary format is usually referred to as an *offset binary* and its value is equal to

$$I_0 = I_u - I_{\mathrm{offset}}, \tag{5.24}$$

where: $I_{\mathrm{offset}} = 2^{n-1}$,

I_u is the integer when the binary number is treated and evaluated as a unsigned binary number. Table 5.2 depicts the integer value of an n-bit signed binary number using Eq. (5.12) for different values of n. Another binary

TABLE 5.2. Integer Values of n-bit Unsigned Binary Numbers for Different Number of Bits

n-bit binary	Integer I_0 $n = 8$	$n = 12$	$n = 16$
11–11111	127	2047	32767
- - - - - - -			
- - - - - - -			
10–00000	0	0	0
- - - - - - -			
- - - - - - -			
00–00000	–128	–2048	–32768

representation that is often used is the *signed binary* representation, in which the most significant bit determines the sign of the number. If the MSB is a 0, then the binary number is a positive quantity; and if it is a 1, the binary number has a negative value. Once again, except for the MSB, each bit (depending on its position) represents a specific integer value as shown below

MSB ↓									LSB ↓	
b_{n-1}	.....	b_7	b_6	b_5	b_4	b_3	b_2	b_1	b_0	← bits
|		|	|	|	|	|	|	|	|	
sign bit		2^7	2^6	2^5	2^4	2^3	2^2	2^1	2^0	← Integer values

$$I_s = (1 - 2b_{n-1}) \sum_{i=0}^{n-2} b_{n-2-i} 2^{n-2-i} = (1 - 2b_{n-1}) \sum_{i=0}^{n-2} b_i 2^i. \tag{5.25}$$

The first summation counts downward starting with $b_{n-2}2^{n-2}$ and the second counts upward starting with $b_0 2^0$. As an example of a signed binary number, consider 11101101, which can be evaluated as $-1(1\times 2^6 + 1\times 2^5 + 0\times 2^4 + 1\times 2^3 + 1\times 2^2 + 0\times 2^1 + 1\times 2^0) = -109$, whereas $01101101 = 1\times 2^6 + 1\times 2^5 + 0\times 2^4 + 1\times{}^2 3 + 1\times 2^2 + 0\times 2^1 + 1\times 2^0 = 109$.

5.6.3 *Analog-to-Digital (A/D) Conversion*

A typical data acquisition system in which the analog signal detected by a hydrophone is converted into digital form is depicted in Fig. 5.16, with a clock controlling the timing of events.

At the leading (or trailing) edge of each clock pulse, the sample and hold (or track and hold) amplifier samples the analog signal and holds the sampled voltage by means of a capacitor. A finite amount of time commonly referred to as the *acquisition time* is required before the sample and hold amplifier can accurately sample the voltage of the analog signal. Once the analog signal is

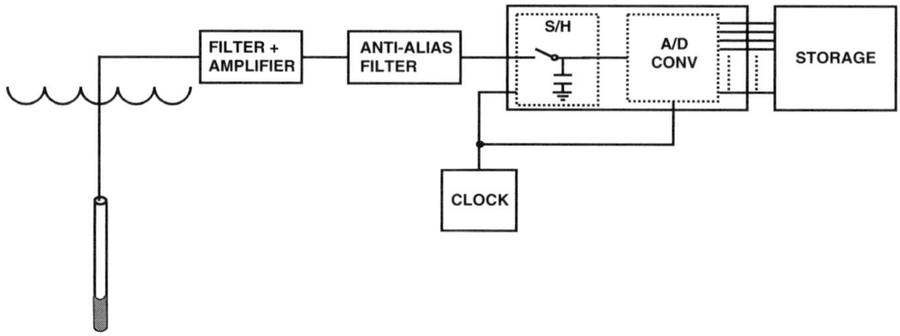

FIGURE 5.16. A typical data acquisition system that converts the analog signal received by the transducer into digital form and stores the digital data. The box labeled S/H is sample-and-hold amplifier.

properly sampled, the A/D converter converts the sampled voltage into digital form that is represented by a binary number. The A/D converter also requires a finite time commonly referred to as the *conversion time* to accurately convert the sampled voltage at its input into digital form. The maximum rate of the clock is restricted by the acquisition and conversion times, so that if t is the sampling interval, then

$$t \geq t_{\mathrm{acq}} + t_{\mathrm{A/D}}, \tag{5.26}$$

where t_{acq} is the sample and hold amplifier acquisition time, and $t_{\mathrm{A/D}}$ is the A/D converter conversion time. Often the sample and hold amplifier, A/D converter, and clock are integrated into a single module or integrated circuit. In this case, the speed of the package or its maximum sampling rate is given in terms of its *throughput*, which is associated with the combined acquisition and conversion times. The output of the A/D converter will be consisting of n bits in some binary format, depending on the capabilities of the A/D converter. Typical values of n are 8, 12, and 16 bits. The larger the number of bits in the A/D converter, the longer the conversion time for each sample. Data acquisition boards for PCs have throughputs of hundreds of MHz for 8-bit A/D conversion, tens of MHz for 12-bit conversion, and up to several MHz for 16-bit conversion.

The A/D converter takes the voltage at the output of the S/H amplifier and converts that voltage into a binary number. There are various types of A/D converters that work on slightly different principles; however, in this section we will assume a generic A/D converter and not be concerned with any specific type of converter. Since a binary number of n bits can represent 2^n different values, the voltage into an A/D converter will be quantized into one of the 2^n values. The specific binary value of a digitized input voltage will depend on the full-scale voltage input and the number of bits that the A/D converter is based on. If $V_{\max}$ is the full-scale or maximum value of a unipolar signal that a unipolar A/D can digitize, then the *resolution* or the discrete incremental voltage between each *quantization* will be

$$\Delta V = V_{\max}/2^n. \tag{5.27}$$

Typical values of $V_{\max}$ are 5, 10, and 20 volts. The voltage value of the least significant bit is equal to ΔV. If the input voltage is zero, then all the bits in the binary output including the LSB will be zero; and if the input voltage is ΔV, the LSB will be 1. For a bipolar signal and a bipolar A/D converter, we need to consider the maximum positive and negative input voltage that the converter can convert in order to calculate the resolution of the converter. Typical maximum values for an A/D converter in the bipolar mode are ±2.5, ±5, and ±10 volts. If we let $V_{\max}$ be the maximum allowable positive input voltage and $-V_{\max}$ be the most negative allowable negative input voltage, then the voltage quantization increment or voltage resolution will be

$$\Delta V = \frac{2V_{\max}}{2^n} = \frac{V_{\max}}{2^{n-1}}. \tag{5.28}$$

When a voltage level is converted into digital form, its value will be rounded off to the nearest quantized level so that the an input voltage can be represented only to an accuracy of ±1/2 ΔV or ±1/2 LSB. The rounding off effect is referred to as the *quantization error*. The effects of the rounding-off of the input voltage is depicted in Fig. 5.17, which shows the sampling time intervals and the voltage quantization levels along with the quantization

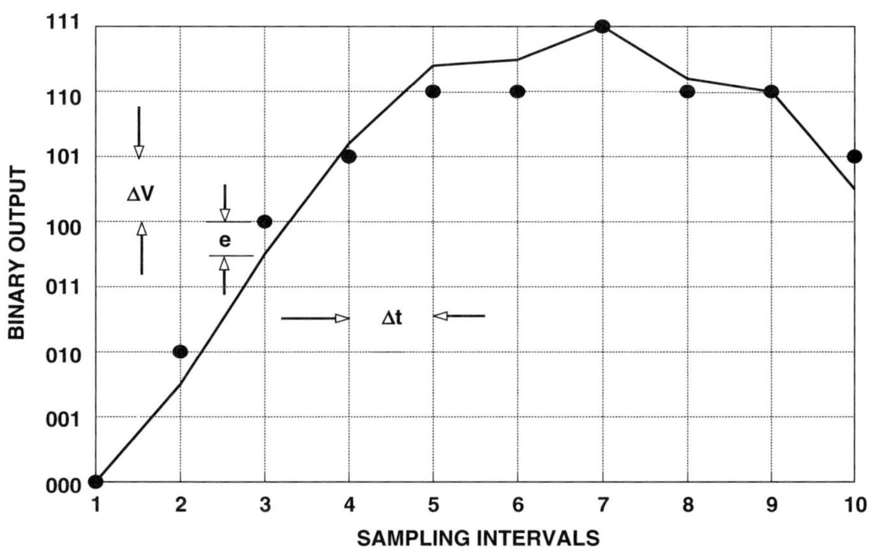

FIGURE 5.17. Example of quantization error in the analog-to-digital conversion process, with an analog signal and the specific level at which the analog signal will be quantized.

error at each time of sample. From the figure we can see that the more bits available in the A/D converter, the more accurate the digitization process will be, since the quantized levels will be smaller as n becomes larger and a better estimate of the input voltage can be made. However, the conversion time of an A/D converter is directly related to the number of bits involved so that the choice of a particular data acquisition card often involves some sort of trade-off between the degree of accuracy desired and the throughput of the data acquisition unit.

Since a bit can have one of two values (0 or 1), the number of different values an n-bit binary number can represent is 2^n. Therefore, an 8-bit binary number can represent $2^8 = 256$ different values varying from 0 to 255, a 12-bit binary number can represent $2^{12} = 4096$ different values varying from 0 to 4095, and a 16-bit binary number can represent $2^{16} = 65536$ different values varying from 0 to 65535. Consequently, an n-bit A/D converter has a theoretical dynamic range of

$$\text{dynamic range (dB)} = 20 \log (2^n) \approx 6\,n, \tag{5.29}$$

which is equal to 48 dB for 8-bit, 72 dB for 12-bit and 96 dB for 16-bit A/D conversion. However, the theoretical limits of dynamic range are never reached because of digital or quantization noise, offset errors, scale factor errors, linearity errors, inaccuracies in the conversion processes, jitter in the clock, variation in circuit parameters caused by temperature effects, etc, so that the LSB is unreliable and that the actual dynamic range is usually about 6 dB less than the theoretical value.

An important part of the data acquisition process is the storage of the digitized voltages and eventual transfer into a PC for analysis. The output of the A/D converter will be in a binary format and will be stored and downloaded into a computer in the same format. The computer will read the downloaded data as integers, and the user must be careful to specify the format (signed, unsigned, 2's complement, etc) of the binary data and ensure that the computer language can handle unsigned integers if the unsigned or offset binary format is used or signed integer if the signed binary format is used. The output voltage value of an A/D converter for the different conversion modes will be

$$\text{unipolar mode}: \quad V = I_u\,\Delta V = I_u\,V_{max}/2^n, \tag{5.30}$$

$$\text{signed binary}: \quad V = I_s\,\Delta V = I_s\,V_{max}/2^{n-1} \tag{5.31}$$

$$\text{offset binary}: \quad V = (I_0 - 2^{n-1})\,\Delta V = (I_0 - 2^{n-1})\,V_{max}/2^{n-1} \tag{5.32}$$

5.6.4 *Data Acquisition Systems*

There are a multitude of different types of data acquisition (DAC) boards that can be plugged directly in a PC EISA, or PCI slots, or a Macintosh

Nubus slot. There are also DAC systems that can be interfaced with a PC via the serial or parallel (printer) port or even a PCMCIA and USB slots commonly present on laptop computers. Variables associated with DAC broads are the number of bits, throughput speed, number of channels, on-board memory space, simultaneous or sequential sampling for multichannel system, and triggering capabilities. The choice of a specific board will depend on the applications envisioned and most often will involve some type of trade-off. It is extremely difficult to have a single general purpose DAC system that will be flexible enough for many different applications.

The issue of bandwidth and sampling rate are two of the most important parameters for the choice of a specific DAC system. The system must have sufficient analog bandwidth and sampling rate to faithfully record the widest band of frequencies for the species of interest. It also makes a tremendous amount of difference if one is interested in recording echolocation signals of dolphins and whales or other types of communications and social sounds. We will learn in Chapter 11 that echolocation signals are of relatively short duration (several hundred μs) in contrast to communication and social sounds that could last for several seconds and even minutes. Echolocation signals can have frequency components up to about 150 kHz, and have repetition intervals of tens of microseconds. Therefore, a transient recording type of DAC would be most ideal for recording echolocation signals. In a transient recording type of board, there is usually a pre-trigger function and sufficient on-board memory to collect several hundred individual signals. A board with a pre-trigger capability will normally continuously digitize incoming signals and the digital output of the A/D converter is fed into temporary memory. When a signal occurs, a certain number of pre-trigger points (digitized points occurring before the onset of the signal) and post-trigger points are transferred to the board's normal memory. Therefore, the dead space between the onset of each echolocation signal is essentially ignored. An example of an echolocation signal recorded with a board having a pre-trigger capability is shown in Fig. 5.18. In most cases, the user can specify the number of pre- and post-triggered points to save in memory. An interesting feature of a pre-trigger function is the ability to capture data that occurred before the onset of the signal that will eventually activate the trigger. In most transient recording type boards, the user can specify the amount of pre-trigger and post-trigger points that are saved into on-board memory. Having a pre-triggering capability helps in minimizing the amount of memory that is needed on the board and ultimately the amount of disc or other memory storage space on the controlling PC. For example, suppose we are sampling echolocation signals at 1 MHz using a 12-bit A/D card with a pre-trigger capability. If we save 30 pre-triggered points and 226 post trigger points, or 256 total points per signal, then for each 1 K word of memory, we can digitize and save four signals. It is not uncommon to have 128, 256, or 512 K word of memory on an ADC board so that even with a 128 K word memory space, we could collect 128 echolocation signals before we would

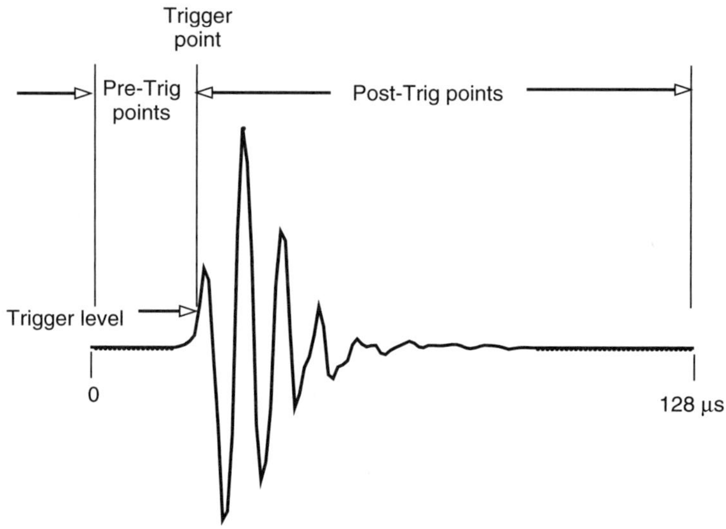

FIGURE 5.18. An example of a dolphin echolocation signal collected with a DAC having a pre-trigger capability.

need to download the data and store it in a disc file. One bit of information that will be lost is the time interval between the onset of each echolocation signal. However, this piece of information can be easily recorded with another board that has a clock that operates at about a rate of 1 ms or greater and a counter that counts the number of ticks of the clock after it has been reset. Therefore, every time an echolocation signal occurs the ADC board flags, the PC, and the count of the counter is read into the computer /'s memory and the counter is then reset, so that the interval between echolocation signals will be encoded by the counter's value each time a signal is detected.

The amount of memory required to store digitized communications and social sounds produced by marine mammals can be much higher than for echolocation signals. Therefore, a DAC board must have a large amount of on-board memory or have very little on-board memory but have the capability to transfer data to computer memory fast enough to maintain a particular throughput. For example, if we are digitizing a dolphin's communication sounds using a 12-bit ADC board operating at a 250 kHz rate, then a 1 s look at the signal would require 250 k words of memory. The amount of memory needed would increase proportionately if we had a longer look at the signals. Social sounds can last from several seconds and up to minutes depending on the species involved. The amount of memory will also increase proportionately if we sample at higher rates. Dead space (digitized data that is not related to the signal of interest) is also an issue in recording social sounds. We can optimize the amount of storage space required for each sound by first passing the input signal through a delay line before it is fed into an A/D

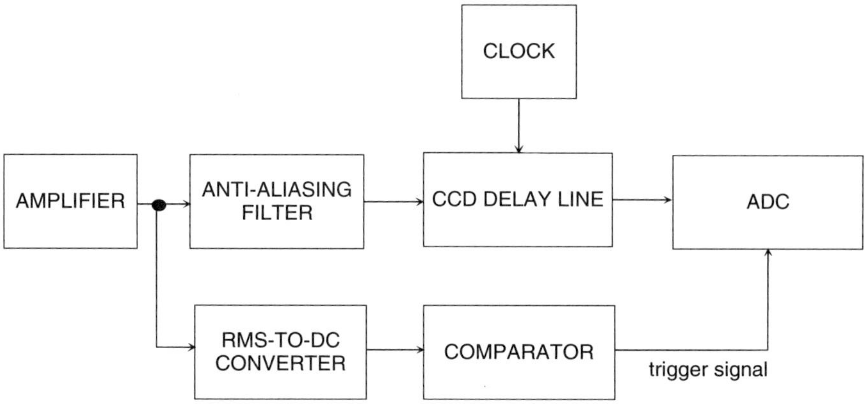

FIGURE 5.19. Functional diagram of a circuit that would allow the pre-examination of a signal before digitizing.

converter, and in the mean time have the signal examined to determine if a sound of interest has occurred. Such a technique is depicted in Fig. 5.19, showing an analog charged coupled device acting as a delay line. Charged coupled devices with bandwidths up to 170 kHz and delay times from hundreds of microseconds to several hundred milliseconds are available. The rms-to-DC converter measures the rms voltage of the signal and converts the voltage to a DC value. The integration time of the rms-to-DC converter can be controlled with a capacitor so that if a long enough integration time is chosen, rms value of the signal for short-duration echolocation signals will be low and will only become large when long duration sounds are being received. With such a circuit working in conjunction with an ADC board, the amount of data that is collected can be minimized since the A/D converter will function only if a social sound is present.

Another popular technique to obtain pre-trigger points consist of using a circular buffer that may be in hardware on a data acquisition board or in a PC that is controlled via software. In order to use a circular buffer within a PC, a data array of N_{total} points is set aside and the A/D converter digitizes continuously and the digitized data are stored into the data. When the end of the buffer is reached, incoming data are then routed to the beginning of the buffer and new data will be written over and replace the old data buffer. This process can continue indefinitely until a trigger signal is received. Upon receiving a trigger signal, the A/D converter will digitized N_{post} and cease digitizing. Therefore, the number of pre-triggered points (N_{pre}) in the circular buffer will be $N_{total} - N_{post}$ and the number of post-trigger points will be N_{post}, where $N_{pre} + N_{post} = N_{total}$. The N_{total} digitized values in the circular buffer are then stored in a disc file starting with the data at the start address corresponding to the trigger address + N_{post} + 1 and ending with the data at the end address corresponding to the trigger address + N_{post}. This process is shown schematically in Fig. 5.20.

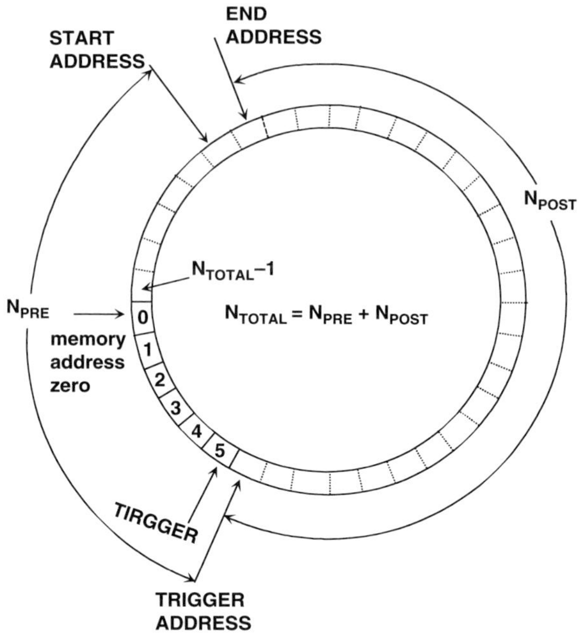

FIGURE 5.20. Schematic of a circular buffer storage scheme.

Although the buffer has a specific beginning address, this is irrelevant since the end address and start trigger address are the relevant addresses. The trigger signal can come from a threshold circuit or could even be done manually by an operator listening to the signals. Au et al. (1999) used a circular buffer technique to measure spinner dolphin sound emissions. Because of the present of snapping shrimp sound, which would cause false trigger signals if a threshold trigger technique was used, the trigger signal was presented by manually depressing a switch after the researcher aurally detected a signal. Upon initiating the trigger signal, 1 s of pre-trigger and 2 s of post-trigger data were collected. In this manner they were able to capture the entire signal. Using a circular buffer and a manual trigger ensures that only N_{total} points will be acquired per signal. Digitizing signals at a high sample rate requires the availability of large amounts of memory, and special precautions must be taken so that digital data are stored only when an acoustic signal is present. For example, if a unit is digitizing continuously for 2 minutes at a 500 kHz sample rate, 120 Mb of memory would be required.

Our discussion of data acquisition systems has so far been restricted to single-channel systems. Often, the use of multi-channel systems is required. There are generally two different types of multi-channels systems, sequential sampling and simultaneously sampling systems. The sequential sampling systems use a single A/D converter and the analog signals from each channel are sampled sequentially using a multiplexer. The throughput of a multi-channel system will depend on the fastest sampling rate available for a single channel and the number of channels being digitized. The throughput will be the fastest sampling rate divided by the number of channels being digitized so that the analog bandwidth and the sample rate per channel will be reduced by a factor of 2 for each channel being digitized. A simultaneous multi-channel board samples all the channels simultaneously and requires the same amount of A/D converters as the number of input channels on the board. Therefore, simultaneous multi-channel boards tend to be more expensive but have wider bandwidths than sequential multi-channel boards.

5.7 Localization with Hydrophone Arrays

There will be many situations in which sounds of marine mammals are recorded with the experimenter not having an idea of the location of the animals. In these situations, it would be good to have a method by which the range and azimuth of a calling animal can be determined acoustically. One good technique in localizing the source of an acoustic signal is to use an array of hydrophones. In this section, we will discuss the technique of localization using a hydrophone array.

5.7.1 The Hyperbola

In order to explain the concepts involved with determining the position of (localizing) a sound source with an array of hydrophones, we should first discuss the geometric and mathematical properties of a hyperbola. Consider the hyperbolas drawn in Fig. 5.21, with two fixed focus points at $x = d_h$ and $x = -d_h$. The definition of a hyperbola is the set of points in a plane such that the difference in the distance between any points in this set and two fixed points (foci) in the plane is a positive constant. From this definition of a hyperbola, we have the following relationship for the hyperbola on the right side of the origin or the space where $x > 0$ in Fig. 5.21, and the two fixed focal points at $x = d_h$ and $x = -d_h$.

$$d_2 - d_1 = \text{constant} = 2a \tag{5.33}$$

or

$$\sqrt{(x + d_h)^2 + y^2} - \sqrt{(x - d_h)^2 + y^2} = 2a. \tag{5.34}$$

Similarly, for the hyperbola on the left side of the origin in the space where $x < 0$,

$$\sqrt{(x - d_h)^2 + y^2} - \sqrt{(x + d_h)^2 + y^2} = 2a. \tag{5.35}$$

Again from the definition of a hyperbola, the point $s(x,y)$ along the hyperbola in Fig. 5.21 will be such that $d_2 - d_1$ is the same no matter where on the

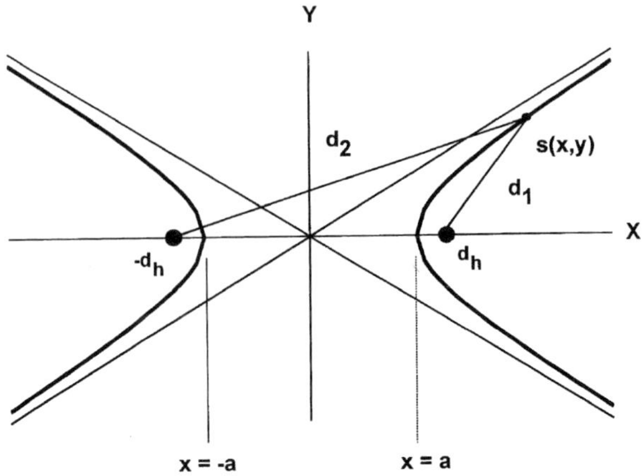

FIGURE 5.21. The geometry of hyperbolas.

hyperbola the point is. By transposing the radicals in Eq. (5.23) to the right-hand side of the equation and squaring both sides twice, we can rewrite the equation of a hyperbola as

$$\frac{x^2}{a^2} - \frac{y^2}{b^2} = 1, \tag{5.36}$$

where $b^2 = d_h^2 - a^2$. Solving for y in Eq. (5.36), we get

$$y = \pm \frac{b}{a} \sqrt{x^2 - a^2}. \tag{5.37}$$

At $x = \pm a, y = 0$, so that the x-intercept or the vertex of the hyperbola is at $\pm a$, as can be seen in Fig. 5.21. The solution for y in Eq. (5.37) is real only when $x^2 \geq a^2$ or $|x| \geq a$. The hyperbola has two asymptotes given by the equation

$$y = \pm \frac{b}{a} x. \tag{5.38}$$

The hyperbola is symmetrical about the line joining the two foci and its eccentricity (wideness) is determined by the ratio of d_h/a.

5.7.2 Sound Localization in a Plane

Consider now the situation in which we have two hydrophones on the x-axis as shown in Fig 5.22, monitoring the sounds of a marine animal. For this geometry, sounds from the animal will arrive at hydrophone 1 before arriving at hydrophone 2, and letting the time difference t_{21} equal the difference in the time of arrival of the acoustic signal at the two hydrophones, we have

$$t_{21} = \frac{d_2 - d_1}{c_0}, \tag{5.39}$$

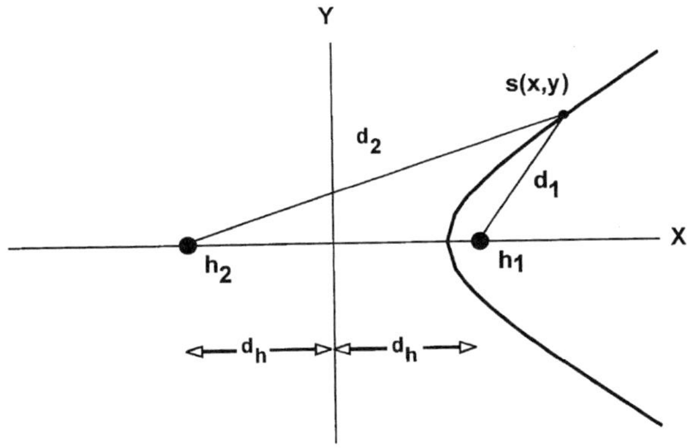

FIGURE 5.22. Geometry of a two-hydrophone array with a sound-producing animal at position $s(x,y)$.

where c_0 is the speed of sound in water. From our discussion on the hyperbola, it follows that the position of the animal cannot be ascertained from only knowing t_{21}. All we know is that the animal's position is somewhere on a hyperbola defined by the position of the two hydrophones and t_{21}.

Suppose we add a third hydrophone along the x-axis. We will now have three time differences and from a knowledge of two of these time differences, for example, t_{21} and t_{31}, where

$$t_{21} = \frac{d_2 - d_1}{c_0}$$
$$t_{32} = \frac{d_3 - d_2}{c_0}$$
$$t_{31} = \frac{d_3 - d_1}{c_0},$$

we can draw four hyperbolas, two associated with t_{21} and two with t_{31}, as shown in Fig. 5.23. However, only two of the hyperbolas, depending on the signs of t_{21} and t_{31}, will be applicable for determining the location of the sound source. If the sound source is on the right side of the y-axis, then $t_{31} < 0$ since $d_3 < d_1$, and $t_{21} > 0$ since $d_2 > d_1$, where dn is the distance from hydrophone n to the sound source. In this case, the hyperbolas associated with $t_{31} < 0$ and $t_{21} > 0$ in Fig. 5.23 will be the appropriate ones. However, if the sound source is on the left side of the y-axis, then $t_{31} > 0$ and $t_{21} < 0$, so that the other two hyperbolas will be the appropriate ones. The sound source will be at one of the intersections of the hyperbolas. However, there will be an ambiguity whether the sound source is above or below the x-axis. This ambiguity is often referred to as a "right–left" or "up–down" ambiguity associated with an array of hydrophones strung out on a line. A third hyperbola can also be drawn from a knowledge of t_{32}; however, it will intersect the other hyperbolas at the same points as if only two hyperbolas were used. The right–left ambiguity of a line array is a

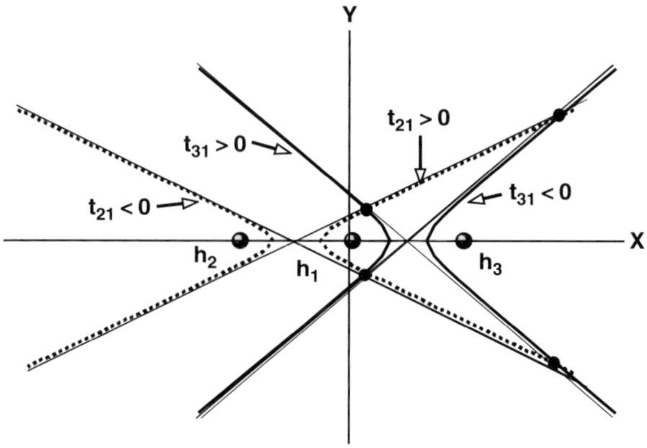

FIGURE 5.23. Geometry of a three-hydrophone array.

characteristic of towed-arrays in which hydrophones are arranged along a line and extended at a distance from the stern of a ship to reduce the noise from the ship. The ambiguity will be present, no matter how many hydrophones are used in the line array.

There are an infinite amount of geometries in which three hydrophones can be arranged in an array that is not in a line. Rearranging the positions of the three hydrophones could eliminate any ambiguity in the location of a sound source. Let us consider the case in which three hydrophones are arranged in a triangle as shown in Fig. 5.24 and the sound source is at a location shown in the figure. In this example, a unique solution can be obtained by knowing the time of arrival difference, t_{12}, t_{13}, and t_{23} between each hydrophone so that d_1, d_2, and d_3 can be determined.

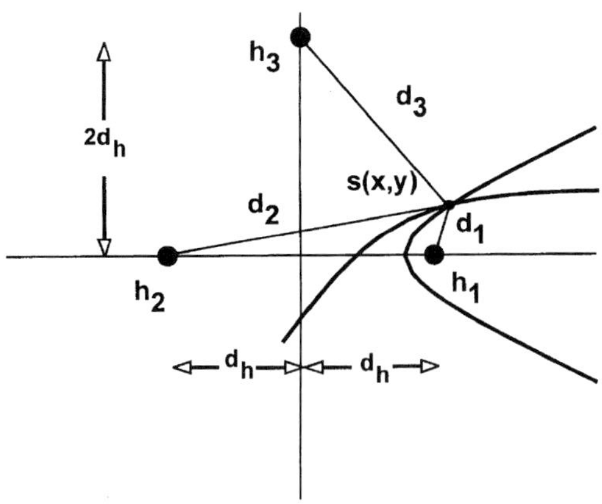

FIGURE 5.24. Geometry of three hydrophones arranged in a triangle localizing a sound $s(x,y)$.

5.7.3 Sound Localization in Three-Dimensional Space

In the previous section, the sound source and the hydrophones were in the same plane and the location of the source can be determined by the intersection of two hyperbolas. When the hydrophones are not in the same plane as the sound source, the situation becomes more complicated. For the two-hydrophone situation depicted in Fig. 5.22, knowing the difference in the arrival time will allow us to localize the sound source on the surface of a hyperboloid as depicted in Fig. 5.25 (the hyperbola of Fig. 5.25 is rotated about the x-axis to form a three-dimensional surface hyperboloid surface). However, if the travel time difference is equal to zero, the hyperboloid reduces to a plane. If the travel time difference is equal to the time for an acoustic signal to travel from hydrophones h_1 and h_2, then the hyperboloid reduces to a line segment along the y-axis. Adding a third hydrophone, as in Fig. 5.23, will cause two hyperboloids formed by the two time of arrival differences to intersect and form two lines, indicating that the source could be anywhere on one of the two intersecting lines. Adding a fourth hydrophone will allow us to determine the intersection of three hyperboloids, which will localize the source to two points, introducing an ambiguity similar to the "left–right" ambiguity present with a line array. Finally, a fifth hydrophone is required in order to localize the sound source to a single point in three-dimensional space. More on four- and five-hydrophone arrays will be addressed in the next section. There is an important exception to sound localization in three-dimensional space, which occurs when there are four hydrophones all in the same plane. Such a situation would arise if a four-hydrophone array is located on a flat bottom at the same depth near the air–water interface. With four hydrophones in the same plane, the sound source can be localized to a solution above and below the plane of the hydrophones (Aubauer, 1995). However, one of the points will be invalid since it will be either below the bottom for a bottom-mounted case or in the air if the four hydrophones are at the same shallow depth. It is important to note that we are assuming that the hydrophones are not configured in a line for either the four- or five-hydrophones arrays.

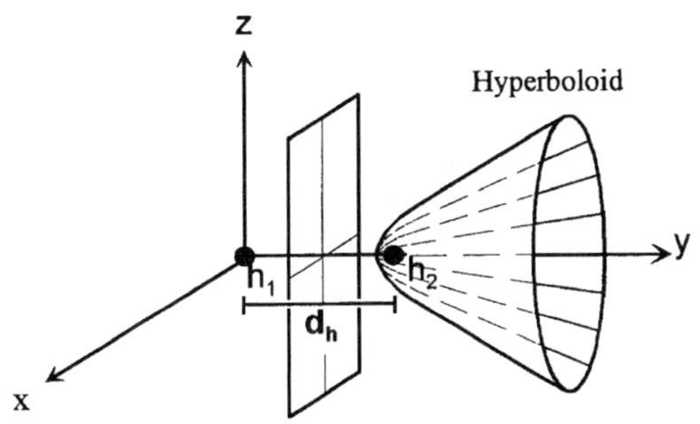

FIGURE 5.25. Geometry of a two-hydrophone array with a sound-producing animal at position $s(x,y,z)$ on the surface of a hyperboloid.

The use of towed arrays in which the hydrophone elements are strung out in a line from the stern of a ship has been used to measure marine mammal sounds (Thomas et al., 1986). The line array is essentially being towed through three-dimensional space. The range and azimuth to an external sound can be determined; however, the sound source can be localized only to a set of points along the rim of a circle that is perpendicular to the line of the array.

5.7.4 *Linear Equation Approach*

5.7.4.1 Three Hydrophone in a Line

The discussion in Sections 5.7.2 and 5.7.3 was presented to provide insight into the underlining geometry involved in localizing a sound source with an array of receivers. However, for hydrophone arrays of certain geometry, the linear equation approach of Spiesberger and Fristrup (1990) can be used. We will first examine the simple two-dimensional case of three hydrophones arranged in a line localizing a sound source in the x–y plane as depicted in Fig. 5.26. In order to use the linear equation approach, one of the hydrophones must be at the origin of the coordinate system (the origin can be placed at the position of any of the three hydrophones). Using the relationship $t = r/c$, where t is the time for an acoustic signal of speed c to travel over a distance r, and Pythagoras' theorem, the distance between the sound source and hydrophone h_1 located at the origin can be expressed as

$$s_x^2 + s_y^2 = c^2 t_1^2. \tag{5.41}$$

Similarly, the distances of hydrophones h_2 and h_3 from the sound source can be expressed as

$$(s_x - x_2)^2 + s_y^2 = c^2 (t_1 + t_{21})^2, \tag{5.42}$$

$$(s_x - x_3)^2 + s_y^2 = c^2 (t_1 + t_{31})^2, \tag{5.43}$$

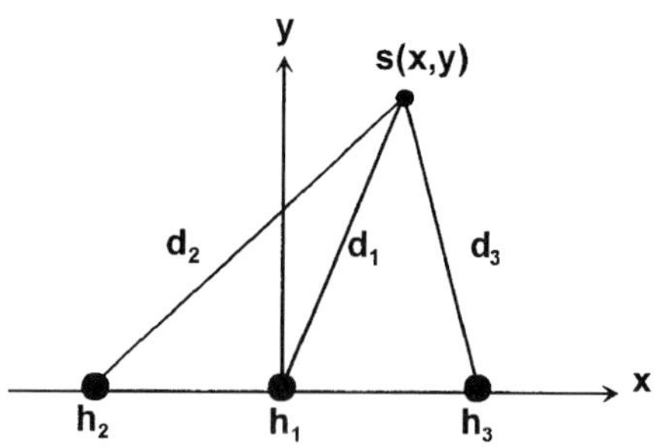

FIGURE 5.26. Geometry of a three-hydrophone line array for localizing a sound source at position $s(x,y)$.

where t_{21} is the difference in the time of arrival of the sound between hydrophones h_2 and h_1 and t_{31} is time of arrival difference between hydrophones h_3 and h_1. Expanding Eqs. (5.42) and (5.43), and subtracting Eq. (5.41) from both equations, we obtain the following equations:

$$\begin{aligned} 2\,x_2 s_x + 2c^2 t_{21} t_1 &= x_2^2 - c^2 t_{21}^2 \\ 2\,x_3 s_x + 2c^2 t_{31} t_1 &= x_3^2 - c^2 t_{31}^2 . \end{aligned} \tag{5.44}$$

Rewriting the equations in matrix format, we have

$$\begin{pmatrix} 2x_2 & 2c^2 t_{21} \\ 2x_3 & 2c^2 t_{31} \end{pmatrix} \begin{pmatrix} s_x \\ t_1 \end{pmatrix} = \begin{pmatrix} x_2^2 - c^2 t_{21}^2 \\ x_3^2 - c^2 t_{31}^2 \end{pmatrix}. \tag{5.45}$$

The system of linear equations is reduced to an $n-1 \times n-1$ coefficient matrix, an $n-1$ column matrix of the unknowns, and an $n-1$ column matrix containing the square of the distance from each hydrophone to the origin subtracted by the square of the difference between the range of the source from each hydrophone and the origin, where n is the number of hydrophones in the array. The solutions to Eq. (5.45) can be obtained using concepts from the theory of determinants. The determinant of the matrix equation is defined as

$$\Delta = \begin{vmatrix} 2x_2 & 2c^2 t_{21} \\ 2x_3 & 2c^2 t_{31} \end{vmatrix} = 4c^2 (x_2 t_{31} - x_3\, t_{21}). \tag{5.46}$$

Cramer's rule can now be used to solve n linear equations in n unknowns, by determinants. This rule states that any unknown is given by a fraction in which the denominator is the determinant formed from the coefficients of the unknown column matrix and the numerator is the same determinant with the column representing the coefficients of the unknown replaced by a column consisting of the elements on the right side of each equation (right side of Eqs. (5.44) or (5.45). Therefore, using Cramer's rule the solutions of Eq. (5.45) can be written as

$$\begin{aligned} t_1 &= \frac{\begin{vmatrix} 2x_2 & x_2^2 - c^2 t_{21}^2 \\ 2x_3 & x_3^2 - c^2 t_{31}^2 \end{vmatrix}}{\Delta} \\ &= \frac{c^2 (x_3 t_{21}^2 - x_2 t_{31}^2) + (x_2 x_3^2 - x_2^2 x_3)}{2c^2 (x_2 t_{31} - x_3\, t_{21})}. \end{aligned} \tag{5.47}$$

$$\begin{aligned} s_x &= \frac{\begin{vmatrix} x_2^2 - c^2 t_{21}^2 & 2c^2 t_{21} \\ x_3^2 - c^2 t_{31}^2 & 2c^2 t_{31} \end{vmatrix}}{\Delta} \\ &= \frac{c^2 (t_{21} t_{31}^2 - t_{21}^2 t_{31}) + (x_2^2 t_{31} - x_3^2 t_{21})}{2 (x_2 t_{31} - x_3\, t_{21})}. \end{aligned} \tag{5.48}$$

The y-coordinate position of the sound source is equal to

$$s_y = \pm\sqrt{c^2 t_1^2 - s_x^2}. \tag{5.49}$$

The ± sign being associated with the "left–right" ambiguity of a line array.

If the hydrophones in Fig. 5.24 are spaced an equal distance a apart, then we have $x_2 = -d_h$ and $x_3 = d_h$, so that Eqs. (5.47) and (5.48) reduce to

$$t_1 = \frac{2d_h^2 - c^2(t_{21}^2 + t_{31}^2)}{2c^2(t_{31} + t_{21})}, \tag{5.50}$$

$$s_x = \frac{c^2(t_{21}^2 t_{31} - t_{21} t_{31}^2) + d_h^2(t_{21} - t_{31})}{2d_h(t_{21} + t_{31})}. \tag{5.51}$$

The solutions for t_1 and s_x contain "poles" and will cause the solutions to explode to infinity when $t_{31} = -t_{21}$. This situation cannot be avoided with a three-hydrophone line array, and there will be locations on the x–y plane in which $t_{31} = -t_{21}$. The sound source does not have to be exactly at these pole locations to create problems. If the sound is close to a pole location, the denominator could be very small in value causing the solution to be unrealistically large. One way to avoid this problem is to have a fourth hydrophone in the same line. Then the solution from a different set of three hydrophones can be calculated.

5.7.4.2 Four Hydrophones in a Plane

We will next consider the case of four hydrophones all in the same plane lying either on the bottom of the ocean or near the surface. Unfortunately, the case of three hydrophones on a plane, but not along the same line, does not lend itself to a linear equation approach. The four hydrophone array can localize sounds in three-dimensional space with a top–bottom ambiguity (two solutions occurring symmetrically above and below the plane of the hydrophones). However, if the hydrophones are lying on a bottom or near the surface, then only one practical solution exists. The geometry for this case is depicted in Fig. 5.27. Using Pythagoras' theorem to calculate the distance from each hydrophone to the sound source, we have the following equations:

$$\begin{aligned} s_x^2 + s_y^2 + s_z^2 &= c^2 t_1 \\ (s_x - x_2)^2 + (s_y - y_2)^2 + s_z^2 &= c^2(t_1 + t_{21})^2 \\ (s_x - x_3)^2 + (s_y - y_3)^2 + s_z^2 &= c^2(t_1 + t_{31})^2 \\ (s_x - x_4)^2 + (s_y - y_4)^2 + s_z^2 &= c^2(t_1 + t_{41})^2. \end{aligned} \tag{5.52}$$

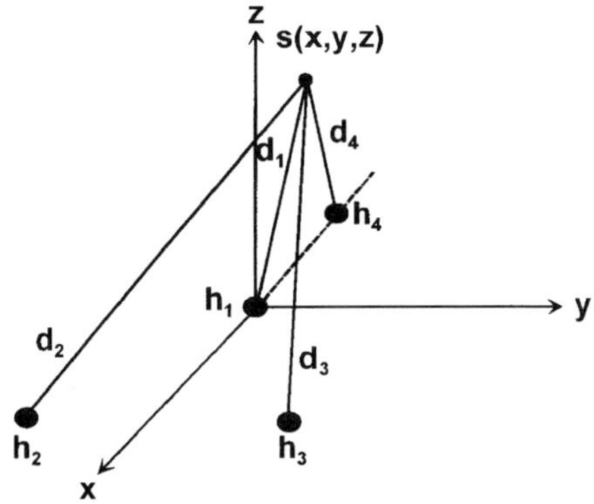

FIGURE 5.27. Geometry of four hydrophones located on a plane localizing a sound source at $s(x,y,z)$.

Subtracting the top equation from the remaining three equations, performing the indicated mathematical operating, and rearranging, we obtain the following set of equations:

$$\begin{pmatrix} 2x_2 & 2y_2 & 2c^2t_{21} \\ 2x_3 & 2y_3 & 2c^2t_{31} \\ 2x_4 & 2y_4 & 2c^2t_{41} \end{pmatrix} \begin{pmatrix} s_x \\ s_y \\ t_1 \end{pmatrix} = \begin{pmatrix} R_2^2 - c^2t_{21}^2 \\ R_3^2 - c^2t_{31}^2 \\ R_4^2 - c^2t_{41}^2 \end{pmatrix}, \tag{5.53}$$

where $R_i^2 = x_i^2 + y_i^2$ is the square of the distance between each hydrophone and the origin. The determinant of the matrix equation can be expressed as

$$\Delta = \begin{vmatrix} 2x_2 & 2y_2 & 2c^2t_{21} \\ 2x_3 & 2y_3 & 2c^2t_{31} \\ 2x_4 & 2y_4 & 2c^2t_{41} \end{vmatrix} \tag{5.54}$$

$$= 8c^2[t_{21}(x_3y_4 - x_4y_3) + t_{31}(x_4y_2 - x_2y_4) + t_{41}(x_2y_3 - x_3y_2)]$$

Using Cramer's rule to solve for t_1, we get

$$t_1 = \frac{\begin{vmatrix} 2x_2 & 2y_2 & R_2^2 - c^2t_{21}^2 \\ 2x_3 & 2y_3 & R_3^2 - c^2t_{31}^2 \\ 2x_4 & 2y_4 & R_4^2 - c^2t_{41}^2 \end{vmatrix}}{\Delta} = \frac{\Delta_{ta} + \Delta_{tb} + \Delta_{tc}}{\Delta}, \tag{5.55}$$

$$\begin{aligned} \Delta_{ta} &= 4(R_2^2 - c^2t_{21}^2)(x_3y_4 - x_4y_3) \\ \Delta_{tb} &= 4(R_3^2 - c^2t_{31}^2)(x_4y_2 - x_2y_4) \\ \Delta_{tc} &= 4(R_4^2 - c^2t_{41}^2)(x_2y_3 - x_3y_2) \end{aligned}$$

Similarly, the solutions for s_x and s_y can be expressed as

$$s_x = \frac{\begin{vmatrix} R_2^2 - c^2 t_{21}^2 & 2y_2 & 2c^2 t_{21} \\ R_3^2 - c^2 t_{31}^2 & 2y_3 & 2c^2 t_{31} \\ R_4^2 - c^2 t_{41}^2 & 2y_4 & 2c^2 t_{41} \end{vmatrix}}{\Delta} = \frac{\Delta_{sxa} + \Delta_{sxb} + \Delta_{sxc}}{\Delta}, \tag{5.56}$$

$$\begin{aligned} \Delta_{sxa} &= 4c^2[(R_2^2 - c^2 t_{21})(y_3 t_{41} - y_4 t_{31})] \\ \Delta_{sxb} &= 4c^2[(R_3^2 - c^2 t_{31})(y_4 t_{21} - y_2 t_{41})] \\ \Delta_{sxc} &= 4c^2[(R_4^2 - c^2 t_{41})(y_2 t_{31} - y_3 t_{21})], \end{aligned}$$

$$s_y = \frac{\begin{vmatrix} 2x_2 & R_2^2 - c^2 t_{21}^2 & 2c^2 t_{21} \\ 2x_3 & R_3^2 - c^2 t_{31}^2 & 2c^2 t_{31} \\ 2x_4 & R_4^2 - c^2 t_{41}^2 & 2c^2 t_{41} \end{vmatrix}}{\Delta} = \frac{\Delta_{sya} + \Delta_{syb} + \Delta_{syc}}{\Delta}, \tag{5.57}$$

$$\begin{aligned} \Delta_{sya} &= 4c^2[(R_2^2 - c^2 t_{21})(x_4 t_{31} - x_3 t_{41})] \\ \Delta_{syb} &= 4c^2[(R_3^2 - c^2 t_{31})(x_2 t_{41} - x_4 t_{21})] \\ \Delta_{syc} &= 4c^2[(R_4^2 - c^2 t_{41})(x_3 t_{21} - x_2 t_{31})]. \end{aligned}$$

Three interesting geometries with three hydrophones spaced at equal distances from a reference hydrophone at the origin of a coordinate system are shown in Fig. 5.28. The hydrophone arrangements are referred to from left to right as *symmetrical star*, *reversed T*, and *rhombus*. The coordinates of the hydrophones in the three symmetrical arrangements are listed in Table 5.3. These coordinates can be substituted directly into Eqs. (5.55)–(5.57) to solve for the coordinate position of a sound source.

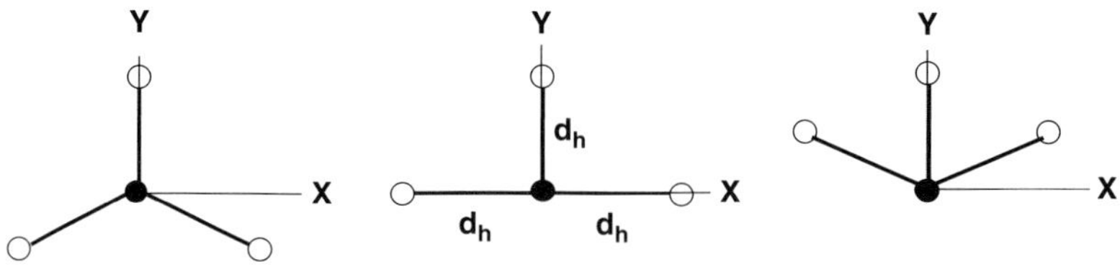

FIGURE 5.28. Symmetrical arrangements of four hydrophones showing a symmetrical star on the left, a reversed T in the middle, and a rhombus on the right. The solid circle is the reference hydrophone at the origin (adapted from Aubauer, 1995).

TABLE 5.3. Coordinates of the Measuring Hydrophones Shown in Fig. 5.28 (From Aubauer, 1995)

	Hydrophone 2			Hydrophone 3			Hydrophone 4		
	x_2	y_2	z_2	x_3	y_3	z_3	x_4	y_4	z_4
symmetrical star	$-\frac{\sqrt{3}}{2}d_h$	$\frac{-d_h}{2}$	0	$-\frac{\sqrt{3}}{2}d_h$	$\frac{-d_h}{2}$	0	0	d_h	0
reversed T	$-d_h$	0	0	d_h	0	0	0	d_h	0
rhombus	$-\frac{\sqrt{3}}{2}d_h$	$\frac{d_h}{2}$	0	$-\frac{\sqrt{3}}{2}d_h$	$\frac{d_h}{2}$	0	0	d_h	0

5.7.4.3 Five-Hydrophone Array

The matrix equation for four hydrophones located on a plane (Eq. (5.53)) indicates a very orderly arrangement where the appropriate equation pertaining to each hydrophone can be simply expressed as

$$\begin{aligned} 2x_i s_x + 2y_i s_y + 2z_i s_z + 2c^2 t_{i1}\, t_1 &= R_i - c^2 t_{i1} 2 \\ R_i^2 &= x_i^2 + y_i^2 + z_i^2 \\ i &= 2, 3, 4, 5 \end{aligned} \tag{5.58}$$

Ri is the distance of the *i*-th hydrophone to the reference hydrophone at the origin. Therefore, the appropriate matrix equation for five hydrophones localizing a sound source in three-dimensional space can be expressed as

$$\begin{pmatrix} 2x_2 & 2y_2 & 2z_2 & 2c^2 t_{21}^2 \\ 2x_3 & 2y_3 & 2z_3 & 2c^2 t_{31}^2 \\ 2x_4 & 2y_4 & 2z_4 & 2c^2 t_{41}^2 \\ 2x_5 & 2y_5 & 2z_5 & 2c^2 t_{51}^2 \end{pmatrix} \begin{pmatrix} s_x \\ s_y \\ s_z \\ t_1 \end{pmatrix} = \begin{pmatrix} R_2^2 - c^2 t_{21}^2 \\ R_3^2 - c^2 t_{31}^2 \\ R_4^2 - c^2 t_{41}^2 \\ R_5^2 - c^2 t_{51}^2 \end{pmatrix}, \tag{5.59}$$

or

$$\mathbf{Am} = \mathbf{b},$$

where **A** is the 4 × 4 coefficient matrix, **m** is the unknown column matrix, and **b** is the also a column matrix. The unknowns can be solved for evaluating determinants and using Cramer's rule. However, this process can be very tedious when dealing with 5 × 5 determinants. There are a variety of numerical software packages such as *Matlab*, which can be used to numerically solve a matrix equation such as Eq. (5.59). Matlab has two methods to solve a matrix equation such as Eq. (5.59). The preferred method is to use a matrix division operation so that $\mathbf{m} = \mathbf{b}/\mathbf{A}$. This method uses an LU factorization approach. The second method is to calculate the inverse of matrix A so that $\mathbf{m} = \mathbf{A}^{-1}\mathbf{b}$, where $\mathbf{A}^{-1}$ is the inverse of matrix A.

5.7.5 Two-Hydrophone Method of Cato

Under some very specific conditions, Cato (1998) showed that a two-hydrophones array would be sufficient to localize a sound source. Assume that there are two hydrophones, H_1 and H_2, located on the x-axis and a sound source $P(x,y)$ at location (x,y) as shown in Fig. 5.29. The intensity received by both hydrophones can be expressed as

$$I_1 = \frac{I_0}{{r_1}^2}, \tag{5.60}$$

where I_0 is the intensity at the source. The source level is then given by

$$\mathrm{SL} = 10 \log I_0 = 10 \log I_1 r. \tag{5.61}$$

We now introduce a variable k such that

$$k = \sqrt{\frac{I_1}{I_2}} = \frac{p_1}{p_2}, \tag{5.62}$$

where p_1 and p_2 are the received acoustic pressures by hydrophone H_1 and H_2. The difference in the received levels between the two hydrophones is merely

$$\mathrm{DSPL} = 10 \log\left(\frac{I_1}{I_2}\right) = 20 \log\left(\frac{r_1}{r_2}\right) = 20 \log(k). \tag{5.63}$$

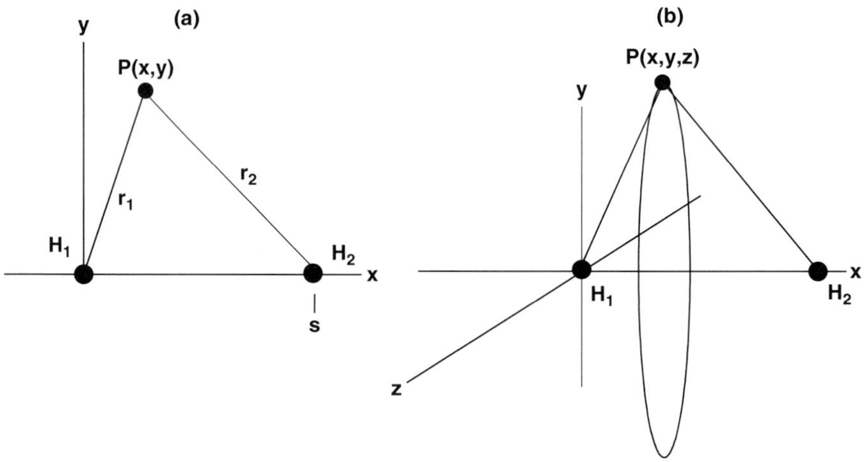

FIGURE 5.29. (a) Geometry for Cato's two hydrophones localization technique, (b) locus of position P of the source in three dimensions determined from differences in levels and arrival time. P lies on the circumference of a circle normal to the plane containing H_1, H_2, and P.

The time of arrival difference for the sound to arrive at hydrophone H_1 and H_2 is

$$\tau = \frac{(r_2 - r_1)}{c_0}, \tag{5.64}$$

where c_0 is the velocity of sound. From Eq. (5.60) and Eq. (5.62), we can express r_1 as

$$r_1 = \frac{c_0\tau}{(k - 1)}. \tag{5.65}$$

This expression for r_1 is useful only if the source is significantly closer to one hydrophone than the other. This can be easily seen by considering the derivative of r_1 with respect to τ and with respect to k

$$\frac{\partial r_1}{\partial \tau} = \frac{c_0}{(k - 1)} = \frac{r_1}{\tau} \to \infty \text{ as } \tau \to 0, \tag{5.66}$$

$$\frac{\partial r_1}{\partial k} = \frac{-c_0\tau}{(k - 1)^2} = \frac{-r_1}{(k - 1)} \to -\infty \text{ as } k \to 1, \tag{5.67}$$

so that r_1 changes very rapidly as $\tau \to 0$ or $k \to 1$. This condition arises as the source approaches the plane bisecting the line between the two hydrophones and in the limit of $\tau = 0$ so that $k = 1$, the source lies in this plane. Also, if the source position is much greater than the separation distance between the two hydrophone, $k \to 1$.

The source level can be determined from Eqs. (5.61) and (5.64) to be

$$\text{SL} = 10 \log\left[\frac{I_1(c_0\tau)^2}{(k - 1)^2}\right], \tag{5.68}$$

where from Eq. (5.63), $k = 10^{\text{DSPL}/20}$. The expression for the source level can now be written as

$$\text{SL} = \text{RL} + 10 \log\left[\frac{(c_0\tau)^2}{(k - 1)^2}\right], \tag{5.69}$$

where $\text{RL} = 10 \log I_1$ is the received level of hydrophone 1.

The position of the source can be estimated by considering the geometry of Fig. 5.29, which indicates that

$$k^2 r_1^2 = r_2^2 = (s - x)^2 + y^2 \tag{5.70}$$

$$\cos\theta = x/r_1, \tag{5.71}$$

where θ is the angle between the horizontal axis and the line between H_1 and $P(x,y)$. Expanding Eq. (5.70) and substituting ${r_1}^2 = x^2 + y^2$, we obtain

$$x = \frac{s}{2} - \frac{r_1^2(k^2 - 1)}{2s}, \tag{5.72}$$

$$\cos\theta = \frac{\left[s^2 - (k^2 - 1)r_1^2\right]}{2r_1 s}. \tag{5.73}$$

Therefore, from Eqs. (5.62), (5.64) and (5.72), the position of the sound source can be determined. Rotation of the plane containing P, H_1, and H_2 about the x-axis does not change the results so that P sweeps out a circle in a plane normal to the x,y plane as shown in Fig. 5.29b.

5.7.6 *One-Hydrophone Localization*

5.7.6.1 Using Direct and Surface Reflections Arrivals

Cato (1998) also considered the situation in which the direct signal and the surface-reflected component of a signal could be separated so that the range and the depth of the source can be estimated. This technique is only good for relatively short signals such as burst pulses and echolocation clicks of dolphins. Consider the geometry shown in Fig. 5.30, showing a dolphin at depth d_S and a single hydrophone at depth d_H. This geometry is exactly the same as in Fig. 5.29a for two hydrophones, except it is rotated by 90°, so that we have the x-axis pointed upward and $s = 2d_H$. For this geometry we can express k as

$$k = \frac{r_2}{r_1} = \frac{p_1}{p_2}, \tag{5.74}$$

where p_1 is the direct signal and p_2 is the surface-reflected signal. The range r_1 is given by Eq. (5.64), where τ is the time delay between the direct and surface-reflected arrivals.

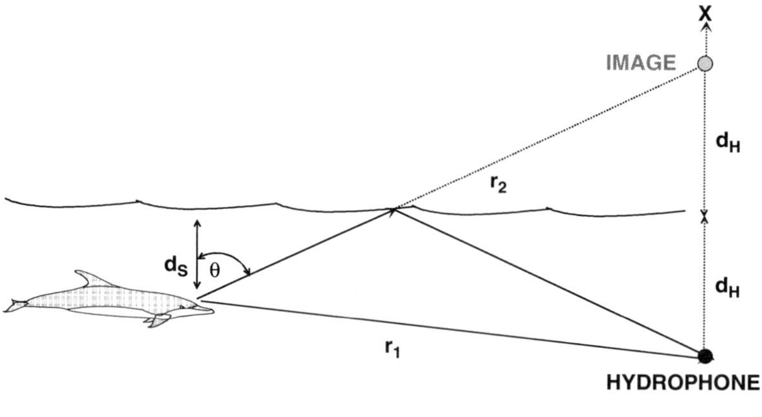

FIGURE 5.30. Geometry for using the direct and surface-reflected arrivals at a single hydrophone.

The range x and the angle θ can be calculated from Eq. (5.72) and (5.73), respectively. The depth of the source can be determined by the expression

$$d_S = d_H - x = \frac{r_1^2(k^2 - 1)}{4d_H}. \tag{5.75}$$

5.7.6.2 Using the Direct, Surface, and Bottom Arrivals

In some special situations where the depth is not large and the bottom is relatively reflected, the time delays between the direct arrival and the surface-reflected component as well as the bottom-reflected component can be used to determine the depth and range of a sound source. Once again, this technique is only good for pulse signals that are short enough that the time of arrival of the direct, surface-reflected, and bottom-reflected components can be determined. Consider the geometry of Fig. 5.31, showing a dolphin, a hydrophone, and both the surface and bottom acoustic paths. The expressions for the different paths are

$$\begin{aligned} r_1 &= \sqrt{x^2 + (d_H - d_S)^2} \\ r_2 &= \sqrt{x^2 + (d_H + d_S)^2} \\ r_3 &= \sqrt{x^2 + (d_{BS} + d_{BH})^2}. \end{aligned} \tag{5.76}$$

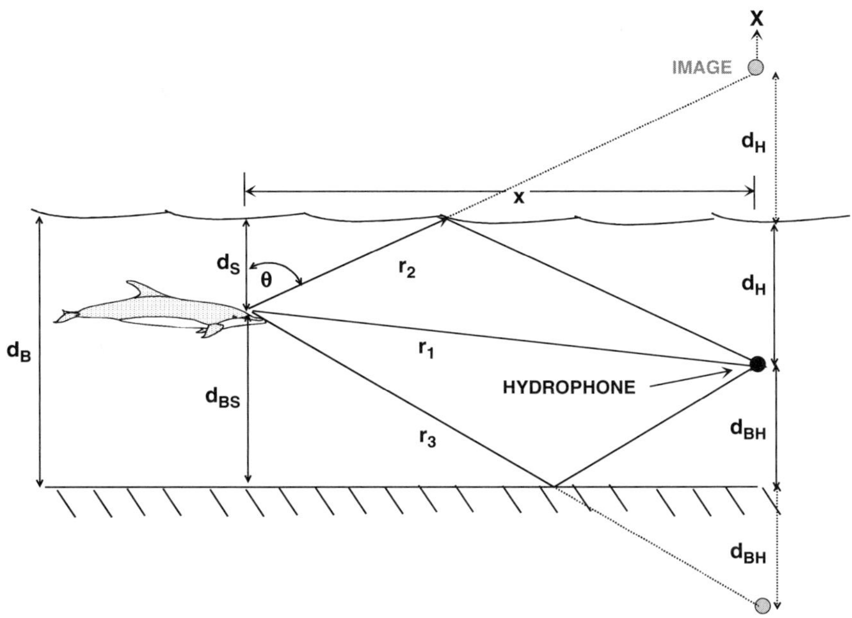

FIGURE 5.31. Geometry for multi-path propagation in shallow water with a flat bottom.

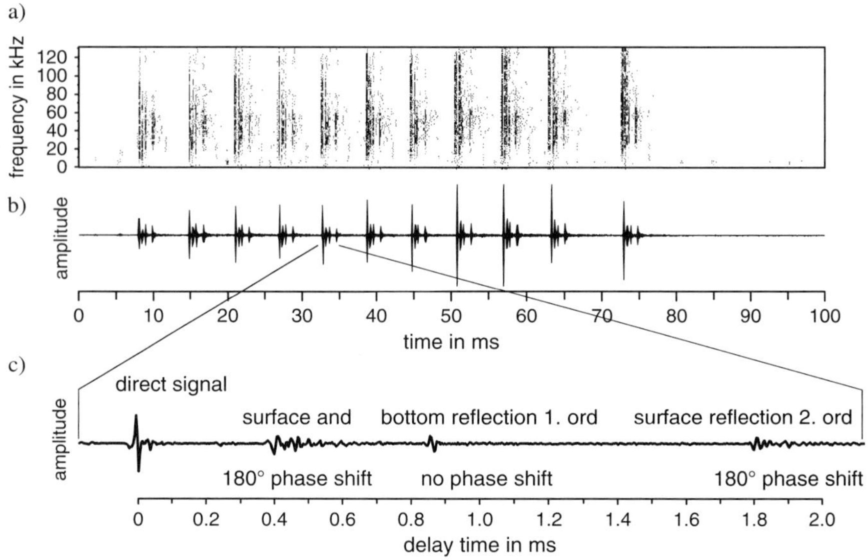

FIGURE 5.32. Spinner dolphin burst pulse signal with multi-path propagation recorded in shallow waters along the Waianae Coast of Oahu (adapted from Aubauer et al., 2000).

If we let τ_{SR} be the time delay between the time of arrival of the direct and the surface-reflected paths and τ_{BR} be the time delay between the time of arrival of the direct and the bottom-reflected paths, then the expression for the range and depth of the source derived by Aubauer et al. (2000) can be expressed as

$$r = \frac{\left(\frac{c\tau_{BR}}{2}\right)^2 + d_B \cdot (d_H - d_B) \; - \; \left(\frac{c\tau_{SR}}{2}\right)^2 \cdot \left(1 - \frac{d_B}{d_H}\right)}{\frac{c\tau_{SR}}{2} \cdot \left(1 - \frac{d_B}{d_H}\right) \; - \; \frac{c\tau_{SR}}{2}}, \tag{5.77}$$

$$d = \frac{\left(\frac{c\tau_{BR}}{2}\right)^2 + \left(\frac{c\tau_{BR}}{2}\right) \cdot r}{d_H - d_B} + d_B. \tag{5.78}$$

An example of a burst pulse sequence for a spinner dolphin is shown in Fig. 5.32, with the spectrogram representation in Fig. 32a and the time domain representations in Fig. 22b and 32c.

5.7.7 *Measurement of Time of Arrival Differences*

One of the most important measurement that has to be done in localizing a calling animal with an array of hydrophones is the measurement of arrival time differences of the sound between the different hydrophones in the array. The exact time an animal will emit a sound is never known, and so a technique is needed to determine the time at which a sound arrives at each hydrophone so that the time of arrival differences can be determined or to directly measure the time of arrival differences of the sound relative to the reference hydrophone at the origin of our coordinate system. One method to determine time of arrival differences that can be implemented with a computer is to determine the cross-correlation between the sound detected by the reference hydrophone and each of the other hydrophones in the array. The cross-correlation function is given by the integral

$$c(t) = \int_{-\infty}^{\infty} x(t)h(t + \tau)\mathrm{d}t. \tag{5.79}$$

This integral compares the waveform $x(t)$ and $h(t)$ for different values of τ. The integral will be maximum when τ is at such a value that the best match between $x(t)$ and $h(t)$ is achieved. If we are working with digital data, the integral can be replaced by a summation. Suppose the digitized signal received by the reference hydrophone is $s_1(i)$ and the digitized signal of another hydrophone is $s_2(i)$, where $i = 1, 2, \ldots, n$, then the cross-correlation function of the two signals can be expressed as

$$c(j) = \sum_{i=1}^{n} s_1(i)s_2(i + j), \tag{5.80}$$

where $j = 1, 2, \ldots, n$, and is the number of samples s_2 is being delayed. The cross-correlation function is determined for a number of j values and the best time estimate of the delay is at the j-values at which the correlation function is maximum.An example of the cross-correlation of the signals received by two receivers for different signal-to-noise ratios are shown in Fig. 5.33. Even when the signal-to-noise ratio is 0 dB, the cross-correlation remains relatively noise-free. In most instances, the cross-correlation is done in the frequency domain using the correlation theorem from Fourier Analysis. The correlation theorem will be discussed in 6.4 of Chapter 6.

Animal calls can have relatively long duration, and using the entire call for localization of specific sections of a call having a high signal-to-noise ratio and distinctive features can be judiciously chosen. An example of a specific portion of a whale's call being used in localizing the whale is shown in Fig. 5.34 in which a narrow band portion of a call was used. An example of the localization purposes may be impractical. However, the entire call does not need to be used. The track of a singing humpback whale is shown in Fig. 5.35, being tracked for about 4-1/2 hours. So far the topic of localization of marine mammals with multi-hydrophone arrays has been treated from an idealized prospective. In the actual implementation of the techniques described in this chapter, there are many practical concerns that need to be addressed. First of all, signals with sharp peaks in their cross-correlation function must be used. Therefore, good

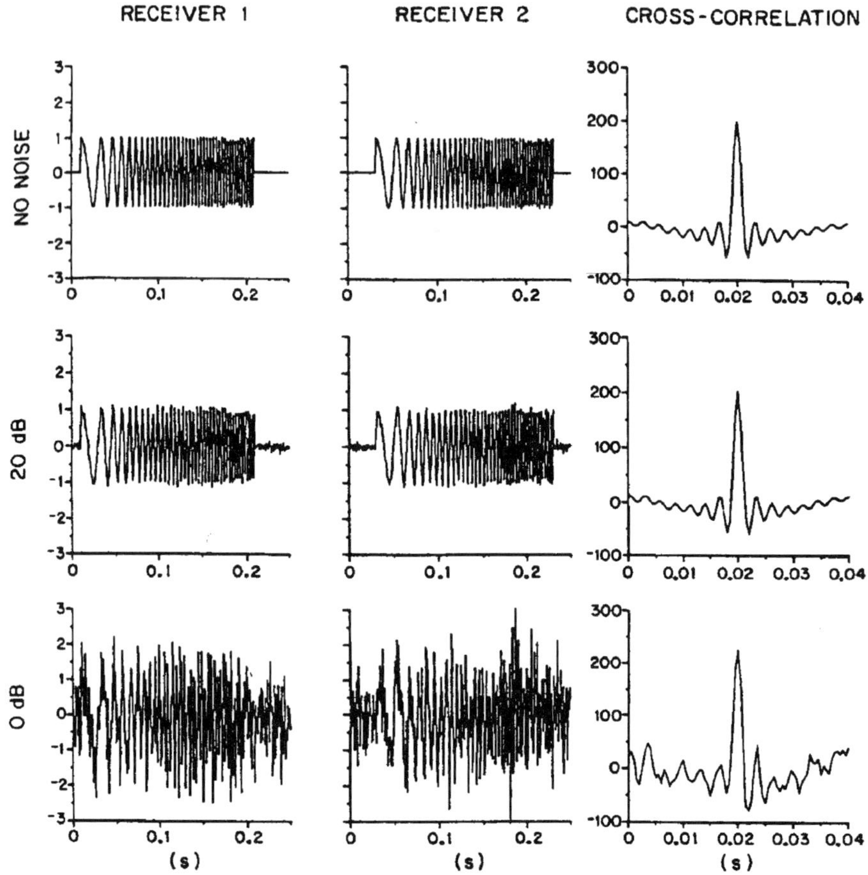

FIGURE 5.33. Examples of the cross-correlation of an animal's call at different signal-to-noise ratios. The right column showing the cross-correlation function has an expanded scale so that its properties can be displayed clearly (adapted from Spiesberger and Fristrup, 1990).

judgment is necessary to pick the specific portions of an animal's call that will be used in the localization calculations. In general, it is best to pick portions of signals that have the highest time-bandwidth products (Spiesberger and Fristrup, 1990). In the real world, multi-path components caused by an acoustic signal reflecting from the bottom often exist. These multi-path signals can make the localization process extremely difficult. If an environment is highly reverberant, it would be best to relocate the localizing array to a location in which multi-paths are not significant. Propagation conditions causing an acoustic signal to bend rather than travel in a straight line will also cause errors in estimating the location of the source. Finally, the presence of noise is always a

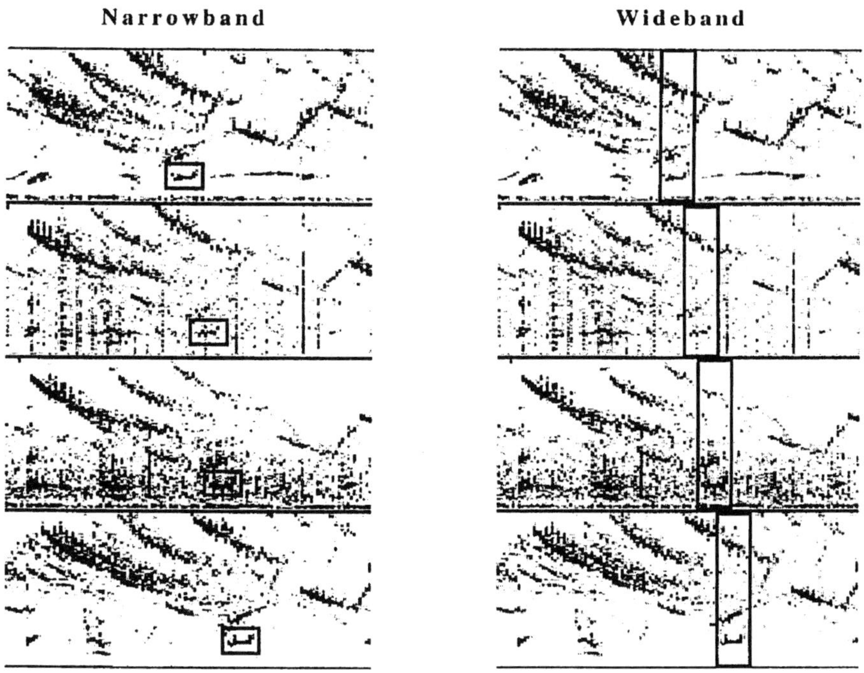

FIGURE 5.34. An example of the sonargram representation of a whale call received by four hydrophones. Frequency is along the vertical axis and time along the horizontal axis. The boxed area in each sonargram was used to localize the whale (adapted from Mellinger, 1995).

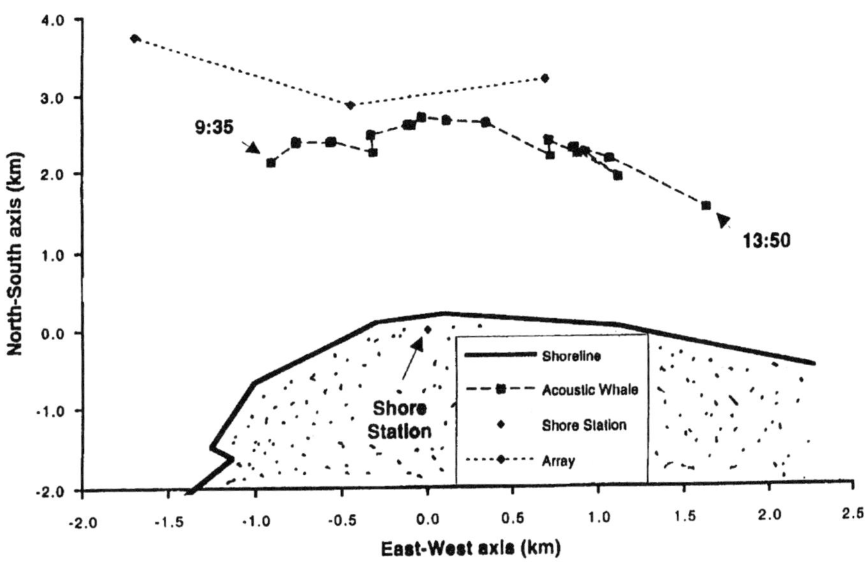

FIGURE 5.35. An example of a singing humpback whale being tracked by a three-hydrophone array with the output of each hydrophone sent to a shore station using sonabuoy transmitters (adapted from Mellinger, 1995).

problem in underwater acoustics and its presence can introduce errors in the determination of time of arrival differences. Therefore, it may not be possible to localize a calling marine mammal to a precise location, but it may only be possible to localize an animal to a certain region of the ocean. Finally, the area over which an array can accurately localize an animal will depend on the size of the area. Accurate localization is generally restricted to distances that are about five times the maximum size of the array (Mellinger, 1995). Some useful references concerning the localization and tracking of whales can be found in Watkins and Schevill (1972), Clark et al. (1986), and Frankel (1994).

References

Aubauer, R. (1995). "Korrelationsverfahren zur Flugbahnverfolgung echoortender Fledermäuse," Fortschr.-Ber. VDI Reihe 17 Nr. 132 (Düsseldorf, VDI-Verlag).

Aubauer, R., Lammers, M. O., and Au, W. W. L. (2000). "One-hydrophone method of estimating distance and depth of phonating dolphins in shallow waters," J. Acoust. Soc. Am. **107**, 2744–2149.

Au, W. W. L., Lammers, M. O., and Aubauer, R. (1999). "A portable broadband data acquistion system for field studies in bioacoustics," Mar. Mamm. Sci. **15**, 526–531.

Cato, D. (1998). "Simple methods of estimating source levels and locations of marine animal sounds," J. Acoust. Soc. Am. **104**, 1667–1678.

Clark, C. W., Ellison, W. T., and Beeman, K. (1986). "Acoustic tracking of migrating Bowhead whales," *IEEE Oceans '86 Conf. Proc.*, Washington DC, pp. 341–345.

Frankel, A. S. (1994). "Acoustic and visual tracking revels distribution, song variability and social roles of humpback whales in hawaiian waters," Ph.D. dissertation, University of Hawaii.

Franz, G. J. (1959). "Splashes as sources of sound in liquids," J. Acoust. Soc. Am. **31**, 1080–1095.

Heindsman, T. E., Smith, R. H., and Arneson, A. D. (1955). "Effect of rain upon underwater noise levels," J. Acoust. Soc. Am. **27**, 378–384.

Huber, D. M. and Runstein, R. A. (1989). *Modern Recording Techniques*, 3rd Edition (Howard W. Sams & Co., Indianapolis, Indiana).

Knudsen, V. O., Alford, R. S., and Emling, J. W. (1948). "Underwater ambient noise," J. Mar. Res. **7**, 410–429.

McGrath, J. R. (1976). "Infrasonic sea noise at the Mid-Atlantic Ridge near 37°N," J. Acoust. Soc. Am. **60**, 1290–1299.

Mellinger, D. K. (1995). "Near-field passive acoustic localization," *Notes from Bioacoustical Oceanography Workshop*, Santa Cruz, CA, August, 1995.

Mellon, R. H. (1952). "The Thermal-Noise Limit in the Detection of Underwater Acoustic Signals." J. Acoust. Soc. Am. **24**, 478–480.

Morris, G. B. (1978). "Depth dependence of ambient noise in the Northeastern Pacific Ocean," J. Acoust. Soc. Am. **64**, 581–590.

Northrup, J. (1974). "Detection of low-frequency underwater sounds from a submarine volcano in the Western Pacific," J. Acoust. Soc. Am. **56**, 837–841.

Piggott, C. L., (1965) "Ambient sea noise at low frequencies in shallow water of the Scotian Shelf," J. Acoust. Soc. Am. **36**, 2152–2163.

Spiesberger, J. L. and Fristrup, K. M. (1990). "Passive localization of calling animals and sensing of their acoustic environment using acoustic tomography," Am. Nat. **135**, 107–153.

Watkins, W. A. and Schevill, W. E. (1972). "Sound source location by arrival-times on a non-rigid three-dimensional hydrophone array," Deep Sea Res. **19**, 691–705.

Thomas, J. A., Fisher, S. R., and Ferm, L. M. (1986). "Acoustic detection of cetaceans using a towed array of hydrophones," Rep. Int. Whal. Comm, No. SC/37/03,**8,** 139–148.

Urick, R. J., Lund, G. R., and Tulko, T. J. (1972). "Depth profile of ambient noise in the deep sea north of St. Croix, Virgin Islands," U.S. Navvy Ord. Lab Tech. Rep. 72–175.

Urick, R. J. (1983). *Principles of Underwater Sound* (McGraw-Hill, New York).

Urick, R. J. (1984). *Ambient Noise in the Sea* (Naval Sea Systems Command, Washington DC).

Wenz, G. M. (1962). "Acoustic ambient noise in the ocean: spectra and sources," Acoust. Soc. Am. **34**, 1935–1955.

6

Fourier Analysis

6.1 The Time and Frequency Domains

Any signal can be represented either in the time domain with its amplitude displayed as a function of time or in the frequency domain with its amplitude displayed as a function of frequency. The time domain representation of a signal is usually referred to as the waveform, or waveshape of the signal. Oscilloscopes are often used to observe the waveforms of signals. The frequency representation of a signal is usually referred to as the frequency spectrum (or just spectrum) of the signal. Spectrum analyzers are often used to observe the spectral characteristics of continuous or long-duration (on the order of several seconds) signals. An example of a 1 v rms, 120 kHz sinusoidal signal observed with an oscilloscope and a sweep frequency spectrum analyzer is shown in Fig. 6.1. The frequency resolution of the spectrum analyzer used to obtain the display was 1 kHz. The frequency analyzer display in Fig. 6.1 is essentially the magnitude of the output of a variable-tuned bandpass filter having a 1 kHz bandwidth, and a variable center frequency that is tuned electronically from 0 to 200 kHz. Since there is energy only at 120 kHz, the output of the filter should be zero for all frequencies except 120 kHz. In actuality, a self-noise voltage of –76 dB re 1 V can be seen in the display of Fig. 6.1. Another popular type of spectrum analyzer often used is the real-time spectrum analyzer, which consists of a bank of many contiguous constant bandwidth filters all tuned to different frequencies. The outputs of the filters can be observed simultaneously so that the spectrum of a signal can be observed in real time. However, real-time spectrum analyzers are usually three to four times more expensive than sweep frequency spectrum analyzers. They also tend to be larger and are limited in the upper frequency to about 100 kHz. The most modern spectrum analyzer will use Fourier transformation techniques to transform time domain signals into the frequency domain via a fast Fourier transform (FFT) computer algorithm implemented on a digital signal processing (DSP) module or in hardware.

W.W.L. Au, M.C. Hastings, *Principles of Marine Bioacoustics*,
DOI: 10.1007/978-0-387-78365-9_6,

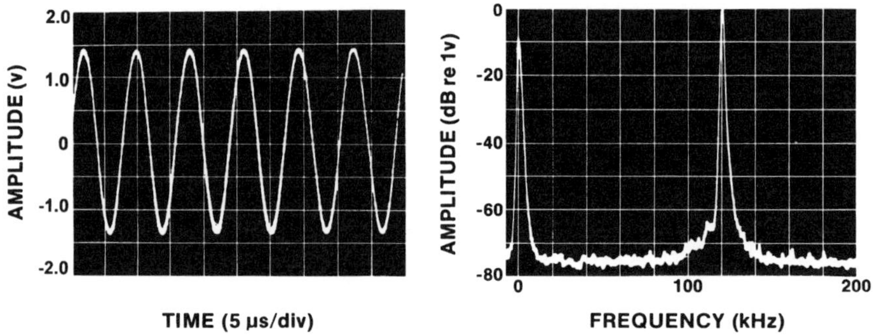

FIGURE 6.1. An example of a sinusoidal signal (1-v rms, 120 kHz) measured with (**a**) an oscilloscope, and (**b**) a spectrum analyzer.

6.2 Foundation of Fourier Analysis: The Fourier Series

We will now proceed on the ambitious task of discussing Fourier analysis in a single chapter. There are volumes written on the topic; some excellent books include those written by Bracewell (1976), Brigham (1988), Kraniauskas (1992), Lyons (1997), and Ramirez (1985). Ramirez (1985) presents a unique and short discussion of Fourier transform, providing the reader with an intuitive understanding of the principles involved. In this chapter, we will concentrate on some important and fundamental concepts involved with Fourier analysis. Our approach will be necessarily abbreviated but mathematically rigorous, and will be more of a "fly through" than a detailed presentation. Our goal here is to be practical and provide the reader with the necessary tools to correctly implement the ideas and concepts found in Fourier analysis. Readers are encouraged to consider the many volumes, especially those referenced above to obtain a deeper presentation of Fourier analysis.

In 1822, Jean Baptist Fourier introduced a series representation of continuous periodic functions that have come to be known as the Fourier series. A function is periodic if $x(t) = x(t + T)$, where T is the period of the signal. In short, a periodic function repeats itself in time continuously from minus infinity to plus infinity and is not physically realizable but can be approximated. For example, if we turned on a signal generator that produces a sine wave, we can approximate the output as a continuous signal by assuming that the function continues into eternity if the generator is left on. The Fourier series of a periodic function can be expressed as

$$x(t) = a_0 + \sum_{n=1}^{\infty} [a_n \cos(n2\pi f_0 t) + b_n \sin(n2\pi f_0 t)], \qquad (6.1)$$

where $f_0 = 1/T$, and the coefficients are related to the following integrals:

$$a_0 = \frac{1}{T} \int_{-T/2}^{T/2} x(t)\,\mathrm{d}t, \tag{6.2}$$

$$a_n = \frac{2}{T} \int_{-T/2}^{T/2} x(t) \cos(n2\pi f_0 t)\,\mathrm{d}t, \tag{6.3}$$

$$b_n = \frac{2}{T} \int_{-T/2}^{T/2} x(t) \sin(n2\pi f_0 t)\,\mathrm{d}t. \tag{6.4}$$

The coefficient a_0 represents the DC (direct current) component of the function $x(t)$. The coefficients a_n and b_n are evaluated for $n = 1, 2, 3, \ldots, \infty$, and the integration is over a period T. These set of equations state that a time domain function can be uniquely described by an infinite sum of cos $(2\pi f_0 t)$ and sin $(2\pi f_0 t)$ waves. Although the summation in Eq. (6.1) is over an infinite extent, the summation will often converge to the desired function after a small number of summations. There are four criteria, known as the Dirichlet conditions, for the existence of the Fourier series: (1) the function must be periodic, (2) the number of discontinuities in a period T must be finite, (3) the number of maxima and minima in a period must be finite, and (4) the function must be integratable in any period, i.e., $\int_0^T |x(t)|\,\mathrm{d}t < \infty$. Any long-duration repeating signal occurring in nature will fulfill the Dirichlet conditions so that they can be represented by the Fourier series.

6.2.1 Even and Odd Functions

Evaluation of the Fourier coefficient can be simplified if the function $x(t)$ is either even or odd. A function is even if $x(-t) = x(t)$ and is odd if $x(-t) = -x(t)$. Examples of an even and an odd function are shown in Fig. 6.2. For an even function,

$$\int_{-T/2}^{T/2} E(t)\,\mathrm{d}t = 2 \int_{0}^{T/2} E(t)\,\mathrm{d}t. \tag{6.5}$$

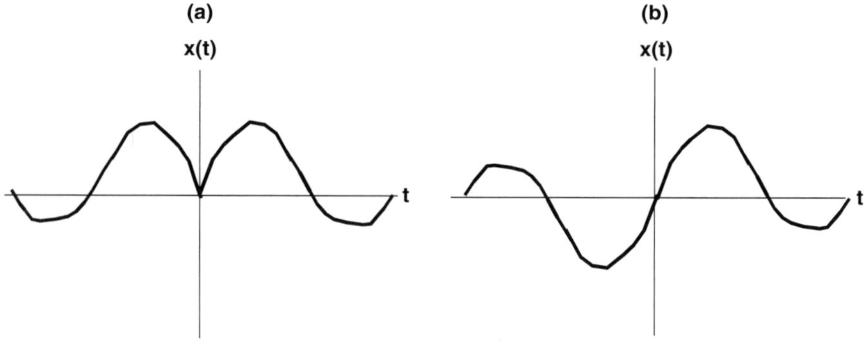

FIGURE 6.2. Example of an even function in (**a**) and an odd function in (**b**).

For an odd function,

$$\int_{-T/2}^{T/2} O(t)\,\mathrm{d}t = 0. \tag{6.6}$$

If $P(t)$ is the product of an even function with an odd function, $P(t)$ will be odd since

$$P(-t) = E(-t)O(-t) = -E(t)O(t) = -P(t). \tag{6.7}$$

If $P(t)$ is the product of an odd function with another odd function, $P(t)$ will be even since

$$P(-t) = O(-t)O(-t) = O(t)O(t) = P(t). \tag{6.8}$$

The product of two even functions will also be even.

If the function $x(t)$ in Eq. (6.1) is even, b_n of Eq. (6.4) will be zero since sin $(n2\pi f_0 t)$ is an odd function and the product of $x(t)$ and sin $(n2\pi f_0 t)$ will be odd and the integral is zero. If the function $x(t)$ is odd, a_0 and a_n will be zero since cos $(n2\pi f_0 t)$ is an even function and the product of an odd $x(t)$ with an even cos $(n2\pi f_0 t)$ will be odd and so the integral for a_n will be zero.

6.2.2 Discrete Spectra from Fourier Series

We will now go through two examples in order to discuss some of the properties of the Fourier series representation of a signal. Consider the triangle signal shown in Fig. 6.3. The triangle signal can be described by the equation

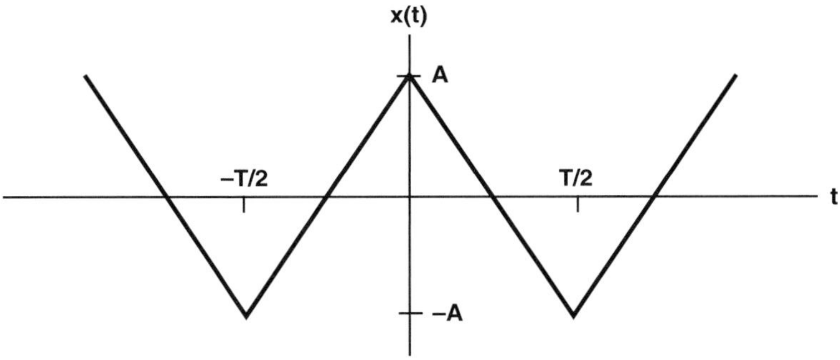

FIGURE 6.3. A triangle signal.

$$x(t) = \begin{cases} A(1 + \frac{4}{T}t) & -\frac{T}{2} < t < 0 \\ A(1 - \frac{4}{T}t) & 0 < t < \frac{T}{2} \end{cases}. \tag{6.9}$$

Since $x(t)$ is an even function, $b_n = 0$; and since the positive and negative excursions are the same, the area under the curve is zero so that $a_0 = 0$. Therefore, only a_n needs to be evaluated.

$$a_n = \frac{2A}{T}\int_{-T/2}^{0}(1 + \frac{4}{T}t)\cos(2\pi n\frac{t}{T})\,\mathrm{d}t + \frac{2A}{T}\int_{0}^{T/2}(1 - \frac{4}{T}t)\cos(2\pi n\frac{t}{T})\,\mathrm{d}t$$

$$= -\frac{16A}{T^2}\int_{0}^{T/2} t\cos(2\pi n\frac{t}{T})\,\mathrm{d}t = \frac{4A}{\pi^2 n^2}[1 - \cos(n\pi)] = \begin{cases} \frac{8A}{\pi^2 n^2} \text{ for } n \text{ odd} \\ 0 \text{ for } n \text{ even} \end{cases}. \tag{6.10}$$

Inserting a_n of Eq. (6.10) into Eq. (6.1), we obtain

$$x(t) = \frac{8A}{\pi^2}\left[\cos(2\pi f_0 t) + \frac{1}{3^2}\cos(3\cdot 2\pi f_0 t) + \frac{1}{5^2}\cos(5\cdot 2\pi f_0 t) + \cdots\right]. \tag{6.11}$$

The Fourier series of the triangular signal contains only odd harmonics of $f_0 = 1/T$. The series converges rapidly to the triangle wave as can be seen in Fig. 6.4a, where the number of terms in the summation of Eq. (6.11) is equal to $(n + 1)/2$, so if we let $n = 3$, two terms are used, etc. From Fig. 6.4a, we see that the Fourier series representation of $f(t)$ converges to $f(t)$ with eight terms. This should not be surprising since a triangle waveform is somewhat similar

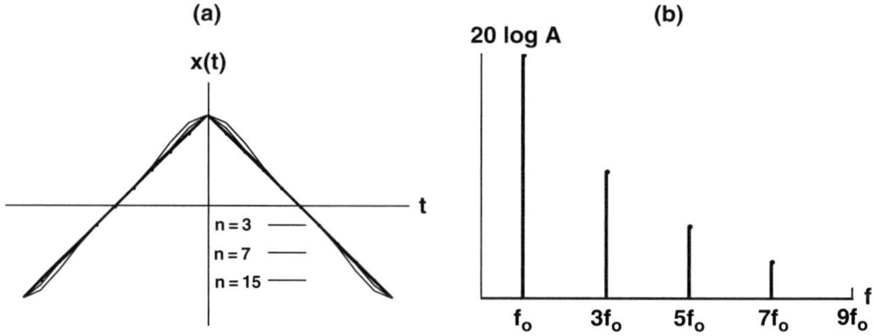

FIGURE 6.4. (**a**) Fourier series results of for $n = 3$, 7, and 15 overlaid on triangle waveform, (**b**) relative amplitude of the amplitude of the spectral lines at odd multiples of f_0.

to a sinusoidal signal. It is also obvious from Fig. 6.4b that the Fourier series produce a spectrum that is discrete in frequency, occurring in this case at every odd multiple of f_0.

We will now consider the square wave in Fig. 6.5a and determine its Fourier series.

The equation for the square wave is

$$x(t) = \begin{cases} -1 & \text{for} \quad -\frac{T}{2} < t < -\frac{T}{4} \\ 1 & \text{for} \quad -\frac{T}{4} < t < \frac{T}{4} \\ -1 & \text{for} \quad \frac{T}{4} < t < \frac{T}{2} \end{cases}. \tag{6.12}$$

As with the case of the triangular wave, the function $f(t)$ is even so that $b_n = 0$, and $a_0 = 0$ because the square wave is symmetrical about the time

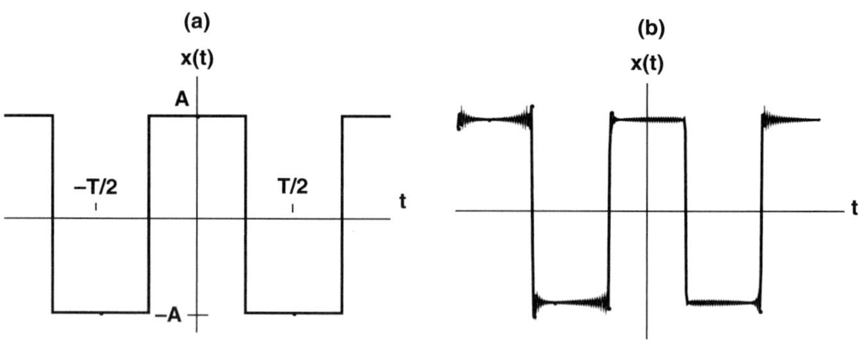

FIGURE 6.5. (**a**) Square wave signal, (**b**) Fourier series representation of the square wave signal with 31 terms corresponding to n = 63.

axis, and the area under the curve for the positive excursions will cancel the corresponding area for the negative excursions. Therefore,

$$a_n = \frac{2A}{T}\left[-\int_{-T/2}^{-T/4}\cos(2\pi n\frac{t}{T})\mathrm{d}t + \int_{-T/4}^{T/4}\cos(2\pi n\frac{t}{T})\,\mathrm{d}t - \int_{T/4}^{T/2}\cos(2\pi n\frac{t}{T})\,\mathrm{d}t\right], \tag{6.13}$$

$$= \frac{2A}{n\pi}\left[-\sin(2\pi n\frac{t}{T})\Big|_{-T/2}^{-T/4} + \sin(2\pi n\frac{t}{T})\Big|_{-T/4}^{T/4} - \sin(2\pi n\frac{t}{T})\Big|_{T/4}^{T/2}\right] = \frac{4A}{n\pi}\sin(\frac{n\pi}{2})$$

where $n = 1,2,3,\ldots, N$. Because of the nature of sin $(n\pi)$, a_n will take on the different values listed in Eq. (6.14).

$$a_n = \begin{cases} 0 & \text{for } n \text{ even} \\ \frac{4A}{n\pi} & \text{for } n = 1, 5, 9, \cdots \\ -\frac{4A}{n\pi} & \text{for } n = 3, 7, 11, \cdots \end{cases}. \tag{6.14}$$

Therefore, the Fourier series of a square wave can be expressed as

$$x(t) = \frac{4A}{\pi}\left[\cos(2\pi\frac{t}{T}) - \frac{1}{3}\cos(3\cdot 2\pi\frac{t}{T}) + \frac{1}{5}\cos(5\cdot 2\pi\frac{t}{T}) - \cdots\right]. \tag{6.15}$$

The Fourier series of the square wave is shown in Fig. 6.5b with 31 terms being summed, corresponding to $n = 63$. The Fourier series representation of the square wave does not converge nearly as rapidly as for the triangular wave. Again this is not surprising since a square wave does not resemble a sine wave at all, so that many terms are needed before the series will take on a square wave appearance. The exact square wave will never be obtained with the Fourier series because of Gibb's phenomenon, which are the oscillations that are present at each point of discontinuity. The Fourier series will represent the original waveform at every point except at the discontinuities. At each discontinuity, there will be an overshoot and the amount of overshoot will be equal to 8.95% of the discontinuity amplitude. The ringing or oscillations at the discontinuities will decrease as the number of term in the series is increased; however, the overshoot will never be eliminated no matter the number of terms used.

6.2.3 *Exponential Form of the Fourier Series*

The Fourier series equation of Eq. (6.1) can be written in an exponential form, a form that will be helpful in our discussion of Fourier Transform. Euler's formula in complex variable theory, relating trigonometric to exponential functions, can be written as

$$\cos\theta = \frac{e^{j\theta} + e^{-j\theta}}{2} \qquad \sin\theta = \frac{e^{j\theta} - e^{-j\theta}}{2j}. \tag{6.16}$$

Substituting the relationships in Eq. (6.16) into Eq. (6.1), we obtain

$$x(t) = a_0 + \sum_{n=1}^{\infty}\left[a_n\left(\frac{e^{j2\pi nf_0t} + e^{-j2\pi nf_0t}}{2}\right) + b_n\left(\frac{e^{j2\pi nf_0t} - e^{-j2\pi nf_0t}}{2}\right)\right] \tag{6.17}$$

$$= a_0 + \sum_{n=1}^{\infty}\left[\frac{(a_n - jb_n)}{2}e^{j2\pi nf_0t} + \frac{(a_n + jb_n)}{2}e^{-j2\pi nf_0t}\right]. \tag{6.18}$$

Let $c_n = (a_n - jb_n)/2$ and $c_{-n} = (a_n + jb_n)/2$ and rewrite Eq. (6.18) as

$$x(t) = a_0 + \sum_{n=1}^{\infty}\left[c_n e^{j2\pi nf_0t} + c_{-n}e^{-j2\pi nf_0t}\right]. \tag{6.19}$$

The summation of the last term in Eq. (6.19) does not change when the signs of the summation index changes to –n and the sum is from –1 to –∞. The coefficient a_0 is the same as c_0, since b_0 is equal to zero, so that we can express Eq. (6.19) as

$$x(t) = \sum_{n=-\infty}^{\infty} c_n e^{j2\pi nf_0t}. \tag{6.20}$$

This is the complex exponential form of the Fourier series. Using Eqs. (6.3) and (6.4) and Euler's formula in Eq. (6.16), the coefficient c_n can be expressed as

$$c_n = \frac{1}{T}\int_{-T/2}^{T/2} x(t)\, e^{-j2\pi nf_0t}\mathrm{d}t. \tag{6.21}$$

For each n, c_n is evaluated to give the magnitude and phase of the harmonic component of $x(t)$ having frequency nf_0. Therefore, the coefficient c_n presents the amplitude of the spectral component of the signal at any frequency nf_0.

6.3 Fourier Transform

In the previous section, we saw that an infinitely continuous periodic signal can be represented by a Fourier series, which provides a discrete spectrum representation of the signal and having values at harmonic intervals associated with the periods, $nf_0 = n/T$. The exponential form of the Fourier series can be extended to take into account non-periodic or aperiodic signals or functions. An aperiodic function can be considered as a function with an infinite period so that T is equal to infinity. As the period T in the Fourier series becomes progressively larger, the interval between harmonics, f_0, becomes progressively smaller. In the limit of T approaching infinity, the frequency components become infinitesimally close, and the discrete spectrum of the Fourier series becomes the continuous spectrum of the Fourier transform. Therefore, Tc_n of Eq. (6.21) merges into a continuum to become a function $S(f)$ of the continuous frequency variable f. The continuous form of Eq. (6.21) can now be written as

$$S(f) = \int_{-\infty}^{\infty} s(t)\, e^{-j2\pi f t} \mathrm{d}t, \tag{6.22}$$

where $S(f)$ is the Fourier transform of the time variable $s(t)$. In this book, the Fourier transform will be denoted in several ways, such as $S(f)$, $\Im[s(t)]$, and $s(t) \Leftrightarrow S(f)$. The inverse Fourier transform is defined by the equation

$$s(t) = \int_{-\infty}^{\infty} S(f)\, e^{j2\pi f t} \mathrm{d}f. \tag{6.23}$$

Equations 6.22 and 6.23 are commonly referred to as the Fourier transform pair. The inverse Fourier transform will be denoted as $\Im^{-1}[S(f)]$. The $\Leftrightarrow$ symbol indicates that the Fourier transform of $s(t)$ is $S(f)$ and the inverse transform of $S(f)$ is $s(t)$. Later in the chapter we will show that inserting Eq. (6.22) into 6.23 will result in $s(t)$ on both sides of the equation. The same Dirichlet conditions with the exception of the periodic criterion applies for the existence of the Fourier transform. Once again, the Dirichlet conditions are satisfied for any natural signal no matter its origin, be it electrical, mechanical, acoustic, cosmic, atomic, etc.

The kernel of the Fourier transform can be expressed as

$$e^{\pm j2\pi ft} = \cos(2\pi ft) \pm j \sin(2\pi ft). \tag{6.24}$$

Although any natural signal will be a real function, $s(t)$, its Fourier transform $S(f)$ will be complex, having a real and an imaginary part. The spectrum of the signal is usually considered to be the absolute value of $S(f)$, or

$$|S(f)| = \sqrt{\mathrm{Re}[S(f)]^2 + \mathrm{Im}[S(f)]^2}, \tag{6.25}$$

where $\mathrm{Re}[S(f)]$ is the real part and $\mathrm{Im}[S(f)]$ is the imaginary part of $S(f)$.

6.3.1 The Impulse or Delta Function

Before proceeding further with our discussion of Fourier transform, it would be very helpful at this time to introduce an important mathematical tool that is often used in Fourier analysis. This tool is called the impulse or delta function $\delta(t)$. Its use will simplify many derivations in Fourier Theory. The delta function is defined as

$$\begin{aligned} &\delta(t - t_0) = 0 \qquad t \neq t_o \\ &\int_{-\infty}^{\infty} \delta(t)\,\mathrm{d}t = 1 \end{aligned} \tag{6.26}$$

The impulse function has an undefined magnitude when $t = t_0$, is zero elsewhere, and has an area (integral) equal to unity. The δ-function can be envisioned as an infinitesimally narrow pulse. As the pulse becomes progressively narrower, its amplitude increases proportionally so that the area is always equal to unity. In the limit of the pulse having zero width, its amplitude approaches infinity. Another property of the δ-function that is useful in Fourier analysis comes from the theory of distribution (Papoulis, 1984) and is

$$\int_{-\infty}^{\infty} \cos(2\pi f t)\,\mathrm{d}f = \int_{-\infty}^{\infty} e^{j2\pi f t}\mathrm{d}f = \delta(t). \tag{6.27}$$

The concepts involved with the δ-function are founded on a rigorous mathematical development, which is beyond the scope of this book.

We can now use the δ-function to examine the Fourier Transform pair of Eqs. (6.22) and (6.23). We can show that given Eq. (6.22), the inverse transform of Eq. (6.23), which is the inverse transform of the forward Fourier transform, is valid. Taking Eq. (6.23) and substituting it into Eq. (6.22), we arrive at the following:

$$\int_{-\infty}^{\infty} S(f)e^{j2\pi f t}\mathrm{d}f = \int_{-\infty}^{\infty} e^{j2\pi f t}\mathrm{d}f \int_{-\infty}^{\infty} s(t')e^{-j2\pi f t'}\mathrm{d}t' = \int_{-\infty}^{\infty} s(t')\mathrm{d}t' \int_{-\infty}^{\infty} e^{j2\pi f(t-t')}\mathrm{d}f'. \tag{6.28}$$

Inserting the relationship given by Eq. (6.27), we obtain

$$\int_{-\infty}^{\infty} S(f)e^{j2\pi f t}\,\mathrm{d}f = \int_{-\infty}^{\infty} s(t')\delta(t - t')\mathrm{d}t' = s(t). \tag{6.29}$$

The dual transform relationship is valid only if $s(t)$ is continuous, and will be valid for functions with a discontinuity only if $s(t)$ is defined as the midvalue at a discontinuity, i.e.,

$$s(t) = \frac{s(t^+) + s(t^-)}{2}. \tag{6.30}$$

Let us now consider the forward and inverse Fourier transform of the δ-function, $s = K\delta(t)$. The Fourier transform of $s(t)$ can be expressed as

$$S(f) = \int_{-\infty}^{\infty} K\delta(t)e^{-j2\pi ft}\mathrm{d}t = Ke^0 = K. \tag{6.31}$$

The inverse Fourier transform of $S(f)$ can be expressed as

$$\begin{aligned} s(t) &= K\int_{-\infty}^{\infty} e^{j2\pi ft}\mathrm{d}f = K\int_{-\infty}^{\infty} \cos(2\pi ft)\,\mathrm{d}f + jK\int_{-\infty}^{\infty} \sin(2\pi ft)\,\mathrm{d}f \\ &= K\int_{-\infty}^{\infty} \cos(2\pi ft)\,\mathrm{d}f = K\delta(t) \end{aligned} \tag{6.32}$$

In deriving Eq. (6.32), we made use of the fact that the integral of $\sin(2\pi ft)$ equals zero because $\sin(2\pi ft)$ is an odd function, and we also made use of Eq. (6.27).

6.3.2 Fourier Transformation of Cosine and Sine Functions

We will now consider the periodic cosine and sine functions making use of the δ-function. This is technically not legitimate since the Fourier transform is applicable to aperiodic signals and not to infinitely continuous (periodic) signals. However, we can take a less rigorous view of what aperiodic means and include very long sinusoidal waves that are much longer than a single period but does not extend to $\pm\infty$. This is akin to turning on a signal oscillator and leaving it on for a very long duration, but not forever. The oscillator output is not infinite in time but only seems to be. Let $s(t) = A\cos(2\pi f_0 t)$, so that its Fourier transform can be expressed as

$$S(f) = A\int_{-\infty}^{\infty} \cos(2\pi f_0 t)e^{-j2\pi ft}\mathrm{d}t = \frac{A}{2}\int_{-\infty}^{\infty} \left[e^{-j2\pi(f-f_0)t} + e^{-j2\pi(f+f_0)t}\right]\mathrm{d}t. \tag{6.33}$$

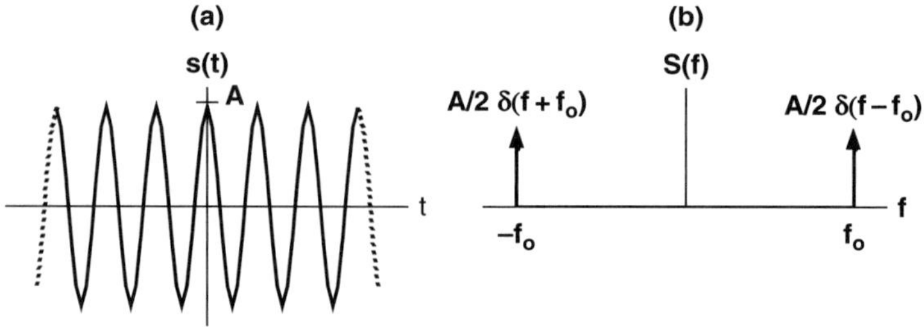

FIGURE 6.6. (**a**) Cos ($2\pi f_0 t$) signal and (**b**) its Fourier transform.

Inserting the relationship of Eq. (6.27), we obtain

$$S(f) = \frac{A}{2}\delta(f - f_0) + \frac{A}{2}\delta(f + f_0). \tag{6.34}$$

The cosine signal and its Fourier transform are shown in Fig. 6.6.

Now consider the sine wave, $s(t) = A\sin(2\pi ft)$. Its Fourier transform is

$$S(f) = A\int_{-\infty}^{\infty} \sin(2\pi f_0 t)e^{-j2\pi ft}\mathrm{d}t = \frac{A}{2j}\int_{-\infty}^{\infty}\left[e^{-j2\pi(f-f_0)t} - e^{-j2\pi(f+f_0)t}\right]\mathrm{d}t. \tag{6.35}$$

From Eq. (6.27), we get

$$S(f) = \frac{A}{2j}\delta(f - f_0) - \frac{A}{2j}\delta(f + f_0). \tag{6.36}$$

The sine signal and its Fourier transform are shown in Fig. 6.7.

The Fourier transforms of the cos and sin function have values only at $\pm f_0$. The amplitude of each component of the Fourier transform for the cosine and sine waves is equal to $A/2$, which is consistent with the fact that the energy in

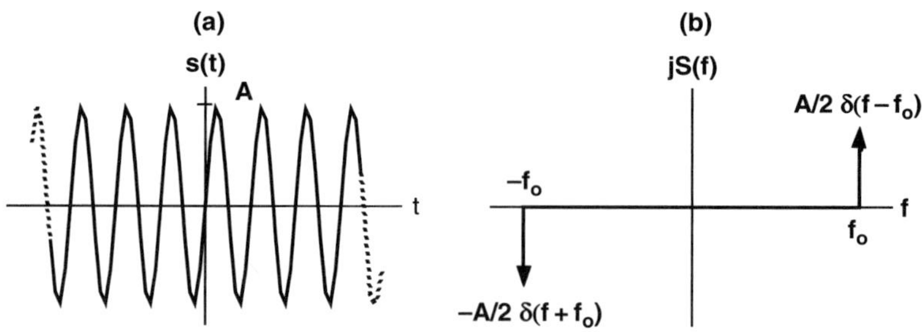

FIGURE 6.7. (**a**) Sin($2\pi f_0 t$) signal and its (**b**) Fourier transform.

the frequency domain is divided equally at $\pm f_0$. $S(f)$ in the frequency domain is purely real for the cos wave and purely imaginary for the sin wave. All natural signals are real, and real even functions will have Fourier transforms that are also real. Similarly, odd real functions will have Fourier transforms that are imaginary. Since the cos function is even, its transform is also real and the sine function is odd so that its transform is imaginary. However, the spectrum of the cos and sine functions is $|S(f)|$ and it will be symmetrical about the origin. Finally, in Figs. 6.6 and 6.7 we see that the Fourier transform also has negative frequencies, a concept that can be very strange to many of us and will be discussed in the following section.

6.3.3 Negative Frequency in Fourier Transform

Since the Fourier transform pair is defined over frequencies and times from minus infinity to plus infinity, we must deal with both negative time and negative frequency. The concept of negative time is not a difficult one since time before a particular event will be negative if we define time zero occurring at the beginning of the event. The same cannot be said about negative frequency. On one hand, we can say just ignore it (as is done in practice) since its not realistic and no harm would be done. On the other hand, it would be good to get a better grip on the notion of negative frequency.

One way to understand negative frequency is to consider the Euler's expression $e^{j2\pi ft} = \cos(2\pi ft) + j\sin(2\pi ft)$. Now take the radius in the polar diagram of Fig. 6.8a and rotate it by making θ increase. As the radius is rotated in the counterclockwise direction, its vertical projection determines the amplitude of the imaginary part of $e^{j2\pi ft}$, plotting out a sine wave. The horizontal projection of the radius determines the amplitude of the real part of $e^{j2\pi ft}$, plotting out a cosine wave. The rate at which the radius is rotated determines the frequency of the cosine and sine waves. Now consider Fig. 6.8b, with the radius being rotated in the clockwise or negative frequency direction. The vertical projection still determines the amplitude of the imaginary part of Euler's formula (the sine wave shown on the right). However, this sine wave has values that are negative to the sine wave in Fig. 6.8a for any given time. The horizontal projection still determines the amplitude of the real part of $e^{-j2\pi ft}$, plotting out the same cosine wave as in Fig. 6.8a. Therefore, positive and negative frequencies can be considered as the direction in which the radius in the polar diagram is rotated. The cosine and sine waves are not changed in shape, but the polarities of the sine waves are reversed for positive and negative frequencies.

6.3.4 More Fourier Transform Examples

At this time it may be helpful to consider some examples of different waveforms and how their Fourier transforms can be computed. These examples

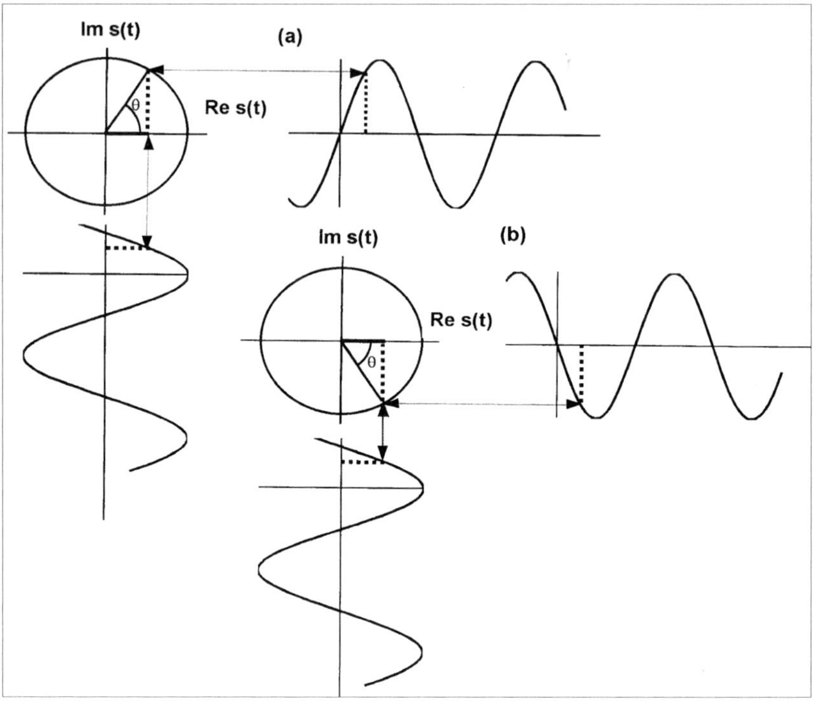

FIGURE 6.8. Representation of the complex function (**a**) $e^{j2\pi ft}$, (**b**) $e^{-j2\pi ft}$.

have been picked with some thought and will have specific applications to different aspects of signal processing of animal signals.

Consider now the exponential function in Fig. 6.9a. The equation for the decaying exponential function can be expressed as

$$s(t) = \begin{cases} \beta \mathrm{e}^{-\alpha t} & t > 0 \\ 0 & t < 0 \end{cases}. \tag{6.37}$$

The Fourier transform of Eq. (6.37) can be expressed as

$$\begin{aligned} \Im[s(t)] &= \int_0^\infty \beta \mathrm{e}^{-\alpha t} \mathrm{e}^{-j2\pi f t} \mathrm{d}t = \beta \int_0^\infty \mathrm{e}^{-(\alpha + j2\pi f)t} \mathrm{d}t = \frac{-\beta}{\alpha + j2\pi f} \mathrm{e}^{-(\alpha + j2\pi f)t} \Big|_0^\infty = \frac{\beta}{\alpha + j2\pi f} \\ &= \frac{\beta}{\alpha + j2\pi f} \cdot \frac{\alpha - j2\pi f}{\alpha - j2\pi f} = \frac{\alpha\beta - j2\pi f \beta}{\alpha^2 + (2\pi f)^2} \end{aligned}. \tag{6.38}$$

The absolute value and phase angle of $S(f)$ can be written as

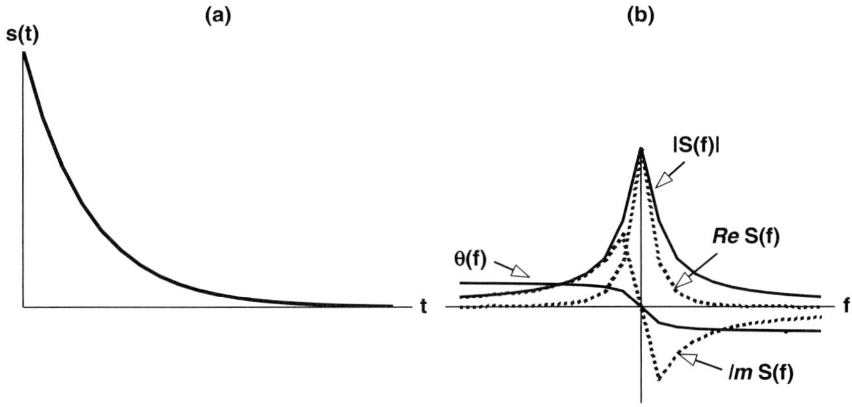

FIGURE 6.9. (**a**) Decaying exponential function in the time domain, (b) the different parts of the Fourier transform in the frequency domain.

$$|S(f)| = \frac{\sqrt{(\mathrm{d}\beta)^2 - (2\pi f\beta)^2}}{\alpha^2 + (2\pi f)^2} \qquad \theta(f) = \tan^{-1}\left(\frac{-2\pi f}{\alpha}\right) \tag{6.39}$$

The absolute value and phase angle along with real and imaginary parts of S(f) are plotted in Fig. 6.9b. This example clearly illustrates that by simply multiplying a time-domain function by the $e^{-j2\pi ft}$ and integrating from $-\infty$ to $+\infty$ the frequency domain function is obtained.

A second example that we will consider is the rectangular pulse shown in Fig. 6.10a. The Fourier transform of such a pulse has an important role in Fourier analysis as we shall see later in this chapter. The Fourier transform of the pulse waveform is

$$S(f) = A\int_{-T}^{T} e^{-j2\pi ft}\,dt = A\left.\frac{e^{-j2\pi ft}}{-j2\pi f}\right|_{-T}^{T} = \frac{A}{\pi f}\left[\frac{e^{j2\pi fTt} - e^{-j2\pi fTt}}{2j}\right] = \frac{A}{\pi f}\sin(2\pi fT) \tag{6.40}$$

$$= 2AT\left[\frac{\sin(2\pi fT)}{2\pi fT}\right]$$

Equation 6.40 is the well-known $\sin(x)/x$ function that occurs so often in physics and engineering that this function is often tabulated in the handbooks of mathematical functions. The Fourier transform of the pulse waveform has zero crossings at intervals of $1/(2T)$ as shown in Fig. 6.10b. The Fourier

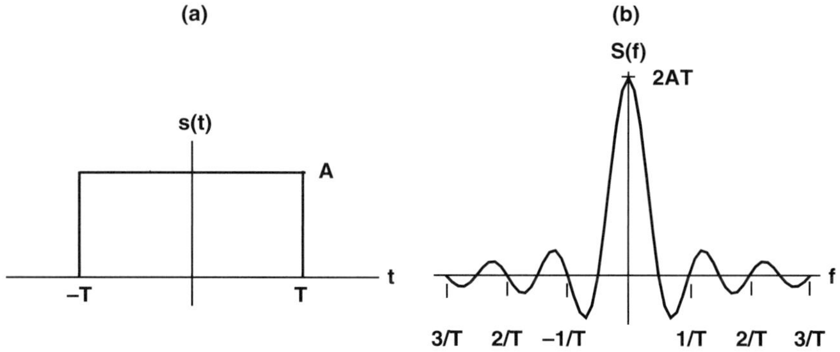

FIGURE 6.10. (**a**) Time domain representation of a pulse, (**b**) Fourier transform of the pulse showing a Sin(x)/x response.

tranform has only a real part; however, the spectrum of the signal is still $|S(f)|$ and the energy is $|S(f)|^2$.

The third example that we will consider is that of a pulsed sine wave that begins and ends at a zero crossing as shown in Fig. 6.11a. Let the frequency of the sine wave be f_0 and its total duration be T, extending from $-T/2$ to $T/2$. The pulse sine wave can be expressed as

$$s(t) = \begin{cases} A\,\sin(2\pi f_0 t) & -\dfrac{T}{2} < t < \dfrac{T}{2}. \\ 0 & \text{elsewhere} \end{cases} \tag{6.41}$$

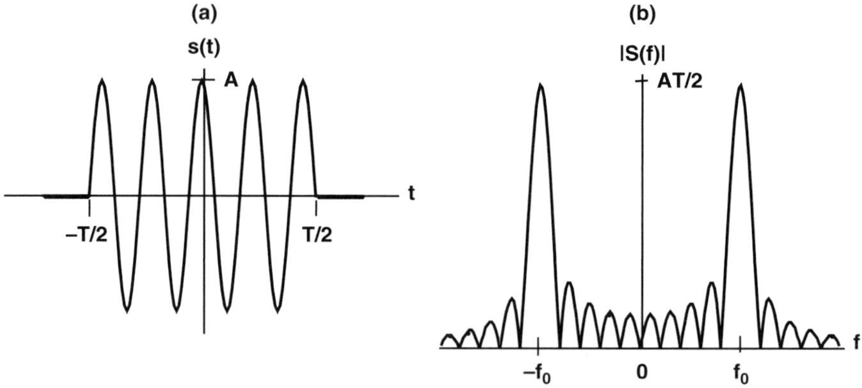

FIGURE 6.11. (**a**) Sine wave pulse in time domain. (**b**) Spectrum of the sine wave pulse in the frequency domain.

The Fourier transform of $s(t)$ is

$$S(f) = A\int_{-T/2}^{T/2} \sin(2\pi f_0 t)\mathrm{e}^{-2\pi f t}\mathrm{d}t = \frac{A}{2j}\int_{-T/2}^{T/2}\left(\mathrm{e}^{-j2\pi(f-f_0)t} - \mathrm{e}^{-j2\pi(f+f_0)t}\right)\mathrm{d}t$$

$$= \frac{A}{2}\left[\left.\frac{\mathrm{e}^{-j2\pi(f-f_0)t}}{2\pi(f-f_0)}\right|_{-T/2}^{T/2} - \left.\frac{\mathrm{e}^{-j2\pi(f+f_0)t}}{2\pi(f+f_0)}\right|_{-T/2}^{T/2}\right] \quad . \quad (6.42)$$

$$= \frac{jAT}{2}\left[\frac{\sin[2\pi(f+f_0)]T/2}{2\pi(f+f_0)T/2} - \frac{\sin[2\pi(f-f_0)]T/2}{2\pi(f-f_0)T/2}\right]$$

The Fourier transform of the pulsed sine wave shown in Fig. 6.11b is a sin $(x)/x$ function centered at $\pm f_o$. As with the pulse signal in Fig. 6.10a, the narrower the sine pulse the broader will be the sin $(x)/x$ function in the frequency domain. Conversely, the wider the sine pulse the narrower will be the sin $(x)/x$ function, i.e., the bandwidth of the spectrum is inversely proportional to the width of the sine pulse. Note also the symmetry in the spectrum about the zero frequency.

6.4 Properties of the Fourier Transform

We will now consider some Fourier transform theorems that will be helpful in analyzing acoustic signals. Most of the derivations are simple and we will attempt to indicate the areas in which the various theorems may be applicable in acoustic signal processing.

6.4.1 Addition or Linearity Theorem

If $x(t)$ and $y(t)$ have the Fourier transforms $X(f)$ and $Y(f)$, respectively, then the sum of $x(t) + y(t)$ has the Fourier Transform $X(f) + Y(f)$. This property can be derived by letting

$$\int_{-\infty}^{\infty} [x(t)+y(t)]\mathrm{e}^{-j2\pi f t}\mathrm{d}t = \int_{-\infty}^{\infty} x(t)\mathrm{e}^{-j2\pi f t}\mathrm{d}t + \int_{-\infty}^{\infty} y(t)\mathrm{e}^{-j2\pi f t}\mathrm{d}t = X(f) + Y(f). \quad (6.43)$$

Often, it is desirable to add noise to a signal when performing a noise masking experiment or to add noise to an echo to test the performance of a signal detector. This theorem states that the noise can be added in either the time or the frequency domain and the results with the same results.

6.4.2 Time and Frequency Scaling

If the Fourier transform of $s(t)$ is $S(f)$, then the Fourier transform of $s(kt)$, where k is a real constant greater than zero, can be determined by letting $t' = kt$ so that

$$\int_{-\infty}^{\infty} s(kt)\,\mathrm{e}^{-2\pi ft}\mathrm{d}t = \int_{-\infty}^{\infty} s(t')\,\mathrm{e}^{-j2\pi f\frac{t'}{k}}\frac{\mathrm{d}t'}{k} = \frac{1}{k}S\left(\frac{f}{k}\right). \tag{6.44}$$

A more general form of Eq. (6.44) with k being either positive or negative is

$$s(kt) \Leftrightarrow \frac{1}{|k|}S\left(\frac{f}{k}\right). \tag{6.45}$$

Multiplying the time variable by a positive constant k causes a stretching of the signal in the time domain, which corresponds to a shift in the Fourier transform to lower frequencies by a factor $1/k$. This type of scaling is found in wavelet analysis where a dilation of a function in the time (left side of Eq. (6.45) causes a corresponding shift to lower frequencies and a compression of the spectrum by $1/k$.

Frequency scaling has a symmetrical property to time scaling, which can be seen by letting the inverse Fourier transform of $S(f)$ be $s(t)$ so that the inverse Fourier transform of $S(kf)$ can be expressed as

$$\int_{-\infty}^{\infty} S(kf)\,\mathrm{e}^{j2\pi ft}\mathrm{d}f = \int_{-\infty}^{\infty} S(f')\,\mathrm{e}^{j2\pi f'\frac{t}{k}}\frac{\mathrm{d}f'}{k} = \frac{1}{|k|}s\left(\frac{t}{k}\right). \tag{6.46}$$

In this case, a dilation in frequency causes a compression of the function in time.

6.4.3 Time and Frequency Shifting

If $s(t)$ has the Fourier transform $S(f)$, then the Fourier transform of $s(t - t_0)$ can be determined by letting $x = t - t_0$, so that

$$\begin{aligned}\int_{-\infty}^{\infty} s(t - t_o)\,\mathrm{e}^{-j2\pi ft}\mathrm{d}t &= \int_{-\infty}^{\infty} s(x)\,\mathrm{e}^{-j2\pi f(x + t_0)}\mathrm{d}x \\ &= \mathrm{e}^{-j2\pi ft_0}\int_{-\infty}^{\infty} s(x)\,\mathrm{e}^{-2\pi fx}\mathrm{d}x = \mathrm{e}^{-j2\pi ft_0}S(f).\end{aligned} \tag{6.47}$$

Therefore, the Fourier transform pair can be expressed as

$$s(t - t_0) \Leftrightarrow e^{-j2\pi f t_0} S(f). \tag{6.48}$$

The function $s(t - t_0)$ represents a signal $s(t)$ being shifted in the positive time axis by a constant t_0. This can be readily seen because the argument of $s(t - t_0)$ is equal to zero only after t increases in value and becomes equal to t_0. Equation 6.48 indicates that if a given signal is shifted in the positive direction by an amount t_0, the amplitude of its Fourier transform does not change from its original value, only a phase change is introduced. Each component of the Fourier transform is delayed in phase by an amount proportional to the frequency f. The shift theorem is self-evident if one considers a sine or cosine wave at the origin. If the sine or cosine wave is shifted by a constant t_0, the phase of the sine or cosine wave with respect to the origin will be shifted by an amount $2\pi(t_0/T)$ since there are 2π radians in each period of a sine or cosine signal, where T is the period of the wave.

If a marine mammal emits a short burst of sound that is detected at a distance R from the animal, the received sound will have a phase change of $e^{-2\pi f\tau}$ from the original emitted signal, where $\tau = R/c$, c being the sound velocity of water. In this case we are assuming ideal propagation an infinite medium.

We can consider the effects of a shift in frequency of a signal by first letting $S(f)$ be the Fourier transform of $s(t)$ and letting $x = f - f_0$ in the frequency-shifted signal $S(f - f_0)$, so that the inverse Fourier transform can be expressed as

$$\int_{-\infty}^{\infty} S(f-f_0)e^{j2\pi f t}df = \int_{-\infty}^{\infty} S(x)e^{j2\pi (x+f_0) t}dx = e^{j2\pi f_0 t}\int_{-\infty}^{\infty} S(x)e^{j2\pi x t}dx = e^{j2\pi f_0 t}s(t). \tag{6.49}$$

The Fourier transform pair can be expressed as

$$s(t)\, e^{j2\pi f_0 t} \Leftrightarrow S(f - f_0). \tag{6.50}$$

6.4.4 Modulation Theorem

In hearing sensitivity experiments with animals, including humans, the sound source is usually turned on and off gradually to avoid the introduction of ringing in the sound projector. Often a sinusoidal signal is modulated by a trapezoidal signal or some other signal that has a rise and fall time, which is much longer than the period of the sinusoidal signal. If an envelope function

$s(t)$ with a Fourier transform $S(f)$ is multiplied with a cosinusoidal signal, the modulated signal $s(t)cos\,(2\pi f_0 t)$ will a Fourier transform that is equal to

$$\int_{-\infty}^{\infty} s(t)\cos(2\pi f_0 t)\mathrm{e}^{-j2\pi f t}\mathrm{d}t = \frac{1}{2}\int_{-\infty}^{\infty} s(t)\mathrm{e}^{-j2\pi(f-f_0)t}\mathrm{d}t + \frac{1}{2}\int_{-\infty}^{\infty} s(t)\mathrm{e}^{-j2\pi(f+f_0)t}\mathrm{d}t \quad . \quad (6.51)$$

$$= \frac{1}{2}S(f-f_0) + \frac{1}{2}S(f+f_0)$$

The spectrum of the amplitude-modulated signal is separated into two equal parts shifted along the f-axis by an amount $\pm f_0$.

6.4.5 Convolution Theorem

The convolution of two functions is an extremely important process in the field of science. In this section, we will take a narrow focus and discuss convolution of an input signal with the transfer function of a system or a device. A transfer function is the response of a system as a function of frequency to an input of varying frequency and the impulse response is the inverse Fourier transform of the transfer function. For electrical devices, the transfer function is determined by inputting a sinusoidal signal of known amplitude and frequency and measuring the response of the system as a function frequency. Transfer function can be determined for almost any system that will accept an input and have a specific output, such as an electronic circuit, a electroacoustic transducer, and almost any type of sensor. The target strength of sonar targets can also be approached from a transfer function point of view. Once the transfer function of a system or device is known, the response of the system can be calculated by convolving the transfer function with any desired input signal. Let $H(f)$ be the transfer function of a device or system and $h(t)$ be its inverse Fourier transform or impulse response of the system, then the response of the system is determined by evaluating the convolution integral expressed as

$$y(t) = \int_{-\infty}^{\infty} x(\tau)h(t-\tau)\mathrm{d}\tau = x(t) * h(t), \qquad (6.52)$$

where $x(t)$ is the input signal and the * denotes the convolution of $x(t)$ and $h(t)$. Technically speaking, the impulse response is the response of a system to an impulse input signal that has frequency components from 0 to infinity. However, in practical situations, any system will cease to respond after the frequency increases beyond its high-frequency cutoff, so that the transfer function does not need to encompass all frequencies but only frequencies up to the system's cutoff frequency.

In order to get a better understanding of the convolution integral we will use a simple, perhaps non-realistic, but illuminating example from Brigham (1988) in Fig. 6.12. The two functions to be convolved are shown in Fig. 6.12a. The steps involved in evaluating the convolution integral consist of (1) changing the variable for t to τ as shown for $x(t)$ in Fig. 6.12b, and folding or taking the mirror image of $h(\tau)$, (2) displacing or shifting $h(-\tau)$ by an amount t to form $h(t - \tau)$ as shown in the right side of Fig. 6.12b, (3) multiplying the shifted function $h(t - \tau)$ by $x(\tau)$, and 4) integrating or finding the area under the product of $h(t - \tau)$ and $x(\tau)$ for different values of the time shift t as shown

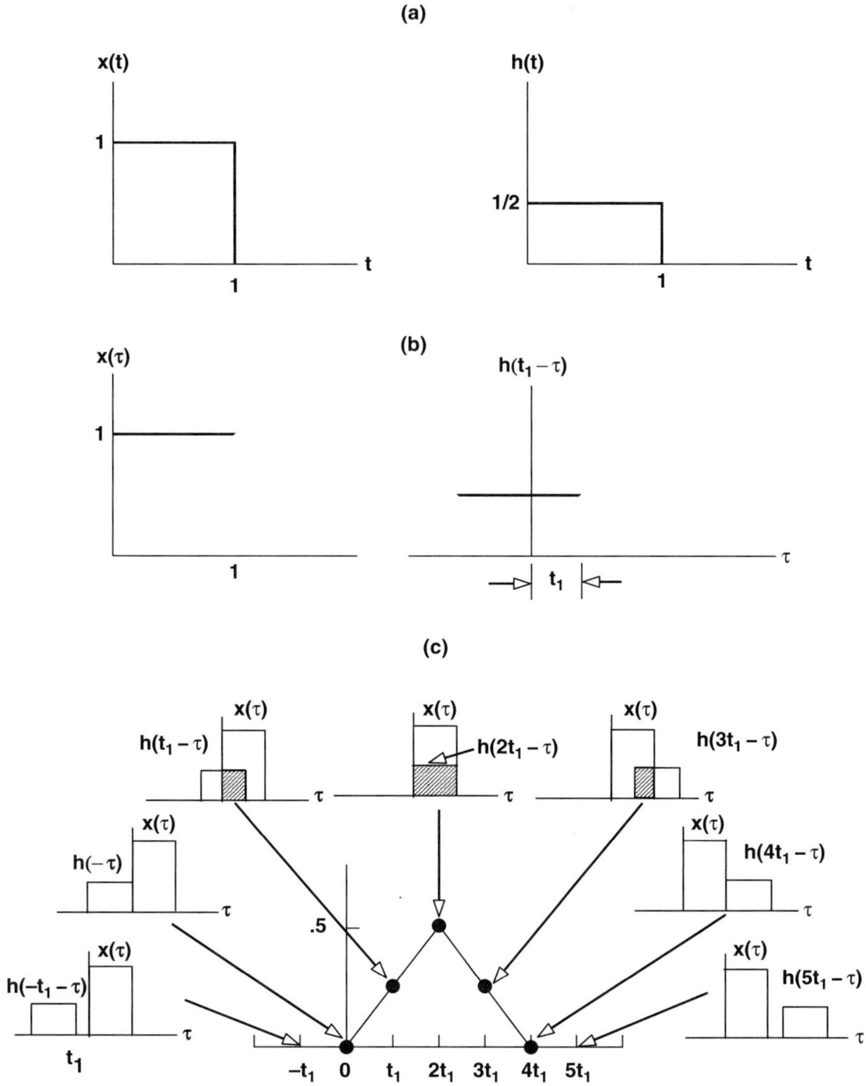

FIGURE 6.12. Graphical example of the convolution integral with t shifting by nt_1.

for various values of t in Fig. 6.12c. The reason for the folding in the convolution integral is easily explained by realizing that the impulse response of a system will vary as a function of time, starting off at time zero+ when the impulse is first applied. Then the response will most likely oscillate and decay as a function of time. The convolution integral is used to determine the response of a system to any signal $x(t)$ and the folding of the impulse response in Fig. 6.12b will ensure that the beginning portion of the signal is multiplied with the beginning portion of the impulse response. As the impulse response is shifted through the signal, as shown in Fig. 6.12c, the results of the multiplication and integration will produce the response of the system as a function of time.

A very important relationship exists between the convolution integral of Eq. (6.52) and its Fourier transform. This relationship, known as the *time-convolution theorem* can be seen by taking the Fourier transform of $y(t)$ in Eq. (6.52),

$$\begin{aligned}\int_{-\infty}^{\infty} y(t)\,e^{-j2\pi ft}dt &= \int_{-\infty}^{\infty}\left[\int_{-\infty}^{\infty} x(\tau)\,h(t-\tau)\,d\tau\right]e^{-j2\pi ft}dt \\ &= \int_{-\infty}^{\infty} x(\tau)\left[h(t-\tau)\,e^{-j2\pi ft}dt\right]d\tau\end{aligned} \quad . \tag{6.53}$$

If we let $\sigma = t - \tau$i then the integral in the brackets becomes

$$\int_{-\infty}^{\infty} h(\sigma)e^{-j2\pi f(\sigma+\tau)}d\sigma = e^{-j2\pi f\tau}\int_{-\infty}^{\infty} h(\sigma)e^{-j2\pi f\sigma}d\sigma = e^{-j2\pi f\tau}H(f) \tag{6.54}$$

Therefore, Eq, 6.53 can now be written as

$$\Im[x(t) * h(t)] = H(f)\int_{-\infty}^{\infty} x(\tau)e^{-j2\pi f\tau}d\tau = H(f)X(f). \tag{6.55}$$

This important result indicates that the convolution integral of Eq. (6.52) does not have to be evaluated directly, rather it can be determined by multiplying the input function $X(f)$ with the transfer function and computing the inverse Fourier transform. This procedure may seem more complicated than merely computing the integral of Eq. (6.52) directly; however, because of the efficiency of fast Fourier transform algorithm, using Eq. (6.55) is actually more time-efficient than evaluating Eq. (6.52) directly.

We now turn to the *frequency convolution theorem* in which two functions represented by their Fourier transforms are convolved, so that

$$Y(f) = \int_{-\infty}^{\infty} X(\nu) H(\nu - f) \mathrm{d}\nu = X(f) H(f). \tag{6.56}$$

We can use the similar steps as in deriving Eq. (6.55) to obtain,

$$x(t) h(t) \Leftrightarrow X(f) * H(f). \tag{6.57}$$

Therefore, multiplying two functions in the time domain is equivalent to convolving two functions in the frequency domain. This theorem plays a significant role in Fourier analysis of very long signals where a portion of the signal is "windowed" and the Fourier transform of the windowed data is calculated. More on this topic will be covered at the end of this chapter.

6.4.6 Correlation Theorem

Another integral that is very important in signal analysis is the correlation integral:

$$c(t) = \int_{-\infty}^{\infty} x(t) h(t + \tau) \mathrm{d}\tau. \tag{6.58}$$

The correlation integral is very similar to the convolution integral; the only difference is argument of the second function is $t + \tau$, instead of $t - \tau$. From a graphical standpoint, the second function is not folded over as in the case shown in Fig. 6.12b. As with the convolution integral, the correlation integral will also have a special relationship to Fourier transformation. Taking the Fourier transform of Eq. (6.58), we obtain

$$\int_{-\infty}^{\infty} c(t) e^{-j2\pi f t} \mathrm{d}t = \int_{-\infty}^{\infty} x(\tau) \left[\int_{-\infty}^{\infty} h(t + \tau) \mathrm{e}^{-j2\pi f t} \mathrm{d}\tau \right] \mathrm{d}t. \tag{6.59}$$

Letting $\sigma = t + \tau$, the integral in the brackets can be expressed as

$$\int_{-\infty}^{\infty} h(\sigma) \mathrm{e}^{-j2\pi f(\sigma - \tau)} \mathrm{d}\sigma = \mathrm{e}^{j2\pi f t} \int_{-\infty}^{\infty} h(\sigma) \mathrm{e}^{-j2\pi f \sigma} \mathrm{d}\sigma = \mathrm{e}^{j2\pi f t} H(f). \tag{6.60}$$

Therefore, Eq. (6.59) can be expressed as

$$C(f) = H(f)\int_{-\infty}^{\infty} x(\tau)e^{j2\pi f\tau}\mathrm{d}\tau = H(f)X(f)^{*}. \tag{6.61}$$

The correlation theorem can now be expressed as

$$c(t) \Leftrightarrow H(f)X(f)^{*}. \tag{6.62}$$

If both functions are the same, then Eq. (6.62) is termed the autocorrelation function; and if both functions are different, then it is termed the cross-correlation function.

The correlation function is very useful in detecting a known signal in noise. Often the signal of interest is known, but is buried in noise. By taking section of the noisy signal and computing the correlation integral, the presence of the signal and location of the signal in time can be "pulled" out of the noise.

6.4.7 *Rayleigh–Parseval Theorem*

This theorem can be stated simply as

$$\int_{-\infty}^{\infty} |s(t)|^2\mathrm{d}t = \int_{-\infty}^{\infty} |S(f)|^2\mathrm{d}f. \tag{6.63}$$

The terms $|s(t)|^2$ and $|S(f)|^2$ relate to energy density and power, one in the time domain and the other in the frequency domain. For example, if $s(t)$ represents voltage $v(t)$, electrical power is equal to v^2/R, or if $s(t)$ represents acoustic pressure $p(t)$, acoustic intensity is equal to $p^2/(\rho c)$. From this prospective, the Rayleigh–Parseval's theorem is self-evident since the energy of a signal in the time domain should also be the same as the energy in the frequency domain. Nevertheless, Eq. (6.65) can be proven mathematically.

For real signals or functions, $|s(t)|^2 = s(t)\cdot s(t)$ so that

$$\int_{-\infty}^{\infty} |s(t)|^2\mathrm{d}t = \int_{-\infty}^{\infty} s(t)s(t)\,\mathrm{d}t = \int_{-\infty}^{\infty} s(t)\int_{-\infty}^{\infty} S(f)\mathrm{e}^{j2\pi ft}\mathrm{d}f\mathrm{d}t. \tag{6.64}$$

Interchanging the order of integration, we have

$$\int_{-\infty}^{\infty} |s(t)|^2 \mathrm{d}t = \int_{-\infty}^{\infty} S(f) \int_{-\infty}^{\infty} s(t) \mathrm{e}^{j2\pi f t} \mathrm{d}t \, \mathrm{d}f = \int_{-\infty}^{\infty} S(f) \, S^*(f) \, \mathrm{d}f, \quad (6.65)$$

which is exactly Eq. (6.63).

6.5 The Discrete Fourier Transform and Fast Fourier Transform

6.5.1 The Discrete Fourier Transform

The Fourier transform pair in Eqs. (6.22) and (6.23) are continuous functions in time and frequency and the integrals extend from $-\infty$ to $+\infty$. Both the continuous nature of the Fourier transform and the integration from $-\infty$ to $+\infty$ make it unsuitable for computations with digital computers. Therefore, a discrete form of the Fourier transform pair must be used in order to perform Fourier analysis with digital computers. There are a series of steps that can be taken to convert the continuous Fourier transform pair into a discrete transform pair that can be used with digital data. We will jump directly to the discrete form of the Fourier transform pair, usually referred to as the *discrete Fourier transform or DFT*. Readers interested in the steps leading from the continuous form to the discrete form can consider the many excellent references listed in the beginning of the chapter. We will merely state that a continuous integral can be approximated by a summation and the infinite nature of the Fourier transform is handled by restricting our computation to N positive points.

An analog signal in the time domain can be quantized or digitized for computation with a digital computer by sampling its amplitude at a fixed time sample interval and converting the amplitude into a digital number, a process known as analog-to-digital conversion (see Chapter 5). If N points are used to describe the signal, then the digitized signal $s(n)$ is the sampled values of $s(t)$ at consecutive time intervals of Δt, where n is the time index with values of 0, 1, 2,. . ., $N - 1$ and the time of the nth sample is equal to $t_n = n \, \Delta t$. The continuous Fourier transform pair of Eqs. (6.22) and (6.23) in its DFT form can be expressed as

$$S(k) = \sum_{n=0}^{N-1} s(n) \, \mathrm{e}^{-2\pi nk/N_1}, \quad (6.66)$$

$$s(n) = \frac{1}{N} \sum_{k=0}^{N-1} S(k) \, \mathrm{e}^{j2\pi nk/N_2}, \quad (6.67)$$

where k is the index for the computed set of discrete frequency components. Its values are $k = 0, 1, 2, \ldots, N-1$ and the frequency of the kth interval is equal to $f_k = k\Delta f$. The frequency interval is equal to one over the product of number of samples, N, of the time domain signal and the sample interval Δt, or

$$\Delta f = \frac{1}{N\Delta t} = \frac{1}{T_{max}}, \tag{6.68}$$

where $T_{max} = N\Delta t$. The discrete frequencies applicable to the DFT are often referred to as frequency bins or the frequency of analysis, $f_{analysis}$. Note, in the DFT equation, that f and t in the exponential term of the continuous Fourier transform has been replaced by the frequency index k and the time index n, respectively.

In the process of developing the discrete Fourier transform pair from the continuous Fourier transform pair, the assumption that $s(n)$ and $S(k)$ are periodic with a period of N was necessary. Brigham (1988) presents for both a graphical and theoretical development of the discrete Fourier transform pair from the continuous Fourier transform pair and those interested in the details are encouraged to look up this excellent reference.

We will now examine another interesting feature of the discrete Fourier transform by showing that for real signals the DFT expression in Eq. (6.66) has a symmetry with respect to frequency so that only $S(k)$ for $k = 0$ to $N/2$ needs to be calculated. Let $k' = (N-k)$ so that as k increases from 0 to $N-1$, k' decreases from N to 1. In other words, the index k counts in the forward direction and k' counts in the backward direction. The DFT can now be written as

$$S(N-k) = \sum_{n=0}^{N-1} s(n)\mathrm{e}^{-j2\pi n(N-k)/N} = \sum_{n=0}^{N-1} s(n)\mathrm{e}^{-j2\pi n}\mathrm{e}^{-j2\pi(-k)/N}. \tag{6.69}$$

Since $\mathrm{e}^{-j2\pi n} = \cos(2\pi n) - j\sin(2\pi n) = 1$ for all integer values of n, Eq. (6.69) becomes

$$S(N-k) = \sum_{n=0}^{N-1} s(n)\mathrm{e}^{j2\pi nk/N} = S*(k). \tag{6.70}$$

Therefore, $|S(N-k)| = |S(k)|$, which indicates that only N/2 frequency values can be determined from the DFT. This means that the maximum frequency component that can be calculated (from Eq. (6.69)) is

$$f_{max} = \frac{N}{2}\Delta f = \frac{1}{2\Delta t} = \frac{\mathrm{sr}}{2}, \tag{6.71}$$

where sr is the digitizing or sampling rate at which an analog signal is converted into a digital signal. Conversely, the sample rate for digitizing data correctly must be greater than or equal to the maximum frequency contained in the data. We discussed in Chapter 5 that a sampling rate equal to twice the maximum frequency is referred to as the *Nyquist rate* or *Nyquist frequency*. Values of the DFT calculated for $k > N/2$ can be considered as the negative frequency portion of the Fourier transform. The frequency corresponding to $k = N/2$ is often referred to as the folding frequency, which is one half the Nyquist frequency.

Although the DFT expression is relatively simple, the number of operation necessary can be very high. There are N complex multiplication and N complex addition for each frequency index k, and k extends from 0 to $N-1$ so that N^2 complex multiplications and additions are necessary to compute the DFT. However, multiplication takes considerably more computer time than addition so that the amount of multiplications is the determining factor in how long it takes to compute a DFT. The amount of complex multiplication can be large. For example, a simple 1,024 point transform would involve approximately 1 million complex multiplications and before the introduction of the fast Fourier transform, it would take tens of minutes to hours to evaluate the DFT. Therefore, the DFT was not commonly used in signal analysis until the fast Fourier transform (FFT) algorithm was introduced in 1966.

6.5.2 The Fast Fourier Transform

The fast Fourier transform is not a new or different version of the DFT, but rather stands for algorithms that can evaluate the DFT in an efficient manner. Its development began in the early 1960 s when Richard Garwin had a need to evaluate Fourier transforms in his study of solid helium. Garwin contacted James W. Tukey and asked him to develop an efficient means of evaluating the discrete Fourier transform. Tukey provided Garwin with an efficient technique to evaluate the summation of the DFT. Then Garwin was led to J. W. Cooley at the IBM Research Center in Yorktown Heights, who programmed Tukey's algorithm in his spare time. The Cooley–Tukey algorithm was introduced to the world in 1965 (Cooley and Tukey, 1965) and it was eventually called the FFT. Since that time, many different algorithms have been developed and these are all lumped under the general heading of FFT. Cooley et al. (1967) has written a historical note on the development of the fast Fourier transform.

The Cooley–Tukey algorithm required that the number of points be limited to 2^γ, where γ is an integer. Therefore, N is restricted powers of 2 such as 2, 4, 8, 16, 32, 64, ..., 2^γ, and an algorithm developed within this constraint is referred to as the radix-2 FFT. Let $W_N = e^{-j2\pi/N}$ in Eq. (6.66) so that the forward DFT can be expressed as

$$S(k) = \sum_{n=0}^{N-1} s_0(n)\, W^{nk}, \tag{6.72}$$

where we have added a subscript to $s(n)$, which will be used to indicate the stages of computation in evaluating the DFT. Therefore, $s_0(n)$ is equal to $s(n)$ since as yet no computation has been performed. We will now consider a simple 4-point FFT so that $N = 4$ and $\gamma = 2$. The indices n and k will be represented as binary numbers, with each bit taking on a value of either 0 or 1 so that for $N = 4$ and $\gamma = 2$ we have

$$n = (n_1, n_0) = 00, 01, 02, 11,$$
$$k = (k_1, k_0) = 00, 01, 02, 11.$$

A compact way of representing n and k is

$$n = 2n_1 + n_0 \quad k = 2k_1, +k_0,$$

where again n_0, n_1, k_0, and k_1 can take on values of 0 and 1 only. Using this type of representation, Eq. (6.72) can be expressed as

$$S(k_1, k_0) = \sum_{n_0=0}^{1} \sum_{n_1=1}^{1} s_0(n_1, n_0)\, W^{(2k_1 + k_0)(2n_1 + n_0)}. \tag{6.73}$$

Nothing has been done so far to reduce the number of multiplications needed to compute the DFT. However, let us now examine the W^{nk} term, noting that $W^{nk} = W^n \mathrm{W}^k$, so that

$$W^{(2k_1 + k_0)(2n_1 + n_0)} = \left[W^{4k_1 n_1}\right] W^{2k_0 n_1}\, W^{(2k_1 + k_0)n_0}. \tag{6.74}$$

The term in the bracket is equal to 1 since for $N = 4$,

$$W^{4k_1 n_1} = \left[\mathrm{e}^{-j2\pi(4/4)}\right]^{k_1 n_1} = \left[\mathrm{e}^{-j2\pi}\right]^{k_1 n_1} = 1^{k_1 n_1} = 1. \tag{6.75}$$

Equation 6.73 can now be written as

$$S(k_1, k_0) = \sum_{n_0=0}^{1} \left[\sum_{n_1=0}^{1} s_0(n_1, n_0)\, W^{2k_0 n_1}\right] W^{(2k_1 + k_0)n_0}. \tag{6.76}$$

We can consider the operation under each summation as a stage of operation, with the first stage being the operation within the brackets. Considering the summation within the brackets, we have

$$s_1(k_0, n_0) = \sum_{n_1=0}^{1} s_0(n_1, n_0) W^{2k_0 n_1}. \tag{6.77}$$

The outer summation can now be written as

$$s_2(k_0, k_1) = \sum_{n_0=1}^{1} s_1(k_0, n_0) W^{(2k_1 + k_0)n_0}. \tag{6.78}$$

From Eq. (6.76), we have for the forward DFT

$$\mathrm{S}(\mathrm{k}_1, \mathrm{k}_0) = \mathrm{s}_2(\mathrm{k}_0, \mathrm{k}_1). \tag{6.79}$$

The set of equations 6.77, 6.78, and 6.79 represents the original Cooley–Tukey (1965) formulation of the FFT algorithm for $N = 4$. These equations are recursive since the second equation is computed in terms of the first equation.

We can see how this formulation reduces the number of multiplication by evaluating Eq. (6.79) to obtain

$$\begin{aligned} s_1(0,0) &= s_0(0,0) + s0(1,0) W^0, \\ s_1(0,1) &= s_0(0,1) + s_0(1,1) W^0, \\ s_1(1,0) &= s_0(0,0) + s_0(1,1) W^2, \\ s_1(1,1) &= s_0(0,1) + s_0(1,1) W^2. \end{aligned} \tag{6.80}$$

W^0 is actually equal to unity, but is kept in this form so that each equation will have the same form in Eq. (6.80). In a similar manner, Eq. (6.78) can be evaluated to give

$$\begin{aligned} s_2(0,0) &= s_1(0,0) + s_1(0,1) W^0, \\ s_2(0,1) &= s_1(0,0) + s_1(0,1) W^2, \\ s_2(1,0) &= s_1(1,0) + s_1(1,1) W^1, \\ s_2(1,1) &= s_1(1,0) + s_1(1,1) W_3. \end{aligned} \tag{6.81}$$

The DFT is given by Eq. (6.79); however, $s_2(k_1,k_0)$ are in bit-reversed order with respect to the desired values $S(k_1,k_0)$, a result of the scrambling in the algorithm. Only four complex multiplications are involved with evaluating Eq. (6.80) and another four in evaluating Eq. (6.81), compared to the $N^2 = 16$ complex multiplication in directly evaluating the forward DFT. Therefore, the number of complex multiplication has been reduced from 16 to 8 for this simple case. The reduction in the number of complex multiplications will be even more for larger values of N.

In a general case where $N = 2^\gamma$, the FFT formulation of Eq. (6.73) will have γ summations and the DFT will be symbolically represented as $S(k_{\gamma-1}, k_{\gamma-2}, \ldots, k_0)$. The γ summations can be evaluated in a similar manner as for the $N = 4$ case. The number of complex multiplications involved in the FFT formulation is $N \log_2 N$. A comparison between the direct or brute force evaluation of the DFT versus the FFT algorithm is shown in Fig. 6.13. From this figure we can see that the FFT algorithm is much more efficient than the direct method of evaluating the DFT. Since the development of the Cooley–Tukey algorithm, a variety of different algorithms have appeared and all of these algorithms are included under the general heading of FFT. There are also algorithms for radix-4, radix-8, etc., and even algorithms that are not based on any radix called "prime-factor" FFT. Readers interested in a more in-depth discussion of different FFT algorithm should consider reading Brigham (1988). FFT computer programs written in almost all conceivable

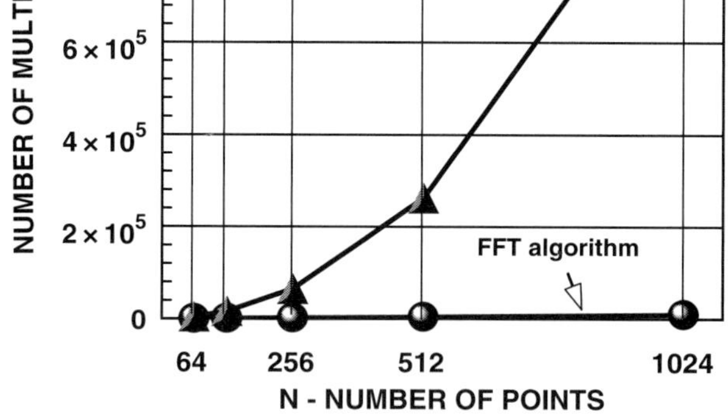

FIGURE 6.13. Comparison of the number of mulitplications by direct calculation and FFT algorithm.

programming languages are readily available and are an integral part of almost all signal processing software. Many digital oscilloscopes have capabilities to Fourier transform the waveform data using the FFT configured in hardware, firmware, or software.

6.5.3 DFT Leakage

There are some subtleties that we should understand when using any FFT program to evaluate a DFT. The first thing to always keep in mind is that we are working in a digital mode where the whole process resolves around quantized data and sometimes we can obtain results that may seem unexpected. For example, if we analyze a sinusoidal signal of frequency f_0, an FFT program will produce correct results only when the input data contain energy precisely at some integer multiple of the Δf, i.e., $f_0 = m\Delta f$, where m is an integer. We can examine this by considering a truncated cosine wave in which the frequency of the wave is an integer multiple of Δf. Since $\Delta f = \text{sr}/N$ and $N = 2^{\gamma}$, then $\text{sr}/N = f_0/m$. Let us consider an example of a cosine wave in which $\text{sr} = 7f_0$, $m = 3$, and $N = 32$. The cosine wave is shown in Fig. 6.14a, where the sample points are clearly denoted, and its spectrum is shown in Fig. 6.14b. What has essentially happened in this example is that the DFT "sees" this signal as if it was an infinitely long or continuous sinusoidal signal. We constructed this example so that $N = 32$ fit exactly an integral multiple of the period of the waveform. In this case, the FFT results are exactly zero except at the frequency of the cosine signal.

We will now consider a case where we sample the cosine signal in such a way that $f_0 \neq m\Delta f$ so that f_0 falls between a Δf interval. All we need to do is multiply our previous sampling rate by some number such as 1.2. We will obtain the following results shown in Fig. 6.15. In this case where the frequency of the input sequence falls between a DFT bin center, energy is leaked into all of the other frequency bins and this effect is referred to as *DFT*

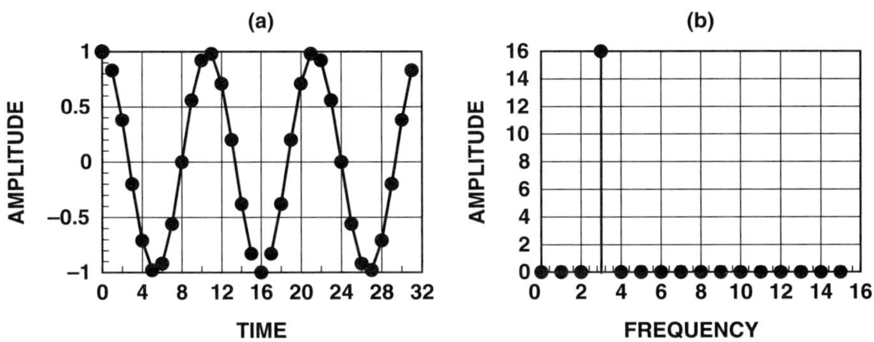

FIGURE 6.14. Cosine signal and its FFT for a case in which NΔt is equal to an integer multiple of the waveform period.

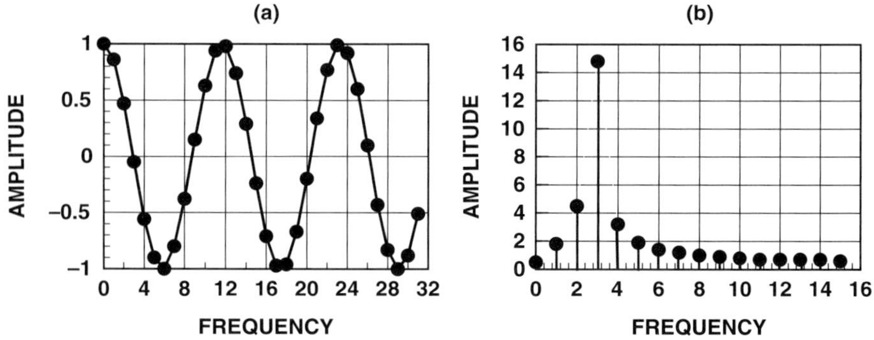

FIGURE 6.15. Cosine signal and its FFT for a case in which $N\Delta t$ is not equal to an integer multiple of the waveform period.

leakage or simply *leakage*. Leakage is an unavoidable consequence of working with long real-world signals that need to be truncated when performing an FFT operation. Therefore, in almost all situations, we must look upon the FFT results as a good approximation of the true frequency characteristics of our signal, but an approximation nevertheless.

The cause of leakage can also be explained in a slightly different but effective manner by considering what is being done when a signal is truncated as in Fig. 6.15. When we truncated the cosine signal, we essentially took an infinitely long signal and multiplied it by a rectangular window of length $N\Delta t$. For a signal starting at $t = 0$, a rectangular window has an amplitude of unity for $0 \leq t \leq N\Delta t$, and an amplitude of zero for all other values of t. The frequency convolution theorem expressed by Eq. (6.57) indicates that multiplication of two functions in the time domain is equivalent to a convolution of the Fourier transform of both functions in the frequency domain. From Fig. 6.10, we know that the Fourier transform of a rectangular function is a $\sin(x)/x$ or sinc function. The frequency convolution operation will be similar to the time convolution operation performed in Fig. 6.12. Therefore, truncating a signal in the time domain by a rectangular window will cause the spectrum of the signal to spread out in a $\sin(x)/x$ manner in the frequency domain and will include the effects of the side lobes shown in Fig. 6.12. Since in many situations truncation of signals cannot be avoided, then the issue becomes one of finding the best way to truncate signals in order to minimize the effects of the sampling window, which brings us to the next section.

6.5.4 FFT Windows

The effects of signal truncation and the resultant leakage in the FFT results can be minimized by applying a data window that forces the amplitude of the input time data sequence to smoothly approach zero at both the beginning and end of the data sequence. The rectangular window does not do this but abruptly truncates the input data at the beginning and end of the sequence. There are many different windowing functions that have been considered over time, all having some advantages and disadvantages over each other. A classic paper of different windows was published by Harris (1976), and the readers are encouraged to read this article for an in-depth discussion on the use of windows. In this book, we will consider just three commonly used windows, besides the rectangular window. These windows are the triangular window, the Hanning (also referred to as the Hann or raised cosine) window, and the Hamming window. The equations for these windows are as follows:

$$\textit{Rectangular window}: \; w(n) = 1, \text{for } n = 0, 1, 2, \ldots, N-1 \tag{6.82}$$

$$\textit{Triangular window}: \; w(n) = \begin{cases} \frac{n}{N/2} & \text{for } n = 0,\ 1,\ 2,\ \ldots N/2 \\ 2 - \frac{n}{N/2} & \text{for } n = N/2 + 1,\ \ldots N-1 \end{cases} \tag{6.83}$$

$$\textit{Hanning window}: w(n) = 0.5[1 - \cos(2\pi n/N)] \text{ for } n = 0, 1, 2, \ldots, N-1 \tag{6.84}$$

$$\textit{Hamming window}: w(n) = 0.54 - 0.46\cos(2\pi n/N) \text{ for } n = 0, 1, 2, \ldots, N-1 \tag{6.85}$$

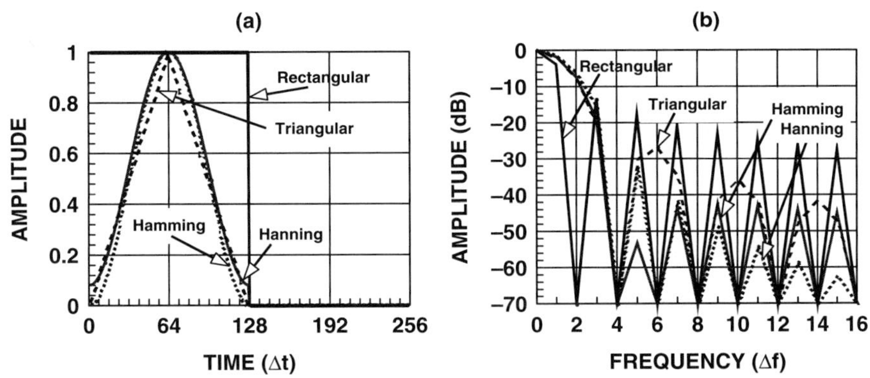

FIGURE 6.16. (**a**) Waveform of the rectangular, triangular, Hamming and Hanning windows, (**b**) spectrum of the various window functions.

The time domain shape of these windows are shown if Fig. 6.16a and their corresponding FFTs are shown in Fig. 6.16b.

The rectangular window is usually used as the measure by which other windows are compared. The rectangular window has the narrowest main lobe; however, it also has many side lobes of relatively high amplitudes, and the first side lobe is only –13 dB below the main lobe peak. The side lobes of the triangular window has lower level side lobes, but its main lobe is almost twice as wide as the main lobe for the rectangular window. As a matter of fact all of the non-rectangular windows have a wider main lobe than that of the rectangular window, making the frequency resolution of non-rectangular windows to be degraded when compared with the rectangular window. However, the important benefits of leakage reduction usually outweigh the loss in FFT resolution. The Hamming window rolls off more rapidly than the Hanning window for the first side lobe; however, the other side lobes do not maintain the rapid roll-off rates as does the Hanning window. Therefore, leakage three or four bins away from the center bin is lower for the Hanning window than the Hamming window. The choice of the specific window to use will definitely depend on the characteristics of the data being examined.

The discussion of FFT window will be helpful in understanding the issue of frequency resolution with computing an FFT. The frequency bin shown in Eq. (6.68) suggests that for a given sampling rate, frequency resolution can be improved by increasing N. This is a common mistake to think that by appending zeros (also known as zero padding) to increase the value of N, the resulting FFT will have greater resolution. The frequency resolution of an FFT will be determined by the main lobe of the windowing function. We saw in Fig. 6.16 that the rectangular window has the narrowest main lobe, but it also has very low side lobe characteristics. Therefore, in the words of Brigham (1988), "FFT resolution cannot be enhanced by appending zeros unless the function is in fact zero-valued over the interval where the zeroes are appended." However, zero-padding will provide smaller values of Δf, and this process is akin to interpolating in the frequency domain.

6.5.5 Digital Convolution

We stated early in this chapter that convolution has an important role in many areas of science and therefore it would be worthwhile to consider an example of evaluating the convolution integral for digital data. Symbolically, the convolution can be expressed as

$$y(m) = x(i) * h(j), \tag{6.86}$$

where $i = 0, 1, 2 \ldots, N_x, j = 0, 1, 2, \ldots, N_h$, and N_x need not be equal to N_h. We will now consider a simple but effective diagrammatic example similar to the graphical example of Fig. 6.12. In this simple example, we will convolve a

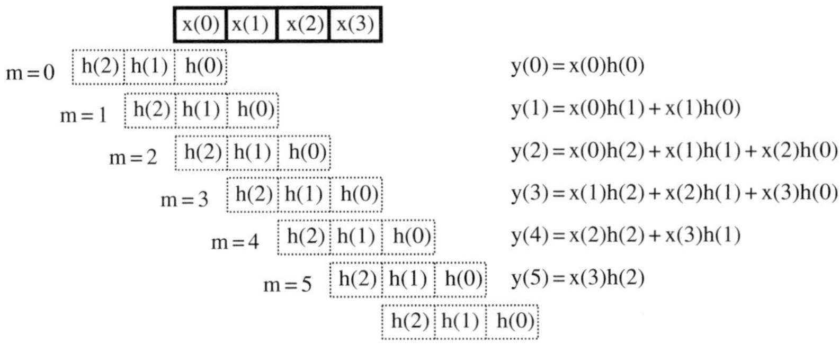

FIGURE 6.17. Diagrammatic example of the convolving digital sequences.

4-point sequence represent by $x(i)$, where $i = 0, 1, 2$, and 3, with a 3-point sequence $h(j-2)$, where $j = 0, 1, 2$. The sequence $x(i)$ is the digital equivalence of $x(\tau)$ in Eq. (6.52), and $h(j-2)$ is the digital equivalence of $h(t-\tau)$ in Eq. (6.52). The two sequences are shown in Fig. 6.17, with the $x(i)$ sequence being the array in the top row, and the $h(j)$ sequence represented by a row array that is being shifted to the right for different values of m. The sifting is similar to the shifting in Fig. 6.12 for different values of t_1. The $h(j)$ sequence starts at index $(j-N_h-1)$, with the index decreasing as j increases to represent the end-to-end flipping of the variable $h(t-\tau)$ of Eq. (6.52). For $m = 0$, $h(0)$ is directly below $x(0)$, and $y(m)$ is the product of $x(0)$ and $h(0)$. When m increases to 1, the $h(j-2)$ array shifts one slot to the right and the corresponding $y(m)$ is equal to the sum of the product of $x(0)h(1)$ and $x(1)h(0)$, and so on, for increasing values of m. This process is similar to the process depicted in Fig. 6.12 for the convolution of two continuous functions. From our digital example, we see that

$$m = 0, 1, 2, \ldots, N_x + N_h - 2. \tag{6.87}$$

The output sequence $y(m)$ of the convolution $x(i)$ and $h(j)$ is defined as

$$y(m) = \sum_{i=0}^{m} x(i)h(m-i) \qquad \begin{array}{rll} x(i) & = 0 & \text{for } i \geq N_x - 1 \\ h(m-i) & = 0 & \text{for } m - i \geq N_{hi} - 1 \\ m & = 0, 1, 2, 3, \ldots, & N_x + N_h - 2 \end{array} \tag{6.88}$$

The number of multiplications involved with evaluating the convolution integral using Eq. (6.88) is equal to the product $N_x N_h$, if the $x(k)$ terms involved with $k \geq N_x - 1$ and the $h(m-k)$ terms involved with $(m-k) \geq N_h - 1$ are ignored.

The convolution integral can also be evaluated by using the digital equivalent of Eq. (6.55), which involves obtaining the inverse FFT of the product of the FFT of $x(i)$ and the FFT of $h(i)$. We can express this process as

$$x(i) * h(i) \Leftrightarrow X(k)H(k), \tag{6.89}$$

where $i = 0, 1, \ldots, N-1$ and $k = 0, 1, 2, \ldots, N/2-1$. When the convolution integral is evaluated using the FFT technique, the number of points for both sequences must be equal. Equation 6.89 involves two FFT operations, each consisting of $N \log_2 N$ complex multiplications, $N/2$ complex multiplications in calculating the product $X(k)H(k)$, and an addition $N \log_2 N$ complex multiplication to calculate the inverse FFT of the product $X(k)H(k)$. Each complex multiplication consists of the four multiplication of real numbers so that the total number real multiplications required to evaluate Eq. (6.87) is equal to $12N \log_2 N + 2N$ or $2N(6 \log_2 N + 1)$.

Evaluating the convolution integral using Eq. (6.88) and Eq. (6.89) will produce identical results. Therefore, it is legitimate to ask which technique would be faster to evaluate. The answer will depend on the length of the $x(i)$ and $h(j)$ sequences in Eq. (6.86). Let us consider an example in which $h(j)$ has 1,024 points. Using the FFT technique of Eq. (6.87), the number of real multiplications required to evaluate the convolution integral will be $2N$ $(6 \log_2 N + 1)$ or 124,928 real multiplications and the number of real multiplications required with Eq. (6.88) is shown in Fig. 6.18 as a function of the number of points in the $x(i)$ sequence. In this example, the FFT technique becomes more efficient than the direct summation technique when $N_x = 128$, and the number of multiplication becomes much less with the FFT technique as N_x increases beyond 128 points.

We will now consider another important characteristic of the discrete convolution integral that arises from the periodic assumption used in deriving the discrete Fourier transform. We will start with the graphical example of

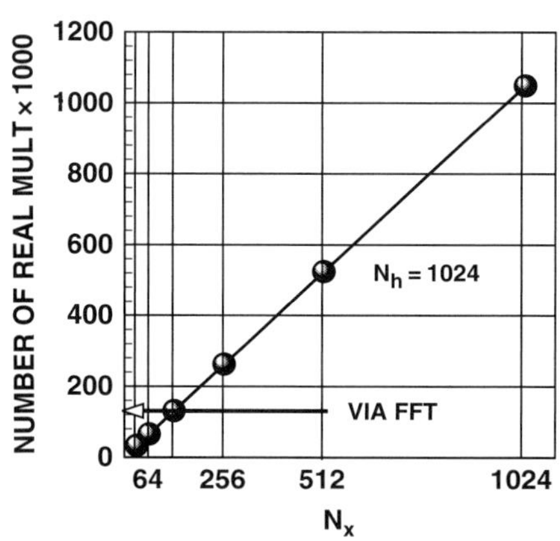

FIGURE 6.18. Number of real multiplications necessary to evaluating the convolution integral using Eq. (6.88), with $N_h = 1024$, for different values of N_x and with the FFT technique.

Fig. 6.12 outlining the steps involved in evaluating the continuous convolution integral of $s(t)$ and $h(t)$ shown in Fig. 6.12. For convenience, part of Fig. 6.12 is redrawn in Fig. 6.19a. The digital representation of $x(t)$ and $h(\tau - t)$ are shown in the left and middle graph of Fig. 16-19b as periodic function with a period of N points. The period is chosen as $N = 9$ and both $x(k)$ and $h(m-k)$ are represented by $P = Q = 6$ samples. The discrete convolution is shown in the right most graph of Fig. 6.19b. The discrete convolution result is also periodic, and this periodicity introduces an overlap in the desired periodic output. Basically, the convolution results of one period interferes with the results of another period and this phenomenon is commonly referred to as an *interference* or *overlap* affect. Therefore, in order for the discrete convolution to accurately approximate the continuous convolution, it is necessary to adjust the period of the signals so that the overlap will not occur. No overlap will occur if the period N is chosen according to the relationship

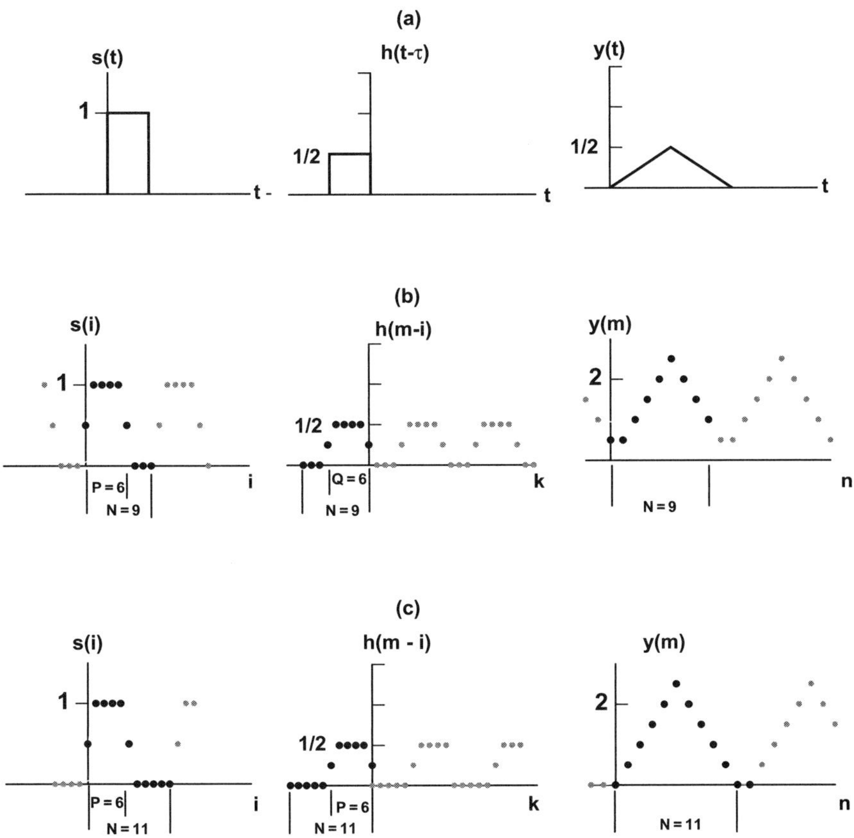

FIGURE 6.19. Relationship between continuous and discrete forms of the convolution integral for finite duration waveforms (adapted from Brigham, 1988).

$$N = P + Q - 1. \tag{6.90}$$

This situation is shown in Fig. 6.19c in which $N = P + Q - 1 = 11$, and there is no overlap in the convolution results shown on the right side of Fig. 6.19c. The length of P and Q can be made longer by zero-padding either of both $x(k)$ and $h(m-k)$. Making $N > P + Q - 1$ will only make the spacing between the discrete convolution results larger and there is no advantage to doing this. Brigham (1988) presented an example of the case in which a large N greater than required by Eq. (6.90) is used. The interference problem will occur for N less than that expressed in Eq. (6.90) regardless if the digital convolution is evaluated directly using Eq. (6.88) or if the FFT method expressed by Eq. (6.89) is used.

The periodicity inherent in the discrete convolution gives rise to what is commonly referred to as *circular convolution*. We will again take an example from Brigham (1988) to illustrate the idea of a circular convolution. Consider Fig. 6.20a that shows two discrete functions that will be convolved. On the left side of Fig. 6.20b is the function $h(m-k)$, in which $h(k)$ appropriately folded and aligned with the $s(k)$ function directly above it. On the right side of Fig. 6.20b is an alternate way of displacing the discrete convolution. The rings represent one period of the two periodic functions: the inner ring being h(m-k) and the outer ring being $s(k)$. The numerical values in each ring correspond to

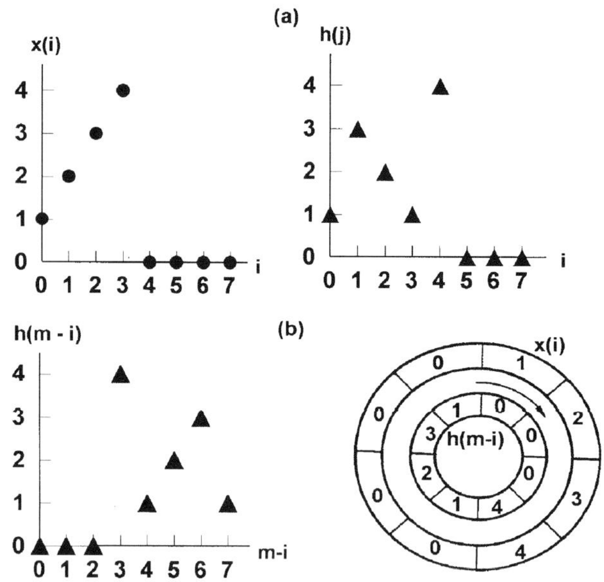

FIGURE 6.20. A graphical example of circular convolution (adapted from Brigham, 1988).

the values of the respective function in Fig. 6.20a, with $h(m-\mathrm{k})$ being the folded function. The appropriate values to be multiplied are adjacent to each other. These multiplied results are then summed around the circle over a single period to obtain a value of $y(m)$ in Eq. (6.88). The inner ring is turned for each shift of m and, as it turns it will eventually return to its original position every eight shifts. Therefore, the same values will be calculated as the inner ring is shifted, giving rise to the results corresponding to the periodic convolution discussed previously. There must be sufficient amount of zero's in either or both rings to avoid any overlapping or interference effects. Therefore, zero-padding of the data is usually recommended to avoid the overlap problem.

6.6 Some Signal Processing Applications

6.6.1 *Beamforming*

In Chapter 3 we discussed electronic beamforming using delay lines as depicted by Fig. 3.18. Each specific delay time τ is associated with a specific angle θ, the major axis of the beam makes with the normal to the array. The output voltage from the M different array element was given by Eq. (3.70) as

$$V_{\theta}(t) = \sum_{m=0}^{M-1} v_m(t - m\tau). \tag{6.91}$$

Beamforming using delay lines is fairly cumbersome as the number of sensor is large. Fortunately, a digital method can be used in beamforming. From the time-shifting property of the Fourier transform expressed by Eq. (6.48), a delay in time is equal to a multiplication of the spectrum by $e^{-j2\pi f\tau}$ so that Fourier transform of Eq. (6.91) can be expressed as

$$V_{\theta}(f) = \sum_{m=0}^{M-1} V_m(f)\, e^{-j2\pi f m\tau}. \tag{6.92}$$

$V_{\theta}(f)$ is the frequency domain relationship for combining the M element outputs with the appropriate delay. The inverse Fourier transform will yield the output voltage $V_{\theta}(t)$ of Eq. (6.92).

The voltage of Eq. (6.92) is valid for a specific value of θ, since from Eq. (3.68), $\tau = (d/c)\sin\theta$, where $d =$ the distance between each element. However, Eq. (6.92) can be used over again for different values of θ which will result in different values for τ. Therefore, the major axis of the beam of an M-element array of hydrophones can be made sweep in angle θ, simulating a planar transducer being mechanically scanned over different values of θ.

If we desire to simultaneously implement the correct delays to combine the M elements outputs for various angles θ_I, we can do so by defining increments of τ_I such that

$$\tau_i = i\left(\frac{d}{M}\right) \qquad i = 0, 1, 2, \ldots, M-1. \tag{6.93}$$

The azimuth angle of the major axis corresponding to each delay τ_I is given by equation

$$\theta_i = \sin^{-1}\left(\frac{\tau_i c}{d}\right) = \sin^{-1}\left(\frac{ic}{M}\right). \tag{6.94}$$

Equations 6.91 can now be written as

$$V_{\theta_i} = \sum_{m=0}^{M-1} V_m(f)\, e^{-2j\pi f(\mathrm{mid}/M)}. \tag{6.95}$$

In Eq. (6.94), an FFT operation is applied to the output of each hydrophone element, with the results multiplied by $e^{-2j\pi f(\mathrm{mid}/M)}$, and finally a summation over the number of element is determined for each azimuth angle θ_i.

6.6.2 *Measuring Ambient Underwater Noise*

The first and foremost principle in measuring ambient noise is to use a hydrophone pre-amplifier combination with a noise floor that is lower than the ambient noise to be measured. As obvious as this principle seems, many researchers have ignored this precaution by using a relatively insensitive but very broadband hydrophone, and have unwittingly measured the noise of the amplifier rather than the ambient noise. If you consider Figs. 6.8 and 6.9, the noise floor of the H-56 hydrophone is about 12 dB below the noise at sea state 0 and would be an appropriate sensor to measure the ambient noise of an indoor pool or a still body of water. However, its frequency range is limited to about 100 kHz. The H-52 in Fig. 6.9 has a much broader bandwidth; however, its noise floor is above the ambient noise typical of sea state 0 and could be used only in situations in which the background noise is rather high, at least 7-10 dB higher than that of sea state 0.

There are several ways in which underwater noise or electronic noise can be measured. Perhaps, the simplest method is to use a spectrum analyzer, which could be a swept frequency instrument or a digital instrument with a built-in FFT analyzer. The major disadvantage of spectrum analyzers is the high cost of such instruments. Perhaps, the most difficult method is to use a variable frequency bandpass filter and measure the rms voltage at the output of the filter for different filter bands. This method is slow and time-consuming.

Another method involves digitizing the noise signal and performing an FFT operation on a PC. On the surface, this technique appears to be fairly straightforward; however, there is an important issue that needs to be considered. This issue has to do with the time over which the spectrum of the noise should be determined. For example, if we are interested in the noise spectrum out to 100 kHz, then our sampling rate must be at least 200 kHz. Assume we actually use a sampling rate that is five times higher than the maximum frequency of interest, so in this example we choose a sampling rate of 500 kHz. That means 1,024 points represent only 2 ms, and 2,048 points represent only 4 ms, and so on. In order to have a time duration of several seconds, the number of points that needs to be sampled will be fairly large and it may be difficult or impossible to calculate the FFT for such large number of points. For instance, 1,024,000 points will represent only 2 s and 2,048,000 points will represent only 4 s worth of data.

Instead of performing an FFT on a very large data array, we can use an averaging process in which the spectrum from consecutive blocks are summed together and divided by the number of block to obtain an averaged spectrum. Each block of digitized data is multiplied by a data window as discussed in Section 6.6.3 before performing an FFT to that block of data. If we let $S_i(k)$ be the FFT of the i-th block of $N-1$ digitized noise voltages, where $k = 0, 1, \ldots, N-1$, then the averaged noise intensity spectrum can be expressed as

$$< N^2(k) > = \frac{|S_1(k)|^2 + |S_2(k)|^2 + |S_3(k)|^2 + \cdots |S_m(k)|^2}{m}, \tag{6.96}$$

where $k = 0, 1, \ldots, N-1$. If the averaged noise density is desired, then Eq. (6.96) should be divided by the bin size or $\Delta f = \mathrm{sr}/N$. If the spectral value for each block is desired, then we would need to "calibrate" the FFT process. One way of doing this is to input N points of a 1 v rms sinusoidal signal that is windowed and then determine the peak values of $|S(k)|$. This peak value should be equivalent to 1 v rms and can be used as a multiplying factor to determine the voltage level of $|S(k)|$ for other signals.

One is often interested in a one-third octave or a one-octave band (or any other band) representation of the noise spectrum. In order to calculate the level within a specific frequency band, we need to sum the energy associated with the energy of each FFT frequency bin that falls within that band. For example, if FFT frequency bins m to $m + 5$ are within the band of interest, then the energy associated with that band will be

$$I_n = |S_1|^2 + |S_2|^2 + \cdots + |S_{m+5}|^2. \tag{6.97}$$

We recommend the use of the preferred American National Standards center frequencies for octave bands of 31.5, 63, 125, 250 Hz, ..., and so on until 16 kHz. The preferred center frequencies then become 31.5, 63, 125 kHz, and so on until 160 kHz. The preferred center frequencies for 1/3 octave bands are 100, 125, 160, 200, 250, 315, 400, 500, 630, and 800 Hz. The next 10-to-1 set starts with 1 kHz and continues by multiplying the first set by 10, 100, and so on. In all likelihood, the first and last FFT bin corresponding to a particular frequency band will not match up with the lower or upper frequency boundary of a band. In such cases, the value of the FFT spectrum for the specific frequency of either the lower or upper frequency can be found by a linear interpolation or some other higher-order interpolation process.

Another consideration that should be taken into account is the lowest center frequency of a one-third or one-octave band representation and the associated bandwidth. From Eq. (5.2) and the definition of the center frequency ($f_0 = \sqrt{f_l f_u}$), the relationship between the center frequency and the width of any band can be expressed as

$$\delta f = f_0 \left(\frac{2^n - 1}{\sqrt{2^n}} \right). \tag{6.98}$$

Since the FFT bin width is a equal to $N\Delta t$, where N is the number of points in a block and Δt is the sample time, the appropriate combination of N and Δt should be chosen so that the FFT bin width is smaller than the bandwidth associated with the lowest center frequency of interest given by Eq. (6.98).

6.6.3 *Time Difference of Arrival Measurements*

In Section 5.7.7 of the previous chapter, we discussed the use of the correlation function as a way to determine the differences in the arrival time of a signal to each element of a hydrophone array. The digital form of the correlation function expressed by Eq. (5-60) is reproduced here in Eq. (6.99) as

$$c(j) = \sum_{i=1}^{n} s_1(i)\, s_2(i + j), \tag{6.99}$$

where $j = 1, 2, \ldots, n$, and the number of samples $s2$ is being delayed. The cross-correlation function is determined for a number of j values and the best time estimate of the delay is at the j-values at which the correlation function is maximum. Equation 6.99 is similar to the digital form of the convolution function of Eq. (6.88) except that $h(m - i)$ is replaced by $s_2(i + j)$ and the length of each signal is usually the same. Figure 6.18 showing the relative number of multiplications necessary to evaluate the convolution integral directly and via the FFT method is applicable in regards to evaluating Eq. (6.98) directly or via the FFT method. Therefore, from Fig. 6.18, one can see that a large saving in computational time can be realized by using the FFT method as expressed by Eq. (6.62).

6.6.4 *Reduction of Noise*

In the process of detecting a narrow band signal in noise, proper application of the FFT can lead to an improvement of the signal-to-noise ratio. Consider the situation depicted by Fig. 6.21a in which a narrow band signal is immerse in broadband noise. The signal-to-noise ratio in this example is –12 dB, making it impossible to detect the signal by examining the time-domain data. If we perform an FFT on the time-domain data, we can expect the sinusoidal signal to be contained within a few bins around the frequency of the signal and for the noise to be spread out across the many frequency bins, thus improving the signal-to-noise ratio. The FFT of a 64 point sample of Fig. 6.21a is shown in Fig. 6.21b. The spectrum is represented in term of the power in the signal. Although a signal-to-noise ratio is much improved in the frequency domain, the signal is still not clearly discernable. However, if the FFT of 512 points is performed, then we obtain the results shown in Fig. 6.21c. The signal is now very obvious in Fig. 6.21c, and thus a gain in the signal-to-noise ratio has been achieved.

The improved signal-to-noise ratio was achieved by evenly spreading out the noise signal into 256 bins. Equivalently, the wideband noise has been filtered by 256 contiguous filters and the noise power contained in the same bin as the signal has been reduced by 1/256 relative to the broadband noise

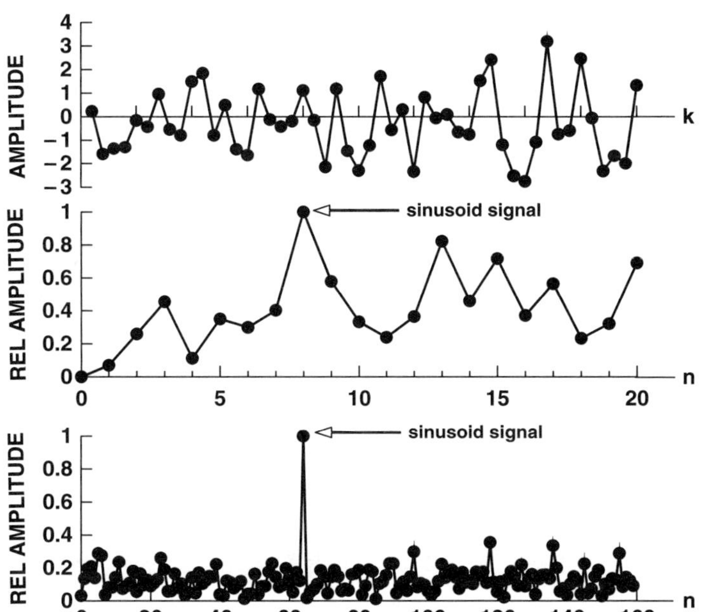

FIGURE 6.21. Example of a narrow band signal buried in noise, where S/N = -12 dB: (**a**) time-domain representation, (**b**) FFT spectrum for N = 64 and (c) FFT spectrum for N = 512.

level. Thus the noise power in a single bin is approximately –24 dB of the broadband noise power. The processing performing an FFT on increasing the number of samples has its limits. Besides the issue of computer memory, as the FFT bins become narrower, the signal may begin to spread out into several contiguous bins and therefore the gain in the signal-to-noise in these bins will be limited.

Another FFT technique that can be used to improve signal-to-noise ratio is the process of averaging successive FFT power spectra. The averaging process will smoothen out and reduce the wild amplitude fluctuations that are typical of noise.

The frequency domain representation of a narrow band signal buried in noise is shown in Fig. 6.22a. The signal is not detectable in the spectrum of 512 points. However, if the noisy spectra are averaged, the signal-to-noise ratio will show a great improvement. Fig. 6.22b shows the averaged results of 64 spectrum of the 512 point FFT results. The noise has been smoothened considerably and the signal is clearly visible above the smoothened noise. Since the power spectra were averaged in 6.23b, the phase of the signal is unknown. The amount of signal-to-noise enhancement for an original S/N of –30 dB or less is approximately $1.5 \log_2 Q$ dB, where Q is the number of successive FFTs that are averaged (Bracewell, 1978). The original spectrum in Fig. 6.23a was –24 dB and the number of successive FFTs average was $64 = 2^6$; therefore, the enhancement due to averaging is $1.5 \log_2 2^6 = 9$ dB. From the previous example, we found that for $N = 512$, there is a 24 dB gain in signal-to-noise over the original time-domain signal. Hence, the processed signal-to-noise ratio is $24 + 9 - 24 = 9$ dB.

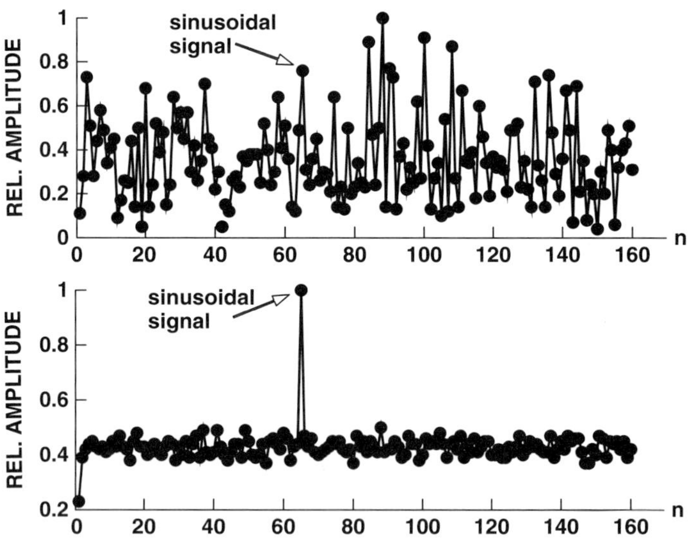

FIGURE 6.22. Example of a signal buried in noise, where S/N = –24 dB: (**a**) FFT spectrum for N = 512, and (**b**) averaged spectrum for 64 successive FFT's with N = 512.

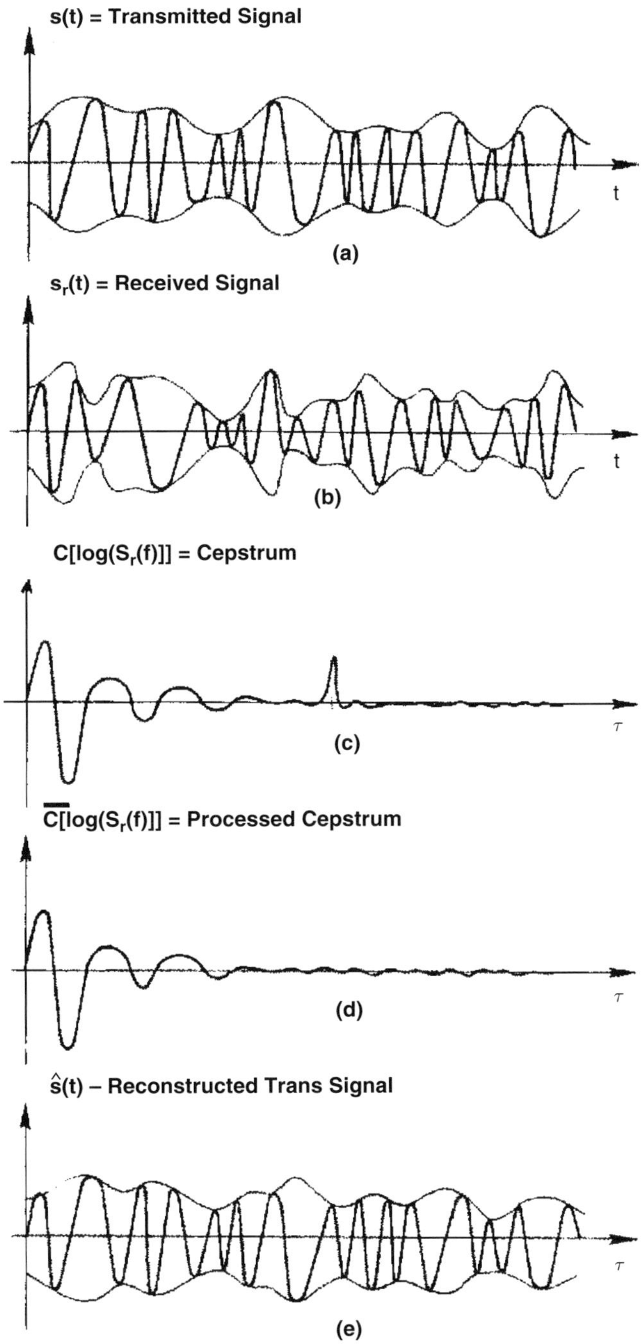

FIGURE 6.23. Example of using a cepstrum process to remove a multipath component in a received signal (from Bingham, 1988).

6.6.5 Ceptrum Analysis: Removal of Multi-paths

Multi-paths occur often in the propagation of underwater acoustic energy. If a hydrophone is close to the surface, then in all likelihood there will be a surface-reflected component in the received signal. There is a technique referred to as the cepstrum signal processing technique that can be useful in detecting the presence of multi-path components in a signal and the eventual removal of the multi-path. The development in this section follows after the presentation of Brigham (1988) who in turn referenced Kemerait and Chiders (1972), Bogert et al. (1963).

Assume that a received signal $s_r(t)$ is given by the expression

$$s_r(t) = s(t) + a_0 s(t + \tau_l), \tag{6.100}$$

where $s(t)$ is the transmitted or direct signal, $s(t + \tau_l)$ is a multi-path component and a_0 is the relative attenuation between the direct and the reflected signal. We perform a Fourier transform on the signal of Eq. (6.100) using the shift theorem as expressed by Eq. (6.68) and then take the natural logarithm of the result to get

$$\begin{aligned} \ln S_r(f) &= \ln[S(f) + a_0 S(f)\, e^{j2\pi f_l \tau}] = \ln[S(f)\,(1 + a_0 e^{j2\pi f \tau_l})] \\ &= \ln S(f) + \ln(1 + a_0 e^{j2\pi f \tau_l}) \end{aligned}, \tag{6.101}$$

where $S(f)$ is the Fourier transform of $s(t)$. The last term of Eq. (6.101) can be rewritten in a series form since

$$\ln(1 + x) = x - \frac{x^2}{2} + \frac{x^3}{3} - \cdots = \sum_{n=0}^{\infty} (-1)^{n-1} \frac{x^n}{n}. \tag{6.102}$$

Substituting Eq. 6.102 into Eq. 6.101, we have

$$\ln S_r(f) = \ln S(f) - \sum_{n=0}^{\infty} (-1)^n \frac{a_0^n}{n} e^{j2\pi f n \tau_l}. \tag{6.103}$$

Now we take the Fourier transform of the ln (spectrum) to obtain

$$C(\tau) = \Im[\ln S_r(f)] = \Im[\ln S(f)] - \sum_{n=0}^{\infty} (-1)^n \frac{a_0^n}{n} \delta(\tau - n\tau_l), \tag{6.104}$$

where the knowledge that the Fourier transform of $e^j 2\pi f n t$ is a delta function is used to obtain the last term of Eq. 6.104. The Fourier transform of the ln of the spectrum is called the *cepstrum*. The first term on the right of the equal sign of Eq. 6.104, $\Im[\ln S_r(f)]$, is the cepstrum of the direct signal and the remaining term is a sequence of impulse functions. Therefore, a cepstrum analysis transforms the multi-path component of the signal into a series of evenly spaced impulse function that can be removed to eventually produce the direct signal by itself.

We will now present an example taken from Brigham (1988) showing the detection and removal of a multi-path component of a signal. Assume a received signal that is similar to the signal described in Eq. 6.100. The signal processing chain consists of the following steps, starting with the signal $s_r(t)$

$$\Im[s_r(t)] \rightarrow \ln() \rightarrow \Im() \rightarrow \text{multi-path removal} \rightarrow \Im^{-1}() \rightarrow e() \rightarrow \Im^{-1}(). \quad (6.105)$$

The logarithm is applied to both the real and imaginary parts of the FFT output. However, the complex FFT output, $[R(f) + j\,I(f)]$ can be expressed in a polar form so that

$$\ln(|S_r(f)|e^{i\theta}) = \ln(|S_r(f)|) + j\theta(f). \quad (6.106)$$

$$\text{where } |S_r(f)| = \sqrt{(R^2(f) + I^2(f)} \text{ and } \theta(f) = \tan^{-1}\left(\frac{I(f)}{R(f)}\right). \quad (6.107)$$

Now the real part of the ln operation is placed in the real part of the FFT and the imaginary part is placed in the imaginary part of the FFT. The FFT operation will result in the *quefrequency* series of Eq.6.107. The impulse response function of Eq. 6.107 will be a $\sin(\tau)/\tau$ function because of the finite duration waveform used in any FFT procedure. Function represents a multi-component that can now be filtered or subtracted out.

An example applying the FFT cepstrum processing procedure for multi-path removal, as outlined by Eq.6.104, is shown in Fig. 6.23. The transmitted signal is shown in Fig. 6.23a, and the received multi-path signal is shown in Fig. 6.23b. The cepstrum analysis of the multi-path signal is shown in Fig. 6.23c. The $\sin(\tau)/\tau$ function at τ_1 is the undesirable reflected component of the signal, which we set to zero. By sequentially applying an inverse FFT, an antilogarithm and another inverse FFT operation, the transmitted signal, can be recovered as shown in Fig. 6.23c.

6.6.6 Digital Equalization of Planar Transducers

Some planar transducers have a relatively broad resonance peak so that the transmit sensitivity does not vary much over approximately half an octave on both sides of the peak. Such a transducer when used as a projector can be digitally equalized to essentially have a relatively flat transmit sensitivity over an octave bandwidth. The equalization process is initialed by feeding a broadband signal $E(f)$ into a projector having an unknown transfer function $T(f)$ resulting in a signal $P(f)$ being projected into the water and received by a broadband hydrophone. The process can be described as

$$E(f)T(f) = P(f). \tag{6.108}$$

If $E(f)$ is a broadband signal with an octave bandwidth, the projected signal received by the hydrophone will in all likelihood not resemble $E(f)$. An equalization filter $G(f)$ can be inserted into the transmission system preceding the projector so that the received hydrophone signal will have the same shape as the original signal. If we let the received hydrophone signal be defined as $U(f)$, we now have

$$E(f)G(f)T(f) = G(f)Pf) = U(f), \tag{6.109}$$

and now $U(f)$ will resemble $E(f)$. The equalization filter is equal to the reciprocal of the transfer function of the transducer,

$$G(f) = \frac{1}{T(f)}. \tag{6.110}$$

The response of the equalization filter can be determined by taking the signal $E(f)$ and dividing by the unequalized received signal $P(f)$, or

$$G(f) = \frac{E(f)}{P(f)e^{j2\pi f\tau}}, \tag{6.111}$$

where the $e^{j2\pi f\tau}$ term comes from the delay-induced phase change as given by Eq. (6.48).

Au and Floyd (1979) presented several examples of the effectiveness of digital equalization of a transducer. They used a sample rate of 500 kHz to generate and project the signals shown in Fig. 6.24. An initial time delay was calculated by determining the time at which the envelope of the cross-correlation function between the input and the received signal reached its peak value. This time may not necessarily correspond to the actual time delay since the actual time delay may fall between a pair of sample points. An improved estimate of the time delay was obtained by fitting a parabola in a least-square error manner through five points about the peak of the cross-correlation function. Once the appropriate delay was found, the equalization filter response was determined by using Eq. 6.111. The prefiltered signals shown in the two examples of Fig. 6.24 are merely the inverse Fourier transform of the original signal $E(f)$ multiplied by the transfer function of the equalization filter given by Eq. 6.110.

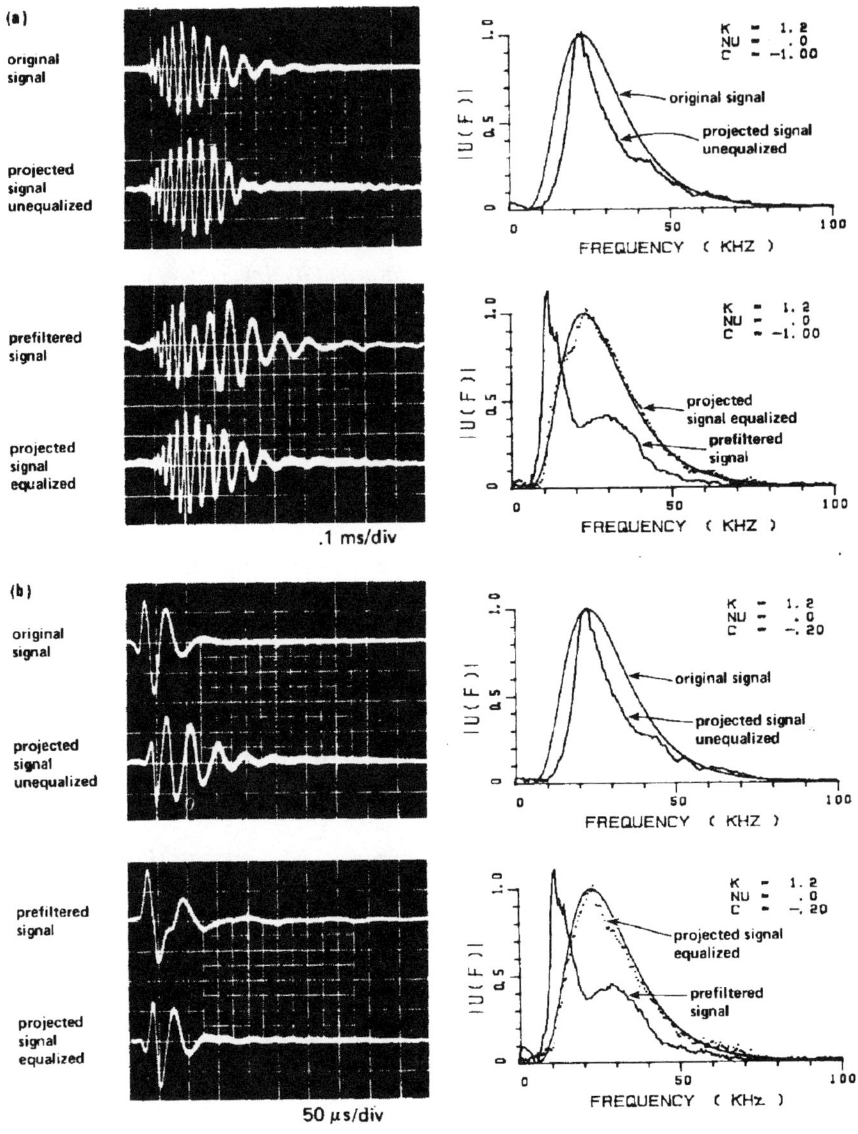

FIGURE 6.24. Examples of the equalization process of a underwater projector.

The equalization process effectively "flattens" out the transducer transmit sensitivity curve. The amount of improvement will be limited to the type of signals that is used. Longer signals with slowly varying instantaneous frequency will function better than short "transient-like" signals. Equalization will also reduce the effective transmit sensitivity of a projector so that the maximum output of the system can be limited. Finally, equalization will work best for transducers with the flattest transmit response.

References

Au, W. W. L. and Floyd, R. W. (1979). "Digital equalization of underwater transducers for the projection of broadband acoustic signals," J. Acoust. Soc. Am. **65**, 1585–1588.

Bogert, B. P., Healy, M. J., and Tukey, J. W. (1963). "The quefrequency analysis of time series for echoes: cepstrum, pseudo autocovariance, cross spectrum and Saphe cracking," in *Time Series Symposium*, M. Rosenblatt, ed. (Wiley, New York), pp. 201–243.

Brigham, E. O. (1988). *The Fast Fourier Transform and Its Applications* (Prentice Hall, Englewood Cliffs, New Jersey).

Bracewell, R. M. (1976). *The Fourier Transform and Its Application* (McGraw-Hill, New York).

Bracewell, R. N. (1978). *The Fourier Transform and its Applications (2nd Edition).* McGraw-Hill, N.Y.

Cooley, J. W. and Tukey, J. W. (1965). "An algorithm for the machine calculation of complex Fourier series," Math. Of Comput., **19**, 297–301.

Cooley, J. W., Lewis, P. A. W., and Welch, P. D. (1967), "Historical notes on the fast Fourier transform," Proc. IEEE, **55**, 1676–1677.

Harris, F. J. (1976). "On the use of windows for harmonic analysis with the discrete Fourier Transform," Proc. IEEE, **66**, 51–83.

Kemerait, R. C., and Childers, D. G. (1972). "Signal detection and extraction by cepstrum techniques." IEEE Trans. Info. Theory, **IT-18**, 745–759.

Kraniauskas, P. (1992). *Transforms in Signals and Systems* (Addison-Wesley,Wokingham, England).

Lyons, R. (1997). *Understanding Digital Signal Processing* (Addison-Wesley, Workingham, England).

Papoulis, A. (1984). *The Fourier Integral and Its Applications*, 2 nd ed. (McGraw-Hill, New York).

Ramirez, R. W. (1985). *The FFT Fundamentals and Concepts* (Prentice Hall, Englewood Cliffs, New Jersey).

7

Auditory Systems of Marine Animals

One of the most important components of bioacoustics has to do with the understanding of hearing and auditory processes in marine organisms. However, to consider hearing in marine animals, it is useful to first discuss the ear of marine mammals. The human auditory system will be used as a model in our discussion of marine mammals, and we will point out specific anatomical structures and characteristics that are different from humans. Again, the discussion of the human auditory system will be brief. There are volumes written on the human auditory system. Two excellent books by researchers that have spent their careers performing psychoacoustic measurements and delving into the complexities of the human auditory system are *An Introduction to Hearing* by Green (1976) and *Fundamentals of Hearing: An Introduction* by Yost (1994).

7.1 Structure of the Human Ear

The human ear consists of three parts: the outer ear, middle ear, and inner ear. A diagram depicting the three parts of the human ear is shown in Fig. 7.1. The outer ear consists of the pinna and the external auditory meatus or canal that leads to the eardrum, or tympanic membrane. The pinna serves as a horn to receive acoustic energy and lead it to the auditory canal. The auditory canal is almost a straight tube that can resonate and cause the acoustic pressure at the eardrum to be higher than at the entrance. A tube closed at one end and open at the other will behave like a quarter-wavelength resonator. The length of the external auditory canal ranges from 2.3 to 2.9 cm, so the wavelength at the resonant frequency will be about 9–12 cm and the resonant frequency will be about 2.7–3.6 kHz. However, the external auditory canal is not a simple tube and has nonrigid walls. The typical resonance frequency of the auditory canal is approximately 3–4 kHz and, at this frequency, the acoustic pressure at the eardrum can be about 12 dB higher than at the entrance to the canal. Since the auditory canal terminates at the compliant tympanic membrane and since the walls of the canal absorb some energy, the resonance is not sharp but quite broad and appreciable gain can be observed between 2 and 6 kHz.

W.W.L. Au, M.C. Hastings, *Principles of Marine Bioacoustics*,
DOI: 10.1007/978-0-387-78365-9_7,

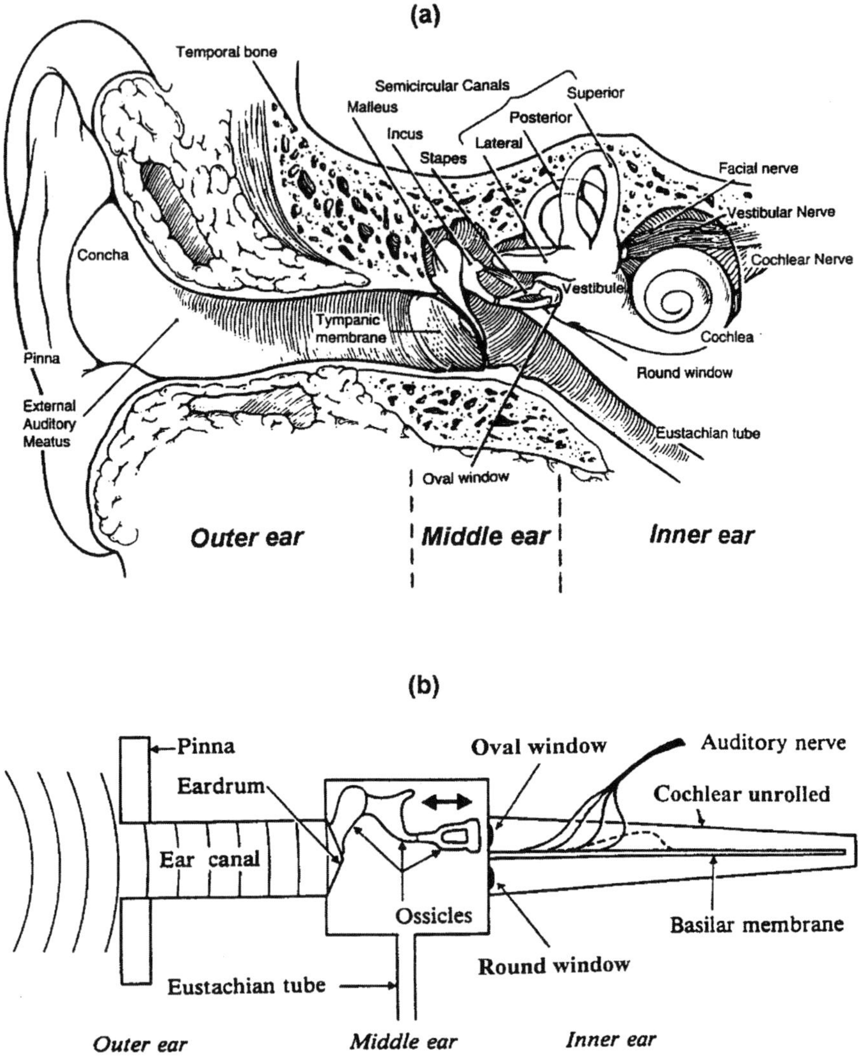

FIGURE 7.1. (**a**) Cross-sectional viewof the human ear, showing the outer, middle, and inner ears, (after Kessel and Kardon, 1979), and (**b**) schematic representation of the ear (after Rossing, 1990).

7.1.1 The Middle Ear

The middle ear starts at the tympanic membrane, which is shaped like a flat cone, lying slightly oblique across the auditory canal with its apex facing inward. Inside of the tympanic membrane is the middle ear cavity, which contains the three middle ear bones, or ossicles: the *malleus* (hammer), the *incus* (anvil), and the *stapes* (stirrup), together with supporting muscles and ligaments. The tympanic membrane is connected to the malleus, which is connected to the incus, which in turn is connected to the stapes. The footplate of the stapes rests on the oval window of the fluid filled cochlea.

Sound enters the ear through the external auditory canal, causing the tympanic membrane to vibrate. The vibration is transmitted via the malleus, incus, and stapes, causing the footplate of the stapes to exert force on the fluid of the inner ear by causing the oval window to vibrate. Figure 7.1b is a schematic of the ear receiving sound at the pinna, which is eventually transmitted to the cochlea via the ear canal, tympanic membrane, ossicles, and the oval window. Therefore, the ossicular chain vibrates the oval window membrane, which causes the fluid in the cochlear to move. In essence, acoustic energy in air has to be transmitted into a fluid medium; however, because of the large impedance difference between air and cochlear fluid (which can be approximated by salt water), the transmission process will be very inefficient. The energy transmission coefficient T for an acoustic signal propagating from one medium into another medium is given by the square of Eq. (4.9)

$$T = \frac{4(\rho_2 c_2/\rho_1 c_1)}{(1 + \rho_2 c_2/\rho_1 c_1)^2}, \tag{7.1}$$

where $\rho_2 c_2$ is the acoustic impedance of sea water, and is equal to approximately 1.5×10^6 Pa-s/m, $\rho_1 c_1$ is the acoustic impedance of air and is equal to 416 Pa-s/m. Inserting the values of the corresponding acoustic impedance, we obtain an extremely low transmission coefficient of 0.001, or a transmission loss of 30 dB. This means that only a small fraction of 0.1% of the acoustic energy in air will be transmitted into water and 99.9% of the energy will be reflected at the interface. Although the calculation just performed is really for an ideal situation of a plane acoustic wave impinging on an infinite boundary, it does indicate that there is a large impedance mismatch involved with the process of sounds entering the auditory canal being coupled into the inner ear. The actual impedances (ratio of pressure to volume velocity) of the outer and inner ear are complex numbers with their magnitudes varying with frequency.

The problem concerning a mismatch of impedances is beautifully compensated for by the ossicles of the middle ear. If we let $F_T = p_T A_T$ be the force, pressure, and the effective area of the tympanic membrane, and assume that this force is transmitted to the oval window via the ossicular chain without loss, then the force at the oval window will be $F_T = p_0 A_0$. Therefore, the ratio of the pressure at the oval window and at tympanic membrane (p_0/p_T) will be equal to A_T/A_0. Measurements have shown that only two-thirds out of a total area of 75 mm^2 of the tympanic membrane is stiffly connected to the malleus so that the effective area of the tympanic membrane is approximately 55 mm^2. Since the area of the footplate of the stapes is about 3.2 mm^2, there is a pressure amplification of 17 or 25 dB between the tympanic membrane and the stapes. The ossicles also act as a lever system. Because the length of the malleus is about 1.3 times longer than the long process of the incus, the force at the tympanic membrane is multiplied by a factor of 1.3 or 2 dB at the stapes. In addition, the tympanic membrane tends to buckle as it moves, causing the malleus to move with about twice the force, adding another 6 dB of gain. Therefore, the 30 dB of loss due to the impedance mismatch is compensated for by a gain of about 33 dB generated by the action of the ossicles. An example of the actual measured pressure gain in a cat is shown in Fig. 7.2. In humans, the pressure between the tympanic membrane and the inner ear increases steadily from 20 Hz to a peak of about 31 dB at 2.5 kHz and decreases to about 10 dB at 10 kHz.

There are two important muscles in the middle ear, the *tensor tympani* and the *stapedius* muscles. Very loud sounds will cause the tensor tympani, which is attached to the malleus just above the point where the manubrium or handle of the malleus joins the tympanic membrane, to tighten the eardrum. At the same time the stapedius muscles, which runs from the posterior wall of the middle ear to the neck of the stapes, will pull the stapes away from the oval window of the cochlea. Activation of these muscles provide about 20 dB or more of protection from very loud sounds. This response to loud sounds, called the *acoustic reflex*, operates between 60 and 120 ms after the sounds arrive at the ear, and therefore is ineffective to impulsive sounds of short duration such as gunshots. Although acoustic reflex is present and provides protection to the ear, hearing loss caused by noise in our environment seems to indicate that too often this protection is not sufficient.

The middle ear cavity is also connected to the *eustachian tube*, which leads to the throat. The eustachian tube allows for equalization of the pressure on the tympanic membrane. If the middle ear cavity was a closed chamber,

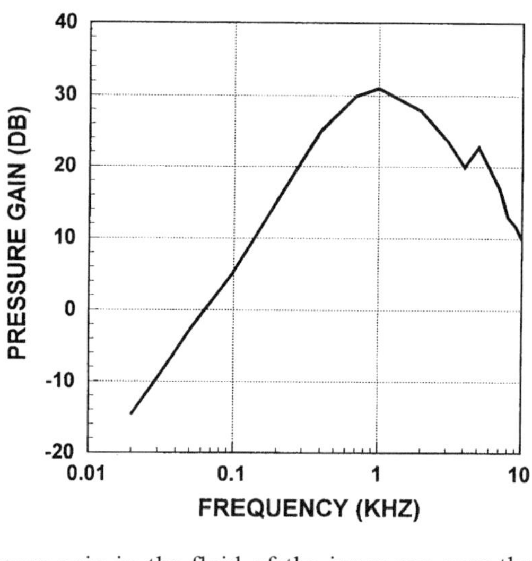

FIGURE 7.2. Pressure gain in the fluid of the inner ear over that at the tympanic membrane as a function of frequency for a cat (after Nedzelnitsky, 1970).

differences in the pressure on both sides of the tympanic membrane will develop when we experience a change in altitude, like in flying in an airplane or diving underwater. This pressure differential will cause the tympanic membrane to move in one direction more than in the other, and if this pressure differential continues to increase we will experience discomfort, pain, and finally damage to our eardrum. However, with the presence of the eustachian tube we can equalize the pressure on both sides of the eardrum by swallowing or blowing with both our nostrils or mouth closed.

There are two other ways in which sounds can be transmitted to the inner ear without going through the ossicles of the middle ear. The inner ear can be stimulated directly by acoustic energy flowing through the middle ear cavity if the ossicles were missing or not functioning. However, the large impedance difference between air and the fluid of the inner ear will make this method highly inefficient. Furthermore, acoustic pressure applied simultaneously on both the oval and round window will inhibit the movement of the basilar membrane, as we shall see in our discussion of the mechanics of the inner ear. Patients with no ossicles typically have a hearing loss of about 60 dB.

Sounds can also enter the inner ear by bone conduction, in which acoustic pressure changes cause the bones of the head to vibrate. This vibration can be transmitted directly into the inner ear. However, there is a large impedance difference between bone and air so that this mode of hearing is not an important part of normal auditory function. Bone conduction can be important if a vibrator is applied to the skull or other parts of the body. In this case, the impedance difference between bone and the fluid of the inner ear is small enough to allow some sounds to be transmitted into the inner ear.

7.1.2 The Inner Ear

The inner ear is divided into three parts: the *semicircular canals*, the *vestibule*, and the *cochlea*, and is found deep within a very dense portion of the skull known as the petrous portion of the temporal bone. A schematic drawing of the labyrinth structure of the inner ear is shown in Fig. 7.3a. The three semicircular canals are not associated with hearing, but affect our sense of balance. The vestibule is the central inner ear cavity and is bounded by the *oval window* on its lateral side. The cochlea is the small snail-shaped structure, which contains the primary auditory organ.

The cochlea is a small, coiled tube of approximately 2–5/7 turns, with a length of only 3.5 cm. Its wall is made up of the hardest bone in the body. Extending along the length of the tube, near the center, are two membranes, the *Reissner's membrane* and the *basilar membrane*, which divides the cochlea into three distinct chambers, as can be seen in the cross-sectional drawing of Fig. 7.3b. The largest chamber is the *scala vestibuli*, the next largest chamber is the *scala tympani*, and the smallest chamber is the *scala media* or cochlear duct, sandwiched between the two larger chambers. The scala media make up less than 10% of the total volume

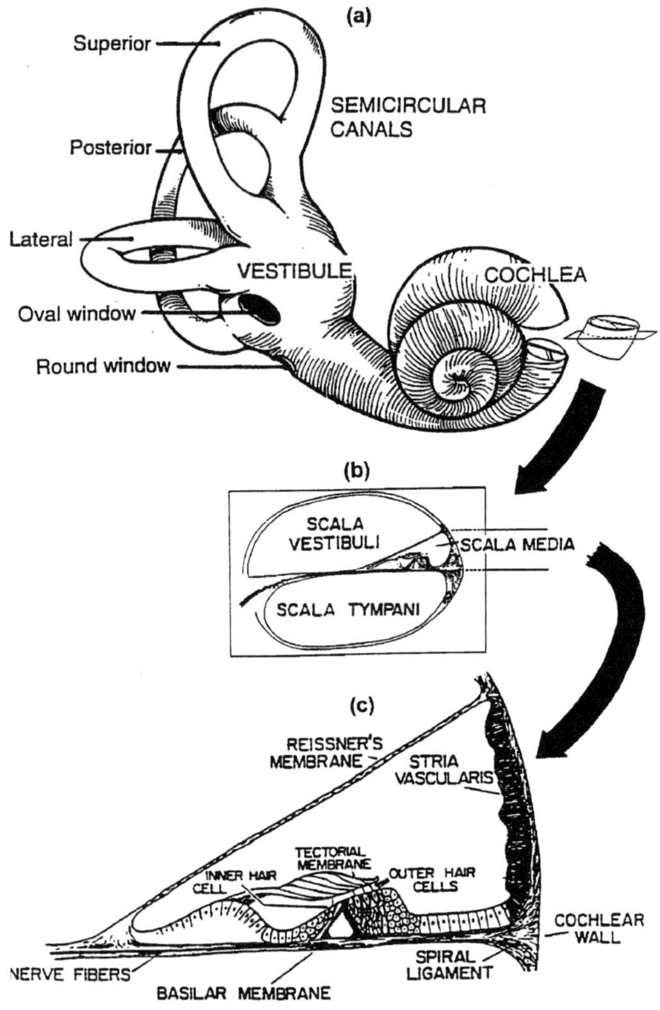

FIGURE 7.3. (**a**) The osseous labyrinth and its landmarks (after Durrant and Lovrinic, 1974); (**b**) cross-section of the cochlea; (**c**) the structure of the scala media (after Green, 1976).

of the entire cochlea and, because it is so small, the cochlea can be thought of as a tube divided lengthwise by a single membrane.

All three chambers of the cochlea are filled with fluid. The scala media is filled with a fluid called endolymph, which is similar to the fluid within cells. The scala vestibuli and scala tympani contain a different fluid called perilymph, which is similar to spinal fluid. The two types of fluids are kept separate by the Reissner's membrane on one side and by the basilar membrane on the other side (see Fig. 7.3c). Two other thin membranes, the oval window and the round window, separate the middle ear from the fluid of the cochlea. The oval window seals one end of the scala vestibuli and the round window seals the scala tympani (see Fig. 7.1b). The scala vestibuli and the scala tympani are connected at the upper end of the cochlea by a small opening near the apex called the *helicotrema*, and fluid from one chamber can flow to the other chamber through this opening (see Fig. 7.1b). The fluid-filled cochlea is essentially sealed so that an inward displacement of the cochlear fluids at the oval window is matched by an outward displacement at the round window. Therefore, when sound vibrations transmitted to the oval window via the tympanic membrane and ossicular chain causes the oval window to be pushed inward by the stapes, the round window is pushed outward by the increased pressure in the inner ear fluid.

Although the volume of the scala media or cochlear duct is only 0.1 of the total volume of the cochlea, within its chamber is the primary auditory sensory organ. A detailed three-dimensional cross-sectional drawing of the sensory organ for hearing, the organ of *Corti*, resting on the basilar membrane is shown in Fig. 7.4. This drawing of the cellular architecture of the organ

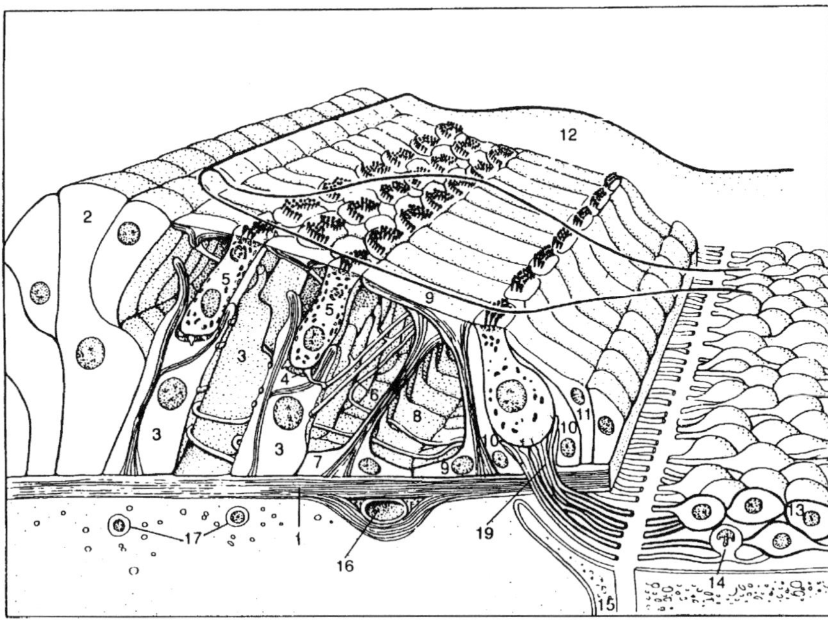

FIGURE 7.4. A three-dimensional drawing of the organ of Corti looking through the tectorial membrane. **1**, Basilar membrane; **2**, Hensen's cells; **3**, Beiter's cells; **4**, endings of spiral afferent fibers on outer hair cells; **5**, outer hair cells; **6**, outer spiral fibers; **7**, outer pillar cells; **8**, tunnel of Corti; **9**, inner pillar cells; **10**, inner phalangeal cells; **11**, border cell; **12**, tectorial membrane; **13**, type I spiral ganglion cell (innervation for inner hair cells); **14**, type II spiral ganglion cell (innervation for outer hair cells); **15**, bony spiral lamina; **16**, spiral blood vessel; **17**, cells of the tympanic lamina; **17**, axons of spiral ganglion cells (auditory nerve fibers); **19**, radial fiber (adapted from Junqueira et al., 1977, and Kelly, 1991).

of Corti is much too detailed for our level of discussion; however, it depicts in a beautiful manner the complexity of this hearing organ. The organ of Corti contains several thousand sensory receptor cells called *hair cells*. There is one row of inner hair cells (IHC) and three rolls of outer hair cells (OHC), organized in a regular and orderly fashion. The single row of inner hair cells contains about 7,000 cells whereas the three rows of outer hair cells have a total of about 24,000 cells. The inner and outer descriptors refer to locations relative to the central axis of the cochlea. The inner hair cells are closer and the outer hair cells are further away from the axis. Each inner hair cell has about 50 hairs (*cilia or sterocilia*), and each outer hair cell has about 70 to 100 cilia. The cilia of the hair cells are arranged in a curved pattern although the cilia of the outer hair cells have pronounced U-shape pattern, and the cilia of the inner hair cells are arranged in a shallow U-shape pattern. The cilia of the inner hair cells are free-standing; they do not connect with any other structures. Approximately 90–95% of the auditory nerve fibers that carry auditory information to the brain are in contact with the inner hair cells. The tops of the cilia of the outer hair cells are in contact with the tectorial membrane.

Motion of the basilar membrane causes the cilia of the hair cells to deflect or bend, which stimulates each hair cell causing it to produce electrical potentials known generally as receptor potentials. Two types of receptor potentials are present with the introduction of an acoustic stimulus; the *cochlear microphonic* potential and the summating potential. The receptor potentials are communicated to the nerve fibers, which sends neural signals to the brainstem. The exact role of the outer hair cells, their receptor potentials, and their interaction between the two types of hair cells is not clear. However, damage to the hair cells and the corresponding elimination of the receptor potentials greatly diminishes the cochlea's ability to perform frequency analysis of sounds and reduces the sensory response for sounds of low and moderate intensity.

We will now consider how sounds enter into the inner ear, causing the basilar membrane to vibrate and stimulate the hair cells. A simplified drawing of the cochlea uncoiled is shown in Fig. 7.5. The cochlea appears as a tapered cylinder divided into two sections by the basilar membrane. Although the cochlea becomes progressively narrower moving from the base or basal end to the apex, the width of the basilar membrane is tapered in the opposite direction, becoming progressively wider moving from the base to the apex. The stiffness of the basilar membrane is maximum at the base end and becomes progressively more flexible to ward the apex of the cochlea, where it is widest. In order to gain an appreciation of how sounds affect the shape of the basilar membrane, it is important to understand that the wavelengths of most sounds are much larger than the length of the basilar membrane. The fluids in the cochlea have a similar composition as sea water, with a sound velocity of about 1,500 m/s. For a high-frequency sound of 10 kHz, the wavelength of the sound will be approximately 15 cm. The cochlear duct is only about 3.5 cm long and therefore, even for a high-frequency sound, the

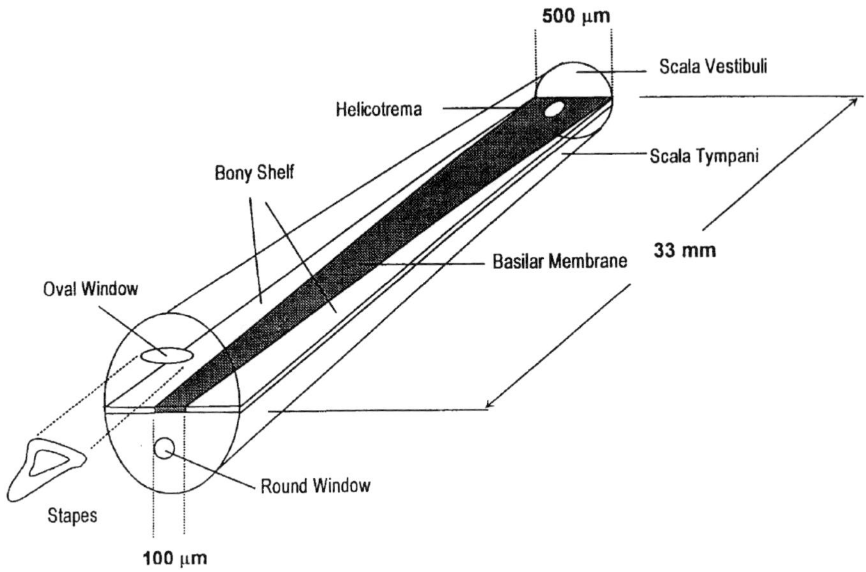

FIGURE 7.5. The major structural features of the uncoiled cochlea.

length of the cochlea is only about 1/5 of a wavelength. At a lower frequency of 1 kHz, the length of the cochlear duct is about 1/50 of the wavelength. This means that the acoustic pressure along the length of the cochlea will have essentially the same value at any instant in time. The duct is never pushed at one end and pulled at the other. Instead, the same physical force is applied throughout the length of the duct as the stapes vibrates and creates a sound field in the fluids of the scala vestibuli. Therefore, sounds in the cochlea can be treated as a pressure difference between scala vestibuli and scala tympani, which is uniform over the entire basilar membrane at any instant in time.

Sounds in the middle ear will cause the stapes to vibrate against the oval window, causing hydraulic pressure waves to be transmitted down the scala vestibuli, inducing ripples in the basilar membrane. The ripples will start at the basal end and propagate toward the apex like a *traveling wave* and cause a vibrational displacement pattern on the basilar membrane that propagates from the basal end to the apex. An example of the vibratory pattern along the basilar membrane at an instant in time is shown in Fig. 7.6.

The figure shows two instantaneous patterns of the traveling wave along the basilar membrane; the bottom figure being a more realistic representation because the basilar membrane is shown attached at its two edges. An example of an instantaneous traveling wave pattern on the basilar membrane is shown in Fig. 7.7 for four successive instances in time. Curve 1 shows the basilar membrane displacement for the first instant in time, curve 2 shows the next instant in time, and so on. These snapshots indicate that one part of the basilar membrane

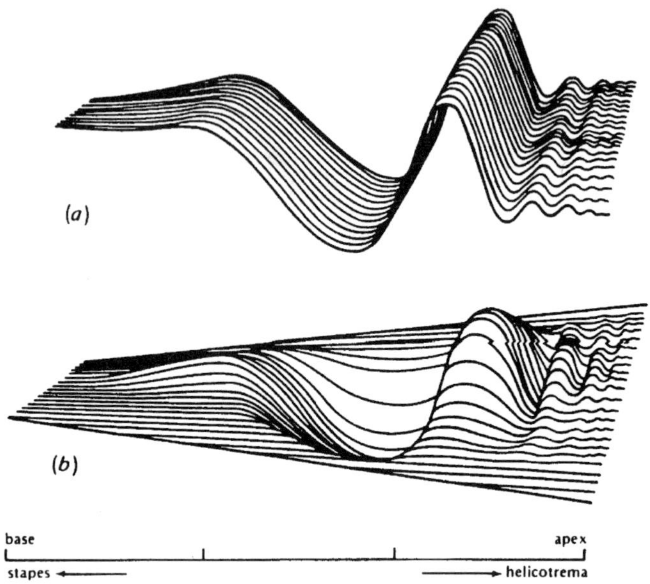

FIGURE 7.6. Instantaneous pattern of a traveling wave along the basilar membrane. (**a**) the pattern that would result if the membrane was ribbon-like. (**b**) a more realistic vibration pattern since the membrane is attached along both edges (adapted from Tonndorp (1960).

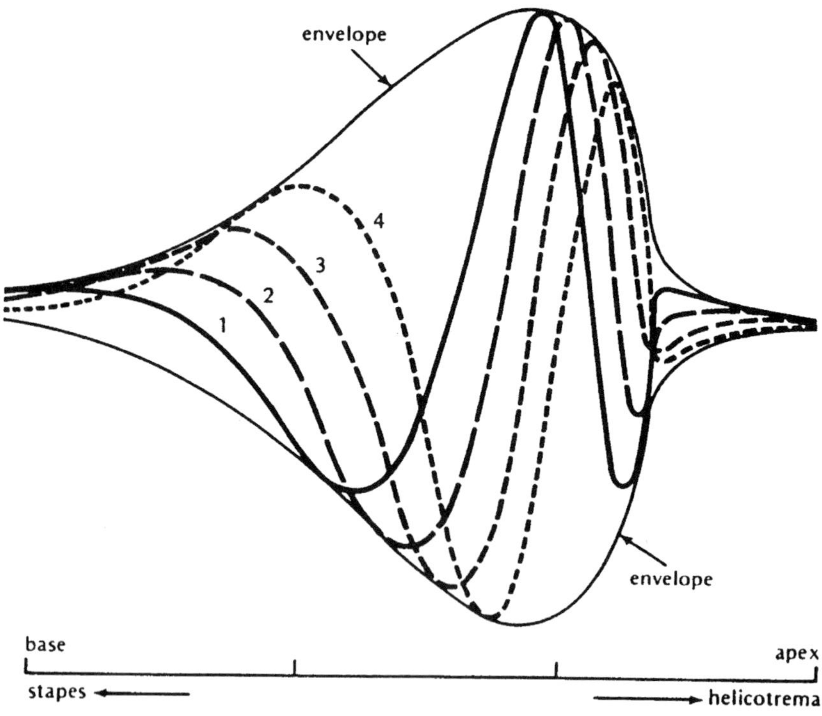

FIGURE 7.7. Four successive instantaneous patterns of a traveling wave. The envelope of the traveling wave is also shown (adapted from Ranke, 1942).

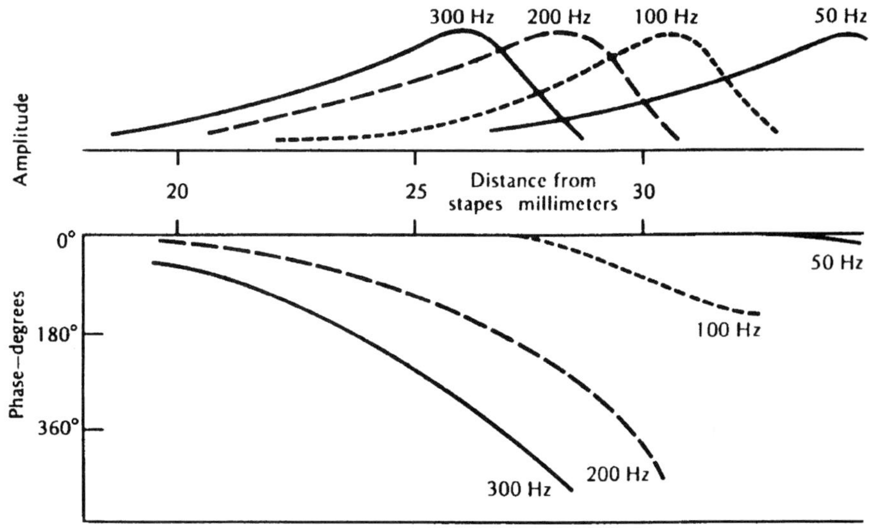

FIGURE 7.8. Envelope of basilar membrane displacement amplitude as a function of distance for different frequencies (after von Bekesy, 1960).

may be displaced toward the scala tympani, while an adjacent part may be deflected in the opposite direction.

The location at which the basilar membrane will have the maximum deflection will depend on the frequency of the sound. The point of maximum basilar membrane displacement for high-frequency sounds will be near the basal end where the width of the membrane is minimal and toward the apex for low-frequency sounds. The relative amplitude of displacement along the basilar membrane for different frequency signals is shown in Fig. 7.7. From the figure, we can see that the basal end of the basilar membrane responds best to high-frequency sounds, but it can also respond to low frequencies. The apical end of the basilar membrane vibrates only to low-frequency stimulation. From Fig. 7.7, we can see that the basilar membrane provides a spectral analysis function in which different frequency signals will cause the basilar membrane to peak at different locations along the membrane and differentially excite hair cells.

The direction of the traveling wave from the basal end toward the apex is not the result of the vibrations being introduced into the cochlear duct at the oval window. The source of vibration can be located at any location along the cochlea, and yet the traveling wave will move toward the apex. The reason for this has to do with the stiffness gradient of the basilar membrane. The stiffness is greatest in the base and decreases toward the apex. The fact that the direction of the wave is always toward the apex no matter where along the cochlear duct the vibration is introduced is called the *traveling-wave paradox*. How far toward the apex a wave will travel depends on the frequency of stimulation; lower frequencies travel further.

7.1.3 Auditory Nervous System

The hair cells of the cochlea are innervated by two separate and distinct systems of nerve fibers. One system is the *afferent fibers* (outgoing) that carry auditory information from the periphery sense organ (organ of Corti) to the brainstem and brain. The other system is the *efferent fibers* (incoming) that typically bring information from higher neural centers to the periphery. The auditory afferent fibers are divided into *radial fibers* or *Type I fibers,* and *outer spiral fibers* or *Type II fibers.* The Type I fibers innervate only inner hair cells and comprise about 75–95% of all the afferent fibers. The nerve fibers enter the cochlea through small openings in the basilar membrane called the *habenula perforata,* as shown in Fig. 7.9. A cat has about 45,000 to 50,000 afferent fibers (compared to about 30,000 for humans), 95% of which innervate the 2,600 inner hair cells. About 20 separate fibers enter though each habenular opening and spread out to the hair cells. The Type II afferent fibers cross the tunnel of Corti below the arch and are aligned along the base of the three rows of outer hair cells. They typically extend along side at least 100 outer hair cells before sending off a small terminal collateral to an outer hair cell in the first row as shown in Fig. 7.10. The fiber then continues on along the base sending off one or two more collaterals to the outer hair cells in the first row. It then extends radially and innervates three of four outer hair cells in the second and third rows. Each Type II afferent fiber typically innervates about 10 outer hair cells. Calculations for humans estimate that an average of eight radial fibers innervate one or two inner hair cells (Yost, 1994). The radial or Type I fibers have a many-to-one connection to the inner hair cells, while the outer spiral or Type II fibers have a one-to-many connection. The radial afferent fibers are shown connecting the inner hair cells to spiral ganglion cells in Fig. 7.4. The axons of

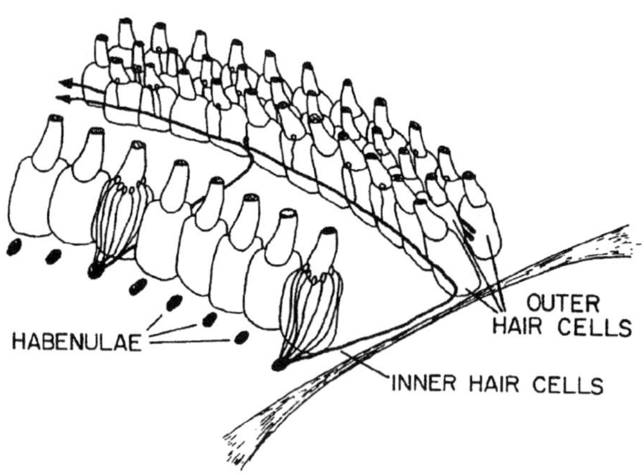

FIGURE 7.9. Typical innervation pattern for the afferent fibers (adapted from Spoendlin, 1970).

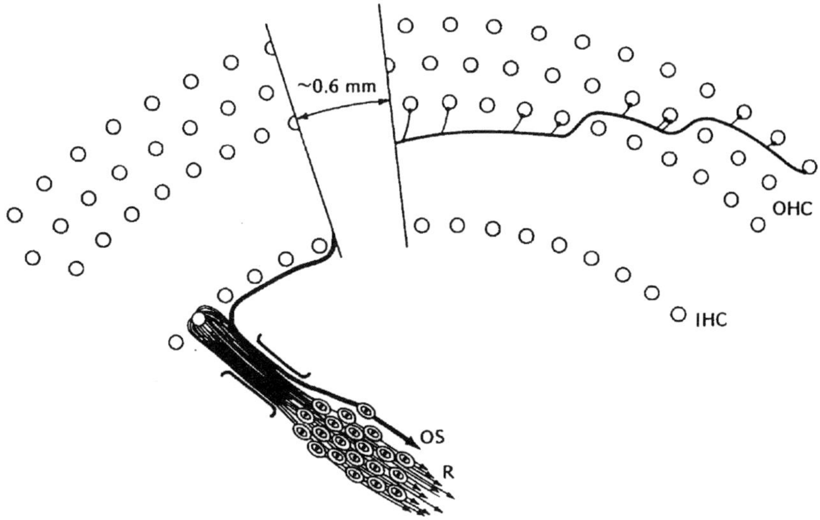

FIGURE 7.10. Horizontal pattern of the afferent innervation of the organ of Corti (adapted from Spoendlin, 1977).

these ganglion cell bodies are the twisted group of fibers that form part of the *eighth (VIII)* cranial nerve. All of the acoustic diVIII nerve terminates on cell bodies located in the *cochlear nucleus*, a collection of cells located in the lower brain stem.

Mechanical deformation of the hair cells causes the hair cells to release a chemical transmitter, which probably initiates a graded electrical potential in the nerve fibers that innervate the base of the hair cell. The neural potential propagates along the nerve fiber to the habenula perforata where the myelinated portion of the fiber is reached. At that point, neural spikes or discharges are produced, which travel along the auditory nerve to the cochlear nucleus. The neural spikes do not vary in amplitude each time the nerves fire. The potential always reaches its maximum amplitude in an all-or-none fashion. The spikes are very short, typically rising to its maximum value in 1–2 ms before returning to the resting state. If a microelectrode is inserted into a nerve fiber, spikes will be observed even without the presence of an auditory stimulus. The neurons have spontaneous activity firing randomly. The lowest sound intensity that gives rise to a minimum of 20% increase in the firing rate is called the threshold for that particular stimulus frequency. As the sound intensity increases above the threshold, the amplitude of the spikes does not change; however, the rate of firing (*spike rate*) will increase. Figure 7.11 illustrates the input-output function for a single auditory nerve fiber, showing the spike rate as a function of the stimulus level. As the stimulus increases above the threshold, the spike rate increases steadily with stimulus intensity until a maximum discharge rate is reached about 30–40 dB above the threshold level.

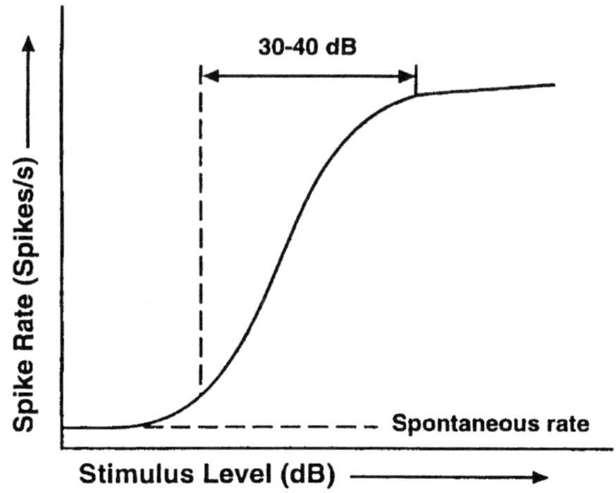

FIGURE 7.11. Input-output function for a single auditory nerve fiber (adapted from Bess and Hume, 1995).

Therefore, a single neuron can encode intensity over only about a 30–40 dB range. Each auditory neuron is tuned to a specific *characteristic frequency* (CF) so that an auditory stimulus at that frequency will have the lowest threshold level when compared with a stimulus of different frequency. Figure 7.12

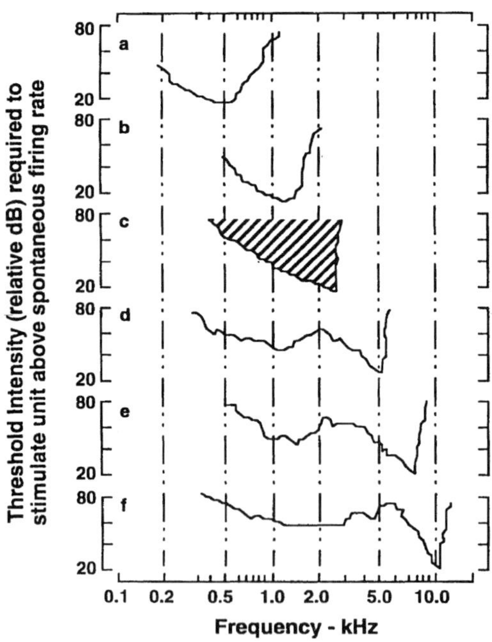

FIGURE 7.12. Tuning curve for six single units with different characteristic frequencies (adapted from Bess and Hume, 1995).

depicts the single-unit frequency threshold curve of six auditory neurons. The vertical axis is the relative intensity in dB required to stimulate the neuron to fire at a rate above its spontaneous rate. The low-frequency portion of the single unit threshold has a shallow slope compared to the steep high-frequency slope. Figure 7.12 clearly indicates that each unit in the graph had a different frequency response and fired best at only a limited range of frequencies. The pattern of discharge reflects the same type of frequency selectivity of the basilar membrane. As the nerve fibers exit the cochlea on the way to the brainstem they maintain an orderly arrangement. Fibers with high characteristic frequencies are located around the perimeter and fibers with low characteristic frequencies make up the core of the cochlear nerve bundle. Thus the auditory nerve is organized so that each characteristic frequency corresponds to a place within the nerve bundle in a similar way as the tuning in the basilar membrane is organized. This mapping of frequency to the place of maximum activity within the auditory system is referred to as *tonotopic organization*. It is a fundamental anatomical characteristic of the auditory system from the cochlea through the auditory cortex.

The auditory nerve fibers also encode temporal information about the auditory stimulus by its firing pattern. Figure 7.13 shows the firing of a neuron to a 1 kHz sinusoidal stimulus. The firings occur at the same point of the sinusoidal stimulus, which in the figure is close to the peak of the stimulus. This type of response is called a *time-* or *phase-locked response*, or a *following response*, because the response appears to be locked to the peak of the stimulus. The neuron may not fire during every cycle of the stimulus waveform; nevertheless, the histogram showing the relative rate at which firing occurs has a peak at the same interval as the period of the 1 kHz stimulus.

7.1.4 The Central Auditory Nervous System

Electrical activity originating in the cochlear branch of the auditory nerve will propagate up toward the cortex through a network of nerve fibers and through various auditory nerve centers. This network of nerve fibers is referred to as the auditory *central nervous system* (*auditory CNS*). The parts of the network that carry information toward the cortex form the ascending or afferent pathways. Nerve pulses can also be sent from the cortex toward the periphery on the descending or efferent pathways. We will briefly discuss the central auditory nervous system, a subject that is extremely complicated and not well known. We will begin our discussion by referring to Fig. 7.14, which is a very oversimplified diagram of the ascending auditory CNS, illustrating the principal connections of the afferent auditory system. The afferent nerve fibers leave the cochlea and travel to the *cochlear nucleus*. As we ascend to higher levels, the picture becomes very complicated with many pathways, or *tracts*, some traveling contralaterally and some remaining ipsilateral. Some fibers may leave one point and go directly to the next, but others may bypass

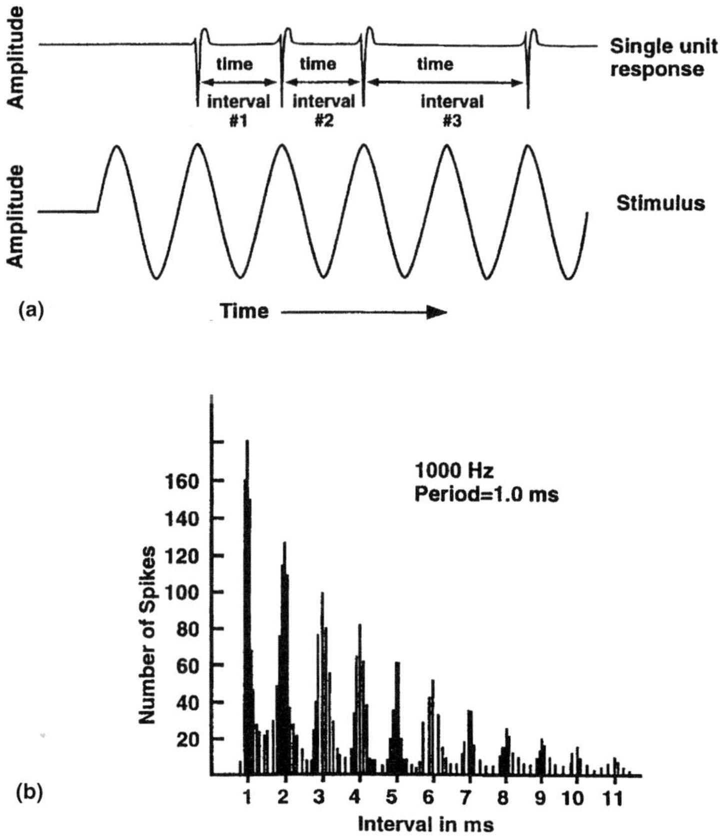

FIGURE 7.13. (**a**) Illustrationof the synchronization of the neuron firing to the stimulus waveform. (**b**) A histogram of the intervals measured in A (adapted from Bess and Hume, 1995).

the next point and travel to a higher location. Some fibers are collateral or branches to one point as the main tract travels past that point to terminate at a higher location.

Neural pulses leaving the cochlea first travel to the cochlear nucleus where the first synapse is made. The axon of all the ganglion cells of the VIIIth nerve terminates in the cochlear nucleus. The cochlear nucleus is the universal terminus of all the VIIIth nerve fibers. Upon entering the cochlear nucleus, each nerve fiber separates and proceeds to three separate regions: the *anteroventral cochlear nucleus*, the *posteroventral cochlear nucleus*, and the *dorsal cochlear nucleus*. In each division of the cochlear nucleus, the cochlea is represented tonotopically. From the cochlear nucleus upward, there are a number of alternate routes until the medial geniculate body is reached. At the geniculate all the auditory fibers synapses and fibers from the medial geniculate body carry auditory information to the cortex.

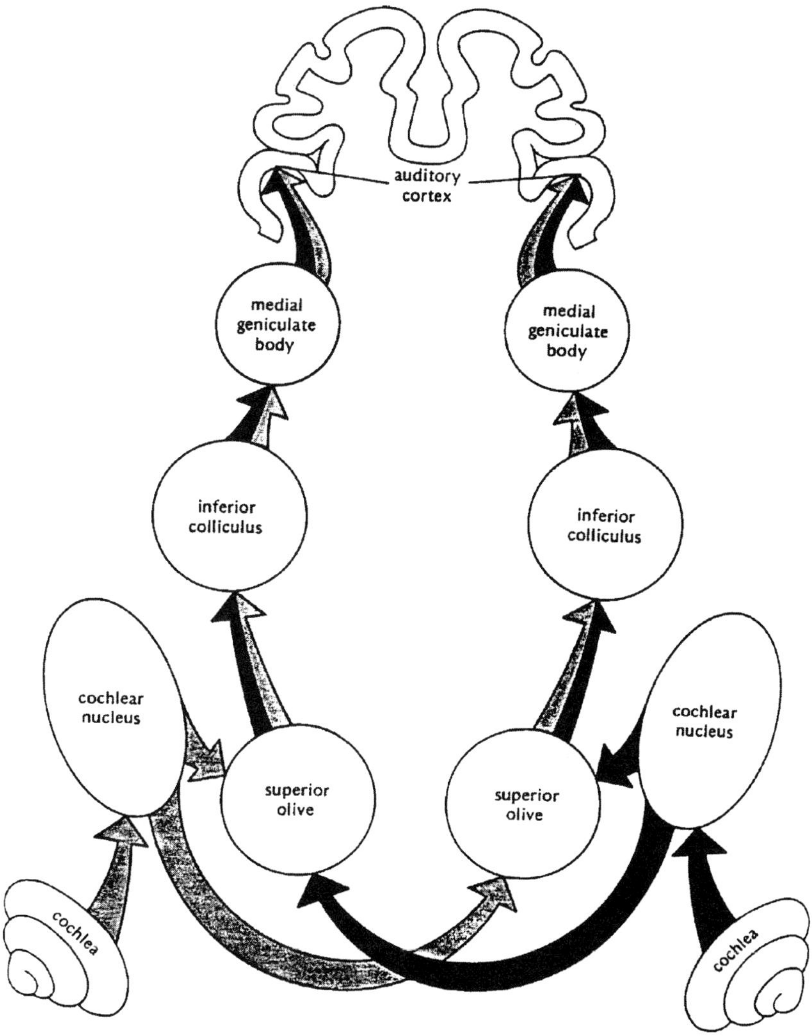

FIGURE 7.14. A simplified diagram of the ascending pathways of the auditory CNS. The black arrows represent input from the right ear and the gray arrow represents input from the left ear (adapted from Yost, 1994).

The *superior olivary complex* is the lowest level where binaural interaction occurs. It is composed on the following nuclei: *lateral superior olive* (LSO), *medial superior olive* (MSO), *traperzoidal body*, and the *preolivary nuclei*. The main nuclei associated with the ascending pathway are the LSO and MSO and the medial nucleus of the trapezoid body (MTB). The MTB is a relay carrying information from the opposite cochlear nucleus to the ipsilateral LSO. The MSO has direct fibers from the cochlear nucleus on both sides and is involved in the localization of low-frequency sounds by means of the difference in the time-of-arrival of sounds at the two ears. The LSO has direct connection from the ipsilateral and indirect ones from the contralateral cochlear nucleus via the MTB. It has a role in sound localization of high-frequency sounds by detecting interaural intensity differences at both ears.

The *inferior colliculus* receives information from both the olive complex and the cochlear nucleus. It seems to be composed of three major areas: the *central nucleus*, the *dorsal cortex of the inferior colliculus*, and the *paracentral nucleus*. Some neurons in the different parts of the inferior colliculus respond only to monoaural stimulus while others respond to bilateral stimulus. The main output of the inferior colliculus is to the *medial geniculate*.

The medial geniculate receives all the afferent fibers that innervate the acoustic parts of the cortex. There are no neurons that cross to the other medial geniculate; all subcortical interactions between the ipsilateral and contralateral sides must take place at lower levels. The outputs of the medial geniculate body go directly into the auditory cortex.

The descending pathways for efferent fibers are also extremely complicated. Some tracts arise in the auditory cortex or in a variety of nuclei and terminate at other nuclei, especially in the olivary complex and the cochlear nucleus. The number of efferent fibers that actually enter the cochlea is relatively small and connect to the outer hair cells. The distribution is not uniform, ranging from 6 to 7 endings per outer hair cell near the base and progressively less toward the apex. The inner row receive efferent innervation through the length of the cochlea whereas the innervation in the second and third rows decrease progressively toward the apex and by the third turn the outer two rows may not be innervated at all. The descending fibers appear to have an inhibitory action on the electrophysiological response of the cochlea although their role is far from being completely understood. Furthermore, both excitatory and inhibitory connections exist in the descending system. It seems that the descending auditory pathways provide control that assists in varying the routing and shape the auditory input.

From this brief discussion, it is obvious the auditory CNS is extremely complex and not well understood. Most of our understanding on the auditory CNS comes from studies with laboratory animals. Therefore, we should not be surprised that our understanding of the auditory CNS of marine animals, especially government-protected marine mammals, will be considerably less than for humans and other laboratory animals.

7.2 The Cetacean Ear

A drawing of the dolphin's ear showing the external auditory meatus, the middle and inner ears is presented in Fig. 7.15. Each middle and inner ears is encased in a bony structure (*Tympanic bulla*), which does not have a bony connection to the skull but is connected by cartilage, connective tissue, and fat (McCormick et al., 1970a). Interested readers should refer to some

(a)

(b)

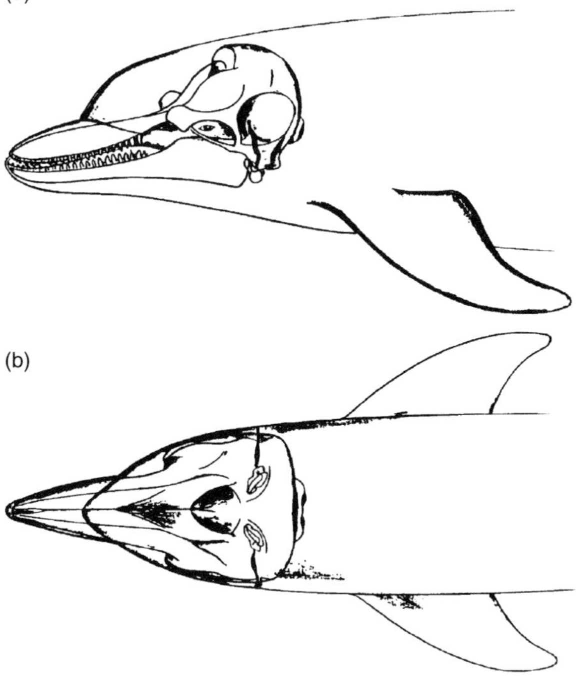

FIGURE 7.15. (**a**) Drawing of a dolphin's skull showing the external auditory meatus (**b**) and the tympanic bulla (**c**) (adapted from Johnson, 1976).

excellent anatomical descriptions of the dolphin ear presented by Wever et al. (1971a, 1971b, 1972), Ridgway et al. (1974), and Ketten (1997).

Unlike other mammals, dolphins and other cetaceans do not have pinnas and their external auditory meatus are not readily visible. The argument for dolphins needing a streamline shape has been used to explain the absence of pinnas. The meatus in most dolphins is like a pinhole with parts of it (nearest the skin surface) composed of fibrous tissue. There is not an air-filled auditory canal similar to terrestrial mammals, but only a narrow channel filled with cellular debris and dense cerumen. There also does not seem to be a direct connection of the external canal to the tympanic membrane. The external meatus with its narrow cross section and occluded structure hardly seems capable of being an acoustic pathway to the middle and inner ears via the tympanic ligament. It is not clear how sound can enter into the meatus and be conducted along its interior to the middle ear. Nevertheless, there have been some who have argued in favor of the external auditory meatus being the primary pathway for sound transmission (Fraser and Purves, 1954, 1960). Since the auditory meatus in cetaceans is so different than all terrestrial mammals, one could justly ask, "if sounds are not conducted in the external auditory meatus, then how does sound enter into the inner ears of dolphins?" An alternate theory of sound

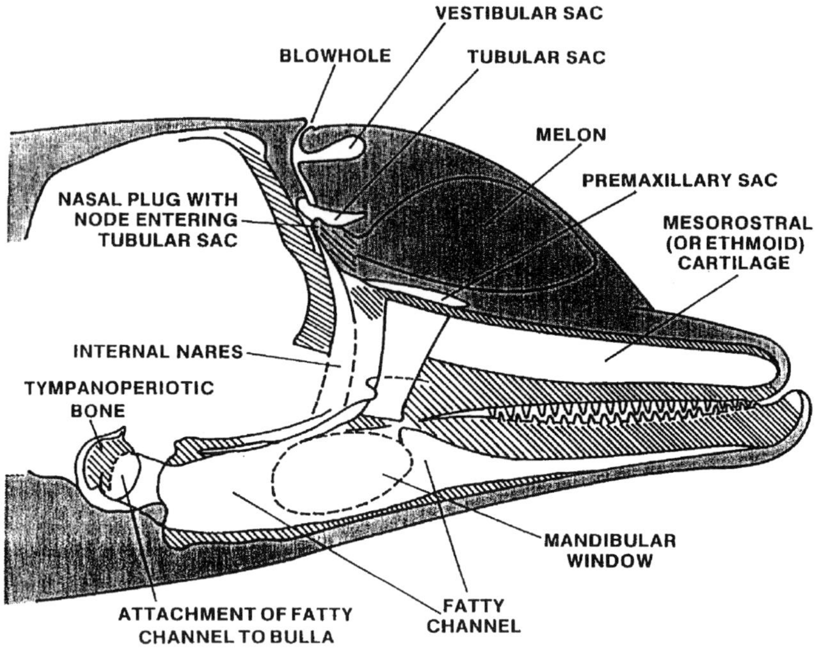

FIGURE 7.16. Schematic of a dolphin's head showing various structures associated with sound reception and sound production (after Norris, 1967a).

conduction in dolphins suggested that the auditory meatus is nonfunctional and that sounds enter the dolphin's head through the thinned posterior portion of the mandible and is transmitted via a fat-filled canal to the tympanoperiotic bone, which contains the middle and inner ears (Norris, 1967a). A schematic diagram of a dolphin's head showing the mandibular window and the fat channel is displayed in Fig. 7.16. The region of the pan bone or "acoustic window" of the mandible is very thin, varying from 0.5 to 3.0 mm in different species of dolphins (Norris, 1967b). The oval-shaped fatty tissue of the acoustic window is almost completely free of muscle strands (Norris, 1970). The interior portion of the pan bone region is filled with a translucent fat, rich in oil, which extends from the pan bone to the bulla. The mandibular fat body is a well-defined subcylindrical mass of fat flaring posteriorly. According to the theory, once sounds enter into this fat body through the pan bone, it may be transmitted directly to the tympanic bulla to which the mandibular fat is attached. Varanasi and Malins (1971, 1972) found that this fatty lipid material had a very low sound absorption characteristic, leading them to coin the term "acoustic fat." Interestingly, this acoustic fat is found only in the mandibular canal and the melon, and nowhere else in an odontocete (Varanasi and Malins, 1971).

Bullock et al. (1967) employed electrophysiological techniques in the study of dolphin hearing and were able to shed some important light on the viability of the two competing theories of sound conduction in dolphins. They measured auditory evoked potentials from the midbrain auditory structure as various types of sound stimuli were played to the dolphins. Using a hydrophone pressed against the head of their subjects, they obtained the largest potentials when the source was placed forward of the pan bone area of the lower jaw window and not at the pan bone. The pattern of sensitivity of an individual Stenella is shown in Fig. 7.17. They also found that when a distant sound source was used and an acoustic shield (10 × 12 cm piece of foam rubber) was held over the sensitive part of the lower jaw and melon, the evoked potential response decreased noticeably. The findings of Bullock et al. (1967) strongly supported Norris's notion that the pathway of sound to the cochlea is via the lower jaw. However, the best point of entry of sounds may be forward of the pan bone area of the lower jaw.

McCormick et al. (1970, 1970) used a different electrophysiological technique to study sound conduction in the dolphin ear. They used the cochlear-potential method, measuring the potential generated by hair cells in the organ of Corti (see Section 2.3). Using a vibrator in contact with the dolphin's skin, they mapped the level of the cochlear potential with the location of the vibrator on the dolphin's body. Their results with four *Tursiops truncatus* are summarized in the schematic of Fig. 7.17, showing the area of greatest sensitivity to sounds. However, their measurements were not fine enough to map out the gradient in sensitivity in the relatively broad area shown in Fig. 7.18. Contrary to the results of Bullock et al. (1967), they found that the upper jaw and most of the skull seem to be acoustically isolated from the ear. When the external auditory meatus was stimulated, the cochlear potentials were 6–14 times less intense than when the most sensitive part of the lower jaw was stimulated. They also found that the cochlear potentials were not affected after the auditory meatus was severed. They concluded that the meatus and tympanic membrane were acoustically nonfunctional. They also concluded that sounds enter the dolphin through the lower jaw.

Although the electrophysiological measurements of Bullock et al. (1967) and McCormick et al. (1970) seemed irrefutable, their conclusions concerning the lower jaw pathway have not been totally accepted. Purves and Pilleri (1973) discounted the findings of McCormick et al. (1970) on the basis that their stimulus levels, which were about 50 dB above the behavioral threshold measured by Johnson (1967), were too high confounding the use of normal auditory pathways. Purves and Pilleri (1973) also argued that the surgical procedure damaged the auditory system. They continued to argue in favor of the external auditory meatus, but without any experimental evidence (Purves and Pilleri, 1983). Johnson (1976) correctly emphasized that the measurements of both Bullock et al. (1967) and McCormick et al. (1970) need to be interpreted with a degree of caution. In both studies, the animal's mouth was open and its larynx connected to a gas-filled tube so that unnatural pockets of gas may have been present to alter the transmission of sound in the dolphin's head. These gas-filled pockets would not be present in a natural state. There are other reasons for caution in examining the results of these two electrophysiological studies. The acoustic conditions for both experiments were less than ideal with the

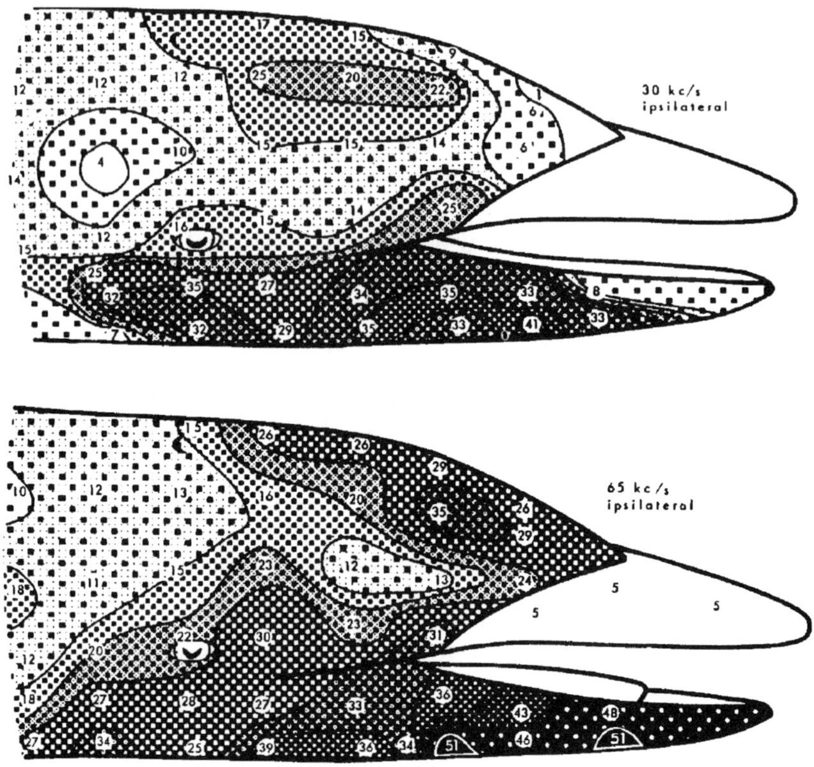

FIGURE 7.17. Pattern of sensitivity of an individual Stenella to the sound produced by a hydrophone pressed against the head surface at the points shown. The numerical values represent attenuation at threshold; therefore, the largest numbers represent greatest sensitivity. Contour lines are drawn at intervals of 5 dB in sensitivity. Recording was from the inferior colliculus (adapted from Bullock et al., 1967).

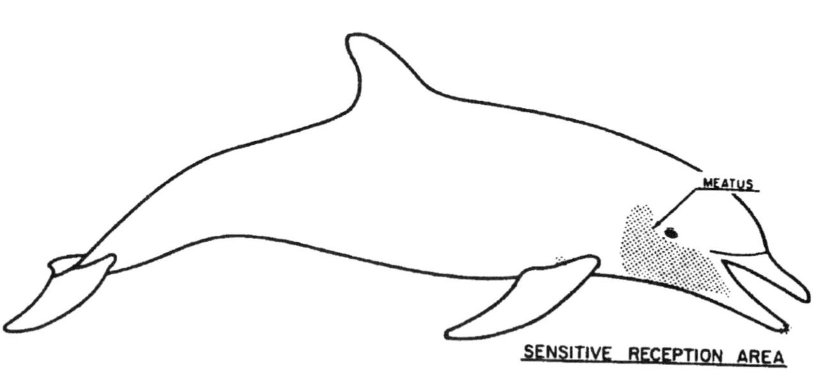

FIGURE 7.18. The three regions of greatest sensitivity to sound for a *Tursiops truncatus* (adapted from McCormick et al., 1970).

subjects confined to small holding tanks, and their heads held near the surface with incisions kept out of the water so that the electrodes could be electrically isolated from each other. The acoustic propagation conditions for such a situation can be extremely variable with SPL changing drastically, on the order of 10 to 20 dB, if the sound source is moved as little as a few centimeters. Furthermore, having the surface of the dolphin's head excited with a hydrophone or a vibrator may set up a totally different vibrational pattern on the skin of the animal than would an acoustic signal approaching from afar. For instance, McCormick et al. (1970) measured large potentials when the vibrator was held against one of the front teeth of the lower jaw. That tooth would probably not experience similar vibration from a sound wave propagating in the water toward the dolphin. Finally, vibrational displacements induced by a hydrophone held against a dolphin's head are probably directed perpendicular to the surface of the head, which would be different from vibrational displacements caused by a sound approaching in a direction in front of a dolphin. Different types of vibrational patterns may be coupled into the inner ear differently.

More recently, Mohl et al. (1999) measured the auditory brainstem evoked potential for a *Tursiops truncatus* that was trained to jump out of the water and beach on a rubber mat. A specially constructed polyurethane suction cup that had a piezoelectric crystal recessed in the cup was used to provide the acoustic click stimulus. There was a layer of water between the crystal and the animal's skin as can be seen in Fig. 7.19. The suction cup projector was constructed

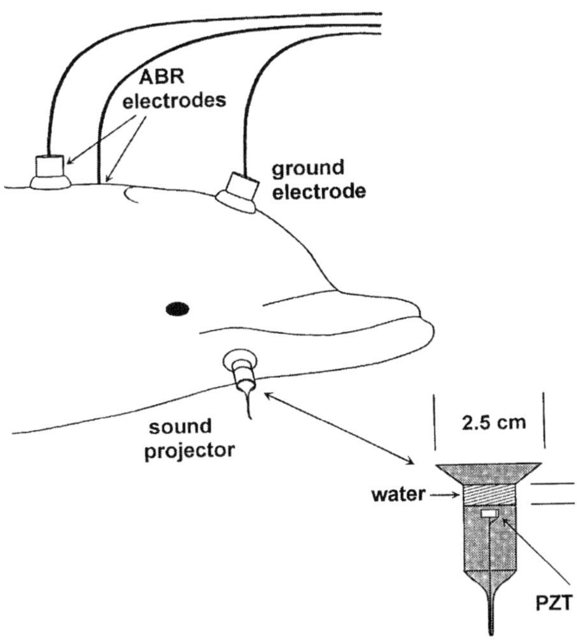

FIGURE 7.19. Electrodes used to measure auditory brainstem evoked potentials and the suction cup projector showing the layer–animal interface (adapted from Mohl et al., 1999).

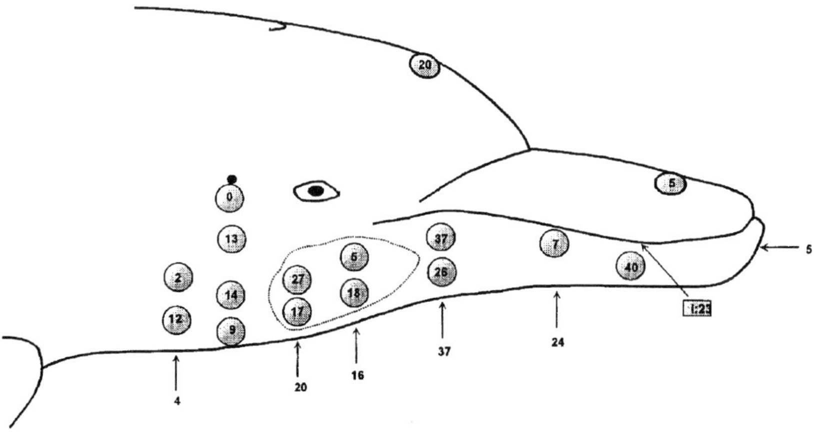

FIGURE 7.20. Pattern of sensitivity of an *Tursiops truncates* to the sound produced by a suction cup hydrophone placed against the head surface at the points shown. The numerical values represent attenuation at the ABR threshold; therefore, the largest numbers represent greatest sensitivity (after Mohl, et al., 1999).

in this manner so that it would simulate sounds traveling through a water–animal interface. Having the experiment done in air also helped to ensure that the stimulus was localized to the spot where the suction cup attached to the animal's skin, since sounds would reflect back into the suction cup at the air interface and not propagate to the animal via some unknown path.

The results of Mohl et al. (1999) are shown in Fig. 7.20. These results are in basic agreement with those of Bullock et al. (1967), with the area of maximum sensitivity being slightly forward of the pan bone area of the lower jaw. The sensitivity at the area of the meatus was extremely low. The results of Bullock et al. (1967) and Mohl et al. (1999) seem to run counter to Norris' hypothesis that sound enters the dolphin's head via the thin pan bone area. However, the fat channel from the pan bone extends forward at an angle and reaches the skin forward of the pan bone in the vicinity where Bullock et al. (1967) and Mohl et al. (1999) obtained good results. Aroyan (1996) studied sound propagation in the head of dolphins by creating a three-dimensional rectangular grid and assigned sound velocity and density values to each grid point that would be appropriate for a dolphin's head. The values were extrapolated from X-ray computer tomography (CT) scans that provided three-dimensional images of a dolphin's head (Cranford et al., 1996). Aroyan (1996) then assumed a plane wave propagating toward the dolphin head and numerically solved the three-dimensional wave equation using a supercomputer and a finite-difference technique. Therefore, at any given point in time, the acoustic energy at each grid point could be determined. He found the incoming acoustic energy converging at the dolphin's bulla with a sound path that entered forward of the pan bone and propagated to and through the pan-bone window

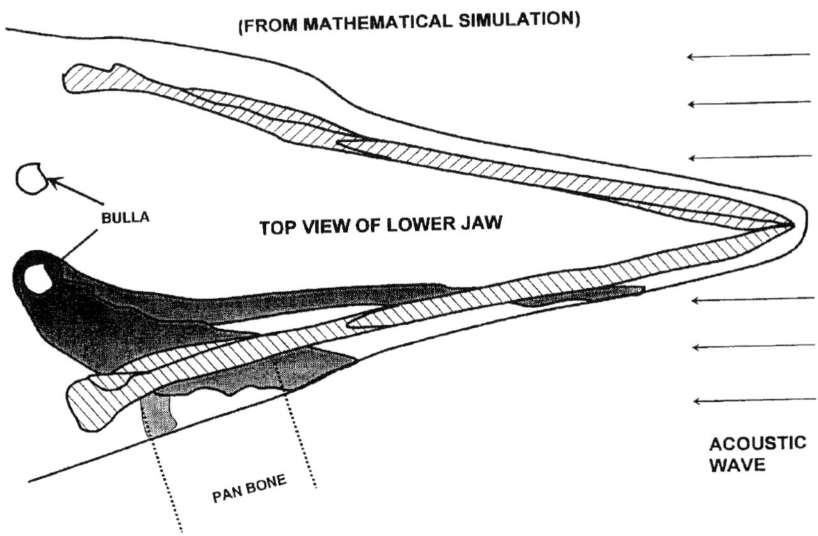

FIGURE 7.21. Numerical simulation results of Aroyan (1997) showing area of high acoustic energy density as an acoustic wave travels through the lower jaw of a dolphin toward the auditory bulla (after Aroyan, 1967).

continuing along the fat channel to the bulla. The results of Aroyan (1996) showing areas of high energy density as an acoustic wave travels toward the bulla are depicted in Fig. 7.21. These results from numeral simulation match the data of Mohl et al. (1999) quite well.

Brill et al. (1977) used a behavioral approach to investigate the role of the lower jaw in sound conduction. They trained a *Tursiops truncatus* to perform a target discriminate task using its sonar. The blindfolded animal was required to discriminate between an aluminum cylinder and a sand-filled ring while wearing a hood that covered the lower jaw. Two hoods, each made of a different type of material, were used. One hood was constructed from 0.16-cm thick gasless neoprene that had a low-acoustic attenuation property and the other hood was constructed from 0.47-cm thick closed cell neoprene that had a high-acoustic attenuation property. The acoustic attenuation property of both materials were measured with simulated dolphin echolocation signals. The gasless neoprene had attenuation values of 2.2 and 1.2 dB for signals with peak frequencies of 35 and 110 kHz, respectively. The closed cell neoprene had attenuation values of 39 and 36 dB for signals with peak frequencies of 55 and 115 kHz. The difference between the animal's performance without the hood and with the low-attenuation hood was insignificant. The difference between the animal's performance without the hood and with the high-attenuation hood was significant. The behavioral results of Brill et al. (1977) are consistent with Norris's theory that the pathway of sound to the cochlea is via the lower jaw.

The results of these three experiments seem to provide overwhelming evidence in support of the theory of sound conduction through the lower jaw, although there is still the question as to where on lower jaw sounds most effectively enter. Norris and Harvey (1974) made acoustic measurements with the head of a recently dead *Tursiops truncatus*, by inserting a small hydrophone into different portions of the head to receive sounds from a projector located away from the head. They found a twofold increase in intensity as the sound traveled from the anterior to the posterior portion of the mandibular fat body, suggesting a structural intensification of received sound similar to the manner in which sound is intensified in the external ear canal of man.

Ketten (1994, 1997) using magnetic resonance imaging has observed a second trumpet-shaped body of lipid that is located between the dorsal extent of the pan bone and the auditory meatus. This funnel-shaped fat channel (lateral fat channel) has its flared end closest to the surface and is aligned almost perpendicular to the surface of the jaw. It becomes progressively narrower toward the middle ear like a megaphone and is aligned toward the middle ear. Ketten (1994, 1997) has hypothesized that the lateral channel may be used in capturing lower frequency signals while the anterior fat channel may be specialized for capturing high-frequency signals associated with echolocation.

The discussion of sound conduction paths way have centered entirely on small odontocetes. It turns out that baleen whales also have occluded external canals that are filled with a homogeneous wax (Ketten, 1992). The proximal end of the canal flares and covers the tympanic membrane, which is shaped like a finger of a glove and this "glove finger" protrudes laterally from the bulla into the auditory canal. The jaws of mysticetes are designed for sieving or gulp feeding and have no evident connection to the temporal bones. It is highly debatable as to how sounds enter into the ears of mysticetes.

7.2.1 Middle Ear

The anatomy of the middle and inner ears of *Tursiops truncatus* and other dolphins was the topic of some excellent studies by Ernest Wever and his colleagues at Princeton University (Wever et al., 1971a, 1971b, 1972; McCormick et al., 1970). Fleischer (1976a,b, 1970) has also studied the structure of the dolphin's ear and has created descriptive analogs of the auditory processes in the dolphin. Interested readers should refer to these excellent works. Only a gross description of the middle and inner ears will be given here in order to gain an appreciation of the suitability of the dolphin's ear to receive high-frequency, low-intensity underwater sounds.

A drawing of the right auditory bulla of *Tursiops truncatus* showing the relationship of its various parts is shown in Fig. 7.22. The ossicular chain consisting of the malleus, incus, and stapes are housed within the bulla. A drawing of the ossicular chain is shown in Fig. 7.23. There is no direct connection between the tympanic membrane and the malleus. The tympanic ligament, which is firmly attached to the bony edge of the dorsal opening

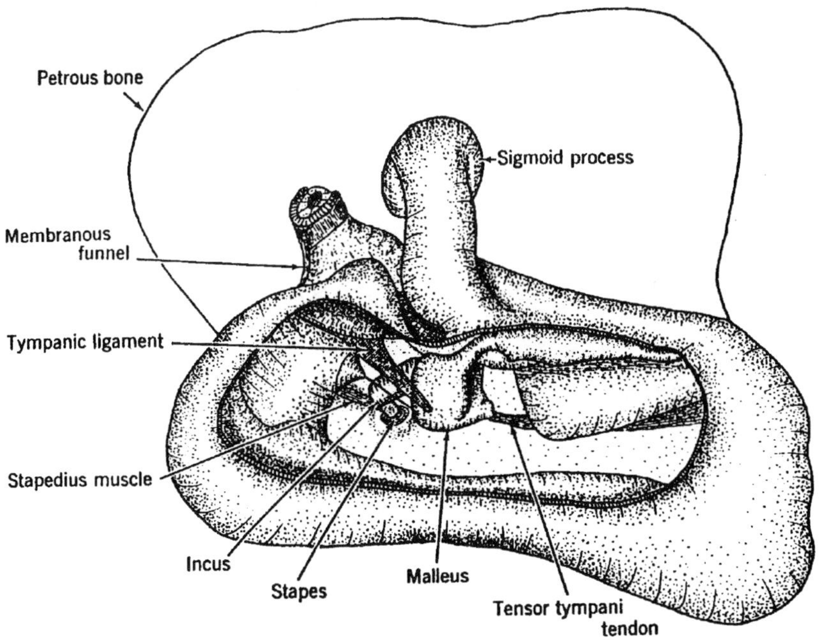

FIGURE 7.22. The right auditory bulla of *Tursiops truncatus* opened on its lateral side to show its contents (adapted from McCormick et al., 1970).

in the tympanic bone, serves mainly as a suspension system for the malleus (McCormick et al. 1970). The incus is connected to the malleus by a saddle-shaped joint. The stapes is connected to the incus and its head surface articulates with the major process of the incus. The footplate of the stapes

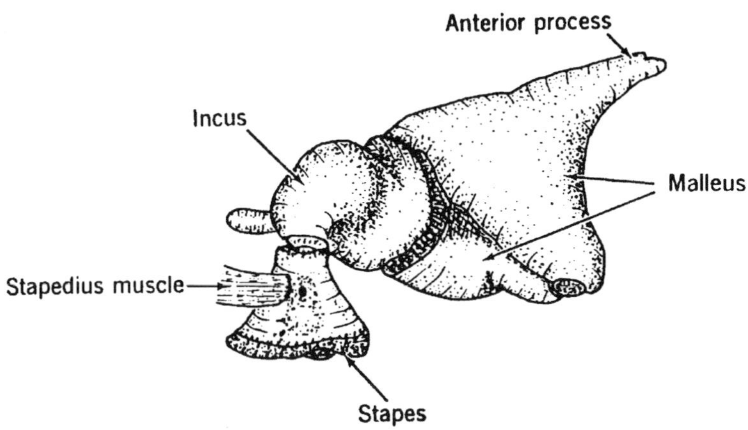

FIGURE 7.23. The ossicles rotated about 70° from their position in the preceding figure, to show the articulation between stapes and incus and the union of incus and malleus (adapted from McCormick et al., 1970).

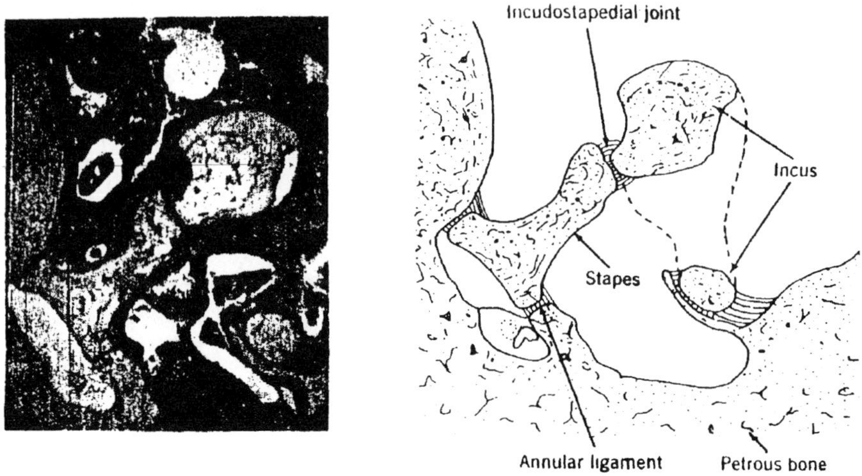

FIGURE 7.24. On the left is a photomicrograph showing the oval window with the footplate of the stapes held by its narrow annular ligament, and portions of the incus. The sketch on the right serves to identify the various parts (adapted from McCormick et al., 1970).

lies on the oval window connecting the middle ear to the cochlea, attached by a narrow annular ligament as shown in Fig. 7.24. After sound arrives at the bulla, propagating through the oval window into the mandibular fat channel, it is not clear how it is coupled into the bulla. Norris (1970) suggested that acoustic energy could possibly propagate through the thin-walled portion of a water- (or blood-) filled bulla into the middle ear without significant losses or perhaps flexural waves could propagate around the bulla and reach the ossicular chain (malleus, incus, and stapes) through the proccessus gracilis of the malleus. McCormick et al. (1970) argued for translational bone conduction and Johnson (1976) favored compressional bone conduction stimulated by pressure waves and not by mechanical bone stimulation. Translational bone conduction would exploit the same stapes-oval window "piston-cylinder" system, which functions in aerial hearing (McCormick et al., 1970). In compressional bone conduction, compressional waves compress the bulla from all sides and causing cochlear fluid to move, exciting the cochlear hair cells as the fluid is pressed out through the oval window. McCormick et al. (1970) argued against compressional bone conduction on the basis that it can only be realized if the bulla is contained in a denser solid and is equally compressed from all sides. However, the bulla is completely physically separated from the skull and is connected by cartilage, connective tissue, and fat. Fleischer (1970) argued for mechanical activation of the ossicular chain, which couples acoustic energy into the cochlear capsule. In the electrophysiological experiments conducted by McCormick et al. (1970, 1970), they found that cochlear potentials were only slightly (4 dB loss) affected by the removal of the malleus after it was severed from the incus. Although the malleus did not have an acoustic role, they concluded that the inner most portion of the ossicular chain was essential for the reception of sounds by the cochlea. A 17 dB drop was observed in the cochlea potential when the ossicular chain was dampened during high-frequency stimulation. McCormick et al. (1970) concluded that the ossicular chain must function as a rigid spring mounted inside the tympanic cavity, and when acoustic energy sets the bulla in motion, the inertia and stiffness of the ossicular chain probably cause the stapes to move relative to the oval window.

7.2.2 The Inner Ear

We will begin our discussion of the inner ears by dividing cetaceans into three different categories or types, following Ketten (1992, 1997), which are based on either the range of frequency they hear or on the type of sounds they produce. Type M animals are Mysticetes, which probably produce sounds of infrasonic to sonic frequencies (about 10 Hz to several kHz but below 20 kHz). Type II animals are dolphins and porpoises, which do not produce either pure tone or fm whistles. They include some of the smallest dolphins and porpoises such as the harbor porpoise (*Phocoena phocoena*), Commerson's dolphin (*Cephalorhynchus commersonii*), Hector's dolphin (*Cephalorhynchus hectori*), finless porpoise (*Neophocaena phocaenoides*), and two slightly larger animals, the Dall's porpoise (*Phocoenoides dalli*) and the pymgy sperm whale (*Kogia breviceps*). As we shall see in a later chapter, these nonwhistling dolphins emit relatively narrow band ($Q \sim 7$–12), high peak frequency (120–140 kHz), long-duration (>120 μs) echolocation signals that have greater than 20 oscillations. The Type I animals emit relatively broad band ($Q < 4$), wide peak frequency (40–140 kHz), short (<70 μs) echolocation signals that have less than about seven oscillations. The definition for Type I and II signals given here is different than the definition used by Ketten (1992, 1997), which was based solely on peak frequency of the echolocation signals and in our estimation is not totally accurate.

Scaled inner schematics of the vestibular and cochlear labyrinths of the Type M, Type I and II odontocetes are shown in Fig. 7.25, along with the corresponding structure for humans. It is obvious from the schematics that the vestibular system in cetaceans is substantially reduced and the semicircular canals are significantly smaller than the cochlea. Less than 5% of the odontocete VIII nerve is devoted to the vestibular fibers, compared to about 30% in terrestrial mammals (Ketten, 1997). The semicircular canal reduction in the mysticetes is less extreme than for odontocetes. The reduced semicircular canals may be an adaptation for life in the ocean, and probably play a much smaller role in balancing than for terrestrial mammals.

The structure of the cetacean cochlea is similar to that of most other mammals, but with some slight variations. The number of cochlea turns for cetaceans are typically slightly over two compared to almost three for humans. The

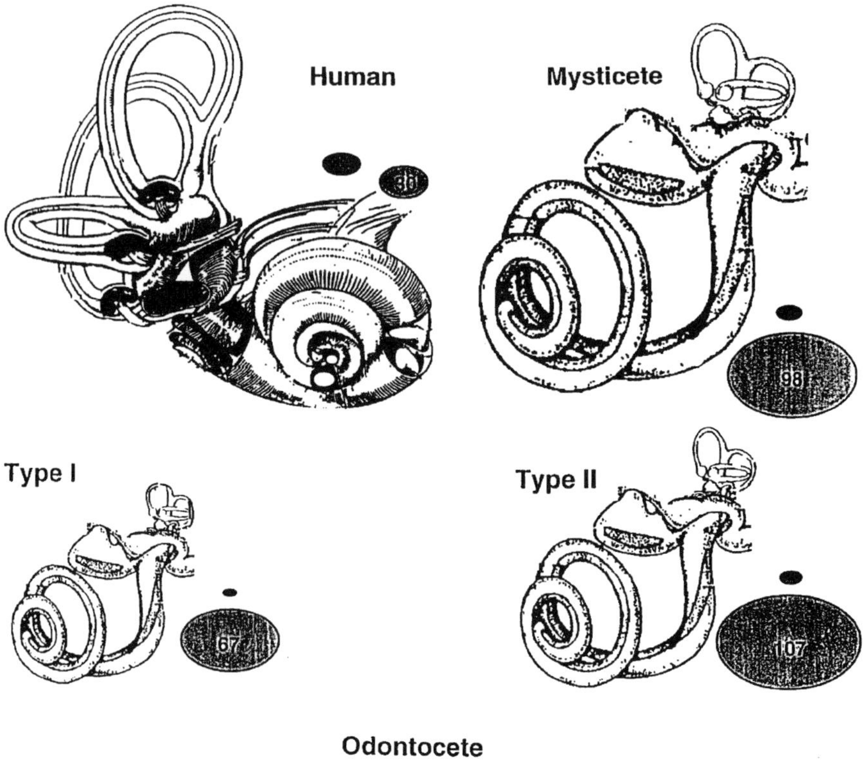

FIGURE 7.25. Scaled inner ear schematics showing the relative volume of the vestibular and cochlear labyrinth in humans, Mysticetes, and Type I and II odontocetes. The gray ovals represent relative cross sections of the vestibular (*dark gray*) and auditory (*light gray*) components of the VIII nerve. The numbers inside each oval are the fiber counts in thousands (adapted from Ketten, 1997).

number of cochlear turns for the Pacific white-sided dolphin (*Lagenorhynchus obliquidens*) is about 1.75. The cochlea is divided along its length by two very thin membranes (basilar membrane and spiral laminae) into two large chambers and is filled with fluid. It begins at the basal end where the oval and round windows are located and coils to a blind end at the apex. Like other mammals, acoustic energy enters the oval window as a compressional wave and causes the basilar membrane to vibrate and deform the acoustic receptor hair cells located in it. The basilar membrane is supported between bony supports along both edges. The width of the basilar membrane is narrow near the basal end and wide near the apical end. Wever et al. (1971b) measured the width of the basilar membrane of four *Tursiops truncatus* and found the width to vary greatly along the cochlea. The basilar membrane is narrow, around 25 μm in width at the basal end and increases in width at a uniform rate over the first 20 mm, and then at a greater rate over the rest of the cochlea. The maximum width is around

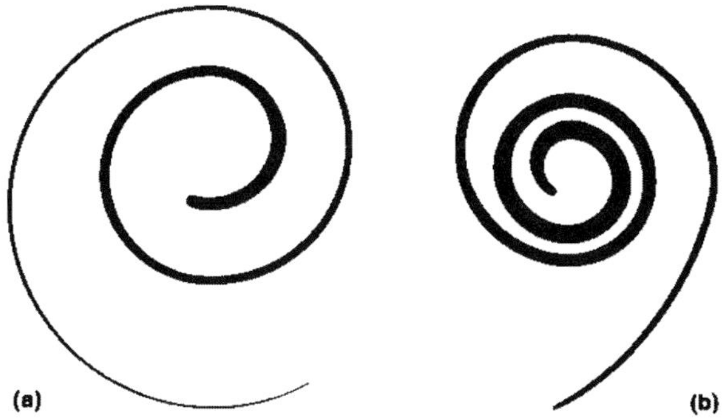

FIGURE 7.26. (a) The rightbasilar membrane of Tursiops truncatus and (b) the left basilar membrane of human (adapted from Wever et al., 1971b).

350 μm near the apical end. A variation in width of 14 times is typical for *Tursiops*. This is approximately twice as much variation as the human basilar membrane, which is a 6.25-fold variation. Scale drawings showing the form and changing width of the dolphin and human basilar membranes are shown in Fig. 7.26. The number of cochlear turns for *Tursiops* is slightly more than two, compared to nearly three for humans. The number of cochlear turns for *Lagenorhynchus obliquidens* is about 1.75, and the variation in width of its basilar membrane is about 11 (Wever et al., 1972). Schematic of uncoiled basilar membranes for representatives of the three types of cetacean inner ears are shown in Fig. 7.27. Also included in the figure are the approximate dimensions of the basilar membrane. The difference between the width of the basal and apical portion of the basilar membrane is about 10 for *Phocoena* and 11 for *Eubalaena*. The thickness of the basilar membrane varies by a factor of 5 from the basal to apical end for odontocetes and only by a factor of about 2.7 for the mysticetes. It should be noted that cetacean basilar membranes are highly differentiated with substantial interspecies differences as can be seen in Table 7.1, which is a modification of a similar table given by Ketten (1994). We did not include the last column of Ketten's table, which gave the peak frequency of sound emissions because of errors in some of her values. Wever et al. (1971b) argued that the most important condition for frequency differentiation in the cochlea is the variation in stiffness of the vibrating structure. Therefore, the variation in width of the basilar membrane and the rigidity of the suspension should contribute to an excellent capability of pitch discrimination for *Tursiops*. They also argued that the narrow width of the basilar membrane near the basal end of the cochlea (25 μm compared to 70 μm for humans) is in consonant with the ability of *Tursiops* to hear very high-frequency sounds. The rigidity of the basilar membrane depends on the amount of membrane stiffening provided by bony luminar support. Ossified outer spiral laminae are found in terrestrial mammals with ultrasonic hearing, but are absent or reduced in animals with lower frequency limits (Pye, 1972). The inner laminae in

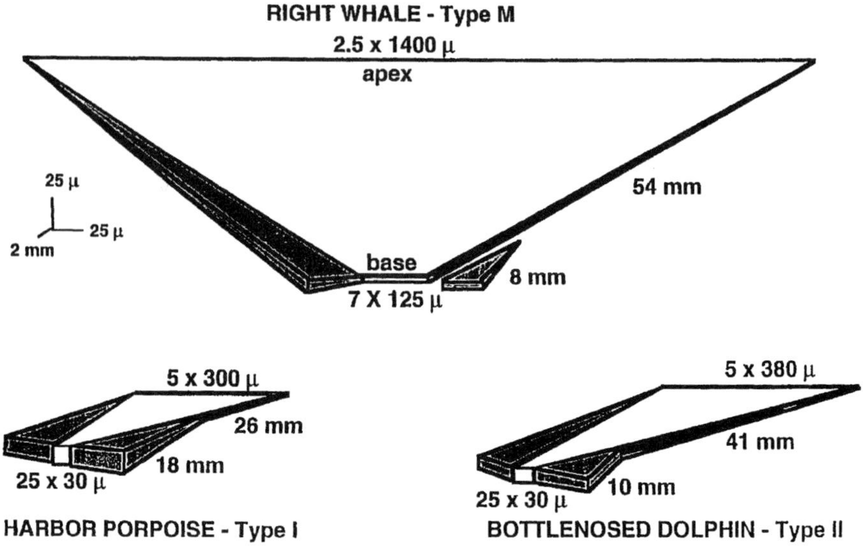

FIGURE 7.27. The basilar membrane and inner and outer spiral laminae (grey) for representative Type I (harbor porpoise), Type II (Bottlenose dolphin), and Type M (right whale species) (adapted from Ketten, 1997). The dimensions are similar but slightly different than those given by Wever et al. (1972).

mysticetes are spongy, fragile structures and the broad, thin membranes are supported laterally solely by a flexible spiral ligament (Ketten, 1994). In odontocetes, the basilar membrane is attached to a rigid bony outer lamina for some portion of the basal turn and is relatively rigid in comparison to most mammals. Ketten (1994) presented an interesting three-dimensional schematic showing the cochlea of three types of cetacean; her schematics are reproduced in Fig. 7.27. Ketten (1994) combined the basilar membrane morphometric data with stiffness distribution estimates based on cochlear type and outer laminar length to estimate the hearing range of several cetacean species. Stiffness gradients for each cochlear type were calculated by modeling basilar membrane segments as simple beams with uniform load partially distributed at one or both ends. These data were then fitted to a species scaled version of the compliance-position function to obtain the species-specific equation for mapping frequency against cochlear spiral length. The estimates of hearing ranges calculated by Ketten (1994) are shown in Fig. 7.29. We shall see in the next chapter that Ketten's estimates for the harbor porpoise and bottlenose dolphin are slightly higher by about 67% than what behavioral hearing studies indicate. The curves for the baleen whales are interesting since there are no behavioral or electrophysiological data on the hearing sensitivity of baleen whales. Ketten's estimate suggests that baleen whales' hearing is about the same frequency range as humans and perhaps even lower.

TABLE 7.1. Basilar Membrane (BM), Cochlear Spiral and Other Measurements (after Ketten, 1994)

Species	Common name	Type	Turns	B. M. Length (mm)	Outer Lamina (mm)	Basal/Apical Width (μm)	Thick (μm)
ODONTOCETI							
I. Geofrensis	bouto	II	1.5	37	–		
P. phocoena	harbor porpoise	I	1.5	26	17.6	30/290	25/5
G. griseus	Risso's dolphin	II	2.5	41	–	40/420	20/5
L. albirostris	white-beaked dolphin	II	2.5	35	7.5	30/360	20/5
S. attenuata	spotted dolphin	II	2.5	37	7.4	40/400	20/5
T. truncatus	bottlenose dolphin	II	2.25 41	10.3	30/370	25/5	sperm whale
L. obliquidens	white-sided dolphin	II	1.75	35			
P. catodon	sperm whale	II	1.75	54	–	–	–
MYSTICETI							
B. acutorostrata	Minke whale	M	2.25	55	–	100/1500	–
B. mysticetus	bowhead whale	M	2.25	61	10	120/1670	7/3.
B. physalus	fin whale	M	–	–	–	100/2200	–
E. glacialis	right whale	M	2.5	50	7	125/1400	7/3
M. novaeangliae	humpback whale	M	2.5	54	–	–	–

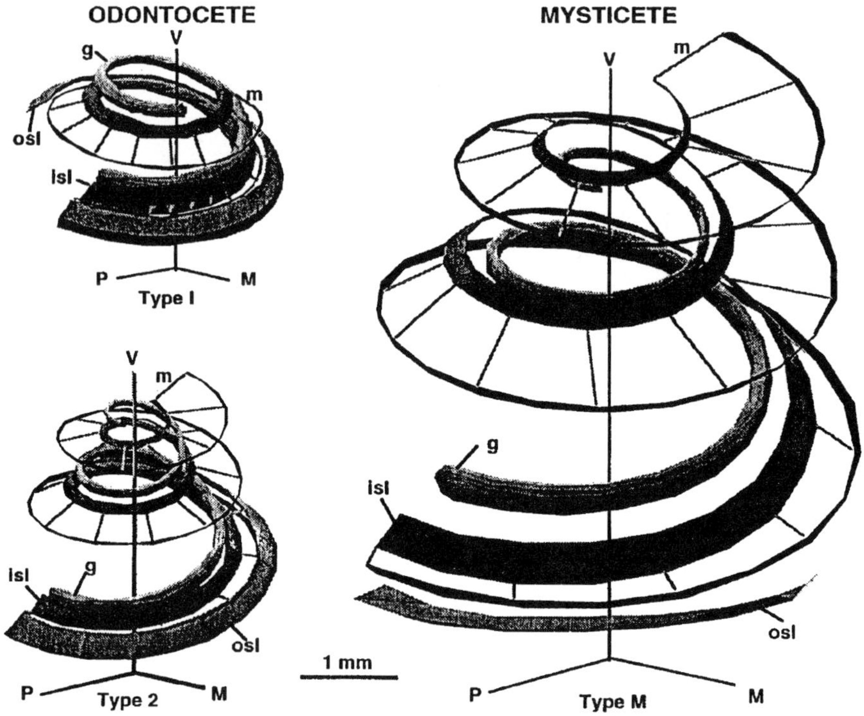

FIGURE 7.28. Basiliar membrane and spiral lamina distribution in Cetecea. The cochlea are shown inverted from *in vivo* orientations. (g) spiral ganlion; (isl) inner osseous spiral lamina; (m) basilar membrane; (osl) outer osseous spiral lamina; (l) lateral; (p) posterior; (v) ventral (adapted from Ketten, 1994).

The structure of the hair cells along the basilar membrane of *Lagenorhynchus* and *Tursiops* were studied by Wever et al. (1971b, 1972), and the number of hair cells were estimated. The number of inner hair cells was approximately 3,451 for *Tursiops* and 3,275 for *Lagenorhynchus*. The number of outer hair cells was approximately 13,933 for *Tursiops* and 12,799 for *Lagenorhynchus*. In comparison, the human ear typically has 3,475 inner hair cells and 11,500 outer hair cells. Therefore, the two species of dolphins and man are on about the same level as regards to the primary receptor elements of the cochlea.

Wever et al. (1972) also estimated the size of the ganglion-cell population associated with the hair cells. They estimated a population on the order of 60,000–70,000 for *Lagenorhynchus* and 95,000 for *Tursiops*. These numbers are considerably greater than man's complement of 30,500 ganglion cells. Comparing the number of ganglion cells to hair cells, *Lagenorhynchus* has about four times more ganglion cells than hair cells, and *Tursiops* about five times. These numbers are again high in comparison to the human

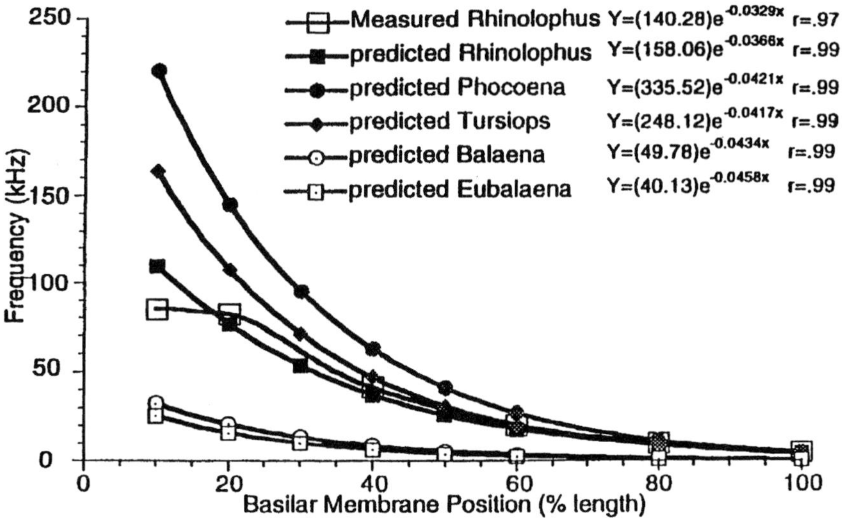

FIGURE 7.29. Frequency vs. cochlear position distributions in the horseshoe bat (*Rhinolphus ferrumequinum*) and four cetaceans. Membrane lengths are normalized as percentages of cochlear length. Curves for the bat are based on published data for membrane dimensions (predicted) and electrophysiological recordings (measured) (adapted from Ketten, 1994).

cochlea, which has about two times more ganglion cells than hair cells. The ganglion cell density for several species of odontocetes, along with comparative values for the horseshoe bat and human, are given in Table 7.2. The data for ceteceans is rather sparse and information is available only for a few species. Nevertheless, it is fairly obvious that the density of ganglion cells per unit length of basilar membrane is much higher in echolocating odontocetes than in other mammals. Wever et al. (1972) suggested that the high ratio of ganglion cells to hair cells in the dolphin might aid in the representation of high-frequency acoustic information and fine details of cochlear events to higher centers of the auditory nervous system.

TABLE 7.2. Ganglion Cell Density for Some Odontocetes Along with Comparative Densities for the Horseshoe, Bat, and Human (after Ketten, 1992)

Species	Type	Total ganglion cells	B.M. length (mm)	Ave density (cell/mm)
Phocoena phocoena	I	66,933	26	2574
Lagenorhynchus obliquidens	II	70,000	35	2000
Stenella attenuata	II	72,506	37	2171
Tursiops truncatus	II	95,000	41	2317
Rhinolophus ferrumequinum	–	15,953	16	997
Homo sapiens	–	30,500	31	973

7.3 The Pinniped Ear

Pinnipeds being amphibious must have good hearing sensitivity in two vastly different media: air and water. Special adaptations and modifications for this complex task from typical terrestrial mammalian or typical aquatic ear would seem to be necessary for efficient operation in both media. The difference in the impedance of water and air, of approximately 3,600, has already been discussed in Section 7.1.1. The middle ears of terrestrial mammals function to compensate for losses attributed to the large impedance difference between air and the fluid of the inner ear. A pinniped would also need some kind of middle ear compensation when hearing in air and may not require any middle ear compensation when hearing underwater. Therefore, it seems that the ear of pinnipeds may be more complex than either the ear of completely terrestrial or completely aquatic animals. A schematic of the ear of a true seal is shown in Fig 7.30 and for our purposes the schematic can also serve as a "generic" pinniped ear applicable for the three pinniped family: phocids, odobenids, and otariids.

The pinniped ear is homologous to the ear of terrestrial mammals, so it is not difficult to envision that it would function in the same manner for airborne sounds as other terrestrial mammals. Plugging the meatus of a pinniped causes the animal's hearing sensitivity to decrease (Mohl and Ronald, 1975). However, such an ear would not seem to be ideal for the reception of underwater sounds. Humans hear underwater sounds by bone conduction and typically experience a 30 dB loss in hearing sensitivity. The middle ear osscicles are essentially

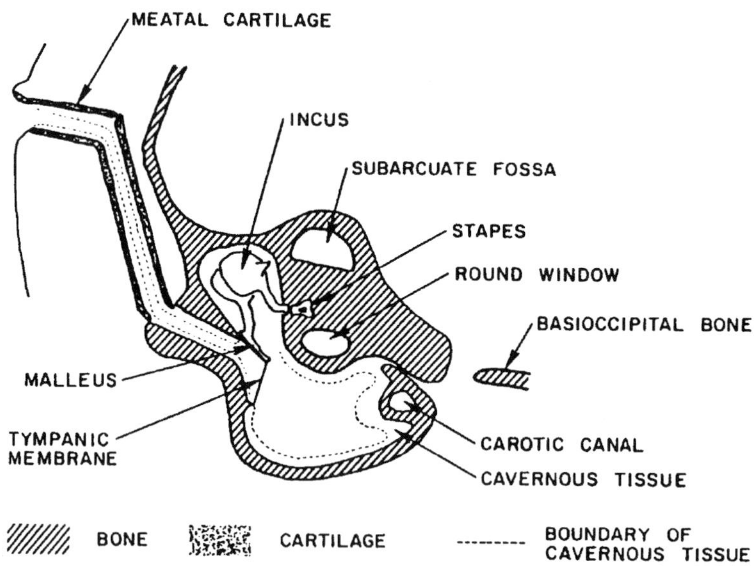

FIGURE 7.30. Schematic of a generic pinniped ear (adapted from Mohl, 1967).

nonfunctional under water. Furthermore, the ability to discriminate frequency of tones and to localize sounds is lost by humans under water. The impedance mismatch between the head and water is slight so that sounds may enter any or all parts of the head and proceed directly to the organ of Corti. Underwater sounds would tend to take other paths through the air-filled middle ear cavity in reaching the inner ear. In shallow depths, the tympanic membrane of pinnipeds is in contact with air in the meatus and the middle ear cavity, which presents an acoustic barrier to underwater sounds. Therefore, it is reasonable to suspect that pinnipeds, like humans, probably hear sounds under water through bone conduction; however, as we shall see in a later chapter, pinnipeds also have directional hearing and frequency discrimination capabilities.

Hearing by bone conduction in pinnipeds may seem very reasonable; however, it is yet to be verified experimentally (Kastak, Personal Comm). Bone conduction can take place in two basic manners or modes. The inertial or translatory mode operates by causing the cochlear capsule to vibrate, which in turn causes the fluid in the cochlea to exhibit motion relative to the organ of Corti. In this mode of bone conduction, the inertia of the ossicular chain also plays a role by restraining the movements of the stapes, causing its motion to lag behind. Therefore, hearing is by the stapedial impulse and by inertia lag of the cochlear fluid (Repenning, 1972). The inertial mode is also referred to as the "resonant" reaction. The second mode of bone conduction, referred to as the "conductive" reaction, involves the cochlear capsule being distorted by the penetrating acoustic energy, producing lymph action in adjusting to changes in the cochlea shape. Both types of bone conduction probably operate simultaneously in contributing to a pinniped's hearing capabilities. A detailed discussion of bone conduction and hearing in pinnipeds has been presented by Repenning (1972); his article is definitely worthwhile reading.

Mohl and Ronald (1975) examined underwater sound conduction in the head of the Harp seal by measuring the cochlear microphonics generated by the hair cells on the basilar membrane in response to sound stimulation. They varied the position of a sound source held gently against the subject's head. The area of maximum sensitivity to underwater sound, found by Mohl and Ronald (1975), are shown in Fig. 7.31. Sound entered the animal's head best at locations below the auditory meatus, providing strong support for hearing by bone conduction in the Harp seal.

A second approach consisted of using sound opaque material to shade specific areas from stimulation while the cochlear microphonics were being measured. This approach confirmed the results shown in Fig. 7.30.

7.3.1 The Outer Ear

True seals (phocids) do not have a pinna while odobenids and otariids have only rudimentary pinnas. The absence or reduced-sized pinnas will not only reduce hydrodynamic drag but also reduce hydrodynamic flow noise

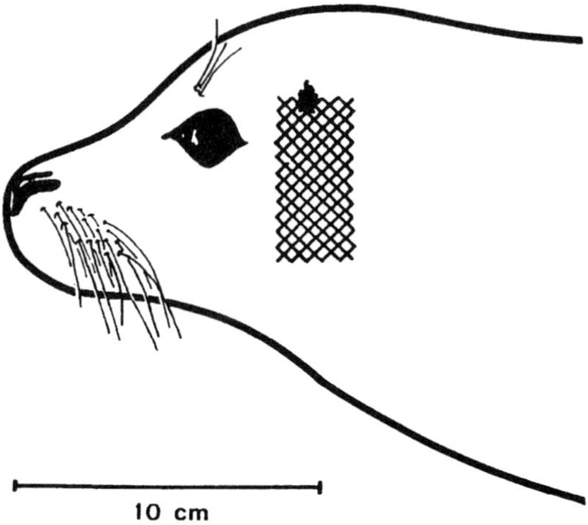

FIGURE 7.31. Area of maximum sensitivity to underwater sound for the Harp seal (adapted from Mohl and Ronald, 1975).

in contrast with conventional pinnas (Mohl, 1967). However, for airborne sounds, the absence of pinnas will affect a mammal's ability to localize front-back difference of received sounds and to minimize the reception of ambient noise. The reduced pinna of otariids may also have a role in closing the external meatus while the animal is under water. The pinna is sculpted in such a way that allows it to fold upon itself and seal the meatus (Huber, 1934).

The external auditory meatus of pinnipeds is long, narrow, and filled with cerumen and wax-covered hair (Ramprashad et al., 1972). The meatus deviates from a straight tubular structure of humans with several bends along its length. It is supported by a meatal cartilage, which is flexible and collapsible and eventually couples to a bony section that leads to the tympanic membrane. From the bony portion, the cartilage-supported section bends dorsally and rostrally, and just before the external orifice it bends laterally again. The meatus opening is controlled by muscles attached to the meatal cartilage, which can serve to close the meatus during diving. According to Mohl (1967), the orifice of a true seal closes to a vertical slit before submergence, but occasionally this is delayed until the animal is well beneath the surface. Since the meatus is only a few millimeters wide and is filled with waxy hair, it seems unlikely that water will enter the ear canal if the animal delays the closing while underwater. Mohl (1967) also argued that the shape of the canal would make it difficult to remove any water, so that water-filling during dives probably does not occur.

7.3.2 *The Middle Ear*

The middle ear bones and middle ear cavity of pinnipeds are similar in shape to terrestrial mammals, but the phocids and odobenids tend to have larger and much denser bones. The ossicles of phocids are much larger than those of other carnivores, except the odobenids (Repenning, 1972). The middle ear bones of phocids are larger and denser than those of otariids. Mohl (1967) found that the ossicles from a harbor seal was about 11 times heavier than the ossicles of a California sea lion of comparable size. The middle ear cavity of phocids is also larger than those of otariids. The tympanic membrane of the harbor seal is also about four times larger in area than that of the California sea lion. The combination of inflated auditory bullae, large ossicles, and large tympanic membranes of phocids are clearly distinctive from those of otariids (Repenning, 1972). However, the vast size difference is not reflected in the difference in the area ratio between the tympanic membrane and the footplate of the stapes: about 27:1 in the seal and 23:1 in the sea lion. These values are well within the range for terrestrial mammals. Listed in Table 7.3 are some typical values for the diameter of the tympanic membrane and oval window for pinnidepds and two other mammals. The tympanic membrane diameter of phocids and odobenids are generally larger than that of humans and about the same order of magnitude as the bear. On the other hand, the tympanic membrane diameter of otariids is about 60% smaller than humans. The shape of the tympanic membrane is also different for otariids and phocids; the ear drums of otariids tend to be rather flat whereas the tympanic membrane of phocids has more of an inflated appearance.

The inside surface of the bony tympanic bulla is covered with a thick layer of cavernous tissue or corpus cavernosum, which is also found in the meatus. The

TABLE 7.3. Tympanic Membrane-Oval Window area Ratios of Several Pinniped Genera, Bear, and Man (adapted from Repenning, 1972)

Genera	Tympanic membrane diameter (mm) A-max	B-min	Oval window diameter (mm) C-max	D-min	Approximate ratio Drum-window area (AXB:CXD)
Homo	7.3	7.2	3.2	1.1	22:1
Ursus	13.0	6.7	1.7	1.0	43.1
PHOCIDAE					
Erignathus	11.7	7.6	2.5	1.6	37:1
Pusa	13.4	10.3	3.2	1.7	22:1
Phoca	12.0	7.9	3.4	1.7	17:1
Mirounga	13.0	10.0	4.1	3.2	10:1
Monachus	7.1	6.4	3.2	1.9	9:1
Leptonychotes	10.0	7.1	3.2	2.4	9:1
ODOBENIDAE					
Odobenus	11.5	7.5	2.6	1.7	20:1
OTARIIDAE					
Arctocephalus	5.7	4.2	1.9	1.0	12:1
Callorhinus	5.7	5.0	2.0	1.2	12:1
Eumetopias	6.4	4.7	1.6	1.6	10:1
Otaria	7.0	5.0	2.0	1.3	7:1
Zalophus	5.7	4.7	2.0	1.3	7:1
Neophoca	6.0	4.5	2.3	1.5	7:1

eustachian tube of pinnipeds is supported by relatively large amounts of cartilage, making the bore relatively small. Since the eustachian tube connects to the middle ear cavity and may be blocked, equalization may be performed by swelling of the cavernous tissue (Mohl, 1967).

Repenning (1972) argued that, for depth from 70 to 100 m, pressure equalization would cause the cavernous tissues to be flooded with blood and be in contact with most or all of both sides of the tympanic membrane so that an acoustic route via the tympanic membrane may be somewhat operable at depths greater than about 70 m. Unlike dolphins, the tympanic bulla of pinnipeds is not acoustically isolated from the rest of the skull, except extensive lacerate foramens separate the bulla from the basioccipital bone. The lacerate foramen may play a role in reducing the coupling of sounds between the two bullas so that time-of-arrival difference and intensity difference cues can be obtained for a directional hearing capability (Mohl, 1967).

7.3.3 The Inner Ear

Very little anatomical work has been done on the inner ear of pinnipeds. The cochlea of phocids has about two and one-half turns. The basal whorl is greatly enlarged in comparison to most mammals, including ortariids and obendenids. The basal whorl of the phocid cochlea is transverse to the orientation of the skull, rather than posterolaterally as in other carnivores including otariids and odobenids (Repenning, 1972). According to Repenning (1972), the basal whorl seems most effective as an adaptation for the reception of sounds from the side of the head in the inertia mode of bone conduction. The enlarged basal whorl of the cochlea is also accompanied by enlarged scalae interior to the cochlea, along with enlarged oval and round windows. In extreme cases like the *mirounga*, the round window is outside the middle ear cavity, through a large foramen at the bulla-mastoid junction just posterior to the auricular foramen and the stylomastoid foramen (Repenning, 1972). The aqueduct of the cochlea of phocids is also relatively wide with a diameter of 2.5 mm (Mohl, 1967). Ramprashad et al. (1972) reported that the number of hair cells in the cochlea of the harp seal (*Phoca groenlandica*) and the ringed seal (*Phoca hispida*) were higher than that of human, but within the range of variations for humans.

7.4 The Sirean Ear

The ears of sireans have not been the subject of much research. Ketten et al. (1992) has written an excellent article on the structure and function of the manatee ear. They used computerized tomography (CT) along with conventional dissection and thin-section microcopy to examine seven ears from one juvenile and three adult Florida manatees (*Trichechus manatus*). The material in this section will be a brief synposis extracted from Ketten et al. (1992); those interested in further details are encouraged to read their article.

Manatees have no pinnae or superficial auditory structures, with the auditory meatus toward the surface being only a dimple, leading to a narrow canal that is packed with black debris and discarded epithelial cells. The role of the external auditory meatus in hearing by the manatee is not clear. Like pinnipeds, manatees probably hear underwater sounds by some form of bone conduction. The tympano-periotics are attached to the inner wall of the cranium and the auditory system seems to be fully mature structurally and may be functional at birth. Consequently, the tympano-periotic complex is disproportionately large in young manatees.

The tympanic membrane is a laterally convex (17×10 mm) ellipse, but is otherwise structurally similar to the tympanic membrane of terrestrial mammals. The averaged membrane area calculated by Ketten et al. (1992) for their subjects was 150 mm^2, which is similar to the membrane area of some of the phocids shown in Table 7.3. The osscicular chain is massive and nearly straight, and is joined together in such a way that there may be two major axes of rotation on a functioning ossicle. The malleus is a thick ovoid. The stapes and incus lie on a medial-lateral line posterior to the malleus. The stirrup-like structure of humans and other animals is replaced by a columnar structure, resembling the columella of reptiles.

The cochlea of *T. manatus* is a multi-turn equiangular spiral that is similar to most other mammals. Some of the properties of the cochlea of adult manatees are given in Table 7.4, along with similar structures for the bottlenose dolphins and humans. The number of turns of the cochlea is less than dolphin and human. There is little difference in the width and thickness of the base and apex of the basilar membrane. The ratio of the basilar membrane width at the base and apex is about 1:3 for *T. manatus* compared to 1:13 for the bottlenose dolphin and 1:6 for human. The thickness ratio at the base and apex is 1.4:1 for *T. Manatus* compared with 5:1 for *Tursiops* and 2:1 for humans. The small difference in the ratio of basilar membrane width and thickness at the base and apex of the cochlea led Ketten et al. (1992) to speculate that the manatee has a narrow range of hearing. However, we shall see in a later chapter that the manatee can hear very well at ultrasonic frequencies. The cochlea structure of *T. manatus* is poorly developed, especially at the basal end. An outer osseous

TABLE 7.4. Morphometry of Adult Manatee, Bottlenose Dolphin, and Human Cochleae (after Ketten et al. (1992).

	Trichechus manatus	Tursiops truncatus	Homo sapiens
Cochlear Canal Length (mm)	35.0	40.0	32.0
Canal Diameter (mm)	7.7	7.5	7.5
Basilar membrane width Base/Apex (um)	200/600	30/400	100/600
Membrane Thickness Basal/Apical (um)	7/5	25/5	10/5
Turns	1.75	2.25	2.50

spiral lamina is missing while the inner osseous spiral lamina is thin and consists mainly of fibrous tissue intermixed with lightly mineralized bone.

7.5 Ears in Fishes

The ears in fishes function very much like ears in mammals and other vertebrates. Fishes localize sound sources, discriminate frequencies, tell predator from prey, and may even use sound for navigation. Little is known, however, about "how" fishes are able to accomplish these tasks. Fishes have two inner ears that lie within the cranial cavity just to the left and right of the brain as shown in Fig. 7.32. On the surface, their inner ear seems quite simplistic, consisting of three semicircular canals and three otolith end organs that appear to function as accelerometers. Fishes have no outer or middle ears and, unlike mammals, their inner ear has no cochlea. However, the basic mechanisms of stimulation of the hair cells in the inner ear and the conversion of acoustic energy to electrical signals compatible with the nervous system, as we described in the first section of this chapter, are the same in fishes as well as all other vertebrates.

7.5.1 Structure of the Inner Ear

Although the exact structure of the ear varies from one species to another (and there are at least 25,000 species), all known inner ears of fish have three semicircular canals and three fluid-filled otolithic end organs – the utricle, saccule, and lagena – as illustrated in Fig. 7.33. The semicircular canals detect angular acceleration (e.g., rotational acceleration of the head) in fishes just as they do in terrestrial mammals. The otolith organs in fish have two roles. First, they serve as vestibular organs and measure the position of the head in the vertical direction relative to gravity and in other directions relative to the acceleration of the body, just as they do in terrestrial vertebrates. Second, they are involved in sound detection. The earliest work suggested that the primary auditory end organs in fishes were the saccule and lagena, but there is a growing body of evidence that now suggests that all three of the otolithic end organs have roles in hearing (Denton and Gray, 1979; Higgs et al., 2004).

Each otolithic organ contains a single, dense, calcareous stone or otolith suspended in a chamber or sac filled with endolymph. On the wall of each sac lies a sensory epithelium, which contains vertebrate Type I and Type II hair cells innervated by branches of the VIIIth nerve as illustrated in Fig. 7.33. In-between the otolith and sensory epithelium is a thin otolithic membrane, which connects to both structures and keeps them in the same position relative to each other. Each hair cell has a bundle of hair-like cilia that project through Swiss-cheese-like holes in the otolithic membrane into the chamber. Because fishes are not much denser than the surrounding water, they move with the oscillatory motion of a sound wave. But the otoliths of the inner ear are four to five times denser than water, so they lag behind the motion of the

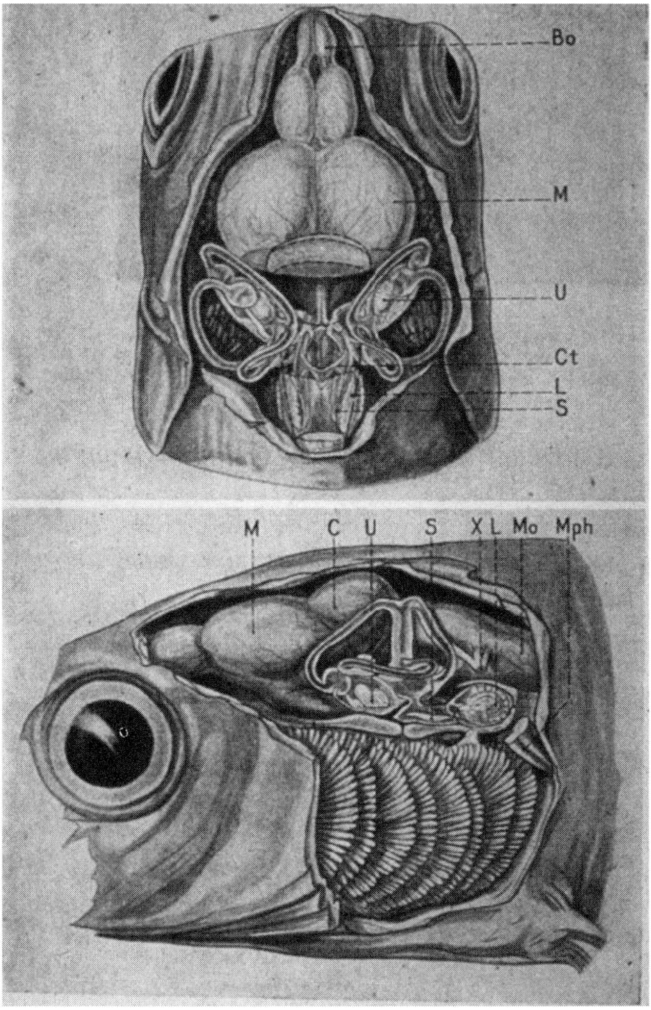

FIGURE 7.32. The inner ear in *Phoxinus laevis* (a minnow) with respect to the brain and other structures in the head (from von Frisch and Stetter 1932). M – medulla; C – cerebellum; U – utricle; S – saccule; L – lagena; X – 10th cranial nerve (not associated with the auditory system).

rest of the body. This creates relative motion between the otolith and sensory epithelium that causes bending of the hair cell ciliary bundles, which results in stimulation of the auditory nerve (van Bergeijk, 1964).

The first direct investigation of the mechanics of the otoliths was that of de Vries (1950), who used X-ray photography to measure the displacements of the otoliths in intact fish heads. Sand and Michelsen (1978) used a laser vibrometry technique to measure the movement of different parts of the saccular otolith in perch (*Perca fluviatilis*) during horizontal vibration of

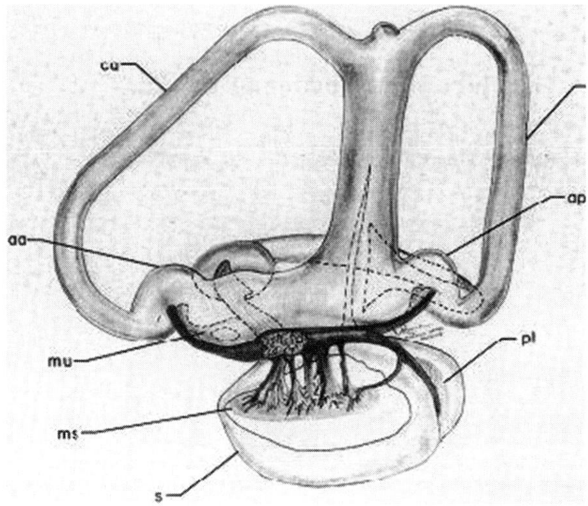

FIGURE 7.33. The right ear of *Salmo salar* (salmon) showing the three semicircular canals, three otolith end organs: mu, utricular epithelium; s, saccule; ms, saccular epithelium; pl lagena (redrawn from Retzius 1881 by Platt and Popper 1981).

the fish along its longitudinal axis. These *post mortem* experiments required the brain to be removed, fluids to be drained from the saccule, and the saccular otolith to be exposed and painted to increase reflectivity. Fluid moved back into the saccular chamber (its origin was unknown), but removal of the brain may have affected the results.

Each otolithic end organ may have thousands of sensory hair cells, resulting in the whole inner ear having tens or even hundreds of thousands of hair cells (Popper and Hoxter, 1984). Fishes continue to produce sensory hair cells throughout much of their lives (Lombarte and Popper, 1994, 2004; Higgs et al., 2003). The number of hair cells increases with growth of the fish. In addition, there is evidence that fishes, unlike mammals, can replace sensory hair cells that have been damaged as a result of exposure to certain drugs (Lombarte et al., 1993) (although there have been no studies to determine if fishes can replace sensory cells that have died as a result of exposure to sound).

The apical ciliary bundles of Type I and Type II sensory hair cells have a single kinocilium and at least 40 stereocilia as shown in Fig. 7.34. The maximum depolarizing receptor potential is achieved when the ciliary bundle bends in the direction of the kinocilium from the stereocilia. A hyperpolarizing potential results from bending in the opposite direction away from the kinocilium. Bending in any other direction generates a receptor potential that is proportional to the cosine of the direction of the sound source (Hudspeth and Corey, 1977). In addition, on the sensory epithelia the hair cells are organized in groups having best sensitivity in a particular direction (i.e., toward the kinocilium from the stereocilia). Figure 7.35 shows these directions of best

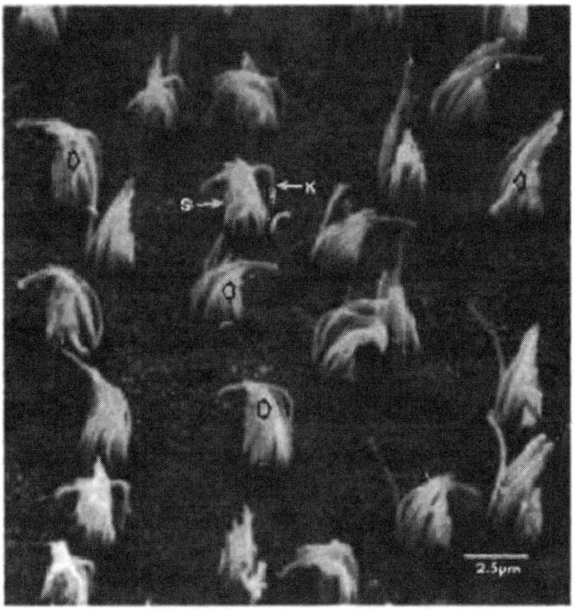

FIGURE 7.34. Scanning electron microscope image of the ciliary bundles protruding from the sensory epithelium in the lagena of *Carassius auratus* (goldfish). The left and right arrows indicate the direction of maximum sensitivity for each hair, which is bending toward the kinocilium (K) from the stereocilia (S) (after Rogers et al. 1988).

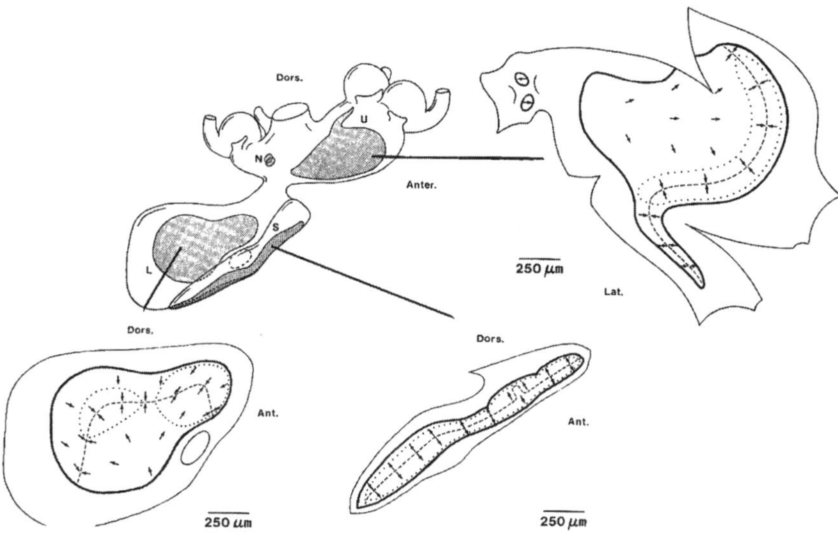

FIGURE 7.35. Exploded views of the sensory epithelia in the left ear of *Carassius auratus* show the organization of hair cells based on the location of the kinocilium with respect to the stereocilia. Arrows indicate the direction of maximum receptor response. M – macula neglecta; U – utricle; L – lagena; S – saccule (after Platt 1977).

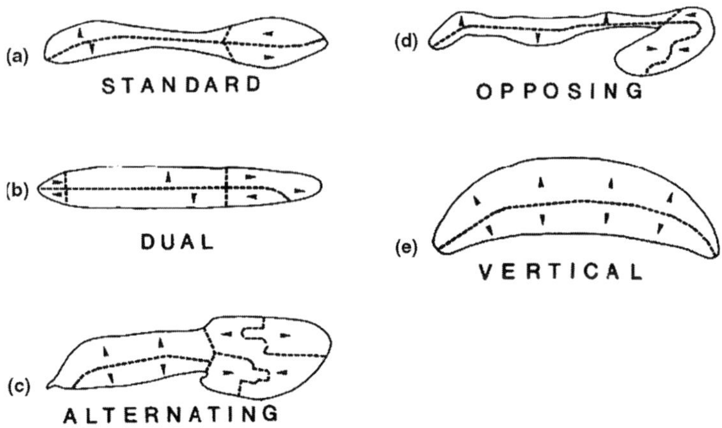

FIGURE 7.36. Hair cell orientation maps on the saccular sensory epithelium in five teleost fishes: (**a**) salmoniform, (**b**) gadiform, (**c**) moray eel; (**d**) anabantid; and (**e**) mormyrid (also characteristic of otophysans, including catfish and goldfish). The arrowheads indicate the direction that would be most excitatory (i.e., movement toward the kinocilium) for hair cell groups bounded by dashed lines. Anterior is to the right and dorsal is up (after Popper and Coombs 1982).

sensitivity in the left ear of a goldfish (*Carassius auratus*). Figure 7.36 displays characteristic hair cell orientation patterns in the saccule of five different species. Arrays of differently oriented hair cells from the otolithic organs in both ears may provide detailed information about the direction and pathway of the motion of impinging sound (Schuijf and Buwalda, 1980; Popper et al., 1988; Rogers et al., 1988).

7.5.2 Frequency Discrimination

Just as mammals, fishes also have the ability to discriminate different frequencies; however, the mechanism for frequency discrimination remains unknown. Only a few studies have been completed in this area. Enger (1981) investigated the possibility of a tonotopic organization similar to that of the basilar membrane in mammals and in the saccule of cod fish. He exposed fish to intense narrowband tones and then used scanning electron microscopy to examine the saccular sensory epithelium for damage to hair cells as illustrated in Fig. 7.37. He found a gross mapping with frequency along the length of the epithelium, with a preference for higher frequencies toward the rostral end of the epithelium and lower frequencies toward the caudal end as shown in Fig. 7.38, but no one has duplicated his results. Cox et al. (1986) conducted a similar experiment on goldfish and found some evidence of tonotopic organization on the saccular epithelium, but their stimulus tones were so intense that hair cell damage was generally widespread.

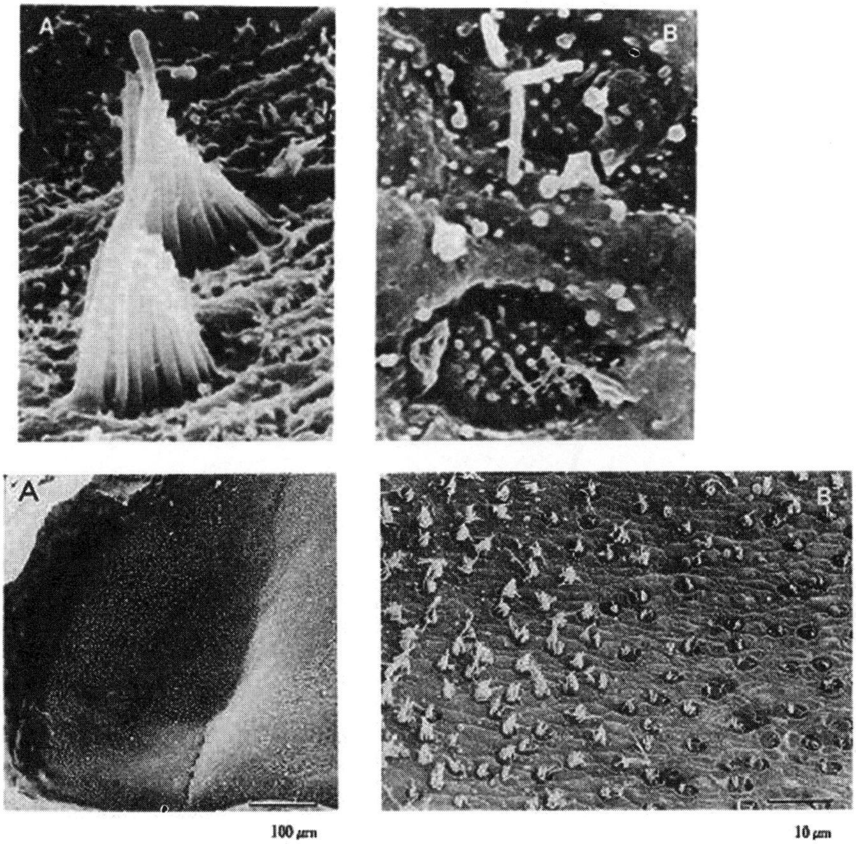

FIGURE 7.37. Enger (1981) exposed cod to intense tonal sounds and found hair cell destruction in different regions of the saccular epithelium whose locations depended on the frequency of the tone. *Top*: Ablated ciliary bundles of underlying hair cells due to overstimulation by intense sound. *Bottom*: Image of saccular epithelium showing a region with hair cell destruction resulting from exposure to single frequency.

7.5.3 Auxiliary Structures and Acoustic Pathways to the Ear

Fishes do not have an outer or middle ear; but because they are acoustically coupled with water, the motion associated with a sound wave directly stimulates their inner ears by generating relative motion between the otoliths and the sensory epithelia. This pathway for sound to the inner ear is called the "direct" path. Many fish species also possess auxiliary structures, which are believed to enhance audition. The most common of these is the swim bladder, a gas-filled chamber (or chambers) located in the abdomen and primarily used to regulate buoyancy. The two chambers of the goldfish swim bladder are easily seen in the X-ray image shown in Fig. 7.39. The swim bladder

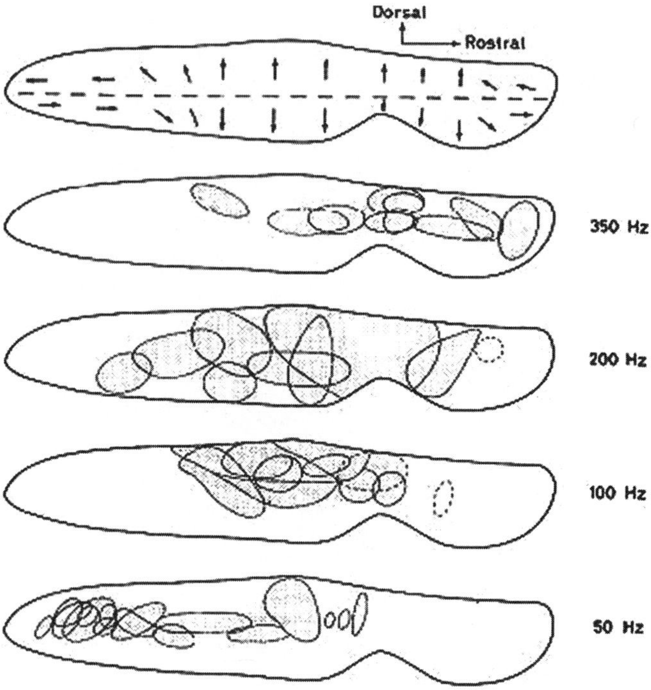

FIGURE 7.38. Enger (1981) found evidence of tonotopic organization in the saccule of cod by exposing fish to intense tones and then examining the sensory epithelium for spatially selective hair cell damage.

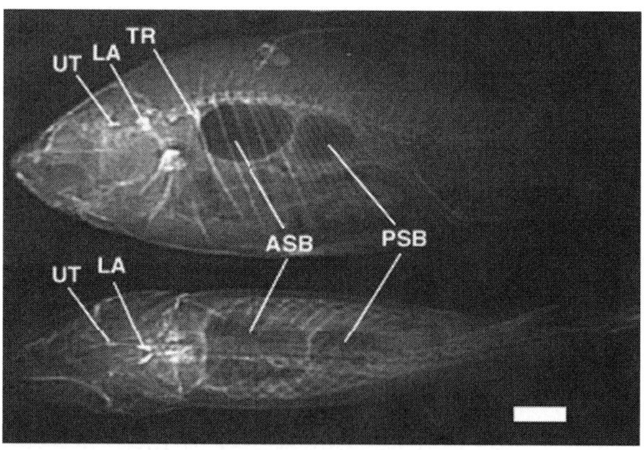

FIGURE 7.39. Lateral (*top*) and dorsal (*bottom*) *in vivo* X-ray images of *Carassius auratus* showing the anterior (ASB) and posterior (PSB) chambers of the swim bladder, the tripus Weberian ossicle (TR) connecting to the anterior swim bladder, and the otoliths in the utricle (UT) and lagena (LA) of the inner ear. The bar indicates a length of 1 em.

oscillates in response to sound pressure, similar to the tympanic membrane in the middle ear of mammals, and radiates acoustic energy, which also may stimulate the hair cells in the otolithic organs of the inner ear. This pathway for sound to reach the inner ear is called the "indirect" path.

Several researchers have investigated the mechanics of the swim bladder. In a *post-mortem* procedure, Popper (1974) used a probe microphone inserted into the anterior chamber of the goldfish (*Carassius auratus*) swim bladder to measure the sound pressure level (SPL) within the swim bladder during acoustic stimulation. Results showed that the swim bladder had a flat response from 50 to 2,000 Hz, which corresponds to the auditory bandwidth in goldfish. Clarke et al. (1975) used a laser light scattering approach to examine the swim bladder of the goldfish. These tests were performed *post mortem*, with the swim bladder partially exposed. Results of this method showed a sharp decrease in swim bladder response above 1,000 Hz. A change in response was also observed due to *post-mortem* aging effects. More recently, Finneran and Hastings (2000, 2004) used a noninvasive ultrasonic vibration measurement system originally developed by Rogers and Hastings (1989) to investigate the *in vivo* response of the swim bladder in goldfish. They found swim bladder resonance frequencies and bandwidths correlating with the most sensitive region of hearing.

Some species possess more elaborate adaptations; for example, Cyprinads and other members of the series Otophysi possess not only a two-chambered swim bladder but also a series of modified vertebrae, the Weberian ossicles, which mechanically couple the anterior swim bladder to the fluids of the inner ear. In this way, the Weberian ossicles function somewhat like the bones of the middle ear in mammals. Fish with Weberian ossicles have lower auditory thresholds and broader auditory bandwidth. Figure 7.40 illustrates the locations of the ear, swim bladder, and Weberian ossicles in the peripheral auditory system of an Otophysan fish. The largest ossicle, the tripus, connects the anterior swim bladder chamber to the second ossicle, the intercalarium, which connects anteriorly to the scaphium and, in some species, the claustrum (Fig. 7.41). The scaphium and claustrum, when present, are embedded in the walls of the sinus impar, a fluid-filled canal that projects rostrally and ultimately communicates with the sinus endolymphaticus, transverse canal, and the saccule (Alexander, 1962; Evans, 1925). The histology slide in Fig. 7.42 shows the interconnection between the transverse canal and saccules.

Expansion of the swim bladder causes forward motion of the ossicles and fluid flow into the saccular chamber, while compression causes fluid flow out of the saccule. This creates additional relative motion between the saccular otolith and its sensory epithelium. Vaitulevich and Ushakov (1974), Vaitulevich (1979), and Altman et al. (1984) used holographic measurement techniques to examine the behavior of the swim bladder in a Cyprinad fish, the carp (*Cyprinus carpio*). These experiments required removal of the lateral wall of the fish and exposure of the swim bladder, but regions of maximum vibration in isolated swim bladders were found to coincide with the approximate location of the attachment of the tripus. This specialized pathway for sound to the inner ear operates in addition to

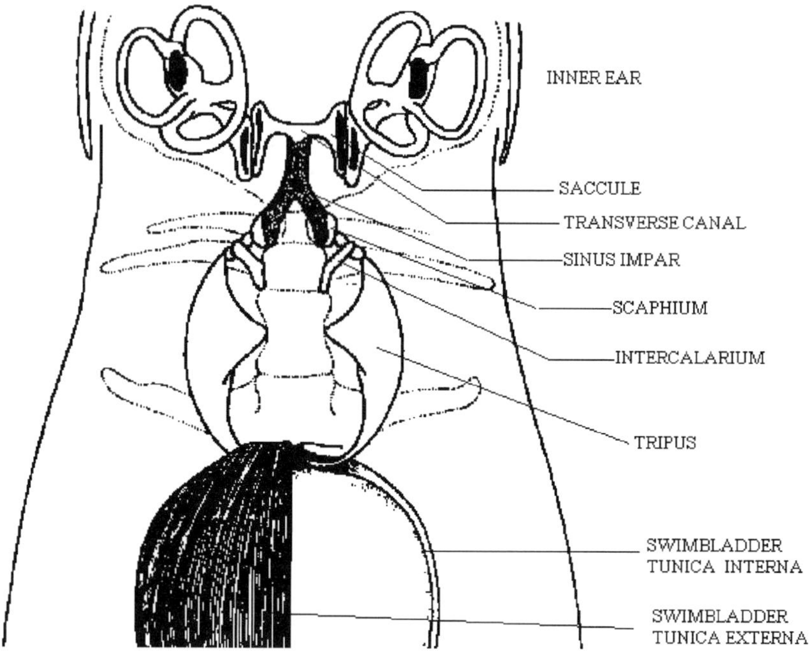

FIGURE 7.40. The Weberian ossicles in Otophysan fish connect the outer wall of the swim bladder (tunica externa) to the fluids of the inner ear via the sinus impar and transverse canal. Motion of the swim bladder rotates the tripus, which in turn moves the interclarium, scaphium and claustrum (not visible), and finally the wall of the sinus impar. The oscillatory motion of the wall of the sinus impar travels rostrally and then laterally through the transverse canal into the saccules (modified from Chardon and Vandewalle 1991).

the indirect path. Thus, Finneran and Hastings (2000) referred to it as the "Weberian path" to distinguish it from the direct and indirect paths.

Several other types of fish have anatomical specializations that play a role in hearing. These include projections of the swim bladder to the inner ear or other air chambers in close proximity to the ear. In clupeids, the utricle plays a major role in hearing (Blaxter et al., 1981). Their utricle has three separate sensory epithelia and is attached to a gas-filled bulla that extends from the swim bladder as shown in Fig. 7.43. Transmission of sound through the bulla pushes fluid into the utricle, which creates additional relative motions between the otolith and sensory epithelia. Mann et al. (1998) speculated that the utricle might play a role in the detection of ultrasound by a clupeid fish, the American shad (*Alosa sapidissima*).

Much progress has been made in understanding the structure and function of the fish auditory system; however, many questions still remain regarding exactly how sound reaches the fish ear and the nature of the coupling between the ear and various accessory structures. Few authors have directly investigated the mechanics of the swim bladder, auxiliary structures such as the Weberian

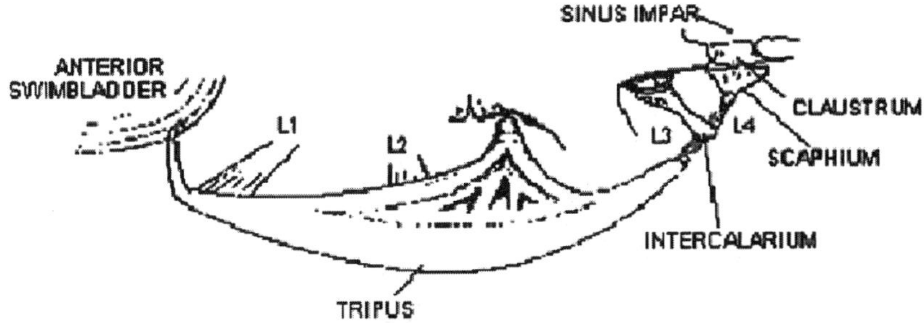

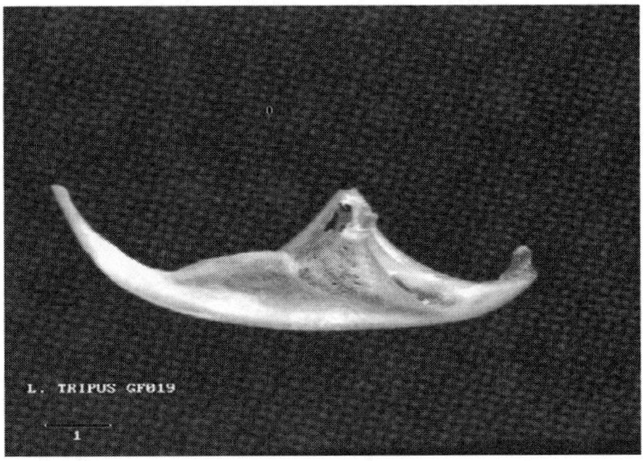

FIGURE 7.41. Detailed view of the Weberian ossicles (top – modified from Krumholz 1943) and photograph of the tripus from a goldfish (bottom; scale bar is 1 mm).

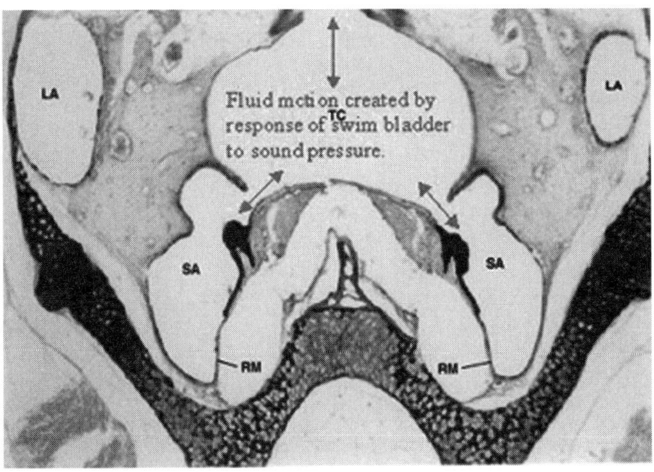

FIGURE 7.42. Transverse slice through the goldfish labyrinth showing fluid connection between the transverse canal (TC) and saccule (SA). The fluid motion drives the saccular otoliths (black cross section at the tip of red arrowhead). (LA – lagena, RM – release membrane).

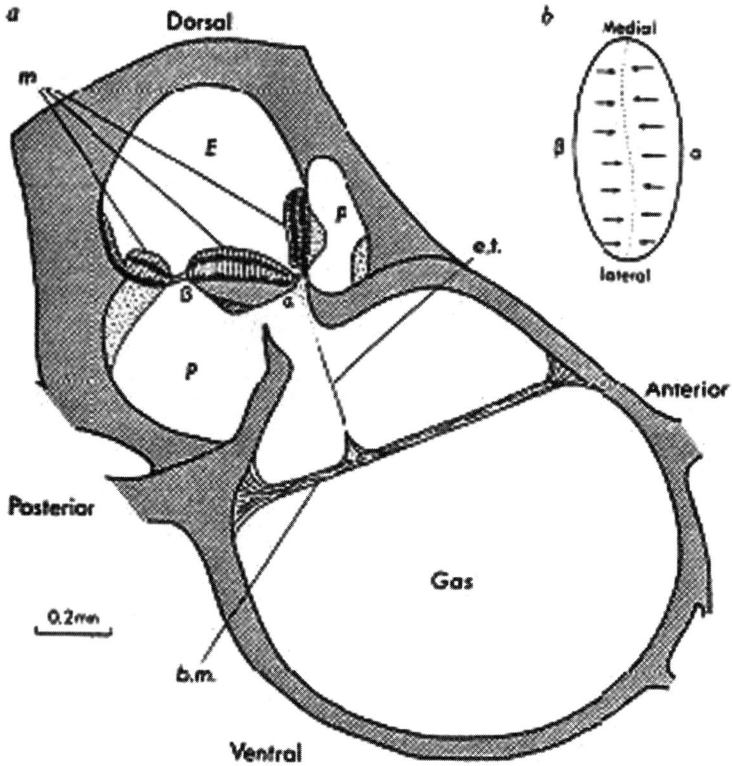

FIGURE 7.43. Cross section of the utricle in the ear of the sprat: sound-induced vibrations of the gas cause vibration of the bulla membrane (b.m.), which oscillates fluid (perilymph – P) through the opening to the sensory epithelia (m) (from Denton and Gray 1979).

ossicles, or the otoliths. Of the works that have been done, the majority has relied upon intrusive or *post-mortem* procedures or have been performed in complex acoustical environments. New technology, which we review in Chapter 13, has been developed to overcome some of these shortfalls.

References

Alexander, R.McN. (1962). "The structure of the Weberian apparatus in the Cyprini," Proc. Zool. Soc. London **139**, 451–473.

Altman, J. A., Butusov, M. M., Vaitulevich, S. F., and Sokolov, A. V. (1984). "Responses of the swimbladder of the carp to sound stimulation," Hear. Res. **14**, 145–153.

Aroyan, J. L. (1996). "Three dimensional numerical simulation of biosonar signal emission and reception in the common dolphin," Ph.D. dissertation, U. C. Santa Cruz.

Bess, F. H. and Hume, L. E. (1995). *Audiology: The Fundamentals* (Williams & Wilkins, Baltimore, MD).

Blaxter, J. H. S., Denton, E. J., and Gray, J. A. B. (1981). "Acousticolateralis system in clupeid fishes," in *Hearing and Sound Communication in Fishes*, W. N. Tavolga, A. N. Popper, and R. R. Fay, eds. (Springer-Verlag, New York), pp. 39–59.

Brill, R. L., Sevenich, M. L., Sullivan, T. J., Sustman, J. D., and Witt, R. E. (1977). "Behavioral evidence for hearing through the lower jaw by an echolocating dolphin (*Tursiops truncatus*)," Mar. Mamm. Sci. **4**, 223–230.

Bullock, T. H., Grinnell, A. D., Ikezono, E., Kameda, K., Katsuki, Y., Nomoto, M., Sato, O., Suga, N., and Yanagisawa, K. (1967). "Electrophysiological studies of central auditory mechanisms in cetaceans," Zeitschrift fur Vergleichednde Phys. **59**, 117–316.

Chardon, M. and Vandewalle, P. (1991). "Acoustic-lateralis system," in *Cyprinid Fishes*, I. J. Winfield and J. S. Nelson, eds. (Chapman and Hall, New York), pp. 332–352.

Clarke, N. L., Popper, A. N., and Adinmann, J. Jr. (1975). "Laser light-scattering investigations of the teleost swimbladder response to acoustic stimuli," Biophys. J. **15**, 307–318.

Cox, M., Rogers, P. H., Popper, A. N., Saidel, W. M., and Fay, R. R. (1986). "Frequency regionalization in the fish ear," J. Acoust. Soc. Am. **79**(Suppl. 1), p. S80.

Cranford, T. W., Amundin, M., and Norris, K. S. (1996). "Functional morphology and homology in the odontocete nasal complex: implications for sound generation," J. Morphol. **227**, 223–275.

Denton, E. J. and Gray, J. A. B. (1979). "The analysis of sound by the sprat ear," Nature **282**, 406–407.

deVries, H. L. (1950). "The mechanics of the labyrinth otoliths," Acta. Oto.-laryngol. **38**, 262–273.

Durrant, J. D. and Lovrinic, J. H. (1974). *Bases of Hearing Science* (Williams and Wilkens, Baltimore, MD).

Enger, P. S. (1981). "Frequency discrimination in teleosts—central or peripheral?" In *Hearing and Sound Communication in Fishes*, W. N. Tavolga, A. N. Popper, and R. R. Fay. eds. (Springer-Verlag, New York), pp. 243–255.

Evans, H. M. (1925). "A contribution to the anatomy and physiology of the air-bladder and Weberian ossicles in Cyprinidae," Proc. Royal Soc. London B. **97**, 545–576.

Finneran, J. J. and Hastings, M. C. (2000). A mathematical analysis of the peripheral auditory system mechanics in the goldfish (*Carassius auratus*), J. Acoust. Soc. Am. **108**, 3035–3043.

Finneran, J. J. and Hastings, M. C. (2004). "A continuous-wave ultrasound system for displacement amplitude and phase measurement," J. Acoust. Soc. Am. **115**, 3202–3209.

Fleischer, G. (1976a). "Hearing in extinct cetaceans as determined by cochlear structrue," J. Paleontol. **50**, 133–152.

Fleischer, G. (1976b). "On bony microstructures in the dolphin cochlea, related to hearing," N. Jb. Geol. Palaont., Abhl, **151**, 161–191.

Fleischer, G. (1970). "Morphological adaptations of the sound conducting apparatus in echolocating mammals," in *Animal Sonar Systems*, R. G. Busnel and J. F. Fish, eds. (Plenum Press, New York), pp. 795–797.

Fraser, F. C. and Purves, P. E. (1954). "Hearing in Cetaceans," Bull. British Mus. (Nat. Hist.), **2**, 103–116.

Green, D. (1976). *An Introduction to Hearing* (Erlbaum Assoc., New York).

Higgs, D. M., Rollo, A. K., Souza, M. J., and Popper, A. N. (2003). "Development of form and function in peripheral auditory structures of the zebrafish (*Danio rerio*)," J. Acoust. Soc. Am. **113**, 1145–1154.

Higgs, D. M., Plachta, D. T. T., Rollo, A. K., Singheiser, M., Hastings, M. C., and Popper, A. N. (2004). "Development of ultrasound detection in American Shad (*Alosa sapidissima*)," J. Exp. Biol. **207**, 155–163.

Huber, E. (1934). "Anatomical notes on pinnipedia and cetacea," in *Marine Mammals: Contribution to Palaeotology*, Packard, Kellogg and Huber, eds. (Carnegie Institute, Washington DC), pp. 105–136.

Hudspeth, A. J. and Corey, D. P. (1977). "Sensitivity, polarity, and conductance change in the response of vertebrate hair cells to controlled mechanical stimuli," Proc. Natl. Acad. Sci. **74**, 2407–2411.

Johnson, S. C. (1967). "Sound detection thresholds in marine mammals," in *Marine bio-Acoustic*, W. Talvoga, ed. (Pergamon Press, New York), pp. 247–260.

Johnson, S. C. (1976). "Dolphin audition and echolocation capacities," in *Dolphin Cognition and Behavior: a Comparative Approach*, R. J. Schusterman, J. A. Thomas and F. G. Wood, eds. (Lawrence Erlbaum Associates, New Jersey), pp. 115–136.

Junqueira, L. C., Carneiro, J., and Contopoulos, A. (1977). *Basic Histology*, 2 nd ed. (Lange Medical Publications, Los Altos, CA).

Kelly, J. P. (1991). "Hearing," in *Principles of Neural Science*, 3rd ed., E. R. Kandel, J. H. Schwartz, and T. M. Jessell, eds. (Elsevier, New York), pp. 471–497.

Kessel, R. G. and Kardon, R. H. (1979). *Tissue and organs: A Text-Atlas of Scanning-Electron Microscope* (W. H. Greeman, San Francisco, CA).

Ketten, D. R. (1992). "The marine mammal ear: specializations for aquatic audition and echolocation," in *The Evolutionary Biology of Hearing*, D. Webster, R. Fay and A. Popper, eds. (Springer-Verlag, New York), pp. 717–754.

Ketten, D. R. (1994). "Functional analyses of Whale Ears: adaptations for underwater hearing," IEEE Oceans '94 Proceedings 13–16 September 1994, Brest, France: 264–270.

Ketten, D. R. (1997). "Structure and function in Whale Ears," Bioacoustics, **7**, 103–135.

Ketten, D. R., Odell, D. K., and Domning, D. P. (1992). "Structure, function, and adaptation of the manatee ear," in *Marine Mammal Sensory Systems*, J. Thomas, R. Kastelein and A. Supin, eds. (Plenum Press, New York), pp. 77–95.

Krumholz, L. A. (1943). "A comparative study of the Weberian ossicles in North American Ostariophysine fishes," Copeia **1943(1)**, 33–40.

Lombarte, A., Yan, H. Y., Popper, A. N., Chang, J. S., and Platt, C. (1993). "Damage and regeneration of hair cell ciliary bundles in a fish ear following treatment with gentamicin," Hear. Res. **64**, 166–174.

Lombarte, A., and Popper, A. N. (1994). "Quantitative analyses of post-embryonic hair cell addition in the otolithic end organs of the inner ear of the European hake, *Merluccius merluccius* (Gadiformes, Teleostei)," J. Comp. Neurol. **345**, 419–428.

Lombarte, A., and Popper, A. N. (2004). "Quantitative changes in the otolithic organs of the inner ear during the settlement period in European hake *Merluccius merluccius*," Mar. Ecol. Prog. Ser. **267**, 233–240.

McCormick, J. G., Wever, E. G., Palin, J. and Ridgway, S. H. (1970a). "Sound conduction in the dolphin ear," J. Acoust. Soc. Am. **47**, 1417–1427.

McCormick, J. G., Weaver, E. G., Ridgway, S. H., and Palin, J. (1970b). "Sound reception in the Porpoise as it relates to Echolocation," in *Animal Sonar Systems*, R. G. Busnel and J. F. Fish, eds. (Plenum Press, New York), pp. 449–467.

Mohl, B. (1967). "Hearing in seals," in *The Behavior and Physiology of Pinnipeds*, R. J. Harrison, R. C. Hubbard, R. S. Peterson, C. E. Rice, and R. J. Schustman, eds. (Appleton-Century-Crofts, New York), pp. 172–195.

Mohl, B. and Ronald, K. (1975). "The peripheral auditory system of the Harp seal, *Pagophilus groenlandicus*, (Erxleben, 1777)," Rapp. P.-v. Rèun. Cons. Int. Explor. Mer. **169**, 516–523.

Mohl, B., Au, W. W. L., Pawloski, J. L., and Nachtigall, P. E. (1999). "Dolphin hearing: relative sensitivity as a function of point of application of a contact sound source in the jaw and head region," J. Acoust. Soc. Am. **104**, 3421–3424.

Nedzelnitsky, V. (1970). "Sound pressures in the basal turn of the cat cochlea," J. Acoust. Am. **67**, 1676–1670.

Norris, K. S. (1967b). "The echolocation of marine mammals," *The Biology of Marine Mammals*, H. T. Andersen, ed. (Academic Press, New York), pp. 391–423.

Norris, K. S. (1970). "Peripheral sound processing in odontocetes," in *Animal Sonar Systems*, R. G. Busnel and J. F. Fish, eds. (Plenum Press, New York), pp. 495–507.

Norris, K. S. and Harvey, G. W. (1974). "Sound transmission in the porpoise head," J. Acoust. Soc. Am. **56**, 659–664.

Pickles, J. O. (1972). *An Introduction to the Physiology of Hearing* (Academic Press, San Diego, CA).

Platt, C. (1977). "Hair cell distribution and orientation in goldfish otolithic organs," J. Comp. Neurol. **172**, 283–297.

Platt, C. and Popper, A. N. (1981). "Fine structure and function of the ear," in *Hearing and Sound Communication in Fishes*, W. N. Tavolga, A. N. Popper, and R. R. Fay, eds. (Springer-Verlag, New York), pp. 3–38.

Popper, A.N. (1974). "The response of the swim bladder of the goldfish (*Carassius auratus*) to acoustic stimuli," J. Exp. Biol. **60**, 295–304.

Popper, A. N. and Coombs, S. L. (1982). "The morphology and evolution of the ear in actinopterygian fishes," Am. Zool. **22**, 311–328.

Popper, A. N. and Hoxter, B. (1984). "Growth of a fish ear: I. Quantitative analysis of sensory hair cell and ganglion cell proliferation," Hear. Res. **15**, 133–142.

Popper, A. N., Rogers, P. H., Saidel, W. M., and Cox, M. (1988). "Role of the fish ear in sound processing," in *Sensory Biology of Aquatic Animals*, J. Atema, R. R. Fay, A. N. Popper, and W. N. Tavolga, eds. (Springer-Verlag, New York), pp. 687–710.

Purves, P. E. and Pilleri, G. E. (1983). *Echolocation in Whales and Dolphins* (Academic Press, New York).

Purves, P. E. and Pilleri, G. E. (1973). "Observations on the ear, nose, throat, and eye of Platanista indi, Invest. on Cetacea, **V**, 13–57.

Pye, J. D. (1972). "Bimodal distribution of constant frequencies in some hipposiderid bats (Mammalia: Hipposideridae)," J. Zool. **166**, 323–335.

Ramprashad, F., Money, K. E., and Ronald, K. (1972). "The harp seal, *Pagophilus groenlandicus* (Erxleben, 1777). XXI. The structure of the vestibular apparatus," Can. J. Zool. **50**, 1357–1361.

Ranke, O. F. (1942). "Das Massenverhaltnis zwischen Membran und Flussigkut im Innenohn," Akust. Z. **7**, 1–11.

Repenning, C. A. (1972). "Underwater hearing in seals: functional morphology," in *Functional Anatomy of Marine Mammals*, R. J. Harrison, ed. (Academic Press, London), pp. 307–331.

Retzius, G. (1881). *Das Gehörorgan der Wirbelthiere Vol. 1* (Samson and Wallin, Stockholm).

Ridgway, S. H., McCormick, J. G., and Wever, E. G. (1974). "Surgical approach to the dolphin's ear," J. Exp. Zool. **177**, 265–276.

Rogers, P. H. and Hastings, M. C. (1989). "Noninvasive Vibration Measurement System and Method for Measuring Amplitude of Vibration of Tissue in an Object Being Investigated," U. S. Patent No. 4,819, 643.

Rogers, P. H., Popper, A. N., Hastings, M. C., and Saidel, W. M. (1988). "Processing of acoustic signals in the auditory system of bony fish," J. Acoust. Soc. Am. **83**, 338–349.

Rossing, T. D. (1990). *The Science of Sound* (Addison-Wesley, Reading, Mass).

Sand, O. and Michelsen, A. (1978). "Vibration measurements of the perch saccular otolith," J. Comp. Physiol. **123**, 85–89.

Schuijf, A. and Buwalda, R. J. A. (1980). "Underwater localization – a major problem in fish acoustics," in *Comparative Studies of Hearing in Vertebrates*, A. N. Popper and R. R. Fay, eds. (Springer-Verlag, New York), pp. 43–77.

Spoendlin, H. (1970). "Structural basis of peripheral frequency analysis," in *Frequency Analysis and Periodicity Detection in Hearing*, R. Plomp and G. F. Smoorenburg, Eds. (A. W. Sythoff, Leiden, The Netherland), pp. 2–37.

Spoendlin, H. (1977). "The afferent innervation of the cochlea," in *Evoked Electrical Activity in the Auditory Nervous System*, R. F Naunton and D. Fernandez, Eds. (Academic Press, New York), pp. 21–41.

Tonndorp, J. (1960). "Shearing motion in scala media of Cochlear models," J. Acoust. Soc. Am. **32**, 238–244.

Vaitulevich, S. F. (1979). "Characteristics of the wall of the swimming bladder in the carp *Cyprinus carpio* during sonic stimulation," Zhurnal Evol. Biokhim. Fiziol. **15**, 196–201.

Vaitulevich, S. F. and Ushakov, M. N. (1974). "Investigations of the vibrations of the swim bladder of the carp (*Cyprinus carpio*) by the method of holographic interferometry," Biofizika **19**, 528.

van Bergeijk, W. A. (1964). "Directional and nondirectional hearing in fish," in *Marine Bioacoustics*, W. N. Tavolga, ed. pp. 281–303.

Varanasi, U. and Malins, D. C. (1971). "Unique lipids of the porpoise (Tursiops gilli): differences in triacylglycerols and wax esters of acoustic (Mandibular and Melon) and blubber tissues," Biochimica et Biophysica Acta, **231**, 415–417.

Varanasi, U. and Malins, D. C. (1972). "Triacylglycerols characteristic of porpoise acoustic tissues: molecular structures of diisovaleroylgly cerides," Science, **176**, 926–927.

Varanasi, U., Feldman, H. R., and Malins, D. C. (1975). "Molecular basis for formation of lipid sound lens in echolocating cetaceans," Nature, **255**, 340–343.

von Bekesy, G. (1960). *Experiments in Hearing* (McGraw Hill, New York).

von Frisch, K. and Stetter, H. (1932). "Untersuchungen über den Sitz des Gohörsinnes bei der Elritze," Z. vergl. Physiol. **17**, 686–801.

Wever, E. G., McCormick, J. G., Palin, J., and Ridgway, S. H. (1971a). "The Cochlea of the Dolphin, Tursiops truncatus: General Morphology," Proc. Nat. Acad. Sci. USA, **67**, 2371–2375.

Wever, E. G., McCormick, J. G., Palin, J., and Ridgway, S. H. (1971b). "The Cochlea of the Dolphin, Tursiops truncatus: The Basilar Membrane," Proc. Nat. Acad. Sci. USA, **67**, 2707–2711.

Wever, E. G., McCormick, J. G., Palin, J., and Ridgway, S. H. (1972). "Cochlea Structure in the Dolphin, Lagenorhynchus obliquidens," Proc. Nat. Acad. Sci. USA, **69**, 657–661.

Yost, W. A. (1994). *Fundamentals of Hearing: An Introduction* (Academic Press, San Diego, CA).

8

Experimental Psychological and Electrophysiological Methodology

In the previous chapter we discussed the anatomical aspects of the ears of marine mammals and now we turn toward psychological and electrophysiological techniques to determine the characteristics and capabilities of the auditory system. We will not attempt an exhaustive treatment of psychological methodology since volumes have been written on this subject and it would be beyond the scope of this book and of our expertise. Our treatment will be specifically related to experiments with aquatic animals. Most of our knowledge on the auditory capabilities of marine mammals and fishes has come from controlled laboratory-type psychophysics and psychoacoustic experiments. Psychophysics involves psychological experiments to determine the relationship between physical stimuli and sensory response. In psychoacoustics, the stimuli are acoustic in nature and experiments in this field involved controlled behavioral testing procedures, following established psychological testing procedures to determine the relationship between an acoustic stimulus and the responses of an organism. Most of us have already been exposed to a psychoacoustic testing procedure when we had our hearing tested in school, at a doctor's office, or in a hospital. Our hearing was tested by sitting at a desk, wearing headphones, or in a soundproof booth with a laboratory technician monitoring our responses through a window. The instruction was very simple: raise a hand when you hear a tone. If we raised a hand at the onset of a tone, there would be a silent period until the tone was presented again, but at a reduced level. This continued until we no longer heard the tone. Then the technician would switch to a different frequency and the procedure would start all over again. What we experienced in this hearing test is a form of a psychoacoustic procedure that can be applied to almost any species of animals.

8.1 Psychoacoustics Procedures

Psychoacoustics experiments generally fall into two basic categories. The first category has to do with the detection of the presence or absence of a stimulus, and experiments in this category are generally referred to as detection

W.W.L. Au, M.C. Hastings, *Principles of Marine Bioacoustics*,
DOI: 10.1007/978-0-387-78365-9_8,

experiments. The subject is trained to respond in a particular manner if a stimulus is present and in a different manner if the stimulus is absent or cannot be perceived. The stimulus can be presented in a noiseless environment or in the presence of different types of masking noise. The stimulus is progressively made weaker until the subject performance in detecting the stimulus approaches the chance level. The second type of experiments has to do with how well one stimulus can be discriminated from other stimuli, and are generally referred to as discrimination experiments. An example of a discrimination task is one in which a subject is presented with a stimulus and is required to classify the stimulus as belonging to category A or B. The difference in the properties of the stimulus will progressively be made smaller until the subject's performance in classifying the stimulus approaches the chance level.

8.1.1 Stimulus Control and Operant Conditioning

One of the major considerations in performing any sensory sensitivity experiments with animals is the need to establish and maintain stimulus control of the subject's behavior. Establishing stimulus control implies that a stimulus exerts control over an animal's behavior and introducing any changes in the characteristics of the stimulus will result in some measurable change in the animal's behavior, usually the performance accuracy of the animal. For example, in a hearing experiment, the presence of a stimulus will evoke a certain measurable behavioral response by the animal while the absence of the target will evoke a different response. Dolphins and other marine mammals in the United States are generally trained by operant conditioning techniques (Skinner, 1961) to respond to the controlling stimulus in a particular manner and thus receive a positive reinforcement, usually a food reward. Often a secondary reinforcement (bridge tone) in the form of an underwater tone or an air-borne whistle sound precedes the primary food reinforcement. Improper responses to the controlling stimulus generally result in an absence of any reinforcement, or a time out. Once stimulus control has been established, an experimenter can test an animal's acuity in detecting, discriminating, recognizing, and classifying various kinds of stimuli using a variety of different psychological testing procedures.

The classical or Pavlovian procedure in which an electric shock is paired with an acoustic stimulus have been experimented with by Russians scientists to study sound reception by dolphins and other marine mammals (Schusterman, 1980). After repeated pairing of a shock with the stimulus, a highly stereotyped response such as changes in galvanic skin condition, changes in respiratory rate, changes in heart rate, or some other conditioned reflexes, is elicited. This technique is still used with fishes and will be discussed further at the end of this chapter.

8.1.2 Yes/No Response Paradigms in Detection Experiments

Psychophysical testing of the sensory capabilities of a subject can be narrowed down to two basic components. The first component has to do with how the subject is to respond to either the presence of the stimulus or its absence. There are a wide variety of ways a subject can respond such as touching a paddle, producing an audible sound, moving to specific locations, etc. The second component has to do with how the stimulus is manipulated and may be based on the subject's performance on the previous trial. There are a wide variety of different response paradigms that can be used to determine the sensitivity of any auditory process associated with animals.

One of the simplest response paradigms involves trials with a single observational or testing interval in which the stimulus will either be present or absent and the subject must respond either stimulus-present or stimulus-absent. If the subject perceives the stimulus, it responds in a specific manner such as touching a specific response paddle; and if the stimulus is not perceived, the subject responds in another manner such as touching a different paddle. This response paradigm is called the *Yes/No* response procedure (Green and Swets, 1966; Gescheider, 1976). In many marine mammal psychophysical experiments, this procedure has been incorrectly referred to as the two-alternative forced-choice procedure. However, in human psychophysics this procedure is known as the Yes/No procedure. Since the human psychoacoustic field has been with us much longer and is more established than animal bioacoustics, it would be prudent to follow the terminology established in the human hearing field.

A variation of the Yes/No response is the *Go/No-go* response in which the subject takes a prescribed course of action such as leaving a station to touch a response paddle, or making an audible or detectable sound when the stimulus is present. The go/no-go procedure is perhaps the most simple and straightforward procedure to study acoustic sensory perception in marine mammals (Schusterman, 1980). The procedure involves a subject moving to a particular location or stationing device to await the presentation of a stimulus. Some common stationing devices include a hoop through which a subject can extend its head, a bite plate on which a subject can position on, and a chin cup or support bar on which a subject can rest some part of its body on. A trial begins with the subject moving to its station and orienting toward the direction of the stimulus source. Upon presentation of the stimulus, the animal leaves the station and touches a response paddle if it perceives the stimulus (go response) for a detection experiment or if the stimulus is the standard, in a discrimination task. Upon striking the response paddle for a correct response, a bridge tone is presented conveying to the animal that it had made a correct detection and will be receiving a food reward. The subject has a definite fixed time period to indicate a go response. Any response after that time period is considered a no-go response. If the stimulus is absent or is not the standard stimulus, the subject remains at station (no-go response) for a short fixed time period until the experimenter presents a bridge or secondary reinforcement tone or sound indicating a correct response. The bridge tone signals the animal that it had made a correct rejection and that it can now leave its station and collect a food reward. Instead of leaving station and touching a paddle, the animal can be trained to emit an audible or detectable sound for a go-response and be silent for a no-go-response. Schusterman et al. (1972) measured the audiogram of a California sea lion by using the conditioned vocalization technique. Ridgway and Carder (1997) required a beluga whale to whistle if it heard a tone or be silent if it did not hear a tone in an auditory sensitivity measurement experiment. In hearing experiments, the stimuli are usually presented by means of transducers, whereas in echolocation detection experiments with dolphins the stimuli are the echoes returning from a target that the dolphin echolocated.

8.1.2.1 Binary Decision Matrix for Yes/No Detection Experiments

In experiments involving the detection of a stimulus and a yes/no or go/no-go response procedure, there are four possible responses: two correct responses and two wrong responses. A correct detection (hit) is one in which the subject responds correctly to the presence of the stimulus. A correct rejection occurs when the subject responds correctly to the absence of the stimulus. A miss occurs when the subject fails to report the presence of a stimulus and a false alarm occurs when the subject indicates that the stimulus is present when it is not. When considering an experiment involving many trials, we can refer to the dolphin's performance in terms of conditional response probabilities. $P(\mathrm{Y/sn})$ is the conditional probability of a correct detection which is the probability of responding "yes" given the presence of the stimulus (sn). It is often referred to as the "hit" rate. $P(\mathrm{Y/n})$ is the conditional probability of a false alarm, which is the probability of responding "yes" given that the stimulus is absent and only noise (n) is present. $P(\mathrm{N/sn})$ is the conditional probability of a miss, which is the probability of responding "no" given the presence of the stimulus. $P(\mathrm{N/n})$ is the conditional probability of a correct rejection, which is the probability of responding "no" given that the stimulus is absent and only noise is present. The binary decision matrix relating the stimulus condition to possible responses is depicted in Fig. 8.1. The different conditional probabilities are related according to the two equations,

$$P(\mathrm{Y/sn}) + P(\mathrm{N/sn}) = 1. \tag{8.1}$$

$$P(\mathrm{Y/n}) + P(\mathrm{N/n}) = 1. \tag{8.2}$$

		RESPONSE	
		Yes	No
STIMULUS	Present	P(Y/sn)	P(N/sn)
	Absent	P(Y/n)	P(N/n)

FIGURE 8.1. The binary decision matrix relating the stimulus condition to possible responses.

If we define the probability of the stimulus being present as $P(\mathrm{sn})$ and the probability of the stimulus being absent as $P(\mathrm{n})$, then the probability of a correct response can be expressed as

$$P(\mathrm{C}) = P(\mathrm{Y/sn})P(\mathrm{sn}) + [1 - P(\mathrm{Y/n})]P(\mathrm{n}). \tag{8.3}$$

In most experiments there are equal number of stimulus present and stimulus absent trials so that $P(\mathrm{sn}) = P(\mathrm{n}) = 0.5$, and Eq. (8.3) can be reduced to

$$P(\mathrm{C}) = 0.5\{P(\mathrm{Y/sn}) + [1 - P(\mathrm{Y/n})]\}. \tag{8.4}$$

Often the four conditional probabilities are expressed as $P(\mathrm{D})$, $P(\mathrm{FA})$, $P(\mathrm{CR})$, $P(\mathrm{M})$ to denote detection, false alarm, correct rejection, and miss probabilities, respectively.

8.1.2.2 Receiver-Operating Characteristics Curve

There are only two degrees of freedom in the stimulus-response matrix of Fig. 8.1 so that all the information in the matrix can be plotted on a two-dimensional graph referred to as the "receiver-operating characteristic" or ROC graph. On this graph the probability of detection $P(\mathrm{Y/sn})$ is on the ordinate, and the probability of false alarm $P(\mathrm{Y/n})$ is on the abscissa as depicted in Fig. 8.2. Points falling along the major diagonal represent chance performance since on the diagonal $P(\mathrm{Y/sn}) = P(\mathrm{Y/n})$, and for this condition Eq. (8.4) indicates that the probability of a correct response will be 0.50. In a two alternative task, change performance is equivalent to $P(\mathrm{C}) = 0.50$. The major diagonal represents the poorest a subject can perform short of making

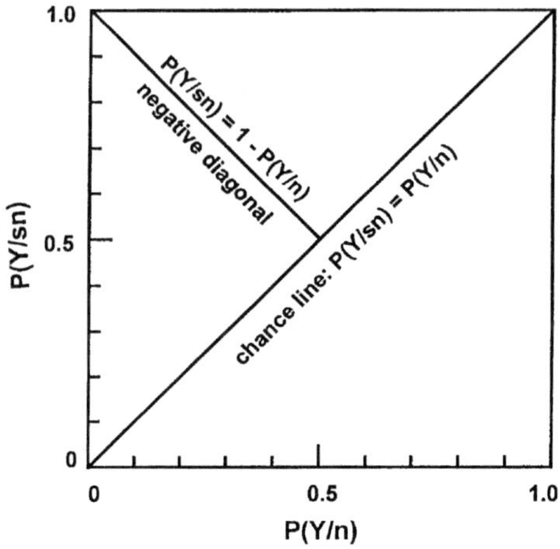

FIGURE 8.2. Graph display for plotting performance data in a ROC format.

deliberate error. Points to the right of the major diagonal represent situations in which a subject made deliberate errors.

In conducting a detection experiment, bias by the subject can be manifested either as a position reference in a Yes/No or as a preference for responding or not responding in a Go/No-go procedure (Schusterman, 1980). A bias in favor or a "no" response will result in an increased threshold while a bias in favor of a "yes" response will result in a decreased threshold determination. Therefore, the bias of the subject can affect the measured threshold. More on this topic will be presented in the section on signal detection theory. The minor diagonal in Fig. 8.2 is the line from the upper left-hand corner of the ROC graph to the mid-point of the major diagonal. Along the minor diagonal

$$P(\mathrm{Y/sn}) = 1 - P(\mathrm{Y/n}). \tag{8.5}$$

Inserting Eq. (8.5) into Eq. (8.1) for points along the minor diagonal, we arrive at

$$P(\mathrm{Y/n}) = P(\mathrm{N/sn}). \tag{8.6}$$

Therefore, along the minor diagonal, the probability of a false alarm is equal to the probability of a miss, a situation representing zero bias, since the two possible types of error (false alarms and misses) would be equally probable. In typical marine mammal psychophysical experiments in which the number of stimulus present and absent trials are the same or balanced and the amount of reward or reinforcement for correct responses is also the same, there is a tendency for animals to exhibit a low false alarm rate, which is the same as having a conservative bias. Schusterman (1974) reported on this phenomenon using the data from acoustic and visual experiments with the California sea lion (*Zalophus californianus*), the harp seal (*Pagophilus groendlandicus*), the harbor seal (*Phoca vitulina*), the Amazon River dolphin (*Inia geofrensis*), and the Atlantic bottlenose dolphin (*Tursiops truncatus*). ROC curves for auditory experiments with some of these animals are shown in Fig. 8.3. The *Zalophus* was involved in an underwater hearing experiment, the *Pagophilus* was involved in an aerial hearing experiment, and the *Inia* and *Tursiops* were involved in echolocation experiments. The ROC curves indicate that these animals tend to exhibit low false alarm rates that were nearly constant, regardless of the difficulty of the task. Even as the correct detection rates continued to decrease to very low levels, the false alarm rates in graphs a–c did not increase. However, the *Tursiops* in graph d did increase its false alarm rate slightly as its detection rate fell below 0.5. In the parlance of signal detection theory, these animals were behaving as Neyman–Pearson observers, keeping a constant false alarm rate while trying to maximize their hit rate (Egan, 1975). According to Schusterman (1974), the need to establish stimulus control in the early training phase and to maintain stimulus control over the course of the experiment is the primary reason for the low false alarm rates. Having a low false alarm rate will result in a low estimate of the sensory sensitivity being measured as will be seen when we discuss signal detection theory.

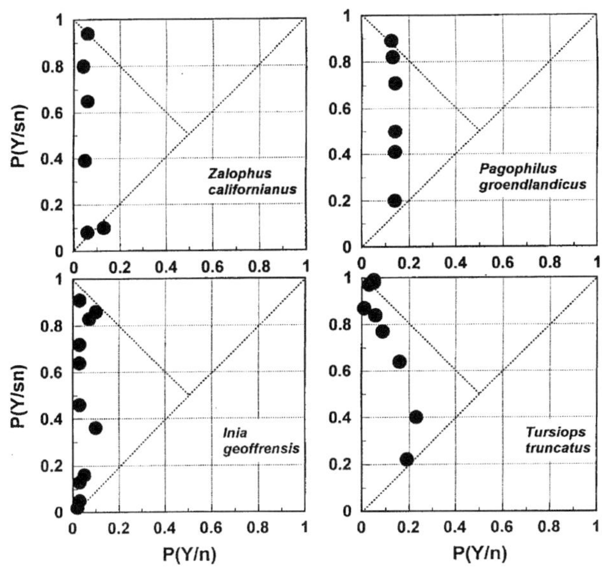

FIGURE 8.3. ROC curves for several different species of marine mammals performing in various acoustic experiments. Curves a–c were redrawn from the data in Schusterman (1974) and the data for curve d came from Au and Synder (1980).

8.1.2.3 The Forced Choice Procedure

The *two alternative forced choice* (2AFC) paradigm involves two intervals within a trial in which the stimulus can be present in either the first or second interval. The subject could touch a specific paddle or produce a specific audible sound if it perceived the stimulus in the first interval or another paddle or another specific audible sound if the stimulus occurs in the second interval. The subject must decide or make a choice in each trial whether the stimulus is in the first or second interval, hence the name forced choice. This technique for measuring a subject's detection sensitivity is generally uncontaminated by the subject's bias (Green and Swets, 1966; Gescheider, 1976), because the stimulus will always be present in one of two intervals during a trial and the subject is required to chose one of the intervals most likely to contain the stimulus. A typical sequence in a 2AFC is shown in Fig. 8.4. In the absence of any response bias, the subject chooses the interval containing the largest sensory stimulus. Since the subjects bias is not a factor in such a procedure, the proportion of correct responses $P(\mathrm{C})$ can be used as a measure of the subject's sensitivity. However, as the stimulus approaches the threshold level, effects of bias can be the subject may resort to guessing and even develop a bias toward one of the two intervals. The examination by a type of ROC curve in which a hit is considered to occur if the subject correctly reports on the presence of a stimulus in the first interval, and a false alarm is considered to occur if the subject incorrectly reports that a stimulus was present in the first interval. Green and Swets (1966) also developed the following formula to take into account the effect of the subject guessing

$$P^*(\mathrm{Y/sn}) = \frac{P(\mathrm{Y/sn}) - P(\mathrm{Y/n})}{1 - P(\mathrm{Y/n})}, \tag{8.7}$$

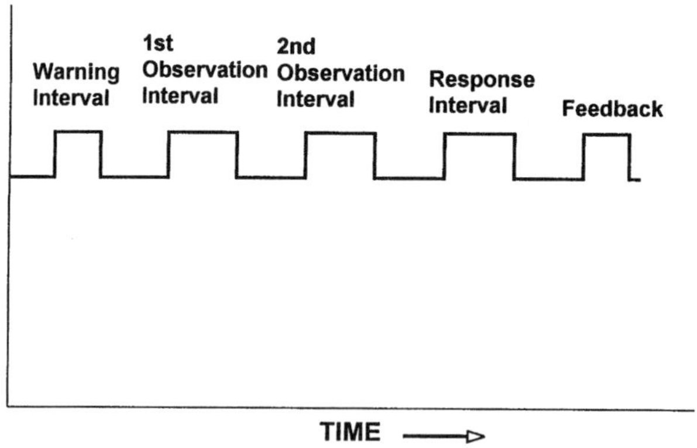

FIGURE 8.4. Sequence in a two-alternative forced choice procedure.

where $P^*(Y/sn)$ is the hit rate corrected for guesses by the subject. The two-alternative force choice procedure is also been called the two-interval forced choice procedure (2IFC). In human sensory research, two or four intervals have been commonly used, but work has also been done with as may as eight intervals (Green and Swets, 1966).

The 2AFC sequence shown in Fig. 8.4 suggests that the acoustic stimulus is being played by a single sound source or sound sources at a single location. Variations to this procedure can be made by having two source at two distinct locations. Suppose in a hearing sensitivity experiment we have an animal stationed between two sound sources as depicted in Fig. 8.5, and there are two sound projectors at the same distance away from the animal at two widely separated angles. During a single trial, one of two sound projectors will project a signal while other remains silent and the animal must tell the experimenter which projector emitted the signal. Therefore, there will be a stimulus present on each and every trial and the subject must choose the correct location from which the stimulus is being presented. Such a procedure would tend to reduce the tendency of the subject to be biased; however, two sound sources are required. If such a procedure is used in noise-masking experiments using two projectors, the experimenter should take the precaution of not projecting both sources simultaneously since it would be difficult to estimate the noise being received by the subject from the two sources.

Although the 2AFC procedure is relatively simple to implement, it has not been a popular procedure for researchers studying the auditory sensitivity of

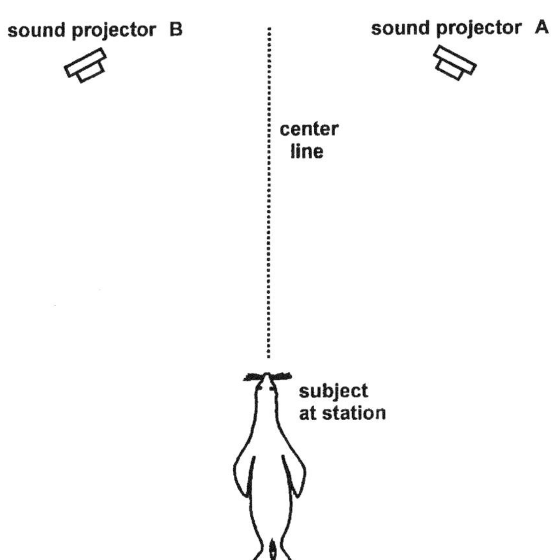

FIGURE 8.5. Geometry of a 2AFC experiment using two spatially separated sound projectors.

marine mammals. Renaud and Popper (1975) used such a technique in measuring the underwater sound localization capabilities of a bottlenose dolphin. They had two sound projectors spaced equal distance from a mid-line and the dolphin was required to be stationed on a bite plate that was aligned with the mid-line between the two projectors, similar to the geometry of Fig. 8.5. During a trial, only one of the projector would emit a signal and the dolphin was required to indicate which projector emitted the signal by touching the paddle on the same mid-line as the sound-emitting projector. The animal's performance was determined as a function of the angular separation between the transducers. Moore and Popper (1975) used a similar technique to determine the sound location capabilities of a California sea lion. Vel'min et al. (1975) in the former Soviet Union used such a 2AFC procedure to study the sensitivity of dolphins to click signals. Two transducers separated by a dividing net as depicted in Fig. 8.6 was used. The animal was required to swim to the same side of the net as the active transducer. The dividing net could be

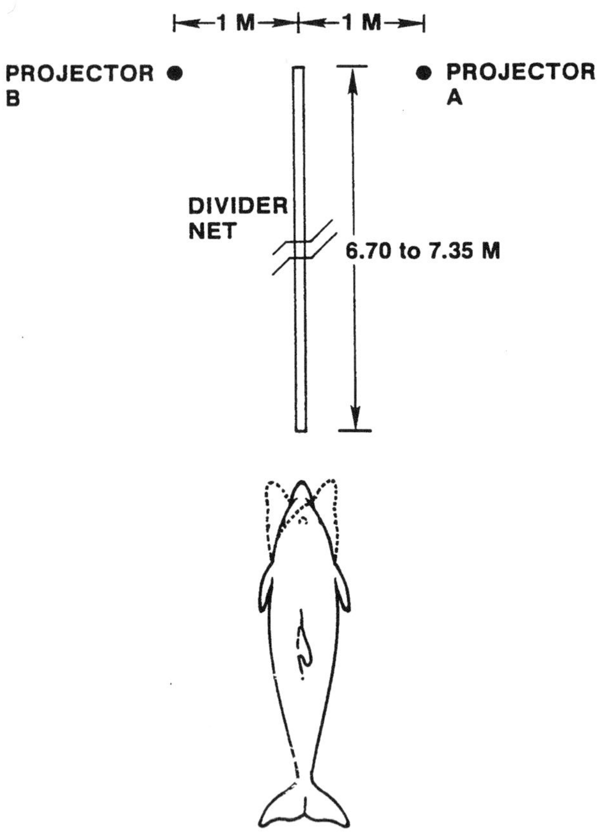

FIGURE 8.6. Experimental configuration using the 2AFC procedure by Vel'min and his colleagues.

extended to different ranges so that the experimenters could determine the minimal distance at which a specific choice was made.

8.1.3 Psychoacoustics Testing Procedures

After reliable stimulus control is achieved, some sort of psychophysical testing procedure is used to estimate the animal's absolute or differential sensitivity. The two most widely used techniques for studying acoustic perception and echolocation in dolphins have been the *modified method of constant stimuli* and the *up/down staircase* or *titration method.* The method of constant stimuli is a nonadaptive method in which the stimulus value is selected before testing begins and remains the same throughout the session. This method can be very arduous since after a session, there may not be any indication of the animal's sensitivity range and what the next stimulus level should be. The subject may also become accustomed to a stimulus level especially if a particular level is used for many consecutive sessions, and the subject may need some brief retraining when the stimulus condition changes. This is especially true in dolphin echolocation experiments in which the target range is a variable. The target range affects not only the level of the echoes but also the time required for a signal to leave an echolocating dolphin, arrive at the target, and return as an echo. A better technique is the modified method of constant stimuli in which a range of stimulus levels are used in a session. Each session is divided into several blocks of equal number of trials, each block having a specific stimulus level. The stimulus values are preselected and usually assigned to each block on a random basis. In either case, the stimulus presentation schedule (stimulus present or absent in a detection task, standard stimulus on the right or left in a forced choice experiment) is randomized in some fashion. An experimental session usually starts off with a block in which the animal can easily perform the detection or discrimination task at a high level of accuracy. The detection or discrimination task is gradually made progressively more difficult by manipulating the stimuli (i.e., decreasing the signal-to-noise ratio in a detection task) in blocks. Behavioral performance accuracy as a function of the stimulus value is recorded and the threshold is usually taken as some interpolated stimulus value associated with an arbitrary level of performance, typically 75% or 70% correct response for all trials or 50% correct detection involving only stimulus-present trials in a detection task.

The up/down staircase method is an adaptive method in which the stimulus values are chosen while testing is in progress. The stimulus value is progressively made weaker or more difficult to discriminate as the animal correctly detects or identifies the stimulus. The process continues until the animal makes an error; then the stimulus value is progressively stepped up or made easier to discriminate, until the animal once again responds correctly to the stimulus. An example of how the signal-to-noise ratio may be varied (by varying an attenuator used to control the signal level) in a detection experiment using the staircase testing procedure is shown in Fig. 8.7. Note that

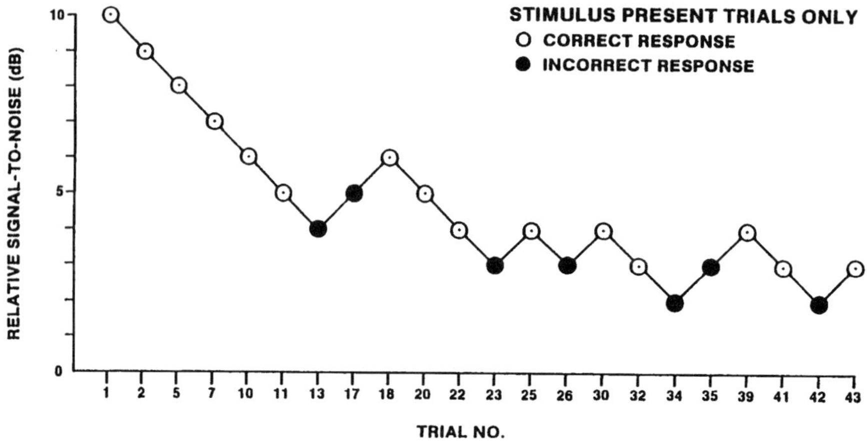

FIGURE 8.7. An example of variation in the signal-to-noise ratio in an experiment using the staircase testing procedure. Only stimulus present trials are shown since the signal-to-noise ratio is changed only after stimulus-present trials.

Fig. 8.7 only displays the results of stimulus-present trials. Usually, an equal number of stimulus-present and absent trials randomly distributed is used during a session. The stimulus-absent trials or "catch" trials are inserted to keep the animal from prospecting or guessing but do not affect the level of the stimulus. The catch trials are also used to calculate the false alarm level. Often the results of a session maybe discarded if the false alarm level exceeds an arbitrary predefined level. Having a relatively constant false alarm level is important if the results will be compared over many sessions, since the thresholds can vary significantly if the false alarm levels vary.

An estimate of the animal's sensitivity is obtained by averaging the stimulus values at all the reversal points, which are the local maxima and minima or peaks and valleys. In the example of Fig. 8.6, the threshold signal-to-noise ratio is

$$\begin{aligned} \text{Threshold S/N} &= \frac{6+4+7+6+7+6+8+6+8+7}{10} \\ &= 6.5 \pm 1.2 \text{ dB}. \end{aligned} \tag{8.7}$$

The threshold obtained with the up/down staircase is equivalent to the 50% correct detection threshold obtained with the method of constant stimuli (Levitt, 1970). The standard deviation given in Eq. (8.7) is understood to be the standard deviation associated with the variation of the value of a signal or noise attenuator. Strictly speaking, the signal-to-noise ratio is a ratio of power so that its standard deviation is not symmetrical about its mean when it is expressed in dB, but is symmetrical about its mean when expressed as a dimensionless ratio. However, for the sake of convenience, this point is often ignored or overlooked. A staircase threshold that is equivalent to the 70% correct detection threshold can be obtained by modifying the staircase procedure slightly. Instead of decreasing the stimulus value after each correct detection, the stimulus value is changed after two consecutive correct detection response. The up/down staircase method and the modified method of constant stimulus yield similar threshold estimates with the same variability, although constant stimulus method tends to produce less bias (Simpson, 1988).

8.1.4 Signal Detection Theory

In this section, a discussion of "signal detection theory" (SDT) will be presented, motivated by the fact that we can gain important insights into the process of detecting signals in noise. Some researchers have also applied the ideas of SDT to dolphin and pinneped bioacoustics, and so it is important to have a basic understanding SDT. Only a brief discussion of SDT will be presented in this section. Interested readers should refer to some excellent books on the subject written by Green and Swets (1966), Swets (1964), and Egan (1975). In considering this section, it is important to always keep in mind that we are discussing a theoretical model using the concept of an ideal observer.

8.1.4.1 Weakness of Classical Psychophysical Threshold

The discussion on measuring detection threshold presented in Sections 8.1.2 and 8.1.4 did not take into account nonsensory biasing factors such as motivation and expectation. Yet these response biases can affect the measurement of a subject's sensitivity in a detection or discriminating task. The response criterion or bias adopted by a subject can encompass a continuum that is affected by such factors as the cost of different types of errors (misses or false alarms), expectation of stimulus condition, and motivation. Toward one end of this continuum, a subject can adopt a conservative criterion and respond to the presence of a stimulus only when he or she is absolutely sure that the stimulus is present. A subject adopting such a cautious or conservative response bias will commit few false alarm errors, but will also have a low probability of detection score. On the other end of this continuum, a subject can adopt a very liberal or lax criterion and respond that a stimulus is present, with only a weak notion or hint that it is. This same subject now adopting a very liberal response bias will commit many false alarm errors but will also have a high probability of detection. For example, instructing human subjects to respond "stimulus present" if they have only the slightest hint that the stimulus is present will bias the subject to be liberal in reporting on the presence of a stimulus. Conversely, instructing a subject to respond "stimulus present" only if he or she is absolutely sure that the stimulus is present will cause the subject to be conservative. Using a procedure in which a subject is penalized for committing false alarms will tend to make the subject conservative and, conversely, giving a subject a greater reward for making correct detections than correct rejections will tend to make the subject liberal in reporting "stimulus present." Another effective technique to shape the response criterion of a subject in a psychophysical experiment is to vary the probability of a stimulus being present in a trial, a technique vividly demonstrated by Tanner et al. (1956). The following is an example taken from Gescheider (1976), showing how the response criterion of a subject can be manipulated by varying the probability of stimulus present, $P(s)$, in a given trial while keeping the stimulus level constant. A summary of the experiment is given in Table 8.1 and the corresponding ROC curve (lower curve) is drawn in Fig. 8.8. The subject's performance, $P(\underline{Y}/\text{sn})$ and $P(\text{Y}/\text{n})$, is given in the Yes-response column, with $P(\text{N}/\text{sn})$ and $P(\text{N}/\text{n})$ given in the No-response column of Table 8.1. The ROC curves indicate that when the probability of stimulus occurrence decreased from 0.5 to 0.3 and 0.1, the subject became progressively more conservative by decreasing the false alarm errors, but with a corresponding decrease in the probability of correct detection. As the probability of stimulus occurrence increased to 0.7 and 0.9, the subject became progressively more liberal, increasing the probability of detection with a corresponding increase in the probability of false alarm. When the probability of stimulus occurrence was 0.5, the subject's performance fell on the minor diagonal, indicating a nonbiased performance. The subject sensitivity was probably constant for each stimulus level, but the performance varied substantially as the nonsensory factor (variation in stimulus occurrence probability) caused the subject's response criterion to vary. This example shows that the same subject having the same sensitivity to a stimulus can have quite different probability of detection scores depending on his or her response criterion. We will return to a discussion of response bias manipulation, but for marine mammals, after considering the elements of signal detection theory.

TABLE 8.1. Signal detection experiment in which the stimulus probability is manipulated from session to session (after Gescheider, 1976)

P(s)	No. of stimulus trials	No. of no stimulas trial	Stimulus condition	Response Yes	Response No
0.90	180	20	Present	0.99	0.01
			Absent	0.95	0.05
0.70	140	60	Present	0.91	0.09
			Absent	0.64	0.36
0.50	100	100	Present	0.69	0.31
			Absent	0.31	0.69
0.30	60	140	Present	0.36	0.64
			Absent	0.09	0.91
0.10	20	180	Present	0.05	0.95
			Absent	0.01	0.99

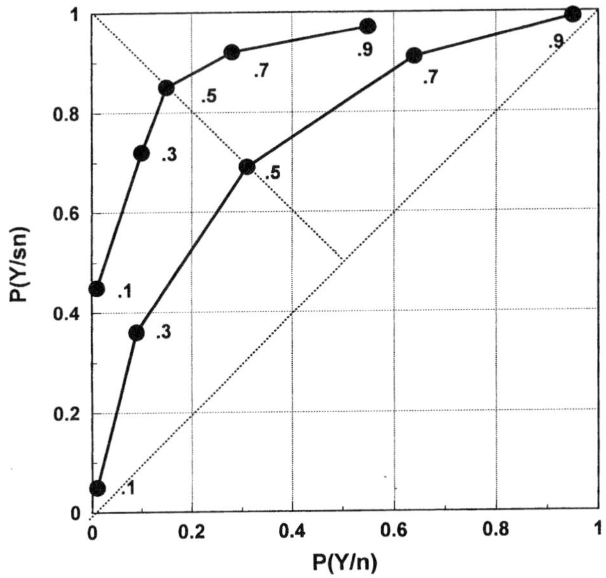

FIGURE 8.8. ROC curves for an experiment in which the probability of stimulus occurrence is varied between sessions and for two constant stimulus level. The lower curve is for a weak stimulus level and the upper curve is for a strong stimulus level. Each data point corresponds to a different probability of stimulus occurrence (after Gescheider, 1976).

8.1.4.2 Elements of Signal Detection Theory

The "theory of signal detectability" (TSD) originally devised by Peterson et al. (1954) to study the detection of electronic signals in noise, provides a way to separate the two aspects of a subject's performance: sensitivity and response bias. Since the pioneering work of Peterson et al. in the field of radar signal detection, many have applied TSD to psychophysics, among the first being Tanner and Swet (1954), and the TSD nomenclature was gradually changed to "signal detection theory" (SDT). SDT provides a theoretical framework from which the process of detecting a signal in noise can be studied and understood. We shall see that the performance of various mammals in detecting a signal in noise can be explained by the principles embedded in SDT and thus provide a means to understand signal detection processes in animals.

SDT holds that there is no true sensory threshold of the type discussed in Section 8.1.4. Rather, sensation is a continuously varying quantity and is zero only when the stimulus intensity is zero. Therefore, the detection of a stimulus in the presence of interference and noise is essentially a decision process and the threshold lies in the subject's decision-making process. Since noise is a random process, the problem is statistical in nature and there is always a finite probability of making an error. In the decision process the subject makes an observation (x) and then makes a decision about the observation. The subject must decide whether x is due to a signal added to the background noise or to the noise alone. When the signal level is much greater than the noise level, the decision is relatively easy. When the signal level is weak and close to the noise level, the decision becomes difficult and the subject will make many errors. If we assume that the interfering noise has a normal or Gaussian distribution and that the signal plus noise is also normal with the same variance as the noise, the detection of a signal in noise can be represented graphically in Fig. 8.9 by two probability distributions, one describing the random variation of the noise $f_N(x)$ and the other describing the signal plus noise $f_{SN}(x)$. The decision process for the case represented by the top-panel is relatively easy, while the process for the middle panel is moderately difficult, and the process for the bottom panel is very difficult. Justification for using the Gaussian assumption has been discussed in detail by Green and Swets (1966). From a theoretical perspective, the "central limit theorem" in statistics indicates that the sum of independent random variables will approach a Gaussian distribution. Since sensory events are often considered as composed of a multitude of similar, smaller events, which are essentially independent, the central limit theorem could be invoked to justify the Gaussian assumption for the net effects (Green and Swets, 1966). From a practical perspective, the Gaussian assumption allows one to derive results that would otherwise be difficult if not impossible to derive (Green and Swets, 1966).

In the decision process, the subject makes an observation x, which consists of a sample from one of the two distributions in Fig. 8.9 and must decide on the correct distribution. The ordinate of $f_{SN}(x)$ gives the probability density or likelihood of x occurring when a signal is present and the ordinate of $f_N(x)$ is the probability density or likelihood of x occurring when only noise is present.

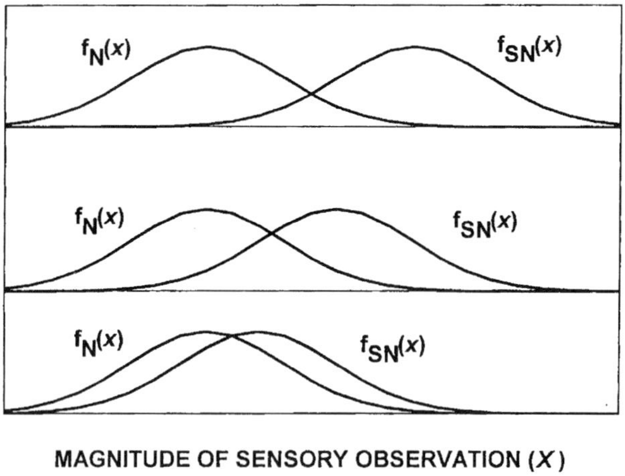

FIGURE 8.9. Probability density functions for three different signal-to-noise situations.

Peterson et al. (1954) analyzed the decision process involving the detection of a signal in noise and introduced a function called the likelihood ratio, which is defined as

$$l(x) = \frac{f_{\mathrm{SN}}(x)}{f_{\mathrm{N}}(x)}. \tag{8.8}$$

They concluded that a receiver which calculates the likelihood ratio for each receiver input is the optimal or ideal receiver for detecting the signals in noise. The decision process can be described by the diagram in Fig. 8.10 showing the two Gaussian probability density functions, one associated with noise alone and the other with signal plus noise. The ideal observer establishes a particular value of $l(x)$ as a cutoff point, or criterion (β), so that if for any particular observation x the likelihood ratio is greater than β, the observer chooses signal plus noise, and if the likelihood ratio is less than β, the observer chooses noise alone. If the value of x corresponding to the criterion β is equal to k, we can define β as

$$\beta = \frac{f_{\mathrm{SN}}(k)}{f_{\mathrm{N}}(k)} = l(k). \tag{8.9}$$

equals the probability of detection. The cross-hatched area to the right of k, under the noise only probability density function $f_{\mathrm{N}}(x)$ equals the probability of false alarm. Let us assume that both the hatched area to the right of k, under the signal plus noise probability density function, $f_{\mathrm{SN}}(x)$ and $f_{\mathrm{N}}(x)$ have a Gaussian distribution with the same standard deviation, σ_{N}, so that these two probability density distributions can be expressed as

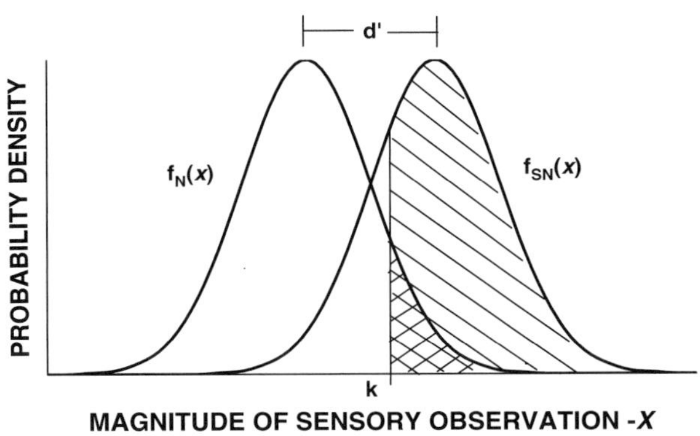

FIGURE 8.10. Probability density function of the noise and signal plus noise in decision process.

$$f_{\mathrm{SN}}(x) = \frac{1}{\sqrt{2\pi}\,\sigma_{\mathrm{N}}} \mathrm{e}^{-\frac{x-M_{\mathrm{SN}}}{2\sigma_{\mathrm{N}}^2}},$$

$$f_{\mathrm{N}}(x) = \frac{1}{\sqrt{2\pi}\,\sigma_{\mathrm{N}}} \mathrm{e}^{-\frac{x-M_{\mathrm{N}}}{2\sigma_{\mathrm{N}}^2}} \qquad (8.10)$$

where M_{SN} and M_{N} are the mean values of the signal plus noise and the noise alone. The probaabilities $P(\mathrm{Y/sn})$ and $P(\mathrm{Y/n})$ can be calculated from the equations

$$P(\mathrm{Y/sn}) = \int_k^\infty f_{\mathrm{SN}}(x)\mathrm{d}x = \Phi\left[\frac{k - M_{\mathrm{SN}}}{\sigma_{\mathrm{N}}}\right],$$

$$P(\mathrm{Y/n}) = \int_k^\infty f_{\mathrm{N}}(x)\mathrm{d}x = \Phi\left[\frac{k - M_{\mathrm{N}}}{\sigma_{\mathrm{N}}}\right] \qquad (8.11)$$

where Φ is the area under the normal distribution curve found tabulated in any textbook on statistics. If k is shifted toward the right (subject becoming more conservative), $P(\mathrm{Y/sn})$ and $P(\mathrm{Y/n})$ will both decrease in value. If k is shifted toward the left (subject becoming more liberal), $P(\mathrm{Y/sn})$ and $P(\mathrm{Y/n})$ will both increase in value. A pictorial description of the effects of shifting the criterion k for an ideal situation assuming Gaussian distributions of equal variance for the probability density functions is shown in Fig. 8.10 for two different signal-to-noise conditions. The expressions in Eq. (8.10) were used to generate the values in the ROC curves of Fig. 8.10. As the criterion point is shifted to the right, both $P(\mathrm{Y/sn})$ and $P(\mathrm{Y/n})$ become larger. Comparing the ROC curves in Fig. 8.7 and the theoretical ROC curves in Fig. 8.9, we can see that human subjects' performance change in a similar manner as predicted by the SDT for varying criterion.

Another parameter that is important in SDT is the measure of signal detectibility, d', shown in Figs. 8.8 and 8.9, and is associated with the sensitivity of the subject. The signal detectability d' is equal to the difference between the means of the $f_{\mathrm{SN}}(x)$ and $f_{\mathrm{N}}(x)$ distributions ($M_{\mathrm{SN}} - M_{\mathrm{N}}$) expressed in standard deviation units of the noise distribution, σ_{N}, so that

$$d' = \frac{M_{SN} - M_N}{\sigma_N}. \qquad (8.12)$$

The locations of the $f_{\mathrm{SN}}(x)$ and $f_{\mathrm{N}}(x)$ distributions are entirely a function of the stimulus and noise intensities so that d′ is a pure index of stimulus detectability and is independent of the location of the subject's threshold level or criterion. For the situation in which the signal is completely known, Peterson et al. (1954) have shown that d' is related to the signal-to-noise ratio according to the relationship

$$d' = \sqrt{2\frac{E}{N}}, \tag{8.13}$$

where E is the energy in the signal and **N** is the noise spectral density (rms noise in a 1-Hz band or rms noise per Hz). Since d' in Eq. (8.13) is the sensitivity associated with an optimal receiver, it represents the best sensitivity possible. Any other receiver will have a lower sensitivity than the optimal receiver.

The signal detectability d' can be calculated for a subject from $P(\mathrm{Y/sn})$ and $P(\mathrm{Y/n})$ since both of these probabilities are equal to the area under the $f_{\mathrm{SN}}(x)$ and $f_{\mathrm{N}}(x)$ distribution curves, respectively, as expressed by Eq. (8.11). By definition, d' is a measure of how far apart the two distributions are in standard deviation units i.e., how many standard deviations the means of the two distributions are apart. The distance between the k-value and the means of the noise and the signal plus noise distributions can be also be measured in terms of standard deviation units or z-scores. To determine how far the k-value is from the mean of the noise distribution, we look up in a table giving the values of the normal distribution function and determine the z-score that corresponds to the area under the normal curve which is equal to the false alarm rate or P(Y/n) to obtain z_{N}. Similarly, to determine how far the k-value is from the mean of the signal plus noise distribution, we look up the z-score corresponding to the area under the normal curve, which is equal to the hit rate or $P(\mathrm{Y/sn})$ to obtain z_{SN}. A simple rule for calculating d', which works for any k-value is (Snodgrass, 1972)

$$d' = z_{\mathrm{N}} - z_{\mathrm{SN}} \tag{8.14}$$

Value of d' can also be obtained directly for any combinations of $P(\mathrm{Y/sn})$ and $P(\mathrm{Y/n})$ by using a table provided by Elliot (1964).

The criterion β can also be computed from a pair of hit and false alarm rates using the z-scores associated with each probability. From Eq. (8.9), β is equal to the ratio of ordinates of the normal distributions $f_{\mathrm{SN}}(k)$ and $f_{\mathrm{N}}(k)$. The ordinate of the $f_{\mathrm{N}}(k)$ distribution can be estimated as the ordinate value given in the table of the normal distribution, which corresponds to the false alarm rate. Likewise, the ordinate of the $f_{\mathrm{SN}}(k)$ distribution can be estimated as the ordinate value of the normal distribution that corresponds to the hit rate. Snodgrass (1972) provides tables containing the area under the normal curve, which is equal to $P(\mathrm{Y/sn})$ or $P(\mathrm{Y/n})$, the corresponding z-scores and the value of the ordinate of the normal curve at k = x. The criterion conveys the bias of the subject in a yes/no task. Values of $\beta > 1$ indicate a liberal subject, and the greater β exceeds 1, the more liberal the subject is. Values of $\beta < 1$ indicate a conservative subject and when $\beta = 1$, the subject is unbiased or has a neutral criterion.

8.1.4.3 Applying SDT to Marine Mammals

Most of the bioacoustics research performed on marine mammals have used the traditional psychological testing and threshold determination procedures discussed in Section 8.1.4. However, there is a scattering of research results published from the SDT prospective. Some publications contain results obtained from traditional psychoacoustic threshold determination procedures, but with results plotted as ROC curves along with calculated values of d' and β.

Schusterman and his colleagues were the first to apply the elements of SDT to study the auditory sensitivity of marine mammals (Schusterman and Johnson, 1975; Schusterman et al., 1975; Schusterman, 1976). They experimented with two techniques to vary the response bias of their subjects. The first technique involved a variation in the stimulus probability in the same manner as was done in Table 8.1. They used three probabilities (3, 0.5, and 0.7) of the stimulus being present to vary the response criterion of a California sea lion in a auditory detection task. The top left panel of Fig. 8.11 shows the ROC curves obtained by Schusterman and Johnson (1975) for three different signal levels. For each signal level, the left most point corresponded to a probability of 0.3 that the stimulus will be present on any trial. The next point to the right corresponded to a 0.5 stimulus probability and the right most point corresponded to a stimulus probability of 0.7. Also shown in the panel are the d' curves for an ideal detector that corresponded to the averaged d' obtained by the animal. The ROC curves suggest that the animal's sensitivity was fairly constant as its response criterion was manipulated by varying the stimulus probability. The response of the sea lion to changes in the stimulus probability was similar to the response of the human subject depicted in Table 8.1.

Schusterman et al. (1975) also manipulated the response criterion of marine mammal subjects by varying the payoff matrix for correct responses (Fig. 8.12). They used three different payoff schedules: the 1:4 payoff schedule awarded the subject four fishes for every correct rejections compared with one fish for every correct detection, the 1:1 payoff schedule awarded one fish for each correct rejection and correct detection, and the 4:1 payoff schedule awarded one fish for each correct rejection and four fishes for correct detection. The upper right panel shows the ROC curves obtained by Schusterman et al. (1975) for a California sea lion. For each signal level, the left most point corresponded to the 1:4 payoff, the middle point for the 1:1 payoff, and the right most point corresponded to the 4:1 payoff matrix. The bottom left panel shows the ROC curve for an Atlantic bottlenose dolphin as the payoff

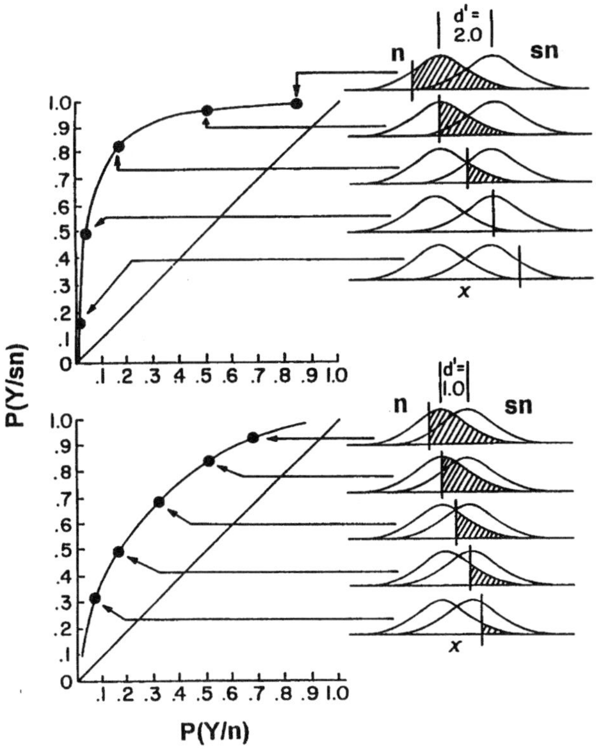

FIGURE 8.11. The relation between ROC curves and the theoretical noise and signal plus noise distributions. Variations in the subject's criterion results in different points along the same ROC curve while variation of signal-to-noise ratios produces different ROC curves (adapted from Gescheider, 1976).

matrix was varied (Schusterman et al., 1975). These results indicate that the subject's decision criterion can be manipulated rather readily by varying the food payoff matrix. The performance results of both the sea lion and the bottlenose dolphin exhibited relatively constant sensitivity as their decision criteria were manipulated. Au and Turl (1984) used the variable payoff scheme of Schusterman to vary the decision criterion of an echolocating dolphin. Their results are shown in the bottom right panel of Fig. 8.11. They used an addition payoff condition (8:1) in which the dolphin was rewarded with eight pieces of fish for every correct detection and only one piece of fish for every correct rejection. Except for the 8:1 payoff condition with the two larger targets, the dolphin's detection sensitivity was fairly constant as its decision criterion shifted from being conservative to relatively liberal.

The SDT approach to hearing studies has not been widely adopted by scientists working with marine mammal acoustics. There are several reasons for this lack of acceptance and use of the SDT approach. First, having a specific

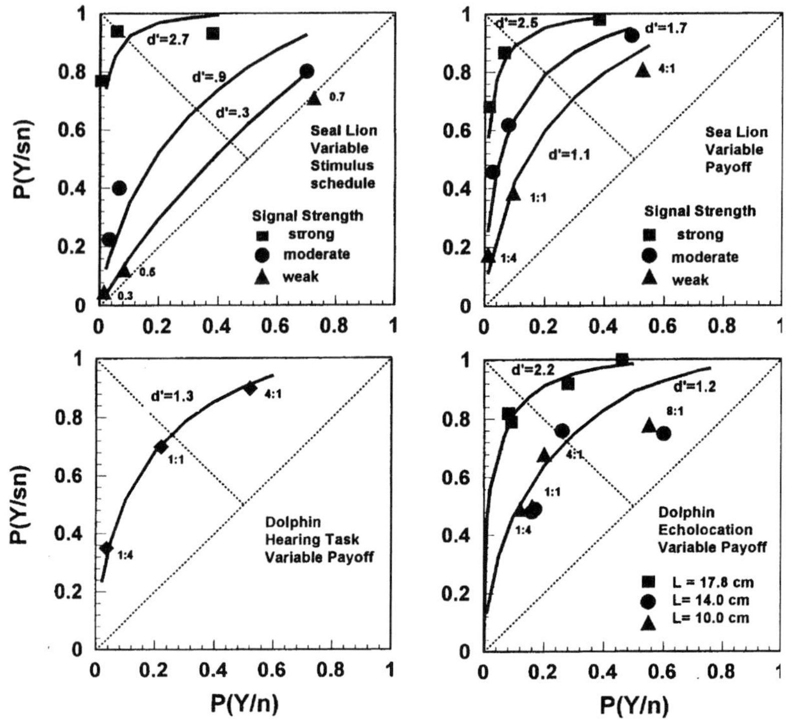

FIGURE 8.12. ROC curves for some marine mammals with the animals' criterion being manipulated by varying the stimulus probability and the payoff matrix (after Schusterman and Johnson, 1975; Schusterman et al., 1975; Au and Turl, 1984).

threshold value either in terms of a sound pressure level or signal-to-noise ratio for passive acoustic experiments or a target range or a signal-to-noise ratio for an echolocation experiment is much easier to relate to and appreciate than a d' value. Secondly, if an audiogram is desired or some echolocation parameter has to be determined as a function of a variable that can have many values, it is more efficient to use a staircase testing procedure to obtain thresholds at different frequencies than to manipulate a subject's bias while obtaining data for several isosensitivity curves. Nevertheless, an understanding of the concepts of SDT is important in order to appreciate how bias and non-sensory factors can affect the performance of subjects.

8.2 Psychoacoustics Discrimination Experiments

We will now consider psychoacoustics experiments in which we desire to learn how well a subject can discriminate between various features of acoustic stimuli in order to infer the properties of the subject's auditory system. Psychoacoustic procedures for discrimination experiments can be divided into several general categories. These categories are not mutually exclusive and there is often a fine line as to what category an experimental procedure belongs. Also, within a category there may be differences in procedure for testing echolocation in dolphins compared with hearing in dolphins and non-echolocating aquatic animals.

8.2.1 Relative Magnitude Difference

In this category, we wish to test how well the subject can determine the relative magnitudes of various acoustic variables. Common variables in hearing studies are amplitude, duration, and frequency of stimuli. The subject is queried as to which projector in a two-projector experiment emitted the higher amplitude, or longer duration, or higher frequency signal, etc. If only one sound projector is used, the subject is asked which interval in a two-alternative forced choice paradigm described by Fig. 8.4 has the higher (or lower) amplitude, or longer duration (or shorter), or higher (or lower) frequency signal. If two sound projectors shown in Fig. 8.5 are used, the experimenter must be careful of any possible constructive and destructive interferences of the two signals presented simultaneously. Even if the signals do not interact with each other, there is a possibility that the presence of multiple signals may interfere with the subject's ability to discriminate whatever difference is being tested. The effect of interference can be eliminated by projecting one stimulus at a time from the two transducers. If two sound projectors are used in a sequential manner, it may be prudent to also randomize which projector will be turned on first, to discourage the subject from locking on to the specific projector that is always turned on first (or second) as the task becomes difficult.

The experiment of Vel'min et al. (1975) in which a dolphin was required to discriminate the intensity difference in click signals projected by two projectors is an example of an auditory experiment in which subjects were required to discriminate the relative magnitude of stimuli. The experimental configuration was similar to the one depicted in Fig. 8.6. Both transducers emitted a click signal at a repetition rate of 30 Hz, with one signal being in higher intensity than the other. The dolphin was required to swim toward the transducer that emitted the higher-intensity signal. In other experiment using a similar experimental setup as depicted in Fig. 8.6, Vel'men and Dubrovskiy (1975) projected a pair of clicks separation by a time τ from one projector and another pair of clicks separated by a time $\tau + \Delta\tau$. The dolphin was required to swim to the same side of the net of as the transducer that projected the click pair having a separation time of $\tau + \Delta\tau$ (the click pair with the larger separation time).

The experiments of Jacobs (1972), Herman and Arbeit (1972), and Thompson and Herman (1975) can also be envisioned as discrimination experiments requiring the subject to specify the relative magnitude of acoustic stimuli. In all three experiments, dolphins were tested on their capability to discriminate constant frequency (CF) or pure tone signals from frequency-modulated (FM) signals. In the experiment of Jacobs (1972) and Herman and Arbeit (1972), a single projector was used and the animal was required to indicate whether a CF or FM signal was emitted during a trial by pressing appropriate paddles. In the experiment of Thompson and Herman (1975), two projectors were used sequentially in a trial so that if one projector projected a CF signal, the other would project an FM signal. The dolphin was required to touch the paddle closest to the projector that emitted the CF signal. These experiments can be viewed as asking the dolphin to specify which stimuli had the largest variation in frequency.

Echolocation experiments in this category could ask similar kinds of questions as the hearing experiments. The dolphin may be asked which of two targets is closer (or further), which is larger or has a stronger (or weaker) echo, etc. The experimenter must take the precaution of spacing the targets sufficiently far apart so that the dolphin can ensonify only one target at a time. The echolocation range difference experiment of Murchison (1980) is a good example of this category of discrimination experiment. The dolphin was trained to station on a chin cup, echolocate two spheres spatially separated by an angle of 40°, and indicate which sphere was closest to it. The animals range difference acuity was tested for three conditions in which the range of the closest target was 1, 3, and 7 m from the tip of the chin cup.

8.2.2 Standard Versus Non-Standard Stimuli

The discrimination procedure in which a subject is trained to accept a particular acoustic stimulus or a class of stimuli as the "standard" and all other stimuli as "non-standard" is perhaps the most widely used procedure with marine mammals. This procedure can be carried out in several ways. In a single sound projector experiment, a single stimulus can be presented in a trial and the subject must indicate whether the stimulus was a standard or non-standard stimulus. There can also be two observational intervals in a trial in which a standard is played in one interval and a non-standard is played in the other interval. The subject must then indicate which of the two intervals contained the standard stimulus. If two widely spaced projectors are used, the standard signal can be emitted from one projector and the non-standard signal from the other projector. The animal's task would be to indicate which of the two projectors emitted standard signal. The experiments of Jacobs (1972), Herman and Arbeit (1972), and Thompson and Herman (1975) can also be classified in this category of discrimination experiments if the CF signals are considered as standard and all FM signals as non-standard.

The wall thickness and material discrimination experiment of Evans and Powell (1967) using echolocating dolphins is an example of a standard versus non-standard echolocation experiment. A circular copper disk with a specific thickness was used as the standard target. Comparison or non-standard targets consisted of copper discs of 30-cm diameter but different wall thickness and aluminum and brass discs. The standard and a single non-standard target were presented simultaneously separated by a distance of 50-cm (center to center) and the blindfolded dolphin was required to determine the location of the standard, swim to it, and touch it. Another example of a standard versus non-standard discrimination experiment was conducted by Evans (1973) in a target strength discrimination with an *Inia geofrensis* and *a Tursiops truncatus*. The animals were trained to accept a chloroprene cylindrical target having a target strength of –19 dB as the standard target and all other chloroprene cylinders having a different target strength as a non-standard or comparison target. The standard and comparison targets were presented simultaneously and the blindfolded echolocating dolphins had to report the position of the standard target.

8.2.3 Same–Different Stimuli

A third method in testing the discrimination capabilities of a subject consists of asking the question whether two stimuli are same or different. This procedure can be used in a listening experiment using one or two sound projectors. If one sound projector is used, the stimuli will need to be presented sequentially in a similar manner as the 2AFC choice sequence shown in Fig. 8.4 and the subject must indicate whether the stimuli are the same or different. If two sound projectors are used, the stimuli can be played sequentially first from one projector and then from the other projector. The subject may respond in a number of ways such as touching a specific paddle to indicate a same response and another paddle to indicate a difference response, or remaining at station to indicate a same response and leaving station to strike a paddle to indicate a difference response, etc. The properties of the two stimuli can be made to approach each other until the subject can no longer discriminate the two stimuli.

The same–difference testing procedure has not been used extensively by the scientists studying marine mammal audition. One study using the same–difference methodology was conducted by Johnson et al. (1988) using a bottlenose dolphin. Click signals were presented in pairs with one of the clicks being larger than the other. The s_1 signal was defined as the click pair with the first click having the larger amplitude. The s_2 signal was also a double click signal but the second click had the larger amplitude. A single transducer was used so that there were two observation intervals separated by a quiet interval. The observation intervals were 1.5 s and the silent interval was 0.5 s in duration. The s_1 signal was always projected in the first observational interval followed by either another s_1 signal or an s_2 signal. A Go/No-go response paradigm was used and the animal was required to remain at station (No-go response) if the s_1 signal was played during both observational intervals. If the s_2 signal was played during the second observational period, the dolphin left its station and touched a response paddle (Go response).

An echolocation experiment using the same–difference procedure should have two targets presented simultaneously, sufficiently separated in azimuth to allow the dolphin to echolocate each target separately. The properties of the two targets can be made to approach each other until the subject can no longer discriminate between them. The experiment of Evans (1973) in which a –19 dB target was used as a standard and all other were non-standard targets could have been conducted using the same–difference methodology. The –19 dB target would have to be paired with another –19 dB target for a "same-targets" trial and a comparison target for a "different-targets" trial. The target presentation schedule could be based on a modified method of constants or an up-down staircase scheme. For a modified method of constants using blocks of ten trials, five of the trials in each block would use a matching –19 dB target and five of the trials would involve a specific comparison target. The specific comparison target in any ten-trial block should be chosen randomly. For an up-down staircase method of target presentation and for each correct different-targets trial, the comparison target with the next closest target strength to the –19 dB target would be used for the next different-targets trial. If the dolphin made an incorrect response for a different-targets trial, the comparison target with the next furthest target strength would be used for the next different-targets trial. The position of the –19 dB target should be randomized for each different-targets trail.

In the hypothetical same–different experiment using the targets of Evans (1973), the –19 dB target was used as an ad hoc standard. However, if the subject truly grasps the same–difference concept, an experiment testing for one discrimination capability can easily shift into testing another discrimination capability. For example, when using Evans' targets, the experiment is to determine the target discrimination capabilities of echolocating dolphins. Upon completion of the target strength discrimination task, cylinders with different wall thickness can be used to determine the subjects' capability in discrimination wall thickness, etc., with the minimum amount of training. Therein lies one of the strengths of the same–different procedure in considering the discrimination capabilities of marine animals.

8.2.4 Matching-to-Sample

Another technique to determine the discrimination capabilities of animals is the *matching-to-sample* procedure. A simple way to explain a matching-to-sample experiment is to consider a visual shape discrimination experiment in which three planar objects are positioned side by side as in Fig. 8.13. The middle shape is taken as the sample and the subject must decide which of the test objects (the left or right one) matches the sample irregardless of the size, color, texture, and orientation of the test objects. A matching-to-sample procedure in which the sample is presented and then removed followed by the presentation of two or three test objects, one having the same shape as the sample, is called a *delayed* matching-to-sample task. The delay is associated

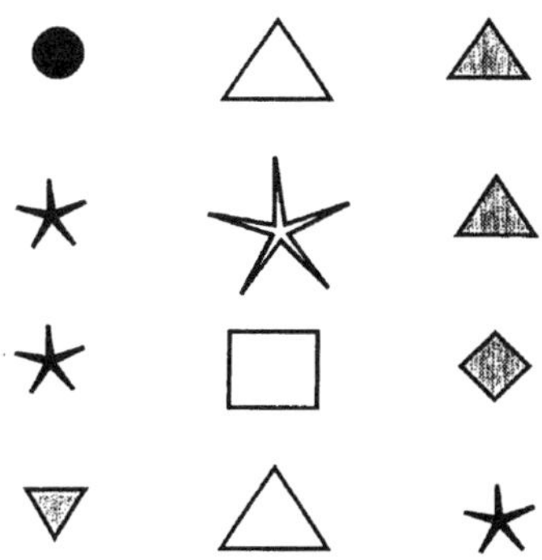

FIGURE 8.13. Example of a visual matching-to-sample task.

with the time between the removal of the sample and the presentation of the test objects. Since as we have previously stated, it is problematic to present sound simultaneously because of possible interference effects, an acoustic matching-to-sample would most likely fall in the delayed category. An acoustic matching-to-sample task can be configured in a number of ways. For example, a single transducer can be used in which the first signal in a trial is taken as the sample. After the cessation of the sample sound, two other test sounds, one matching the sample, are played sequentially and the subject must indicate which interval contained the match. If two transducers are used, one could be used to present the sample at the beginning of a trial and after a short delay the test sounds are played sequentially, one matching the sample, by both transducers. If three transducers are used, one can be used to project the sample signal only and the test signals are projected sequentially by the other two transducers.

The acoustic matching-to-sample procedure, to the best of our knowledge, has not been used to determine the discrimination capabilities of aquatic animals. Thompson and Herman (1975) conducted an acoustic experiment using a procedure that had some characteristics of a matching-to-sample task. They projected several different sounds (as many as 8) sequentially to a dolphin, and after a period, a sample sound was presented. The bottlenose dolphin had to indicate whether or not any of the test sounds matched the sample sound. However, their primary motivation in this procedure was to test the dolphin's auditory memory rather than its discrimination capabilities.

A variety of acoustic discrimination tasks can be performed using a matching-to-sample procedure including the discrimination of amplitude difference, frequency difference, amplitude modulation depth, and frequency modulation rates. If the frequency discrimination capability is of interest, a set of frequencies could be selected apriori to be used as the sample in a given trial. One of the test sounds would have the same frequency as the sample but with a very different amplitude, and the other test sound would be of a different frequency but with the same amplitude as the matching test sound. On the next trial, the sample sound could be another frequency from the set of possible sample frequencies. The frequency of the nonmatching comparison test sound can be gradually made closer to the frequency of the sample sounds. In using such an experimental procedure, a researcher will eventually be able to determine the frequency resolution capability as a function of frequency of a test subject.

The experiment of Roitbalt et al. (1990) is an example of using the matching-to-sample procedure. However, their objectives were not to determine the limits of a dolphin echolocating discrimination capabilities but to demonstrate a dolphin's ability to comprehend the matching-to-sample procedure in an echolocation modality and to study the animal's decision-making process. A target set consisted of a large hollow PVC tube (10 dm O.D., 25 cm length, 0.3 cm wall), a small hollow PVC tube (7.5 cm O.D., 15 cm length, 0.3 cm wall), a solid aluminum cone (10 cm base, 10 cm length), and a 5-cm water-filled sphere. Since the targets differed in material composition, shape, dimensions, and structure, this experiment could be considered a multidimensional discrimination experiment. The experiment was conducted with four arrays of targets, a sample array and three comparison or test arrays. The sample array was located 4.8 m directly ahead of the animal's station and the comparison arrays were suspended from a bar located 3.9 m from the dolphin's hoop station, as is depicted in Fig. 8.14. A trial consisted of the blindfolded dolphin inserting its head into the observational aperture while a sample target from the sample array was gently lowered into the water. The dolphin echolocated the sample target until a screen was raised to block its echolocation signals. Next, a target from each comparison array was lowered into the water, one of the targets matched the sample and two were different. The acoustic screen was again lowered and the dolphin echolocated the targets from the comparison arrays to determine which was the matching target. The animal responded by touching one of the three response paddles to indicate the location of the matching target.

The matching-to-sample procedure like the same–different procedure can be used to test for different types of discrimination capabilities with a minimum of retraining. For example, a matching-to-sample procedure could also be used to test the target strength discrimination capabilities of dolphins using same targets of Evans 1973. In this case, the sample array would consist of the five targets used by Evans and one comparison target would always be the –19 dB target and the other comparison target would be the matching target. After this task, cylindrical targets of different wall thicknesses could be used to test the dolphin's capabilities to discriminate wall thickness. One again, if the dolphin truly understands the matching-to-sample concept, a minimum of training would be required to switch from

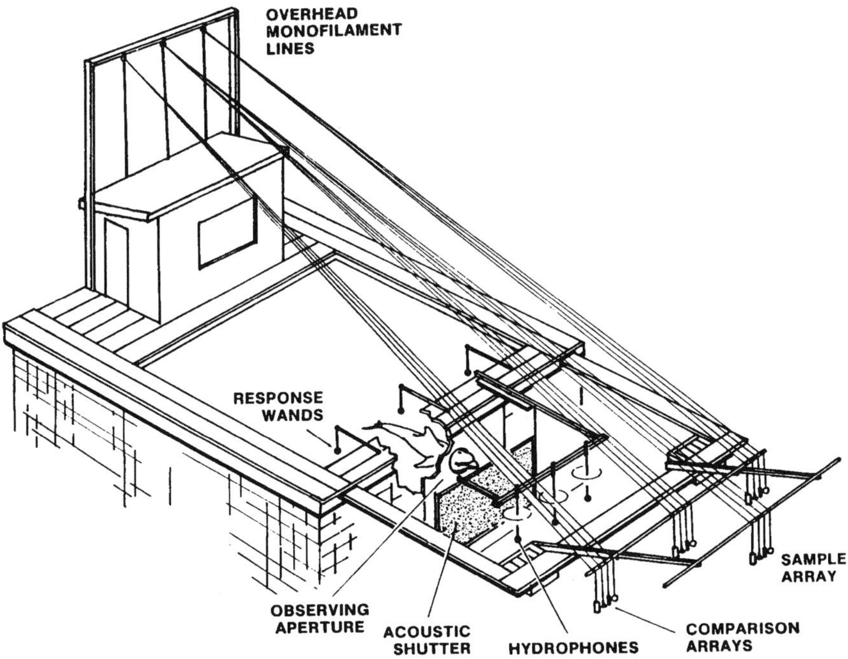

FIGURE 8.14. Diagram of the matching-to-sample apparatus in the echolocation experiment of Roitbalt et al. (1990).

the target strength discrimination task to the wall thickness discrimination task. However, using the same–different technique may be simpler than the matching-to-sample technique to study discrimination capabilities in echolocating dolphins, because fewer targets can be used and only two target locations are necessary.

8.2.5 Probe Technique in Echolocation

All of the procedures previously described to measure the discrimination capabilities of echolocating dolphins compared the behavioral responses with corresponding manipulation of a specific target parameter (i.e., size, material, shape, wall thickness, etc.). These methods do not convey any information as to how dolphins classify targets or echo parameters. The probe technique can provide insights to the dimensions along which some sensory quality is represented in an echolocating dolphin. The technique requires the establishment of an extremely stable discrimination between two different classes of targets and then interspersed with trials in which novel targets are used in "probe" trials. The number of probe trials per session should necessarily be small, or the novel targets will no longer be considered novel.

An example of using the probe technique is the experiment of Hammer and Au (1980), in which a dolphin was trained with two sets of different-sized cylinders. One set of cylinders composed of hollow aluminum cylinders controlled response to a specific paddle and another set of cylinders composed of coral rock embedded in degassed epoxy controlled response to another paddle. The animal was then presented with probe trials of different-sized hollow cylinders of glass, bronze and steel, and the response of the animal was observed. Hammer and Au (1980) found that the dolphin classified glass most like aluminum and steel least like aluminum, regardless of target size. In a typical 64-trial session, a maximum of only eight-probe trials with two novel targets were used.

8.3 Psychoacoustics Techniques for Fish Hearing

Since the 1950s three techniques, coupled with either the descending method of limits or the staircase psychophysical procedure as described in Section 8.1.4, have been used to study hearing capabilities in fish. The first technique, conditioned approach for food, was also used prior to 1950, but had several pitfalls. In this method, fishes are trained to approach a particular area on hearing a stimulus by receiving a food reward for the correct response. It is difficult to know, however, the sound pressure level detected by the fish because it is free to move about in the tank or test area. In addition, the response is often ambiguous near threshold because of the spatial inconsistency of the stimulus.

The instrumental avoidance conditioning technique was often used in the 1960 s and 1970 s. In this method, fish are trained to cross a barrier in a tank on hearing a sound to avoid getting an electric shock (Horner, Longo and Bitterman 1961; Tavolga and Woodinsky 1963). As with the conditioned approach for food, the fish is free to move about; this makes for a non-uniform stimulus. To overcome this deficiency, Weiss (1966) minimized the size of the shuttle box to a $4'' \times 4'' \times 8''$ tank with a rho-c rubber divider. He cut a small passage hole in the rubber divider so the fish could quickly move from one side to the other. The stimulus was generated by running two sound projectors located at each end of the tank in phase (i.e., push-pull mode). This created a fairly uniform sound pressure level from side to side and up and down the small chamber as illustrated in Fig. 8.15.

One of the problems with the conditioned approach for food and instrumental avoidance conditioning techniques is the amount of time it takes to train a fish. Weiss (1966) reported a one-hour training period per day for 25 days and that not all the specimens were trainable. To reduce the training period and avoid methods with negative reinforcement using shock, Yan and Popper (1991) developed a positive reward operant conditioning paradigm that was used to measure the auditory sensitivity in goldfish (*Carassius*

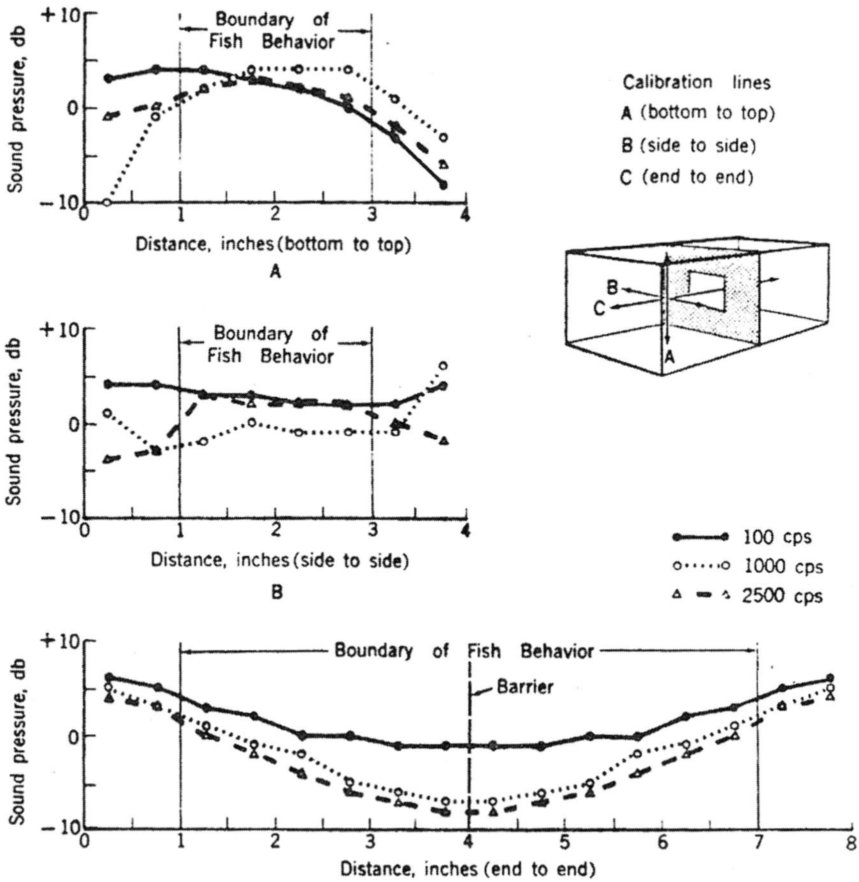

FIGURE 8.15. Limiting the volume of the shuttle box for instrumental avoidance conditioning and driving it with sound projectors operating in-phase and located at each end produces a uniform stimulus (Weiss 1966).

auratus), and later in the oscar (*Astronotus ocellatus*), and cichlid fish (Yan and Popper 1992). Positive operant conditioning is difficult to achieve with fish because of food delivery problems underwater and maintaining motivation of the test subject over rather lengthy test sessions. They overcame these difficulties by developing a fixed feeding station that could be easily recognized by the fish. It had an observation paddle and a report paddle covered with a piezo film, which required very little force to activate the reward. The observation paddle had food visible and served to keep the fish in one spot during presentation of the signal. The fish were deprived of food for 16–23 hours prior to testing to maintain motivation throughout the trials. They also used light as a reinforcing stimulus and a 45–90 second blackout period as a timeout. This set-up reduced the training period to 2–7 days with a 3–5 day average. Then thresholds could be determined at the rate of only two per day.

Because of the long training period and the difficulty in keeping the fish in one location for presentation of the stimulus, most scientists have used the conditioned inhibition of respiration and heart rate via electric shock for psychophysical studies. This was first demonstrated in fish by Otis et al. (1957) with a light stimulus. In this classical conditioning technique, the termination of a conditioned stimulus (CS) is repeatedly associated with the delivery of an electric shock or electric stimulation of some portion of the animal's anatomy. After a sufficient number of pairings of the CS and shock, the CS will elicit a slowing of heart rate or inhibited respiration or both. Conditioning disappears with repeated presentations of the CS and must be reinforced. Otis et al. (1957) delivered a shock through the water by using two sheets of aluminum foil placed on opposite sides of the tank as the electrodes, but it is usually just applied across the tail of the animal. In this technique the fish is restrained as shown in Fig. 8.16, and respiratory movements can be mechanically transmitted by a lever arm. A thermistor positioned in front of the mouth of the fish can also be used to monitor respiration via temperature fluctuations. To monitor heart rate, an electrode must be inserted in the cardiac cavity and another electrode inserted in a neutral area on the other side of the body.

The conditioned cardiac response in a flatfish as recorded by Chapman and Sand (1974) is shown in Fig. 8.17. The classically conditioned suppression of respiration or heart beat is a fast and reliable method for studies of hearing in fish. In goldfish the suppression can be conditioned to tones in about 10 trials. It has been replaced only recently in studies of hearing sensitivity by recording of the auditory brainstem response or ABR as described in the following sections.

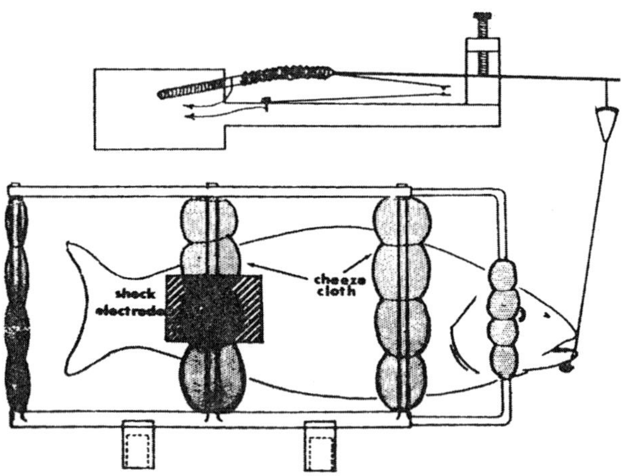

FIGURE 8.16. Fish restrained for classical conditioning (from Fay 1969). The lever attached to the lower lip transmitted respiratory movements.

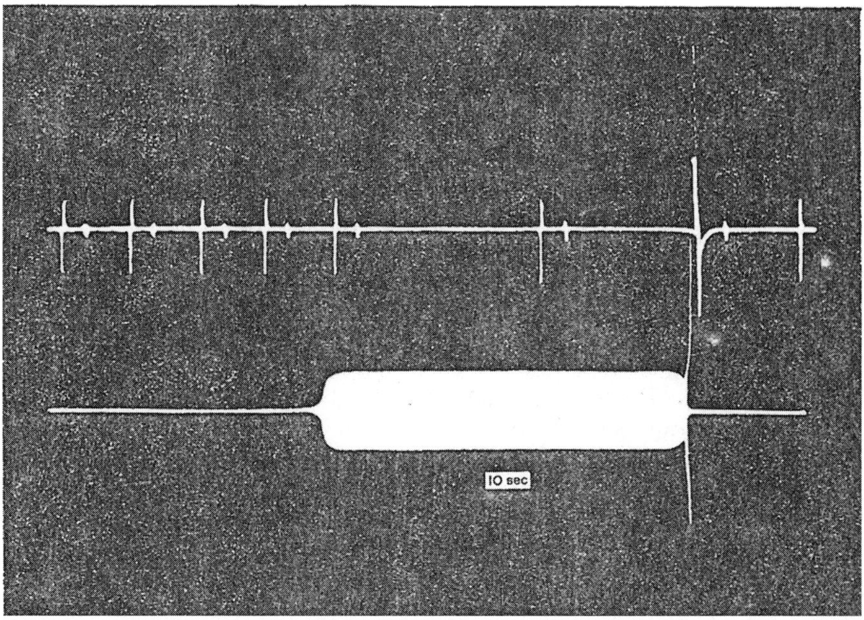

FIGURE 8.17. Conditioned cardiac response (*upper trace*) to pure tone stimulus (*lower trace*) at 200 Hz and 110 dB re 1 μPa (adapted from Chapman and Sand, 1974).

8.4 Electrophysiological Techniques for Marine Mammals

The hearing capabilities of animals can also be assessed by measuring the evoked potential response of the nervous system along the auditory pathway from the inner ear to the central auditory nervous system. An auditory evoked potential (AEP) is an electrical response of the nervous system produced by an external acoustic stimulus that can be measured as voltage by measuring electrodes. There are a variety of electrophysiological techniques that can be used with animals. Some of these techniques are extremely invasive and would eventually lead to the demise of the subject while other techniques are totally non-invasive. Bullock et al. (1968) were the first to use electrophysiology to examine the auditory processing of acoustic signals by dolphins. They inserted electrodes in the brain stem of their subjects and measured the evoked potentials produced by various types of acoustic stimuli. The work of Bullock et al. (1968) is a landmark that deserves serious and careful examination. Two excellent review articles by Ridgway (1980, 1983) of electrophysiological experiments on hearing in odontocetes are also worthwhile reading to gain an appreciation of early evoked potential experiments with dolphins.

In this section we will present a brief overview of auditory evoked potential techniques that can be performed in a non-invasive manner to highlight what is involved in performing electrophysiological measurements on marine mammals. The excellent textbook by Supin et al. (2001) is a must for those interested in a detailed discourse on evoked potential techniques to study marine mammals hearing. For large baleen whales, the use of evoked potential techniques may be one of the few ways in which we will be able to gain an understanding of their hearing capabilities. Whales often strand themselves on beaches or are caught in fishing wares and a team of scientists properly equipped with both acoustics equipment and evoked potential measurement instruments and ready to react quickly may be able to study the auditory sensitivity of a large whale.

Auditory evoked potentials can be measured by placing electrodes on the surface of an odontocete's head. The electrodes can be thin, needle-shaped wires on the order of 0.1–0.3 mm diameter inserted subdermally 3–5 mm into the skin as used by Supin and Popov (1993) and Supin et al. (1993). A dolphin's skin is relatively insensitive to these small subdermally inserted wires. Surface electrodes in the form of a 1-cm cup secured by adhesive tape have been used by Supin and Popov (1995). Dolphin et al. 1995) used a 2.4-cm silvered disc embedded in 4 cm latex suction cups that were filled with conductive electrode gel as electrodes. Small 1-cm gold-plated metallic cups used for human EEG measurements have also been embedded in latex suction cups and used as electrodes to measure AEP from dolphins and small whales. A schematic of a killer whale (*Orcinus orca*) wearing surface electrodes and remaining relatively still in one location is shown in Fig. 8.4I. The neural signal was differentially amplified between suction cups 1 and 2. Suction cup 3 served as a ground electrode.

8.4.1 Auditory Evoked Potentials Caused by Brief Signals

Auditory evoked potentials are relatively low-amplitude signals varying from about 10 μv to less than 1 μv, depending on the strength of the acoustic stimulus. Signals at these levels are normally immersed in noise and are not detectable; however, through a process of averaging, signals can be "pulled" out of the noise. If the distance between the acoustic source and the subject is relatively fixed, the neural signals will occur repetitively at the same time after the brief acoustic signal is projected toward the subject. Therefore, the neural signal will be synchronized to the acoustic signal so that the acoustic signal or a trigger signal related to the acoustic signal can be used to trigger an averaging computer. In its simplest form, a data acquisition system can be configured to digitized *n* number of points when it receives a trigger signal and store the data in an acquisition buffer. The data in the acquisition buffer are then examined for any artifacts. Artifacts are manifested by large voltage values (much larger than any AEP) that may be caused by movement of the subjects, muscle twitches and spasm, eye movements, etc. If artifacts are present, the data in the acquisition buffer are discarded. If no artifacts are present, the data in the acquisition buffer are added to the data in the accumulation buffer in which is stored the sum of all the previous blocks of n points. The neural signal evoked by the acoustic stimulus will be essentially the same for each projection of the acoustic signal so that the process of accumulating the data will cause that portion of the data containing the signal to become progressively larger while the noise that is random does not increase in value as rapidly. This process of digitizing n points and adding the results to the accumulation buffer can continue until signal-to-noise ratio is sufficiently high enough that the evoked potential can be clearly seen. The increase in signal-to-noise ratio will be the $\sqrt{m}$, where m is the number of averages taken. In this simple scheme, data acquisition cannot take place during the time that the content of the accession buffer is being added to the accumulation buffer, limiting the repetition rate of the acoustic signal.

A faster method of performing real-time averaging involves configuring two acquisition buffers, each of ***n***-points, and a third n-point buffer to hold the accumulated data. When a trigger is received, the signal is digitized and the values stored one of the acquisition buffer. While the content of the most recent acquisition is being examined and eventually transferred and added to the content of the accumulation buffer, the other acquisition buffer is available to receive any digitized data. The two acquisition buffers are used in a "ping-pong" fashion to store the most recent acquisition of n-points. There are special auditory evoked potential instruments that use special high-speed digital signal processing chips (DSP) to perform this averaging procedure in real time.

Auditory evoked potentials are classified into different categories depending on the delay or latency between the presentation of the acoustic stimulus and the occurrence of the evoked potential. Three different latency regimes are generally used to categorize AEP: short-latency brain stem AEP occurring within 10 ms, mid-latency AEP occurring between 10 and 50 ms, and long-latency occurring over 50 ms after the onset of a brief acoustic stimulus. Examples of the three different types of AEPs are shown in Fig. 8.4 II. The short-latency AEP are referred to as auditory brain stem responses or *ABR*, and are caused by the onset or beginning portion of a brief acoustic signal. Popov and Supin (1990a) presented an interesting illustration showing the effects of electrode placement on the received ABR signals; their illustration is reproduced in Fig. 8.4III. According to Popov and Supin (1990a), ABRs for dolphins can be best measured with an active electrode placed about 6 cm posterior of the blow hole. The mid-latency AEP is thought to originate from the auditory cortex and is referred to as the auditory cortical response (*ACR*) by Supin and Popov (1990). ACRs can best be measured with an active electrode placed about 25 cm posterior of the blow hole along the midline of the dolphin as illustrated in The ABR signal is actually a summation of several signals, denoted by roman numerals, which are generated in different portions of the auditory system. Click-evoked dolphin ABR like human ABR consist of seven waves that occur within 10 ms. Earlier studies in human auditory electrophysiology related wave I to the acoustic VIIIth nerve, wave II to the cochlear nucleus and trapezoid body, wave III to the superior olive complex, wave IV to the lateral lemniscus, wave V to the inferior colliculus, wave VI to the medial geniculate body of the thalamus, and wave VII to the medial geniculate body (Spehlmann, 1985). However, doubts have been raised about such simple wave-to-point correlation. Furthermore, it is not yet proven whether ABR waves come from the auditory brainstem nucleus or from axon bundles between them (Ridgway, 1983).

The dolphin ABRs are larger than most mammals and are about 10 times larger than for humans (Ridgway, 1983). The ABR pattern shown in Figs. 8.18 and 8.19 is consistent with the ABR patterns measured for other species of dolphins such as *Delphinis delphis* (Ridgway, 1983), *Sotalia fluviatilis* Supin et al. (2001), *Inia geoffrensis* (Popov and Supin, 1990b). Wave IV is typically the largest wave reaching amplitudes up to 10 V, but typically around 3 μV. The amplitude and latency of the waves in an ABR signal are related to the intensity of the auditory stimulus, with the amplitude becoming smaller and the latencies increasing as the stimulus intensity is reduced. The effect of stimulus intensity on the amplitude and latency of ABR signals are shown in Fig. 8.20. The ABRs were evoked by a click signal with a center frequency of 66-kHz. The repetition rate of the stimulus also has an effect on the amplitude and latency of ABR waves as can be seen in Fig. 8.20. The ABR amplitude does not vary linearly with stimulus amplitude, except when the stimulus amplitude is within about 30 dB from the animal's ABR threshold as can be seen in Fig. 8.21b. As the repetition rate increases, the ABR amplitude and latency decreases.

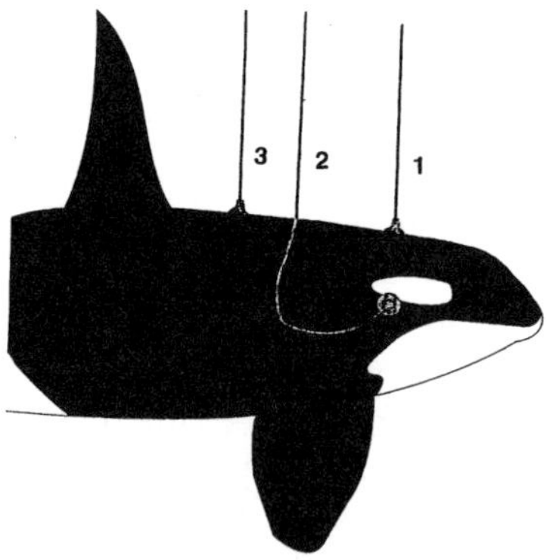

FIGURE 8.18. Schematic of a killer whale (*Orcinus orca*) wearing suction cup electrodes and remaining relatively still in one location (adapted from Szymanski et al., 1999).

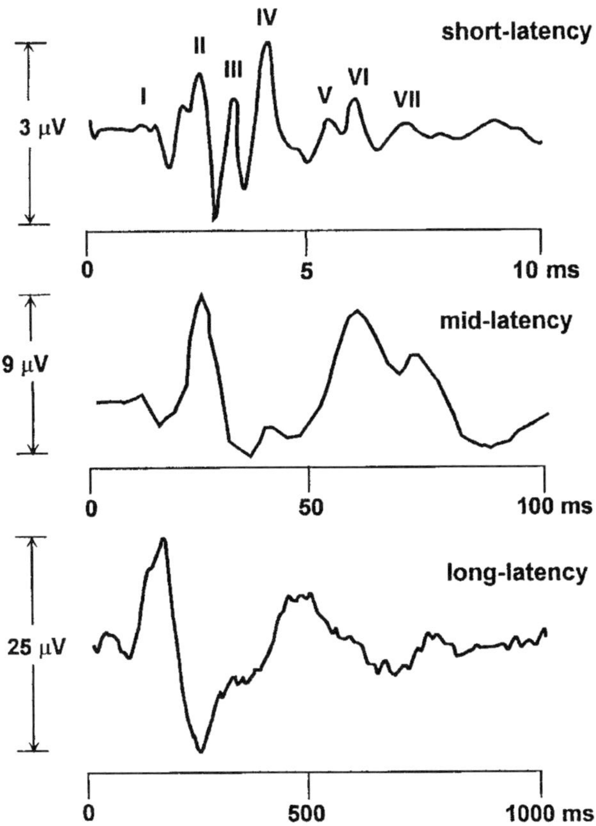

FIGURE 8.19. Examples of three different types of AEPs recorded from bottlenose dolphins. The short- and long-latency AEPs are from Ridgway (1983) and the mid-latency AEP was obtained from Supin and Popov (1990).

The approximate sensitivity of a mammal to sounds at a particular frequency can be obtained by measuring ABRs. As the intensity of the stimulus decreases, the amplitude of the ABR also decreases until the point that it can no longer be distinguished from the noise. The level of the acoustic stimulus at that point will be a rough estimate of the animal's hearing sensitivity at a particular frequency. However, there are several critical issues involving the use of ABR to estimate the hearing sensitivity. First, ABRs are evoked by short transient acoustic stimuli that are broadband or by the onset of a tone burst. Therefore, it is impossible to specify the exact frequency that the auditory system is responding to. Supin et al. (1993) presented an illuminating graph (reproduced in Fig. 8.20) to illustrate this point. Although the acoustic stimulus may be relatively long, the effect part of the signal that will evoke an ABR will be the beginning portion of the signal. Secondly, the hearing sensitivity shown in behavior audiogram is normally measured in terms of the rms

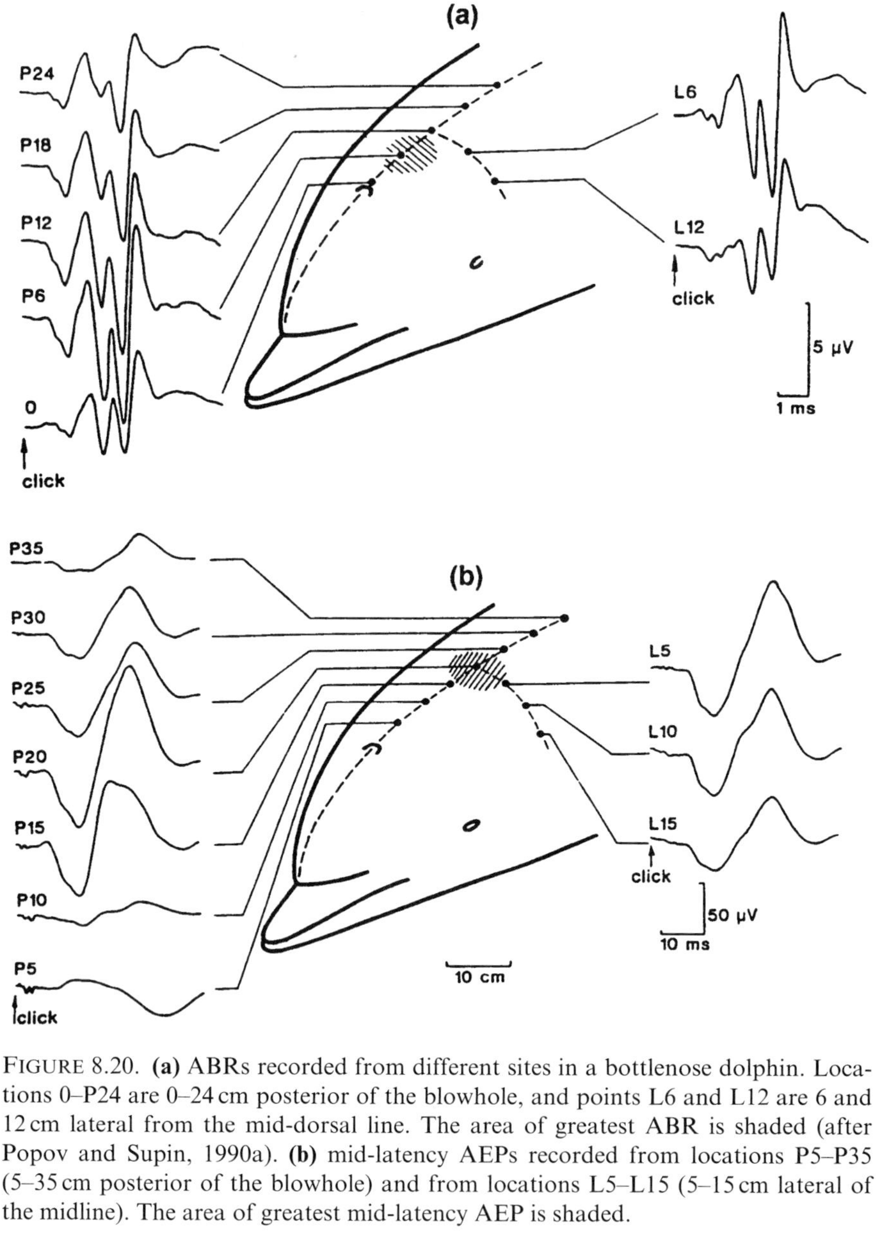

FIGURE 8.20. **(a)** ABRs recorded from different sites in a bottlenose dolphin. Locations 0–P24 are 0–24 cm posterior of the blowhole, and points L6 and L12 are 6 and 12 cm lateral from the mid-dorsal line. The area of greatest ABR is shaded (after Popov and Supin, 1990a). **(b)** mid-latency AEPs recorded from locations P5–P35 (5–35 cm posterior of the blowhole) and from locations L5–L15 (5–15 cm lateral of the midline). The area of greatest mid-latency AEP is shaded.

value of the signal at the listener's threshold. How the rms value of a long tone is related to the peak-to-peak value, or even the rms value of a transient signal, or the effective part of the signal in Fig. 8.4 V is anyone's guess. The rms value of a short transient may be as much as 15 dB less than a sine wave having the same peak-to-peak amplitude. Third, hearing sensitivity is usually measured

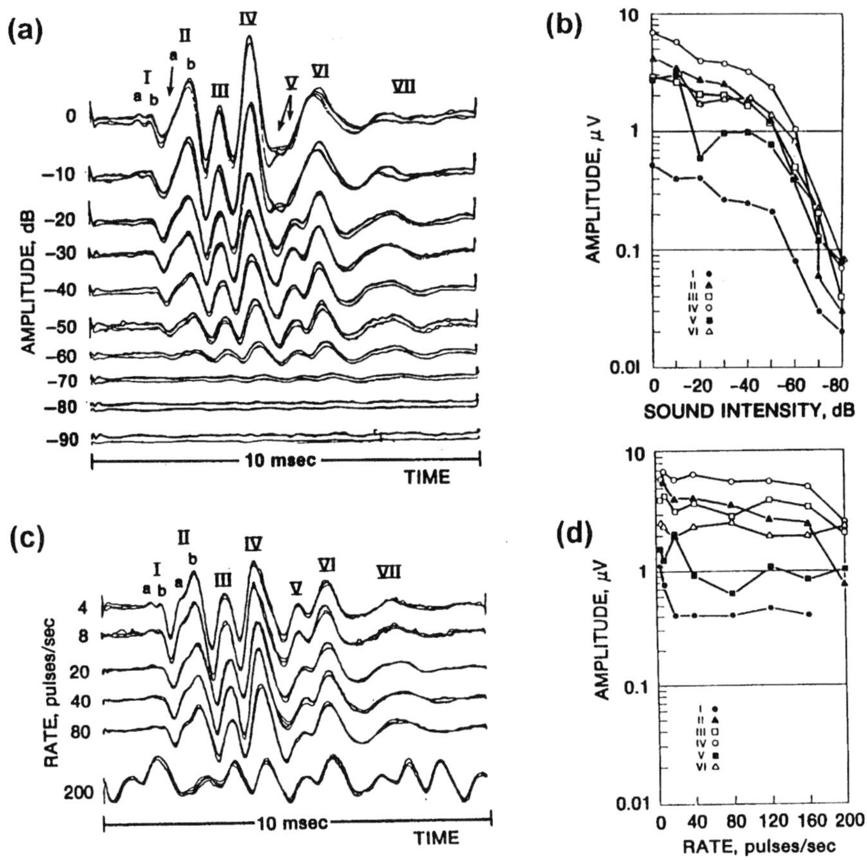

FIGURE 8.21. **(a)** Series of dolphin ABRs showing the decrease in amplitude and increase in latency as the intensity of the acoustic stimulus is decreased. **(b)** The amplitude of *Waves* I–VI as a function of the decreasing stimulus intensity. **(c)** Change in the latency of *Waves* I–VI as a function of decreasing stimulus intensity. **(d)** Series of ABRs illustrating decreasing amplitude and increasing latency with increasing stimulus repetition rate (after Ridgway, 1983).

with a long tone that is longer than the integration time of the auditory system, whereas ABRs are evoked by only the onset of the signal. The hearing threshold of mammals will progressively improve as the duration of a tonal signal is increased until the integration time of the auditory system is reached. After that, the auditory system is operating in a steady state mode and any further increase in the duration of the signal will not improve the sensitivity of the subject. Finally, the use of ABRs to estimate hearing sensitivity can only be used for relatively high frequency stimuli. Wave I arrives between 1 and 2 ms of the stimulus onset, and any stimulus with frequency of less than about 1 kHz will have a period greater than the arrival time of wave I.

Although there are some problems in using ABRs to estimate the absolute hearing sensitivity of subjects, the technique is still useful in getting some rough estimates. Measuring the relative difference in the ABR response of subjects as the properties of the acoustic stimuli are manipulated can be very effective in studying the auditory system. Alexander Supin and his colleagues at the Severtsov Institute of Evolutionary Morphology in Moscow, Russia, have been at the forefront in the use of electrophsyiological techniques to study the auditory processes in dolphins. Among some of the auditory processes in dolphins they have studied with the ABR technique include frequency selectivity, interaural time-delay, interaural intensity difference, and location of the auditory window.

One final comment about the use of ABRs in study hearing in dolphins should be made. Through the use of ABR, we have gained an appreciation of the rapid reaction of the dolphin auditory system, not only in the cochlea but also in the central nervous system (CNS) as well. According to Ridgway (1983), transmission time from wave I to Wave IV is likely a function of CNS brainstem transmission time, and is considerably faster in *Tursiops* than in man or cat and is very similar to that of the rat. Despite having a longer nerve pathway because of a larger brain, the brainstem transmission time in the dolphin is equal to or faster than that in much smaller-brained species. Some comparative latency values of ABR waves are shown in Fig. 8.23. Mid- and long-latency AEPs can be used to study the response of the auditory system to tone bursts and low frequency signals. However, the amplitude and latency of mid- and late latency AEPs tend to vary considerably from individual to individual and may not be as reliable and consistent as ABRs. Most of the studies using evoked potentials to determine the auditory properties of odontocetes have been conducted by measuring ABRs rather than ACRs.

8.4.2 Envelope-Following Responses

Auditory-evoked potentials recorded from the human scalp have been shown to follow the periodicity of pure-tone sinusoidal signals for frequencies up to about 1 kHz (Moushegian et al., 1983) and have been termed the frequency-following responses. Scalp-recorded responses have also been found to follow the envelope of continuous sinusoidally amplitude-modulated stimuli in humans (Richards and Clark, 1984) and in gerbils (Dolphin and Mountain, 1993), using carrier frequencies considerably higher than the pure-tone phase locking capability of the frequency-following response. Continuous sinusoidal amplitude-modulated (SAM) signals produced by amplitude modulating a high-frequency sinusoidal carrier with a low-frequency sine wave can be expressed as

$$s(t) = \sin(2\pi f_c t)[1 + m\sin(2\pi f_{\mathrm{mod}} t)], \tag{8.15}$$

where $s(t)$ is the SAM signal, t is time, m is the modulation depth, f_{mod} is the modulation frequency ,and f_c is the carrier frequency. An example of a SAM for a modulation depth of 100% ($m = 1$) is shown in the time and frequency domains in Fig. 8.24. The frequency spectrum has three tonal components, at f_c and at $f_c \pm f_{mod}$. A modulation depth of 100% will produce sidebands that are equal in amplitude and 6 dB below the value of the carrier signal at f_c. Auditory-evoked potentials from the mammalian auditory system will follow the waveform of the amplitude modulation, giving rise to the nomenclature enveloped following response (EFR). The neurons of the auditory system are actually responding to the carrier frequency but are firing at the rate of the modulation frequency causing ear to behave like an AM radio, demodulating the AM signal to recover the energy at f_{mod}. The evoked response to AM signal contains significant energy at the modulation frequency. There are inherent nonlinearities in the transduction of acoustic, mechanical energy to neuroelectrical energy in the cochlea and at higher centers along the auditory pathway modulation, or envelope frequency of the stimulus (Dolphin and Mountain, 1993).

Amplitude-modulated signals can also be produced by summing two sinusoidal or tonal signals of different frequencies so that

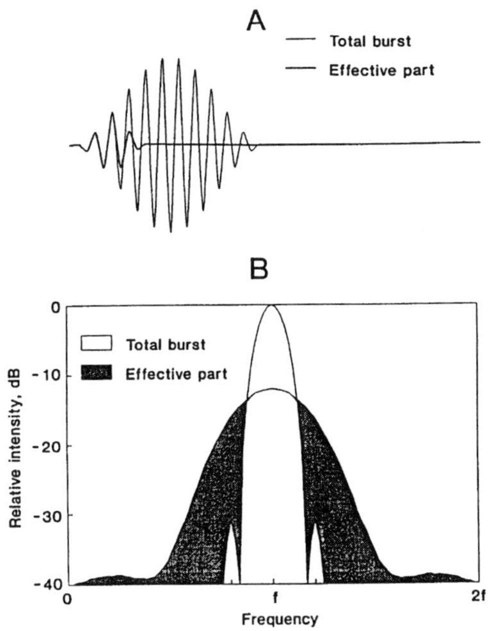

FIGURE 8.22. Schematic representation of difference between spectra of a tone burst and its effective part. (a) Waveform of a cosine-enveloped tone burst and its arbitrary isolated initial part. (b) Frequency spectra of the burst and its initial part (adapted from Supin et al., 1993).

$$s(t) = \sin(2\pi f_1 t) + \sin(2\pi f_2 t), \tag{8.16}$$

where f_1 and f_2 are the frequencies of sinusoid 1 and 2, respectively. A two-tone signal (TT) contains energy only at the frequencies of the individual tones; however, the amplitude of the signal will be modulated with a period corresponding to the absolute difference in the frequency between the two tones, or $f_{2,1} = |f_2 - f_1|$. An example of a two-tone stimulus and its frequency spectrum is shown in Fig. 8.25. Popov et al. (1995) found that AEPs measured with TT and SAM stimuli had comparable amplitudes when the modulation depth was 100%. The use of continuous amplitude-modulated stimuli in conjunction with EFR measurements have been performed on dolphins by Dolphin et al. (1994) and Dolphin (1996) at low frequencies ($<$ 10 kHz) and by Popov et al. (1995) for high frequencies (64–128 kHz). Three different species, an Atlantic bottlenose dolphin, a beluga whale, and a false killer whale, were used in the study of Dolphin et al. (1995). The animals were trained to wear suction cup electrodes and station themselves within a hoop or on a biteplate at a depth of 1 m. The electrodes were silver discs 2.4-cm diameter imbedded in custom-designed latex suction cups. Evoked responses

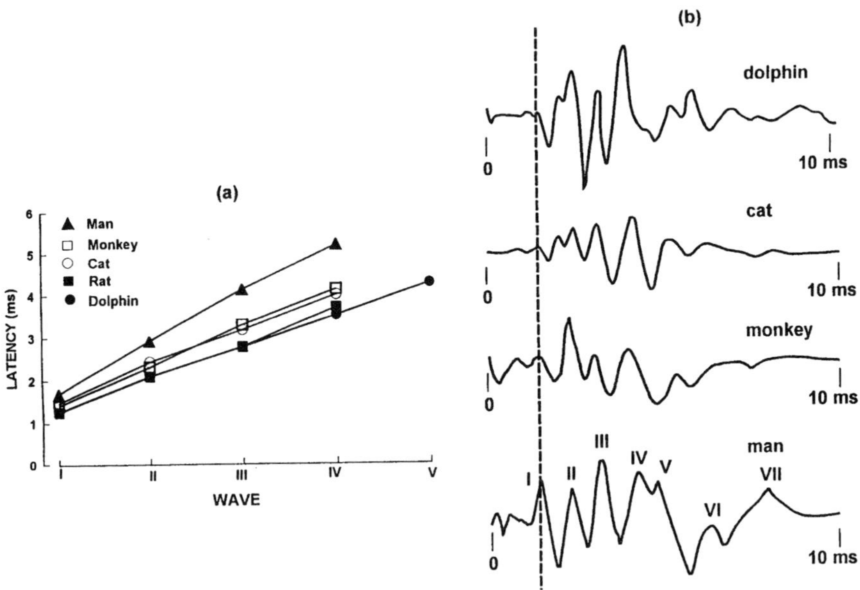

FIGURE 8.23. **(a)** Comparison of latencies of the first five ABR waves in the dolphin and in other species. **(b)** Comparison of a typical dolphin ABR with one from a cat, monkey, and man. The vertical dashed line connects presumably similar components, the auditory nerve deflection, *Wave* I in the human record (adapted from Ridgway, 1983).

were recorded differentially from the scalp source between the parietal (non-inverting) just posterior to the blowhole and the mastoid (inverting). A ground electrode was placed either on the melon or on the trunk region. A J-9 sound projector was located 1 m from the animals' lower jaw.

An example of a two-tone signal and the corresponding evoked response in the time and frequency domains are depicted in Fig. 8.26. The two-tone signal has energies only at f_1 and f_2. The evoked potential response was averaged over 350 presentations of the stimulus in 164 ms blocks. Although the stimulus does not have any energy at the difference or envelope frequency, the presence of the envelope component is obvious in the averaged evoked potential response. The evoked response also contained the two-tones stimulus and several distortion product components, one at $2f_1 - f_2$, along with a 60 Hz electrical power component. As the amplitude of the stimulus is decreased, the amplitude of the EFR will also decrease so that a subject's threshold can be estimated when the EFR response is no longer observed.

One of the initial questions concerning the use of EFR has to do with the most relationship between the frequencies of the carrier and modulated signals called the modulation rate transfer function (MRTF). This was determined for dolphins by modulating the primary or carrier tone at frequencies between 17 and 4019 Hz and measuring the response magnitude and phase at f_{mod} or $f_{2,1}$.

Examples of the modulation rate transfer function for *Tursiops* using both TT and SAM signals of different carrier and primary frequencies are shown in Fig. 8.27. From the figure, one can see that there are fairly broad areas in which the magnitude of the evoked potential responses does not vary much

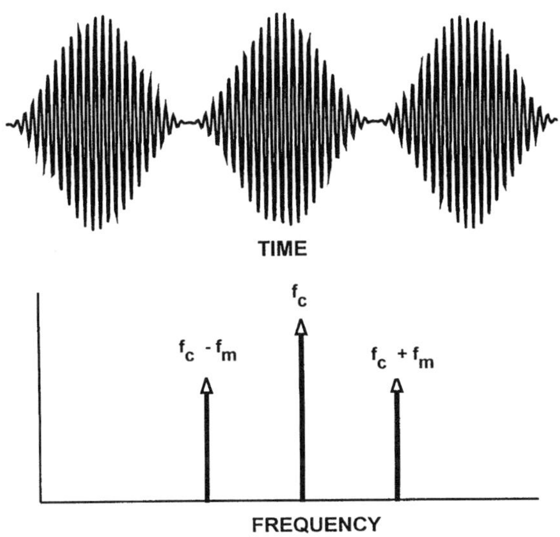

FIGURE 8.24. The waveform and frequency spectrum of a sinusoidal amplitude-modulated signal.

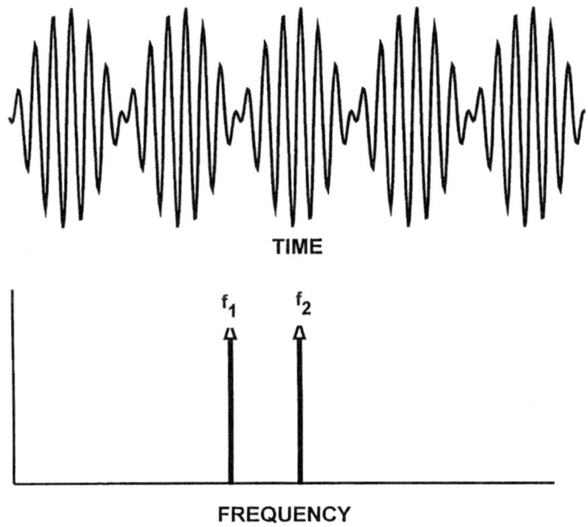

FIGURE 8.25. An example of a two-tone stimulus and its frequency spectrum.

with modulation or difference frequency from 20 Hz to the high-frequency roll-off. In all of the cetacean studied by Dolphin et al. (1995), the cutoff frequency for the modulation rate transfer function was in the 1–2 kHz range. A high-frequency corner was located at approximately 1 kHz in the *Pseudorca*, 1.5 kHz in the *Tursiops*, and 2 kHz in the *Delphinapterus*. There were also sharp nulls in the MRTF, which for *Tursiops* fell between 300 and 400 Hz. Popov et al. (1995) also measured the MRTF for a bottlenose dolphin, but for a much higher frequency of 90 kHz, as shown in Fig. 8.4XI. They obtained a similar cutoff frequency of approximately 1 kHz for a *Tursiops*, as shown in Fig. 8.28. The MRTF measured by Popov et al. (1995) was bimodal with peaks at 600 Hz and 1 kHz, and a broad null between 200 and 400 Hz and a sharp null at 800 Hz. The MRTF roll-off for frequencies greater than 1 kHz.

The envelope-following response technique using SAM and TT signals has several advantages over the ABR technique. The auditory system is responding to the carrier frequency, which is clearly defined. The rms value of the stimulus can be easily determined, which makes it simple to compare the thresholds obtained by this technique and behaviorally. Finally, the hearing sensitivity of the subject can be tested to relatively low frequencies, on the order of hundreds of Hertz. The averaging process involved in measuring EFR is very different than for short clicks and tone burst signals. The stimulus is best generated digitally with a buffer holding the digital representation of the signal that is fed to a digital-to-analog converter. When the end of the buffer is reached, the addressing of the buffer loops around to the beginning and the process begins again. The data in the buffer must be configured in such a way the end of a signal corresponds to end of the buffer, so that when the pointer of the buffer loops around to the beginning the process is seamless as if the signal

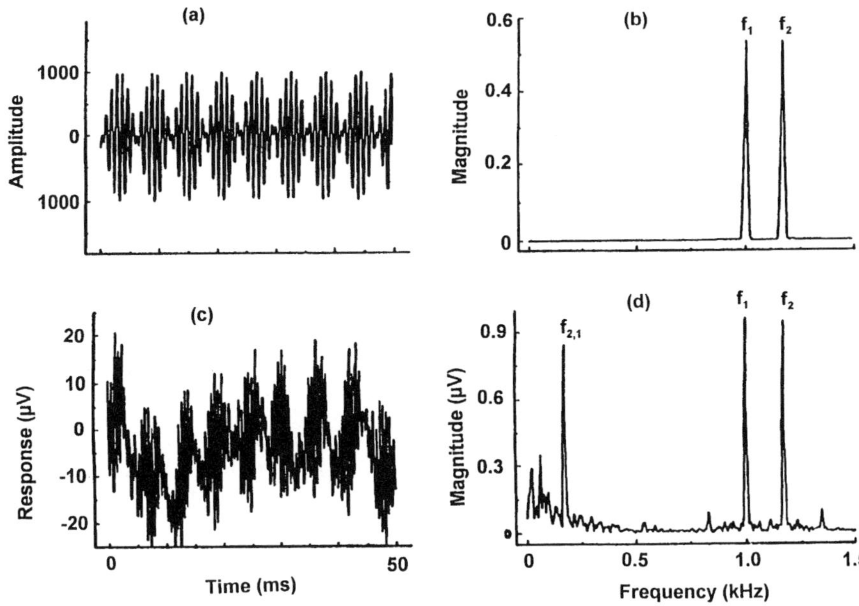

FIGURE 8.26. **(a)** Example of the two-tone amplitude-modulated acoustic stimulus in the time domain, **(b)** frequency domain representation of the two-tone signal, **(c)** the evoked potential response in the time domain, **(d)** the evoked potential response in the frequency domain (after Dolphin et al., 1995).

is a continuous signal. Every time the pointer is at the beginning of the buffer, a trigger signal is generated and is used to synchronize the averaging process. In this manner, the evoked potential is averaged synchronously and eventually after many averages, the evoked potential generated by an audible signal will be clearly above the noise. As an example, we can consider the signals used by Dolphin et al. (1995). Stimuli were digitized at a sample rate of 50 kHz and stored in a 8,192-point buffer, providing a signal of 164 ms duration. The EFR digitized at a sampling rate of 25 kHz and the data stored in a 4,098-point buffer. Each time a pointer looped to the start of the output signal buffer, a trigger signal was generated to start the data acquisition process and collect 4,098 points. Therefore, the signal generation and data acquisition processes were in sync so that the acquired data could be averaged.

Supin et al. (2001) used a modified envelope-following response technique than Dolphin et al. (1995). Instead of using a continuous SAM or TT signal, Supin et al. (2001) used a SAM tonal burst of 20 ms at any desired repetition rate so that averaging process is much easier. The averaging process can be synchronized with the onset of the 20 ms SAM signal. The role of integration time is not taken into account in the technique of Supin et al. (2001) because of the shortness of the stimulus so that their results may be different than if continuous SAM signals were used.

8.5 Electrophysiological Techniques for Fishes

Much of our knowledge of hearing in fish has been gained through invasive electrophysiological recordings of microphonics and single units in the eighth nerve. Single-unit recording is used to measure the discharge patterns of a single nerve fiber. Enger (1967) published the first audiogram for herring (*Clupea harengus*), based on the recordings of single units at frequencies from 30 to 4000 Hz. The fishes were anesthetized in with tricaine methane sulfonate (MS-222 Sandoz), clamped in a holder, curarized, and then artificially respirated by running sea water through a tube inserted in its mouth. In this way they could be kept alive for at least 24 hours for the recording of neurophysiological signals. Micropipettes filled with 4 M NaCl were used for the recording of single-unit activity. In the fish species studied so far, single units show tuning to a particular frequency. Although the tuning is not as sharp as the spatial tuning of the basilar membrane in mammals, it is enough to construct an audiogram as shown in Fig. 8.29. Fay (1978, 1984) used single unit recordings extensively to study temporal, spectral, and directional characteristics of hearing in goldfish.

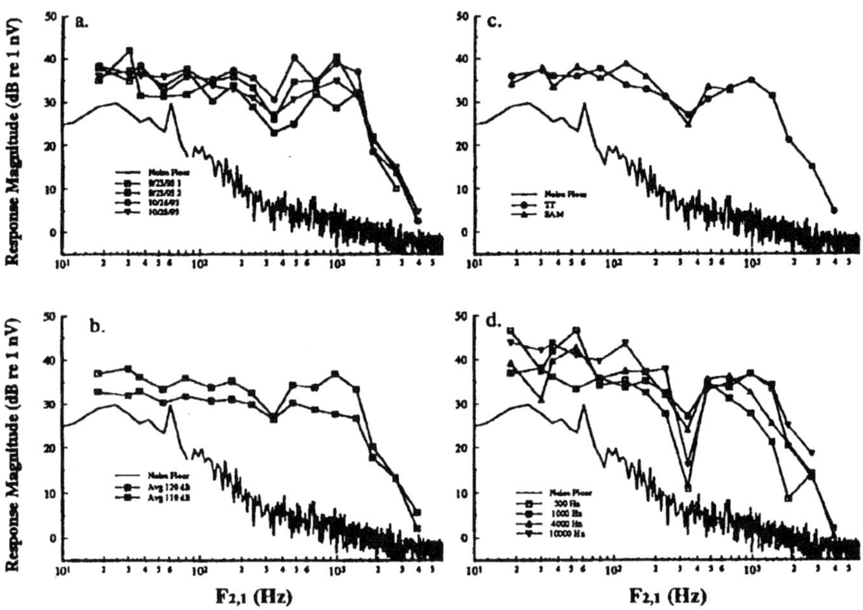

FIGURE 8.27. The modulation rate transfer function obtained for *Tursiops*. (a) Four MRTF obtained during four different experimental sessions. The session dates are indicated in the legend. Stimuli were two-tone signals projected at 130 dB. (b) The averaged MRTF of (a) along with the MRTF obtained with a 120 dB signal. (c) MRTF obtained using SAM and TT stimuli with $f_1 = f_c = 1$ kHz. (d) MRTF obtained using TT stimuli at 130 dB and f_1 of 0.5, 1.0, 4.0, and 10.0 kHz (adapted from Dolphin et al., 1995).

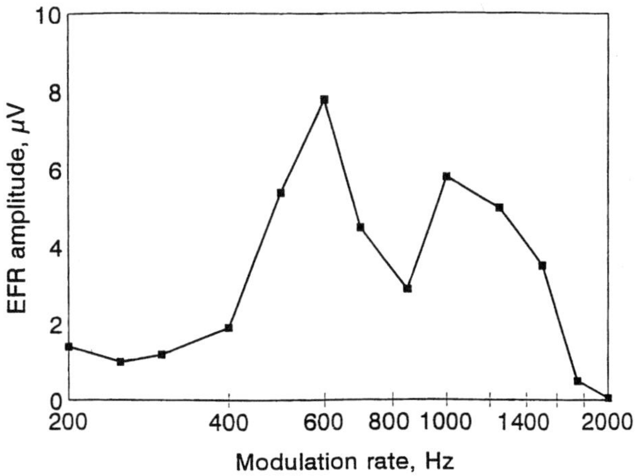

FIGURE 8.28. Modulation rate transfer function for a *Tursiops* for a AM signal with a carrier of 90 kHz (adapted from Popov et al., 1995).

Saidel and Popper (1987) constructed audiograms for two anabantid species, the blue gourami (*Trichogaster trichopterus*) and the kissing gourami (*Helostoma temincki*), by measuring saccular microphonic potentials. This procedure is invasive and very similar to the recording of single units as described above, except that the electrode is placed near the medial aspect of the saccule. Because of opposing hair cell groups within the same epithelium, saccular microphonics occur at twice the stimulus frequency (Furukawa and Ishii 1967).

Since 1998 (Kenyon et al. 1998), recording of auditory evoked potentials (AEPs) in fishes has been used in most all studies of hearing. Traditionally

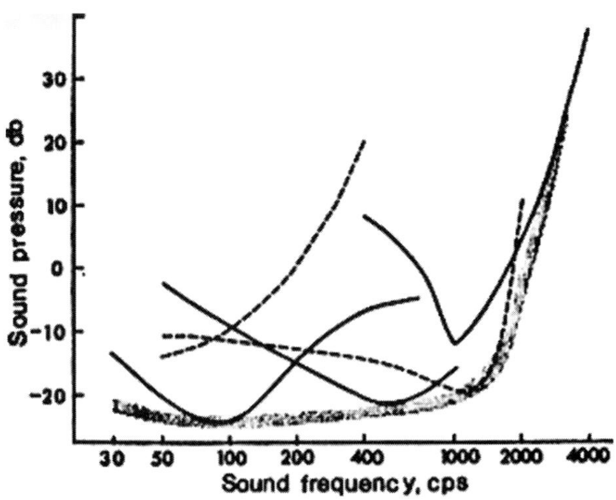

FIGURE 8.29. Herring audiogram reconstructed from single unit recordings (from Enger, 1967). The *dotted lines* are for clarity.

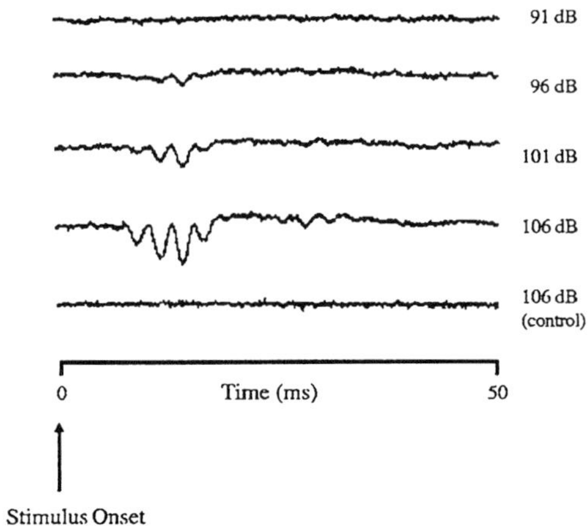

FIGURE 8.30. ABR recording from spot (*Leiostomus xanthurus*) at a test frequency of 300 Hz; the threshold is judged to be 96 dB re 1 μPa (adapted from Ramcharitar et al., 2006).

this method has been described as the ABR (auditory brainstem response) technique. However, these potentials arise from several centers in the auditory pathway and similar to saccular microphonics, most often contain a signal at twice the stimulus frequency. So the more general term, auditory evoked potential (AEP) is applicable. Typically, pure tone bursts of 20-50 ms duration are used to elicit an AEP. Recordings are made similarly as described previously for marine mammals, either with small surface or subdermal electrodes. Fig. 8.30 illustrates AEPs for the spot (*leiostomus xanthurus*) as recorded by Ramcharitar et al. (2006). The auditory threshold is defined as the minimum sound pressure level at which an AEP is evoked. This method has revolutionized hearing studies in fish because it is noninvasive, can be done on wild animals, and can be done in the field as well as in the laboratory. However, techniques and testing configurations vary greatly from one lab to another, so data are not directly comparable. In addition, the relationship between AEP thresholds and behavioral audiograms is not understood. This needs to be examined before AEP thresholds can be used to infer acoustic behavior of fishes in the wild.

References

Au, W. W. L. and Snyder, K. J. (1980). "Long-Range Target Detection in Open Waters by an Echolocating Atlantic Bottlenose Dolphin (Tursiops truncatus)," J. Acoust. Soc. Am., **68**, 1077–1084.

Au, W. W. L. and Turl, C. W. (1984). "Dolphin biosonar detection in clutter: variation in the payoff matrix," J. Acoust. Soc. Am., **76**, 955–957.

Bullock, T. H., Grinnell, A. D., Ikezono, E., Kameda, K., Nomoto, K., Sato, O., Suga, N., and Yanigisawa, K. (1968). "Electrophysiological studies of central auditory mechanisms in cetaceans," Z. Vergl. Physiol. **59**, 117–156.

Chapman, C. J. and Sand, O. (1974). "Field studies of hearing in two species of flatfish *Pleuronectes platessa* (L.) and *Limanda limanda* (L.) (Family Pleuronectidae)," Comp. Biochem. Physiol. **47A**, 371–385.

Dolphin, W. F. and Mountain, D. C. (1993). "The envelope following response (EFR) in the Mongolian gerbil to sinusoidally amplitude-modulated signals in the presence of simultaneously gated pure tones," J. Acoust. Soc. Am. **94**, 3215–3226.

Dolphin, W. F., Chertoff, M. E., and Burkard, R. F. (1994). "Comparison of the envelope following response in the Mongolian gerbil using two-tone and sinusoidally amplitude modulated tones," J. Acoust. Soc. Am. **96**, 2225–2234.

Dolphin, W. F., Au, W. W. L., Nachtigall, P. E., and Pawloski, J. (1995). "Modulation rate transfer functions to low-frequency carriers in three species of cetaceans," J. Comp. Physiol. A. **177**, 235–245.

Dolphin, W. F. (1996). "Auditory evoked responses to amplitude modulated stimuli consisting of multiple envelope components," J. Comp. Physiol. A. **179**, 113–121.

Egan, J. P. (1975). *Signal Detection Theory and ROC Analysis* (Academic Press, New York).

Elliot, P. B. (1964). "Table of d′." in *Signal Detection and Recognition by Human Observers*, J. A. Swets, ed. (John Wiley & Sons, New York).

Enger, P. S. (1967). "Hearing in herring," Comp. Biochem. Physiol. **22**, 527–538.

Evans, W. E. (1967). "Discrimination of different metallic plates by an echolocating delphinid," in *Animal Sonar Systems: Biology and Bionics*, R. G. Busnel, ed. (Laboratoire de Phsiologie Acoustique, Jouy-en-Josas, France), pp. 363–383.

Evans, W. E. (1973). "Echolocation by marine delphinids and one species of fresh-water dolphin," J. Acoust. Soc. Am. **54**, 191–199.

Fay, R. R. (1969). "Behavioral audiogram for the goldfish," J. Aud. Res. **9**, 112–121.

Fay, R. R. (1978). "Coding of information in single auditory-nerve fibers of the goldfish," J. Acoust. Soc. Am. **63**, 136–146.

Fay, R. R. (1984). "The goldfish ear codes the axis of acoustic particle motion in three dimensions," Science **225**, 951–954.

Furukawa, T. and Ishii, Y. (1967). "Neurophysiological studies on hearing in goldfish," J. Neurophysiol. **30**, 1377–1403.

Gescheider, G. A. (1976). *Psychophysics: Method and Theory*, (John Wiley & Sons, New York).

Green, D. M. and Swets, J. A. (1966). *Signal Detection Theory and Psychophysics* (R. E. Krieger Publishing Co., Huntington, New York).

Hammer, C. E., Jr. and Au, W. W. L. (1980). "Porpoise echo-recognition: an anlysis of controlling target characteristics," J. Acoust. Soc. Am. **68**, 1285–1293.

Herman, L. M. and Arbeit, W. R. (1972). "Frequency difference limens in the bottlenose dolphin 1-70 kHz," J. Aud. Res. **12**, 109–120.

Horner, J. L., Longo, N. and Bitterman, M. E. (1961). "A shuttle-box for fish and a control circuit of general applicability," Am. J. Psychol. **74**, 114–120.

Jacobs, D. W. (1972). "Auditory freqeuncy discrimination in the Atlantic bottlenose dolphin, *Tursiops truncatus* Montagu: a preliminary report," J. Acoust. Soc. Am. **53**, 696–697.

Johnson, R. A., Moore, P. W. B., Stoermer, M. W., Pawloski, J. L., and Anderson, L.C. (1988). "Temporal order discrimination within the dolphin critical interval," in Animal Sonar: Processes and Performance. P. E. Nachtigall and P. W. B. Moore, eds., Plenum Press, New York, pp. 317–321.

Kenyon, T. N., Ladich, F., and Yan, H. Y. (1998). "A comparative study of hearing ability in fishes: the auditory brainstem response approach," J. Comp. Physiol. A **182**, 307–318.

Levitt, H. (1970). "Transformed up-down methods in psychoacoustics," J. Acoust. Soc. Am. **49**, 467–477.

Moore, P. W. B. (1975). "Underwater localization of click and pulsed pure-tone signals by the California sea lion (*Zalophus californianus*)," J. Acoust. Soc. Am. **57**, 406–410.

Moushegian, G., Rupert, A. L., and Stillman, R. D. (1983). "Scalp recorded early responses in man to frequencies in the speech range," Electroencephalogr. Clin. Nenrophysiol. **35**, 665–667.

Murchison, A. E. (1980). "Maximum detection range and range resolution in echolocating bottlenose popoises (*Tursiops truncates*)," in *Animal Sonar Systems*, R. G. Busnel and J.F. Fish, eds. Plenum Press, NY. pp. 43–70.

Otis, L. S., Cerf, J. A., and Thomas, G. J. (1957). "Conditioned inhibition of respiration and heart rate in the goldfish," Science **126**, 263–264.

Peterson, W. W., Birdsall, T. G., and Fox, W. C. (1954). "The theory of signal detectability," IRE **PGIT 4**, 171–212.

Popov, V. V. and Supin, A. Y. (1990a). "Electrphysiological studies of hearing in some cetaceans and a manatee," in *Sensory Systems of Aquatic Mammals*, R. A. Kastelein, J. A. Thomas, and P. E. Nachtigall, eds. (De Spil. The Netherland), pp. 405–415.

Popov, V. V. and Supin, A. Y. (1990b). "Localization of the acoustic window at the dolphin's head," in *Sensory Systems of Aquatic Mammals*, R. A. Kastelein, J. A. Thomas, and P. E. Nachtigall, eds. (De Spil. The Netherland), pp. 417–426.

Popov, V. V., Supin, A. Y., and Klishin, V. O. (1995). "Frequency tuning curves of the dolphin's hearing: Envelope-following response study," J. Comp. Physiol. A **178**, 571–577.

Ramcharitar, J. U., Higgs, D. M., and Popper, A. N. (2006). "Audition in sciaenid fishes with different swim bladder-inner ear configurations," J. Acoust. Soc. Am. **119**, 439–443.

Renaud, D. L. and Popper, A. N. (1975). "Sound localization by the bottlenose porpoise *Tursiops truncates*," J. Exp. Biol. **63**, 569–585.

Rickards, F. W. and Clark, G. M. (1984). "Steady state evoked potentials to amplitude-modulated tones," in *Evoked PotentialsII* , R. H. Nodar and C. Barber, eds. Butterworth, Boston. pp. 163–168.

Ridgway, S. H. (1980). "Electrophysiological experiments on hearing in odontocetes," in *Animal Sonar*, P. E. Nachtigall and P. W. B. Moore, eds. (Plenum, New York), pp. 483–493.

Ridgway, S. H. (1983). "Dolphin hearing and sound production in health and illness," in *Hearing and Other Senses: Presentations in Honor of E. G. Wever*, R. R. Fay and G. Gourevitch, eds. (The Amphora Press, Gronton, CT), pp. 247–296.

Ridgway, S. H. and Carder, D. A. (1997). "Hearing deficits measured in some *Tursiops truncatus* and discovery of a deaf/mute dolphin." *J. Acoust. Soc. Am.* **101**, 590–593.

Roitbalt, H. L., Penner, R. H., and Nachtigall, P. E. (1990). "Matching-to-sample by an echolocating dolphin," J. Exp. Psych: Anim. Beh. Proc. **16**, 85–95.

Saidel, W. M. and Popper, A. N. (1987). "Sound reception in two anabantid fishes," Comp. Biochem. Physiol. **88A**, 37–44.

Schusterman, R. J. and Johnson, B. W. (1975). "Signal probability and response bias in California sea lions," Psychol. Rec. **25,** 39–45.

Schusterman, R. J., Balliet, R. F., and Nixon, J. (1972). "Underwater audiogram of the California sea lion by the conditioned vocalization technique," J. Exper. Analy. Behav. **17**, 339–350.

Schusterman, R. J., Barrett, B., and Moore, P. (1975). "Detection of underwater signals by a California sea lion and bottlenose porpoise: variation in the payoff matrix," J. Acoust. Sec. Am. **57**, 1526–1632.

Schusterman, R. J. (1974). "Low false-alarm rates in signal detection by marine mammals," J. Acoust. Soc. Am. **55**, 845–847.

Schusterman, R. J. (1976). "California sea lion underwater auditory detection and variation of reinforcement schedules," J. Acoust. Soc. Am. **59**, 997–1000.

Schusterman, R. J. (1980). "Behavioral methodology in echolocation by marine mammals," in *Animal Sonar Systems*, R.-G Busnel and J. F. Fish, eds. (Plenum, New York), pp. 11–41.

Simpson, W. A. (1988). "The method of constant stimuli is efficient," Per. Psych. **44**, 433–436.

Skinner, B. F. (1961). *Cumulative Record*, (Appleton-Century-Crofts, New York).

Snodgrass, J. G. (1972). *Theory and Experimentation in Signal Detection* (Life Science Assoc., Baldwin, New York).

Spehlmann, R. (1985). *Evoked Potential Primer: Visual, Auditory, and Somatosensory Evoked Potentials in Clinical Diagnosis*,(Butterworth, Boston).

Supin, A. Y. and Popov, V. V. (1990). "Frequency-selectivity of the auditory system in the bottle-nose dolphin, *Tursiops truncatus*," in *Sensory Systems of Aquatic Mammals*, R. A. Kastelein, J. A. Thomas, and P. E. Nachtigall, eds. (De Spil. The Netherland), pp. 385–93.

Supin, A. Y. and Popov, V. V. (1993). "Direction-dependent spectral sensitivity and interaural spectral difference in a dolphin: evoked potential study," J. Acoust. Soc. Am. **93**, 3490–3495.

Supin, A. Ya, Popov, V. V. (1995). " Temporal resolution in the dolphin's auditory system revealed by double-click evoked potential study," J. Acoust. Soc. Am. **97**, 2586–2593.

Supin, A. Y., Popov, V. V., and Klishin, V. O. (1993). "ABR frequency tuning curves in dolphins," J. Comp. Physio. A. **173**, 649–656.

Supin, A. Ya, Popov, V. V., and Mass, A. M. (2001). *The Sensory Physiology of Aquatic Mammals* (Kluwer Academic Publishing, Boston).

Szymanski, M. D., Bain, D. W., and Henry, K. R. (1995). "Auditory evoked potentials of a killer whale (*Orcinus orca*)," in *Sensory Systems of Aquatic Mammals*, R. A. Kastelein, J. A. Thomas, and P. E. Nachtigall, eds. (De Spil. The Netherland), pp. 1–10.

Szymanski, M. D., Bain, D. E., Kiehl, K., Pennington, S., Wong, S., and Henry, K. R. (1999). "Killer whale (/Orcinus orca/): Auditory brainstem response and behaviorial audiograms," J. Acoust. Soc. Am. **106**, 1134–1141.

Swets, J. A. (1964). *Signal Detection and Recognition by Human Observers* (John Wiley & Sons, New York).

Tanner, W. P., Jr. and Swets, J. A. (1954). "A decision-making theory of visual detection," Psych. Rev. **61**, 401–409.

Tanner, W. P., Jr., Swets, J. A., Green, D. M. (1956). *Some General Properties of the Hearing Mechanism,* University of Michigan: Electronic Defense Group Technical Report No. 30.

Tavolga, W. N. and Wodinsky, J. (1963). "Auditory capacities in fishes: Pure tone thresholds in nine species of marine teleosts," Bull. Amer. Mus. Nat. Hist. **126**, 179–239.

Thompson, R. K. R. and Herman, L. M. (1975). "Underwater frequency discrimination in the bottlenose dolphin (1-140 kHz) and the human (1-8 kHz)," J. Acoust. Soc. Am. **57**, 943–948.

Vel'min, V. A. and Dubrovskiy, N. A. (1975). "On the analysis of pulsed sounds by dolphins," Dokl. Adak. Nauk. SSSR **225**, 470–473.

Vel′min, V. A., Titov, A. A., and Yurkevich, L. I. (1975). "Time summation of pulses in the bottlenose dolphin," in Morskiye mtekopitayusheiye. Mater. 6-go Vses. soveshch. poizuch. morsk. mtekopitayshchikh, Part 1. Kiev: Naukova Dumka, pp. 77–80.

Weiss, B. A. (1966). "Auditory sensitivity in goldfish (*Carassius auratus*)," J. Aud. Res. **6**, 321–335.

Yan, H. Y. and Popper, A. N. (1991). "An automated positive reward method for measuring acoustic sensitivity in fish," Behav. Res. Meth., Instru. & Compu. **23**, 351–356.

Yan, H. Y. and Popper, A. N. (1992). "Auditory sensitivity of the cichlid fish *Astronotus ocellatus* (Cuvier)," J. Comp. Physiol. A **171**, 105–109.

Part II
Acoustics of Marine Animals

9

Hearing in Marine Animals

Part I consisted mainly of the necessary background materials that are needed tools in order to make bioacoustic measurements, perform psychoacoustic experiments, and discuss topics in marine bioacoustics. In the second part of this book, we will examine the marine animals themselves and discuss hearing and sound production mechanisms, characteristics and function of sounds.

9.1 Hearing in Dolphins

All of our knowledge of hearing in cetaceans comes from studies with small odontocetes. The most studied dolphin is the Atlantic bottlenose dolphin, *Tursiops truncatus*. This species also happens to be the most common species in oceanariums, aquariums, and theme parks. Despite the amount of research performed with *Tursiops*, there are still many gaps in our knowledge of auditory processes. Much of the auditory research with dolphins has followed along the lines of human auditory research, but at a much slower pace and with considerably less intensity.

9.1.1 Hearing Sensitivity

9.1.1.1 Sensitivity to Continuous Tones

In a pioneering study, Johnson (1967) performed a carefully controlled psychophysical experiment to measure the underwater auditory sensitivity of a bottlenose dolphin to continuous wave (cw) tones as a function of frequency. A cw tone is a signal with a long enough duration to produce a steady-state response by the subject, i.e., the subject's sensitivity would not improve if the signal duration was longer. The dolphin was required to station in a sound stall and respond to the presence or absence of a 2-sec tone signal. A go/no-go response paradigm was used in conjunction with the up/down staircase psychophysical testing procedure. The dolphin audiogram measured by Johnson is shown in Fig. 9.1, along with a human audiogram measured by Sivian and White (1933). Johnson's results indicate that a bottlenose dolphin can hear over a wide frequency range, between 75 and

W.W.L. Au, M.C. Hastings, *Principles of Marine Bioacoustics*,

DOI: 10.1007/978-0-387-78365-9_9,

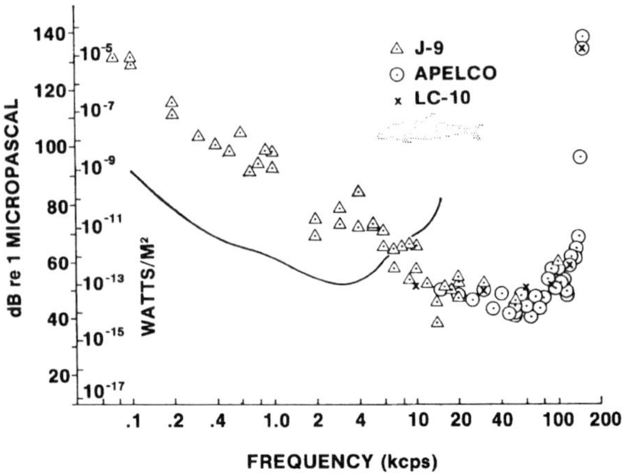

FIGURE 9.1. Auditory sensitivity of a bottlenose dolphin and human subjects. The *solid curve* is the human audiogram of Sivian and White (1933) plotted against the right ordinate in watts/cm^2. The left ordinate is sound pressure level in dB (re 1 :Pa) in water (adapted from Johnson, 1967).

150 kHz, and has good sensitivity (within 10 dB of the maximum sensitivity) between approximately 15 and 110 kHz. The upper frequency limit of hearing cuts off very sharply, on the order of 495 dB per octave, which is far greater than filter (Johnson, 1968b).

The only meaningful way thresholds in air and water can be compared is on the basis of intensity, hence the human audiogram was plotted in watts per cm^2 by Johnson (converted to mks units here). The reason for this is the difference in the acoustic impedance (ρc) between water and air and the role of impedance on acoustic pressure, where $\rho$is the density of the medium and c is the sound velocity of the medium. The relationship between impedance, pressure, and intensity for a plane wave is

$$I = \frac{p^2}{\rho c}. \tag{9.1}$$

From Eq. (9.1) we can see that for a given intensity (or energy) the sound pressure level will be very different in air and underwater. The dolphin and human audiograms in Fig. 9.1 are similar in shape, with the dolphin's shifted to higher frequencies by a factor of 9. The data also indicate that the maximum hearing sensitivity of the dolphin underwater and human in air are similar.

Johnson's auditory sensitivity results probably represent a very conservative estimate of the bottlenose dolphin's capability. The animal was conditioned against committing false alarm errors. A 90-sec "timeout," which included the removal of the response levers from the tank, was taken every time the dolphin committed a false alarm error. This type of experimental procedure probably caused the dolphin to adopt a relatively conservative response pattern in which it would choose the go response only if it was fairly sure that a signal was present.

Since the experiment of Johnson (1967) with *Tursiops truncatus*, the auditory sensitivity of many other small cetacean species have been measured. Audiograms have been measured for the harbor porpoise, *Phocoena phocoena* (Andersen, 1970a), killer whale, *Orcinus orca* (Hall and Johnson, 1971; Bain and Dallheim, 1992), Amazon River dolphin, *Inia geoffrensis* (Jacobs and Hall, 1972), beluga or white whale, *Delphinapterus leucas* (White et al., 1978), Pacific Bottlenose dolphin, *Tursiops gilli* (Ljungblad et al., 1982), false killer whale, *Pseudorca crassidens* (Thomas et al., 1988), Chinese river dolphin, *Lipotes vexillifer* (Wang et al., 1992), Risso's dolphin, *Grampus griseus* (Nachtigall et al., 1995), and Tucuxi, *Sotalia fluviatilis* (Sauerland and Dehnhardt, 1998). The results of auditory sensitivity measurements of different cetacean species are depicted in Fig. 9.2. The data of Johnson (1967) for *Tursiops* was smoothed to avoid confusion in separating the various curves. The curve for *Orcinus orca* is from Bain and Dallheim (1992) and is the average from two animals. This curve came from unpublished data and the investigators used an unusual reinforcement procedure. They not only reinforced the subjects for correct responses, but also for misses which would tend to cause the subject to be very conservative (it need not respond to the presence of the signal in order to receive a reinforcement). However, one whale responded to the presence of signals up to 104 kHz and another to 120 kHz, which indicates that the animals could hear at least signals up to these frequencies. Previous data from Hall and Johnson (1971) indicated a high-frequency limit of 35 kHz for the killer whale. Apparently, that particular whale experienced high-frequency hearing losses. Only the low-frequency and high-frequency of the *Grampus* data are plotted since the animal was masked by the ambient noise of Kaneohe Bay for the mid-frequencies. In all of these studies with the exception of Andersens, the up/down staircase psychophysical procedure was used to measure the animals' threshold. Andersen used the method of constant stimuli to test the *Phocoena's* hearing sensitivity. Thomas et al. (1988) and possibly Hall and Johnson (1971) were the only investigators that did not observe a "time out" when the animal committed a false alarm error. Ljungblad et al. (1982) used a rather unusual modification to the staircase procedure in measuring the audiogram of the *Tursiops gilli.* The signal level after each correct signal-present trial was reduced by 5 dB until the animal missed the signal. The signal level was then increased by 15 dB for the next signal-present trial. It seems that with this procedure, the animal would not be motivated to detect a difficult signal since the signal level would be raised by 15 dB if it missed the stimulus, making the task considerably easier. The relatively low sensitivity of the *Tursiops gilli* for low frequencies may be a reflection of the unusual testing procedure rather than the actual sensitivity of the dolphin.

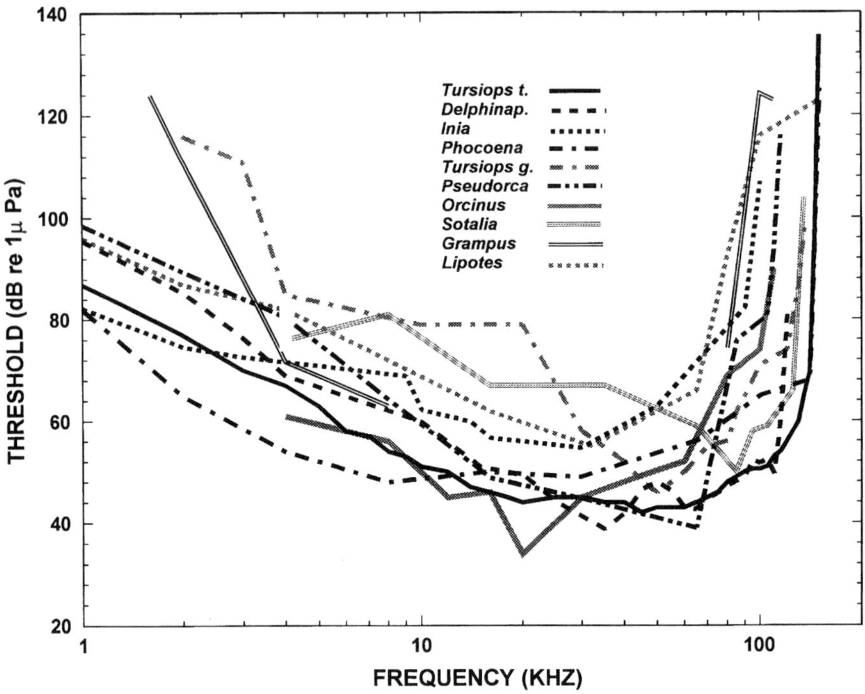

FIGURE 9.2. Audiograms for different cetacean species.

An interesting feature of the data in Fig. 9.2 is how similar all the different audiograms appear in shape, levels of maximum sensitivity, and high-frequency limits. This seems quite remarkable in light of the different sizes of the various animals and the different habitats they live in. The relative sizes of all the odontocetes represented in Fig. 9.2 are depicted in Fig. 9.3. All of the animals tested could hear high-frequency sounds beyond 100 kHz, considerably higher than for most mammals. Perhaps with the exception of the sperm whale, all echolocating odontocetes have very sensitive hearing that extends to very high frequencies.

The audiograms depicted in Fig. 9.2 can be divided into three frequency regions. The first region is the low-frequency region in which the sensitivity gradually improves with frequency at a rate of approximately 10–15 dB per octave (with the exception of the *Tursiops gilli*). The second region is the mid-frequency where the sensitivity forms a "U"-shaped trough, or dip in which the maximum sensitivity is reached. The third region is the high-frequency cutoff region where the sensitivity decreases rapidly with frequency (a minimum of 100 dB per octave, depending on the specific animal).

A summary of some important properties of the different audiograms depicted in Fig. 9.2 is given in Table 9.1. In the table, the frequency of best hearing is arbitrarily defined as the frequency region in which the auditory

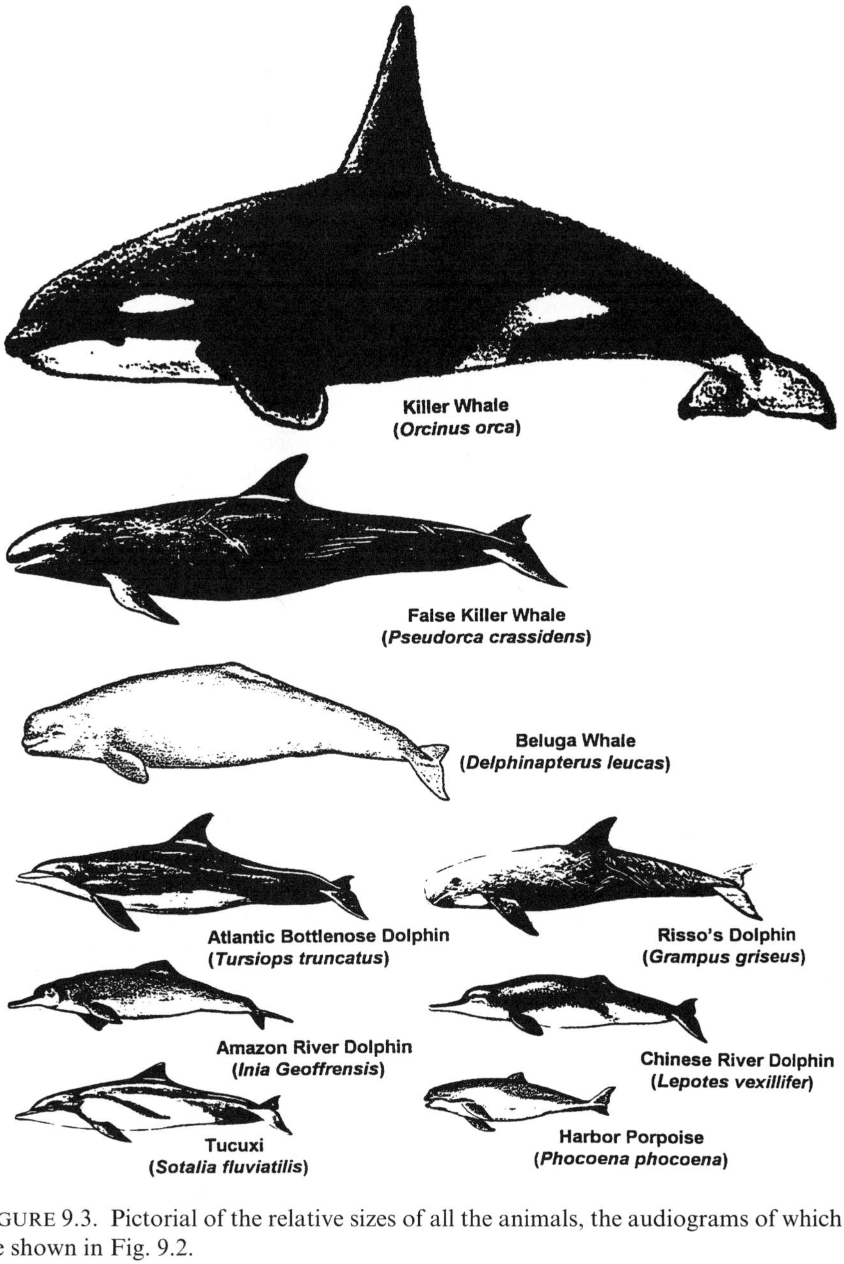

FIGURE 9.3. Pictorial of the relative sizes of all the animals, the audiograms of which are shown in Fig. 9.2.

sensitivity is within 10 dB of the maximum sensitivity depicted in each audiogram of Fig. 9.2. With the exception of the *Orcinus* and the *Lipotes*, the maximum sensitivity of the rest of the species represented in Fig. 9.2 and Table 9.1 are very similar, and within experimental uncertainties, especially

TABLE 9.1 Some Important Properties of the Audiograms Plotted in Fig. 9.2

Species	Maximum sensitivity (dB re 1 :Pa)	Frequency of best hearing (kHz)	Upper frequency limit (kHz)
Tursiops truncatus	42	15–110	150
Phocoena phocoena	47	3–70	150
Orcinus orca	34	15–30	120
Inia geoffrensis	51	12–64	100
Delphinapterus leucas	40	11–105	120
Tursiops gilli	47	30–80	135
Pseudorca crassidens	39	17–74	115
Lipotes vexillifer	55	15–60	100
Grampus griseus	–	–	100
Sotalia fluviatilis	50	35–50	135

for the audiograms obtained with the staircase procedure using relatively large step sizes of 5 dB or greater. At the frequency of best hearing, the threshold for *Orcinus* is much lower than for the other animals. It is not clear whether this keen sensitivity is a reflection of a real difference or a result of some experimental artifacts. The data of Table 9.1 also indicate that *Tursiops truncatus* and *Delphinapterus leucas* seem to have the widest auditory bandwidth.

9.1.1.2 Sensitivity to Pulse Tones

The phenomenon of temporal auditory summation in which the auditory sensitivity to a tone increases as the duration of a tone pulse increases is well known in human audition (Plump and Bouman, 1959; Zwislocki, 1960). The ear behaves like an integrator with an "integration time constant" (Zwislocki, 1960). Energy is summed over the duration of a tone pulse until the pulse is longer than the integration time constant. As the duration of a pulse tone is made longer, more energy is contained within the pulse and, therefore, the ear is better able to perceive the pulse tone. For tone pulses with duration shorter than the integration time, the amount of energy increases by 3 dB for each doubling of the duration. Hughes (1946) found that the change in the absolute hearing threshold as a function of the duration of a single tone pulse could be described by the equation

$$I/I_{\infty} = 1 + (\tau/t), \tag{9.2}$$

where I = threshold intensity of a tone pulse of duration t; I_{∞} = threshold intensity for t; ∞ and τ is a constant. For signals longer than τ, the threshold is essentially constant and equal to I_{∞}.

Johnson (1968a) studied temporal auditory summation in a dolphin, using the same animal, experimental configuration, and procedure as in his cw audiogram measurement (Johnson, 1967). He found a similar phenomenon to humans, of increasing auditory sensitivity with increasing duration of the pulse tone. The results of Johnson's measurements are shown in Fig. 9.4a and 9.4b for tone frequencies of 0.25, 1, 4, 20, 40, and 100 kHz. The dolphin's threshold shifted downward as the pulse duration increased and gradually leveled off as the duration approached and exceeded the integration time constant. The dashed curves represent Eq. (9.2) fitted to the data with a minimum least square error, at each frequency except 100 kHz. The data of Fig. 9.4 varied in a manner consistent with Eq. (9.2). The dolphin's threshold shifted at approximately 3 dB per doubling of duration in the time region of steepest decent.

The time constants used in the curve fit are plotted in Fig. 9.5, along with the time constants obtained in experiments with humans. Johnson (1968a) plotted the time constant as a function of frequency on a semi-log plot. Here, a log–log graph is used so that the time constant at 100 kHz can be included on the graph. Johnson (1968a) did not include the time constant at 100 kHz because he found that the transient response of the J-9 used to obtain the data of Fig. 9.4a caused the signal to have a broad spectrum. The LC-10 had a better transient response but still introduced unwanted frequency spreading, particularly above 100 kHz. Nevertheless, the 100 kHz data obtained with the LC-10 is included in Fig. 9.4 in order to get an estimate of the dolphin's time constant at 100 kHz, which is close to typical peak frequencies of sonar signals used by *Tursiops truncatus*. The estimate of the time constant at 100 kHz seems to be consistent with the rest of the data. From Fig. 9.5, we can see that the dolphin's time constant rises to a peak at frequencies between 1 and 4 kHz, and then decreases rapidly with frequency. It is interesting to note that the dolphin's time constants associated with frequencies between 0.50 and 10 kHz closely match those obtained for human subjects. Johnson (1968a) concluded that the porpoise integrates pure-tone acoustic energy in the same way as humans.

9.1.1.3 Hearing at Depth

All of the audiograms for dolphins have been measured near the water surface so the effects of ambient pressure are negligible. However, many species of dolphins swim to deep depth and the effects of external pressure may affect the animals' hearing sensitivity. In order to examine the effects of pressure on the hearing sensitivity of odontocetes, Ridgway et al. (2001) trained two beluga whales to station on a bite plate connected to a test platform that could be lowered to different depths. They used a unique response paradigm requiring each subject to emit a whistle whenever it heard a pure-tone signal and incorporated an extended reinforcement schedule so that the whale would be reinforced after many trials.

Hearing sensitivity was measured at depths of 5, 100, 200, and 300 m. The stimulus signal was a 500 ms tone, which was turned on at random intervals. Ridgway et al. (2001) found that the hearing sensitivity of the two beluga whales did not change with depth, suggesting that sound is conducted through the head tissue of the whales directly to the inner ears without going through the middle ear even though all the components of a typical

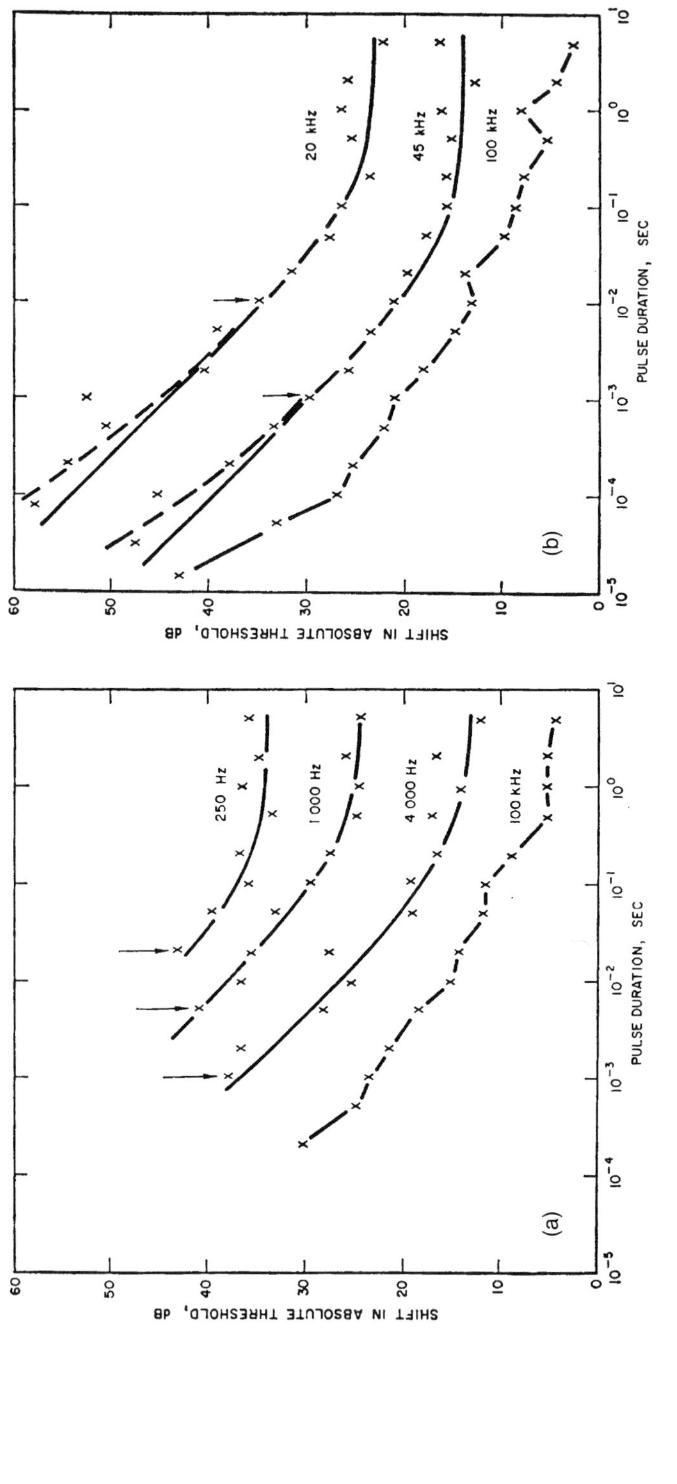

FIGURE 9.4. Absolute thresholds vs. pulse duration for the data obtained with a J-9 projector **(a)** and with an LC-10 hydrophone **(b)**. The *arrows* indicate the shortest pulses used in least-square fit of the data to Eq. (9.2) (adapted from Johnson, 1968a).

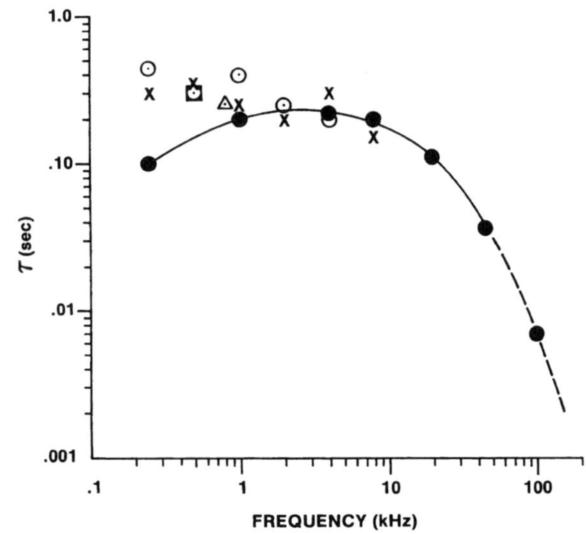

FIGURE 9.5. Time constants vs. frequency for the dolphin and humans. The *open circles* and *crosses* are data from Plump and Bouman (1959), the *square* from Blodgett et al. (1958), the *triangle* from Hamilton (1957), and the *closed circles* are for the dolphin (after Johnson, 1968a).

mammalian ossicular chair are present. The middle ear of dolphins is extremely complex and its role in hearing and how it functions is unclear. The ossicular chain is stiffened and tightly bound together with sheaths and annular ligaments, and the ossicles are denser and more massive than for terrestrial mammals of similar size (Ketten, 2000).

9.1.1.4 Temporary Threshold Shift

When the mammalian auditory system is exposed to a high level of sound for a specific duration, the hair cells in the cochlear begin to fatigue and will not immediately restore to their normal shape, the animal's hearing sensitivity decreases. If the exposure is below some critical energy flux density limit, the hair cells will eventually return to their normal shape so that the hearing loss is temporary and the effect is termed "temporary threshold sift" in hearing sensitivity or TTS. If the sound exposure exceeds a particular energy flux density limit, the hair cells in the cochlear become permanently damaged and will eventually die so that the hearing loss is permanent, a condition referred to as "permanent threshold shift" in hearing or PTS.

TTS experiments have been conducted with three species of odontocetes (*Tursiops truncatus, Pseudorca crassidens, Delphinapterus leucas*) using both behavioral and electrophysiological techniques (Schlund et al., 2000; Finneran et al., 2000 and Nachtigall et al., 2003). Schlund et al. (2000)

conducted a behavioral experiment to measure the shift in the masked underwater hearing thresholds in five bottlenose dolphins and two white whales. The fatiguing stimuli were 1-s tones at 0.4, 3, 10, 20, and 75 kHz. San Diego Bay has a large population of snapping shrimp and the noise produced by these organisms probably mask the hearing of the dolphins and white whales at some frequencies. Therefore, the term masked temporary threshold shift (MTTS) was used by Schlund et al. (2000). They found that levels between 192 and 201 dB re 1 μPa induced a 6 dB or larger MTTS. An exception occurred at 75 kHz, where one dolphin exhibited an MTTS after exposure only 182 dB and the other dolphin did not show any shift after exposure to maximum levels of 193 dB, and at 0.4 kHz, where no subjects exhibited shifts at levels up to 193 dB. The shifts occurred most often for test frequencies above the fatiguing stimulus. At the conclusion of the study, all thresholds were at baseline values. Finneran et al. (2000) also conducted a behavioral experiment to measure masked underwater hearing thresholds in a bottlenose dolphin (*Tursiops truncatus*) and a white whale (*Delphinapterus leucas*) caused by a single underwater impulsive sounds produced from a seismic water gun. Hearing thresholds were measured at 0.4, 4, and 30 kHz. MTTSs of 7 and 6 dB were observed in the white whale at 0.4 and 30 kHz, respectively, approximately 2 min following exposure to single impulses with peak pressures of 160 kPa, peak-to-peak pressures of 226 dB *re* 1 μPa, and total energy fluxes of 186 dB *re* 1 $\mu Pa^2 \cdot s$. Thresholds returned to within 2 dB of the pre-exposure value, approximately 4 min after exposure. No MTTS was observed in the dolphin at the highest exposure conditions: 207 kPa peak pressure, 228 dB *re* 1 μPa peak-to-peak pressure, and 188 dB *re* 1 $\mu Pa^2 \cdot s$ total energy flux.

Nachtigall et al. (2003) used a band of continuous random noise with frequencies between 4 and 11 kHz as the fatiguing stimulus to induce TTS in a bottlenose dolphin in Kaneohe Bay, Hawaii. Behaviorally determined hearing thresholds for a 7.5-kHz tone for an Atlantic bottlenose dolphin (*Tursiops truncatus*) were obtained following exposure to the fatiguing stimulus. A fatiguing noise stimulus at the energy flux density of 213 dB *re* 1 μPa^2 s caused temporary threshold shifts of 18.8, 12.0, 2.0, 10.4, and 12.4 dB for different sessions in which TTS was induced. The fatiguing noise stimulus had an equivalent to a noise pressure level of 179 dB *re* 1 μPa. Threshold determination took at least 20 min. Recovery was examined 360, 180, 90, and 45 min following exposure and was essentially complete within 45 min. Although Kaneohe Bay also has a large population of snapping shrimp, the 7.5 kHz test tone was low enough in frequency where the hearing sensitivity of the bottlenose dolphin was probably not masked by the ambient background noise between 4 and 11 kHz.

These TTS experiments with the bottlenose dolphins and white whales indicate their auditory system is susceptible to TTS as in any mammal auditory system. However, the auditory systems of the beluga and the bottlenose dolphin apparently have the capability to recover relatively rapidly from TTS. Clearly, more research is needed to understand how the auditory system of odontocetes are affected by anthropogenic noise.

9.2 Spectral Analysis Sensitivity

One of the fundamental characteristics of the human auditory system is its ability to derive spectral information from sounds. Our ability to hear one sound in the presence of other sounds and to discern individual harmonics from complex tones are examples of the ears' frequency resolution capability. The idea that our auditory frequency analyzer could be thought of as a bank of overlapping, linear, bandpass filters was suggested as early as 1863 by Helmholtz (Paterson and Moore, 1986). It was not until Fletcher (1940), in his famous article, suggested several ways in which the width of the bandpass filters could be estimated. The first technique is commonly referred to as the critical ratio technique and the second one is referred to as the critical bandwidth technique. These techniques developed to study the humans auditory system can also be applied to study the auditory system of dolphins.

9.2.1 Critical Ratio

In the critical ratio technique, the auditory sensitivity of subjects to tones masked by broadband white noise is measured. Fletcher hypothesized that: (a) a pure tone would be masked only by a narrow band of noise surrounding the frequency of the tone, (b) at threshold of hearing, the noise power was equal to the signal power. From the second hypothesis, the following equation would be applicable at the masked threshold:

$$I_t = N_0 \times \Delta f, \tag{9.3}$$

where

I_t = Intensity of the tone at threshold in μPa^2
N_0 = Noise spectral density in $\mu Pa^2/Hz$
Δf = Bandwidth of the auditory filter at the tone frequency.

Therefore, the bandwidth of the auditory filters can be indirectly estimated by performing masked threshold measurements. The bandwidth is the ratio of the threshold intensity to noise spectral density, and the bandwidths derived from this technique are referred to as critical ratios. The shape of the filters are assumed to be rectangular with infinitely steep low and high pass cutoffs. The critical ratio can be expressed in dB as

$$\mathrm{CR} = 10\ \mathrm{Log}\ (\Delta f). \tag{9.4}$$

Johnson (1968b) used the critical ratio technique of Fletcher (1940) to measure the filter bandwidth of the auditory system of *Tursiops truncatus*. The masked threshold for 15 different frequencies at six different noise levels were measured, using the same animal, experimental configuration, and

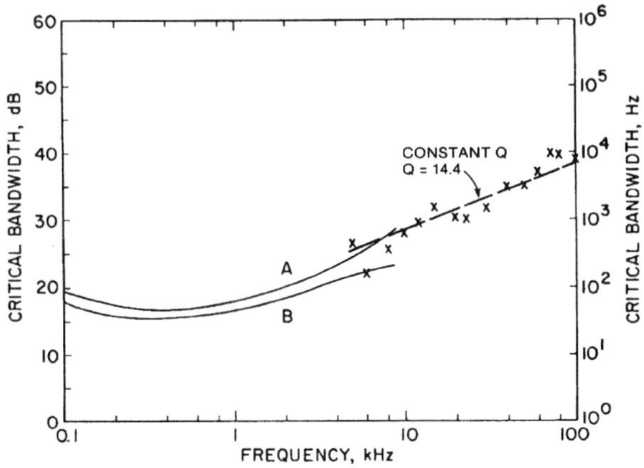

FIGURE 9.6. Critical ratios for the dolphin (X) and for human (**A**) monaural (French and Steinberg, 1947) and (**B**) binaural (Hawkins and Stevens, 1950) (adapted from Johnson, 1968b).

procedure as in his cw audiogram measurement (Johnson, 1967). The critical ratio results measured by Johnson is shown in Fig. 9.6, along with monaural and binaural results for humans. The dolphin's critical ratio seems like an extension of the human critical ratio for higher frequencies. The dolphin's critical ratio increases almost proportionately with frequency, suggesting that the dolphin's auditory system may be modeled by a bank of constant-Q filters. The Q or quality factor of a bandpass filter is defined as the ratio of the center frequency to the bandwidth as expressed by the equation

$$Q = F_0/\Delta f. \tag{9.5}$$

The effectiveness of noise in masking pure tones can be determined by plotting the shift in threshold (difference between threshold in quiet and threshold in noise) as a function of the effective level (difference between the threshold in quiet and total noise energy in one critical band).

Johnson (1968b) compared the effectiveness of noise in masking the *Tursiops* with the data obtained in human experiments as illustrated in Fig. 9.7. The human data are from the studies of French and Steinberg (1947) and Hawkins and Stevens (1950). Except for the 5 kHz data, the dolphin's results compare favorably with the human data, showing a linear shift in threshold as a function of the masker level.

Au and Moore (1990) also measured the critical ratio of an Atlantic bottlenose dolphin. The animal's masked threshold was determined using a staircase technique and a step size of 1 dB. Time outs were not used when the animal committed false alarm errors.

The results of Au and Moore (1990) along with the results of Johnson (1968b) for *Tursiops* are shown in Fig. 9.8. Also included in the figure are the

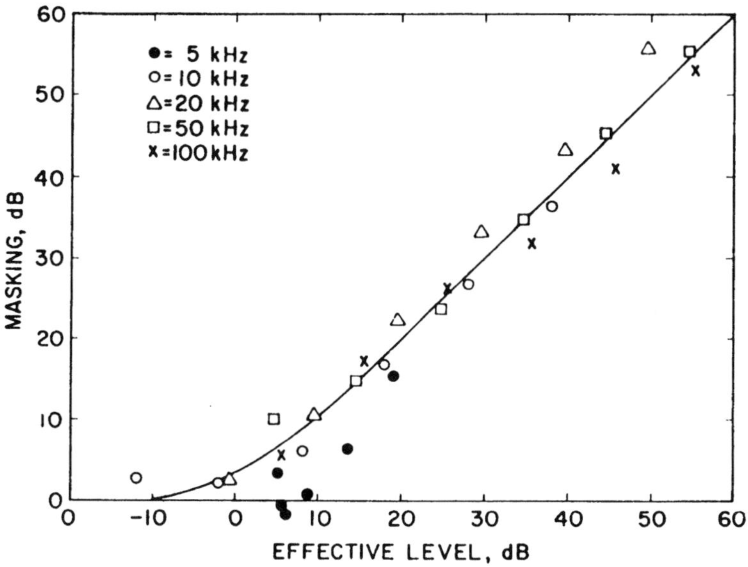

FIGURE 9.7. Plot of the shift in threshold resulting from the presence of masking noise vs. the effective level of the masker. The *solid curve* gives results from experiments with human subjects (adapted from Johnson, 1968b).

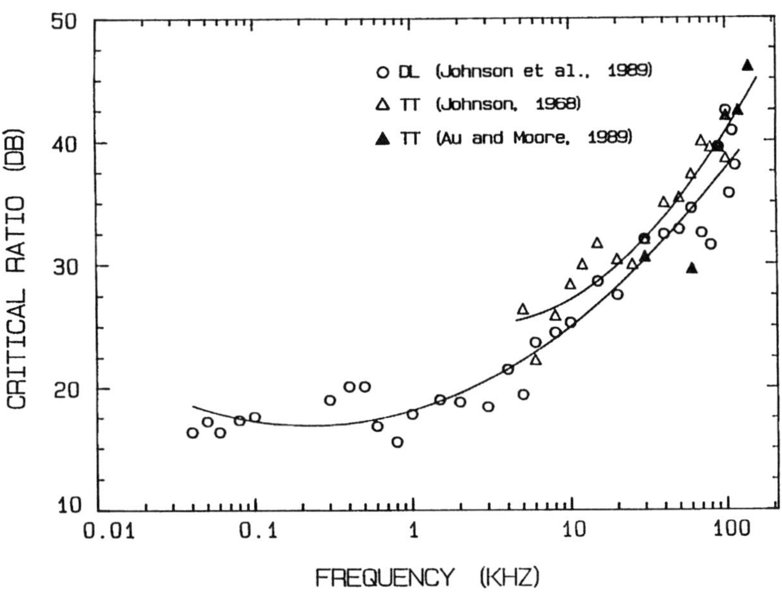

FIGURE 9.8. Critical ratios for two Atlantic bottlenose dolphins and a beluga (adapted from Johnson, 1968b; Johnson et al., 1989; Au and Moore, 1990).

critical ratios for a beluga (*Delphinapterus leucas*), which was measured by Johnson et al. (1989). They used a staircase technique with 5-dB stepsize and one noise level to obtain their results. The solid curves are second-order polynomial curve fits to the combined data of Johnson (1968) and Au and Moore (1990) for *Tursiops* and for the *Delphinapterus leucas* (Johnson et al., 1989). The data indicate that the critical ratios of the beluga are slightly narrower than for the bottlenose dolphin at low frequencies and approach that of the dolphin at high frequencies.

9.2.2 Critical Bandwidth

Fletcher (1940) also devised a more direct method to measure the width of the auditory filter. In this second technique, the bandwidth of the masking noise is varied and the masked threshold determined for each bandwidth. For noise bandwidths that are narrower than the auditory filter at the test frequency, the masked threshold should increase proportionately as the width of the noise filter increases. When the width of the noise becomes equal to or greater than the auditory filter, the masked threshold should no longer increase as the noise filter increases since the auditory filter cannot receive any more noise. Therefore, the masked threshold should be a constant with respect to the bandwidth of the noise. In this technique, the masked threshold is plotted as a function of the noise bandwidth. Two straight lines are fitted to the data, one fitted to the varying portion and the other to the constant portion of the masked threshold data. The width of the auditory filter is the intersection of the two lines fitted to the data. The filter bandwidths determined by this technique are referred to as "critical bandwidths." As in the critical ratio technique, the shape of the filters is also assumed to be rectangular.

Au and Moore (1990) used masking noise of different bandwidths to measure the critical bandwidth of a *Tursiops* at frequencies of 30, 60, and 120 kHz. The masked threshold was determined as a function of the noise bandwidth using the staircase procedure and a go/no-go response paradigm. The masked thresholds as a function of the noise bandwidth for the three test frequencies are displayed in Fig. 9.9. Also included in Fig. 9.9 are the intersecting lines that were matched to the data in a minimum least square error manner. The bandwidth associated with each point of intersection is the critical bandwidth and represents an estimate of the animal's auditory filter bandwidth at the test frequencies. The data from the critical ratio and critical bandwidth measurements with the same dolphin are depicted in Fig. 9.10. The solid lines are second-order polynomial curve fits of the data. The relationship between the critical ratio/critical bandwidth can be seen in the figure. The critical bandwidth is approximately 12 dB (16 times) greater than the critical ratio at 30 kHz and 60 kHz and 3.5 dB (2.6 times) at 120 kHz. Critical bandwidths for humans are typically 3 dB (2.5 times) greater than critical ratios (Scharf, 1970).

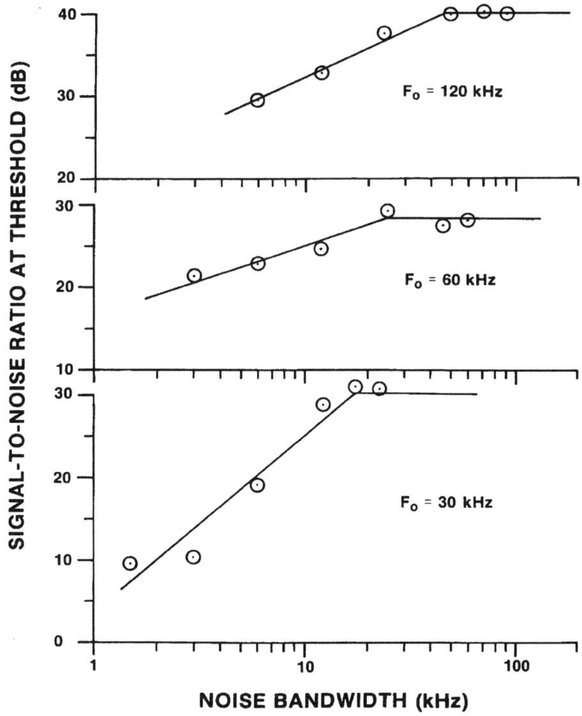

FIGURE 9.9. Masked hearing threshold as a function of noise bandwidth for three test frequencies (adapted from Au and Moore, 1990).

The dolphin's critical bandwidth did not increase with frequency as rapidly as the critical ratio. There are also large differences between the critical bandwidth and the critical ratio. The reason for this is not known. One possibility is that the masking noise used was too high causing the auditory system to react in a nonlinear fashion. Another possibility is that Fletcher's assumption for humans that signal and noise power are equal at the detection threshold may not be valid for the dolphin because of increased efficiency of the dolphin in detecting signals in noise. Larger critical ratios would result if the signal power was smaller than the noise power at threshold.

9.2.2.1 Masking by a Pure Tone

Another technique that can be used to study the frequency analysis property of the dolphin auditory system involves masking of a pure tone signal by another pure tone. Johnson (1971) continued his auditory research on *Tursiops* by conducting a tone-on-tone masking experiment using the same experimental apparatus, procedure, and dolphin as in his original cw auditory sensitivity study. A 70 kHz tone was used as the masking tone with two masking levels, 40 and 80 dB, above the animal's threshold. The results of the

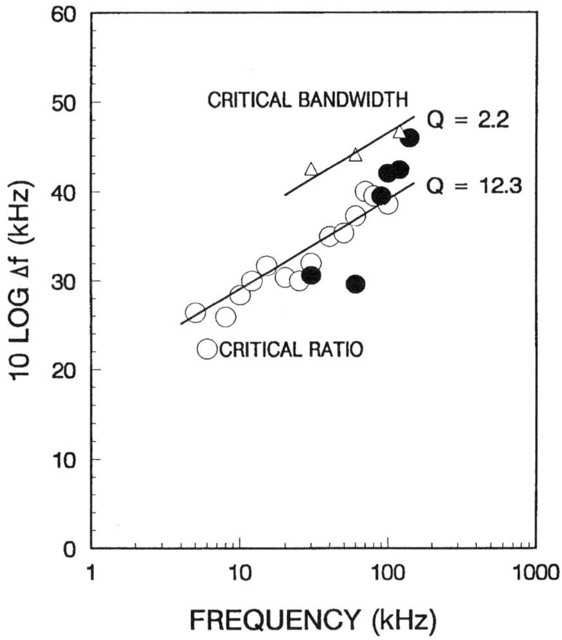

FIGURE 9.10. Critical ratio/bandwidth for a *Tursiops truncatus* (adapted from Au and Moore, 1990).

tone-on-tone masking experiment are shown in Fig. 9.11, with the threshold shift from the unmasked condition plotted as a function of the frequency of the test tone. The curves indicate how many dBs the intensity of the signal at a particular frequency must be raised above its absolute threshold to be detectable in the presence of the 70 kHz masking tone. Above the curves, both masker and signal are audible. Below the curves, only the masker is audible.

Most of the masking occurred when the signal frequency approached the masker frequency with maximum masking occurring when the signal and masking frequencies were almost equal. When the signal frequency equaled the masker frequency, there was a dip in the amount of masking, as can be seen in Fig. 9.11. This phenomenon also occurs with humans (Wengel and Lane, 1924) and is the result of the signal and masker interacting to produce beats. The dolphin's masking results resembled tone-on-tone masking results for humans, with the exception that the dolphin's results are relatively symmetrical about the masking frequency compared to the asymmetric nature of the human results. Signals both lower and higher in frequency than the masking frequency are masked equally well in the dolphin. With humans, lower-frequency tones are more effective in masking higher frequencies than high-frequency tones are in masking low frequencies (Wengel and Lane, 1924).

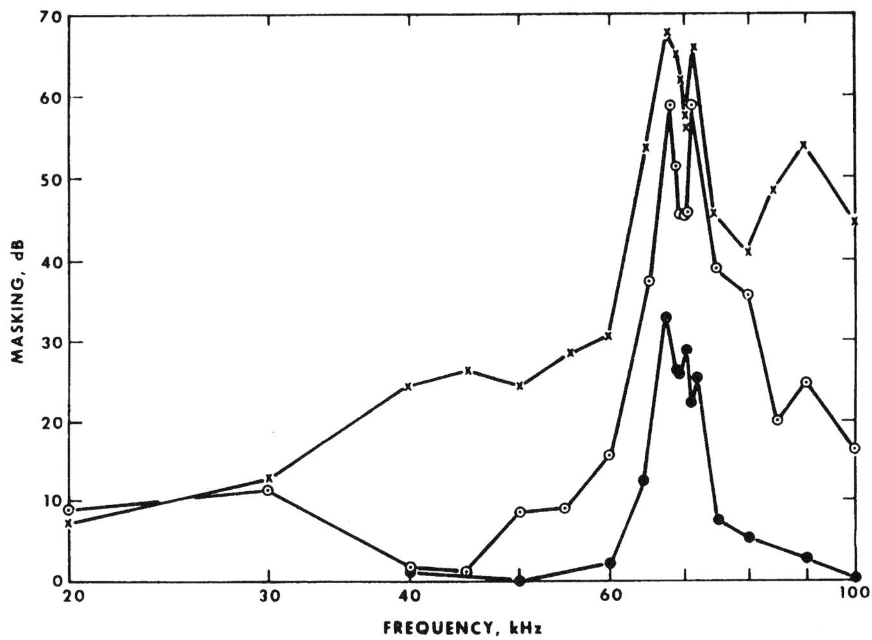

FIGURE 9.11. Threshold shift (masking) from threshold in quiet due to the presence of the masking tone for various test frequencies. The *crosses* give the first run data, and the points in *open circles* give the second run data at 80-dB masking level. The *closed circles* give the data taken at the 40-dB masking level (adapted from Johnson, 1971).

Johnson (1971) also estimated the minimal intensity difference that the dolphins can perceive from the magnitude of the dips in the 70 kHz masking curves of Fig. 9.11. Assuming that at 70 kHz the animal's threshold is determined by its ability to resolve beats in the tone, the following relation was derived

$$\Delta L = 20\ Log \frac{1 + 10^{-M/20}}{1 - 10^{-M/20}}, \tag{9.6}$$

where ΔL is the variation in sound-pressure level accompanying beats and M is the difference in the level of beating tones. Both ΔL and M are in dBs. From Eq. (9.6), the values of ΔL for the three curves of Fig. 9.11 are 1.0, 0.35, and 2.0 dB for runs 1 and 2 at the 80 and 40 dB masker levels, respectively.

9.2.3 *Frequency Discrimination*

The ability to distinguish one frequency from another is an important characteristic in detecting and processing acoustic signals as well as sonar echoes. The frequency discrimination capability of the bottlenose dolphin was studied by Jacobs (1972), Herman and Arbeit (1972), and Thompson and

Herman (1975). In all three studies, the dolphin was required to discriminate a constant frequency (CF) pure tone signal from a frequency-modulated (FM) signal having the same center frequency as the CF signal. Jacobs (1972) and Thompson and Herman (1975) used a single projector, and during a trial the dolphin was required to indicate whether a CF or FM signal was emitted by pressing one of two paddles. Herman and Arbeit (1972) used two spatially separated projectors and presented sounds sequentially during a trial. A CF signal was transmitted from one projector followed by a FM sound from the other projector, or vice versa. The dolphin was required to touch the paddle closest to the projector that transmitted the CF signal. In all three studies, the amount of frequency modulation was decreased in a staircase procedure until the animals could not discriminate the FM signal from the CF signal.

The difference limen (DL) is the difference between the upper and lower frequency of the FM signal at the animal's threshold. The relative difference limen is the Weber fraction ($\Delta F/F$), where F is the center frequency of the FM signal. The results of the three frequency discrimination experiments with *Tursiops truncatus* are shown in Fig. 9.12, with the relative difference limen in

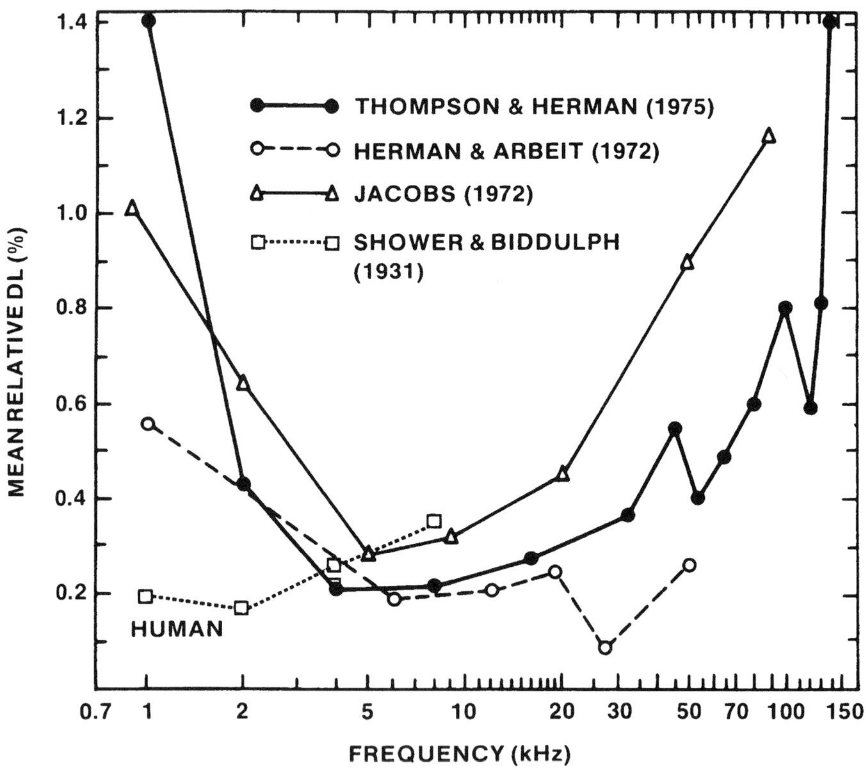

FIGURE 9.12. Acoustic frequency discrimination capability of *Tursiops truncatus* and human.

percent plotted as a function of frequency. The data of Herman and Albeit and Thompson and Herman were obtained with the same animal. Their results, if averaged, would be a good representation of the dolphin's difference limen. The data plotted in Fig. 9.12 indicate that the dolphin's frequency discrimination capability is similar to humans, but shifted to higher frequencies by a factor of 5–9. The dolphin's best discrimination occurred between 2 and 55 kHz, with relative DLs ranging from 0.2% to 0.4%. Relative DLs remained within 1% for all frequencies between 1 and 140 kHz. According to Fay (1974), the frequency discrimination capability of the bottlenose dolphin is superior to any other vertebrate studied except man.

9.3 Directional Hearing Capability

9.3.1 Receiving Beam Patterns

A directional hearing capability is an important property of a sonar system operating in either the passive or active mode. Having narrow transmission and reception beams allows the dolphin to localize objects in a three-dimensional volume, spatially separate objects within a multi-object field, resolve features of extended or elongated objects, and to minimize the amount of interference received. The amount of ambient noise from an isotropic noise field and the amount of reverberation interference received is directly proportional to the width of the receiving beam. The effects of discrete or partially extended interfering noise or reverberant sources can be minimized by simply directing the beams away from the sources.

A common method of measuring the beam pattern of a hydrophone is to rotate the hydrophone about an axis and measure the voltage produced by a stationary sound source, as depicted in Fig. 9.13a. If the hydrophone is too

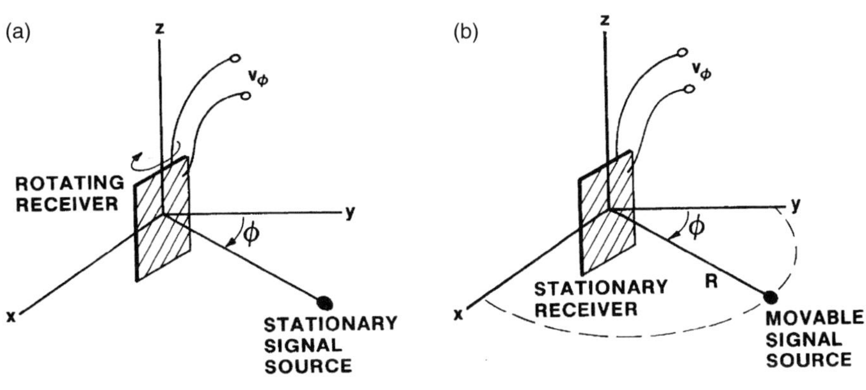

FIGURE 9.13. The geometry of two common methods of measuring the beam pattern of a hydrophone. (**a**) The acoustic source is fixed and the hydrophone is rotated about one of its axis. (**b**) The hydrophone is fixed and a source is moved in a circle about the hydrophone.

large or is integrated within the structure of a ship or some other structure, the beam pattern can be determined by fixing the position of the receiving hydrophone and measuring the SPL produced by a sound source as it is moved along the arc of a circle, as depicted in Fig. 9.13b. Unfortunately, the receiving beam pattern of a dolphin cannot be measured with either technique, unless an electrophysiological technique such as the evoked potential technique using waterproof surface electrodes is used. Instead, the masked hearing threshold as a function of the azimuth about the animal's head needs to be measured. The relative masked hearing threshold as a function of azimuth is equivalent to the received beam pattern since the receiving beam pattern is the spatial pattern of hearing sensitivity.

The receiving beam pattern of a *Tursiops truncatus* was measured in both the vertical and the horizontal planes by Au and Moore (1984). We measured the dolphin's masked hearing threshold as the position of either the noise or signal sources varied in their angular position about the animal's head. The dolphin was required to voluntarily assume a stationary position on a bite plate constructed out of a polystyrene plastic material. The noise and signal transducers were positioned along an arc, with the center of the arc located approximately at the panbone of the animal's lower jaw. A schematic drawing of the experimental apparatus is shown in Fig. 9.14. In order to measure the dolphin's receiving beam in the vertical plane, the animal was trained to turn to its side before biting the specially designed vertical bite plate stationing device, as shown in Fig. 9.14. A go/no-go response paradigm was used, and the animal's masked threshold was determined with an up-down staircase psychophysical procedure. For the measurement in the vertical plane, the position of the signal transducer was fixed directly in line with the bite plate and position of the noise transducer along the arc. The level of the noise was varied in order to obtain the masked threshold. A threshold estimate was considered complete when at least 20 reversals (10 per session) at a test angle had been obtained over at least two consecutive sessions, and if the average reversal values of the two sessions were within 3 dB. After a threshold estimate was achieved, the noise transducer was moved to a new azimuth over a set of randomly predesignated azimuths. As the azimuth about the dolphin's head increased, the hearing sensitivity of the dolphin tended to decrease, requiring higher levels of masking noise from a transducer with its acoustic output was held constant. Masked thresholds were measured for different angular located at that azimuth, to mask the signal from a source located directly ahead of the animal.

The receiving beam patterns in the vertical plane are plotted for signal frequencies of 30, 60, and 120 kHz, in Fig. 9.15a. The data points of the beam patterns in Fig. 9.15a were fitted with a cubic spline function (Forsythe et al., 1977) in order to interpolate values between points. The beam patterns redrawn as smooth curves based on a cubic spline interpolation process are shown in Fig. 9.15b. The radial axis of Fig. 9.15 represents the difference in dB between the noise level needed to mask the test signal at any azimuth and

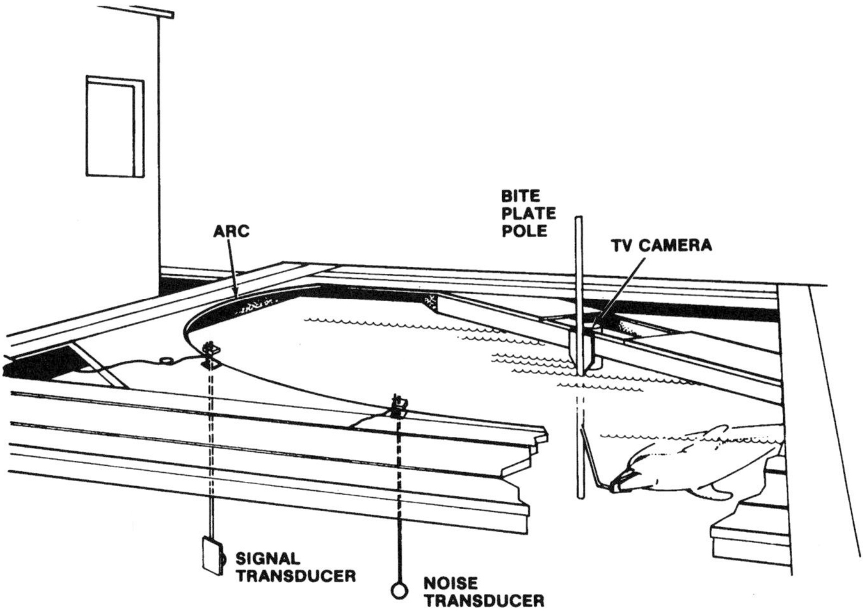

FIGURE 9.14. Experimental configuration for the receive beam pattern measurement of a bottlenose dolphin. The animal was trained to turn to its side and position on the vertical bite plate for the measurement in the vertical plane. For the measurement in the horizontal plane, a different bite-plate assembly was used (adapted from Au and Moore, 1984).

the minimum noise level needed to mask the test signal at the azimuth corresponding to the major axis of the vertical beam. The shape of the beams in Fig. 9.15 indicates that the patterns were dependent on frequency, becoming narrower, or more directional as the frequency increased. The beam of a planar hydrophone also becomes narrower as frequency increases. The 3-dB beamwidths were approximately 30.4°, 22.7°, and 17.0° for frequencies of 30, 60, and 120 kHz, respectively. There was also an asymmetry between the portion of the beam above and below the dolphin's head. The shape of the beams dropped off more rapidly as the angle above the animal's head increased than for angles below the animal's head, indicating a more rapid decrease in the animal's hearing sensitivity for angles above the head than for angles below the head. If the dolphin receives sounds through the lower jaw, as was discussed in Chapter 8, the more rapid reduction in hearing sensitivity for angles above the head may have been caused by shadowing of the received sound by the upper portion of the head structure. There is a slight peculiarity in the 60-kHz beam, which shows almost the same masked threshold values for 15° and 25° elevation angles.

The radial line passing through the angle of maximum sensitivity is commonly referred to as the major axis of the beam. The major axis of the vertical

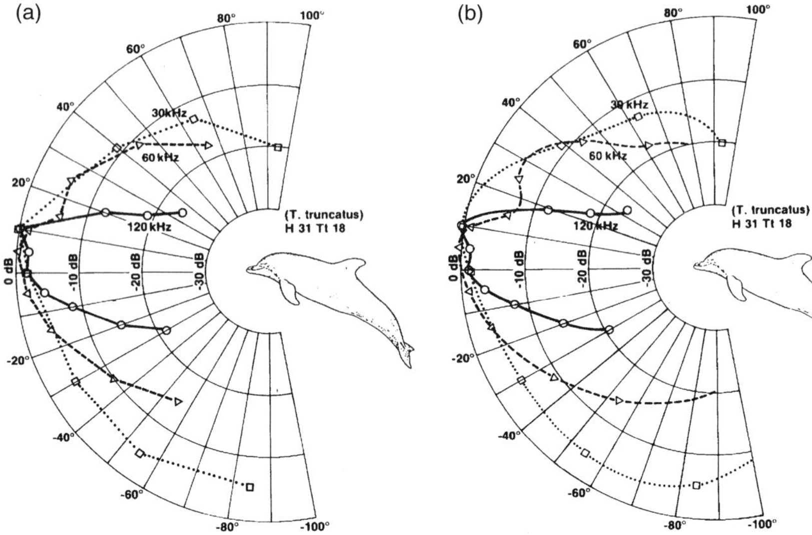

FIGURE 9.15. (**a**) Receive beam patterns in the vertical plane for frequencies of 30, 60, and 120 kHz. The relative masked thresholds as a function of the elevation angle of the masking noise source are plotted for each signal frequency. (**b**) The smoothened beam patterns obtained by fitting the data with a cubic spline function (adapted from Au and Moore, 1984).

beams in Fig. 9.15 is elevated between 5° and 10° above the reference axis. The 30- and 120-kHz results show the major axis at 10°, while the 60-kHz results showed the major axis at 5°. It will be shown in Chapter 11 that the major axis of the received beam in the vertical plane is elevated at approximately the same angle as the major axis of transmitted beam in the vertical plane.

In the horizontal beam pattern measurement, two noise sources were fixed at azimuth angles of ±20°. The level of the noise sources was also fixed. The position of the signal transducer was varied from session to session. Masked thresholds were determined as a function of the azimuth of the signal transducer, by varying the signal level of the signal transducer in a staircase fashion. A threshold estimate was considered completed when at least 20 reversals at a test angle had been obtained over at least two consecutive sessions, if the average reversal values were within 3 dB of each other. After a threshold estimate was determined, the signal transducer was moved to a new azimuth over a set of randomly predesignated azimuths. Two noise sources were used in order to discourage the dolphin from internally steering its beam in the horizontal plane. If the animal could steer its beam, it would receive more noise from one of the two hydrophones, and therefore not experience any improvement in the signal-to-noise ratio. The masking noise from the two sources was uncorrelated but equal in amplitude.

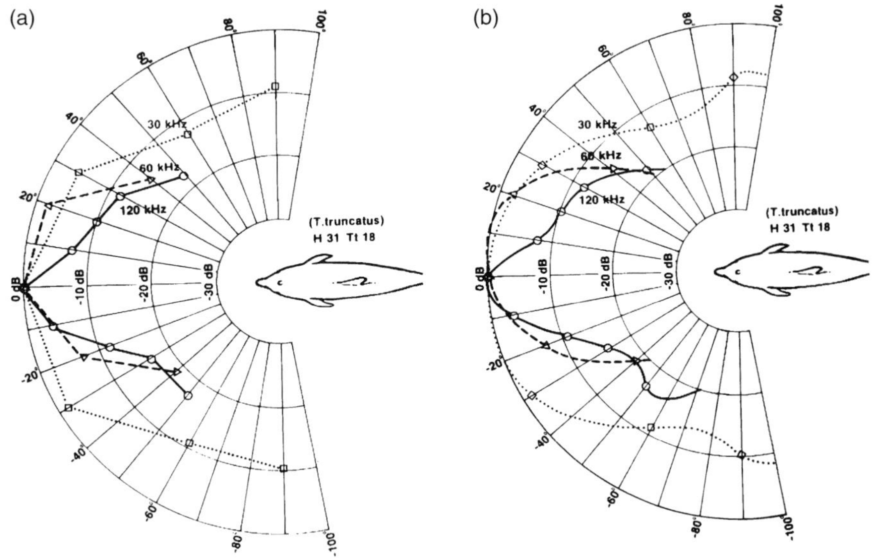

FIGURE 9.16. (**a**) Receive beam patterns in the horizontal plane for frequencies of 30, 60, and 120 kHz. The relative masked thresholds as a function of the angle of the signal source are plotted for each signal frequency. (**b**) The smoothened beam patterns obtained by fitting the data with a cubic spline function (adapted from Au and Moore, 1984).

The received beam patterns in the horizontal plane for frequencies of 30, 60, and 120 kHz are shown in Fig. 9.16a. The beam patterns are redrawn as smooth curves based on Fig. 9.16b. The radial axis represents the difference in dB between the signal level at the masked threshold for the various azimuths and the signal level of the masked threshold for 0° azimuth (along the major axis of the horizontal beam). The horizontal receiving beams were directed forward, with the major axis being parallel with the longitudinal axis of the dolphin. The beams were nearly symmetrical about the major axis. Any asymmetry was within the margin of experimental error involved in estimating the relative thresholds. The horizontal beam patterns exhibited a similar frequency dependence as the vertical beam patterns, becoming narrower or more directional as frequency increased. The 3-dB beamwidths were 59.1°, 32.0°, and 13.7° for frequencies of 30, 60, and 120 kHz, respectively.

Zaytseva et al. (1975) measured the horizontal beam pattern of a dolphin by measuring the masked hearing threshold as a function of azimuth. Their beamwidth of 8.2° for a frequency of 80 kHz was much narrower than the 13.7° for a frequency of 120 kHz. The difference in beamwidth is even larger if the results of Au and Moore (1984) are linearly interpolated to 80 kHz. We calculated an interpolated beamwidth of 25.9° at 80 kHz, which was considerably greater than the 8.2° obtained by Zaytseva et al. (1975). The difference in beamwidth measured by Zaytseva et al. and Au and Moore may be attributed to the use of only one noise source by Zaytseva et al., compared to the two noise sources used by Au and Moore. With a single masking noise source in the horizontal plane, there is a possibility of the animal performing a spatial filtering operation by internally steering the axis of its beam in order to maximize the signal-to-noise ratio. Another possibility is that Zaytseva et al. did not use a fixed stationing device. Rather, the dolphin approached the signal hydrophone from a start line, always oriented in the direction of the signal hydrophone. The animal responded to the presence or absence of a signal by either swimming or not swimming to the hydrophone. In such a procedure, it is impossible to control the orientation of the animal's head with respect to the noise masker so that the dolphin could move its head to minimize the effects of the noise.

9.3.2 Directivity Index

The directivity index was defined in Section 3.3.6 as a measure of the sharpness of the beam or major lobe of either a receiving or transmitting beam pattern. For a receiving hydrophone, it is the ratio of the power received by an omnidirectional receiver to that received by a directional receiver in the same isotropic noise field given by Eq. (3.68) as

$$DI = 10 \log\left(\frac{P_0}{P_D}\right), \tag{9.7}$$

where P_0 is the total acoustic power received by an omnidirectional hydrophone and P_D is the total acoustic power received by a directional hydrophone. Using the coordinate system shown in Fig. 3.18, the directivity index of a transducer was derived (Eq. (3.72)) as

$$DI = 10 \log \frac{4\pi}{\int_0^{2\pi} \int_{-\pi/2}^{\pi/2} \left[\frac{p(\theta,\phi)}{p_0}\right]^2 \sin\theta \, d\theta \, d\phi}. \tag{9.8}$$

Although the expression for directivity index is relatively simple, using it to obtain numerical values can be quite critical unless transducers of relatively simple shapes (cylinders, lines, and circular apertures) with symmetry about one axis are involved. Otherwise, the beam pattern needs to be measured as a function of both θ and φ. This can be done by choosing various discrete values of θ and measuring the beam pattern as a function of φ, a tedious process. Equation (9.10) can then be evaluated by numerically evaluating the double integral with a digital computer. Fortunately, directivity index measurements for simple planar transducers shaped as circles, rectangles, and ellipses have

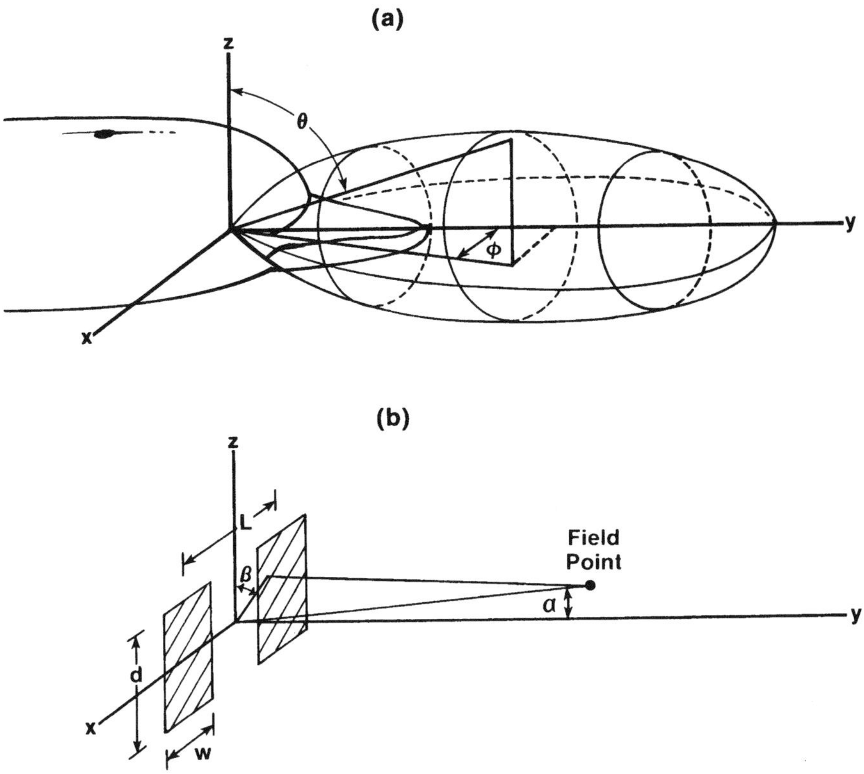

FIGURE 9.17. Geometry for directivity index calculations (a) for dolphin data, (b) for two-element rectangular array model.

been performed, and nomograms relating directivity index to the beamwidth exist (see Bobber, 1970).

The directivity indices associated with the dolphin's beam patterns in Figs. 9.15b and 9.16b were estimated by Au and Moore (1984), using the geometry shown in Fig. 9.17a. We assumed that the overall three-dimensional beam pattern was a product of the beam pattern in the vertical and horizontal planes, namely,

$$\left(\frac{p(\theta,\phi)}{p_0}\right)^2 = \left(\frac{p(\theta)}{p_0}\frac{p(\phi)}{p_0}\right). \tag{9.9}$$

The expression of Eq. (9.8) was then evaluated numerically using a digital computer and a two-dimensional Simpson's 1/3 rule algorithm (McCormick and Salvadori, 1964). The results of the numerical evaluation of Eq. (9.8) using the smoothed beam patterns of Figs. 9.15b and 9.16b are plotted as a function of frequency in Fig. 9.18. DIs of 9.4, 15.3, and 20.6 dB were obtained for frequencies of 30, 60, and 120 kHz, respectively. A linear curve fitted to the computed DIs in a least-square-error manner is also shown in the figure The equation of the line is

$$\mathrm{DI(dB)} = 16.9 \log f(\mathrm{kHz}) - 14.5 \tag{9.10}$$

The results of Fig. 9.18 indicate that the dolphin's receive directivity index increased with frequency in a manner similar to that of a linear transducer (Bobber, 1970). The expression in Eq. (9.10) is only valid for frequencies at which DI(dB) is greater or equal to 0.

Although the directivity index expressed by Eq. (9.10) is for a *Tursiops*, it can also be used to estimate the directivity index of other dolphins by applying an appropriate correction factor, so that Eq. (9.10) can be rewritten as

$$\mathrm{DI(dB)} = 16.9 \log f(\mathrm{kHz}) - 14.5 + \mathrm{CF(dB)}, \tag{9.11}$$

where CF(dB) is a correction factor taking into account different head sizes. The directivity index of a planar circular plate is proportional to its diameter, so if we let d_T be the diameter of the head of a *Tursiops* at the location of the blowhole and d_D be the diameter of the head of a different species of dolphin, then the correction factor can be expressed as

$$CF(dB) = 20 \log(d_D/d_T). \tag{9.12}$$

The correction factor will be positive for a dolphin with a larger head and negative for a dolphin with a smaller head than *Tursiops*.

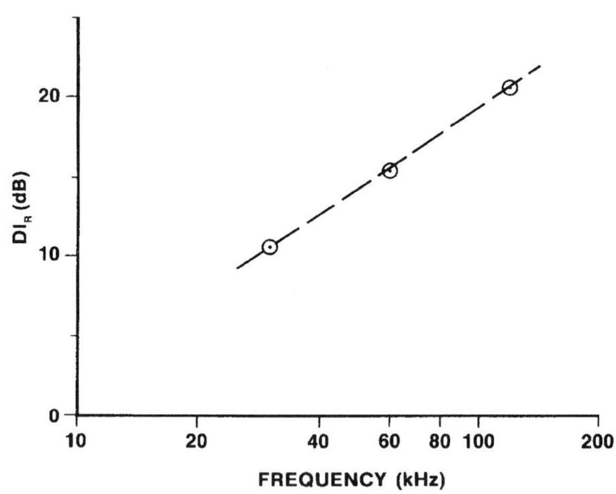

FIGURE 9.18. Receiving directivity index as a function of frequency for a *Tursiops truncatus*.

9.3.3 Sound Localization

The ability to localize or determine the position of a sound source is important in a sonar in order to resolve echoes from closely spaced targets or different portions of an extended target and to determine the relative position of targets within the sonar beam. The capability to localize sounds has been studied extensively in humans and in many vertebrates (see Fay, 1988). Lord Rayleigh, in 1907, proposed that humans localized in the horizontal plane by using interaural time differences for low-frequency sounds and interaural intensity differences for high-frequency sounds. Consider the geometry of Fig. 9.19a with a sound approaching a dolphin at an angle *N*. If the frequency of the sound is high enough so the wavelength is smaller than the dimension of the dolphin's head, the sound reaching the furthest ear will be shadowed by the head and will have different intensity in each ear. Lower frequency sounds with longer wavelengths will merely diffract around the head with minimum loss, so the interaural intensity difference cue will be minimal and interaural time difference (or phase difference) cues must be used for localization. The distance between the pan bone areas on either side of the mandible is approximately 12 cm, the wavelength of an 11 kHz signal in water.

The sound localization acuity of a subject is normally defined in terms of a minimum audible angle (MAA). It is defined as the angle subtended at the subject by two sound sources, one being at a reference azimuth at which the

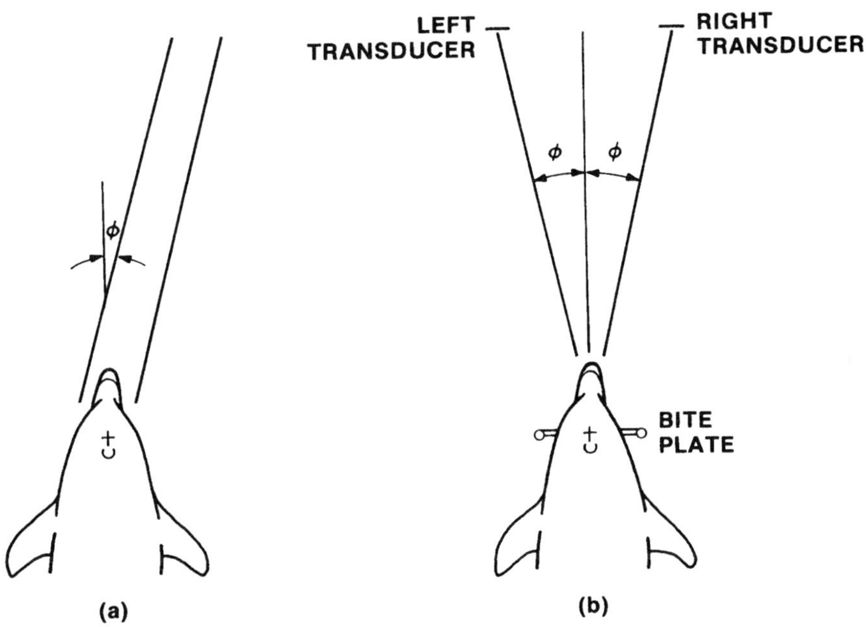

FIGURE 9.19. (**a**) Geometry for sound localization of a distant source. (**b**) Geometry for the minimum audible angle measurement in the horizontal plane.

subject can just discriminate the sound sources as two discrete sources (Mills, 1958). If the sound sources are separated symmetrically about a midline, the MAA is one half the threshold angular separation between the sound sources. If one of the sound sources is located at the midline, then the MAA is the same as the threshold angular separation between the two sources.

Renaud and Popper (1978) examined the sound localization capabilities of a *Tursiops truncatus* by measuring the MAA in both the horizontal and the vertical planes. During a test trial, the dolphin was required to station on a bite plate facing two transducers positioned at equal angles from an imaginary line running through the center of the bite plate (Fig. 9.19b). An acoustic signal was then transmitted from one of the transducers and the dolphin was required to swim and touch the paddle on the same side as the emitting transducer. The angle between the transducers was varied in a modified staircase fashion. If the dolphin was correct for two consecutive trials, the transducers were moved an incremental distance closer together, decreasing the angle between the transducer by 0.5°. After each incorrect trial, the transducers were moved an incremental distance apart, increasing the angle between the transducers by 0.5°. This modified staircase procedure allowed threshold determination at the 70% level as discussed in section 8.1 the Psychophysical Procedures. The threshold angle was determined by averaging a minimum of 16 reversals. A randomized schedule for sound presentation through the right or left transducer was used.

The localization threshold determined in the horizontal plane as a function of frequency for narrow band pure tone signals is shown in Fig. 9.20. The MAA had a U-shaped pattern, with a large value of 3.6° at 6 kHz, decreasing to a minimum of 2.1° at 20 kHz and then slowly increasing in an irregular fashion to 3.8° at 100 kHz. MAA for humans vary between 1° and 3.7°, with

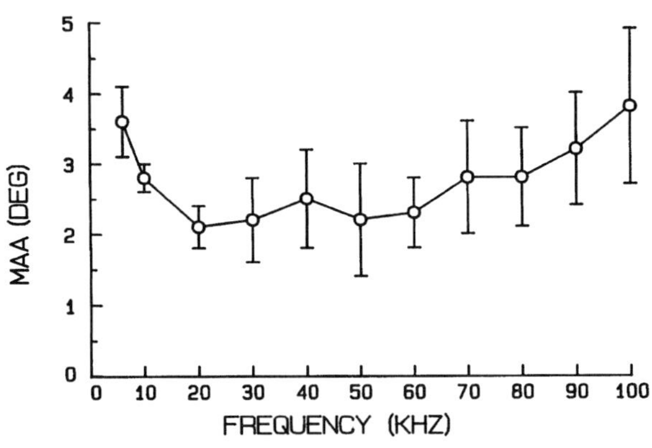

FIGURE 9.20. Localization threshold determined in the horizontal plane as a function of frequency. Shown are the mean ± one standard deviation for seven determination per frequency. The animal faced directly ahead at 0° azimuth (adapted from Renaud and Popper, 1978).

the minimum at a frequency of approximately 700 Hz (Mills, 1972). The region where the MAA decreased to a minimum in Fig. 19.20 (about 20 kHz) may be close to the frequency at which the dolphin switches from using interaural time difference cues to interaural intensity difference cues. The MAAs in the horizontal plane for the bottlenose dolphin were smaller than the MAAs (in the horizontal plane) of 3.5°°at 3.5 kHz and 6° at 6 kHz for a harbor porpoise measured by Dudok van Heel (1962), and 3° at 2 kHz measured by Andersen (1970b) also for a harbor porpoise.

In order to measure the MAA in the vertical plane, the dolphin was trained to turn to its side (rotate its body 90° along its longitudinal axis) and to bite on a sheet of plexiglass, which was used as the stationing device. The experimental geometry was similar to Fig. 9.20b, but with the dolphin turned to its side. The results of the MAA measurements in the vertical plane as a function of frequency are shown in Fig. 9.21. The MAA in the vertical plane varied between 2.3° at 20 kHz and 3.5° at 100 kHz. These results indicate that the animal could localize in the vertical plane nearly as well as in the horizontal plane. These results were not expected since binaural affects either interaural time or intensity should not be present in the vertical plane. However, the dolphin's ability to localize sounds in the vertical plane may be explained in part by the asymmetry of the acoustic pathways in the vertical plane of the dolphin's head as shown in Fig. 9.20. Sounds can enter the dolphin's head along the bottom midline of the lower jaw. Both intensity and time difference cues may be present for such unsymmetrical sound pathways. The asymmetry in the vertical receiving beam patterns of Fig. 9.15 can also contribute to the dolphin's ability to localize in the vertical plane. If the dolphin was able to rock its jaws slightly in the vertical plane, it would hear larger intensity variations for a sound approaching from a direction above the jaw than if the sound approached from below the jaw.

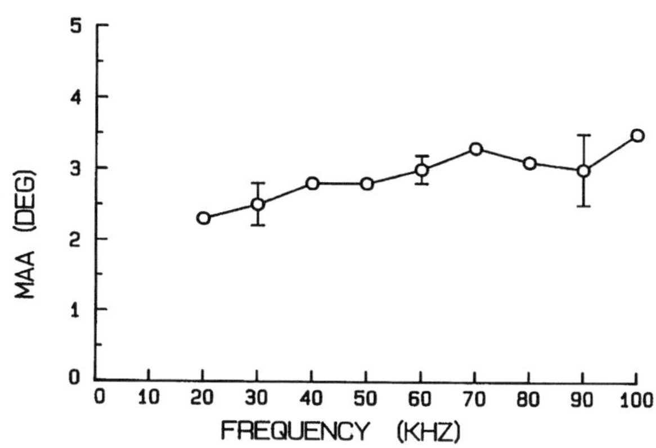

FIGURE 9.21. Localization threshold determined in the vertical plane as a function of frequency. Standard deviations are indicated for 30, 60, and 90 kHz vertical data (seven sessions each). The dolphin's azimuth was 0° (adapted from Renaud and Popper, 1978).

Renaud and Popper (1978) also determined the MAA for a broadband transient signal or click signal, having a peak frequency of 64 kHz and presented to the dolphin at a repetition rate of 333 Hz. The MAA in the horizontal plane was found to be approximately 0.9° and 0.7° in the vertical plane. It is not surprising that a broadband signal should result in a lower MAA than a pure tone signal of the same frequency as the peak frequency of the click. The short onset time and the broad frequency spectrum of a click signal should provide additional cues for localization.

9.4 Hearing in Pinnipeds

Pinnipeds, being amphibious animals, must hear in air and underwater. Therefore, it is important to understand how to compare the audiograms obtained in air with those obtained underwater. In Section 9.1.1, we stated that the only meaningful way thresholds in air and water can be compared is on the basis of intensity (which is directly related to energy and power) because the mammalian auditory system is basically an energy detector as is evident from temporal summation studies (see Section 9.1.1). Therefore, the important parameter is the amount of acoustic energy or power that is being delivered to the auditory system and how that acoustic energy is translated to pressure. Equation (9.1) gives the relationship between the intensity and pressure for a situation in which the sound field is not affected by much boundary reflections and is propagating outward from a source. This is the simplest situation; however, in some tanks where there is a considerable amount of reflection and the sound field at the position of the subject is affected by these reflections, the intensity would have to be measured rather than calculated using Eq. (9.1).

The intensity scale can be expressed in watts/m^2 as was done for Fig. 9.1 or in dB units where the reference is either watts/m^2 or μwatts/m^2. Another factor that must be considered is the how sound pressure levels are expressed for airborne and underwater sounds. Airborne sounds are usually expressed in term of dB re 20 μPa and has the notation dB SPL. The reference for underwater sound is 1 μPa. The acoustic impedance, ρc, in Eq. (9.1) is equal to about 416 Pa s/m for air and about 1.5×10^6 Pa s/m for salt water. If we let SPL be the underwater sound pressure level in dB and dB SPL be the in-air sound pressure level, then the acoustic intensity in dB can be expressed as

$$I_{\mathrm{DB}} = \mathrm{SPL} - 121.7[\mathrm{dB\ re\ 1\ \mu watts/m^2}] \tag{9.13}$$

for underwater sound. The intensity in dB for airborne sound can be expressed as

$$I_{\mathrm{DB}} = \mathrm{dB\ SPL} - 60[\mathrm{dBre1\mu watts/m^2}]. \tag{9.14}$$

Both expressions were derived from the relationship in Eq. (9.1).

9.4.1 Hearing Sensitivity

Mohl (1968) was to first measure the audiogram of a pinniped. He measured the audiogram of a common or harbor seal (*Phoca vitulina*) both underwater and in air. For the underwater measurement, the sound transducer was 1.8 m from the seal's station, which consisted of a rod (key) connected to a switch. Both the transducer and the stationing key were at a depth of 0.8 m. The seal was required to touch the key that controlled the stimulus. For the in-air measurement, a loudspeaker was mounted on one edge of a floating platform and the seal's station was on the opposite edge, 1 m from the loudspeaker. The seal was trained to stick its head out of the water and touch a key connected to a switch, which also controlled the stimulus. On a signal-present trial, the stimulus would turn on and remain on as long as the seal was in contact with the key. Therefore, in both the underwater and in-air experiment, the seal controlled stimulus onset as well as stimulus duration, which averaged about half a second. If the stimulus was present, the seal was trained to touch a specific key; and if the stimulus was absent, it was trained to touch another key. For every correct response, the seal was rewarded with small fish or pieces of fish. For incorrect responses, the seal was "punished" with a blast of air in its face from nozzles mounted just behind the response keys. A method of constant stimulus was used with an equal number of stimulus-present and absent trials in each sessions.

The seal's underwater audiogram was measured first and the results are shown in Fig. 9.22 along with the in-air human audiogram from Sivian and White (1933). The seal's audiogram had a similar shape as dolphin audiograms shown in Fig. 9.2, with a slowly increasing hearing sensitivity as frequency increased until a maximum sensitivity was reached, after which the sensitivity decreased rapidly with frequency. The harbor seal had a

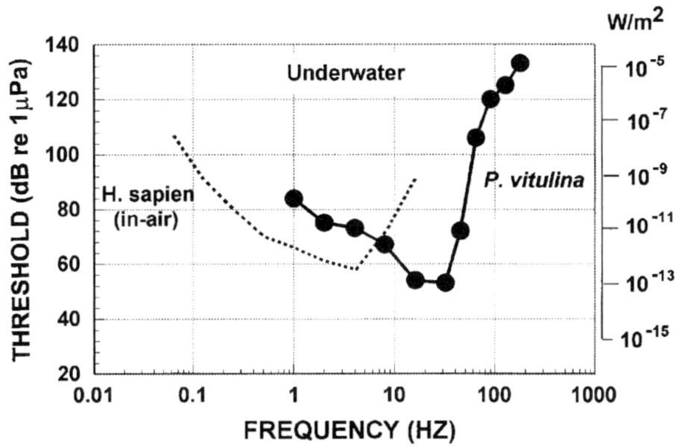

FIGURE 9.22. Underwater audiogram of a harbor seal (after Mohl, 1968), and a human in air plotted against the right ordinate in Watts/m^2 (adapted from Sivian and White, 1933).

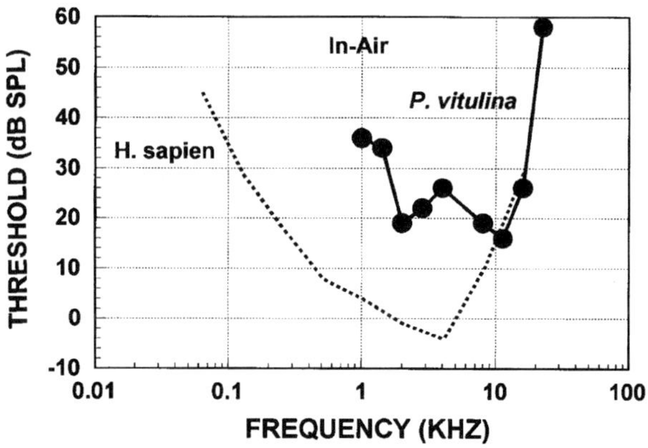

FIGURE 9.23. In-air audiogram of a harbor seal (after Mohl, 1968), and a human (adapted from Sivian and White, 1933).

maximum sensitivity at about 32 kHz and could hear out to 180 kHz, although its sensitivity decreased rapidly for frequencies beyond about 50 kHz. The best sensitivity also matches the best sensitivity of the human audiogram.

The harbor seal in-air audiogram is shown in Fig. 9.23 along with the human audiogram of Sivian and White (1933). The seal did not hear as well in-air when compared to humans. The best in-air sensitivity was about 24 dB less sensitive than for humans. The upper limit of in-air hearing for the harbor seal was similar to that of a human.

The in-air and underwater audiograms of the harbor seal can be compared to each other by converting the sound pressure level threshold values into intensity using Eq. (9.1). The results of converting the sound pressure thresholds into intensity are shown in Fig. 9.24. Both audiograms are similar in shape

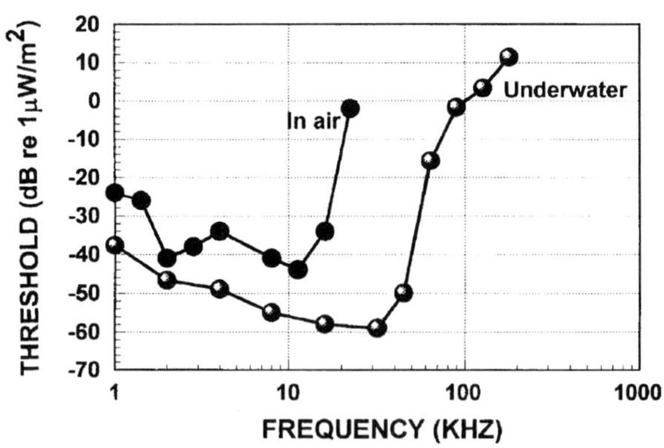

FIGURE 9.24. Sound intensity at the threshold of the harbor seal (adapted from Mohl, 1968).

with a slope of about 5 dB/octave at the lowest frequencies. At frequencies above the area of maximum sensitivity, the thresholds increase at about 65 dB/octave; and between 64 and 90 kHz, the underwater audiogram leveled off and the threshold increased at a rate of 12 dB/octave up to 180 kHz (Mφhl, 1968).

The results of Fig. 9.24 clearly show that the harbor seal could hear much better under water than in air. The underwater sensitivity was better and the upper frequency limit was about 10 times higher under water than in air. At the lower frequencies, below the area of maximum sensitivity of the in-air audiogram, the seal's sensitivity underwater was about 15 dB better than in air. The conspicuous dip in the in-air audiogram at 2 kHz may have been caused by a quarter-wave resonance in the long and narrow meatus of the seal (Mohl, 1968).

Since the experiment of Mohl (1968), the audiograms of several other pinnipeds have been measured. The underwater audiogram of a California sea lion (*Zalophus californianus*) was determined by Schusterman et al. (1972). Terhune and Ronald (1972) measured the underwater hearing sensitivity of a harp seal (*Pagophilus groenlandicus*) and a ringed seal (*Pusa hispida*) (Terhune and Ronald, 1975). Schusterman and Moore (1978) measured the underwater audiogram of a Northern fur seal (*Callorhinus ursinus*). The hearing sensitivity data for these pinniped studies have been tabulated by Fay (1988) in his psychophysics databook. The underwater hearing sensitivity of the Hawaiian monk seal (*Monachus schauinslandi*) was determined by Thomas et al. (1988). Finally, Kastak (1996) measured both the underwater and in-air sensitivity of a California sea lion, a harbor seal, and an elephant seal. The underwater hearing sensitivity of various pinniped species are shown in Fig. 9.25. The lightly marked curves on the left at low frequency

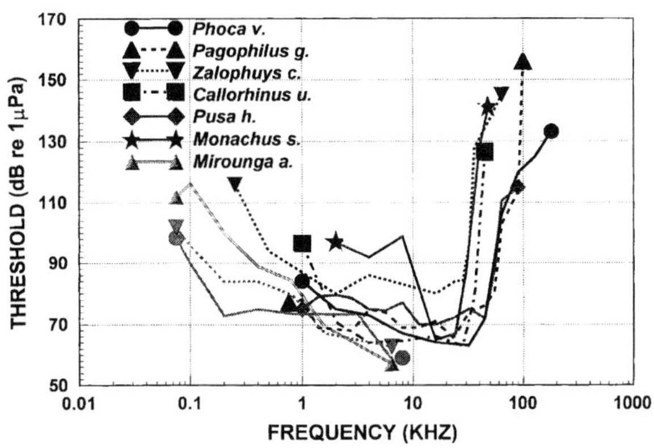

FIGURE 9.25. Underwater hearing sensitivity of different pinniped species. Data taken from Mohl (1968) and Kastak (1996) for *Phoco vitulina.*, Schusterman et al. (1972), and Kastak (1996) for *Zalophus californianus*, Terhune and Ronald (1972) for *Pagophilus groenlandicus*, Terhune and Ronald (1975) for *Pusa hispida*, Schusterman and Moore (1978) *Callorhinus ursinus*, Thomas et al. (1988) for *Monachus schauinslandi* and Kastak (1996) for *Mirounga angustirostris*.

came from the study of Kastak (1996). Two *Zalophus* were used by Kastak; only the data from the more sensitive animals are shown.

The audiograms depicted in Fig. 9.25 are similar to typical mammalian audiograms. There is a low-frequency region that increases in sensitivity with frequency at a rate of approximately 10 dB per octave, a high-sensitivity dip at mid frequencies, and a high-frequency region in which the sensitivity decreases rapidly with frequency. The low-frequency sensitivity curves of Kastak have some overlapping frequencies with other measurements. Kastak's data for *Zalophus* overlapped in frequency with the data of Schusterman et al. (1972), and seem to suggest that the animal used by Schusterman et al. (1972) may have suffered from some sort of hearing loss since its sensitivity was on the order of 10–20 dB worse than the animal used by Kastak. The data of Kastak for the harbor seal were obtained at frequencies that overlapped the frequencies used by Mohl (1968). At some frequencies, the data from both animals were very similar and at other frequencies there were about a 10-dB difference.

A summary of some important properties of the different audiograms shown in Fig. 9.25 is given in Table 9.2. As with odontocetes in Table 9.1, the frequency of best hearing is arbitrarily defined as the frequency region in which the auditory sensitivity is within 10 dB of the maximum sensitivity depicted in each audiogram of Fig. 9.25. The data shown in Fig. 9.25 and Table 9.2 indicate that with the exception of *Zalophus*, the maximum underwater hearing sensitivity of the different species is very similar, differing at most by only 4–7 dB. The data also indicate that pinnipeds do not hear quite as well as odontocetes and their upper frequency limits are generally less than 100 kHz.

In-air audiograms have been determined for harbor seal (Mohl, 1968), the harp seal (Terhune and Ronald, 1971), the California sea lion (Schusterman,

TABLE 9.2 Some Important Properties of the Audiograms Plotted in Figs. 9.25 and 9.26

	Underwater			In-air		
Species	Max sens dB (re 1 μPa)	Freq-best hearing (kHz)	Upper freq (kHz)	Max sens dB (re 1 μPa)	Freq-best hearing (kHz)	Upper freq (kHz)
Harbor seal *Phoco vitulina*	63	2–46	180	16	1.0–16	23
Harp seal *Pagophilus groenlandicus*	66	1–45	100	29	1.5–15	32
California sea lion *Zalophus californianus*	57		64	31	4.0–27	32
Northern fur seal *Callorhinus ursinus*	63	2–33	45	7	1.8–20	32
Ringed seal *Pusa hispida*	70	2–50	90			
Hawaiian monk seal *Monachus schauinslandi*	65	12–27	46			

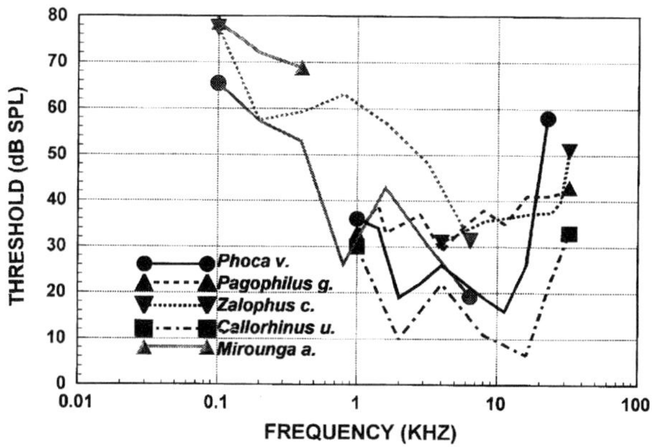

FIGURE 9.26. In-air audiogram of different pinniped species. Data taken from Mohl (1968) and Kastak (1996) for *Phoca virtulina*, Terhune and Ronald (1971) for *Pagophilus groenlandicus*, Schusterman (1974) and Kastak (1996) for *Zalophus californianus*, Schusterman and Moore (1980) for *Callorhinus ursinus*, and Kastak (1996) for *Mirounga angustirostris*.

1974), and the northern fur seal (Schusterman and Moore, 1980). These audiograms are depicted in Fig. 9.26, along with the low-frequency sensitivity threshold measured by Kastak (1996). One interesting feature of Fig. 9.26 is the hump in the audiogram of all the pinnipeds but *Zalophus*. The data of Kastak (1996) and Mohl (1968) for *Phoca* show the hump at slightly different frequencies; however, we should keep in mind that Kastak used a head set on his subjects, and Mohl (1968) had a speaker that was approximately 1 m from his subject. Mohl (1968) attributed the hump to resonance that occurred within the meatus of the harbor seal. It is possible that by using a headset, the resonance frequency could be affected. The in-air hearing of the *Zalophus* used by Schusterman (1974) did not exhibit a reduced sensitivity as it did for underwater hearing.

A summary of some important properties of the different audiograms shown in Fig. 9.26 is given in Table 9.2. The data clearly indicate that the northern fur seal has the most sensitive in-air hearing of the four pinnipeds, and can also hear to frequencies up to 32 kHz. The *Zalophus* and the *Pagophilus* did not exhibit good hearing sensitivity in air. The upper limit of hearing denoted in Table 9.2 is not very accurate, but merely reflects the highest frequency at which the animals' sensitivity were measured. The underwater audiogram indicates that the northern fur seal is as sensitive to sounds as the harbor seal. The northern fur seal in-air hearing sensitivity is more acute than that of the harbor seal and the harp seal. These threshold values do not support the assumption of Schusterman et al. (1972) that Phocidae generally have better hearing abilities than Otariidae. We shall see in the next sections that the northern fur seal also has a lower critical ratio than the ringed and harp seals.

9.4.1.1 Hearing at Depth

The role of the middle ear of pinnipeds in hearing not only underwater but at depths that pinnipeds typically dive to is not very clear. Pinnipeds need some kind of middle ear compensation when hearing in air, but may not require any middle ear compensation when hearing underwater. Therefore, the ears of pinnipeds may be more complex than either the ears of completely terrestrial or completely aquatic animals. In Section 7.3.2 we pointed out the presence of a thickened mucosa on both sides of the tympanic membrane that can be engorged with blood when an animal is submerged, allowing sounds transmission along a pathway identical to that of terrestrial mammals (Repenning, 1972). In an attempt to resolve the role of the middle ear at depth, Kastak and Schusterman (2002) conducted a hearing experiment with a California sea lion at depths of 10, 50, and 100 m. They used a stationing apparatus with a bite plate that could be lowered to any desired depth. The stationing platform had a projecting hydrophone directly forward of the bite plate along with a video camera, a response paddle, trial light and a receiving hydrophone, similar to the stationing platform used by Ridgway et al. (2001) for testing the hearing of two beluga whales at depth. As in the Ridgway et al. (2001) study, Kastak and Schusterman (2002) conducted multiple go/no-go trials during a single dive procedure. The sound levels were adjusted according to a up-down staircase schedule.

The results of the hearing sensitivity measurements at depths of 10 and 50 m are shown in Fig. 9.27, along with the results of Schusterman et al. (1972) obtained in a small pool with the animal at a depth of 1 m. The data obtained at 10 m closely matched the results obtained at the surface by Schusterman et al. (1972); however, the threshold at 50 m was significantly

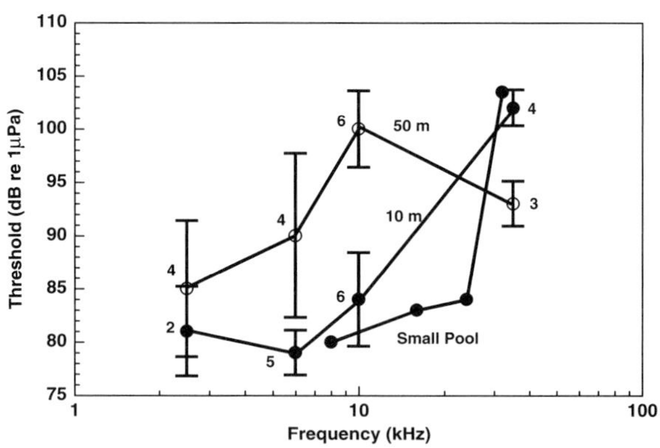

FIGURE 9.27. The mean and standard deviation of a California sea lion audiogram at 10 and 50 m depth. The numbers of sessions for each data point is shown next to the mean values. The small pool data are from Schusterman et al. (1972) (adapted from Kastak and Schusterman, 2002).

lower than that at the surface, except at 35 kHz where there was a cross-over. These results clearly suggest that the middle ear of *Zalophus californianus* is functional underwater and at depth. The same could be said for all seal lions since the anatomy of the ear is very similar in all species. Kastak and Schusterman (2002) argued for a similar study be done with the true seals or phocids because they have larger middle-ear spaces, more extensive cavenous tissues, and are deeper divers.

9.4.2 *Spectral Analysis Sensitivity*

9.4.2.1 Critical Ratio

Underwater masked hearing threshold experiments have been conducted with the harp seal by Terhune and Ronald (1971), the ringed seal by Terhune and Ronald (1975), the northern fur seal by Moore and Schusterman (1987). Southall et al. (2000) measured the critical ratio at low frequencies for three species of pinnipeds, the northern elephant seal, harbor seal, and a California sea lion. The critical ratios of various pinnipeds are plotted in Fig. 9.28. The data suggest that the northern fur seal may have the narrowest critical ratio in marine mammals. The critical ratio of the fur seal is smaller than that of the bottlenose dolphin over its entire range of hearing sensitivities and of the six

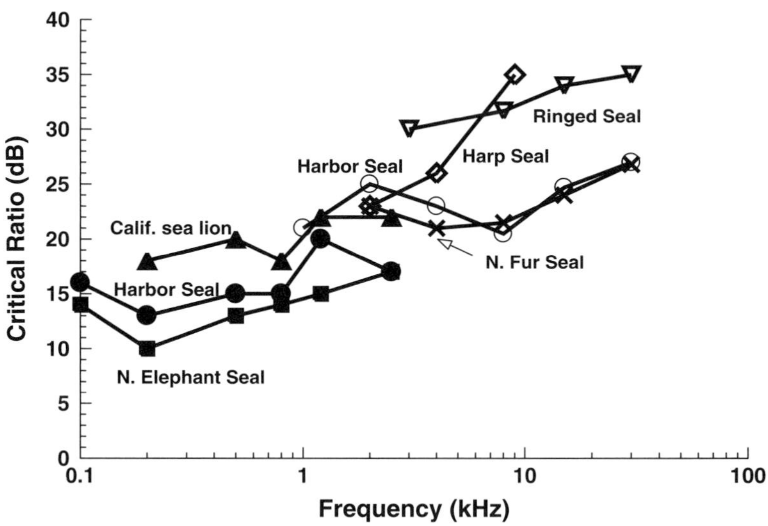

FIGURE 9.28. Critical ratio for a six pinnipeds and three other mammals. The *solid symbols* are from Southall et al. (2000). The other data were taken from Terhune and Ronald (1971) and Turnbull and Terhune (1990) for the harbor seal, Terhune and Ronald (1971) for the harp sea, Terhune and Ronald (1975) for ringed seal, and Moore and Schusterman (1987) for the northern fur seal (adapted from Southall et al., 2000).

pinnipeds species depicted in Fig. 9.25. The results also indicate that the ringed seal has the widest filter bandwidth of the six pinnipeds. Although the hearing sensitivities are relatively similar, the masked threshold sensitivities are different. This suggests that there are wide variations in the filter bandwidths within the auditory system of different pinnipeds and how well different species can handle masking noise and discriminate different tones will also vary widely. One note of caution should be interjected here in comparing the results between species. There is the issue of individual variations within a species, which could be the major reason for the differences found in Fig. 9.28. Unfortunately, because of high expenses involved with marine mammal research, most auditory research tends to be one-subject experiments.

The overall shape of the curves in Fig. 9.28 is consistent with the critical ratio results obtained with other mammals including humans. There is a general trend in the critical ratio (and the filter width) increasing with frequency, displaying a constant-Q quality over much of the frequency range. The results for the harbor seal results from Southall et al. (2000) seem somewhat consistent with the results of Turnbull and Terhune (1990) at around 1 kHz, but the two curves diverge at about 1.5 kHz. Again, natural variations between individuals may be the major factor in the differences.

Southall et al. (2003) expanded their investigation by measuring the critical ratio of the northern elephant seal, the harbor seal, and the California sea lion in air. The measurements were conducted in a hemianechoic chamber (Eckel Industries) mounted on a semi-isolated cement foundation. The chamber measured $3.3 \times 2.3 \times 2.2$ m in dimensions, with an experimental control room immediately adjacent to the test area. The chamber with its double-paneled, 8-gauge stainless steel outer walls provided 40–60 dB of noise attenuation of external noise. The in-air critical ration results of Southall et al. (2003) are shown in Fig. 9.29, along with the underwater critical ratio from their earlier study (Southall et al., 2000).

The in-air and underwater results shown in Fig. 9.29 are very similar, which they should be. The auditory filtering process in a mammal begins at the level of the basilar membrane and progresses into the higher auditory centers (see Section 7.1.2). Therefore, once sound enters into the cochlear, the filtering process is the same whether or not the ear is in air or underwater. Ambient pressure should have negligible effects on the basilar membrane. For the results shown in Fig. 9.29, the same animals were used in both measurements, with at least 2 years between the two experiments. There is some spread to the data for each subject that could be attributed to temporal changes in the auditory system of the subject or even to difference in the methodology. The underwater measurements were done in a small pool with the animals close to the surface (1 m), so that reflections from the water surface and the walls of the pool would make the sound pressure level fluctuate in time because of movement of the water surface and slight difference in the animal's position with respect to the transducer both within a trial and between trials.

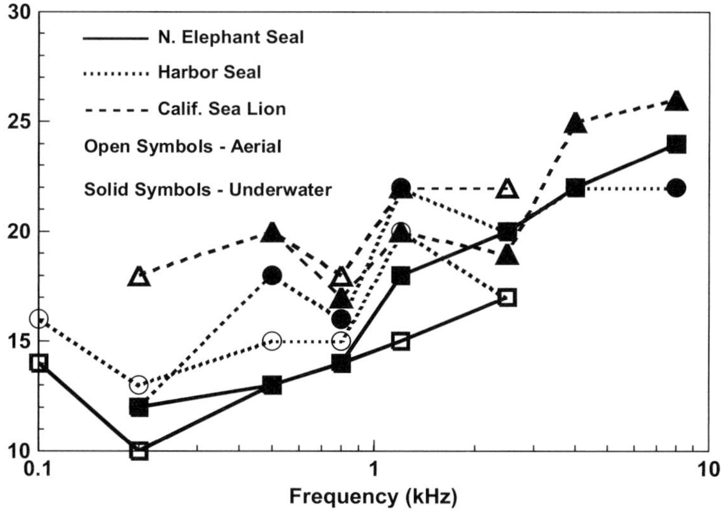

FIGURE 9.29. Critical ratio vs. frequency of three species of pinniped measure in air (Southall et al., 2003) and underwater (Southall et al., 2000).

9.4.2.2 Critical Bandwidth

Southall et al. (2003) also measured the critical bandwidth of three pinnipeds by varying the band of the noise in an analogous manner, as Au and Moore (1990) with the bottlenose dolphin, discussed earlier in this chapter. The technique of varying the bandwidth of the noise in order to measure the auditory filter width at a particular frequency is usually referred to as the direct method in contract to the indirect method of measuring the critical ratio using broadband noise of a fixed width. They performed the critical bandwidth experiment in air using the same chamber used to measure the critical ratio. The results for the northern elephant seal, the harbor seal and the California sea lion are shown in Fig. 9.30, along with the critical ratio of each species measured in air and underwater.

The critical bandwidth technique usually results in a wider estimate of the auditory filter width than the critical ratio technique in most mammals, and the results of Southall et al. (2003) were consistent with this general pattern. The critical bandwidth was on the average 5 times greater than the critical ratio for the northern elephant seal, 9.4 times greater for the harbor seal, and 5.2 times greater for the California sea lion. In contrast, the critical bandwidth of a bottlenose dolphin measured by Au and Moore (1990) was between 2.6 and 16 times wider than the critical ratio results. The results of Southall et al. (2003) suggest that pinnipeds may be more efficient in detecting a tonal signal in noise than humans whose critical bandwidth is typically 2.5 times greater than critical ratios (Scharf, 1970).

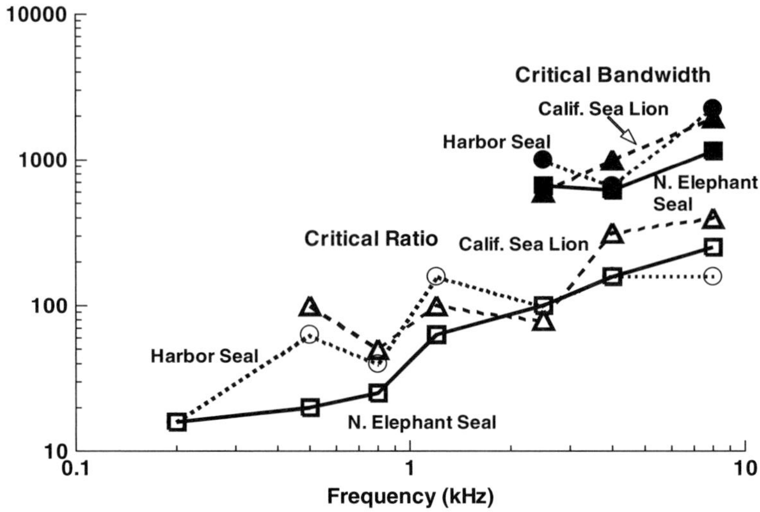

FIGURE 9.30. Critical bandwidth of three species of pinniped measured in air using the band widening technique and critical ratio of the same species measured both in air and underwater (adapted from Southall et al., 2003).

9.4.2.3 Frequency Discrimination

The frequency discrimination capabilities of a harbor seal have been determined by Mohl (1967). The seal was presented with a sound that was either a constant frequency or one in which the frequency alternated between f and $f + \Delta f$ at a rate of once every other second. The seal's task was to touch a specific key if the signal had a constant frequency and another specific key if the signal changed frequencies. Schusterman and Moore (1978) measured the frequency discrimination capabilities of a California sea lion, and Terhune and Ronald (1976) studied the ringed seal. In Schusterman and Moore's experiment, the sea lion was presented with a standard tone of 1-s duration followed by a second tone of 1-s duration that was either the same or lower in frequency than the standard tone. The sea lion was trained to bark when the two tones had a different frequency. In the experiment of Terhune and Ronald, the ringed seal had to discriminate between a pure tone pulse and a tone pulse that was frequency modulated at a rate of 2.5 Hz. The data from these three studies are depicted in Fig. 9.31, along with the data for *Tursiops truncatus* obtained by Thompson and Herman (1975).

The data in Fig. 9.31 indicate that the harbor seal could discriminate frequency differences the best of the three pinnipeds tested. The California sea lion's frequency discrimination accuracy was better than that of the ringed seal at frequencies below 30 kHz at which point there was a cross-over in accuracy and the ringed seal became better than the sea lion for frequencies greater than 30 kHz. For frequencies between 4 and about

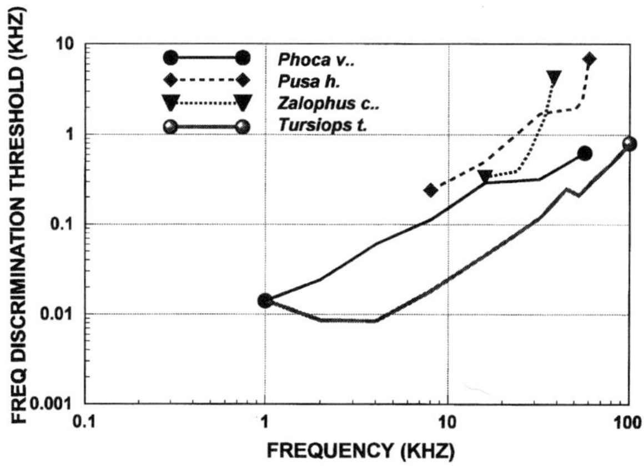

FIGURE 9.31. Frequency discrimination capabilities of three different pinniped species and a bottlenose dolphin. The pinniped data were taken from Mohl (1967) for *Phoca vitulina*, Schusterman and Moore (1978) for *Zalophus californianus*, and Terhune and Ronald (1976) for the *Pusa hispida*. The *Tursiops* data were taken from Thompson and Herman (1975).

30 kHz, the bottlenose dolphin could discriminate frequency differences that were between 1/3 and 1/7 of the frequency differences discriminated by harbor seal.

9.4.3 *Intensity Discrimination*

The intensity difference discrimination capabilities of a California sea lion was determined by Moore and Schusterman (1976). The sea lion was required to place his head on a head stand that positioned him 3.1 m directly in front of the transducer at a depth of about 0.48 m. The animal was trained to identify a 16-kHz tone as the louder or softer of a stimulus pair. The data was analyzed in terms of signal detection theory and d' values. They found that a d' of 1 corresponded to an intensity difference of approximately 3 dB. Unfortunately, this is the only intensity difference data for pinnipeds.

9.4.4 *Sound Localization*

The sound localization capabilities of a wide range of terrestrial mammals have been studied (see Fay, 1978); however, sound localization has been studied for only four species of marine mammals: two pinnipeds and two dolphin species. Mohl (1964) determined an MAA of 1.5° at the 75% correct response level for a 2 kHz pure-tone burst signal. Gentry (1967) measured MAAs of 15° (at the 63% correct response level) for a 3.5 kHz and 10° for a 6.5 kHz pure tone burst signal using a California sea lion. Terhune (1974) measured the sound localization capabilities of a harbor seal in air and underwater using a click train, noise bursts, and a 10-kHz puretone pulse. The harbor seal did best with the click train, with a MAA of 9° underwater and 3° in air. It is not surprising that the MAA in air was 1/3 the MAA measured in water by Terhune (1974). The speed of sound in air is about 4.5 times slower than in water, and if the harbor seal was using time of arrival difference cue, then the difference in the time of arrival will be about 4.5 times greater for a given angular separation of the sound source in air. Moore (1975) used a click signal having a peak frequency of 9 kHz with a California sea lion and obtained an MAA of 9° at the 75% correct response level and 6° at the 63% level. Moore and Au (1975) measured the MAAs of a California sea lion over a frequency range from 2 to 16 kHz in one octave increments. The results of all the underwater MAA measurements are depicted in Fig. 9.32.

The results of Moore and Au (1975) experiment with *Zalophus* indicated that the animal had difficulties localizing the sound source for frequencies between about 1.5 and 7.5 kHz. The results also indicated that the lower frequencies (0.5 and 1 kHz) were easier to localize than the higher frequencies (4, 8, and 16 kHz). Our data resemble human MAA data presented by Mills (1972) in which there is a mid frequency region where localization acuity is relatively poor. Mills (1972) suggested that the low-frequency region where localization is optimal, humans use interaural time difference or interaural phase difference cues. For the high-frequency region, Mills argued that interaural intensity difference cues were used by humans. The frequency region where performance is degraded corresponds to the region in which there is a transition from the use of time or phase cues to intensity cues.

Moore and Au (1975) performed additional experiments in order to gain a better understanding of how the sea lion used interaural time difference and interaural intensity difference cues. In the first experiment, one transducer

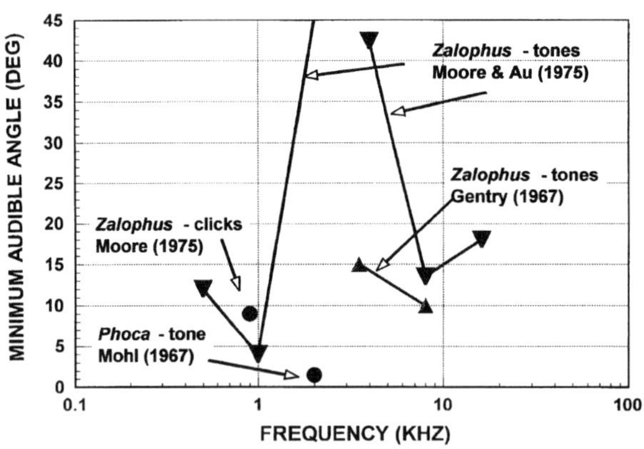

FIGURE 9.32. Minimum audible angle as a function of frequency for different studies.

was located directly ahead of the animal at 0° and the second transducer was kept at an angular separation of 20°. The *Zalophus* localization performance was determined as a function of frequency. The results of this experiment are shown in Fig. 9.33, and basically confirmed that in the frequency region between about 1.5 and 7.5 kHz the sea lion's localization acuity was poor, suggesting that neither interaural time difference or intensity difference cues were available to the animal.

In Chapter 8, we discussed the likelihood that pinnipeds hear underwater mainly by some kind of bone or tissue conduction. The conventional wisdom has been that with bone or tissue conduction, the mammalian head structure is transparent to sound so that intensity difference cues would not be present (Reysenbach de Haan, 1957). In order to address this issue, Moore and Au (1975) took a skull of a sea lion and mounted two miniature LC-10 hydrophones in a manner that placed the acoustic centers against the bulla as shown in Fig. 9.34. We then mounted the skull in an rotatable apparatus with a J-9 transducer positioned in the midsagittal plane 61 cm directly ahead of the skull as shown in Fig. 9.34. The axis of rotation of the skull corresponded to the mid-point between the acoustic centers of the LC-10 hydrophones. It should be pointed out that any inferences concerning the method by which a live sea lion localizes underwater sound based solely on results of the skull measurements is totally unwarranted, since a submerged sea lion skull holds only slight resemblance to a living sea lion head covered with flesh and skin. However, if the trends observed through behavioral studies show close similarities with trends associated with the skull measurements, then speculations on localization mechanisms based on these similarities may have heuristic value.

The results of the skull measurement are shown in Fig. 9.35, with the intensity difference between the two hydrophones plotted against the angle of the skull for different frequencies. It is quite obvious from the results that

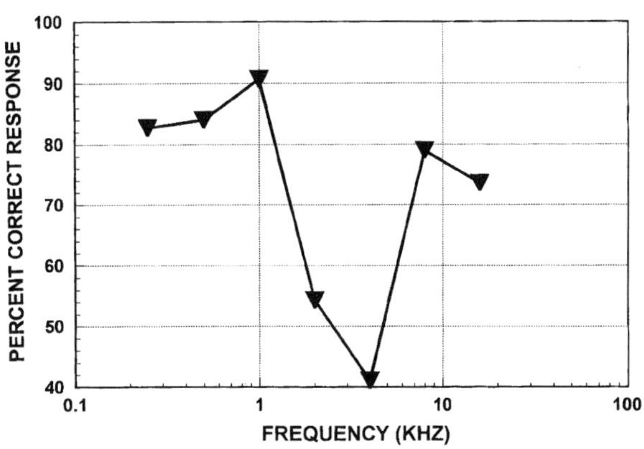

FIGURE 9.33. Localization performance vs. frequency for a transducer angle of 20°.

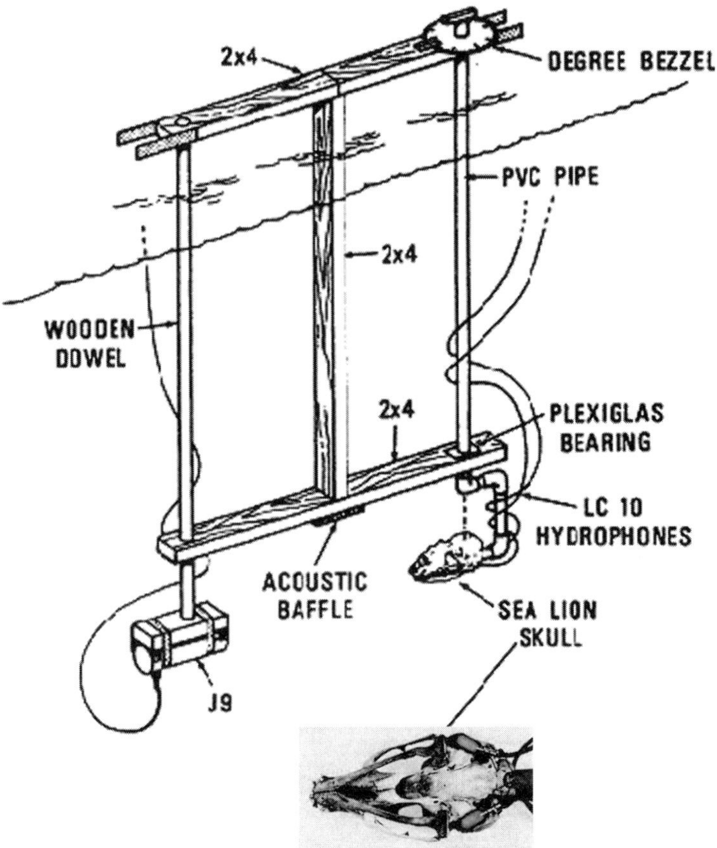

FIGURE 9.34. Schematic of mounting apparatus used for the skull measurement (adapted from Moore and Au, 1975).

the skull was not acoustically transparent and considerable intensity differences can exist between the bullas of a sea lion for frequencies greater than 2 kHz. The MAAs for 4, 8, and 16 kHz are depicted by the open circles in Fig. 9.35. It is interesting to note that the MAAs corresponded to similar intensity differences, between 2.7 and 3.4 dB. The data of Moore and Schusterman (1976) indicated that a *Zalophus* intensity difference discrimination threshold was approximately 3 dB, which matched the skull data extremely well. The skull data also showed that at an angle of 20°, the intensity difference was greater for the 8 kHz signal than for the 16 kHz signal. This matched the animal's performance data shown in Fig. 9.32.

One of the more interesting features of the data relates to the 8- and 16-kHz results. For angles below 30°, the 16-kHz signal had lower intensity difference than the 8-kHz signal. However, a cross-over point occurred at 28°,

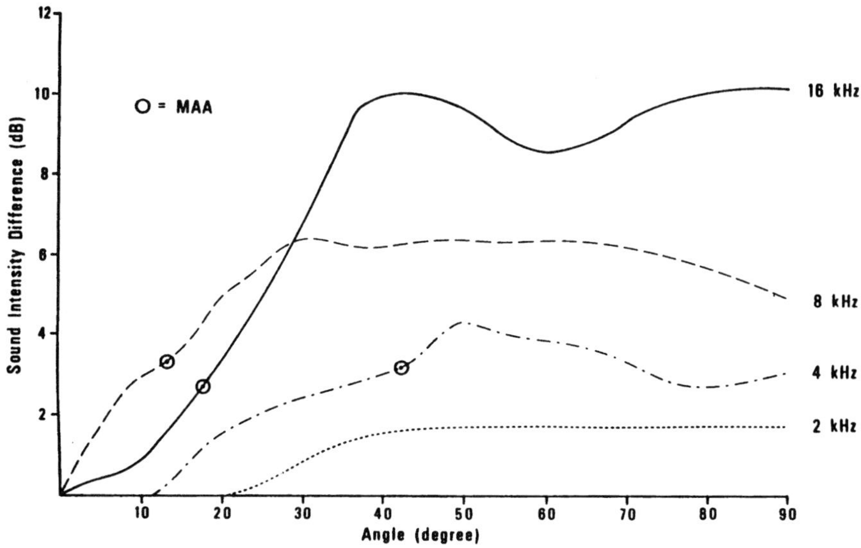

FIGURE 9.35. Sound intensity difference between skull-mounted hydrophone as a function of skull azimuthal angle for different frequencies. *Circles* indicate sea lion MAA values at the respective frequencies obtained from the animal behavioral data (adapted from Moore and Au, 1975).

after which the intensity difference for the 16-kHz signal became greater. This observation motivated further animal behavioral measurements at extended angles between 30 and 45° using an 8-kHz and 16-kHz signal. The results of this additional animal test are shown in Fig. 9.361 in which the localization

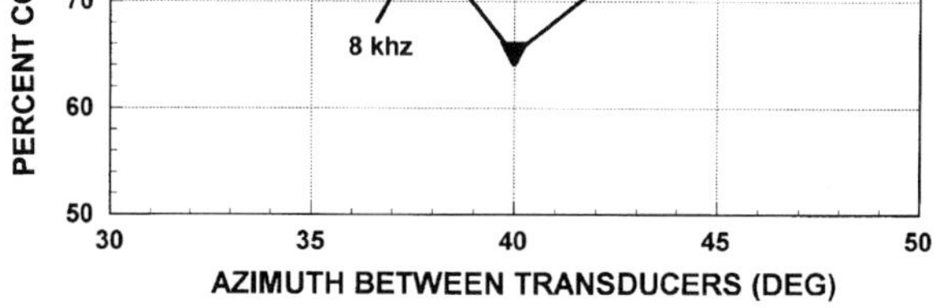

FIGURE 9.36. Localization performance as a function of transducer separation angle for 8 and 16 kHz.

performance was plotted against the angular separation of the transducers for 8- and 16-kHz signals. From the data, we can see that a cross-over in performance also occurred but at approximately 36.5° angular separation. The skull data of Fig. 9.32 indicated a possible cross-over at approximately 28°, only 9° from the observed behavioral cross-over.

Pinniped airborne calls are typically broadband so that from a pinniped's perspective, the ability of localize broadband sound is important. Therefore, Holt et al. (2004) examined the aerial localization of broadband noise by three species of pinnipeds: the northern elephant seal, harbor seal, and California sea lion. They used a broadband noise stimulus that is relatively flat between 2 and 15 kHz and conducted the experiment in the same hemianechoic chamber as Southall et al. (2003). The subjects were required to station in a chin cup and the stimulus was presented to speakers either to the right or left of the center line and the subjects were required to touch the paddle that was on the same side of the center line as the activated speaker. The results of Holt et al. (2004) are presented in a different format than the original manuscript in Fig. 9.37 using data provided by the investigators. A sixth-order polynomial curve fit of the data is shown by the solid line. Holt et al. (2004) reported MAA of 3.6° for the northern elephant seal, 4.2° for the harbor seal, and 4.2° for the California sea lion. The curves shown in Fig. 9.37 indicate MAA that are very similar to those reported by the investigators.

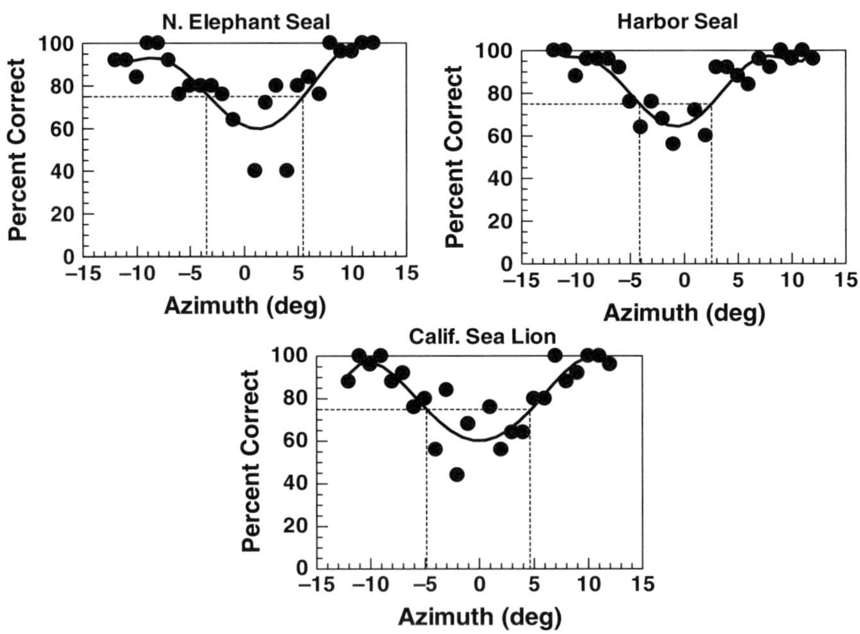

FIGURE 9.37. Results from the sound localization of Holt et al. (2004) using broadband noise as the stimuli.

The results suggest that head size alone does not account for small MAA differences between the three subjects, since the northern elephant seal had the largest head and the largest MAA. Head diameters of 30.8 cm were measured for the northern elephant seal compared to 16.0 cm for the harbor seal and 16.6 cm for the California sea lion. Heffner and Heffner (1992) have shown that sound localization capability is inversely proportional with the width of the visual field. The aerial results obtained by Holt et al. (2004) are much smaller than the underwater results obtained by Gentry (1967) and Moore and Au (1975) for frequencies greater than about 2 kHz. One could conclude that the higher sound velocity in water may constrain the localization capability of amphibious mammals such as phocids and otariids, forcing the animals to place a heavy reliance on vision.

9.5 Hearing in Manatees

The underwater hearing sensitivity of the West Indian manatee (*Trichecus manatus*) was measured by Gerstein et al. (1993). Two 7-year old male manatees were trained to place their heads into a hoop facing a 1.5 m away. The center of the rectangular hoop was at a depth of 1.5 m and two PVC response paddles, one associated with signal-present and the other with signal-absent responses were located equal distance from the stationing hoop. A pure-tone signal was then played to the animal and if it could hear the tone, the animal would respond by touching the signal-present paddle and if a signal could not be heard, the animal would respond by touching the signal-absent paddle. Correct responses were reinforced with monkey chow biscuits, carrots, and assorted aquatic plants. The signal level was decreased systematically in a staircase fashion after correct detection trials and increased after the animal missed or failed to respond to a signal. A USRD J-9 transducer was used for frequencies between 0.1 and 46 kHz and a USRD J-13 was used for frequencies between 15 and 100 Hz.

The audiogram of one of the subjects (Gerstein et al., 1999) is shown in Fig. 9.38 along with the audiograms of a California sea lion and a harbor seal. There are two curves for the harbor seal, the high-frequency one from 1 to 100 kHz is from Mohl (1967) and the low-frequency one is from Kastak (1996). Only the low-frequency data of Kastak (1996) is shown for the sea lion. The data indicated that the manatee has a maximum sensitivity of about 50 dB re 1μPa between 16 and 18 kHz. The manatee hearing seemed to be approximately 10 dB more sensitive than the sea lion and the harbor seal. The highest frequency measured by Gerstein et al. (1999) was 46 kHz, a frequency close to the upper limit of hearing for the manatee.

The manatee audiogram also indicated that these animals can detect low-frequency underwater sounds down to 15 Hz. The hump in the sensitivity at 400 and 500 Hz may have been caused by the animal switching over from hearing the sound to feeling the sound (Gerstein et al., 1999). Whatever

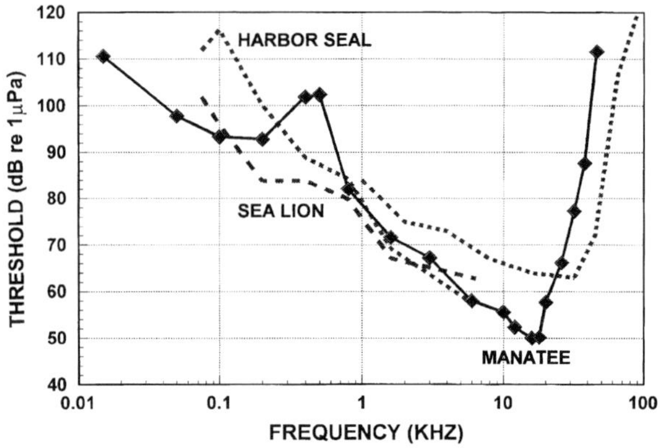

FIGURE 9.38. Audiogram of a manatee (from Gerstein et al., 1999), harbor seal (from Mohl, 1967 and Kastak, 1996), and a California sea lion (from Kastak, 1996).

technique was used to detect low-frequency sounds below 1 kHz, the data indicated that the manatee was on the whole more sensitive than the sea lion and the harbor seal, especially for frequencies below 100 Hz.

9.6 Hearing in Fishes

While only 125 species of marine mammals are known to exist, over 30,000 species of fish live in the fresh and salt waters of the world. The taxonomic class, Osteichthyes (bony fishes), with more than 25,000 species, is the largest of all vertebrate groups (Popper and Fay, 1993). Most all auditory systems that have been studied out of this class are species in the subdivision Telostei of the subclass Actinopterygii. There are over 20,000 freshwater and marine teleost species, and hearing sensitivity varies widely depending on the structure of the peripheral auditory system and neuroanatomy. This diversity is illustrated in Fig. 9.39, which displays audiograms based on acoustic pressure for 50 different species taken from the data compiled by Fay (1988). The solid lines are for species considered to be hearing specialists and the dotted lines for hearing generalists. As described in Chapter 7, hearing specialists have more sensitive and/or broader bandwidth hearing capability because of specialized anatomy that enhances stimulation of the inner ear by acoustic pressure in addition to direct stimulation by acoustic particle motion. Below 2 kHz, these species have more sensitive underwater hearing than most marine mammals.

Based on sound pressure level, lowest thresholds for hearing specialists are about 50–60 dB re 1 μPa from 200 to 1,500 Hz (with useful hearing range from below 100– 2,000 Hz), while lowest thresholds for those fishes without a swim bladder or with no obvious connection between the swim bladder and inner ear are 65–110 dB re 1 μPa from 50 to 500 Hz. For the latter, hearing sensitivity is better described in terms of acoustic particle motion.

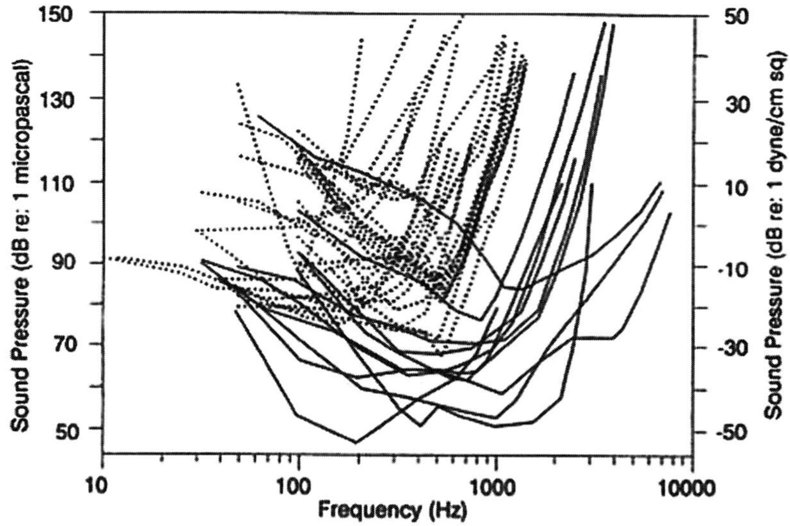

FIGURE 9.39. Behavioral sound pressure audiograms for over 50 species of fish taken from an extensive review by Fay (1988). *Dark lines* are considered to be hearing specialists and *dotted lines* are hearing generalists (from Popper and Fay, 1999).

One group of hearing specialists, the clupeids (e.g., herring, anchovy, shad), which have an air chamber (auditory bulla) adjacent to their utricle, has several species that have been shown to detect ultrasound up to about 200 kHz (Mann et al., 1997, 1998). The only other species reported to detect high-intensity ultrasound is the cod (*Gadus morhua*). Astrup and Møhl (1993) measured an average sound pressure threshold of 194 dB re 1 μPa at 38 kHz for 15 cod. This threshold is about 40–50 dB higher than thresholds reported for American shad by Mann et al. (1997, 1998). It is postulated that these species evolved to detect echolocation clicks as they are prey for many marine mammals. Figure 9.40 shows auditory thresholds for American shad (*Alosa sapidissima*) measured by Mann et al. (1998) relative to hearing thresholds for *Tursiops* over the same frequency band. Not all clupeids with an auditory bulla detect ultrasound. Higgs et al. (2004) postulated that the reason for this was due to differences in the structure of the sensory epithelia in the utricle; however, further study is needed to understand the relationship between utricular anatomy and the capability of certain clupeids to detect ultrasound.

9.6.1 Influence of the Lateral Line at Low Frequencies

In addition to the inner ear, at frequencies below 200 Hz the lateral line system also detects acoustic particle motion. This mechanosensory system is made up of series of neuromasts, each of which contains a number of hair and supporting cells covered by a gelatinous mass called a cupula as shown in

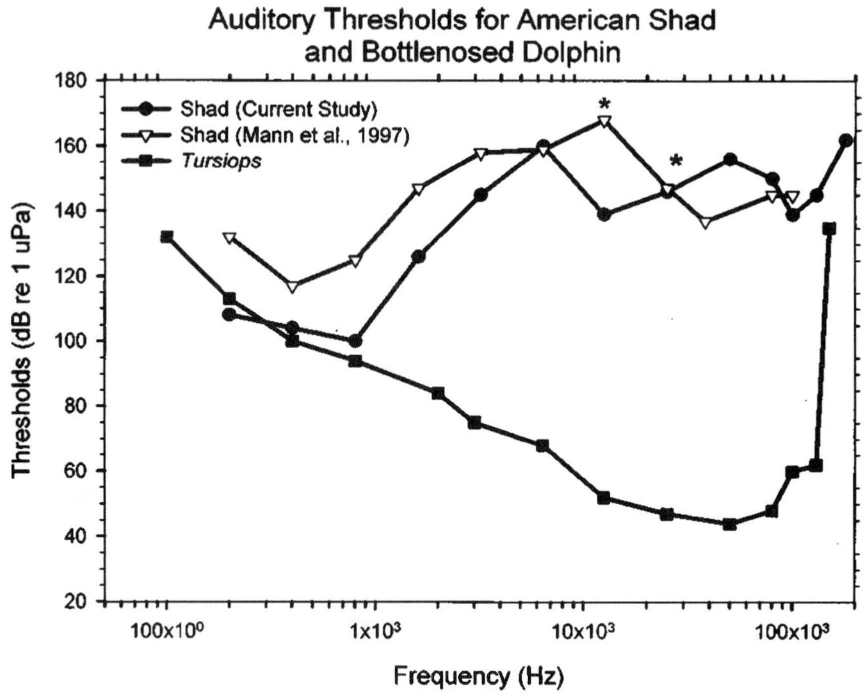

FIGURE 9.40. Sound pressure thresholds for American shad (*Alosa sapidissima*) and the bottlenose dolphin (*Tursiops truncatus*) as reported by Mann et al. (1998).

Fig. 9.41a. Neuromasts are found over the body of most teleost species, either free-standing on the surface or inside canals that open to the surrounding water through pores usually spaced a few millimeters apart (Fig. 9.41b). The combined inner ear–lateral line system is referred to as the *octavolateralis* system, as both are hair-cell-based and overlap functionally in terms of the

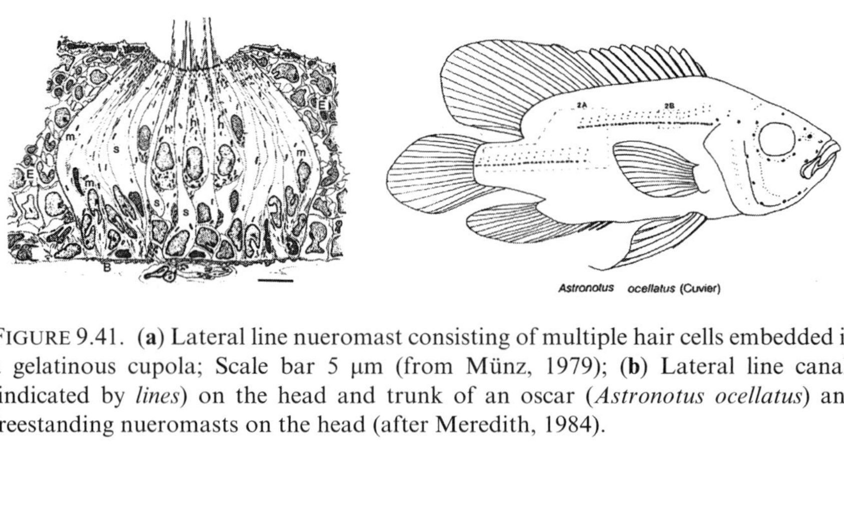

FIGURE 9.41. (**a**) Lateral line nueromast consisting of multiple hair cells embedded in a gelatinous cupola; Scale bar 5 μm (from Münz, 1979); (**b**) Lateral line canals (indicated by *lines*) on the head and trunk of an oscar (*Astronotus ocellatus*) and freestanding nueromasts on the head (after Meredith, 1984).

signals detected and peripheral mechanics. A major difference between the two is range of detection. The lateral line detects hydrodynamic signals within a few body lengths of the fish, while the inner ear detects sound waves from greater distances. Thus the lateral line could be considered analogous to the sense of touch rather than auditory in nature (Biaun et al., 2002).

9.6.2 Hearing Sensitivity

Because fishes may detect both acoustic particle motion and pressure, have two receptor systems, and generally hear best at frequencies below 1,000 Hz, measurement of auditory sensitivity is a major challenge, especially in aquaria and laboratory test chambers that are smaller relative to the wavelength of sound. Auditory sensitivities below 200 Hz measured by either standard behavioral or electrophysiological methods are based on signal detection by both lateral line and inner ear, because the two receptors cannot be differentiated. A few studies have blocked function of the lateral line using sodium cobalt and determined low-frequency sensitivity of the inner ear. But no method is currently available to block the reception of the inner ear while leaving the lateral line with normal function.

These issues, in addition to the normal biological variability and error associated with psychophysical and electrophysiological measurements, have created relatively large variations in the sensitivities reported for the same species by different researchers. For example, Fig. 9.42 shows three audiograms for the goldfish (*Carssius auratus*) measured in three different laboratories. Enger (1966) and Fay (1969) both used a loud speaker in air above a water tank. This type of set-up produces little or no particle motion in the water. But even though the stimulus was similar in each case, Enger used reward conditioning with food and Fay used classical conditioning of respiration. Offutt (1968) used an underwater J-9 transducer and classical heart rate conditioning.

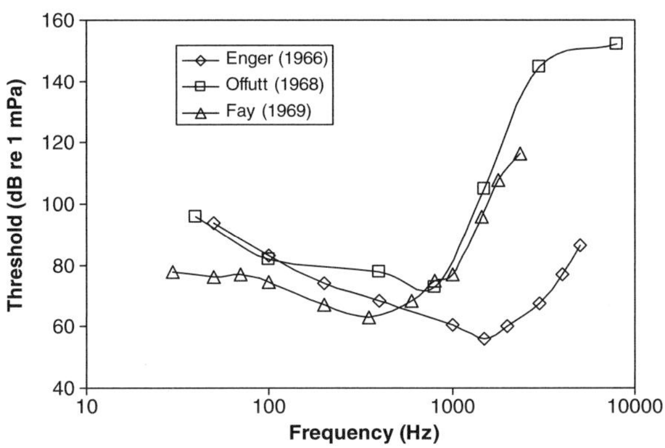

FIGURE 9.42. Goldfish (*Carassius auratus*) sound pressure thresholds measured by three different researchers. Inconsistency among laboratories is caused by differences in stimulus and psychophysical methods.

9.6.2.1 Pressure vs. Particle Motion

All teleost fishes detect acoustic particle motion directly with the inner ear at frequencies higher than about 200 Hz, and with the ear and lateral line at frequencies below 200 Hz. However, hearing sensitivity is most often reported in terms of sound pressure levels. For fish species that do not have a special connection between the swim bladder (or other gas-filled structure) and the inner ear, it is not known how much the swim bladder response to sound pressure enhances hearing sensitivity. Yan et al. (2000) tried to test the effect of deflating the swim bladder on hearing sensitivity in two hearing generalists, the blue gourami (*Trichogaster trichopterus*) and the oyster toadfish (*Opsanus tau*) and found no effect; however, they conducted the experiment in a very shallow tub with the head of the fish at the water surface where it is physically impossible to generate a radiated sound field from the swim bladder. So their results are not conclusive. For hearing specialists, which have a special anatomical connection (e.g., Weberian ossicles) between the swim bladder and inner ear or projection of the swim bladder close to the ear, stimulation of the inner ear occurs primarily via response of these gas-filled structures to pressure. This has been demonstrated by several researchers who have measured threshold shifts between 5 and 50 dB after deflation of the swim bladder or surgical removal of one of the Weberian ossicles in goldfish (*Carassius auratus*) and bullhead catfish (*Ictalurus nebulosus*) (Ladich and Wysocki, 2003; Kleerekoper and Roggenkamp, 1959; Poggendorf, 1952). The amount of threshold shift increases with frequency, because the swim bladder and Weberian apparatus provide the largest stimulus to the inner ear above about 500 Hz.

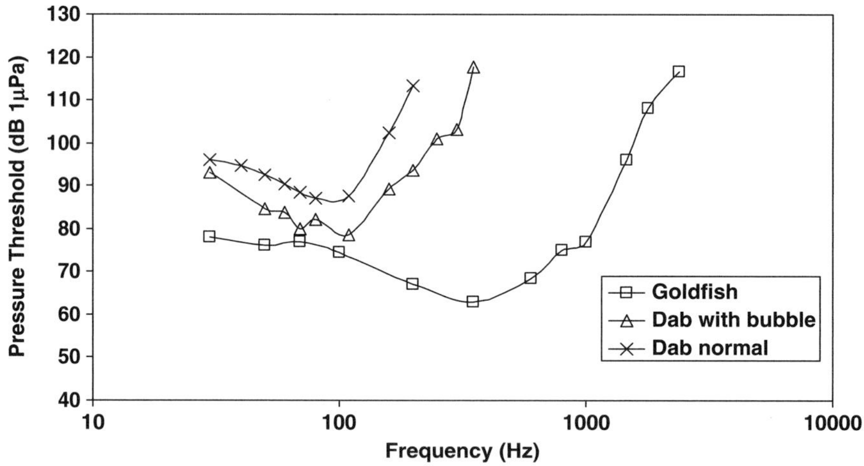

FIGURE 9.43. Hearing sensitivity of a flatfish, the dab (*Limanda limanda*) from Chapman and Sand (1974) compared with that of the goldfish (*Carassius auratus*) from Fay (1969). Placing an air bubble in close proximity to the dab's ear improved hearing sensitivity by about 10 dB, especially at higher frequencies.

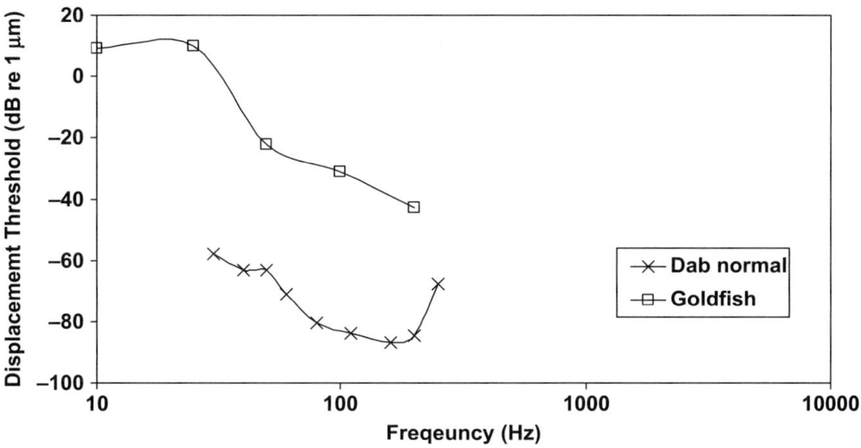

FIGURE 9.44. When thresholds are in terms of acoustic particle motion, fishes without swim bladders or special anatomical structures to transfer swim bladder motion to the inner ear appear to have lower thresholds than hearing specialists. The goldfish (data from Fay and Patricoski, 1980) is less sensitive to acoustic particle displacement than the dab (data from Chapman and Sand, 1974).

Figures 9.43 and 9.44 illustrate how misleading reporting auditory thresholds in terms of pressure can be for species that have no swim bladder or no special anatomical pathway connecting the inner ear to a gas-filled structure. Figure 9.43 displays the results of an experiment performed by Chapman and Sand (1974) to determine if proximity to a gas bubble would enhance hearing sensitivity in a fish without a swim bladder. They measured normal sound pressure thresholds in a flatfish, the dab (*Limanda limanda*), and then measured them again with a small balloon placed externally near its head. The results indicated that the presence of the balloon reduced thresholds by about 10 dB and slightly increased the audible frequency range. Figure 9.44 shows auditory thresholds based on acoustic particle displacement in dB re 1 micron for the dab and goldfish. It is obvious that the goldfish is much less sensitive to particle displacement than the dab.

9.6.2.2 Temporary Threshold Shift

Whether a fish is sensitive to sound pressure or particle motion is very important in determining what types of anthropogenic sounds could cause a hearing impairment and in measuring the appropriate hearing threshold. However, there are no data on temporary threshold shift (TTS) in terms of acoustic particle motion. Recent studies have confounded the problem by reporting "no TTS" in hearing generalists when exposed to a predominantly diffuse pressure field and TTS measured in terms of sound pressure thresholds. Based on sound pressure thresholds, TTS has been found in species known to be hearing specialists, including *Carassius auratus* (Popper and Clarke, 1976; Smith et al., 2004) and fathead minnows (*Pimephales promelas*) (Scholik and Yan, 2001, 2002). Smith et al. (2004) tested hearing in goldfish and tilapia (*Oreochromis niloticus*), a species with no special adaptation for stimulation by acoustic pressure. They found that goldfish had a 5-dB TTS after only 10 minutes of exposure to band-limited noise (0.1–10 kHz, approximately 170 dB re 1 μPa overall sound pressure level), and that goldfish with a 3-week exposure experienced a 28-dB threshold shift that took over 2 weeks to recover. Similarly, Scholik and Yan (2001) demonstrated that fathead minnows did not recover from TTS within 14 days after a 24-hour exposure to white noise from 0.3 to 2.0 kHz with an overall spectral sound pressure level of 142 dB re 1 μPa.

Not surprisingly, both Smith et al. (2004) and Scholik and Yan (2002) reported no TTS (based on sound pressure thresholds) in two species of generalists, tilapia and bluegill sunfish (*Lepomis macrochirus*). In their tests the fatiguing sound was a pressure field with negligible particle motion (typical of a sound speaker in air above a water tank containing the fish) and auditory thresholds before and after exposure were based on sound pressure. Figure 9.45 shows the results of the TTS study by Smith et al. (2004). Care must be taken in extrapolating these results because of the nature of the fatiguing stimulus and TTS based on sound pressure thresholds. This is especially true since TTS in hearing generalists was found following exposure to emissions from a small airgun array (Popper et al. 2005) and a low frequency active sonar transducer (Popper et al. 2007).

9.6.2.3 Hair Cell Damage

A few studies have shown that exposure to high sound pressure levels can destroy sensory cells in fish ears. No study, however, has yet examined the relationship between destruction of hair cells and hearing loss in fishes, or between the destruction of hair cells and exposure to high levels of acoustic particle motion. Enger (1981) and Hastings (1995) found that some sensory cells lost their ciliary bundles while looking for evidence of frequency discrimination in the ears of Atlantic cod fish (*Gadus morhua*) and goldfish (*Carassius auratus*), respectively. Using scanning electron microscopy, Enger found that some sensory hair cells were destroyed after 1–5-hours exposure to pure tones at frequencies from 50 to 400 Hz with a sound pressure level of 180 dB re 1 μPa (rms). This is 100–110 dB above the threshold in the most sensitive hearing frequency range for cod fish. In this study, Enger used a waveguide instrumented with a sound projector at each end to produce an exposure that had negligible particle velocity. Hastings (1995) reported similar damage to auditory hair cells in goldfish (*Carassius auratus*) exposed to continuous tones having levels of 189, 192, and 204 dB re 1 μPa (peak) at 250 Hz and 197 dB re 1 μPa (peak) at 500 Hz for approximately 2 hours. These levels are about 120–140 dB above the sound pressure thresholds for goldfish at these frequencies. This study also used a waveguide that allowed

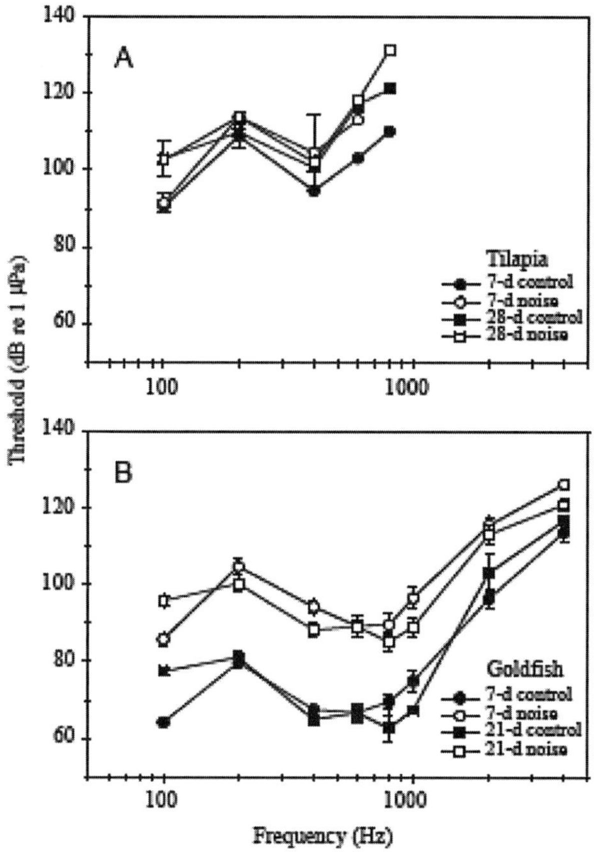

FIGURE 9.45. Smith et al. (2004) found TTS in goldfish exposed to band-limited noise (0.1–4 kHz) for 0 (control), 7, or 21 days at an overall level ranging from 164 to 170 dB re 1 µPa .Thresholds returned to baseline levels after 14 days. They found no TTS in tilapia, a species that has no special anatomy to enhance the detection of acoustic pressure.

exposure to sound with negligible particle velocity. Four fish were exposed at each set of conditions, and damage was found to correlate with sound pressure level at a 95% confidence level. In addition, goldfish exposed to 182 dB re 1 µPa (peak) at 500 Hz had no hair cell damage (approximately 120 dB above auditory threshold at a frequency in their most sensitive range).

In another study using scanning electron microscopy, Hastings et al. (1996) demonstrated similar effects in the ears of the oscar (*Astronotus ocellatus*). Small regions of sensory cells in the article and/or legend in four out of five animals were damaged after 1 hour of continuous exposure to a 300-Hz pure tone at 180 dB re 1 µPa (peak), while no damage occurred after 1-hour exposure to the same tone applied for 20% of each minute. Damage was not evident in animals after 1 day, but was evident in the animals that were kept alive for 4 days following exposure. These results suggest that damage from exposure to sound takes time to become visually apparent and is related to received acoustic energy since damage did not occur for pulse durations limited to 20% of one minute.

McCauley et al. (2003) investigated the effects of exposure to the sounds of a single seismic air gun on an Australian fish, the pink snapper (*Pagrus auratus*). Fish were in a cage and exposed to several air gun emissions at different distances in shallow water. The animals were kept alive for different time intervals after exposure. After the animals were sacrificed, their ears were examined for signs of damage using scanning electron microscopic techniques very similar to those used by Hastings et al. (1996). The results showed damage to about 15% of the sensory hair cells in the caudal end and 2.7% overall for the saccular epithelium in several of the animals as shown in Fig. 9.46, after a post-exposure period of 58 days. Because this test took place in shallow water about 9-m deep, however, the pink snapper, a hearing generalist, was no doubt exposed to particle motion higher than would be expected at the same sound pressure level in a free field. Future testing for TTS and sensory hair cell damage in fishes, especially with real anthropogenic sources, should include measurement of both acoustic pressure and particle motion during sound exposure in an effort to understand the cause of hearing loss and hair cell damage in fish with no special anatomical pathway for the detection of acoustic pressure.

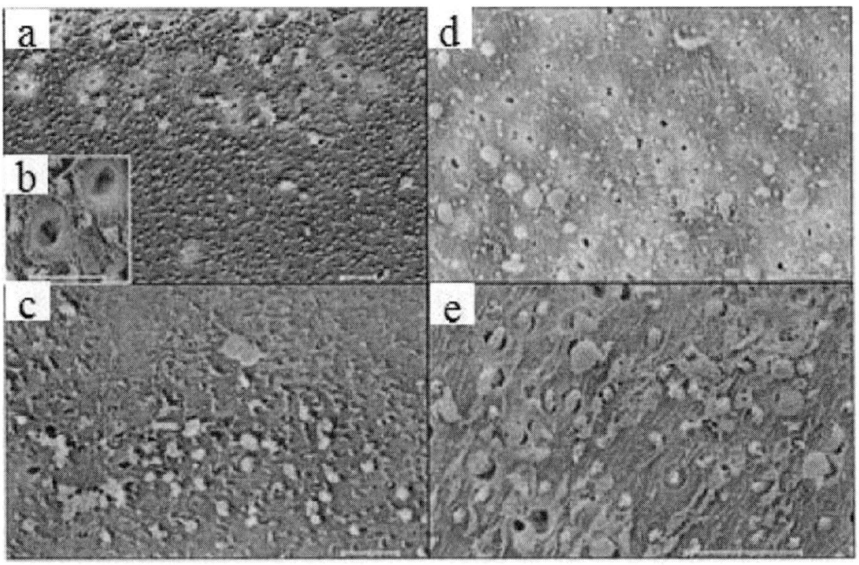

FIGURE 9.46. (**a**, **b**, **c**) Epithelia from exposed fish 18 h after exposure to the air gun; the photographs show numerous holes and "blebbing"; d, e: Photographs from saccular epithelia 58 days after exposure showing damage. Scale bars: a: 20 mm; b: 2 mm; c, d, e: 20 mm (adapted from McCauley et al., 2003).

9.6.3 Spectral and Temporal Analysis

Fish display many of the same abilities for spectral and temporal analysis as found in mammals. Critical masking ratios (CR) increase with signal frequency by about 3 dB per octave, which results in auditory filters with constant Q but with bandwidths slightly larger than found in other vertebrates.

Critical bandwidth data are available for only a few species, but the form of these functions is similar to humans and other vertebrates. Figure 9.47 displays critical bands in cod (*Gadus morhau*) as determined by Hawkins (1981).

9.6.4 Directional Hearing

Several studies have shown that many species can determine the direction and swim toward a sound source (Richard, 1968; Myrberg et al., 1969; Chapman and Johnstone, 1974; Schuijf and Buwalda, 1980); however, "how they do it" is still not fully understood. Localization mechanisms based on the phase difference between the direct signal and indirect signal radiated from the swim bladder (de Munck and Schellart, 1987), and elliptical orbits of hair cells with respect to the otoliths (Schellart and de Munck, 1987) have been proposed.

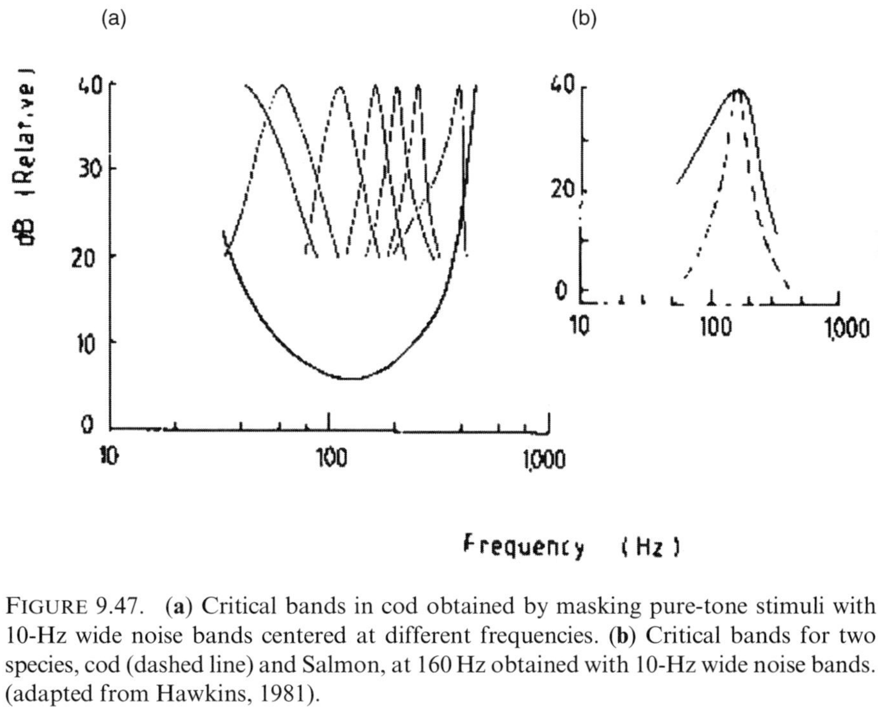

FIGURE 9.47. (**a**) Critical bands in cod obtained by masking pure-tone stimuli with 10-Hz wide noise bands centered at different frequencies. (**b**) Critical bands for two species, cod (dashed line) and Salmon, at 160 Hz obtained with 10-Hz wide noise bands. (adapted from Hawkins, 1981).

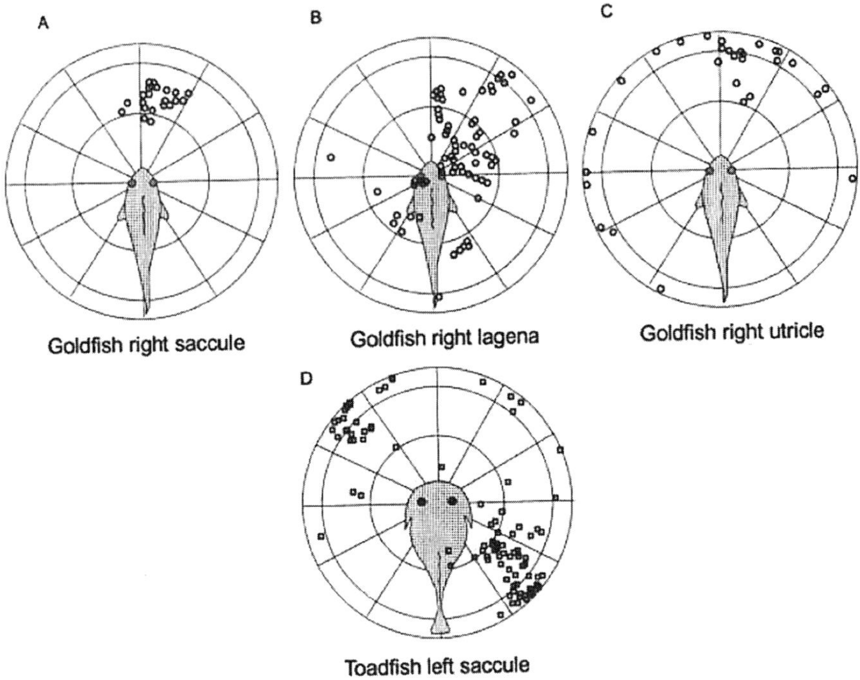

FIGURE 9.48. Distributions of best direction vectors determined by recording responses of single units in different branches of the eighth nerve.

Hair cell orientation patterns indicate that the inner ear is very likely to be involved in sound localization. Experiments on haddock (*Melanogrammus aeglefinus*), perch (*Perca fluviatilis*), goldfish (*Carassius auratus*), and toadfish (*Opsanus tau*) have shown that the ears are responsive to stimulus direction (Enger et al., 1973; Sand, 1974; Fay, 1984; Fay and Edds-Walton, 1997). By measuring microphonic potentials as the animal was rotated 360° relative to the sound source, Sand (1974) found that the anterior part of the saccule in perch responded to both horizontal and vertical stimulation while the posterior region responded best to stimulation from the vertical direction.

Enger et al. (1973), Fay (1984), and Fay and Edds-Walton (1997) used a vibration table in air to linearly accelerate the whole body of the fish along selective axes to create relative motion between the otoliths and sensory epithelia in the auditory end organs. Enger et al. (1973) simultaneously measured saccular microphonics in haddock and found that different sensory cells had different axes of maximal sensitivity. Fay (1984) and Fay and Edds-Walton (1997) made extracellular recordings of single units.

References

Andersen, S. (1970a). "Auditory Sensitivity of the Harbor Porpoise (*Phocoena phocoena*)," Invest. Cetacea **2**, 255–259.

Andersen, S. (1970b). "Directional Hearing in the Harbor Porpoise (*Phocoena phocoena*)," Invest. Cetacea **2**, 260–263.

Astrup, J. and Møhl, B. (1993). "Detection of intense ultrasound by the cod *Gadus morhua*," J. Exp. Biol. **182**, 71–80.

Au, W. W. L. and Moore, P. W. B. (1984). "Receiving Beam Patterns and Directivity Indices of the Atlantic Bottlenose Dolphin (*Tursiops truncates*)," J. Acoust. Soc. Am. **75**, 255–262.

Au, W. W. L. and Moore, P. W. B. (1990). "Critical Ratio and Critical Bandwidth for the Atlantic Bottlenose Dolphin," J. Acoust. Soc. Am. **88**, 1635–1638.

Bain, D. and Dallheim, M. E. (1992). Personal Communications.

Blodgett, H. C., Jeffress, L. A., and Taylor, R. W. (1958). "Relation of masked threshold to signal-duration for various interaural phase-combinations," Am. J. Psychol. **71**, 283–290.

Bobber, R. J. (1970). *Underwater Electroacoustic Measurements* (U.S. Govern. Printing Office, Washington, DC).

Braun, C. B., Coombs, S., and Fay, R. R. (2002). "Multisensory interactions within the octavolateralis systems: What is the nature of multisensory integration?" Brain, Behav. and Evol. **59**, 162–176.

Chapman, C. J. and Sand, O. (1974). "Field studies of hearing in two species of flatfish *Pleuronectes platessa* (L.) and *Limanda limanda* (L.) (Family Pleuronectidae)," Comp. Biochem. Physiol. **47A**, 371–385.

Chapman, C. J. and Johnstone, A. D. F. (1974). "Some auditory discrimination experiments on marine fish," J. Exp. Biol. **61**, 521–528.

de Munck, J. C. and Schellart, N. A. M. (1987). "A model for the nearfield acoustics of the fish swimbladder and its relevance for directional hearing," J. Acoust. Soc. Am. **81**, 556–560.

Dudok van Heel, W. H. (1962). "Sound and Cetacea," Nether. J. Sea. Res. **1**, 407–507.

Enger, P. S. (1966). "Acoustic threshold in goldfish and its relation to the sound source distance," Comp. Biochem. Physiol. **18**, 859–868.

Enger, P. S. (1981). "Frequency discrimination in teleosts – central or peripheral?" in *Hearing and Sound Communication in Fishes*, W. N. Tavolga, A. N. Popper and R. R. Fay, eds. (Springer-Verlag, New York), pp. 243–255.

Enger, P. S., Hawkins, A. D., Sand, O. and Chapman, C. J. (1973). "Directional sensitivity of saccular microphonic potentials in the haddock," J. Exp. Biol. **59**, 425–433.

Fay, R. R. (1969). "Behavioral audiogram for the goldfish," J. Aud. Res. **9**, 112–121.

Fay, R. R. (1974). "Auditory frequency discrimination in vertebrates," J. Acoust. Soc. Am. **56**, 206–209.

Fay, R. R. (1978). "Coding of information in single auditory-nerve fibers of the goldfish," J. Acoust. Soc. Am. **63**, 136–146.

Fay, R. R. (1984). "The goldfish ear codes the axis of acoustic particle motion in three dimensions," Science **225**, 951–954.

Fay, R. R. (1988). *Hearing in Vertebrates: A Psychophysics Databook* (Hill-Fay Assoc., Winnetka, Illinois).

Fay, R. R. and Edds-Walton, P. L. (1997). "Directional response properties of saccular afferents of the toadfish, *Opsanus tau*," Hear. Res. **111**, 1–21.

Fay, R. R. and Patricoski, M. L. (1980). "Sensory mechanisms for low frequency vibration detection in fishes," in *Abnormal Animal Behavior Prior to Earthquakes, II*, R. Buskirk, ed. (U.S. Geological Survey Open File Report 80–453), pp. 63–91.

Fay, R. R. and Popper, A. N. (eds.). (1999). *Comparative Hearing: Fish and Amphibians*, Spring Handbook of Auditory Research, Vol. 11, (Springer New York), 438 p.

Finnerana, J. J., Schlundt, C. E., Carder, D. A., Clark, J. A., and Young, J. A., Gaspin, J. B., and Ridgway, S. H. (2000). "Auditory and behavioral responses of bottlenose dolphins (*Tursiops truncatus*) and a beluga whale (*Delphinapterus leucas*) to impulsive sounds resembling distant signatures of underwater explosions," J. Acoust. Soc. Am. **108**, 417–431.

Fletcher, H. (1940). "Auditory Patterns," Rev. Mod. Phys. **12**, 47–65.

Forsythe, F. E., Malcolm, M. A., and Moler, C. B. (1977). *Computer Methods for Mathematical Computations* (Prentice-Hall, Englewood Cliffs, NJ).

French, N. R. and Steinberg, J. C. (1947). "Factors Governing the Intelligibility of Speech Sound," J. Acoust. Soc. Am. **19**, 90–119.

Gentry, R. L. (1967). "Underwater Auditory Localization in the California Sea Lion (*Zalophus californianus*)," J. Aud. Res. **7**, 187–193.

Gerstein, E. R., Gerstein, L. A., Forsythe, S. E., and Blue, J. E. (1993). "Underwater Audiogram of a West Indian Manatee (*Trichecus manatus*)," *Proc. Tenth Bien Conf. Biol. Mar. Mamm.*, (Galveston, Texas), P. 53 (A).

Gerstein, E. R., Gerstein, L. A., Forsythe, S. E., and Blue, J. E. (1999). "The Underwater Audiogram of the West Indian Manatee (*Trichechus manatus*)," J. Acoust. Soc. Am. **105**, 3575–3583.

Hall, J. D. and Johnson, C. S. (1971). "Auditory Thresholds of a Killer Whale," J. Acoust. Soc. Am. **51**, 515–517.

Hamilton, P. M. (1957). "Noise masked thresholds as a function of tonal duration and masking band width," J. Acoust. Soc. Am. **29**, 506–511.

Hastings, M. C. (1995). "Physical effects of noise on fishes," Proceedings of INTER-NOISE 95, The International Congress on Noise Control Engineering, Vol. II, pp. 979–984.

Hastings, M. C., Popper, A. N., Finneran, J. J., and Lanford, P. J. (1996). "Effect of low frequency underwater sound on hair cells of the inner ear and lateral line of the teleost fish *Astronotus ocellatus*," J. Acoust. Soc. Am. **99**, 1759–1766.

Hawkins, A. D. (1981). "The hearing abilities of fish," in *Hearing and Sound Communication in Fishes*, W. N. Tavolga, A. N. Popper and R. R. Fay, eds. (Springer-Verlag, New York), pp. 109–137.

Hawkins, J. E. and Stevens, S. S. (1950). "The Masking of Pure Tones and Speech by White Noise," J. Acoust. Soc. Am. **22**, 6–13.

Heffner, R. S. and Heffner, H. E. (1992). "Hearing in Large Mammals: Sound-localization Acuity in Cattle (*Bos Taurus*) and Goats (*Capra hircus*)," J. Comp. Psychol. **106**, 107–113.

Herman, L. M. and Arbeit, W. R. (1972). "Frequency Difference Limens in the Bottlenose Dolphin: 1-70 kHz," J. Aud. Res. **12**, 109–120.

Higgs, D. M., Plachta, D. T. T., Rollo, A. K., Singheiser, M., Hastings, M. C., and Popper, A. N. (2004). "Development of ultrasound detection in American Shad (*Alosa sapidissima*)," J. Exp. Biol. **207**, 155–163.

Holt, M. M., Schusterman, R. J., Southall, B. L., and Kastak, D. (2004). "Localization of aerial Broadband Noise by Pinnipeds," J. Acoust. Soc. Am. **115**, 2339–2345.

Hughes, J. W. (1946). "The Threshold of Audition for Short Periods of Stimulation," Proc. Roy. Soc. (London) **b133**, 486–490.

Jacobs, D. W. (1972). "Auditory Frequency Discrimination in the Atlantic Bottlenose Dolphin (*Tursiops truncates*) Montagu: A Preliminary Report," J. Acoust. Soc. Am. **53**, 696–698.

Jacobs, D. W. and Hall, J. D. (1972). "Auditory Thresholds of a Fresh Water Dolphin (*Inia geoffrensis*) Blainville," J. Acoust. Soc. Am. **51**, 530–533.

Johnson, S. C. (1967). "Sound Detection Thresholds in Marine Mammals," in *Marine Bio-Acoustics*, W. Talvoga, ed. (Pergamon Press, New York), pp. 247–260.

Johnson, S. C. (1968a). "Relation Between Absolute Threshold and Duration of Tone Pulse in the Bottlenosed Porpoise," J. Acoust. Soc. Am. **43**, 757–763.

Johnson, S. C. (1968b). "Masked Tonal Thresholds in the Bottlenosed Porpoise," J. Acoust. Soc. Am. **44**, 965–967.

Johnson, S. C. (1971). "Auditory Masking of One Pure Tone by Another in the Bottlenose Porpoise," J. Acoust. Soc. Am. **49**, 1317–1318.

Johnson, S. C., McManus, M. W., and Skaar, D. (1989). "Masked Tonal Thresholds in the Belukha Whale," J. Acoust. Soc. Am. **85**, 2651–2654.

Kastak, D. A. (1996). "Comparative Aspects of Hearing in Pinnipeds," Ph.D. dissertation, U. C. Santa Cruz.

Kastak, D. and Schusterman, R. J. (2002). "Changes in auditory sensitivity with depth in a free-diving California se lion (*Salophus californianus*)," J. Acoust. Soc. Am. **112**, 329–333.

Ketten, D. R. (2000). "Cetacean Ears," in *Hearing by Whales and Dolphins*, W. W. L. Au, A. N. Popper, and R. R. Fay, eds. (Springer-Verlag, New York), pp. 43–108.

Kleerekoper, H. and Roggenkamp, P. A. (1959). "An experimental Study on the Effect of the Swimbladder on Hearing Sensitivity in *Ameiurus nebulosus nebulosus* (LeSueur)," Can. J. Zool. **37**, 1–8.

Ladich, F. and Wysocki, L. E. (2003). "How Does Tripus Extirpation Affect Auditory sensitivity in Goldfish?" Hearing Res. **182**, 119–129.

Ljungblad, D. K., Scoggins, P. D., and Gilmartin, W. G. (1982). "Auditory Thresholds of a Captive Eastern Pacific Bottlenosed Dolphin, (*Tursiops* spp.)," J. Acoust. Soc. Am. **72**, 1726–1729.

Mann, D. A., Lu, Z. and Popper, A. N. (1997). "A clupeid fish can detect ultrasound," Nature **389**, 341.

Mann, D. A., Lu, Z., Hastings, M. C., and Popper, A. N. (1998). "Detection of ultrasonic tones and simulated dophin echolocation clicks by a teleost fish, the American shad (*Alosa sapidissima*)," J. Acoust. Soc. Am. **104**, 562–568.

McCauley, R. D., Fewtrell, J., and Popper, A. N. (2003). "High intensity anthropogenic sound damages fish ears," J. Acoust. Soc. Am. **113**, 638–642.

McCormick, J. M. and Salvadori, M. G. (1964). *Numerical Methods in FORTRAN* (Prentice-Hall, Englewood Cliffs, NJ).

Mills, A. W. (1958). "On the Minimum Audible Angle," J. Acoust. Soc. Am. **30**, 237–246.

Meredith, G. E. (1984). "Peripheral configuration and central projections of the lateral line system in *Astronotus ocellatus* (cichlidae): a nonelectroreceptive teleost," J. Comp. Neurol. **228**, 342–358.

Mills, A. W. (1972). "Auditory Localization," in *Foundations of Modern Auditory Theory*, Vol 2, J.V. Tobias, eds. (Academic Press, New York), pp. 303–348.

Mohl, B. (1964). "Preliminary Studies on Hearing in Seals," Vidensk. Medd. Cansk Naturh. Foren. **127**, 283–294.

Mohl, B. (1967). "Frequency Discrimination in the Common Seal," in *Underwater Acoustics*, V. A. Albers, ed. (Plenum Press, New York), pp. 43–45.

Mohl, B. (1968). "Auditory Sensitive of the Common Seal in Air and Water," J. Aud. Res. **8**, 27–38.

Moore, P. W. B. (1975). "Underwater Localization of Click and Pulsed Pure-tone Signals by the California Sea Lion (*Zalophus californianus*)," J. Acoust. Soc. Am. **57**, 406–449.

Moore, P. W. B. and Au, W. W. L. (1975). "Underwater Localization of Pulsed Pure Tonesby the California Sea Lion (*Zalophus californianus*)," J. Acoust. Soc. Am. **58**, 721–727.

Moore, P. W. B. and Schusterman, R. J. (1976). "Discrimination of Pure-tone Intensities by the California Sea Lion," J. Acoust. Soc. Am. **60**, 1405–1407.

Moore, P. W. B. and Schusterman, R. J. (1987). "Audiometric Assessment of Northern Fur Seals (*Callorhinus ursinus*)," Mar. Mamm. Sci. **3**, 31–53.

Münz, H. (1979). "Morphology and innervation of the lateral line system in *Sarotherodeon niloticus*," Zoomorphologie **93**, 73–86.

Myrberg, A. A., Jr., Banner, A., and Richard, J. D. (1969). "Shark attraction using a video-acoustic system," Mar. Biol. **2**, 264–276.

Nachtigall, P. E., Pawloski, J., and Au, W. W. L. (2003). "Temporary threshold shifts and recovery following noise exposure in the Atlantic bottlenosed dolphin (*Tursiops truncatus*)," J. Acoust. Soc., Am. **113**, 3245–3429.

Nachtigall, P. E., Au. W. W. L., Pawloski, J. L., and Moore, P. W. B. (1995). "Risso's Dolphin (*Grampus griseus*) Hearing Thresholds in Kaneohe Bay, Hawaii," in *Sensory Systems of Aquatic Mammals*, R. A. Kastelein, J. A. Thomas, and P. E. Nachtigall, eds. (De-Spiel, Woerden, The Netherlands), pp. 49–53.

Offutt, G. C. (1968). "Auditory response in the goldfish," J. Aud. Res. **8**, 391–400.

Paterson, R. D. and Moore, B. C. J. (1986). "Auditory Filters and Excitation Patterns as Representations of Frequency Resolution," in *Frequency Selectivity in Hearing*, B. C. J. Moore, ed. (Academic Press, New York), pp. 123–177.

Plump, R. and Bouman, M. A. (1959). "Relation between Hearing Threshold and Duration for Tone Pulses," J. Acoust. Soc. Am. **31**, 749–758.

Poggendorf, D. (1952). "Die absoluten Hörschwellen des Zwergweises (*Amiurus nebulosus*) und Beiträge zur Physik des Weberschen Apparatus der Ostariophysan," Z. Vergl. Physiol. **34**, 222–257.

Popper, A. N. and Clarke, N. L. (1976). "The auditory system of goldfish (*Carassius auratus*): effects of intense acoustic stimulation," Comp. Biochem. Physiol. **53A**, 11–18.

Popper, A. N. and Fay, R. R. (1993). "Sound Detection and Processing by Fish: Critical Review and Major Research Questions," Brain Behav. Evol. **41**, 14–38.

Popper, A. N., Smith, M. E., Cott, P. A, Hanna, B. W., MacGillivray, A. O., Austin, M. E., and Mann, D. A. (2005). "Effects of exposure to seismic airgun use on hearing of three fish species," J. Acoust. Soc. Am. **117**, 3958–3971.

Popper, A. N., Halverson, M. B., Kane, E., Miller, D. D., Smith, M. E., Stein, P., and Wysocki, L. E. (2007). "The effects of high-intensity, low-frequency active sonar on rainbow trout," J. Acoust. Soc. Am. **122**, 623–635.

Renaud, D. L. and Popper, A. N. (1978). "Sound Localization by the Bottlenose Porpoise (*Tursiops truncates*)," J. Exp. Biol. **63**, 569–585.

Repenning, C. A. (1972). "Underwater hearing in seals: Functional morphology," in *Functional Anatomy of Marine Mammals Vol. 1*, R. J. Harrison, ed. Academic, London, pp. 307–331.

Reysenbach de Haan, F. W. (1957). "Hearing in Whales," Acta Oto-Laryngol. Suppl. **134**, 1–114.

Richard, J. D. (1968). “Fish attraction with pulsed low-frequency sound,” J. Fish. Res. Board Can. **26**, 1441–1452.

Ridgway, S. H., Carder, D. A., Kamolnick, T., Smith, R. R., Schlundt, C. E., and Elsberry, W. R. (2001). “Hearing and Whistling in the Deep Sea: Depth Influences Whistle Spectra but Does not Attenuate Hearing by While Whales (*Delphinapterus leucas*), (Odontoceti, Cetacea).

Sand, O. (1974). “Directional sensitivity of microphonic potentials from the perch ear,” J. Exp. Biol. **60**, 881–899.

Sauerland, M. and Dehnhardt, G. (1998). “Underwater Audiogram of a Tucuxi (*Sotalia fluviatilis guianensis*), J. Acoust. Soc. Am. **103**, 1199–1204.

Schaft, B. (1970). “Critical bands,” in *Foundation of Modern Auditory Theory. Vol 1*, J.V. Tobias, ed. (Academic Press, NY.), pp. 159–202.

Schellart, N. A. M. and de Munck, J. C. (1987). “A model for directional and distance hearing in swimbladder-bearing fish based on the displacement orbits of the hair cells,” J. Acoust. Soc. Am. **82**, 822–829.

Schlund, C. E., Finneran, J. J., Carder, D. A., and Ridgway, S. H. (2000). “Temporary shift in masked hearing thresholds of bottlenose dolphins, *Tursiops truncatus*, and white whales, *Delphinapterus leucas*, after exposure to intense tones,” J. Acoust. Soc. Am. **107**, 3496–3408.

Scholik, A. R. and Yan, H. Y. (2001). “Effects of underwater noise on auditory sensitivity of a cyprinid fish,” Hear. Res. **152**, 17–24.

Scholik, A. R. and Yan, H. Y. (2002). “The effects of noise on the auditory sensitivity of the bluegill sunfish, *Lepomis marochirus*,” Comp. Biochem. Physiol. A **133**, 43–52.

Schuijf, A. and Buwalda, R. J. A. (1980). “Underwater localization – a major problem in fish acoustics,” in *Comparative Studies of Hearing in Vertebrates*, A. N. Popper and R. R. Fay, eds. (Springer-Verlag, New York), pp. 43–77.

Schusterman, R. J., Balliet, R. F., and Nixon, J. (1972). “Underwater Audiogram of the California Sea Lion by the Conditional Vocalization Technique,” J. Exp. Anal. Behav. **17**, 339–350.

Schusterman, R. J. (1974). “Auditory Sensitivity of the California Sea Lion to Airborne Sound,” J. Acoust. Soc. Am. **56**, 1248–1251.

Schusterman, R. J. and Moore, P. W. B. (1978). “The Upper Limit of Underwater Auditory Frequency Discrimination in the California Sea Lion,” J. Acoust. Soc. Am. **63**, 1591–1595.

Schusterman, R. J. and Moore, P. W. B. (1978). “Underwater Audiogram of the Northern Fur Seal (*Callorhinus ursinus*),” J. Acoust. Soc. Am. **64**, S87 (Abstract).

Schusterman, R. J. and Moore, P. W. B. (1980). “Auditory Sensitivity of Northern Fur Seals (*Callorhinus ursinus*) and a California Sea Lion (*Zalophus californianus*)to Airborne Sound,” J. Acoust. Soc. Am. **68**, S6 (Abstract).

Sivian, L. J. and White, S. D. (1933). “On Minimum Audible Sound Fields,” J. Acoust. Soc. Am. **4**, 288–321.

Smith, M. E., Kane, A. S. and Popper, A. N. (2004). “Noise-induced stress response and hearing loss in goldfish (*Carassius auratus*),” J. Exp. Biol. **207**, 427–435.

Southall, B. L., Schusterman, R. J., and Kastak, D. (2000). “Masking in Three Pinnipeds: Underwater, Low-frequency Critical Ratios,” J. Acoust. Soc. Am. **108**, 1322–1326.

Southall, B. L., Schusterman, R. J., and Kastak, D. (2003). “Auditory Masking in Three Pinnipeds: Aerial Critical Ratios and Direct Critical Bandwidth Measurements,” J. Acoust. Soc. Am. **114**, 1660–1666.

Terhune, J. M. and Ronald, K. (1971). "The Harp Seal, *Pagophilus groenlandicus* (Erxleben, 1771), X. The Air Audiogram, Can. J. Zool. **49**, 385–390.

Terhune, J. M. and Ronald, K. (1972). "The Harp Seal, *Pagophilus groenlandicus* (Erxleben, 1771), III. The Underwater Audiogram, Can. J. Zool. **50**, 385–390.

Terhune, J. M. (1974). "Directional Hearing of a Harbor Seal in Air and Water," J. Acoust. Soc. Am. **56**, 1862–1865.

Terhune, J. M. and Ronald, K. (1975). "Underwater Hearing Sensitivity of Two Ringed Seals (*Pusa hispida*)," Can. J. Zool. **50**, 565–569.

Terhune, J. M. and Ronald, K. (1976). "The Upper Frequency Limit of Ringed Seal Hearing," Can. J. Zool. **54**, 1226–1229.

Thomas, J., Chun, N., Au, W., and Pugh, K. (1988). "Underwater Audiogram of a False Killer Whale (*Pseudorca leucas*)," J. Acoust. Soc. Am. **84**, 936–940.

Thompson, R. K. R. and Herman, L. M. (1975). "Underwater Frequency Discrimination in the Bottlenose Dolphin (1-140 kHz) and the Human (1-8 kHz)," J. Acoust. Soc. Am. **57**, 943–948.

Turnbull, S. D. and Terhune, J. M. (1990). "White noise and pure tone masking of pure tone thresholds of a harbor seal listening in air and underwater. (Phoca vitulina)," Can. J. Zool. **68**, 2090–2097.

Wang, D., Wang, K., Xiao, Y., and Sheng, G. (1992). "Auditory Sensitivity of a Chinese River Dolphin (*Lipotes vexillifer*), in *Marine Mammal Sensory Systems*, J. A. Thomas, R. A. Kastelein, and A. Ya. Supin, eds. (Plenum Press, New York), pp 213–221.

Wengel, R. L. and Lane, C. E. (1924). "The Auditory Masking of One Pure Tone by Another and its Probable Relation to the Dynamics of the Inner Ear," Phys. Rev. **23**, 266–285.

White, M. J. Jr., Norris, J., Ljungblad, D., Baron, K., and di Sciara, G. (1978). "Auditory Thresholds of Two Beluga Whales (*Delphinapterus leucas*)," HSWRI Tech. Rep., No. 78-109 (Hubbs Marine Research Institute, 1700 S. Shores Road, San Diego, CA).

Yan, H. Y., Fine, M. L., Horn, N. S., and Colón, W. E. (2000). "Variability in the role of the gasbladder in fish audition," J. Comp. Physiol. A **186**, 435–445.

Zaytseva, K. A., Akopian, A. I., and Morozov, V. P. (1975). "Noise Resistance of the Dolphin Auditory Analyzer as a Function of Noise Direction," BioFizika **20**, 519–521.

Zwislocki, J. (1960). "Theory of Temporal Auditory Summation," J. Acoust. Soc. Am. **32**, 1046–1060.

10

Emission of Social Sounds by Marine Animals

Social sounds of marine mammals are usually studied with a spectrographic analyzer that determines the "instantaneous" frequency and relative amplitude of a signal as a function of time, with the information usually plotted as a spectrogram. A spectrogram is a two-dimensional time- frequency representation of a signal with one axis representing time and the other orthogonal axis representing frequency. The relative amplitude of the signal at any specific time and frequency is represented by various shades of gray (the darkest shade representing the maximum amplitude) for a black-and-white display or by different colors, with a color scale usually plotted somewhere on the spectrogram. Most, if not all, modern spectrograph analyzers use a short-term Fourier transform technique to determine the frequency content of the signal. In the short-time Fourier transform technique, the signal is divided into fixed blocks of N points. Each block is passed through a windowing function and an FFT operation is performed on it. The blocks of data are not usually contiguous but are overlapping so, except for the first block, each block will have some data that are common to the previous block. A schematic of how data are subdivided into blocks is depicted in Fig. 10.1, starting with an array of data points that are subdivided into blocks of N point each with some overlapping between adjacent blocks. The data in each block are multiplied by a window function to ensure that the values of the data at both ends of the block will be zero. An FFT is performed on each block of N points. The Hanning and the Hamming windows are two of more popular windows, although there are many different windows (see Chapter 6). The short-term FFT technique is only valid if the frequency of the signal within each block does not change; it is assumed that the frequency of a signal within a block is constant. Therefore, some care must be taken in choosing the analyzer block size. The frequency and time resolution of each block is dependent on the sample rate and the size of the block, or $\Delta f = \mathrm{sr}/N$, where Δf is the frequency resolution, sr is the sample rate, and N is the size of each block. The frequency resolution can only be improved by increasing the number of points in a block, and this can be done at the expense of the time resolution since a longer block will decrease the time resolution. Therefore, there is an inherent frequency–time resolution trade-off that cannot be

W.W.L. Au, M.C. Hastings, *Principles of Marine Bioacoustics*,
DOI: 10.1007/978-0-387-78365-9_10, © Springer Science+Business Media, LLC 2008

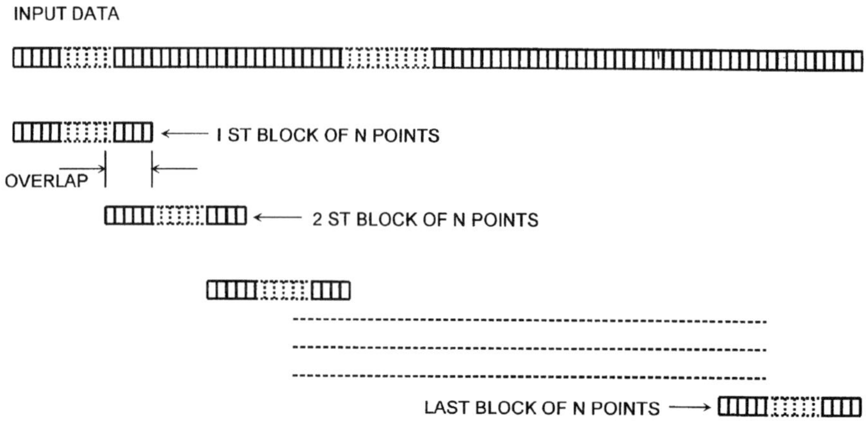

FIGURE 10.1. A schematic showing how the data are subdivided into overlapping blocks of N points.

circumvented. The time corresponding to each block can be chosen as the time corresponding to the beginning, ending, or mid-point of each block. Overlapping is done so that there is a smooth transition from the frequency in one block to the frequency in an adjacent block of data. Unfortunately, there is not a general rule of thumb as to the amount of overlap that is necessary to properly analyze a signal; the amount of overlapping will depend on the properties of the signals being examined. In presenting data in the spectrogram form, it is important to specify the sample rate and the number of points per block so that the frequency and time resolution of the spectrogram can be determined. In the pre-digital era, the bandwidth of the analyzing filter would customarily be listed.

Many of the sounds emitted by marine mammals will have a pulse-like or burst-like property so it would be wise at this time to consider how pulse signals appear on a spectrogram. Watkins (1966) examined the use of spectrograms to analyze pulse signals and wrote a very enlightening article on the topic. Let us assume that we have N repetition of a signal $s(t)$ and the interval between the beginning of one signal and the beginning of the next signal is τ. The burst or pulse $s(t)$ can be any type of signal, such as a sine wave burst, a damped sinusoid, or a transient signal. Let us further assume that the Fourier transform of $s(t)$ is equal to $S(f)$. A signal of N repetitions of $s(t)$ can be expressed as

$$x(t) = s(t) + s(t - \tau) + s(t - 2\tau) + \cdots + s(t - N\tau). \tag{10.1}$$

In this notation, $s(t - N\tau)$ means that the variable takes on the values of $s(t)$ after a delay of $N\tau$. From the shift theorem in Fourier analysis, we know that $\Im(s - \tau) = e^{-i2\pi\tau f}S(f)$. Therefore, the Fourier transform of Eq. (10.1) can be expressed as

$$\Im[x(t)] = S(f)\left[1 + N e^{-i2\pi\tau f}\right]. \tag{10.2}$$

We made use of the fact that $e^{-i2\pi\tau f} = e^{-i4\pi\tau f} = \cdots = e^{-i2N\pi\tau f}$ to obtain Eq. (10.2.) Taking the absolute value of both sides of Eq. (10.2), we have

$$|\Im[x(t)]| = |S(f)|\sqrt{(N^2 + 1) + 2N\cos(2\pi\tau f)}. \tag{10.3}$$

There are two terms on the right side of Eq. (10.3), the first being the absolute value of the spectrum of the pulse multiplied by a term that varies in a sinusoidal fashion and will cause ripples to appear in the overall spectrum of the signal $x(t)$. The minima in the spectrum will occur every time the cosine term is equal to –1, which will occur whenever $2\pi\tau f = m2\pi$. The maxima in the spectrum will occur every time the cos term is equal to +1, which will occur whenever $2\pi\tau f = (m-1)2\pi$, where $m = 1, 2, \ldots$. Therefore, frequencies of the peaks and nulls in the spectrum will be

$$f_{\mathrm{peak}} = (m-1)/\tau. \tag{10.4}$$

$$f_{\mathrm{min}} = m/\tau. \tag{10.5}$$

These two equations indicate that the peaks will be spaced apart at a frequency interval of $1/\tau$,and so will the nulls. This will cause the spectrum to have a rippled pattern to it, and such a rippled pattern can induce time-separation pitch (TSP) in humans. In other words, many humans listening to a stimulus with a rippled spectrum can match a tone exactly to the frequency spacing of the ripple (Small and McClellan, 1963).

It would be useful at this time to consider some specific examples. Following Watkins (1966), let us consider a 1-kHz sinusoidal pulse of two cycles occurring at a repetition rate of 166 Hz ($\tau = 6$ ms) and continuing indefinitely. Two repetitions of the signal are shown in Fig. 10.2a, along with the spectrum for only two repetitions, and the spectrogram showing the spectral property of the signal. The next example is a signal consisting of a series of single 1-kHz sine pulses separated by the duration of a single cycle (1 ms). The results for this example are shown in Fig. 10.2b. From these two examples, we can see that the further apart the pulses are in the time domain the closer together the ripples will be in the frequency domain. The lines in the spectrogram, except at the fundamental frequency of 1 kHz, appear to be harmonics. In reality, these lines represent the peaks of the ripples in the frequency domain and these peaks are separated by the repetition rate that is equal to a frequency of $1/\tau$. Therefore, whenever lines of this sort appear on in a spectrogram, there is a strong possibility that the signal has a pulse character to it.

The next example taken directly from Watkins (1966) involves a situation in which both the pulse-tone and the repetition rate are progressively and simultaneously lowered, producing a sloping frequency pattern along with converging ripple bands. The signal consists of a series of sine pulses,

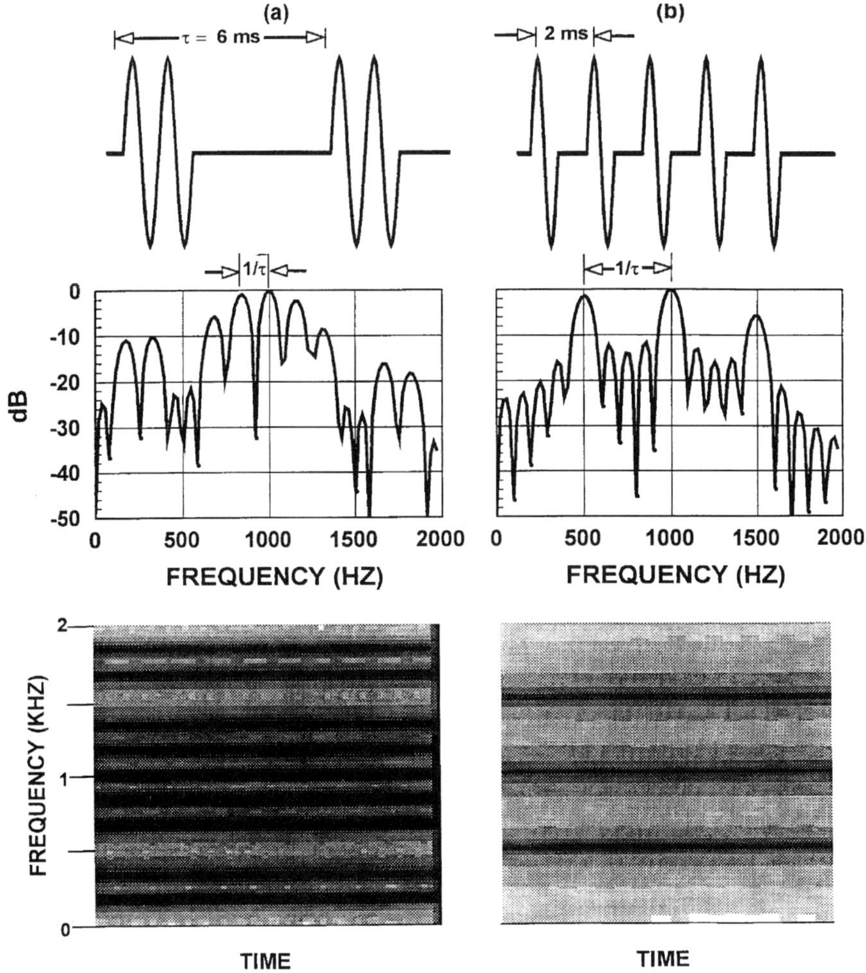

FIGURE 10.2. (**a**) A two-cycle 1 kHz tone burse pulsed at a rate of 166 pps. (**b**) A single cycle 1 kHz single pulse pulsed at a rate of 500 pps so that there is a one-cycle on and one-cycle off interval. Sample rate was 25 kHz and each block consisted of 1,024 points.

each eight cycles in duration with about ten cycles between pulses, and the tone starts with a frequency of about 2.2 kHz that gradually drops to about 0.5 kHz. The spectrogram of this signal is shown in Fig. 10.3 The ripples surrounding the fundamental for time less than 1.2 s indicate the pulse nature of the signal, and eventually the pulse intervals become long enough to be observed directly on the spectrogram for time greater than about 1.3 s.

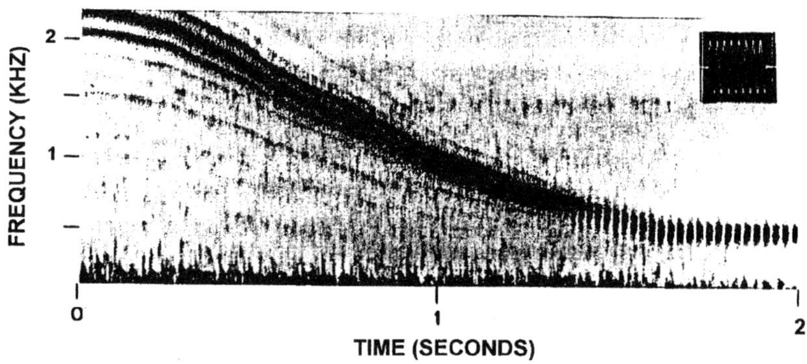

FIGURE 10.3. Eight-cycle pulse at changing frequency separated by about ten cycle intervals. The repetition rate is progressively lowered as the tone frequency drops. The analyzing bandwidth is 200 Hz (adapted from Watkins, 1966).

10.1 Social Sound Emissions by Odontocetes

Sound emissions by odontocetes (toothed whales and dolphins) can be classified into two broad categories of frequency-varying continuous tonal sounds referred to as whistles and broadband clicks (Evans, 1967), including burst pulse sounds. Whistles and burst sounds can be categorized as social sounds. Whistles appear to be used for intraspecific communications (Herman and Tavolga, 1980). Most of the research on social sounds has been performed on whistles and very little work has been done on burst pulse sounds. Perhaps, the main reason for this state of affairs is that the fundamental frequency of whistles is mainly in the sonic or audible range and are therefore easily recorded by off-the-shelf audio tape and DAT recorders. Later in this chapter, we will show that burst pulse sounds tend to be broadband and have significant ultrasonic frequency components and therefore require equipment that has frequency bandwidth larger than standard audio equipment. In our discussion of sound emissions, we will limit ourselves to the sounds that are produced internally by the animals and not consider the sounds produced by an animal striking an object or the water surface with any part of its body. Social sounds are those that are used by odontocetes in a social context and could be, but not necessarily, used for communications. In this chapter, we will focus more on the characteristics of the emitted sounds and not so much on the purpose or behavioral context under which certain sounds are produced.

The general practice for those studying social sounds of dolphins is to refer to these sounds as vocalizations. This terminology probably began with describing human voice sounds and has carried over in acoustic research with terrestrial animals. The term vocalization means to give voice and refers to the fact that voice sounds are produced by the vocal cords in the larynx of humans and other animals. This is an unfortunate choice of terminology, because research on odontocetes using widely different techniques and instrumentation has produced data that clearly indicate that sounds are produced within the nasal system and not in the larynx. Some of these experiments have produced data that seem to specifically rule out the larynx as a source of sound production. Diercks et al. (1971) used an array of contact hydrophones to measure a dolphin sonar signal and obtained results that placed the sound source at a depth of 1.5–2.0 cm from the surface in the vicinity of the nasal plug. Norris et al. (1971) and Dormer (1979) used high speed x-ray motion pictures to observe movements of the laryngeal and nasal region associated with sound production in live phonating *Tursiops truncatus*, *Tursiops gilli*, and *Stenella longirostris*. Hollien et al. (1976) used soft tissue spot x-rays to observe the larynx and melon of two phonating *Tursiops truncatus*. All of these investigators observed movements in the nasal system that correlated with sound production and did not see any correlated movements in the larynx. According to Dormer (1979), "the mechanism of sound production in porpoises consists of a series of muscular pumps, valves, and compliant sacs, with the sound being generated at the nasal plug." He observed the movement of the left nasal plug to be precisely associated with whistle production and speculated that the right nasal plug is associated with click production. Mackay and Liaw (1981) used an ultrasonic Doppler motion detector to measure vibrational motion in *Tursiops truncatus* and *Delphinus delphis* while the dolphins were phonating. The motion detector emitted a continuous wave of 2 MHz sound and could detect movements smaller than 10 μm. They observed that the nasal plug and the air sacs all vibrated in synchrony with the production of sound. They also observed that the nasal diverticular on the right side vibrated with clicking sounds all the time, but only some of the time on the left side. The vestibular sac was steadily inflated while the animal produced clicks. They indicated that the vibration of the nasal plugs probably is the originator of the clicking or buzz sound. Gross movements and rapid vibrations in the larynx were not observed when the animals emitted clicks and whistles. However, motion in the larynx was detected with each breath.

Ridgway et al. (1980) measured the muscle activities (using electromyographic techniques) and air pressures associated with sound production in five *Tursiops truncatus*. They found that the anterior internus, posterior internus, diagonal membrane muscle, and nasal plug muscles (all associated with the nasal system) fired just before and during the production of sounds, whereas the hyoepiglottal muscle and the intercostal muscles (associated with the larynx) did not. They also found that pressure increased in the nares and premaxillary sacs prior to sound production and dissipated after the sound. There were no pressure changes in the trachea during sound production. An example of muscle activities and pressure events in the nasal system associated with whistles is given in Fig. 10.4. An example of muscle activities

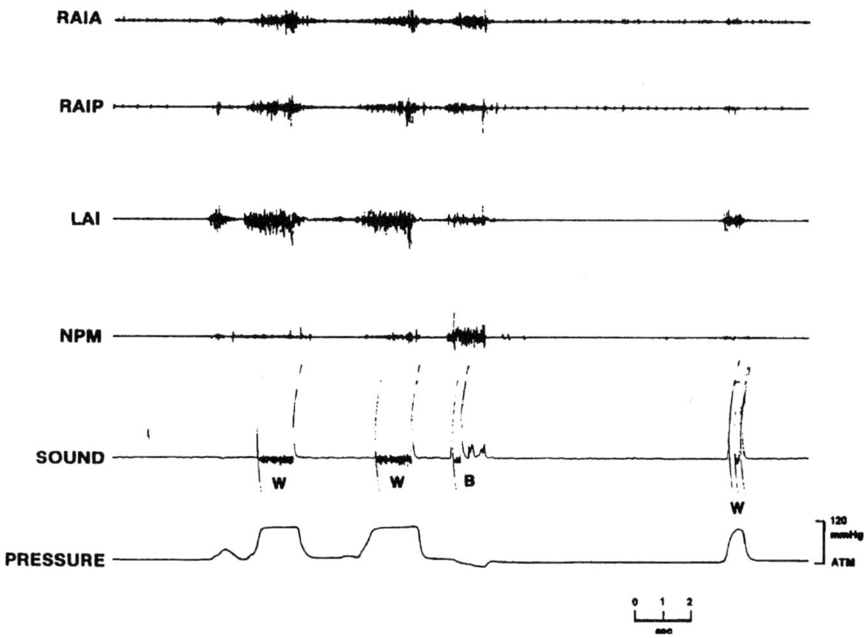

FIGURE 10.4. EMG, sound, and pressure events in the nasal system of *Tursiops* – right anterior internus muscle, anterior part (RAIA), right anterior internus muscle, posterior part (RAIP), left anterior internus muscle (LAI), nasal plug muscle (NPM), whistles (W) and blows (B) (adapted from Ridgway et al., 1980).

associated with clicks and whistles are shown in Fig. 10.5. Note in both figures that simultaneous muscle, pressure, and sound events were recorded. Amundin and Andersen (1983) measured the muscle activity of the nasal plug muscle and air pressure in the bony nares of a *Phocoena* and a *Tursiops*

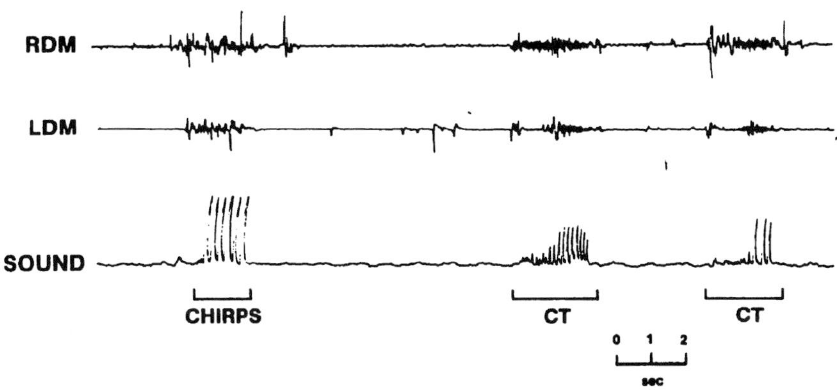

FIGURE 10.5. EMG of the right and left diagonal membrane muscles during pulsed sounds described as chirps and click train (CT) (adapted from Ridgway et al., 1980).

truncatus and obtained results similar to Ridgway et al. (1980). Ridgway and Carder (1988) measured air pressure in the nasal cavities and in the trachea adjacent to the larynx associated with sound production in an echolocating *Delphinapterus leucas*. They found that the intranasal pressure increased markedly but intratracheal pressure remained unchanged during the emission of echolocation clicks by the beluga. All of the measurements cited in this section suggest that the larynx of dolphins is not involved in the production of either click or tonal sounds.

The proponents of a laryngeal sound production mechanism based their arguments mainly on anatomical considerations and experiments with dead animals (Lilly and Miller, 1961, Purves, 1967; Purves and Pilleri, 1986; Blevins and Parkins 1973; Schenkkan, 1973).

There is a complete lack of experimental evidence with live dolphins that would implicate the larynx as the source of sound production. Therefore, in light of the overwhelming evidence pointing toward the nasal system of dolphins as the location of sound production, and lack of evidence to support the hypothesis of sounds having a laryngeal source, it is not proper to refer to sound emissions by dolphins as vocalizations. This terminology simply carries the wrong connotation; that is that sounds are produced in the larynx of dolphins by some kind of vocal tract mechanism. As an example of such misuse, consider the following statement from the abstract of a review article by Tyack (1997), a well-known marine mammalogist, "Diving marine mammals may not be able to rely upon involuntary voice cues for individual recognition, but rather may require vocal learning to maintain a stable signature as their vocal tract changes shape with increasing pressure during a dive."

The controversy on the site of sound generator by dolphin has become a non-issue by the recent data obtained by Cranford (2000). Cranford and his colleagues inserted a laryngioscope through the blowhole to visually examine the "monkey lips" while a dolphin produced clicks and whistles. They also simultaneously recorded the sounds the animal produced. Good correspondence was found between the movements of the monkey lips and the observations of transmitted signals. In their presentation at the Acoustical Society of America meeting at Newport Beach, California, Cranford et al. (2000) observed movements of the left monkey lips with the emission of whistle signals and movement of either the left or right monkey lips with the production of clicks.

10.1.1 Whistles

It is important to start this section by recognizing the fact that although many odontocetes produce whistle sounds, there are a few that do not. Among the odontocete species that never or rarely whistle are the harbor porpoise (*Phocoena phocoena*), Dall's porpoise (*Phocoenoides dalli*), Commerson's dolphin (*Cephalorhynchus commersonii*), Hector's dolphin (*Cephalorhynchus *hectori*), finless porpoise (*Neophocaena phocaenoides*), pygmy sperm whale (*Kogia* sp.), and the sperm whale (*Physter catadon*). Other porpoises in the family *Phocoenidae* and dolphins in the genus *Cephalorhynchus,* as well as a few other dolphins, do not emit whistles. Riverine dolphins also do not seem to emit whistles. However, all odontocetes studied to date do emit clicks and burst pulse sounds.

Dolphin whistles have traditionally been recorded with standard commercial tape recorders, cassette recorders, DAT recorders, and the audio track of video recorders. All of these instruments tend to have bandwidths that are limited to the audio range with upper frequency limits between 15 and 22 kHz. Although the frequency extent of the fundamental component of dolphin whistles often falls between 5 and 15 kHz (Herman and Tavolga, 1980), some species like the white-beak dolphin (*Lagenorhynchus albirostris*) produce whistles that may extend beyond 20 kHz. A spectrogram of a series of white-beaked dolphin whistles is shown in Fig. 10.6. Note that some of the whistles extend to frequencies of approximately 35 kHz. The signal in this example was recorded with a Racal instrumentation tape recorder.

Furthermore, whistles emitted by dolphins with the fundamental component well within the sonic range may have higher order harmonics that extend considerably beyond the ultrasonic frequency range (Lammers and Au, 1996). An example of such a whistle signal with higher order harmonics that extend close to 80 kHz is shown in Fig. 10.7a for a spinner dolphin (*Stenella longirostris*). These harmonics are an integral part of a whistle, and their presence indicates that the whistle signal being recorded deviates from a pure sine wave. Electrical engineers have traditionally used the amplitudes of the higher order harmonics as a measure of the amount of distortion in a signal when compared to a pure sine wave. An example of the waveform of the signal shown in Fig. 10.7a at the location of the dashed line occurring at 0.83 s is shown in Fig. 10.7b. If the whistles were measured with an instrument

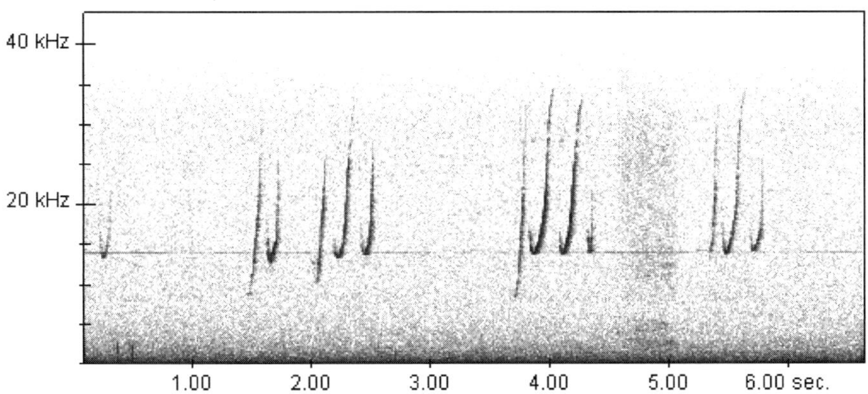

FIGURE 10.6. A spectrogram of whistles emitted by a white-beaked dolphin (courtesy of Marianna Rassmusen).

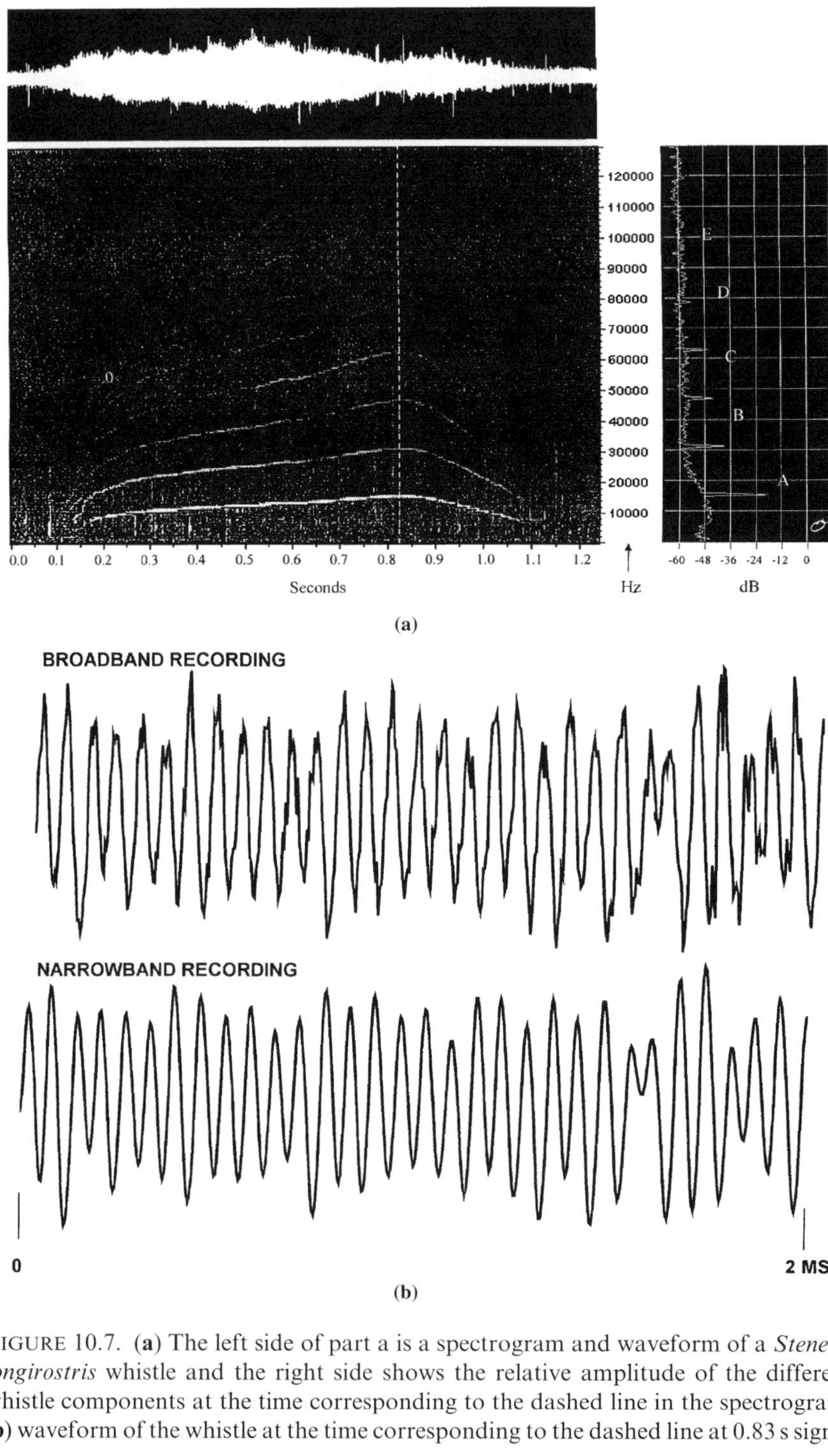

FIGURE 10.7. (**a**) The left side of part a is a spectrogram and waveform of a *Stenella longirostris* whistle and the right side shows the relative amplitude of the different whistle components at the time corresponding to the dashed line in the spectrogram, (**b**) waveform of the whistle at the time corresponding to the dashed line at 0.83 s signal recorded by a broadband and a narrowband system limited to 22 kHz.

having a high-frequency cutoff at 22 kHz, the waveform of the signals would appear as in Fig. 10.7b. One can clearly see by the cutoff measurement at 22 kHz, the appearance of the waveform is more sinusoidal. If band-limited versions of signals are used in a playback experiment, dolphins may readily recognize that the signals are not natural.

At our current level of knowledge, it is difficult to obtain a clear understanding of what the orientation of a dolphin has to be in order to measure undistorted whistle signals. Whistle signals are probably emitted in the form of a beam that has different width depending on the frequency of the signal. Higher frequency signals will have narrower beams than lower frequency signals. Therefore, the higher order harmonics may not propagate off axis from a dolphin as readily as the fundamental frequency. The animal's head structure may function like a low-pass filter, with the corner frequency decreasing as a function of the angle from the beam axis.

There are many different types of whistles that dolphins emit, both in captivity and in the wild. The traditional method (one still used by many to establish different categories and to determine the repertoire of whistle sounds) consists of visual inspection of spectrograms with particular emphasis on the shape of the whistle contour on the spectrograms. This technique is very subjective and has produced highly variable results among researchers. Therefore, research in call classification has been plagued with nomenclature difficulties and the lack of any standardization in techniques and classification categories. Some of the calls can be very complex and can include combinations of different individual structural units. In some of the earlier studies, considerable emphasis was placed on the linguistic possibilities of these sounds, so some investigators made very fine distinctions between similar contours. The findings of many of the earlier pre-1983 publications are included in an excellent review by Herman and Tavolga (1980) and some of these articles will be briefly mentioned here. Caldwell and Caldwell (1965) observed 5 different contours from over 1,400 whistles from four common dolphins, whereas Dreher and Evans (1964) identified 17 different contours among 9 captive *Tursiops truncatus*; 16 in a group of five *Tursiops gilli*, 19 in a group of wild common dolphins, and 7 in a pod of about 12 short-finned pilot whales. Dreher and Evans (1964) also reported 23 different contours for 3 juvenile *Tursiops* although they stated that some of the contours classified as being unique may have been minor variations of one another. An example of the contour-type categorization for four difference species from the research of Dreher (1966) are reproduced in Fig. 10.8. In this figure, subjectivity comes into play when one is trying to decide if a particular contour is really unique to all the other contours or just has some minor variations to another contour, and in determining what specific category a particular contour belongs to. One can be too strict in accepting contours into established categories and therefore come up with more contours than if one is relatively lax in assigning a contour to a category. Classification of sound repertoire has also suffered from nomenclature difficulties involved

Contour	Animal: Atlantic Bottlenose		Pacific Bottlenose		Delphinus Bairdi		Pilot Whale	
1	USED	1	USED	6	USED	3	USED	1
2	USED	2			USED	8		
3	USED	3	USED	1	USED	1	USED	2
4	USED	4	USED	3	USED	7		
5	USED	5	USED	6	USED	5	USED	6
6	USED	6	USED	3	USED	6	USED	5
7	USED	7	USED	6	USED	17		
8	USED	8	USED	2	USED	2		
9	USED	9					USED	4
10	USED	10			USED	9		
11	USED	11	USED	6				
12	USED	12	USED	6	USED	4	USED	3
13	USED	13			USED	14		
14	USED	14						
15	USED	15						
16	USED	16						
17			USED	6				
18			USED	5				
19			USED	4				
20			USED	5				
21			USED	5				
22			USED	4				
23			USED	2				
24					USED	11		
25					USED	18		
26					USED	13		
27					USED	16		
28					USED	16		
29					USED	12		
30					USED	15		
31					USED	10		
32	USED						USED	6

FIGURE 10.8. Examples of cetacean whistle contours published by Dreher (1966).

with naming of a sound without sufficient spectral and structural information to allow for consistency between researchers (Herzing, 1988). However, if very broad general categories are used, then whistles can be lumped into six, perhaps seven different general broad contours (Bazua-Duran, 1997). Additional categories entail minor variations within the six general broad categories shown in Fig. 10.9. The six fundamental categories are as follows.

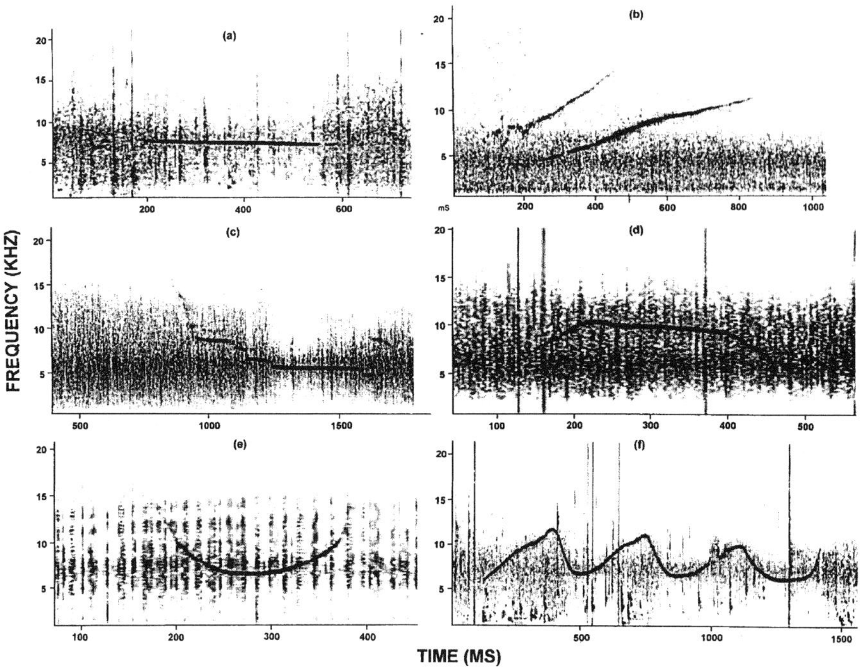

FIGURE 10.9. Examples of the spectrograms of some of the basic fundamental contour shapes, (**a**) constant frequency, (**b**) upsweep, (**c**) downsweep, (**d**) convex or hill, (**e**) concave or valley, (**f**) sinusoidal or multiple (courtesy of Carmen Bazua-Duran).

1. **Constant frequency** – signals where the frequency changed by about 25% or less over the total duration of the signal (see Fig. 10.9a). In this category, the frequency of the whistle does not really remain constant throughout its duration, but has a minimum amount of frequency change. Taruski (1979) was the first to use this category to describe the whistles of pilot whales.
2. **Upsweep** – whistles that are frequency modulated with the instantaneous frequency increasing over time and do not have any large inflection points (see Fig. 10.9b).
3. **Downsweep** – whistles that are frequency modulated with the instantaneous frequency decreasing over time and do not contain any large inflection points (see Fig. 10.9c).
4. **Concave** or **Hill** – whistles that are frequency modulated with the instantaneous frequency initially increasing with time, followed by an ending portion with the instantaneous frequency decreasing with time (see Fig. 10.9d).
5. **Convex** or **Valley** – whistles that are frequency modulated with the instantaneous frequency initially decreasing with time, followed by an ending portion in which the frequency increases with time (see Fig. 10.9e).

6. **Sinusoidal** or **Multiple** – whistles that are frequency modulated with more than one repetition of a hill or a valley and the contour appearing somewhat like a sinusoidal signal with at least two inflection points (see Fig. 10.9f). Categories 2–6 are based on the slope of the whistle and the number of inflection points, where an inflection point is defined as a point at which the slope of the contour reverses in direction, i.e., an upsweep changing to a downsweep (negative inflection) or a downsweep changing to an upsweep (positive inflection). Driscoll (1995) has added another category based on time and the presence of some frequency modulation. She refers to this category as **Chirp,** and signals in this category have short duration (less than or equal to 300 ms) and are frequency modulated. The contour shape may actually resemble some of the shapes of the other categories, but the main difference is that signals in this category are of relatively short duration. Caldwell and Caldwell (1970) were the first to describe chirps they observed in whistles produced by bottlenose dolphins. Any whistle with duration less than 300 ms automatically falls in another category.

Some researchers consider contours with slight variations from the fundamental contour shape as belonging to a separate category. For example, the sinusoidal or multiple contour of Fig. 10.8 may have one, two, three, or more positive humps, and each contour may be categorized into different classes. On the other hand, another researcher may consider these as being in one category. An upsweep may have a slight inflection and be considered to be in a different category as an upsweep without any inflections. Since there are no general, acceptable guidelines that the majority of researchers can agree on, considerable differences can exist between different groups of researchers. Therefore, the contour shape categorization process has a large subjective element to it.

A more quantitative method of categorizing whistle signals involves a determination of various parameters that can characterize whistles. Figure. 10.10 is a schematic of a generic spectrogram. Some of the different parameters associated with this spectrogram trace include the following: (a) starting frequency, (b) ending frequency, (c) minimum frequency, (d) maximum frequency, (e) number of inflection points, (f) frequency of inflection points, (g) duration, (h) break in contour or number of steps, and (i) the presence of harmonics. These parameters are then fed to different statistical software programs for analysis. The results of this quantitative method generally classify whistles into the six contour categories of Fig. 10.9 (Bazua-Duran, 1997). Steiner (1981) studied the species-specific differences in pure tonal whistles of five Western North Atlantic Dolphin species using a multivariate discriminant analysis and parameters a, b, c, d, e and h. The discriminant analysis easily discriminated the whistles of the long-finned pilot whale (*Globicephala melaena)* and the Atlantic bottlenose dolphin (*Tursiops truncatus)* from whistles of the Atlantic white-sided dolphin (*Lagenorhysnchus acutus*), Atlantic spotted dolphin (*Stenella plagiodon*), and spinner dolphin

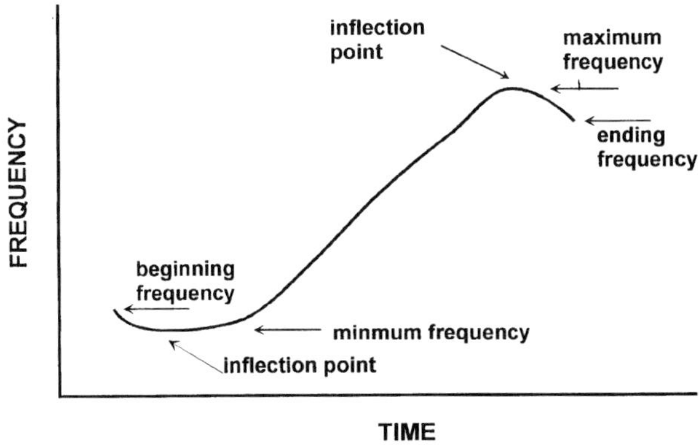

FIGURE 10.10. A generic spectrogram with some of the different parameters of the contour labeled.

(*Stenella longirostris*). The analysis did not discriminate between whistles of *L. acutus*, *S. longirstris*, and *S. plagiodon*. Wang et al. (1995) compared the whistles of seven odontocete species, *Tursiops truncatus*, *Lagenorhysnchus obscurus* (dusky dolphin), *Stenella frontalis* (Atlantic spotted dolphin), *Stenella longirostris*, *Stenella attenuata* (pan-tropical spotted dolphin), *Inia geofrensis*, and *Sotalia fluviatilis* (tucuxi). All the recordings were done in the wild. Wang et al. (1995) used the parameters a-h except for f and i, and included two other parameters that were not clearly defined; beginning sweep and end sweep. These parameters were analyzed by IBM PC computer statistical software called PC-SAS and STATVIEW (a Macintosh statistical program). They also used a similar discriminate analysis program as Steiner (1981) in order to make their results somewhat compatible. They found that species-specific differences existed in the whistles and the degree of differences correlated with taxonomic relations, habitats, and body length. Species with close taxonomical relations had similar whistles. Pelagic species had whistles of higher frequency and greater frequency modulation than coastal species, and riverine species had much lower frequency range and lower modulation than oceanic species. Another interesting result from their study is the strong correlation between the maximum frequency of the whistles with body length as shown in Fig. 10.11. The longer the body length, the lower the maximum frequency.

The contour of spectrograms can also be used "directly" in categorizing whistles. However, care must be taken to account for slight differences in the duration of a call. Whistles may have the same starting, ending, minimum, and maximum frequency, but slightly different durations and may therefore be classified into a different category by a statistical algorithm. Buck and Tyack (1993) used a non-uniform time dilation or time warping technique developed for human speech recognition (see Itakura, 1975) to essentially normalize the duration of whistles by either stretching or compressing the

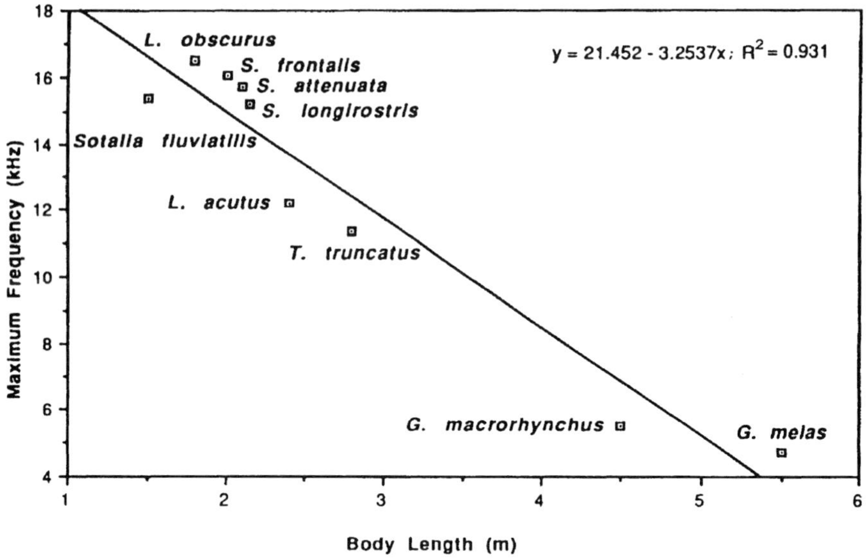

FIGURE 10.11. Maximum whistle frequency as a function of body length for different odontocete species (adapted from Wang, et al., 1995).

time axis of a contour. The warping was constrained so that it could not stretch or compress the time axis by a factor greater than two, and the number of points of the unknown contour $s(n)$ must be the same as the reference contour $r(n)$. Assume that $s(n)$ is of length N so that $n = 0,\ldots, N-1$ and $r(m)$ is of length M with $m = 0,\ldots, M-1$. A warping function $w(n)$ of length N was determined such that

$$D(s,r) = (\frac{1}{N}) \min_{w} \left\{ \sum_{n=0}^{N-1} (s[n] - r[w[n]])^2 \right\}. \tag{10.6}$$

The first and last values of $w[n]$ were constrained to ensure the warped version of $s[n]$ had the same length as $r[n]$, i.e., $w[0] = 0$ and $w[N-1] = M-1$. The following constraint on $w[n]$ limited the maximum and minimum dilation rate,

$$w[n+1] - w[n] = \begin{cases} 0, 1, 2, & \text{if } w[n] \neq w[n-1] \\ 1, 2 & \text{if } w[n] = w[n-1]. \end{cases} \tag{10.7}$$

An example of two contours before time dilation and after time dilation is shown in Fig. 10.12. A typical statistical analysis program would have probably placed the undilated whistles into two different categories and the dilated whistles into the same category. Two recognition experiments were performed by Buck and Tyack using three dolphin whistles, each from five dolphins of the Sarasota population using an "out-of-set" recognition algorithm. In the out-of-set recognition experiment, a library of

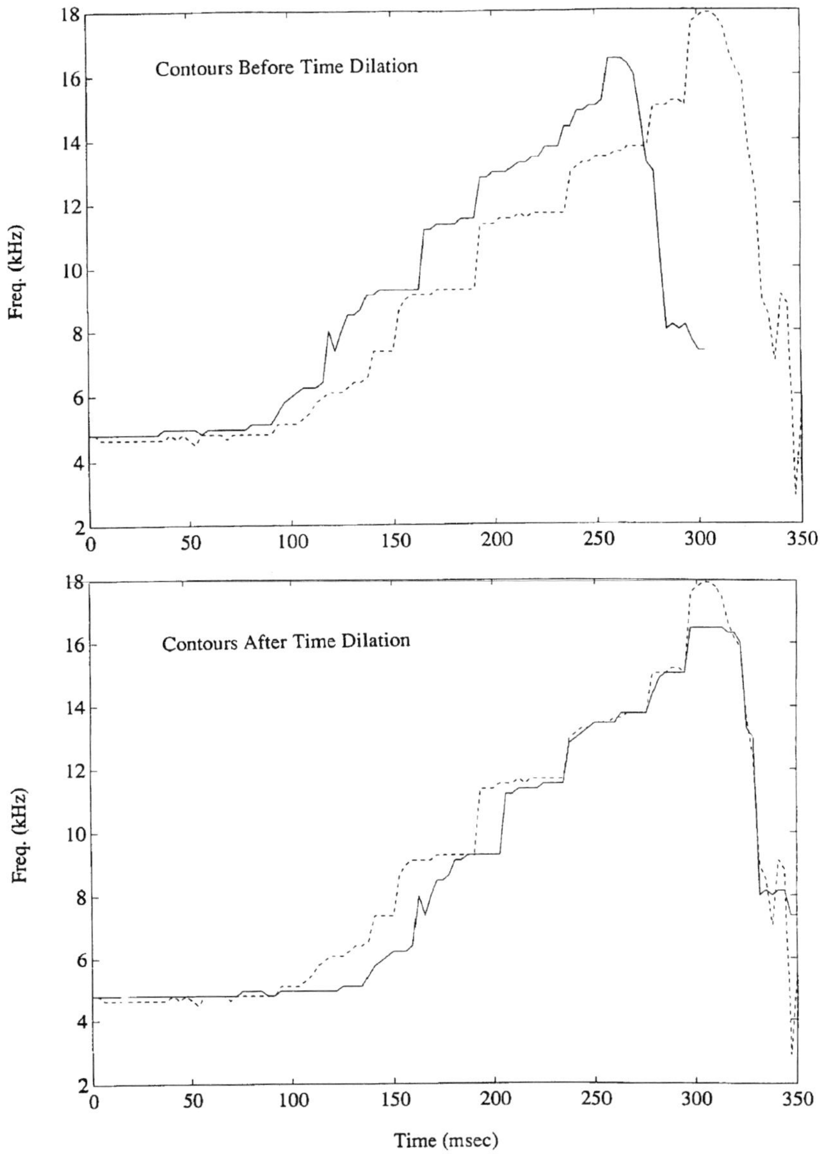

FIGURE 10.12. Two contours before time dilation (*top*) and same two contours after time dilation (*bottom*). (adapted from Buck and Tyack, 1993).

"known" signals was used to identify a set of unknown signals. Each unknown signal was compared against all the known reference signals, and a preferential listing of the reference signals was generated in the order of increasing distance. The unknown signal was assigned to the nearest reference signal. Each of the two experiments used a dictionary of five reference signals chosen as typical of the five different dolphins. The set of unknown signals was obtained by choosing three more signals at random from a larger library for each of the same five dolphins, giving them 15 whistles to be identified. The out-of-set algorithm was able to identify correctly 15 out of 15 single-loop whistles and 14 out of 15 central loops for multiple-loop whistles, and has been referred to as the contour similarity technique (CS technique) by McCowan (1995).

She tested the two techniques of normalization by analyzing both simulated and red signals and McCowan (1995) took a similar approach as Buck and Tyack (1993) in the sense of using a time dilation or warping to compare the contour shape of spectrograms. She linearly divided each contour into 19 parts to obtain 20 points and determined the peak frequency at each point. Dividing the time axis in this way amounts to normalizing the time axis, since every contour will be represented by 20 points and the time is essentially replaced by a point. Applying the linear time normalization technique of McCowan on the contours of Buck and Tyack shown in the top panel of Fig. 10.12, we obtained the contours shown in Fig. 10.13. McCowan's technique also produced contours that were fairly similar to each other. McCowan also applied a frequency normalization procedure in which the 20 peak frequencies for a normalized time contour were normalized by multiplying each frequency value by a normalization constant (each signal would use a different normalization constant). The frequency normalization used by McCowan could also be achieved by dividing the frequency value of each contour by the maximum of the 20 peak frequencies. The technique of time and frequency normalization has been referred to as the contour similarity technique (CS technique) by McCowan.

McCowan and Riess (1995) tested the two techniques of normalization by analyzing both simulated signals and 20 whistles, each from five captive

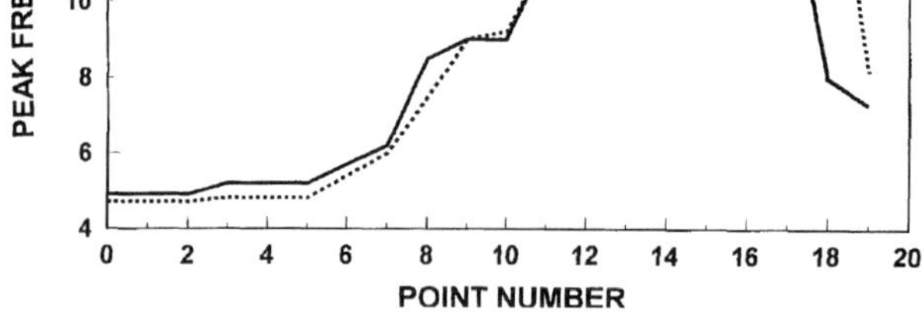

FIGURE 10.13. Results of applying the linear time normalization of McCowan (1995) to the contours of Buck and Tyack (1993) shown in the top panel of Fig. 10.12.

dolphins. She subjected the actual frequency time-normalized data to a K-means cluster analysis (Hartigan, 1975), and also generated a rectangular Pearson product-moment correlation matrix to obtain an index of similarity between the signals. A principal components analysis (Dixon et al., 1990) was conducted on the correlation matrices to reduce the number of collinear variables. Factors with eigenvalues greater than 1.0 were used in subsequent analyses. The data set that was subjected to both time and frequency normalization was tested with the K-means clutter analysis. Her results indicated that the technique grouped actual frequency data together according to the number of shared actual frequencies and grouped correlation coefficients successfully according to signal contour. The normalized time and frequency data were grouped together according to contour shape.

The use of artificial neural networks is a technique that deserves more attention in regards to classifying, categorizing, and discriminating call types. This topic will be considered in Chapter XII.

Many studies of social sounds have been done with captive dolphins. While valuable information can be obtained from acoustic measurements on captive animals, one cannot be all to sure if and how these sounds are used in the wild. Having a group of animals in a tank merely represents an artificial society that is in one way similar to a society in the wild. On the other hand, recording of sounds in the wild can be difficult, especially if broadband recordings are wanted and the identity of the phonating individual is also desired. Listed in Table 10.1 is a representative list of whistle recordings performed with wild dolphins.

10.1.2 Social Sounds: Signature Whistles

Dolphins are known to develop individually distinctive whistles that are called signature whistles, a term introduced by Caldwell and Caldwell (1965). From the acoustic recordings of 126 captive bottlenose dolphins, Caldwell et al. (1990) found that signature whistles made up about 94% of each individual's whistles. Most of the whistles were recorded while the

TABLE 10.1. A Representative List of Whistle Studies Involving Wild Dolphins

Species	Reference
Pilot whale (*Glophicephla melaena*)	Taruski (1979)
Atlantic Bottlenose Dolphin (*Tursiops truncatus*)	Wang et al. (1995)
	Schultz and Corkeron (1994)
	Smolker et al. (1993)
	dos Santos et al. (1995)
Spinner dolphin (*Stenella longirostris*)	Driscoll (1995)
Atlantic spotted dolphin (*Stenella frontalis*)	Herzing (1996)
Killer whale (*Orcinus orca*)	Ford and Fisher (1983)
Beluga whale (*Delphinapterus leucas*)	Sjare and Smith (1986)
Narwhal (*Monodon monoceros*)	Ford and Fisher (1978)

individual dolphins were isolated. The signature whistles were distinctive among individuals and each individual produced a stereotyped signature whistle that was stable over many years. Most of the Caldwells' research has been with captive dolphins, and while one can learn a great deal from these animals, there was always the nagging question if wild dolphins behave acoustically in similar manners as captive dolphins. Readers interested in an detailed discussion of signature whistles should consider the review by Tyack (1997).

Signature whistles have also been observed in wild bottlenose dolphins. Dr. Randy Wells and his team in Sarasota, Florida, have been involved in an annual capture–release program for over 25 years in which animals are captured, measured, aged, sexed, sampled, marked, and their whistles recorded by means of a hydrophone in a suction cup before the animals are released (Wells, 1991). Sayigh et al. (1990) analyzed a total of 398 recordings of 134 known individuals in the Sarasota bottlenose population and obtained signature whistle results that were similar to the captive-animal work of Caldwell et al. (1990). Some signature whistles can be very stereotyped while others can exhibit variable features. Some signature whistles show variations in the number of repetitive elements, called loops by Caldwell et al. (1990). They can also have features that vary in both time and frequency. In Fig. 10.14a are two examples of signature whistles with repetitive loops, one from a captive bottlenose dolphin and another from a wild dolphin in the Sarasota population. In this example, the whistles have an introductory loop designated as "I", from one to three repeated central loop designated as "C," and a terminal loop designated as "T." Usually, there are several discrete whistle sounds that are separated by silence. In Fig. 10.14b are examples of signature whistles from a specific bottlenose dolphin in the wild that do not contain repetitive loops, but have variations in time and frequency. These whistles all have a basic pattern of frequency modulation, with a central segment of relatively constant frequency of 5 kHz, along with higher order harmonics. The variation in the time duration of the constant frequency portion is obvious. Sayigh (1992) analyzed the whistle contours from 83 wild dolphins from the Sarasota population and found that 73% had repeated loops while only 27% did not. Although the recordings used by Sayigh (1992) were from wild dolphins, the animals were temporarily captured and hauled onto a boat. This procedure introduces an uncertainty as to the degree these recorded signals represent typical signals used in the wild.

Smolker et al. (1993) studied the use of signature whistles by a population of wild *Tursiops* in the waters off Peron Peninsula in Western Australia, focusing on mother–infant pairs involving infants up to 4 years in age. Bottlenose dolphins exist in a fission–fusion society in which individuals typically associate in temporary parties that change frequently as individuals join and leave in a fluid manner, and all members of the social network are never found in a single group (Wells et al., 1980, 1987). Therefore, signature

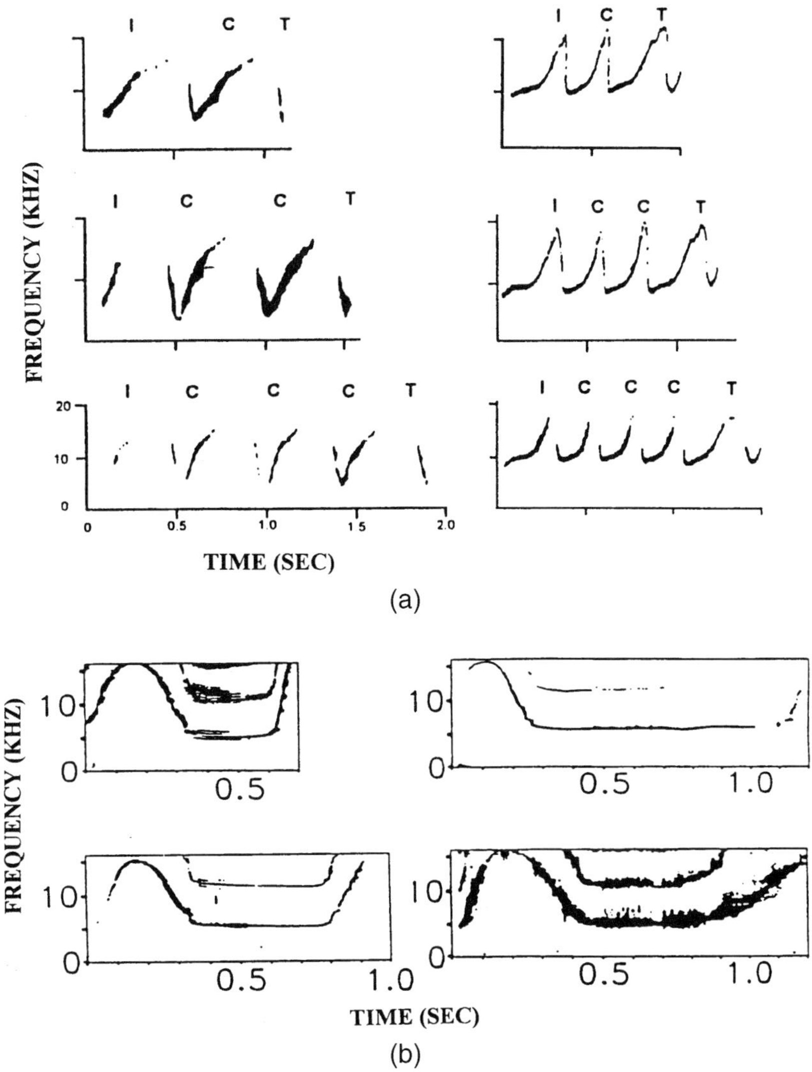

FIGURE 10.14. (**a**) Examples of signature whistle with varying number of repetitive sum units called loops by Caldwell et al. (1990). The whistles in the left column are from a juvenile male in captivity and the whistles in the right column are from an adult female recorded in the wild (**b**) Examples of a signature whistle without repetitive loops from an adult female dolphin in the wild (adapted from Tyack, 1997).

whistles can be very useful to locate individuals and to identify other individuals using acoustic cues. Smolker and her colleagues found that signature whistles occurred primarily when mother–infant pairs were separated and the whistles were repetitive and were generally concentrated toward the later stages of the separation involving the process of reunion. They found that the mothers did not appear to whistle as frequently as the infants during the separation. They hypothesized that infant whistles may function to facilitate reunions by conveying information to the mother concerning the infant's desire to reunite and/or its location . They also hypothesized that the infant whistles could induce a cooperative response from the mother including approach, slowing to allow the infant to catch up or to whistle. Another interesting finding of their study is the good correlation between the separation distance between mother–infant pairs and the amount of whistling done by the infant. The histogram in Fig. 10.15 depicts the percentage of acoustic events that contained whistles as a function of the approximate separation distance between mother and infant. When the separation was greater than 100 m, the emitted sounds of the infants were always whistles. As the range decreased, the percentage of whistle events also decreased. Whistles rarely occurred when infants were together with their mothers, but did occur during "infant position breaks" involving separation distance up to about 5 m. Smolker et al. (1993) concluded that highly individualized signature whistles may be useful in a fission–fusion society in which individuals join and leave

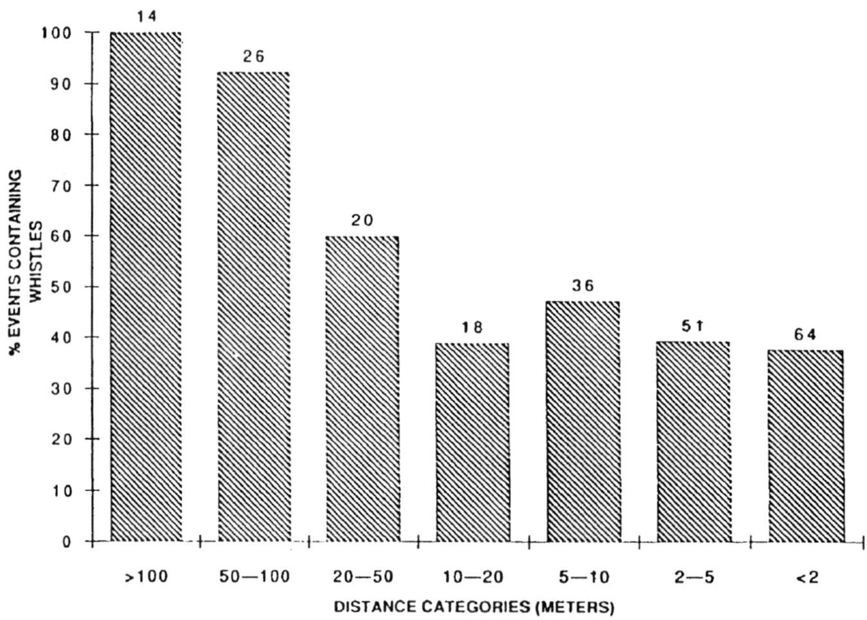

FIGURE 10.15. The percentage of separations during which infants whistled as a function of maximum distance. Distance classes are infant position break (<2 m, 2–5 m), close separation (5–10 m, 10–20 m), and far separation (20–50 m, 50–100 m, >100 m). Numbers above each bar are the number of events within each class (adapted from Smolker et al., 1993).

temporary groups in a fluid manner, yet maintain long-term association with particular individuals.

Recording the social sounds from specific individual dolphins in the wild can be an extremely difficult matter and it may require years of recording to obtain sufficient information on specific animals. Dr. Denise Herzing has spent over 12 seasons obtaining simultaneous underwater video and audio recordings of a population of Atlantic spotted dolphins (*Stenella frontalis*) along the sand banks and adjacent deep waters off Grand Bahama Island in the Bahamas. One hundred and fifty five individual spotted dolphins have been identified by photographs and video of dorsal fins, flukes, and constellation of spots. Whistling dolphins were easily identified since bubbles were often expelled from the blowhole during sound emissions as shown in Fig. 10.16. The simultaneous emission of a bubble stream and whistles has also been used by McCowan and Reiss (1995) to identify whistling dolphins.

One type of whistle contour observed by Herzing (1996) was a predominate and unique repetitive whistle that was consistently observed for up to ten years for some dolphins. These types of whistles were considered signature whistles. Whistles typically had a fundamental frequency between 4 and

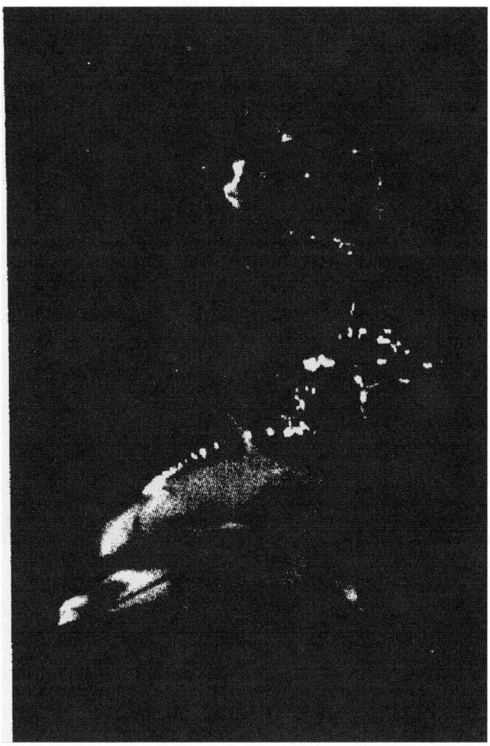

FIGURE 10.16. Bubbles from the blowhole of a specific identifiable dolphin while emitting whistles (adapted from Herzing, 1996).

18 kHz and durations from 0.5 to 8 s. Signature whistles were observed in three behavioral contexts:

1. *Mother/calf reunions*. Mothers produced their own signature whistles after calves departed from them. After the mother whistled, the calves rejoined her.
2. *Alloparental care*. Older male and female spotted dolphins in an association with younger dolphins produced their own signature whistles prior to the retrieval of younger dolphins.
3. *Courtship*. Female and male spotted dolphins were observed repetitively broadcasting their own signature whistle during attempted courtship and mating activity.

Examples of some signature whistles of *Stenella frontalis* from Herzing's (1996) work in the Bahamas are shown in Fig. 10.17. The spectrograms labeled A1, A2, A4 are signature whistles of mothers, and the corresponding signature whistles of the calves are shown below in the spectrograms labeled B1, B2, B4, and B5.

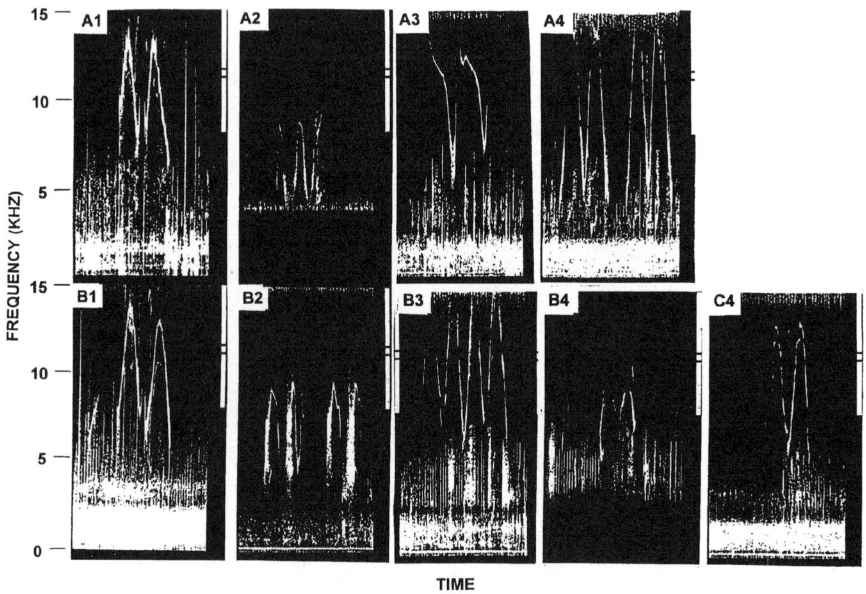

FIGURE 10.17. Spectrograms of signature whistles of wild *Stenella frontalis*. A1 is the signature whistle of Nippy, a young adult female, and B1 is the signature whistle of Nassau, a 3-year-old female calf of Nippy. A2 is the signature whistle of Mugsy, a young adult female, and B2 is that of her 2-year-old male calf, Mel. A3 is the signature whistle of a juvenile female named Trimy and B3 is that of a juvenile female named Katy. A4 is the signature whistle of Luna, an old adult female, B4 is the signature whistle of Diamond, her 4-year-old female calf and B5 is of Latitude, a 3-year-old male calf (after Herzing, 1996).

The spectrograms of Fig. 10.17 clearly indicate that some *Stenella frontalis* infants produce signature whistles that closely matche the signature whistle of their mother. The spectrograms of A1 and B1, A4 and C4 are very similar in shape. This type of correspondence was not found in the signature whistle studies with *Tursiops truncatus* (Caldwell and Caldwell, 1979; Sayigh, 1992). With the bottlenose dolphins, most calves appear to develop signature whistles that are quite different from those of their parents. In a study of 14 infants born in captivity, Caldwell and Caldwell (1979) found that only one dolphin produced a signature whistle similar to its parents. In a study of the signature whistles of nine bottlenose calves, only one calf developed a signature whistle somewhat similar to that of her mother. A variation on the signature whistle that Herzing (1996) has recorded is the excitement call. This call combines low-frequency burst pulse sounds, which are discussed in the next section, with high-frequency signature whistles. This type of call had durations between 2 and 30 s. Dolphins producing excitement calls were usually highly erratic in their swimming behavior and emitted streams of bubbles. Excitement calls were also observed during periods of distress, frequently emitted by calves under three years of age. Cause of distress behavior varied from intraspecific social behavior to the presence of humans in the water. After the start of a distress call, an older dolphin would often calm the youngster down by a slight pectoral flipper brush on the flank of the calling calf. Distress calls rarely exceeded 30 s before a conspecific was successful in calming the calling youngster down.

The signature whistle hypothesis has been challenged by McCowan and Reiss (1995). Using the contour similarity technique discussed in Section 10.1.1, they analyzed the whistles of ten adult dolphins in captivity from three different social groups. The normalized spectrogram data were subjected to a K-cluster analysis and a discriminant analysis. They determined whistle categories within social groups and across social groups. Their results were inconsistent with the signature whistle hypothesis, since their subjects not only produced many different whistle types but also shared several whistle types within as well as across social groups. Needless to say, much more research needs to be done to resolve the signature whistle issue. Again, we would like to stress the importance of obtaining good reliable data from wild dolphins in a natural social setting in order to resolve the signature whistle issue.

10.1.3 Burst Pulses

Burst pulse sounds are another major category of sound emissions by odontocetes. All odontocetes produce burst pulse sounds. For dolphins and small whales, burst pulse sounds are characterized by a high repetition rate (greater than about 300 pulses per second) or low interpulse intervals (less than about 3 ms). To the human ear, burst pulse sounds resemble squawks, blats, squeaks, cracks, snaps, bleats, barks, groans, and moans, and have been described as such in the literature. Humans and perhaps dolphins probably hear the time-separation pitch associated with the interclick interval or repetition rate that was discussed in the beginning of this chapter. Another way of describing the auditory sensation coming from listening to burst pulse sounds is to describe the process as listening to the repetition rate of the pulse train.

Researchers have suggested that burst pulse sounds are an important part of dolphin sound emissions (see Herman and Tavolga, 1980), yet these sounds have not been studied as much as whistles. The heavy emphasis on studying whistles has led many to suppose that whistles are the primary mode of communication in dolphins and small whales. Part of the reason for this state of affairs can probably be attributed to the wide band nature of burst pulse sounds compared to the frequency of the fundamental components of whistles, which are mainly in the human auditory range. Burst pulse sounds can have frequency components that extend beyond 100 kHz (Lammers and Au, 1996). An example of such a burst pulse sequence is depicted in Fig. 10.18. Therefore, special wide band recording instruments and wide band hydrophones must be used to capture the entire spectrum of these pulses. The burst pulse sound of Fig. 10.18 from a wild Hawaiian spinner dolphin has no frequency components in the audio range and only frequency components in the ultrasonic range.

There is a fine region of demarcation between echolocation clicks and burst pulse sounds, and one type of sound may merge into the other during a sound emission bout. It would not be surprising if we eventually learn that burst pulse sounds and echolocation clicks are produced by the same mechanism. The distinction between the two types of pulse sounds can be found in the interclick intervals and intensity. However, determining the source intensity or pressure of a sound emission requires a knowledge of the range between the hydrophone and the emitter, a task that is not simple in the wild or for that

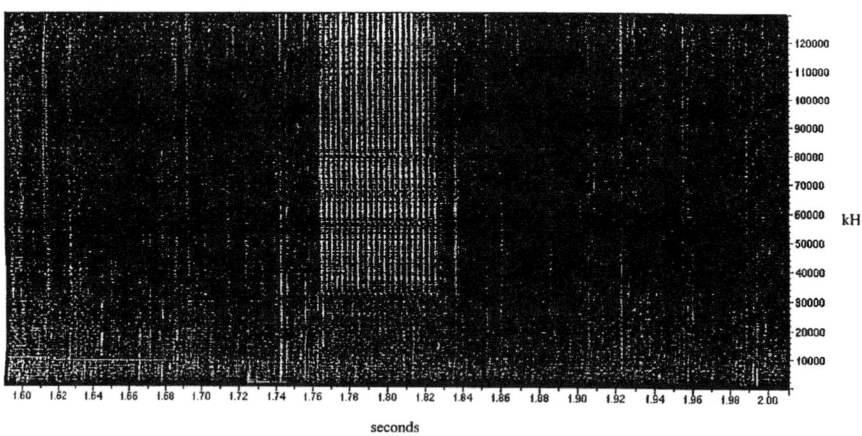

FIGURE 10.18. Spectrogram of the burst pulse signal of a Hawaiian spinner dolphin having only ultrasonic frequency components and no audio frequency components (adapted from Au et al., 1998).

matter, in a tank if the dolphin is swimming rapidly while emitting the pulses. Therefore, interclick interval is probably the best method to use in determining whether a pulse train is an echolocation click train or a burst pulse sound. We will see in the next section that when a dolphin is echolocating, the interclick intervals are usually longer than the two-way travel time for the signal to travel from the animal and back plus a lag time. Lag time is defined as the difference in time between the two-way travel time and the interclick interval. Evans and Powell (1967), using a simultaneous audio-video system with an echolocating bottlenose dolphin that swam toward a target, found an almost constant mean lag time of 15.4 ms for target ranges of 1.4–0.4 m. Only when the animal approached within 0.4–0.03 m of the target did the mean lag time decrease to a minimum of 2.5 ms. However, it is doubtful that at these short ranges a dolphin can do meaningful echolocation since the animal's transmitter is on the top portion of its head and sounds enter through the lower jaw so that it would be very difficult to align both portions of the dolphin's head toward a very close target.

Burst pulse sounds were observed by Au et al. (1987) while measuring the beam pattern of echolocation signals used by the beluga whale (*Delphinapterus leucas*). At that time we did not recognize that we were recording burst pulse sounds that occurred at the end of an echolocation click train whenever they were observed. We referred to these signals as Mode 3 echolocation signals

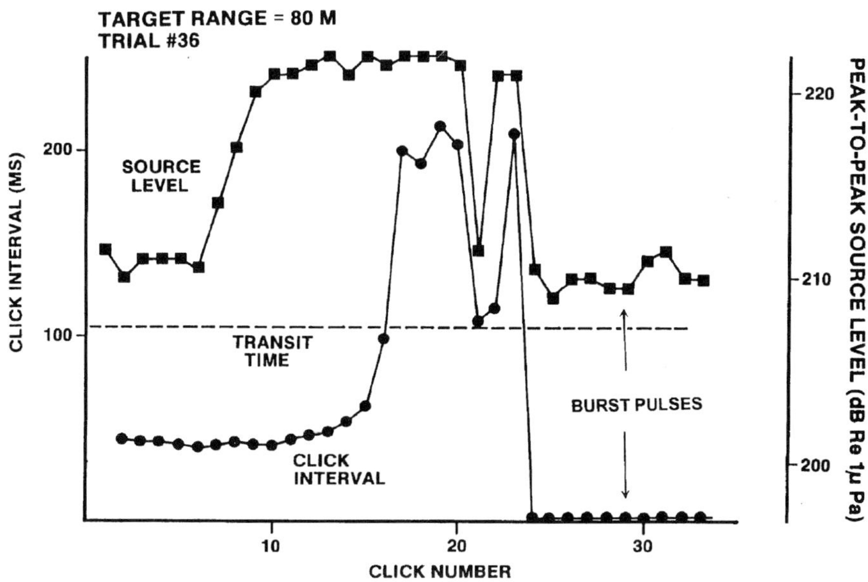

FIGURE 10.19. An example of the variations in the click interval and source level in a single trial. The left ordinate represents the click interval and the right ordinate represents the source level. The horizontal dashed line is the two-way travel time for a signal to travel from the animal to the target and back (adapted from Au et al., 1987).

when they should have been labeled as burst pulse sounds. There were weeks in which these sounds would not be detected and then prolonged periods in which they would be present, at the terminal phase of a echolocation trial, and they seemed to signify the completion of the echolocation task. We found that the properties of the burst pulse sounds were much different than those of the echolocation signals. The average interclick intervals were very low, 1.7 ± 1 ms, and the amplitudes were between 12 and 20 dB below the amplitude of the higher echolocation signals. An example of the difference in the echolocation signals and the burst pulse signals can be seen in Fig. 10.19, which shows both the amplitude and interclick interval during the click train. We also found that the burst pulse signals were highly directional. Examples of the vertical and horizontal beam patterns of burst pulse signals in comparison to the beam patterns of echolocation signals are shown in Fig. 10.20. The beluga whale was required to station in a bite bar and a vertical array of hydrophones at about 1.9 m from the beluga's blowhole was used to measure the vertical beam A horizontal array was used to measure the horizontal beam. The beam. patterns in Fig. 10.20 strongly suggest that the same source mechanism is involved in the production of burst pulse and echolocation signals. The data also indicate that

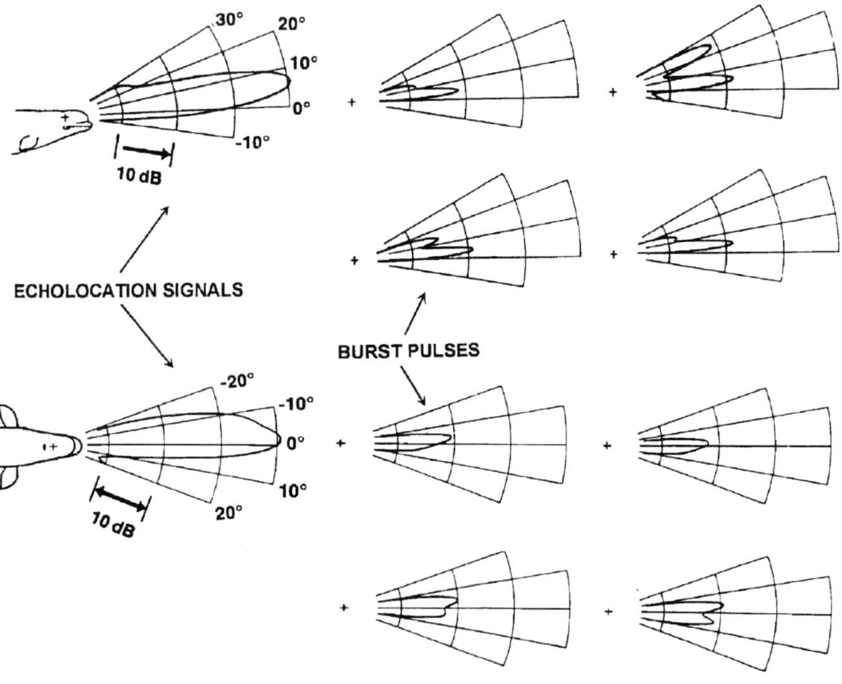

FIGURE 10.20. Examples of the beam pattern of burst pulse sounds in the vertical and horizontal planes from a single click train. The beam patterns in the vertical and horizontal planes for the highest amplitude echolocation click in a trial are also shown. The outer amplitude scale for all the polar plots is 223 dB re 1 μPa (after Au et al., 1987).

in order to obtain accurate source level and frequency information when recording burst pulses, a hydrophone must be aligned along the major axis of a dolphin's beam. We will see in the section on echolocation signals that the waveform of clicks and pulse signals will be distorted if the signals are not measured along the major axis of the transmission beam.

Burst pulse sounds that are associated with different behaviors have been reported by many investigators. Herzing (1988) has summarized much of the literature on burst pulse sounds published before 1988. For the bottlenose dolphin, high-intensity, broad-band cracks or pops have been observed in alarm and fright, in agonistic, aggressive, and head-to-head open mouth encounters often accompanied by head nodding, shaking, and arching. Squawks and squeals have been observed in head-to-head and agnostic and open mouth encounters and in play. Buzzes and yelps in precopulatory behavior have also been reported. Cracks or pops have been observed for *Phocoena phocoena* in an alarm context. Pulsed calls are also common in *Orcinus orca* (Ford, 1985) and in the beluga whale (Sjare and Smith , 1986). It is common to encounter burst pulse sounds in the Hawaiian spinner dolphin (Norris et al., 1985; Lammers and Au, 1996). Blats and squeaks from pilot whales, squawks, and squeaks from the Atlantic spotted dolphin and in narwhals has been reported. The use of pulsed codas in sperm whales have also been reported by Watkins and Schevill (1977).

Despite the many observations and reports of burst pulse sounds in the literature, there is a paucity of quantitative description of these sounds. Herzing (1988) performed a quantitative analysis of a particular type of burst pulse sounds and squawks in captive bottlenose dolphins. The various parameters considered were squawk duration, repetition rate, separation between harmonic bands, intersquawk interval, maximum frequency, distribution of sound energy, changes in repetition rate within squawks, frequency modulation by whistle-like contour within squawks, number and presence before, after, or overlap of other types of sound emissions, including eight categories of whistles and two categories of other pulsed sounds, and whether the squawk was discrete in time or continuous with echolocation. A principal components analysis was performed with repetition rate, maximum frequency, a continuity measure of squawks relative to surrounding echolocation, duration, and a singularity variable of the squawk. Descriptive statistics were computed for the duration, intersquawk interval, and maximum frequency.

Squawks were found to occur separately and continuously with echolocation, occurring either immediately before or immediately after echolocation. Those that occur separately have been viewed as being emotionally induced social sounds. Discrete squawks were found more often than continuous squawks and were used more often in direct confrontational contexts. Discrete squawks were also more frequently found in sequences that increased or decreased in activity. Continuous squawks were found in sequences that had multiple changes in intensity or behavior or that showed no change. The repetition rate of squawks showed a consistent relationship to the degree of frequency modulation of the squawk (low repetition rate associated with simple modulation and high repetition rate with complex modulation). The principal components analysis showed that two heavily weighted independent factors emerged, the repetition rate and the continuity aspect of the squawk with echolocation. Duration of squawk and of the intersquawk interval were highly correlated. As squawks became longer, the intersquawk interval became proportionately longer. Whistles were commonly found overlapping squawks or at the beginning or end of squawks. Herzing (1988) recordings were band limited to 22 kHz, so that some of her results must be considered with caution since she may have missed many burst pulse sounds that have only ultrasonic components.

The long-term research of Herzing (1996), observing primarily the Atlantic spotted dolphin (*Stenella frontalis*) but also *Tursiops truncatus*, has been very instrumental in our understanding of how various types of sound emissions

TABLE 10.2. Sound Emissions and Associated Underwater Behavior of *Stenella Frontalis* and *Tursiops Truncatus* (after Herzing, 1996)

Call type	Spectral description	Behavioral context	Species, age class, and sets
Signature whistles	Frequency modulated (FM) whistle 4–10 kHz	Mother/calf reunion	Spotted dolphins all age classes Female/ male offsprings
	0.5–8 s Bubbles emitted from blowhole occasionally	Alloparental care	Spotted dolphins Juvenile, young, and old adults-female/ male
Excitement calls	Burst pulsed sounds with overlapping signature whistles: 4–18 kHz 2–30 s duration Bubbles emitted from blowhole	Distress or excitement	Spotted dolphins All age classes Most frequently, calves female/male
Genital buzz	High repetition rate clicks 1.2–2.5 kHz 8–2000 clicks/s	Courtship	Spotted dolphins Juvenile, young, and old adult males
	6–20 s duration	Discipline	Intra and Interspecific Adult, juvenile Females/males
Squawk	Broad-band burst-pulse sounds 0.2 –12 kHz	Agonistic and aggressive	Intra and Interspecific All age classes Females/males
	0.2–1.0 s duration 200–1200 clicks/s	Sexual play	Intra and Interspecific All age classes Females/males
Scream	Overlapping FM whistles 5.8 –9.4 kHz 2.5 –4.0 s duration	Agonistic and aggressive	Intra and Interspecific Juvenile, young, and old adult males
Bark	Burst-pulse sounds 0.2 –2.0 kHz 0.3 –1 s duration	Agonistic and aggressive	Intra and interspecific Juvenile, young, and old adult males
Synchronized squawks	Burst-pulse sounds 0.1–15 kHz Main energy 0.1–2.2 kHz 0.9 –1.0 s duration	Agonistic and aggressive	Intra and interspecific Juvenile, young, and old adult males

are used in a wild dolphin society. The results of her simultaneous video and audio recordings of these wild dolphins in the shallow sand banks of the Grand Bahama Island are summarized in Table 10.2. The frequency range for the burst pulses in Table 10.2 should be regarded with caution, since the audio portion of the video was band limited to 22 kHz.

The following description of pulse burst sounds and the context in which they are produced and tabulated in Table 10.2 comes directly from Herzing (1996). Genital buzzes were observed in courtship and disciplining. Genital inspection and buzzing of a female by a male approaching from below often occurred. When mothers were unsuccessful in retrieving their calves, they swam inverted and buzzed the genital area of their offspring. Mothers eventually made physical contact with their rostra to the calf's flank. Squawks were recorded during bouts of agonistic and aggressive behavior and during sexual play. Squawks were commonly recorded in the following contexts:

(1) Intraspecific agonistic/aggression displays between spotted dolphins with head-to-head and open-mouth posturing and including body charges and tail slaps to the head. Aggression was frequently observed between male groups and between unreceptive/unapproachable females and courting males.
(2) Interspecific agonistic/aggression encounters between spotted and bottlenose dolphins. Groups of spotted male dolphins would face off with solitary or small groups of bottlenose dolphin and chase them as a solidified unit, often emitting synchronized squawks. Intraspecific sexual play between spotted dolphins.
(3) Interspecific sexual play/courtship between spotted and bottlenose dolphins in which adult male bottlenose dolphins copulated with young female spotted dolphins and when adult female bottlenose dolphins solicit copulation with young male spotted dolphins.

Barks, which are low-frequency burst pulse sounds, often occurred with screams, which are overlapping frequency-modulated whistles and were recorded during highly escalated agonistic/aggressive interspecific and intraspecific encounters. Synchronized squawks are highly coordinated burst pulse sounds that were recorded during highly escalated aggressive activity in which male spotted dolphins coordinated their swimming behavior and posture in agonistic/aggressive fights with other groups of spotted dolphins and in escalated fights with bottlenose dolphins.

Nonwhistling dolphins like *Phocoena phocoena* must adjust the repetition rate of their click signals in order to produce communication or social sounds. The clicks produced by *Phocoena* are often discussed in terms of having a low-frequency (LF) component at about 2 kHz and a high-frequency (HF) component between 120 and 140 kHz as if there were two separate signals. In reality, there is only one signal; the high-frequency "rides" on the low-frequency component, as can be seen in Fig. 10.21. However, if band-limited

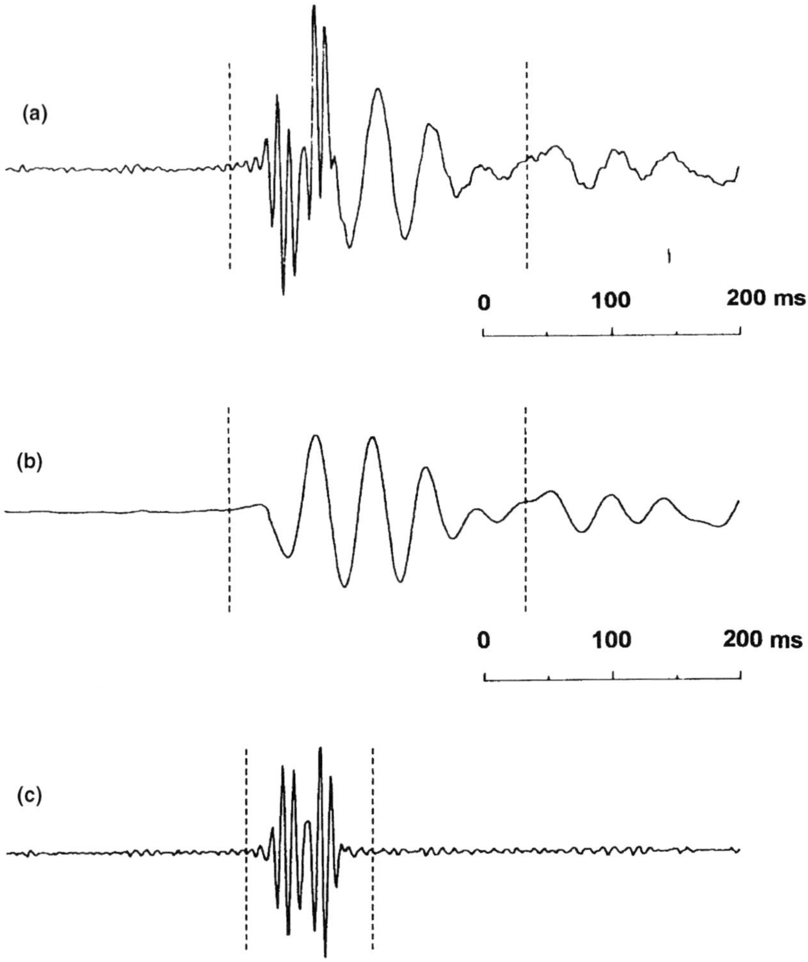

FIGURE 10.21. (**a**) An example of a *Phocoena* click signal having low- and high-frequency components, (**b**) same click signal after low-pass filtering, (**c**) same click signal after high-pass filtering (after Kamminga and Wiersma, 1981).

equipment is used, then only the LF portion will be observed as in Fig. 10.21b. Conversely, if the recording system has a high-pass filter, then only the HF portions will be observed as in Fig. 10.21c. It is also possible that when *Phocoena* emits high-amplitude echolocation signals, the LF portion may be depressed or totally absent. The HF portion of the signal is 6–10 dB higher than the LF portion.

Amundin (1991) studied the social sounds produced by five subadult males and one subadult female *Phocoena* in a pool and described these sounds for different behavioral contexts. The first type of repetition rate pattern

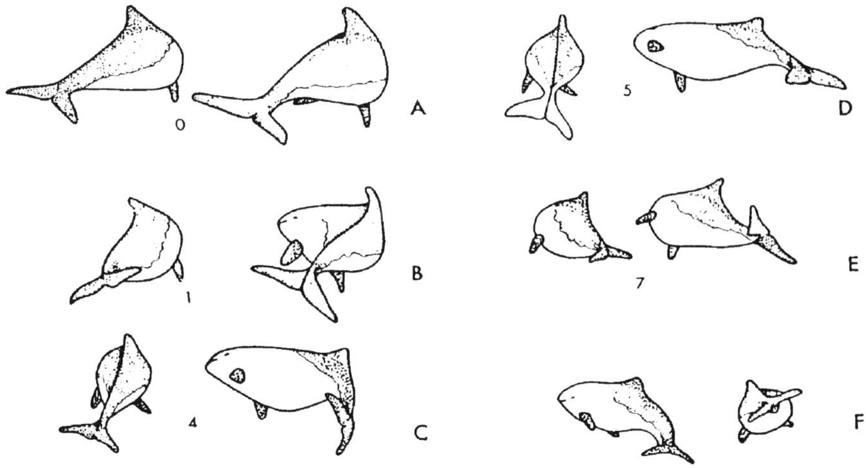

FIGURE 10.22. "Sideward turn threat" seen in connection with a specific call. The porpoise to the right points its rostrum toward the head region of the other by quickly turning sideward (starting in frame **B**) with the anterior part of the body. The other animal reacts by avoidance (starting in frame **C**). The pictures were drawn from an 8-mm film sequence, filmed at 18 frames/s. The numeric values indicate the frame number (adapted from Amundin, 1991).

described was the *sideward turn threat call*. An example of the social context of this type of call is depicted in Fig. 10.22 showing-six frames (A–F) advancing in time. Two porpoises are shown swimming side by side in frame A and are swimming apart by frame F. The call during this agonistic behavior between two subadult males is shown in Fig. 10.23a, as calling porpoise did a quick sideward turn of the anterior part of its body pointing its rostrum directly at the other during the sound emission. The latter reacted by a prompt avoidance that started in frame C. The second signal type was the *push threat call* with a repetition rate between 750 and close to 1,000 pulses as shown in Fig. 10.23b. The repetition rate increased in the beginning of the call and stayed relatively constant in the middle and decreased toward the end of the call. The duration ranged from about 0.5 s to over 1.5 s.

The repetition rate pattern in the beginning sometimes resembled that of the sideward turn threat call, but was always longer than the sideward turn threat call. This call was heard in agonistic encounters between the subadult males, 1, 2, and 3. The aggressive animal chased after another in full speed, often making contact by nodding with its rostrum against the other's back.

The third type of call is the *S-display sound*. This sound consisted of a train of clicks with a very low repetition rate (<10 Hz) interrupted with a burst of somewhat higher repetition rate and lower intensity. The repetition rate pattern of S-display sounds is shown in Fig. 10.23c.

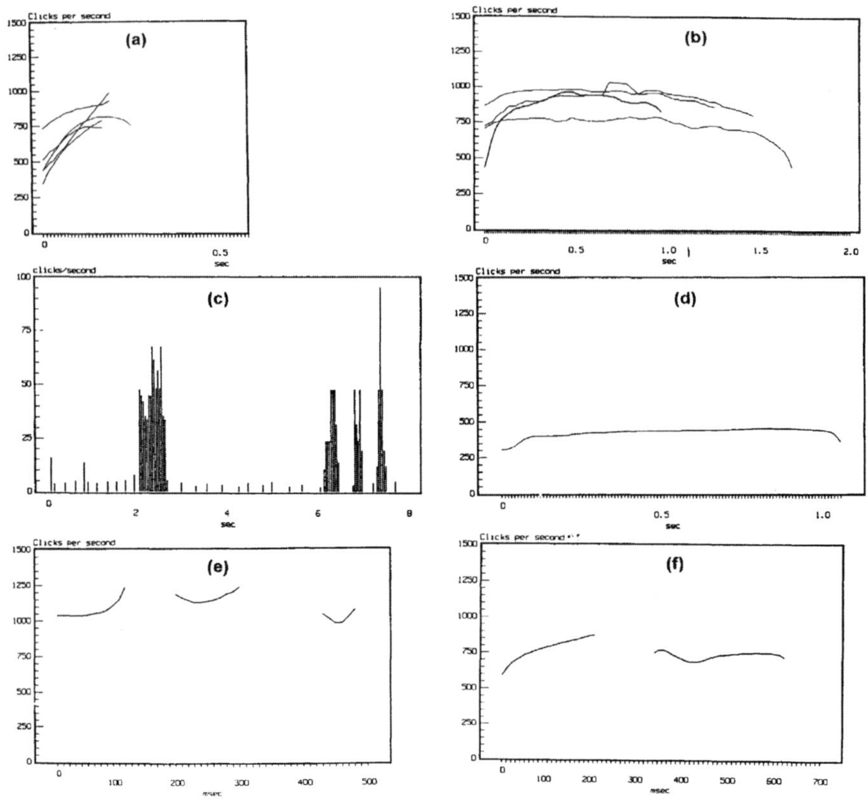

FIGURE 10.23. Repetition rate pattern of the (**a**) sideward threat call, (**b**) the push threat call, (**c**) S-display sound, (**d**) distress call, (**e**) signal of dominance, and (**f**) signal of pain (adapted from Amundin, 1991).

The total duration could be up to over 10 s. This call was so named because the subadult males that would make this call would flex their bodies in an S-shaped posture as shown in Fig. 10.24. The head would angle downward by making the contour of the skull behind the blowhole more evident. This sound was made in a sexual context with the tail initially pointing downward and was then slowly bent upward while the animal obtained a full erection. During this display the animal would glide slowly through the water.

The fourth type of call was the *distress call*. This call had a long, but varying duration with a rather even click repetition rate as can be seen in Fig. 10.23d. This call to the human ear resembles that of a sheep bleating and was heard from animal of both sexes in situations where they obviously experienced discomfort. It could be heard with an animal in a stretcher during transport. Almost all newly caught animals emitted this call, sometimes more or less continually for the first couple of days after arrival to a new tank.

FIGURE 10.24. The S-display observed in connection with the S-display sound (adapted from Amundin, 1991).

Amundin (1991) observed an incident after a heavy thundershower; the three subadult males emitted this distress call in chorus while swimming in a tight formation in an outdoor pool. Often one started the call and the others would join in, suggesting that social facilitation might have occurred. In another instance, two of the males were separated for the first time by a net frame lowered in the pool and they emitted this distress call while swimming at high speeds in tight circles.

The fifth call was the *signal of dominance*, which is shown in Fig. 10.23e. Spectrograms of these short calls have been published previously by Busnel and Dziedzic (1967). This call was heard in food competition situations in which adult females intimidated younger males in order to gain access to fish given to them (Busnel and Dziedzic, 1967). The sixth call was the *signal of pain*, which was recorded when an animal accidentally was inflicted pain. The first call in Fig. 10.23f was 200-ms long, which showed an increase in repetition rate from 600 to 800 c/s. The second signal was almost 300-ms long and showed a complex modulation in the click repetition rate.

Non-whistling odontocetes can also emit pulses in codas like the sperm whale (*Physter catadon*), which has long been named the "carpenter fish" by Navy sonar operators, because the clicks sounded like hammer blows on nails. (It is not entirely certain that sperm whales do not whistle. Most investigators attribute the emission of clicks by sperm whales to an echolocation function; however, Watkins and Schevill (1977) have been able to detect certain patterns within click trains that suggest sperm whales may use clicks in a social context. The pattern of clicks in a coda was usually about 0.5–1.5 s in duration and was repeated from 2 to 6 or more times, almost exactly at variable intervals of a few seconds or minutes. The spectra of the clicks within a coda were similar, and successive codas had the same click repetition patterns and similar frequency emphasis. The codas appeared to be unique to individuals over the periods of observation. Watkins and Schevill (1977) heard codas in the following situations: (1) from whales that met underwater, when two separate groups came together, and when individuals approached each other; (2) as exchanges between whales that were not far apart, whereas distant whales did not exchange codas; (3) as sometimes eliciting silence in other more distant sperm whales; (4) in response to unusual sounds, such as low-frequency rumbling from underwater landslide or starting and stopping of ships' engines, or the passage of a noisy aircraft; and (5) only from submerged sperm whales. Examples of a nine-click and a seven-click coda are shown in Fig. 10.25a and b, respectively. The nine-click coda was repeated 11 times in about 1 min, and the seven-click coda was repeated 7 times in 31 s. A more complicated seven-click coda that is repeated 3 times in 6.5 s is shown in Fig. 10.25c. More than one coda pattern could often be heard within 1 or 2 min, sometimes alternating and overlapping to demonstrate that more than one whale was involved.

Codas may serve as unique individual identifiers since they were easily separated by human listeners and presumably other sperm whales could easily identify different coda patterns. According to Watkins and Schevill (1977), codas from different whales within acoustic contact were remarkably varied. This suggest to them that the sperm whales were using codas as purposefully chosen identifications sufficiently varied from all other whales in an area over at a particular time period.

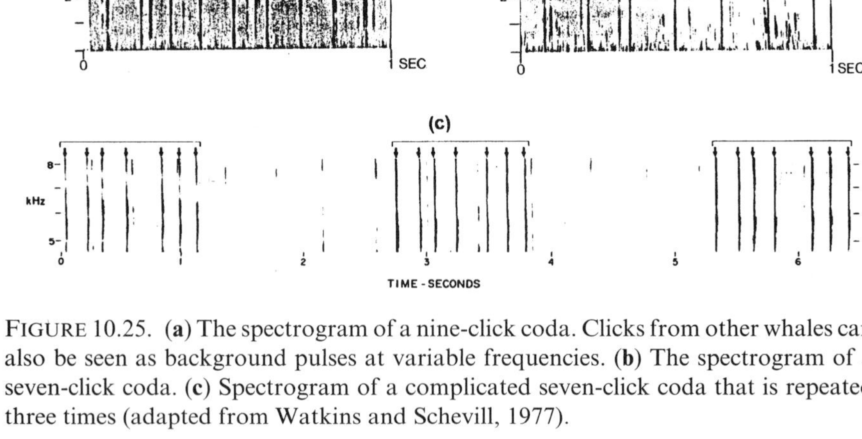

FIGURE 10.25. (**a**) The spectrogram of a nine-click coda. Clicks from other whales can also be seen as background pulses at variable frequencies. (**b**) The spectrogram of a seven-click coda. (**c**) Spectrogram of a complicated seven-click coda that is repeated three times (adapted from Watkins and Schevill, 1977).

10.1.4 Geographic Difference and Dialect

In our discussion of dialect, it would be proper to make a distinction between geographical difference and dialect. Geographical differences are associated with widely separated populations that do not normally mix. Dialect is best reserved for sound emission differences on a local scale among neighboring populations, which can potentially intermix (Grimes, 1974). Geographic variations are generally considered to result from acoustic adaptations to different environments, or else to be a functionless by-product of isolation and genetic divergence caused by isolation (Ford, 1991). The functional significance of dialects is controversial, with some maintaining that dialects are epiphenomena of song learning and social adaptation, whereas others believe that they play a role in assortative mating and are of evolutionary significance (Ford, 1991). According to Nottebohm (1969) and others, dialects occur in many species of birds in order to serve as a means of reducing gene flow between breeding populations, thereby allowing for more efficient and rapid adaptation to local-specific conditions. A true dialect is defined as consistent differences in the calls or songs of neighboring populations of a particular species or between populations with individuals that have the potential for interbreeding. Dialects are known to occur in many species of birds but appear to be very rare in mammals. Krebs and Kroodsma (1980) have an excellent review of dialect formation in birds. According to Lemon (1966), at least three conditions are necessary for dialects to be formed: (a) some isolation of the population should exist either through behavioral differences or geographical barriers, (b) a means of transmitting calls or songs from generation to generation, and (c) sources of variations within calls so that individuals of a species residing in different areas may diverge in the use of different sound emission types.

Killer Whale The dialects of resident killer whales (*Orcinus orca*) in the coastal waters of British Columbia and Washington have been studied over a prolonged period by Dr. John Ford and his colleagues (Ford and Fisher, 1983; Ford, 1984; Ford, 1991). Those with a special interest of this subject should refer to the excellent and detailed description provided by Ford (1991). Ford and his colleagues used photographic identification techniques, keying on unique natural markings on the whales to identify 16 pods or stable kin groups of 232 resident killer whales. Acoustic exchanges within pods were dominated by repetitious, discrete calls. Pods produced between 7 and 17 discrete calls and call repertoires differed significantly among pods. The 16 pods are divided into 4 clans, each clan having a unique repertoire of distinctive calls. Three of the clans comprising the northern (N) community are designed by the letters **A**, **C**, and **R**. The southern (S) pod consists of a single clan, **J**. A map of the study area and the extent of the northern and southern ranges are shown in Fig. 10.26. Pods within a clan shared several call types, but no sharing took place among clans. Shared calls within a clan often contained structural variations specific to each pod or group of pods in the

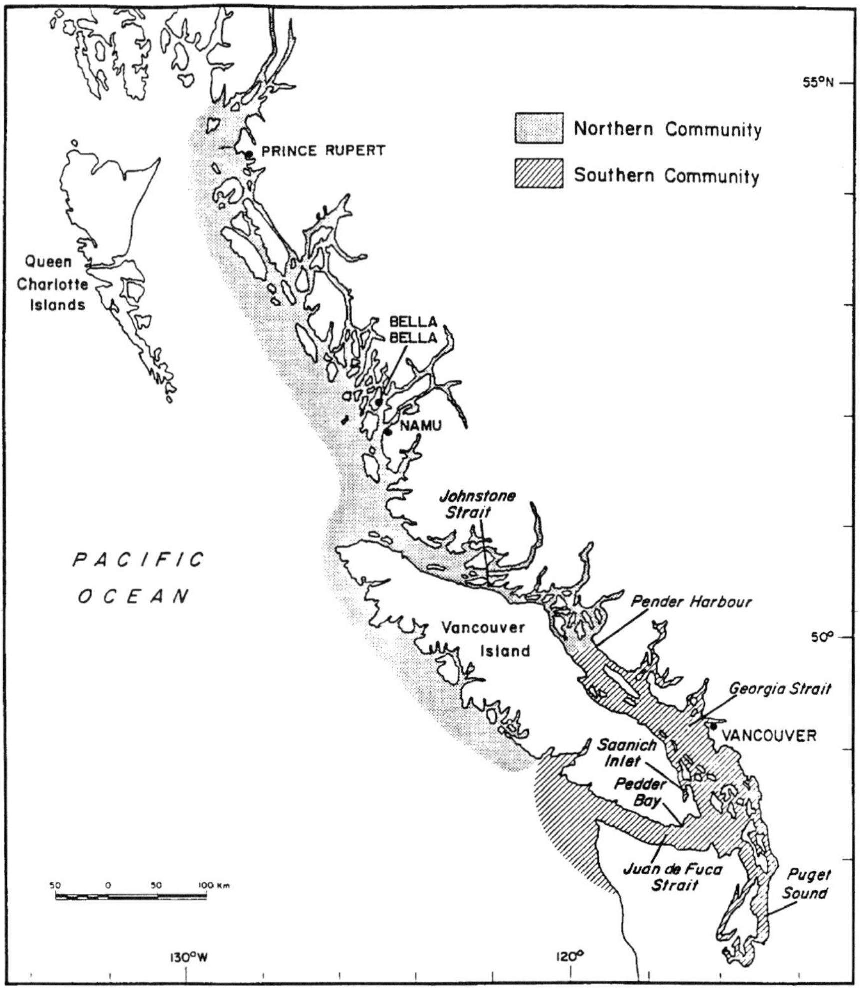

FIGURE 10.26. Map of the known distribution of the northern and southern communities of resident killer whale pods (adapted from Ford, 1991).

clan. Differences in acoustic behavior formed a system of related pod-specific dialects within the call tradition of each clan. Ford (1991) has proposed that each clan comprised related pods that have descended from a common ancestral group, and pod-specific repertoires probably serve to enhance the efficiency of acoustic communications within the group and act as behavioral indicators of pod affiliation.

Killer whale calls are typically made up of rapidly emitted pulses, which to the human ear have a tonal quality (Ford, 1991). Many calls have several abrupt shifts in pulse repetition rate, allowing them to be divided into different segments or parts. The acoustic relationship of the 16 pods is illustrated in

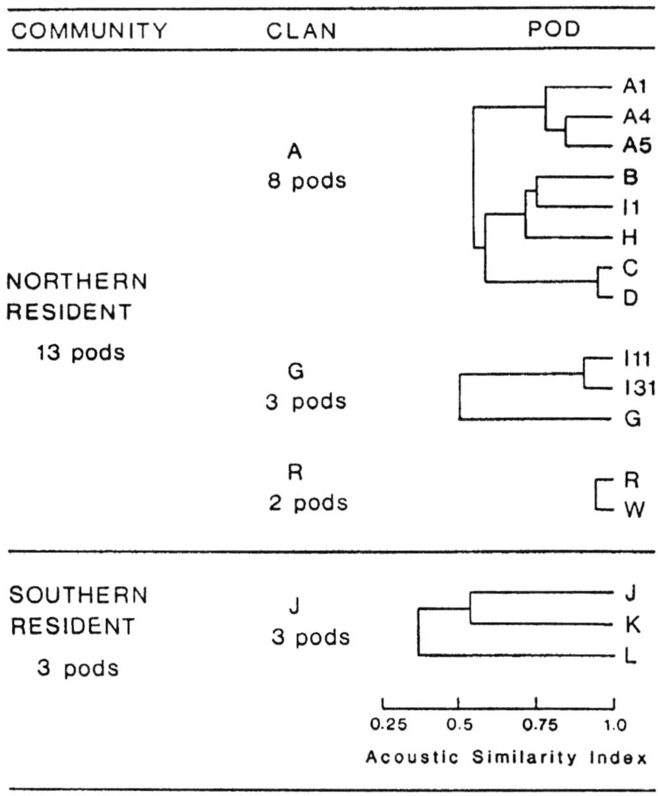

FIGURE 10.27. Summary of acoustic relationship of resident pods in British Columbia. The two resident communities had exclusive ranges, while clans had exclusive call traditions. Pods in the three northern community clans often associated with each other. All pods within a clan shared calls, yet many also had unique calls (adapted from Ford, 1991).

Fig. 10.27. Pods and clans in the figure were defined by the characteristics of their calls. Although all pods belonging to a clan shared a number of calls, these calls were often rendered in consistently different forms by different pods. Also, certain pods produced calls that were not used by the rest of the clan. Such variations produced a set of related group-specific dialects within the call tradition of each clan. Pods that shared the majority of calls in their repertoire and formed distinct acoustic subgroups are referred to as a subclan by Ford (1991) as depicted in Fig. 10.27. Clan A consisted of 90 whales divided into eight pods, A1, A4, A5, B, C, D, H, I1, all of which shared a portion of their call repertoires. The 19 call types for clan A are summarized in Table 10.3. All eight pods shared a minimum of four call types, N3, N7, N8, and N12 where the letter N refers to the northern clan and the numbers refer to variations in a call type by the different subpods. Clan A was divided into

TABLE 10.3. Call Types and Subtypes Produced by Pods Belonging to Clan A in the Northern Resident Community (adapted from Ford, 1991)

	Pod							
Call	**A1**	**A4**	**A5**	**B**	**C**	**D**	**H**	**I1**
N1								
i	x							
ii				x				x
iii					x	x		
iv							x	x
v		x						
N2	x	x	x					
N3	x	x	x	x	x	x	x	x
N4	x	x	x					
N5								
i	x	x	x	x			x	x
ii				x			x	x
N7								
i	x	x	x					
ii	x	x	x				x	x
iii				x				x
iv					x	x		
N8								
i	x	x	x				x	
ii					x	x		
iii				x				x
iv				x				x
N9								
i	x							
ii		x						
iii			x					
N10	x	x	x					
N11								
i	x	x	x	x				
ii				x	x	x	x	
N12	x	x	x	x	x	x	x	x
N13		x	x					
N16								
i				x				
ii					x	x		
iii							x	x
iv								x
N17			x					
N18				x	x			
N19		x						
N20				x	x	x		x
N21				x				
N27	x							
N47	x							
Total	**14**	**14**	**13**	**14**	**9**	**8**	**9**	**13**

NOTE: The clan designation includes pods that share call types.

two major acoustic categories. Furthermore, calls N1, N5, and N11 were produced by all but one or two of the pods. In the table, the different Roman numerals below a call labeled by a number is an example subgroup. Subclan A consisted of pods A1, A4, and A5. Subclan B consisted of pods B, C, D, H, and I1, which shared four calls, N16, N18, N20, and N21, that did not occur in the repertoire of any other pod. However, only N16 was shared by all five pods in subclan B.

Spectrograms of call types N7 and N8 are shown in Fig. 10.28. In this example, we can see that different pods produce similar but different versions of call type N7 and N8. Pods A1, A4, and A5 produced two versions of call type N7 and pod B, and I1 produced two versions of call type N8. All of the spectrograms have two parts except for call type N7ii, iii, and iv, which had three parts. Examples of a call type (N16) of subclan B that contained four parts are shown in Fig. 10.29. From Table 10.3 we can see that pods B, C, D, H, and I1 produced different versions of call type N16.

It is the opinion of Ford (1991) that pod-specific repertories can be retained for periods in excess of 25 years. He also contends that discrete calls generally serve as signals for maintaining contact within the pod and that the use of repertoires of pod-specific calls enhances this function by conveying group identity and affiliation. Part of the rationale for this opinion is the fact that most of the call types are heard by observers, regardless of the type of activity in which the group is engaged, and this suggests that the calls are context-independent. Each pod is capable of producing most or all calls made by the group as a whole. Killer whales, like most other small dolphins, are able to learn and mimic a wide variety of sounds. Even from young age, killer whale infants can selectively learn specific calls, especially the calls of their mothers. Bowles et al. (1988) studied the development of calls of a captive-born killer whale calf and found that it learned and reproduced only the calls of its mother and ignored the calls of other killer whales in the same pool. Ford (1991) also observed killer whales imitating the call types of different pods and even those from other clans; however, these instances were rare, but does show that a capacity for learning and mimicry of acoustic signatures.

Bottlenose Dolphin Research on dialects in dolphins and whales has mainly been with the residential killer whale pods in the coastal waters of British Columbia and Washington State. However, some differences in the structure of whistles of different populations of bottlenose dolphin in the Gulf of Mexico (Wang et al., 1995) suggest that they also may have dialects. Much more research would need to be performed in order to accumulate sufficient evidence to address this issue. However, working with bottlenose dolphins may be much more difficult than killer whales, since bottlenose dolphins exist in free fission–fusion group in which many individual animals tend to come and go at will. Therefore, there does not seem to be a clan-like structure to the bottlenose dolphin society and dialect may be more associated with habitat differences rather than clan and pod membership like the killer whale.

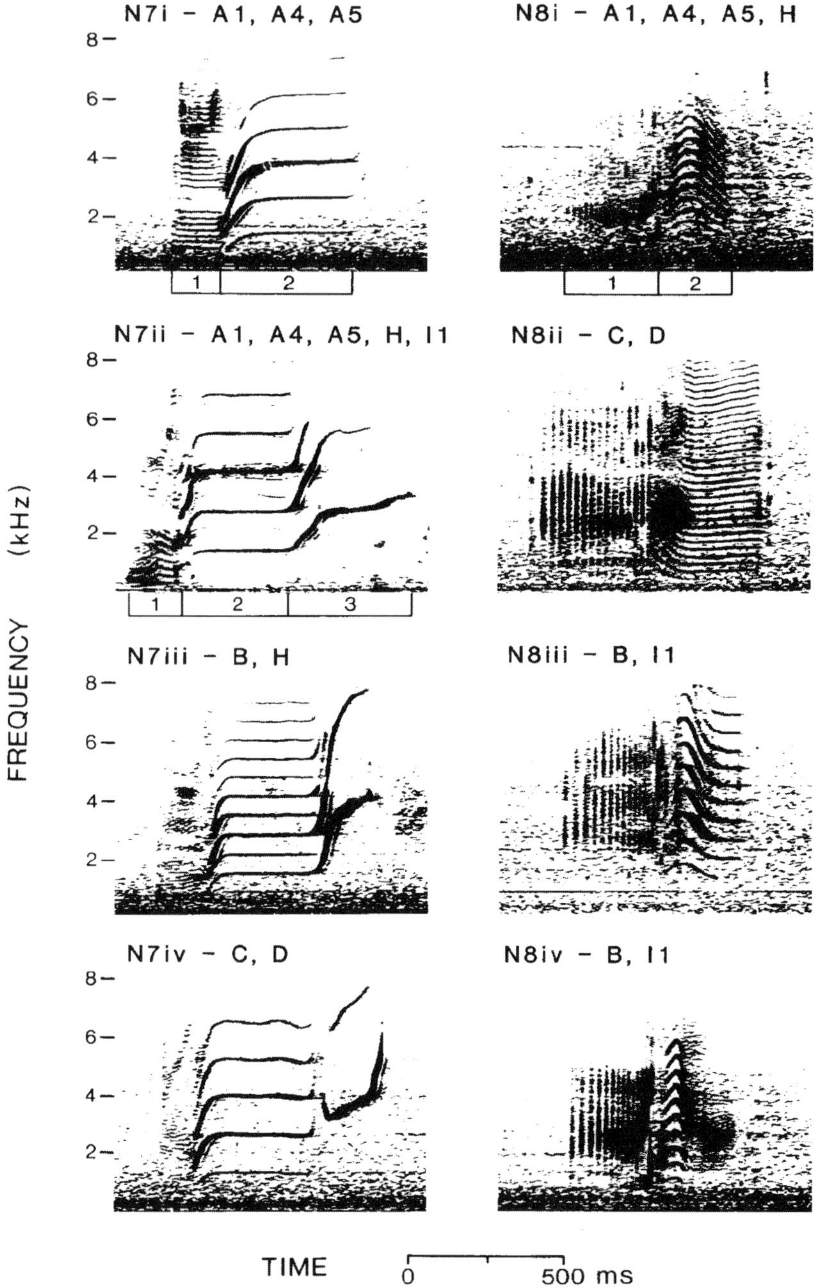

FIGURE 10.28. Spectrograms of call types N7 and N8 for clan A. Above each spectrogram is the subtype identification and the pods that produce the variant, and below certain spectrograms are division marks separating calls into their component parts (adapted from Ford, 1991).

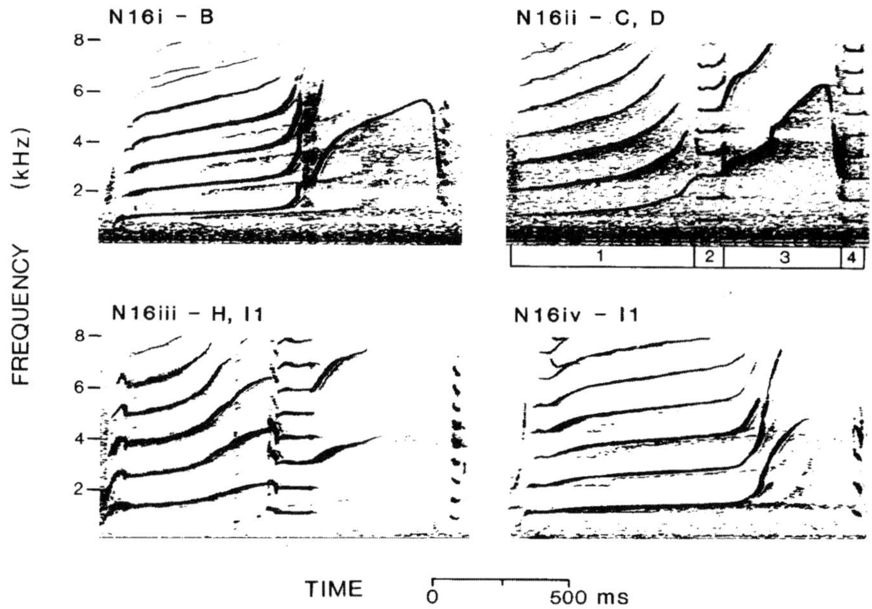

FIGURE 10.29. Spectrogram of variants of call N16 produced by subclan B of Clan A, pods B, C, D, H, and I1 (adapted from Ford, 1991).

Sperm Whales Sperm whales live in a matrilineal family unit where there exists cooperative behavior, including communal care of the young in ways similar to killer whales. The family units are very stable and females live as long as 60–70 years (Whitehead et al., 1991). In a previous section, we discussed the production of codas consisting of short-patterned series of clicks by sperm whales. These codas seem to be the primary means of acoustic communications for these animals (Watkins and Schevill, 1977). Weilgart and Whitehead (1993) postulated that codas primarily function to maintain social cohesion within groups of females following periods of dispersion during foraging. They also emphasized that codas appear to be principally used as communication among members of a group rather than between groups, since they do not seem to propagate beyond about 600 m at the surface and groups are usually separated by several kilometers.

Weilgart and Whithead (1997) conducted a study on the geographical variation of sperm whale codas and the possible existence of dialects. They recorded codas from a number of sperm whales around the South Pacific and in the Caribbean Sea from a 1.25 auxiliary cutter named Balaena. These recordings were made over a year between June 1992 and May 1993. Photographic identifications also allowed them to assign recording sessions to particular groups. Detailed analysis was done on a subsample of 3,644 codas from 86 recording sessions. The codas were first classified according to the number of clicks in the codas, usually between 3 and 12 clicks were contained in the codas. Rare codas with less than 3 or more than 12 clicks were discarded. Then all classified codas were characterized by the intervals between clicks relative to the total length of the coda. A K-means cluster analysis was used to classify the click interval patterns into 10 clusters and 30 coda types. Codas were also classified into four nonmutually exclusive classes: short (<5 clicks), long (>6 clicks), regular (equally spaced interclick intervals), and plus-one (double intervals between the last two clicks) codas. A coda repertoire was constructed for each recording session from both the 30 coda types and 4 coda classes. A hierarchical system used to classify the recording sessions included (1) days, (2) groups, (3) places, (4) areas, and (5) ocean. The repertoires of codas were compared using Spearman and Pearson correlation coefficients.

Weilgart and Whitehead (1997) found that the coda repertoires recorded from the same group on the same or different days were much more similar than those recorded from different groups in the same place. Groups recorded in the same place had more similar coda repertoires than those in the same broad area but different places. Groups from the same area were in turn marginally more similar than those in the same ocean but different areas or different oceans. A multivariate analyses indicated that the coda class repertoires of groups in different oceans and in different areas within the same ocean were statistically significantly different. Weilgart and Whitehead (1997) concluded that strong group-specific dialects were apparent and was overlaid on weaker geographic variation. Sperm whales and killer whales are now the only cetaceans known to have dialects. They join the elephant seal (which will be discussed in Section 1.2.3) as the only nonhuman mammals known to exhibit any acoustic dialect.

10.2 Sound Emissions by Mysticete Whales

There are 11 species of mysticetes, or baleen whales, and sounds have been recorded from all but the pygmy right whale (Clark, 1990). We will not attempt to discuss the sounds produced by all of the mysticetes, nor will we attempt a detailed discussion of the acoustics of any one species. Rather the treatment in this section will be rather broad, and important references will be cited so that those interested in learning more can refer to these references. In the case of baleen whale, it would be appropriate and convenient to discuss sound emissions in terms of vocalizations since there is no evidence as to how the sounds are produced and they could well be produced within the larynx.

The vocalizations of baleen whales can be divided into two general categories: (1) songs and (2) calls (Clark, 1990). The calls can be further subdivided into three categories: (1) simple calls; (2) complex calls; and (3) clicks, pulses, knocks, and grunts (Clark, 1990). Simple calls are often low-frequency, frequency-modulated signals with narrow instantaneous bandwidth that sounds like moans if a recording is speeded up or slowed down, depending on the specific animal. Amplitude modulation and the presence of harmonics are usually part of a simple call, with most of the energy below 1 kHz. Complex calls are pulse-like broadband signals with a variable mixture of amplitude-modulated noise and/or a frequency-modulated fundamental. They sound like screams, roars, and growls, with most of the energy between 500 and 5,000 Hz. Clicks, pulses, knocks, and grunts are short-duration (<0.1 s) signals with little or no frequency modulation. According to Clark (1990), clicks and pulses are very short (<2 ms) signals with frequencies between 3 and 31 kHz, while grunts and knocks are longer (50–100 ms) signals in the 100–1,000 Hz range. Clicks and pulses have not been reported widely and they are the most controversial signals attributed to mysticetes. Songs are defined as "sequences of notes occurring in a regular sequence and patterned in time" (Clark, 1990). Songs are easily discriminated from calls in most instances.

10.2.1 Songs of Mysticete Whales

Four mysticete species have been reported to produce songs; the blue whale, *Balaenoptera musculus* (Alling and Payne, 1991; Alling et al. 1991), the fin whale, *Balaenoptera physalus* (Watkins et al., 1987), the bowhead whale, *Balaena mysticetus* (Ljungblad et al., 1982), and the humpback whale (Payne and Mcvay, 1971). Songs of humpback whales have, without a doubt, received the most attention from researchers. Part of the reason for this is the relative ease for investigators to travel and do research in the summer grounds of humpback whales, especially in Hawaii in the Pacific and Puerto Rico in the Atlantic. An excellent review of the general acoustic behavior of mysticete whales has been written by Clark (1990).

Humpback Whale Songs The list of studies involving humpback whale songs is long and extends from 1971 to the present time. Helweg et al. (1992) have written an excellent review on the current understanding of humpback whale songs. They also included many references that interested parties can refer to. From the many studies of humpback whale songs, we can summarize some general properties of songs and the whales that are doing the singing:

(1) Songs from the north Pacific, south Pacific, and Atlantic populations are different.
(2) Singing peaks during the winter months when humpback whales migrate to warmer waters at lower latitudes.
(3) Whales within a population sing the same basic song in any one year although the song may undergo slight changes during a breeding season.
(4) Changes in songs are not due to forgetting during the summer months, which are mainly non-singing months, since songs recorded early in the winter breeding season are nearly the same as the songs recorded late in the previous breeding season.

(5) Songs from consecutive years are very similar, but songs across nonconsecutive years will have fewer similarities.
(6) Singers are most probably only males, since no females have been observed singing.
(7) Some singing also occurs during the summer and fall.
(8) Singing whales are often alone, although they have been occasionally observed singing in the presence of other humpback whales.
(9) Singers tend to remain stationary; however, they have also been observed singing while swimming.

There are several hypotheses on why humpback whales sing, although the function of songs is still uncertain. Some have proposed that singing might be a sexual advertisement to females in the vicinity (Payne and McVay, 1971; Winn and Winn, 1978; Tyack, 1981), or that songs may establish and/or maintain space between competing males (Winn and Winn, 1978; Frankel, 1994). Songs may play a role in inducing or synchronizing ovulation in females as for canaries, parakeets, and red deer (Helweg, 1989). Other proposals include establishing dominance among males or serving as navigational beacons for whales traveling to breeding grounds (Winn and Winn, 1978). Helweg et al (1992) believe that males sing with messages to both sexes. The message for other males is to "stay away", that is to maintain the spacing between singers. The message for females is one of advertisement. The song would convey location and indication of reproductive fitness, and perhaps identify information such as regional dialect. They suggest that singers maintain a spatially dynamic array through which females pass, and females choose a mate based on some characteristic of a singer's song.

Humpback whale songs have been typically described from a human hearing prospective based on their aural qualities. Payne and McVay (1971) developed a hierarchical model describing songs in terms of themes, phrases, and units. Units are the shortest sounds that seem continuous to humans, and are often repeated numerous times with a high degree of redundancy. Singers often repeat a single type of unit, alternate between two different units, or repeat sequences of multiple units in a stereotypical fashion. A phrase consists of repeated units or repeated sequences of units and is often subdivided into subphrases. Phrases can be heard repeated 20–30 times or more by a singer. Often, one sequence of repeated phrases is directly followed by another repeated sequence of a different phrase. These repeated sets of phrases are called themes, and themes may be composed of phrases that remain the same during multiple repetitions or may gradually change with repetition. Songs consist of 5–7 themes that are produced in a stereotypical order. A song session is often made up of several consecutive songs. Song sessions are usually continuous with essentially no breaks between themes, making it difficult to define the beginning and end of a song. Winn et al. (1971) observed a specific theme called the ratchet theme that often occurred while whales surfaced, and this theme is often designated by many as the beginning of a song. The duration of

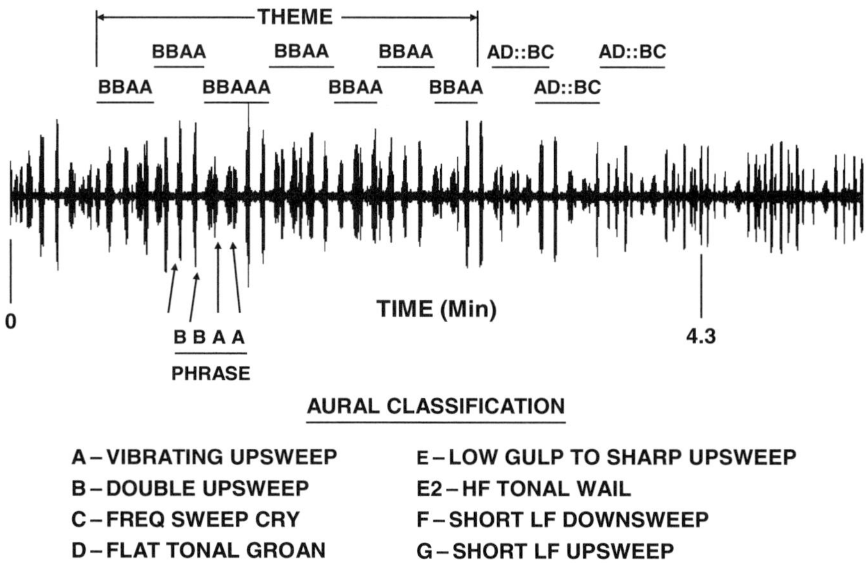

FIGURE 10.30. Example of a segment of a typical humpback whale song. Each letter represents one sound. Each individual sound is called a unit where different letters correspond to aurally distinctive sound units. Repeated groups of units are called phrases. Some phrases consist of repeated groups of subphrases. A theme is a set of repeated phrases. Songs consist of repeated themes within a song.

songs is highly variable with some lasting just minutes, others for several hours, and some as long as the 24-hour song observed by Winn and Winn (1978).

An example of a segment of a humpback whale song is shown in Fig. 10.30, recorded during the February of 2002. This example shows the pulsatile nature of humpback whale songs. Typically, units are categorized by an aural classification process and each category often arrives as a result of some consensus by members of a research team. Unfortunately, this subjective process of categorizing units makes it difficult to have a good standardization of units among humpback whale researchers. The song segment of Fig. 10.30 is organized as phrases and a theme.

The waveform and spectrogram of typical units (described in Fig. 10.30) recorded during the 2001–2002 winter in the water of Maui are shown in Fig. 10.31. The units had durations between 1 and 3 s. The spectrograms can be separated into two groups: those units that have some tonal quality and higher frequency harmonics (Units C, E, E_2 and H) and those that are relatively broadband with rumbles, grunts, or gurgle-like qualities and with weak harmonics (Units A, B, D, F, and G) . The units with some tonal elements and high-frequency harmonics were described as cries or wails. Four distinct themes were found for the nine whales that we recorded. The sequence of units producing each theme is listed in Table 10.4. One common theme often sounds like a ratchet sound that is typically sung as the whale approaches the surface and while on the surface.

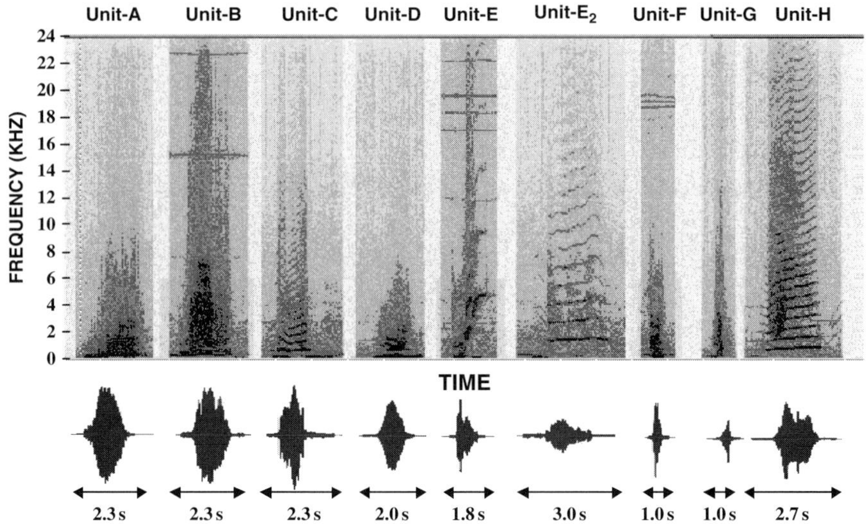

FIGURE 10.31. Spectrogram and waveform representation of the nine units contained in the humpback whale songs during the winter season in Hawaiian waters in 2001–2002. The units were from the recordings of several whales and were chosen because they showed the characteristics of the unit most clearly (adapted from Au et al., 2006).

One of the parameters of humpback whale vocalizations that has not been well defined is the source levels of the song units. In order to address this problem, Au et al. (2006) used a vertical array of hydrophones to localize the position of singing humback whales and to obtain calibrated measurements of their songs. Singing humpback whales often remain relatively stationary in the water column pointing at an angle to the bottom. The geometry of their measurements is shown in Fig. 10.32a, and the locations of three different whales are shown in Fig. 10.32b. The source levels of the three whales are shown in Fig. 10.33 for the nine different units. The results clearly showed that the source levels of the different units were different and that some units have relatively consistent source levels between individual whales (units A, E, D, E2) and some units have very different source levels. The results of this

TABLE 10.4. The general sequence of units producing the four themes observed in the recordings of nine singers. There is considerable variability in a theme associated with the number of or sequence of the units. The units in parenthesis are those that are repeated seemingly randomly from one to seven times, depending on the specific whale (From Au et al., 2006)

Theme	Units
1	B C (B–C) D D A
2	D A D E F D (E-F-D) E A
3*	A F F F A F F E F D (E-F-D) E2 E or A F F F A F F F F E E E [without (E-F-D)]
4**	G G G G H

*The specific number of F units can vary between 2 and 7, and E units between 1 and 3.
**Sometimes the H unit is not emitted.

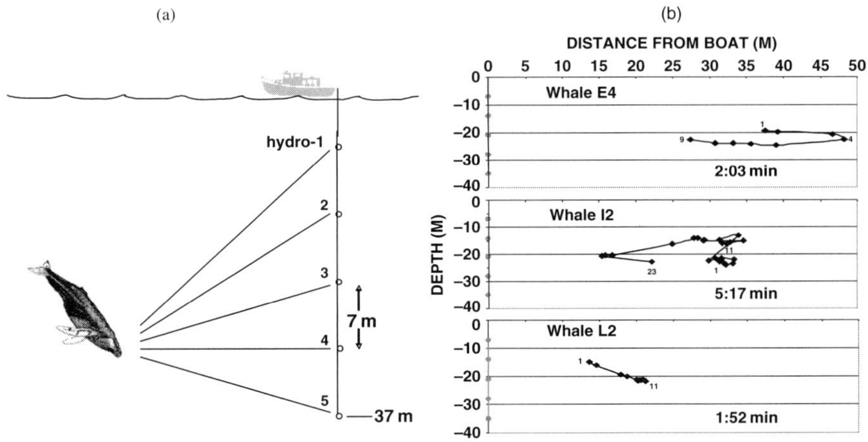

FIGURE 10.32. (**a**) Geometry of the vertical array used to localized singing humback whales, and (**b**) the tracks of three singing humback whales relative to the position of the drifting boat (adapted from Au et al., 2006).

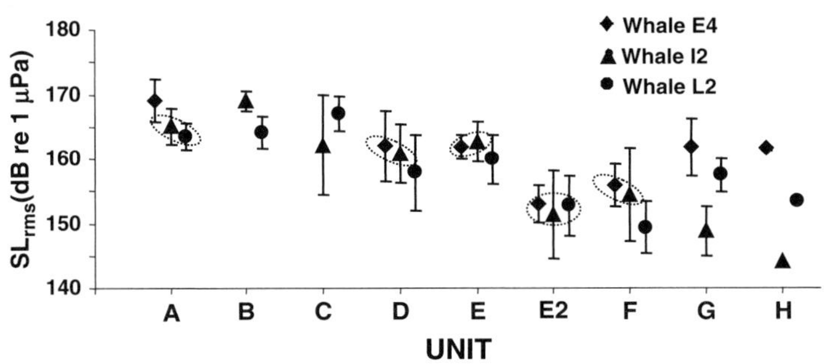

FIGURE 10.33. The mean and standard deviation of the rms source level of the different units for the three whales tracked in Fig. 3. The means that are enclosed within a *dash oval* are not significantly different. All other units are significantly different at $p < 0.05$ (adapted from Au et al., 2006).

study indicate that the source levels for the different units are not the same but can vary as much as 20 dB between the least intense and the most intense units. The higher level units were the longer broadband units with durations of about 2 s that are described as rumbles, grunts, or gurgles, whereas the lower level units were units with some tonal quality or short broadband sounds such as units F and G. In an anthropomorphic sense, we can usually shout much louder than we can sing. It seems that the same tendency may be true for humpback whales.

The rms source levels for the most intense unit emitted by the three whales are shown in Fig. 10.34. Unit A had the highest average source level for Whale E4 with a maximum of 173 dB. The source level remained above 170 dB on

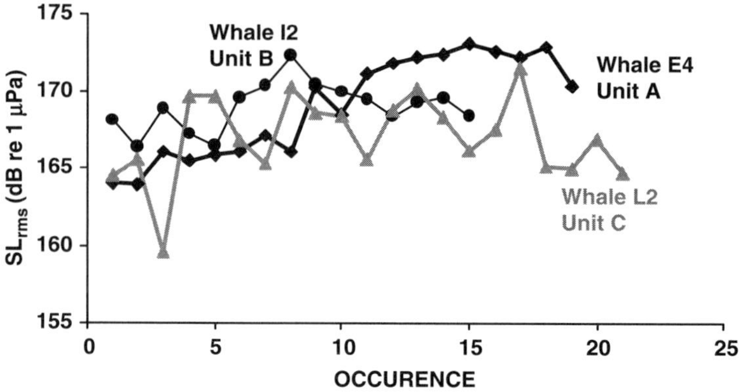

FIGURE 10.34. The rms source level for the unit having the highest intensity for each of the three whales tracked in Fig. 10.32 (adapted from Au et al., 2006).

nine occasions during the period the whale was tracked. Unit B was the most intense unit emitted by whale I2, reaching a level of 172 dB. Six emissions of unit B sounds were equal to or above 170 dB source level. Unit C was the most intense unit emitted by whale L2, with the highest level being 172 dB. However, only four of the twenty seven units were above 170 dB. Although the average source level of unit A emitted by whale E4 was the same as unit B emitted by whale I2, Fig. 10.34 clearly shows that whale E4 emitted higher level signals than whale I2 but had a greater range between the lowest and highest levels, which brought down its average source level for unit A.

There were three units (E, E_2 and H) that had very high-frequency harmonics that extended beyond 24 kHz, which was the upper limit of the TASCAM recorder. The spectrogram and frequency spectrum of Unit-H sounds produced by three different whales are shown in Fig. 10.35. These spectrograms show the highest harmonic levels ever reported. The frequency spectra plots alongside the spectrograms indicate that the higher frequency harmonics do not drop off very rapidly. At a frequency of 10 kHz, the amplitude of the spectra is about –20 dB below peak amplitude for whales E4 and F4, about –22 dB for whale L2. At a frequency of 22 kHz, the amplitude of the spectra is between –38 and –42 dB below the peak amplitude. Although Unit-H was recorded only 11 times total from nine whales, representing the least recorded unit, it was selected to be reported because of its unusually high-frequency harmonic structure.

The source levels measured by four of the five hydrophones in the array for a Unit-H sound from whale E4 and a unit E_2 sound from whale L2 are shown in Fig. 10.36. Units-H and E_2 were specifically chosen because both units were rich in high-frequency harmonics so that the sound field at the fundamental and harmonic frequencies could be easily determined. The closed circles for the fundamental frequency are the estimated source level from each of the four working hydrophones. Specific points were not shown for the higher

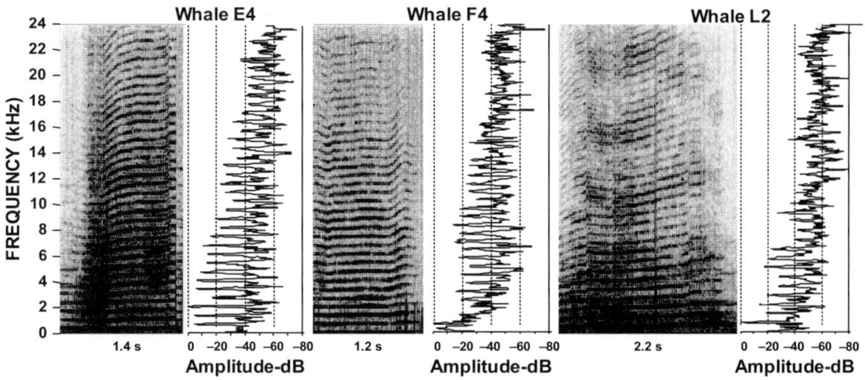

FIGURE 10.35. Spectrogram showing the frequency versus time variation of unit H sounds and corresponding frequency spectra plotting relative amplitude as a function of frequency. The plots depict the high-frequency nature of unit H sounds showing harmonics extending beyond 24 kHz for three different whales (adapted from Au et al., 2006).

harmonics to keep the plots relatively simple and uncluttered. The line connecting the points for the fundamental frequency consists of a third-order polynomial fitted to the four source levels. The polar plots suggest that a beam has started to form in the vertical plane at the fundamental frequency of 750 Hz for whale E4. The same is true for whale L2 having a fundamental

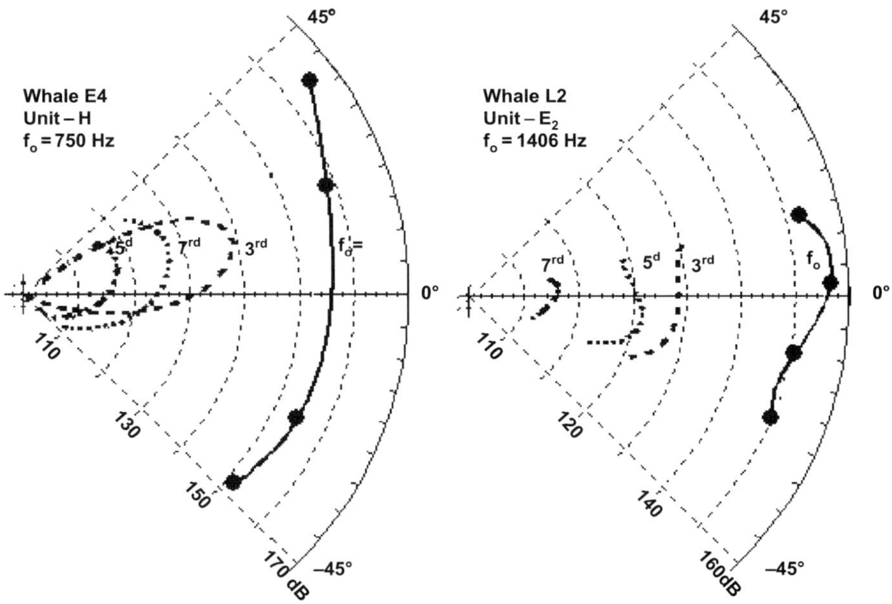

FIGURE 10.36. Polar plot of the rms source level for whale E4 projecting a unit H sound and for whale L2 projecting a unit E_2 sound (adapted from Au et al., 2006).

frequency of 1,406 Hz. As the frequency increases, the beam becomes narrower; however, there is an anomaly between the 5th and 7th harmonic for whale E4, where the beam of the 7th harmonic is slightly larger than for the 5th harmonic. The reason for this anomaly is not known, but may be related to the orientation of the whale. The angle of the whale to the vertical was not ascertained and the array was initially positioned directly in front of the whale; however, the whale could easily pivot about its longitudinal axis to modify the geometry between the array and the whale.

Initially, Au et al. (2006) were surprised to see the beginning of directivity in the sound field at the fundamental frequency of units H and E_2. However, a fundamental frequency of 750 Hz for unit H is associated with a wavelength close to 2 m. The average length of a male humpback whale is approximately 12–14 m (Winn and Reichley, 1985) so that considering the ratio of the linear measurement of the head close to the blowhole from photographs, one can estimate a diameter of roughly 1.5–1.7 m. Here we are assuming that the sound source is in the vicinity of the blowhole (Aroyan et al., 2000). Therefore, for the head diameter to wavelength ratio, we have $d/\lambda \sim 0.75$–0.85. For unit $E_2$and whale L2, the wavelength corresponding to the fundamental frequency of 1,406 Hz is close to 1 m so that $d/\lambda \sim 1.5$–1.7. A linear circular transducer will exhibit a directional characteristic when d/λ is close to unity (Urick, 1983). The fundamental frequencies of other units, such as unit A and B, are relatively low at about 210 Hz. For these units, $d/\lambda \sim 0.21$–0.24 so that the sound has no directivity at the fundamental frequency.

The relevance of various humpback sounds to conspecifics has been studied in playback experiments where different types of sounds are played to a pod of humpback whales and the effects of the sounds on the pod can be observed visually. Tyack (1981) broadcasted both humpback whale songs and winter social sounds in the vicinity of groups of whales being tracked by shore-based theodolite stations. He found a marked difference in the responses of the whales to social sounds and to songs. Of the 16 groups observed during song playback, 14 moved away from the boat, while only two, a singer and a long adult, moved toward the boat. Three of four singers, all five cows/calf/escort pods, and six of seven non-singing adult pods moved away from the boat. When social sounds were played back, there was a pronounced tendency for certain classes of whales to charge the boat. Three of four singers observed and six of eight pods of 1–2 adults charged the playback boat. The whales would swim rapidly on the surface toward the boat and in some cases accelerate to 12 km/hr. A charge typically terminated in a sudden dive near the boat and some of the whales would come within several meters of the underwater speaker. In contrast, both cow/calf/escort pods and pods of 3–4 adults moved away from the boat. Tyack (1981) concluded that winter social sounds made by these surface-active groups induced a joining response, whereas songs produced an avoidance response.

Bowhead Whales Song The second mysticete whale that was reported to produce song is the bowhead whale, *Balaena mysticetus* (Ljungblad et al.,

1982; Cummings and Holliday, 1987). Studies of the bowhead whale sounds have not been nearly as extensive as for the humpback whale. Part of the reason is that these whales live in the Bering, Chukchi and Beaufort Seas close to and in the Arctic, and it is difficult to get access to them without mounting a fairly large-scale expedition. Each year in April–May, about 4,000 bowhead whales migrate eastward from the Bering and Chukchi Seas past Pt. Barrow, Alaska, to their summer grounds in the Canadian Beaufort Sea. Around mid-September, the western return migration occurs (Ljungblad et al. 1982).

Ljungblad et al. (1982) used sonobuoys that were dropped from an aircraft ahead of the positions of migrating whales to record the sounds of bowhead whales in both the spring and fall of 1979. Cummings and Holliday (1987) used a 2.46-km, nearly linear, three-hydrophone array placed on the offshore side of a large grounded ice ridge to record the sound of bowhead whales on their eastward spring migration. Each element of their array was a modified sonobuoy that telemetered the sounds back to a base station. The recordings of Cummings and Holliday (1987) are especially important because they could localize vocalizing whales and obtain estimates of source levels and also gain more insights into the propagation conditions.

Bowhead whales emit a variety of simple and complex sounds that resemble moans to the human ear. They also emit sequential sounds with repeatable phrases that can be classified as songs. These patterned moan sequences were only observed during the spring migration by Ljungblad et al. (1982), while songs have not been noted during summer or autumn (Wursig and Clark, 1993). A bowhead whale song is relatively simple, very short, and unpretentious as compared to the humpback whale. Ljungblad et al. (1982) described the bowhead song as having just one theme with basically only two sounds repeated over and over. Songs are almost continuous, seldom interrupted by more than a second or two and have a more of a singing quality than the humpback whale songs. The songs are always very tonal with clear pitch even though they are produced by pulsive moans. Cummings and Holliday (1987) provided a very different description of the bowhead whale song. They described the songs as sounding like raucous elephant roars and trumpeting, in metrical arrangement. These trumpet-like sounds appeared in discrete repetitions or phases put together to form longer sequences. Braham et al., (1980) also described the bowhead whale sounds as elephant trumpeting; however, his sounds were recorded in the fall and were not associated with songs. Cummings and Holliday (1987) pointed out that the songs they recorded were quite different form the songs described by Ljungblad et al. (1982), and attributed the difference to the bowhead whales changing their songs from year to year. The intersound interval of songs was appreciably longer (8 vs. 1.5 s) and the overall frequency content was also much higher (4–5 kHz vs. 0.8–2 kHz) than those recorded by Ljungblad et al. (1982).

An example of a bowhead whale song is shown in Fig. 10.37. Note the regularity between the intervals and the shape of the spectrogram contour for the different phrases in the songs. The maximum frequency of each moan in

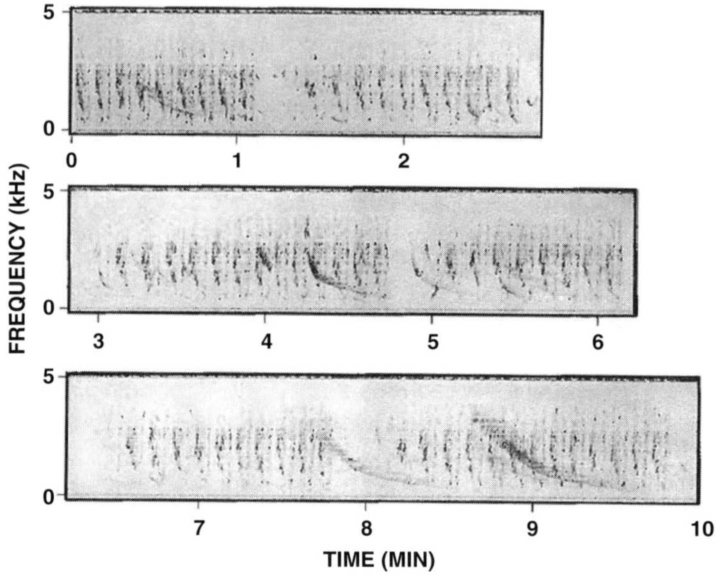

FIGURE 10.37. Continuous time-compressed sound spectrogram of six songs (each composed of repeated phrases) from a bowhead whale. The downward FM warbles are from bearded seals. Analyzing bandwidth was 18.75 Hz (adapted from Cummings and Holliday, 1987).

Fig. 10.37 was about 4 kHz. Cummings and Holliday (1987) found the number of phrases in a song varied from 3 to 20, with a mean of 10 phrases per song. The mean duration of a song was 66.3 s. In ten songs of ten elements each, the mean interval between the start of one phrase to the start of the next phrase was 7 s. The spectrogram in Fig. 10.38 is a close-up example of three nearly identical phrases with a 7 s interval between the start times of the phrases. Generally, only one whale sang at a time; however, as soon as one whale stopped singing, a different whale would begin singing. Cummings and Holliday (1987) felt it was remarkable that virtually only one whale sang at a time even though during a 2-day period, 414 whales passed within acoustical range of the array and an average of 350 song phrases per hour were recorded. They reported that singing was nearly continuous during this time with two diurnal peaks in sound production; a pronounced peak at 0600–0800 and a weaker one at 1600–1800 with a 10-hour interval between the two peaks. Cummings and Holliday (1987) speculated that perhaps the reason for only one whale singing at a time might be some sort of behavioral hierarchy wherein there may be vocal dominance, or, that whales simply chose not to cope with the problems of singling out one another's call in a reverberant acoustic environment.

Source levels were also determined by Cummings and Holliday (1987) for the phrase in each of 33 songs having the highest received level, presumably

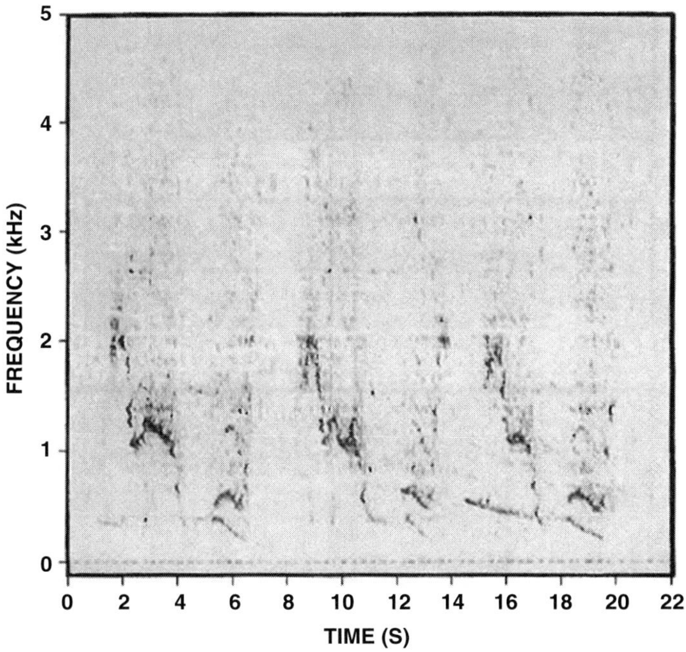

FIGURE 10.38. Spectrogram showing three nearly identical phrases form a bowhead whale song. Analyzing bandwidth was 18.75 Hz (adapted from Cummings and Holliday, 1987).

from different whales. Peak spectrum source levels of these songs were between 158 and 189 dB re 1 μPa, with a median of 177 dB. They estimated the range at which other whales would be able to hear a singer using measured propagation loss values, the median source level, and a median ambient noise level below 2 kHz of 60 dB, to be about 10.7 km. They concluded that under the shallow-water , ice-laden, noisy conditions typical of the area, bowhead whales have a relatively limited underwater sound communication range in comparison to some other species.

Finback Whale Songs Watkins et al. (1987) analyzed 25 years of recordings at a variety of geographic locations on near-surface hydrophones close to whales and on deep hydrophone systems. The data came from a wide variety of locations in the Pacific and Atlantic Oceans, but particularly around Cape Cod, Massachusetts. The sounds from finback whales include single 20-Hz pulses, irregular series of 20-Hz pulses, and stereotyped 20-Hz signal bouts of repetitive sequences of 20-Hz pulses. An example of a 20-Hz pulse from a finback whale is shown in Fig. 10.39 as measured by two hydrophones 250 m apart. Also included in Fig. 10.39 is the spectrogram of two other finback whale pulses. The 20 Hz pulse is in reality a slowly varying FM pulse with each pulse starting at about 23 Hz and dropping to about 18 Hz in about 1 s.

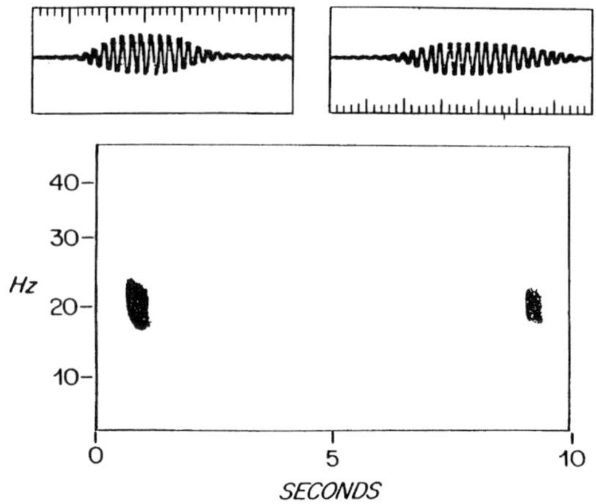

FIGURE 10.39. Time waveform of a finback whale pulse as detected by two hydrophones 250 m apart and spectrogram of two other finback whale pulses. The analysis bandwidth was 5.6 Hz (adapted from Watkins et al., 1987).

The range of frequencies varies from about 25 to 17 Hz. From this spectrogram, one can see the smear in frequencies centered around 20 Hz. The source levels are variable but high with a maximum of about 186 dB re 1 μPa.

Vocalizing finback whales typically behave in a very stereotypical manner. They move little and appear to stay at about 50 m depth. They surface slowly and have only gentle fluke movements during respiration, making 3–7 blows over about 2 min. During their blows, they barely submerge, and after

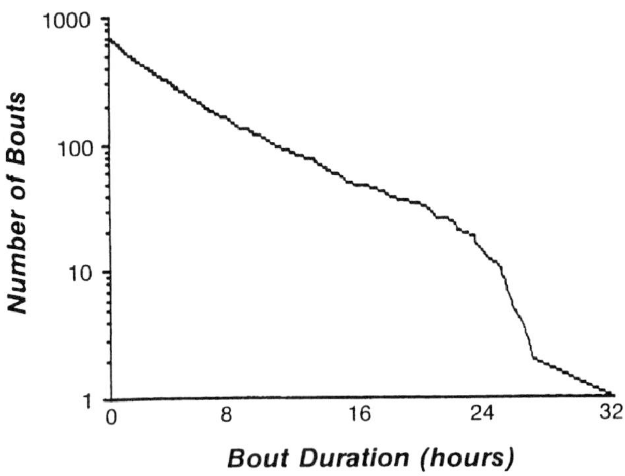

FIGURE 10.40. Plot of all 20-Hz signal bouts over a 5-year period in Bermuda (adapted from Watkins et al., 1987).

blowing they submerge slowly. Generally, only one vocalizing whale was audible from the near-surface hydrophones, and when other vocalizing whales were heard, they were at least 5 km apart. Vocalizing whales seem to be separated from non-vocalizing whales by 200 m or more. When more than one vocalizing whale was audible, the signals were not synchronized although they had similar pulse rates.

The 20-Hz signals are emitted in bouts that can last for hours. A plot of bout duration versus the number of bouts over a 5-year period is shown in Fig. 10.40. In Bermuda, bouts have a seasonal high during the winter months from about September to May and a broad peak between October and December. Only a few bouts were recorded in June and none were noted in July and August. The pulse intervals in a bout were very regular, as can be seen in the two examples of Fig. 10.41, which exhibit a regular doublet pattern

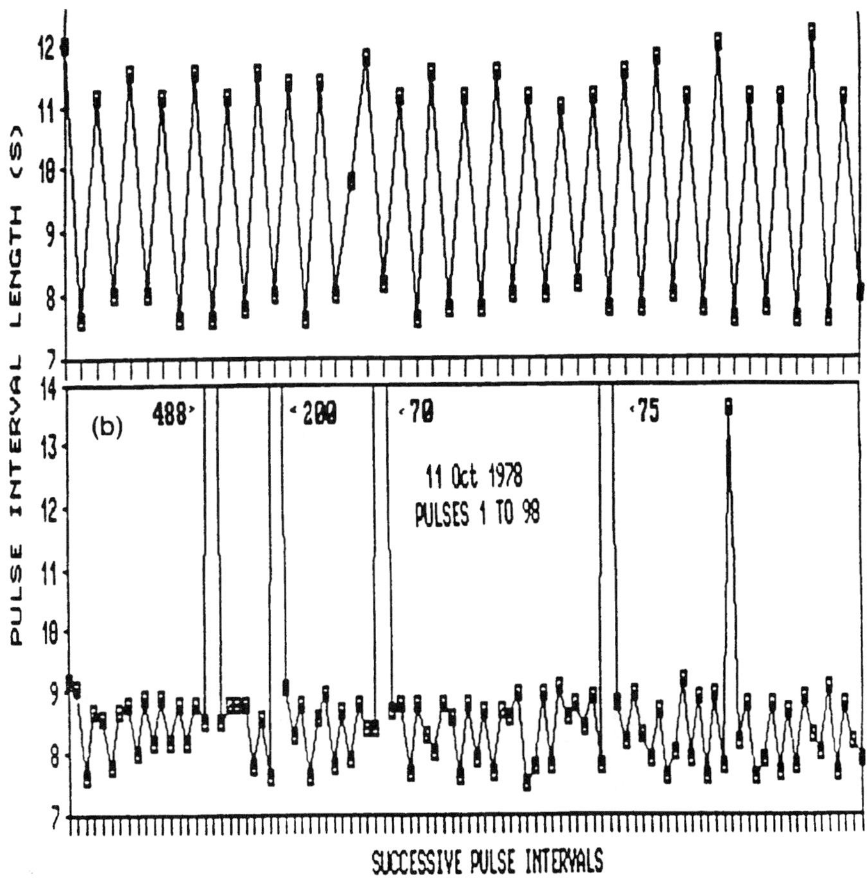

FIGURE 10.41. Pulse interval patterns for a finback whale song (adapted from Watkins et al., 1987), long as 32.5 hours. During a bout, periodic rests averaged about 115 s at roughly 15-min intervals.

common in New England waters. The bottom pattern was taken 20 years after the top pattern. In general, signals are produced in a relatively regular sequence of repetition at intervals ranging from about 7 to 26 s, with bouts that can last and 120 min were observed. There was also some variability in the 20-Hz signals in that they were never exactly replicated.

Blue Whale Songs The research of Alling et al., (1991) has provided some evidence that blue whales (*Balaenoptera nusculus*) sing. The songs consist of four distinct notes lasting about 2 min. The first, second, and fourth notes are pulsive and the third note is a pure tone. Alling et al., (1991) recorded a singer in two different years and the song was basically the same for both years.

McDonald et al (1994) were also able to observe blue whale songs from a seafloor array in the Northeast Pacific. A typical two part blue whale song time series and corresponding spectrogram are shown in Fig. 10.42. The spectrogram of the first part of the two part song had six spectral lines separated by about 1.5 Hz and this type of spectrogram is typically generated by pulses. Cummings and Thompson (1971) previously reported on the pulsive nature of some blue whale moans. The second part of the song was tonal in nature with a slight FM downsweep varying from 19 to 18 Hz in the first 3–4 s. The 18 Hz tone is then carried until the last 5 s when there is an

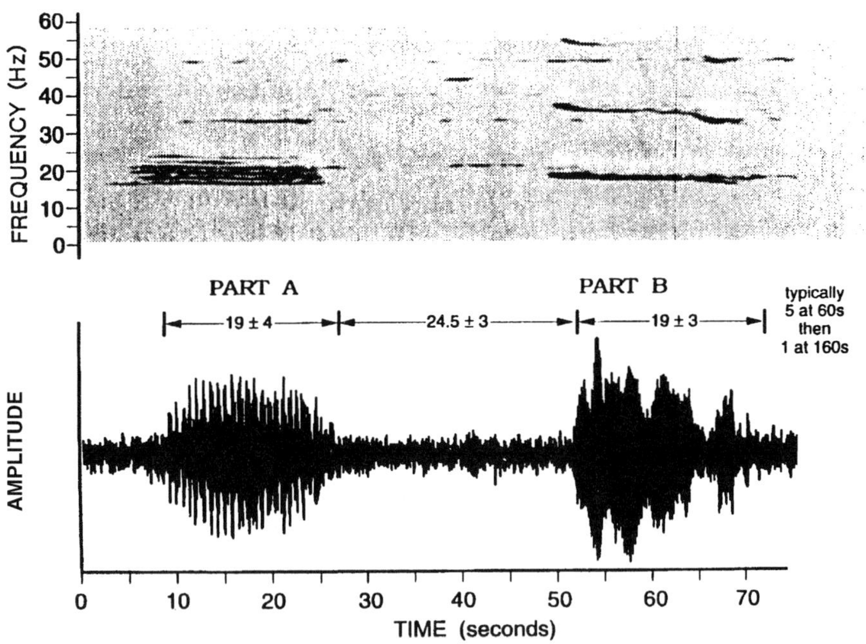

FIGURE 10.42. Time series and spectrogram of a typical blue whale song (adapted from McDonald et al., 1994).

abrupt step down to 17 Hz. The amplitude modulation in the second part of the song was probably caused by multi-path propagation of the signal from the whale to the hydrophones.

There were 134 songs or doublets in the 8-hour long sequences of blue whale songs recorded by McDonald et al. (1994). They all basically followed a pattern of a 19 s pulsive signal followed by a 24.5 s gap and a 19 s monotonic signal. Similar patterns in song intervals were reported by Thompson et al. (1987) where songs occurred in series of two to six with a median repetition interval (breathing time) of 131 s.

10.2.2 Calls of Mysticete Whales

The calls of mysticete whales have been the subject of much research over the past three decades. The U.S. Navy has begun to open its Sound Surveillance System (SOSUS), originally designed for the detection and tracking of submarines in north Atlantic and north Pacific Oceans, to selected and fortunate bioacousticians, providing another avenue to study the acoustics and other facets of whale behavior. Much has been learned about the characteristics of these sounds; however, there is much more to be learned concerning the function or roles of the different calls. One of the problems in studying whale calls is the same as in studying the whistles of dolphins, which is the lack of any standard nomenclature for emitted sounds. Similar calls are often given different names by different researchers. The calls of baleen whales will be discussed with an emphasis on the acoustic properties of the calls. A summary of some of the acoustic properties of the different baleen whales is presented in Table 10.5. One important area of research involves correlating various calls of whales with behavioral patterns. Calls and songs are most likely used for some sort of communications; however, at this time the specific kinds of communication cannot be clearly elucidated. It is extremely difficult to study the context and functions of baleen whale vocalizations. Only on rare occasions have baleen whales been kept in captivity long enough so that controlled acoustic experiments could be done. A stranded juvenile Bryde's was rehabilitated at Marineland in Florida (Edds et al., 1993) and sounds were recorded from it. These situations are far and few in between and the subjects are usually in course of being rescued and slowly nursed back to health for a future release. The sounds of mysticete whales have also been summarized nicely in a book entitled *Marine Mammals and Noise* (Richardson et al., 1995). Some of the information in Table 10.4 has been gleamed from Richardson et al. (1995).

Blue Whale From our discussion of blue whale songs we know that these animals typically produce sounds in the infrasonic frequency range that are strong, low-frequency moans. Cummings and Thompson (1971), off the coast of Chili, measured three part sounds that lasted about 36.5 s, having frequencies between 11.5 and 200 Hz. The major energy peak occurred at about

TABLE 10.5. Characteristics of Mysticete Whales Vocalizations

Species of whales	Signal type	Frequency limits (Hz)	Dominant frequencies (Hz)	Source Level (dB re 1 μPa) at 1 m	References
Blue	FM moans	11.5–200	16–25	188	Cummings and Thompson (1971), Edds (1982)
gs	Songs	16–60	16–60		McDonald et al (1994)
Bowhead	tonal moans	25 – 900	100–400	129–178	Cummings and Holliday (1987); Wursig and Clark (1985)
	pulses	25–3500	152–185		
	songs	20–500	158–189		Cummings and Holliday (1987); Ljungblad et al. (1982)
Bryde's	FM moans	70–245	124–132	152–174	Cummings et al. (1986b); Edds et al. (1993)
	pulsed moans	100–930	165–900	–	Edds et al. (1993)
	discrete pulses	700–950	700–950	–	"
Finback	FM moans	14–118	20	160–186	Watkins (1981); Edds (1988); Cummings and Thompson (1994)
	tonals	34–150	34–150	–	Edds (1988)
	songs	17–25	17–25	186	Watkins (1981)
Gray	pulses	100–2000	300–825	–	Dalheim et al. (1984); Crane (1992)
	FM moan	250–300	250–300	152	Cummings et al. (1968); Dalheim et al. (1984)
	LF-PM-moan	125–1250	<430		" "
	PM pulses	150–1570	225–600		" "
Humpback	simple moans	20–1800	25–30	175	Thompson et al. (1986)
	complex moans	35–360	35–360	175	"
Humpback	grunts (pulse and FM)	25–1900	25–1900		Thompson et al. (1986)
	pulses	25–89	25–80	176	"
	songs	30–8000	120–4000	144–174	Payne and Payne (1985)
	clicks	500–2000	800, 1800	143–154	Stimpert et al. (2007)
Minke	FM tones	60–130	60–130	165	Schevill and Watkins (1972)
	thumps	100–200	100–200	–	Winn and Perkins (1976)
	grunts	60–140	60–140	151–175	"
	ratchets	850–6000	850		Rankin and Barlow (2005)
	boing	1200–5000			

TABLE 10.5. (continued)

Species of whales	Signal type	Frequency limits (Hz)	Dominant frequencies (Hz)	Source Level (dB re 1 μPa) at 1 m	References
Right-N	moans	<400	–	–	Watkins and Schevill (1972)
	scream	500 – 9000			Parks and Tyack (2005)
	warble	500 – 10000			"
	blow	1000 – 10000			"
	upcall	100 – 500			"
	downcall	100 – 500	450		"
	gunshot	100 – 10000			"
Right-S	tonal	30–1250	160–500	–	Cummings et al. (1972); Clark (1982, 1983)
	pulses	30–2200	50–500	172–187	" "
Sei	FM sweeps	1,500–3,500	1,500–3,500	–	Knowlton et al. (1991)

20 Hz, with frequency components extending to about 200 Hz. They estimated these moaning, pulsive sounds to have a source level of about 188 dB re 1 μPa, which would probably represent the most powerful sustained utterance known from whales or any other living source. Edds (1982) measured narrowband moans sweeping from 20 to 18 Hz and lasting about 16 s. Rivers (1995) measured the calls of three blue whales in the waters off Central California and found the calls to be similar to the two part songs reported by McDonald et al. (1994). The first part consisted of a series of pulses that were about 0.7 s apart and the second part was a downsweep with a fundamental frequency shift of 1.2 Hz, on the average.

There is evidence suggesting that the vocalization of blue whales may vary between geographic areas (Cummings and Thompson, 1994; Rivers, 1995). Blue whales recorded off Oregon by McDonald et al. (1994), Mexico (Thompson et al. 1996), and California all emit two-part calls (Cummings and Thompson, 1994). Blue whales from the Gulf of Mexico emit both two- and three-part calls (Thompson et al. 1996), with the standard two parts often separated by a sharply downswept third part. The blue whales off Hawaii also emit a two-part call (Thompson and Friedl, 1982). Blue whales off Chili emitted three sequential AM pulses with no apparent downsweep (Cummings and Thompson, 1971).

Bowhead Whales Most calls of bowhead whales can be categorized as "moans," which are relatively low-frequency, mostly tonal sounds. These moans are typically constant frequency, upsweep, and downsweep FM sounds that last about 2.5 s. The frequencies vary from about 25 to 900 Hz (Cummings and Holliday, 1987). Another common sound is a gargle-like one that is pulse-modulated and extends to about 910 Hz with the peak frequency at about 400 Hz. Cummings and Holliday (1987) measured the peak spectrum source level of 44 moans that varied from 129 to 178 dB re 1 μPa with a mean of 177 dB. The peak spectrum source level of the gargle-like sounds varied from 152 to 169 dB re 1 μPa.

Bryde's Whale The vocalizations of the Bryde's whale (*Balaenoptera edeni*) have been reported by Cummings et al. (1986b) and Edds et al. (1993). Cummings et al. (1986b) reported that Bryde's whales in the Gulf of California produced short downswept FM moans with a mean frequency near 124 Hz with a range of 70–245 Hz. The source levels were estimated to be about 152–174 dB re 1 μPa. Edds et al. (1993) also reported on short moans from an adult-calf pair in the Gulf of California similar to those reported by Cummings et al. (1986). However, they also observed pulsed moans that had durations of 0.7–1.4 s. Pulsed moans were also reported for a captive juvenile at Sea World of Florida that produced a pulsed moan; however, the duration tended to much longer, between 0.5 and 5.1 s. The repetition rate of the pulsed moan varied between 60 and 130 Hz, which was higher than the 20–70 Hz produced by the captive juvenile. Edds et al. (1993) suggested that the pulsed moans they observed may have been related to the presence of an adult-calf pair and the absence of adult-calf pairs for the recording of Cummings et al. (1986).

Finback Whale We have already mentioned that the sounds emitted by finback whales include a single 20-Hz pulse, irregular series of 20-Hz pulses, and stereotyped 20-Hz signal bouts of repetitive sequences of 20-Hz pulses. An example of a 20 Hz pulse from a finback whale was shown in Fig. 10.40 along with the spectrogram of two other finback whale pulses. The 20-Hz pulse is in reality a slowly varying downsweep FM pulse with frequencies varying from 23 to 18 Hz in about 1 s. The frequency sweep of the Pacific finback whale is greater than that of their Atlantic cousins. The major difference in the songs and calls for the finback whale is the pattern or sequence of sounds that are emitted in a song. The same basic 20-Hz pulses are used for both calls and songs.

Finback whales also produce sounds at frequencies up to 150 Hz, which include a 34 Hz and a 75-Hz tone, a 129–150 Hz tone preceding 20-Hz sounds, and downward sweeping FM pulses varying from 118 to 14 Hz (Edds, 1988). These higher frequency calls are heard most often when two or more whales are close together and may be used for social reasons. A low-frequency broadband rumble with energy mainly at 30 Hz is often heard when two finback whales approach each other, during close encounters with a ship or boat and in agonistic encounters between whales.

Gray Whale The gray whale (*Eschrichtius robustus*) produces four basic types of sounds (Cummings, et al., 1968; Dalheim et al., 1984; Crane, 1992). Dalheim et al. (1984) collected 200 hours of underwater recordings of gray whale sounds in Laguna San Ignacio, Baja California (winter breeding grounds of gray whales), and Crane (1992) had 69.8 hours of useful recordings of migrating gray whales passing Monterey and Carmel Bays, California. The most prevalent sounds recorded by Dalheim et al. consisted of transient pulses (S1), which were broadband with energy ranging from <100 to about 2 kHz, with an emphasized band between 300 and 825 Hz (See Fig. 10.43a). These pulses typically occurred in a series or burst ranging from 2 to 30 with an average of 9.4 pulses per series and an average of 1.8 s per series. Each pulse was less than 50 ms long with a pulse repetition rate varying from 2.2 to 14.7 Hz, and an average of 5.9 Hz. The second sound was termed an S2 signal, which had a rapid FM-up- and downsweep that had a metallic aural quality. These signals had a frequency range of 250–300 Hz and an average duration of 0.3 s (see Fig. 10.43b). The third type of sounds, designated as S3, was a low-frequency moan that was pulse-modulated and had some FM present. The sound had a frequency range of 125–1250 Hz and a duration of 1–4 s. Most of the energy was concentrated below 430 Hz (See Fig. 10.43c). The fourth sound was called S4 and had a higher mean frequency (150–1570 Hz) than the S3 calls. Their average energy was centered between 225 and 600 Hz (See Fig. 10.43d). The average duration was less than a second and this signal also appeared to be pulse-modulated.

The sounds produced by gray whales were recorded by Dalheim et al. (1984) under a variety of behavioral activities and environmental conditions. Although they did not attempt to correlate behavior with sound production,

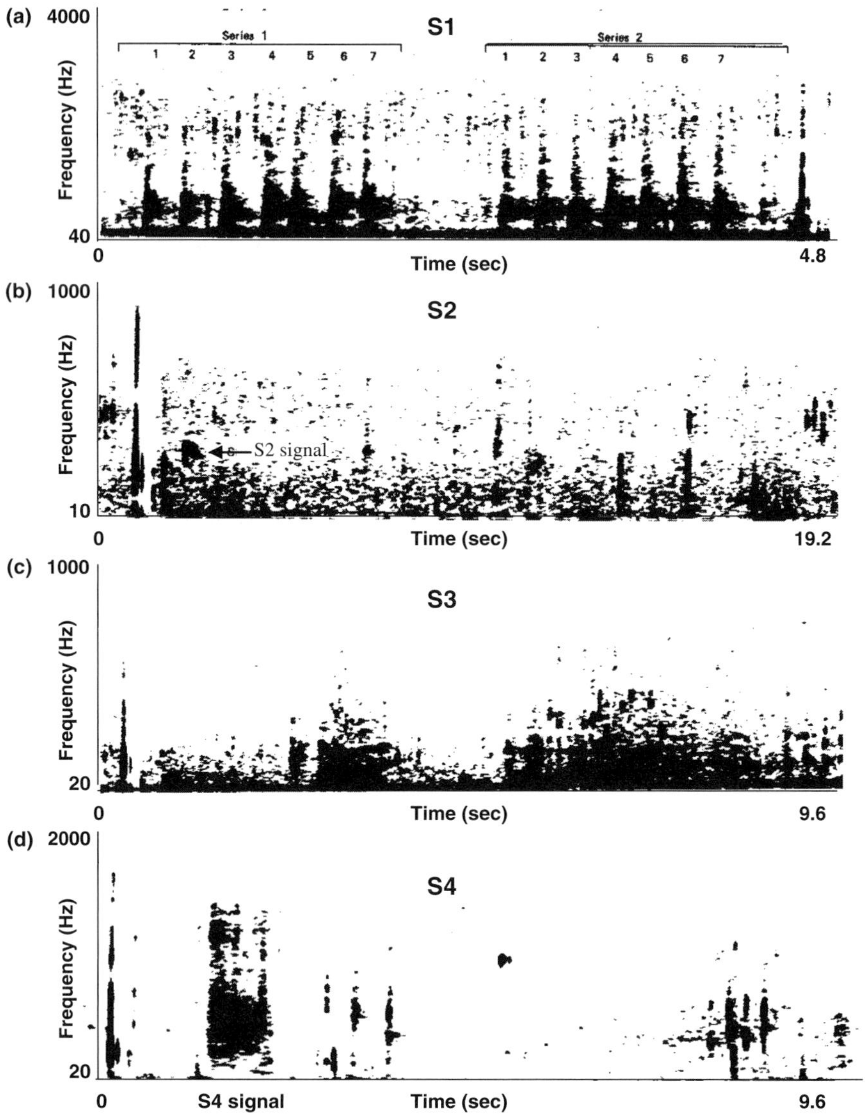

FIGURE 10.43. The four different types of calls of the gray whale (**a**) S1 calls, (**b**) S2 calls, (**c**) S3 calls, and (**d**) S4 calls (adapted from Dalheim et al., 1984).

they noted that the whales appeared to produce more sounds when numerous whales were together in a small area, when whales were on collision courses, when single whales chased cow/calf pairs, when whales were interacting with bottlenose dolphins, when whales were on collision courses with vessels, and when whales were in the presence of increased non-biological noise due to vessels transiting the area.

The gray whale calls recorded by Crane (1992) could also be divided into similar categories as those of Dalheim et al. (1984). However, there are minor differences in the frequencies and duration between both data sets. These differences may reflect differences in the situations under which the recordings were made. Crane's data were recorded as gray whales migrated along the California coast, while the data of Dalheim et al. were obtained after the whales had reached their winter destination.

Humpback Whales The interest in humpback whale song has taken center stage in such a way that there is very little data on other sounds that they emit, especially in the summer feeding grounds. To rectify this situation, Thompson et al. (1986) recorded the sound production and associated behavior of Pacific humpback whales in Southeast Alaskan waters on their summer feeding grounds. Underwater vocalization sounds from humpback whales consisted of moans, grunts, and pulse trains. Sounds were also recorded that were associated with the blow hole and with surface impacts. The moans were prolonged vocalizations that lasted over 400 ms and had frequencies between 20 and 1,800 Hz. The moans seemed to be pulsive in nature with a harmonic structure. Simple moans had a strong fundamental component between about 25 and 30 Hz. Complex moans did not exhibit strong fundamentals, and were often pulse-modulated in a band from 35 to 360 Hz. Both complex and simple moans had an average duration of 800 ms. The estimated source level of moans was approximately 175 dB re 1 μPa. Grunts were shorter vocalizations than moans, with frequencies between 25 and 1,900 Hz, and did not usually have a harmonic structure. Most grunts were pulsed and had an FM upsweep. The estimate source level of grunts was 190 dB re 1 μPa. The vocalization rate of these humpback whale moans and grunts was 1.8–2.3 sounds/min. Pulse trains consisted of low-frequency pulses with energy between 25 and 80 Hz. These pulses were about 300–400 ms in length with a bandwidth of about 50 Hz. Some of the pulse trains sounded like tones, with some having harmonics. Thompson et al. (1986) measured 183 pulse trains and found a 5 pulses per min median frequency of occurrence in 8.5 hours. They also estimated the source level of these low-frequency pulse trains as about 176 dB re 1 μPa.

Minke Whales The calls of the relatively small Minke whale (*Balaenoptera acutorostrata*) of the Antarctic were recorded by Schevill and Watkins (1972). The calls were composed of a single downsweep FM signal starting at about 130–115 Hz and sweeping down to about 60 Hz in about 0.5 s. They measured nine sounds, presumably from one whale, and found that the sounds were not all identical but had a similarity in duration, frequency sweep, and sound pressure level. They also estimated the maximum source level to be approximately 165 dB re 1 μPa. The low-amplitude signals had no harmonic structure and appeared to be nearly sinusoidal. Winn and Perkins (1976) reported that Minke whales also produce upsweep FM, grunts, clicks, thump trains, and ratchets. Thump trains typically last over 1 min and consist of pulses having durations between 50 and 70 ms with most of the energy between 100 and 200 Hz.

A sound that Naval sonar operators have been well familiar with, but without any idea of the source, is the "boing" sound. It has a distinctive "metallic-like" ring, resulting in its boing label. Wenz (1964) was the first to describe the boing sound from the tapes recorded on Navy submarines. This state of affairs continued for several decades, with scientists slowly beginning to suspect minke whales. Gedamke et al. (2001) suggested the minke whales may be the producers of the boing sounds from his analysis of dwarf minke whale calls in the Great Barrier Reef, Australia. It was not until 2005 that Rankin and Barlow (2005), using a towed array, were able to localize boing sounds and visually identify minke whales as the source. Spectrograms of the boing sound from the central and eastern Pacific are shown in Fig. 10.44. The central boing had frequencies between 1.3 and 5.0 kHz, whereas the eastern boing had frequencies between 1.5 and 4.6 kHz. The boing consists of a brief pulse followed by a long call that is frequency- and amplitude-modulated (Rankin and Barlow, 2005). Their data from 128 boings could be grouped into two distinct call types, with non-overlapping pulse repetition rates of 91–93 pulses/s detected east of 138° W, matching Wenz's (1964) description of the San Diego boing, and with repetition rates of 114–118 pulses/s matching Wenz's description of the Hawaii boing. The eastern boing had a longer duration with a mean of 3.6 s than the central boing with a mean of 2.6 s.

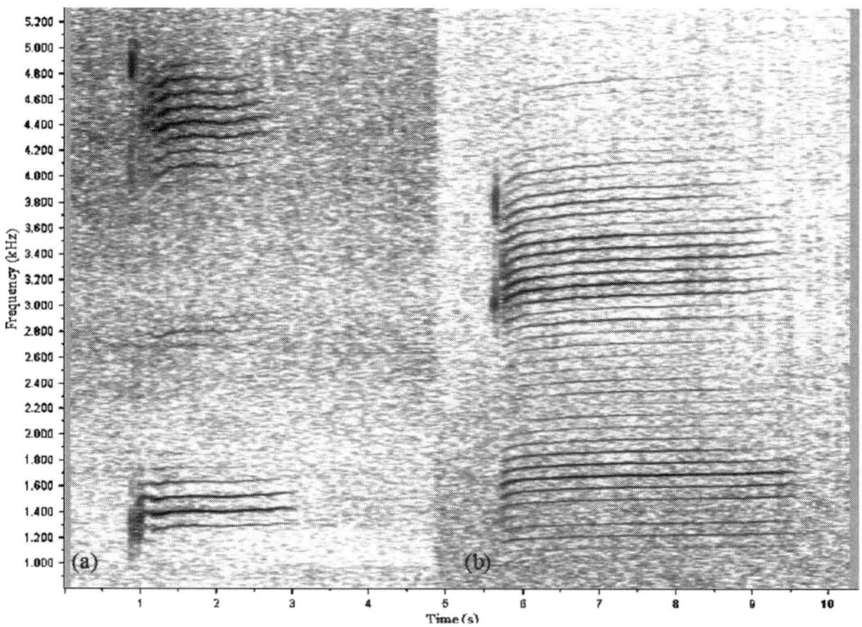

FIGURE 10.44. Spectrogram of (a) central boing and (b) eastern boing sounds (adapted from Rankin and Barlow, 2005).

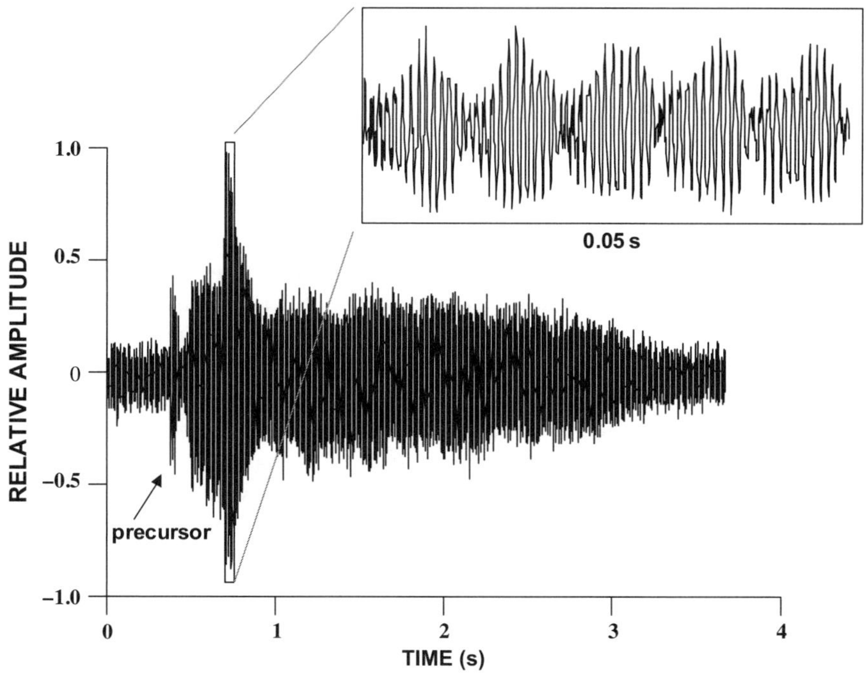

FIGURE 10.45. An example of a boing sound waveform.

The waveform for a boing sound detected by a near-bottom hydrophone at a depth of 5 km approximately 100 km north of Oahu, Hawaii, is shown in Fig. 10.45. Often, but not always, a precursor or two will precede the main boing sound. The expanded view indicates that the boing sound consists of a number of repetitions of a basic waveform and this repetition results in the ripple patterns shown in the spectrogram as was discussed earlier in this chapter.

Right Whales The calls of the southern right whale (*Eubalaena australis*) were first described by Payne and Payne (1971) and Cummings et al. (1972). Clark (1982, 1983) performed a 2-year study in Golfo San Jose, Peninsula Valdes, of the southern right whale acoustic communication. He described their repertoire as a continuum consisting of two functional subdivisions; a set of simple FM, discrete calls associated with resting and transiting whales, and a set of highly variable, intergraded signals associated with groups of active whales. The most common discrete call was an FM upsweep, or "up-call" that was produced by all the whales, including newborn calves. Clark (1983) also found that the southern right whales produced countercalls over many miles and often joined other whales who called back. From his observations, Clark (1983) concluded that the up-call was a contact call, which served to keep animals in acoustic range and bring them together. Clark (1982, 1983) also noted that when the animals were in close, active (often sexually active) groups, they produced a much richer set of acoustic signals. The sounds were typically produced in a rapid series with only one whale in the active group vocalizing at any one time.

The calls of the northern right whale (*Eubalaena glacialis*) seem to be similar to the southern right whale. Parks and Tyack (2005) used single hydrophones, synchronous hydrophone/video recordings and a horizontal array of 15 hydrophones in the Bay of Fundy, Canada, to measure the sounds of northern right whale in surface active groups. Surface active groups were described as two or more whales (up to 35 observed) interacting at the surface. They labeled the calls into the six categories shown in Fig. 10.46, keeping their labels similar to those of Clark (1982). Screams were the most numerous followed by gunshot, blows, upcalls, warbles, and downcalls. Screams were produced mainly in groups of females and males and rarely produced by whales separated from their group. Gunshot sounds were usually recorded from lone right whale males although it is not known if females can produce this sound. Upcalls seemed to be produced by males while swimming into a surface active group for the first time or while alone at the surface while a focal female was on a dive. Warbles were only recorded in the presence of calves in a surface active group, and were pulses with energy at peak frequency of 3 kHz.

Sei Whales Very few recordings of sei whales (*Balaenoptera borealis*) exist. Thompson et al. (1979) reported that sei whales produced a sonic burst of

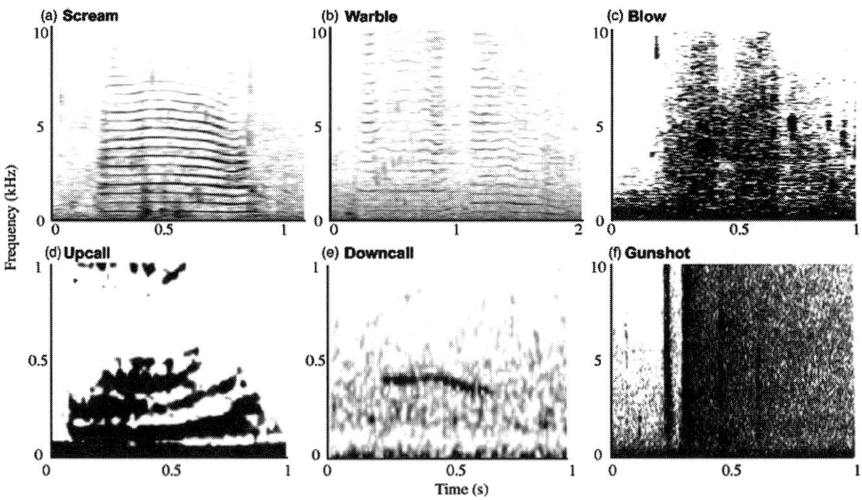

FIGURE 10.46. Spectrograms of six calls of the northern right whale (adapted from Parks and Tyack, 2005).

7–10 metallic-like sounding pulses with energy at a peak frequency of 3 kHz. The train of pulses lasted 0.7 s with each pulse being about 4 ms in duration. Knowlton et al. (1991) reported sei whale sounds that consisted of two phrases of about 0.5–0.8-s duration spaced about 0.4–1 s apart. Each phrase was composed of a series of 10–20 FM sweeps in the range of 1.5–3.5 kHz and lasting about 30–40 ms/sweep.

10.3 Underwater Vocalization of Pinnipeds

Pinnipeds being amphibious animals vocalize in air and underwater. However, some Arctic and Antarctic species spend almost all of their time in the water or hauled out on ice, making it difficult to learn and observe the adaptive significance of various calls. The functions of in-air calls of pinnipeds are much easier to study since the observation of behavior of calling animals and the reactions of animals around them can be easily observed. In-air vocalizations by seals and sea lions functions either to threaten or to attract individuals by identifying the caller as to species, sex, age, location, and probably as an individual (Schusterman, 1978). Vocalizations by mothers and pups play an important function for pinnipeds that are hauled out in a large herd. According to Schusterman (1978), it appears likely that underwater calls may serve the same socially communicative functions as in-air calls. In this section, we will concentrate mainly in the characteristics of pinniped vocalizations underwater. We will not attempt to render a detailed description of the various vocalizations used by the different species of pinnipeds.

10.3.1 Underwater Vocalization of Phocids

Weddell Seal Phocid (hair) seals make a variety of sounds that have been described as grunts, snorts, buzzes, barks, yelps roars, creaks, trills, etc. Most seals do not have a large repertoire of calls. The Weddell seal (*Leptonychotes weddelli*) of Antarctica, whose vocalization structure was first described by Schevill and Watkins (1965a), seems to have the largest repertoire of underwater calls. Thomas et al (1983) classified the calls of Weddell seals into 12 distinct calls with 34 types. Examples of these 12 calls are shown in Fig. 10.47. These calls were given designations according to how they sounded to humans. The 11 T calls were designated after the "trill" quality of the sounds, 3 R for "cricket call", G for "guttural glug", 1 M for "mew", 1 E for "eeyoo", 1 L for "growl", 5 P for "chirp",4 C for "chug", 1 A for "click", 2 Z for "seitz", 2 K for "knock", and 1 H for "teeth chatter". The Weddell seal underwater calls cover a frequency range up to about 15 kHz. From the spectrograms of Fig. 10.47, it seems that most of the calls have a pulse or burst characteristic and even the long-duration trill calls seem to have pulsating components. The

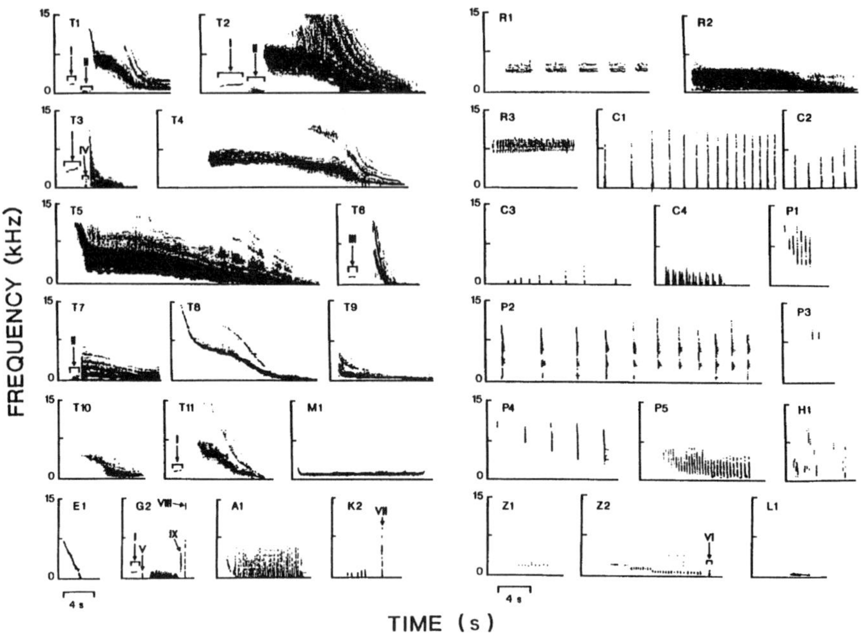

FIGURE 10.47. Spectrogram of call types in the Weddell seal underwater repertoire (adapted from Thomas et al., 1983).

harp seal (*Phoca groenlandica*) of the Arctic also has a relatively wide repertoire of 15 call types (Mohl et al., 1975).

Leopard Seal The leopard seal (*Hydrurga leptonyx*) of the Antarctic has been reported to emit 12 different types of underwater calls, all within the frequency range of 65–4,800 Hz (Rogers et al., 1995). Rogers et al., (1995) made underwater recordings of a male and female leopard seal housed in a concrete saltwater pool at Taronga Zoo, Australia, and also made 110 hours of field recordings of leopard seals calls in the Antarctic. Examples of the 12 different call types of the leopard seal are shown in Fig. 10.48. Comparing the spectrograms of Figs. 10.48 and 10.49, one can see that the calls are very different from those of the Weddell seal, with only a few calls that one can say are relatively similar.

Harp Seal The harp seal (*Phoca groenlandica*) typically gathers into large herds near the ice edge in the northwest and northeast Atlantic. They probably use underwater sounds to find the main herd and also to locate mates (Terhune and Ronald, 1986). They are very vocal during this spring breeding season. Mohl et al. (1975) identified 16 types of harp seal sounds, and Terhune (1994) has since identified another 3 types of calls. Most of their calls have a maximum sound pressure level of 135–140 dB re 1 μPa (Watkins and Schevill, 1979; Terhune and Ronald (1986). Under quiet conditions, a strong harp seal sound might be detectable by another harp seal at a distance

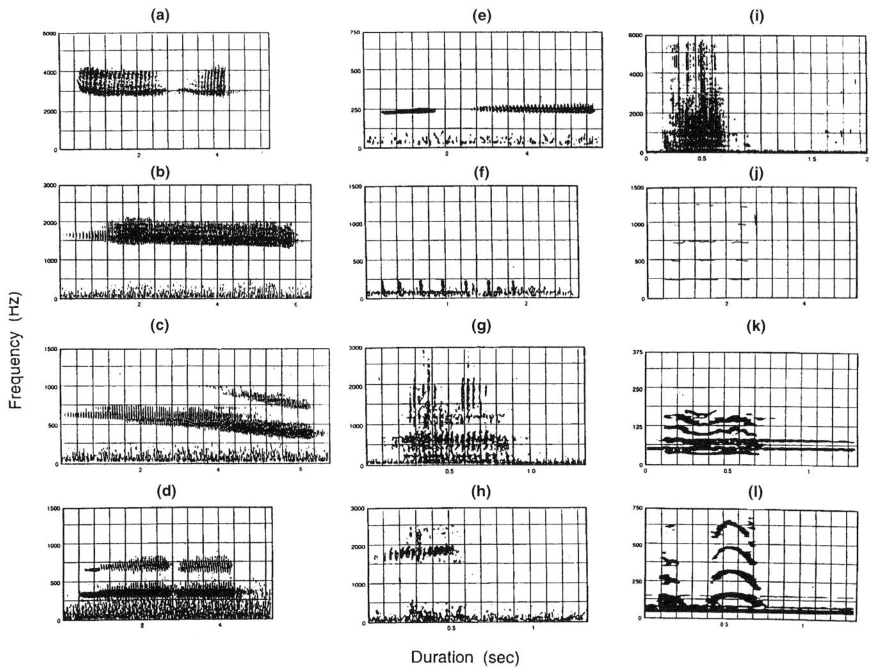

FIGURE 10.48. Examples of 12 call types of the leopard seal (a) high double trill, (b) medium trill, (c) low descending trill, (d) low double trill, (e) hoot with a trill, (f) thump pulse, (g) roar, (h), noseblast, (i) blast, (j) hoot, (k) growl, and (l) snort (after Rogers et al., 1995).

of 2 km (Terhune and Ronald, 1986). A vocalizing herd could be detected with hydrophones out to 30–60 km (Watkins and Schevill, 1979; Terhune and Ronald, 1986).

Gray Seal The gray seal (*Halichoerus grypus*) has about seven types of underwater calls during the breeding season (Asselin et al., 1993). Examples of the gray seal repertoire of calls are shown in Fig. 10.48. The majority of the gray seal calls recorded by Asselin et al. (1993) consisted of guttural sounding "rups" and "rupes" with frequencies between 100 and 3,000 Hz, and low-frequency growls with frequencies between 100 and 500 Hz. Other less common calls were low-frequency clicks as well as loud knocks. The guttural rup was the most common call, that occurred in 47% of the 3,856 recorded vocalizations. The fundamental frequency was relatively low, between 0.1 and 0.4 kHz with a sharp upsweep at the end of the call, which increased from 0.4 to 4.7 kHz. The second most common call was the guttural rup, which accounted for 29% of all the vocalizations. The fundamental frequency of the guttural rupe ranged from 0.1 to 0.4 kHz, with numerous ripples that extended to 3.6 kHz.

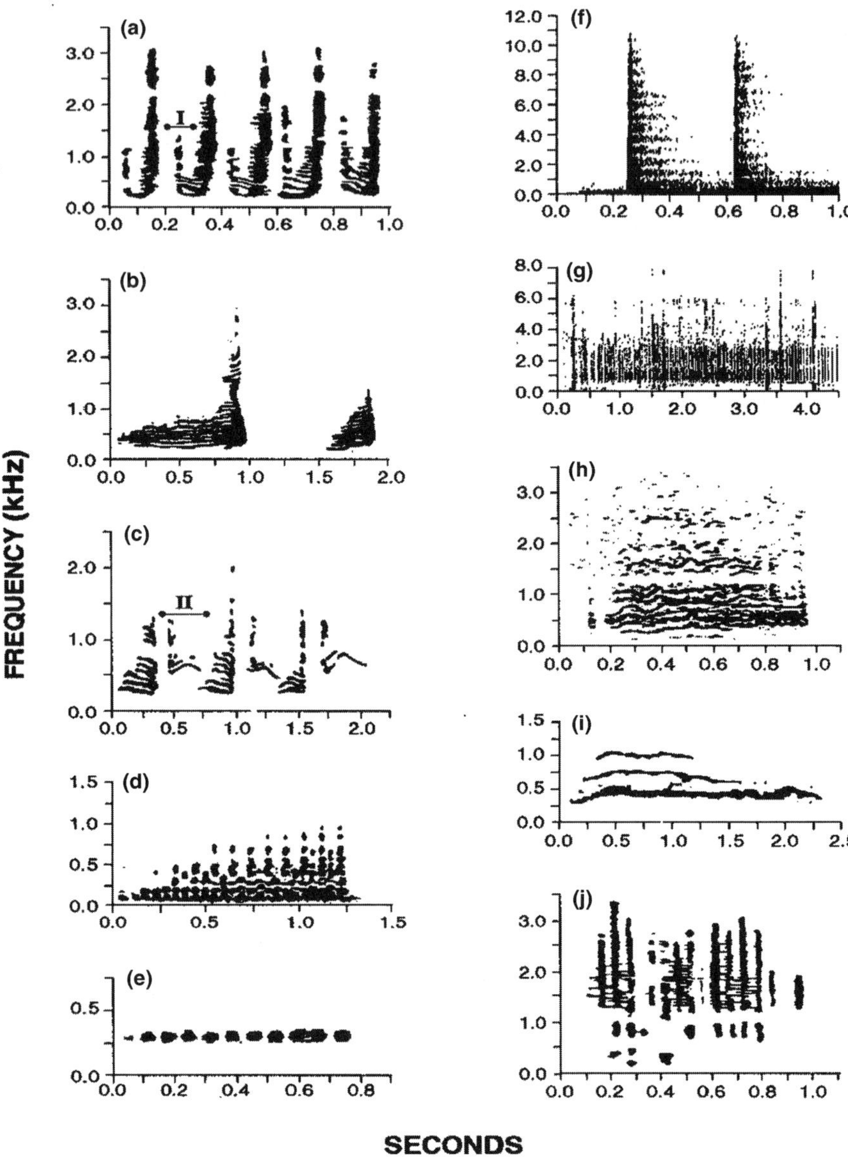

FIGURE 10.49. Examples of underwater call types emitted by gray seals during the breeding season. (a) series of 5 type-1 rups with an auxiliary structure, (b) Series of type-2 rups. (c) Series of 3 type rupes. (d) Growl type 1. (e) Growl type 2. (f) Series of two nocks. (g) Click train with a repetition rate 72 Hz. (h) Click train with a repetition rate o f 125 Hz. (i) Roar. (j) Trot with repetition rate of 13 Hz (adapted from Asselin et al., 1993).

Bearded Seal The bearded seal is the largest of the northern phocids and is distributed throughout the Arctic. Cleator et al. (1989) classified their calls into six types, which included several types of trills, moans, and groans. Trills were by far the most common vocalization of the bearded seal. A dictionary definition of a trill is a rapid reiteration of the same tone or the alternation of two musical tones a diatonic second apart. Trills were mainly downward swept FM calls and can be seen as the FM warble in the spectrogram of the bowhead whale call. Trills can have a starting frequency as high as 4.8 kHz, which can decrease to as low as 130 Hz and last up to almost 3 min. Trills were placed into five different categories. In Alaskan waters and parts of the Canadian Arctic, ascending trills can be heard, though less frequently than the descending trills. Bearded seals also emitted trills in sequences to produce songs. The acoustic propagation characteristics of trills were also examined by Cleator et al. (1989). Sixty three percent of the trills appeared to originate with a 5-km radius around their camp. Trills as far as 20–30 km could also be detected. They concluded that for their ice, weather, and topographic conditions, the bearded seal trill can be heard underwater for a minimum of 25 km.

Harbor Seal The harbor seal (*Phoca vitulina*) apparently produces only five basic types of underwater vocalizations (Hanggi and Schusterman, 1992, 1994). Like the other investigators, Hanggi and Schusterman (1992, 1994) classified the calls according to how they sounded to the human ear. The five vocal types are (a) roar, which had little if any tonal quality; (b) bubbly growl, which also was non-tonal; (c) grunt, which was an assortment of very short calls of 0.1–0.5 s that were generally produced in a series; (d) groan, which was similar to grunts but longer in duration and ranged from 1 to 5.5 s; and (e) creak, which had a tonal quality with harmonics in the frequency range of 700–2000 Hz and sounded like a rusty hinged door being opened. The durations of creaks varied from 0.5 to 6 s.

Remaining Seals Many other seals produce only limited types of underwater vocalizations. The general characteristics of underwater vocalizations of all phocids are tabulated in Table 10.6. It seems that the hooded (*Cystophora cristata*), ribbon (*Phoca fasciata*), crabeater (*Lobodon car cinophagus*), and elephant seals (*Mirounga* sp.) may all have only one type of vocalization. The Ross seal (*Ommatophoca rossi*) and hooded seal (*Cystophora cristata*) emit two types and the ringed seal (*Phoca hispida*) emits three types of underwater calls. The underwater vocalizations of the Hawaiian (*Monachus schauinslandi*) and Mediterranean (*Monachus monachus*) monk seals have not been reported as of yet. The Ross seal produces an interesting underwater siren call, which is shown in spectrogram form in Fig. 10.50 (Watkins and Ray, 1985). This siren call has three to five strong "sideband" harmonics or ripples in the spectrum. The fundamental frequency typically drops from 4 to 1 kHz during the first part of the call and then returns smoothly upward to end at 4.1–4.5 kHz. Another interesting feature of the siren calls was the presence of a secondary siren-like tone whose frequency is higher and changed more rapidly than the primary tone (Watkins and Ray, 1985). The ribbon

TABLE 10.6. Characteristics of Vocalization of Phocid (Hair) Seals

Species	Signal type	Frequency Extent (kHz)	Dominant frequencies (kHz)	Source level (dB re 1 μPa)	References
Bearded seal (1983)	trill, moans, grunts	.02–6	1–2	178	Cleator et al. (1989); Cummings et al.
Crabeater seal	groan	<0.1–8	0.1–1.5	—	Stirling and Sniff (1979)
Gray seal	rups and rupes	.1–3	0.1–3	—	Asselin et al. (1993)
	growl	0.2–1	0.1–1		"
	knocks	<0.1	–11		"
	clicks and roars	0.2–3.5	0.2–3.5		"
Harbor seal	roar	0.4–4	0.4–0.8	—	Hanggi and Schusterman (1994)
	bubbly growl	<0.1–0.4	0.1–0.25	—	"
	grunt, groan	"	—	—	"
	creak	0.7–4	0.7–2	—	"
Harp seal	17 types	<0.1–16	0.1–3	130–140	Mohl et al. (1975); Terhune (1994)
Hooded seal	grunt	0.2–0.4	0.2–0.4	—	Terhune and Ronald (1973)
	snort	0.1–1	0.1–1	—	"
	buzz	1.2–6	1.2	—	"
Leopard seal	12 types (see Fig. 10.42)	0.06–4.8	—	—	Rogers et al. (1995)
Ribbon seal	FM & puff	0.1–7.1	0.1–7.1	—	Watkins and Ray (1977)
Ringed seal	barks, clicks, yelps	0.4–16	<5	95–130	Stirling, 1973; Cummings et al. (1986)
Ross seal	siren	1–4	1–4	—	Watkins and Ray (1985)
Spotted seal	social sounds	0.5–3.5	—	—	Beier and Wartzok (1979)
	clicks	8 – 150	—	—	Renouf et al. (1980)
Weddell seal	34 types	0.1–11.8	—	153–193	Thomas and Kuechle (1982)

seal also produces interesting underwater sounds, including a relatively intense prolonged downward sweep in frequency, and a broadband puffing sound (Watkins and Ray, 1977). The spectrograms of these two sounds are also shown in Fig. 10. 50. The downward FM sound varies in frequency from

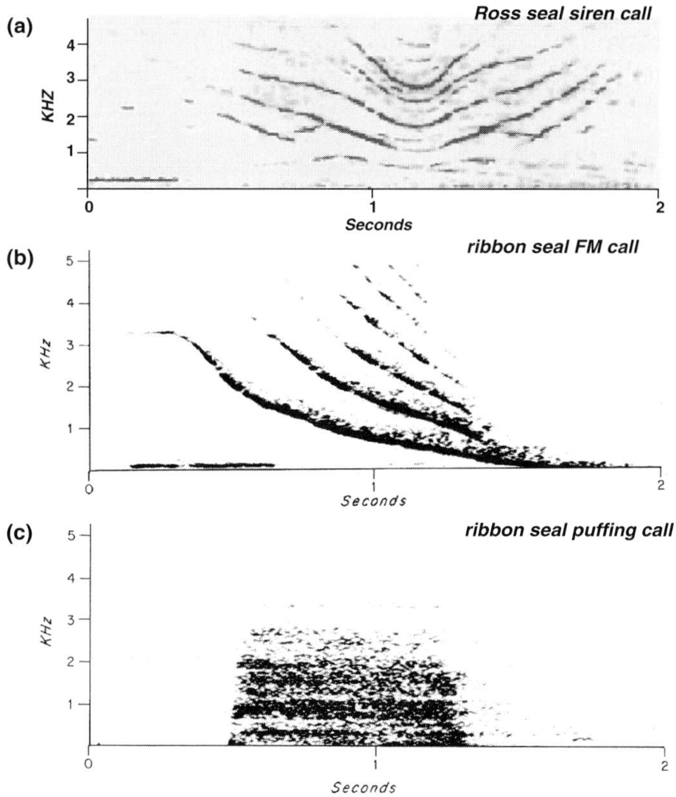

FIGURE 10.50. (a) Siren call of the Ross seal (after Watkins and Ray, 1977; (b) downsweep FM and "puffing" calls of the ribbon seal; (c) analysis filter bandwidth as 45 Hz for these calls (adapter from Watkins and Ray,1977).

7 to 0.1 kHz in downward sweeps of 2–5 kHz. The broadband puffing sound has frequencies below 5 kHz and lasts a little less than 1 s.

Geographical Variations and Dialect Although elephant seals produce a relatively simple pulsive vocalization in air and underwater, they can vary the repetition rates of the pulses to change the audible quality of the sound. Le Boeuf and Perinovich (1974) have measured the difference in the pulse rates of the in-air threat vocalization of male elephant seals in four populations on the islands of the coast of California and Baja California, Mexico, over a 5-year period. The mean pulse rate at Año Nuevo Island increased every year from 1968 to 1972, but still was slower than the mean pulse rate of all southern rookeries even though 43% of the breeding males on Año Nuevo Island came from southern populations. Males at St. Nicholas Island off Southern California emit sound pulses at more than double the rate of the seals at Año Nuevo Island (Le Beouf and Peterson, 1969). Mean pulse rate of males at San Miguel Island and Isla de Guadalupe (408 and 944 km south of

Ana Nuevo Island) were intermediate to the pulse rates used by seals on St. Nicholas and Año Nuevo. Le Beouf and Peterson (1969) suggested that these geographical differences in threat vocalizations resemble local dialects in birds and humans. The vocalizations of other seal species have also been found to have geographical differences. Thomas et al. (1988) reported variations in the underwater vocalization in the Weddell seal in the Antarctic. Weddell seal vocalizations from Davis Station, McMurdo Sound, and Palmer Peninsula, three widely separated locations on Antarctica, were similar but none were identical. One vocalization was unique to Davis Station. At all stations, trills or territorial defense calls were most common and had more types than other calls. Cleator et al. (1989) reported that the bearded seals had a geographically different vocal repertoire that may be indicative of discrete breeding stocks. A one-way ANOVA showed significant differences in start and end frequencies of all trill types from experimental stations in six different locations separated by hundreds of kilometers. Terhune (1994) found differences in the underwater vocalization of the harp seals in the Gulf of St. Lawrence and north of Jan Mayen Island. Each herd had one unique call type and shared an additional 17 call types. Within each herd, there were no differences in the call types and the number of call elements over a 19-year period. From his findings and other separate tagging studies, Terhune (1994) suggested that the two herds were reproductively isolated. Thomas and Golladay (1995) reported that the underwater vocal characteristics of leopard seals recorded from McMurdo Sound and Palmer Peninsula were different. Leopard seals at McMurdo Station only had five calls, compared to nine call types at Palmer Peninsula. At McMurdo Sound, calls had one or two components whereas at Palmer Peninsula some calls had three components. More calls from Palmer Peninsula had FM and most McMurdo Sound call durations were significantly longer.

10.3.2 Underwater Vocalization of Otariids

California sea lion The underwater vocalization of several captive California sea lions (*Zalophus californianus*) under three different sets of conditions was studied by Schusterman et al. (1967). The vocalizations of the animals were recorded when involved with conspecific social interaction, when presented with a mirror underwater, and when fleeing from a human observer. They found that *Zalophus* produce five basic sounds: clicks, barks, whinnies, buzzing, and bang or crack sounds. Clicks were produced in trains typically lasting about 2 s, although many trains lasted as long as 23 s with pauses of less than 0.5 s. Click repetition rates varied within click trains from less than 5 pps (pulses per second) to 70 or 80 pps. An example of a click train emitted by *Zalophus* is presented in Fig. 10.51. The click train was emitted by one sea lion while play-fighting with another sea lion. Schusterman et al. (1967) noted that clicking sounds were never emitted at regular intervals by any of the animals during any of the free-swimming conditions. The second vocalization

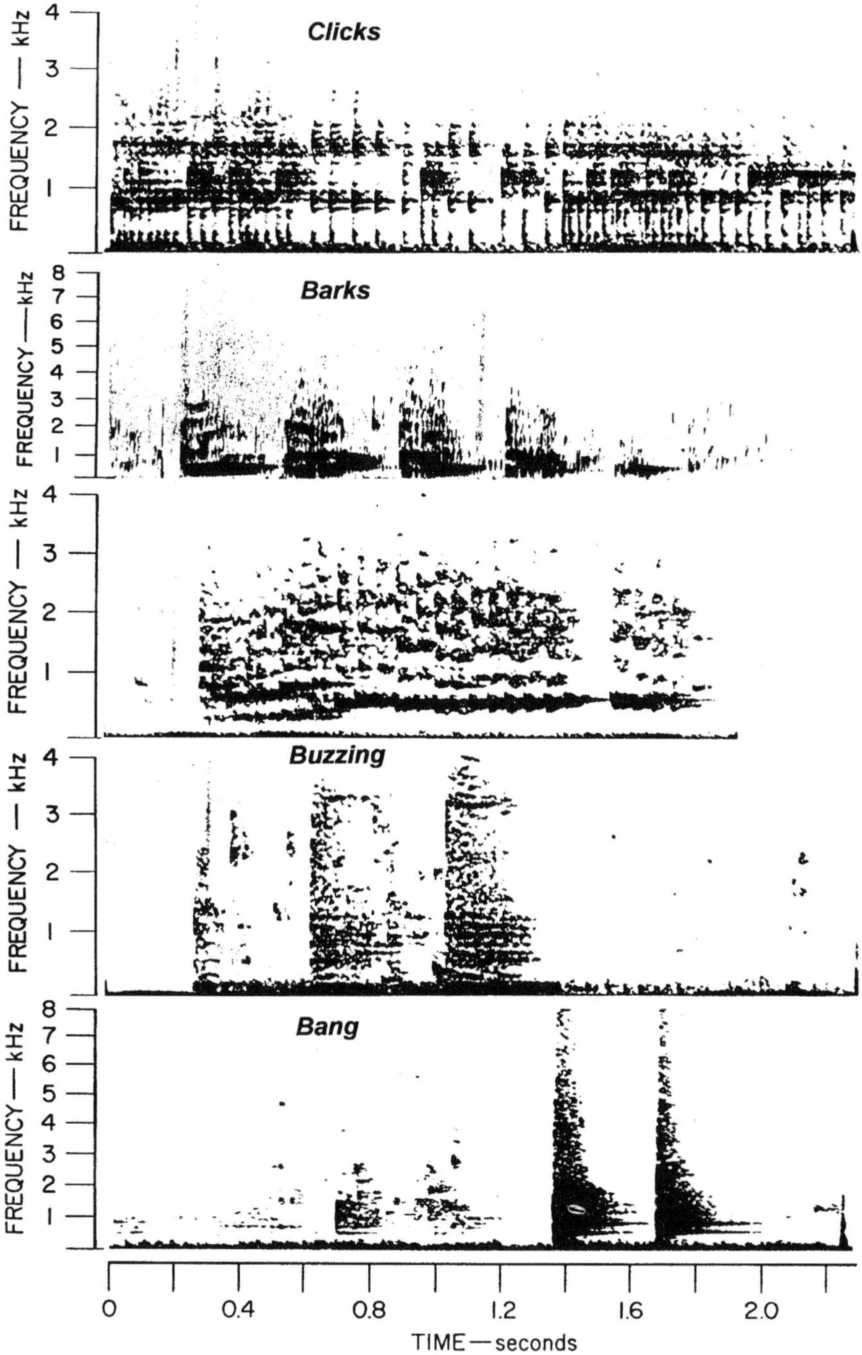

FIGURE 10.51. Five different vocalization by the California sea lion; click train, barks that are preceded by some clicks, whinny, buzzing, and bang or crack (after Schusterman et al., 1967).

type was the bark. This form had most of its energy below 3.5 kHz although some energy could be found out to 8 kHz. A single bark usually had a duration of 200–300 ms with little variation. Barks were sometimes preceded by a click train, as can be seen in Fig. 10.51. The bark in the figure was produced while a 2-year-old male was fleeing from the experimenter. The third call type was a whinny that was frequently produced by a 3.5-year-old female during aggressive encounters. The whinny sound typically lasted about 1.5 s and was repeated three or four times in succession. The spectrogram of a whinny is fairly complex with rapidly occurring pulses and frequency extending between 0.5 and 3.5 kHz. Schusterman et al. (1967) suggested that the whinny sound may be the female counterpart of a male bark; however, they also heard female barking both in air and underwater. The next sound type was a "buzz" sound usually emitted by a sea lion while in a social situation. An example of a buzz is shown in Fig. 10.51. This vocalization may actually be a series of discrete pulses occurring so rapidly that they have a buzzing aural quality. Most of the energy was between 0.5 and 1.5 kHz, with components extending to about 4 kHz. The final sound type was the bang or crack, which was first heard when one of the sea lions was confronted with its mirror image. This sound was repeated several times over a period of days, usually under the same condition. Several bangs are shown in Fig. 10.51. This sound always was associated with extremely rapid swimming and appeared to the ear to be quite loud and mechanical. Most of the energy of bangs was between 0.5 and 1.5 kHz, although frequency components as high as 8 kHz were frequently present.

Schusterman et al. (1967) also observed several sea lions while they emitted underwater clicks. They found that *Zalophus* can produce clicks with their mouth closed and without any emission of bubbles or with their mouth opened with the emission of bubbles. They also found some movement of the area around the throat or larynx while the sea lion vocalized. Such movements were not present when an animal was silent, implicating the laryngeal area as the underwater sound generator. Barking could also take place underwater without the emission of bubbles, indicating that air is recycled within the animal (Schusterman, 1978). Apparently, barking underwater may be more difficult than in air, resulting in the repetition rate of barks underwater to be almost a half that in air. Barks produced in air have a slight directional property with the sound being projected forward. However, the sound intensity is only about 10 dB lower toward the side of the head perpendicular to the forward axis and an additional 6–8 dB in the backward direction. The directional property of sounds emitted underwater has not been investigated.

Walrus Male walrus vocalize intensively underwater during the breeding season (Stirling et al., 1987). The underwater vocalizations tend to be very stereotyped (Ray and Watkins, 1975; Stirling et al., 1987). The sounds produced by a walrus (*Odobenius rosmarus*) in captivity have been described by Schevill et al. (1966). Ray and Watkins (1975) recorded male walruses phonating in the water adjacent to a mixed group of cows, subadults, and juveniles in the Bering Sea and found that the sounds were essentially the same as for the captive animals studied by Schevill et al. (1966). Three basic sound patterns were produced by submerged males: a double pulse followed by a bell sound, a pulse series ending in a coda of seven sounds, and a final pulse series introduced by a triplet and ending with a strong final pulse, which indicated that the animal would soon surface (Ray and Watkins, 1975). Examples of these three types of sounds produced by the walrus are shown in Fig. 10.52. The double pulse followed by a bell tone in Fig. 10.52a was typically emitted shortly after diving. This sequence may be repeated up to five or more times during the entire submergence period. The pulses are relatively low-frequency, extending to only 1.5 kHz. The bell sound had fundamental frequencies between 0.4 and 1.2 kHz. The next sound is the seven-pulse coda shown in Fig. 10.52b, which was the most common repetitive pattern of walrus pulses. They were repeated again and again and at a higher acoustic level than preceding pulses so that they stood out at a distance and could be easily identified amidst a heavy background of pulse sounds. The pulses in the coda are broadband pulses with frequencies extending to 4 kHz. The final pattern shown in Fig. 10.52c was usually emitted just before surfacing. This sound starts with three rapid pulses at a rate of 18 pps followed by 14–20 pulses or more at a lower rate (6 pps), followed still by a sudden shift to a more rapid rate (10 pps), ending with two slower pulses, a 0.5 s pause, and a final higher intensity pulse. It is the contention of Ray and Watkins (1975) that the stereotyped and repetitive, seasonally produced sequences of sound emitted by walrus fit the criteria of a "song" used to describe animal sounds.

Steller sea lion and fur seals Very few underwater acoustic recordings exist for the Steller sea lion (*Eumetopias jubata*) and its related species, the northern and southern fur seals. According to Poulter (1968), both the Steller sea lions and fur seal bulls appear to produce continuous series of under water clicks by slowing the rate at which they produce growls. These clicks are produced both underwater and in air. However, neither species is very vocal under water. Growls, snorts, and bleats are also produced under water by the Steller, South

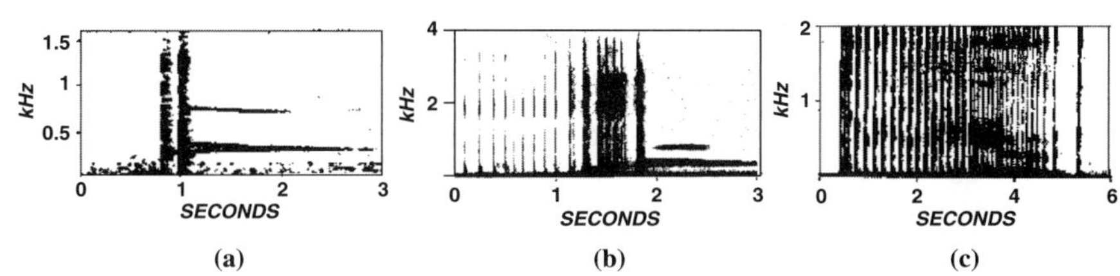

FIGURE 10.52. The three sound types produced by walrus: (1) double pulse followed by a bell sound, (2) a seven-pulse coda terminating a series of pulses and followed by the bell sound, (3) pulse pattern most often used just before surfacing. Analysis filter bandwidth was 22.5 Hz in (**a**), 150 Hz in (**b**), and 75 Hz in (**c**) (after Ray and Watkins, 1975).

American, Australian, and New Zealand sea lions, and bleats are produced by the Northern fur seal (Poulter, 1968). The Juan Fernández fur seal (*Arctocephalus philippii*) has been reported by Norris and Watkins (1971) to produce low-frequency clicks under water.

10.4 Underwater Vocalization of Sirenians

Very little research has been done on the underwater vocalization of manatees and dugongs. Although manatees spend their entire lives in water, very little is known of their underwater vocalizations. Hartman (1979) reported that the West Indies manatee (*Trichechus manatus*) produces high-pitched squeals, chirp-squeaks, and screams underwater. However, manatees are generally very quiet and only make sounds when males are sexually aroused or under conditions of fear, aggravation, and protest (Richardson et al., 1995). Schevill and Watkins (1965b) reported that underwater calls of the West Indies manatees have a fundamental frequency between 2.5 and 5 kHz, although some are as low as 0.6 kHz. The sounds are relatively weak, only about 10–12 dB above the ambient noise at 3–4 m distance. Durations of the sounds were also relatively short, 0.15–0.5 s. This species produced about 10 types of short-duration sounds (0.06–0.4 s) with frequencies between 1 and 12 kHz, which are often accompanied by rapid amplitude modulations (Steel and Morris, 1982). However, Gerstein (pers comm) has detected ultrasonic sounds from the West Indies manatee, and an example of these sounds from an adult male are shown in Fig. 10.53. Most of the signals in the spectrogram resemble a series of short FM pulses, each having a concave or hill structure with harmonics extending to 22 kHz. The

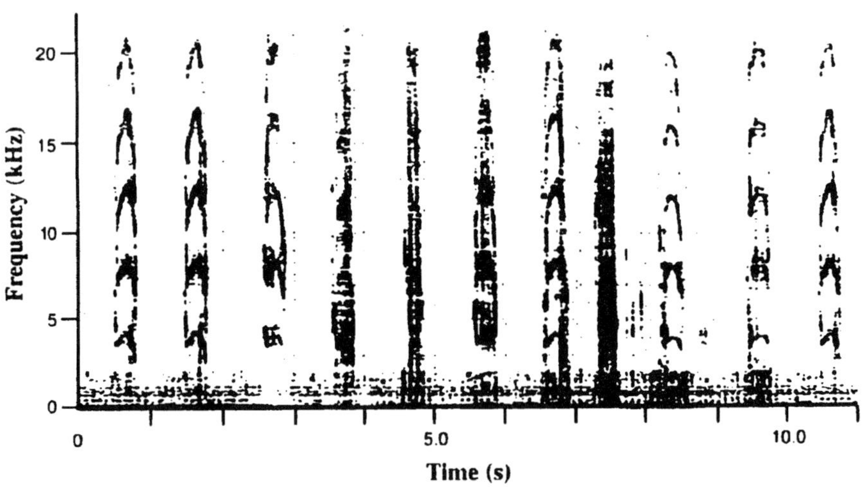

FIGURE 10.53. Example of some ultrasonic vocalizations produced by West Indies manatees (courtesy of Edmund Gerstein).

duration of the FM pulses is approximately 0.3 s. Signals number 4–6 and 8 seem to be broad band pulses with energy extending out to about 22 kHz.

Dugongs (*Dugong dugong*) produce underwater vocalizations that have been described as whistles, chirps, and chirp-squeaks (Nishiwaki and Marsh, 1985). The sounds are in the frequency range of 1–8 kHz and seem to be similar to those produced by manatees.

10.5 Sound Production by Fishes

The topic of sound production by fishes is an extremely difficult subject to adequately address because of the many species of fishes and the limited amount of research performed on them. In this section, we will merely introduce the topic and readers interested in further details are encouraged to consider two excellent review papers by Zelick et al. (1999) and Myrberg and Fuiman (2002). A volume that contains the sound characteristics of the largest collection of fishes is the one written by Fish and Mowbray (1970). In this book the waveforms and spectrograms of over a hundred different fish species are shown.

The complexity of many species is compounded by the fact that teleost fishes produce sound in a wide variety of ways, none of which involves a larynx or syrinx-like structure as used by terrestrial vertebrates. There are two common methods in which fish produce sounds: by stridulation and by manipulation of the muscles on the swim bladder or muscles connected to bones around the swim bladder (Zelick et al., 1999). However, there are many different ways in which stridulation can take place. Some fish grind their pharyngeal jaws, others strum tendons attached to the fourth and fifth pectoral fins, and others by the articulation of incisors or more often by dorsal and ventral patches of mosaic-like teeth located in the pharynx (Fish and Morbray, 1970). A more complex mechanism involves the movement of exceptionally fast muscles connected to the swim bladder. The muscles can be entirely on the swim bladder (intrinsic) or the muscles can insert onto a bone with the other end either on bone or on the swim bladder (extrinsic). An example of an intrinsic muscle configuration is shown in Fig. 10.54. The volume of the swim bladder can be rapidly changed by the contraction of these muscles, which causes a sound to be produced (see Zelick et al. 1999). The gas-filled swim bladder (or gas bladder) in the abdominal cavity may serve as a sound amplifier, although it has other very critical functions as well (Steen, 1970). Sounds produced by stridulation are typically pulsed, broad-band sounds, with the pulses being of short duration on the order of 10–50 ms. The pulse repetition rate and the number of pulses produced in a sonic event vary between species. An example of a sound produced by stridulation is shown in Fig. 10.55 for the silver perch (*Bairdiella chrysura*). The clicks resemble echolocation clicks produced by dolphins, except for the longer duration and lower frequencies. The frequency of stridulation sounds

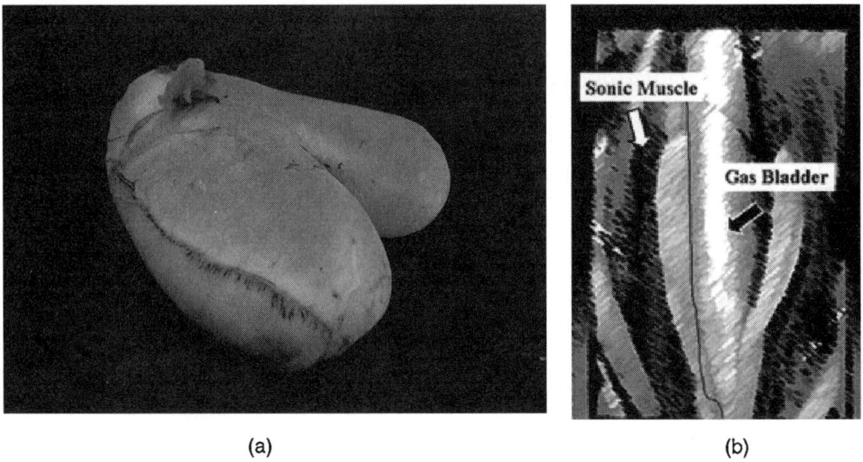

FIGURE 10.54. (**a**) Swimbladder of the toadfish, Opsanus beta. Sonic muscles can be seen on the lateral walls of the swimbladder (courtesy of David Mann), (**b**) Representative sciaenid internal anatomy revealing the gas bladder and sonic muscles of a male weakfish, *Cynoscion regalis* adapted from Gilmore, 2002).

have been correlated with fish size; the larger the fish, the lower the fundamental frequency (Myrberg et al., 1993). The size–frequency relationship can probably be attributed to the resonance properties of the swim bladder. The larger the swim bladder, the lower its resonance frequency, and vice versa.

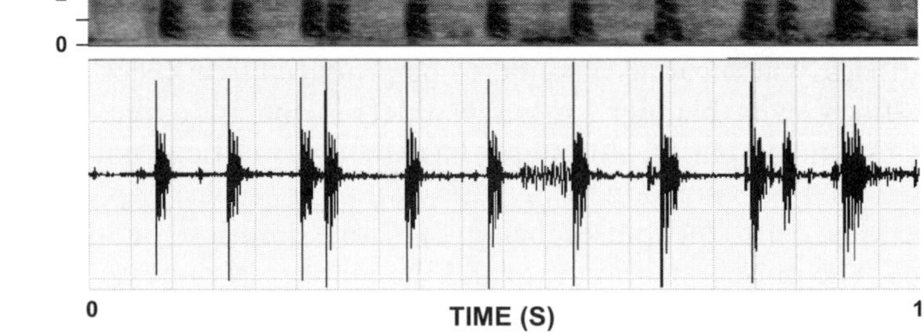

FIGURE 10.55. The waveform and spectrogram of the sounds produced by the silver perch (*Bairdiella chrysura*) (courtesy of David Mann).

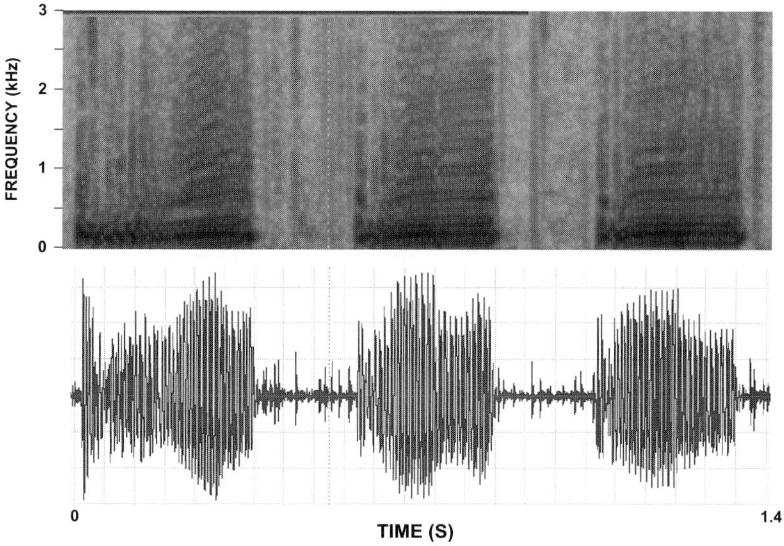

FIGURE 10.56. The waveform and spectrogram of typical sounds produced by the coral reef toadfish (*Sanopus astrifer*) (courtesy of David Mann).

Sounds produced by sonic muscles on the swim bladder tend to be pulsed tonal sounds. These sounds will also be longer than the sounds produced by stridulation. An example of a sound produced by the sonic muscle is shown in Fig. 10.56. Note the higher frequency harmonics in the spectrogram, indicating the tonal nature of the sound. The sounds of fishes with sonic muscles, including most deep fishes, have never been described. Mann and Jarvis (2004) published what may be the first record of a sound from a deep fish from a recording made at the AUTEC test range in the Tongue of the Ocean, Bahamas, from four deep-sea hydrophones.

Fish use sounds in a wide variety of behaviors including aggression, protection of territory, defence, and reproduction (reviewed in Tavolga, 1971; Demski et al. 1973; Zelick et al. 1999). There is also evidence that at least one species of marine catfish (*Arius felis*) uses a form of "echolocation" to identify objects in its environment by producing low-frequency sounds and listening to their reflections from objects (Tavolga, 1976). Data in the literature suggest that it is the temporal pattern of fish sounds, rather than their frequency spectrum, that is most important for acoustic communication by fishes (Wenz, 1964; Spanier, 1979). Most of the sounds are not very intense. Source levels have not been measured for many fish sounds, but it seems fish that use the sonic muscles on the swim bladder produce higher intensity sounds than fish that use stridulation. Toadfish produce sounds with the sonic muscles and their source level is on the order of 140 dB re 1 μPa (Tavolga, 1971), whereas the damselfish uses stridulation and produces sounds that are about 110 dB (Myrberg and Riggio, 1985). These are not very large values, making them difficult to detect in a high ambient noise environment.

Fishes not only produce sounds as individual entities but some also do so in choruses. It was found by Knudsen et al. (1948) that croakers (*Scianidae*) produced choruses. Other studies followed and showed that many species of fish produced a wide variety of sounds in choruses (Cato and McCauley, 2002). The appearance of fish chorusing sounds is dependent on species, time of day, and time of season. An example of the time of day factor can be found in Fig. 10.57, showing the nocturnal nature of fish choruses in the Gulf of Carpentaria in Northern Australia waters. The fish choruses dominated the acoustic scene between 2000 and 0500 hours. The dominant feature of fish chorusing sounds are illustrated by the spectra shown in Fig. 10.58, recorded in Cleveland Bay and Middle Reef in northern Australia (McCauley, 2001). The Cleveland Bay chorus was recorded on February 19, 1989, at 2108 hours and the middle reef chorus was recorded on January 26, 1988, at 2202 hours for curve 54 and on May 28, 1988, at 2022 hours for curve 157. The Cleveland Bay chorus indicates that the sounds can be as much as 30 dB higher than the ambient noise in the absence of the fish. The increase in the ambient noise level is very broad, extending from about 50 to over 5 kHz. The Middle Reef chorus suggests a seasonal variation with the chorusing sounds being considerably louder in January than in May.

This presentation of fish sound production has been kept necessarily brief. Whole volumes can be written on this subject, since there are so many

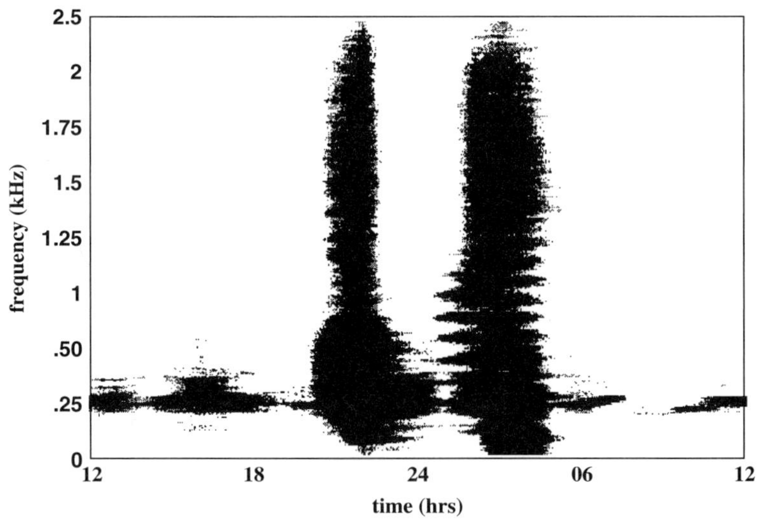

FIGURE 10.57. Example of two fish choruses in the Gulf of Carpentaria in northern Australia. The choruses were produced by the species of *Terapontidae* (adapted from McCauley, 2001).

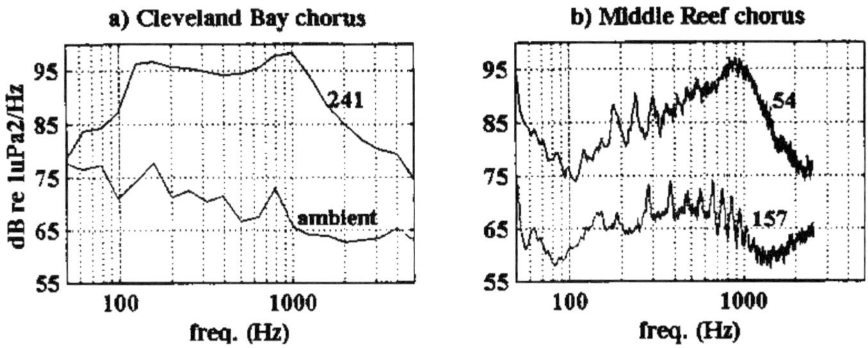

FIGURE 10.58. Spectra of *Terapontidae* choruses in Cleveland Bay measured at 2108 hours on February 19, 1998, and at Middle Reef on January 26, 1988, at 2202 hours for 54 and May 28, 1988, at 2022 hours (adapted from McCauley, 2001).

different species and there are still many more species that have not been examined from an acoustics perspective.

10.6 Sound Production by Snapping Shrimp

One of the most pervasive sources of noise that can interfere with the measurement and recording of underwater bioacoustic signals in shallow coastal waters (depth less than 60 m) is due to snapping shrimps. Snapping shrimps are found throughout the world at latitudes between approximately ±35°. As a community, they produce the loudest continuous biological sounds. Examples of snapping shrimp noise for three locations in the Pacific, Sydney Harbor in Australia, Kaneohe Bay on Oahu, Hawaii, and San Diego Bay in California, are shown in Fig. 10.59. Also shown in the figure is the shallow water noise produced by wind blowing at 22–27 knots and 10–17 knots. Even when compared against the underwater noise created by 27 knot winds, the levels of the snapping shrimp sounds are at least 12–22 dB louder.

The snapping shrimp responsible for producing the high ambient noise levels in Kaneohe Bay belongs to a family of crustaceans, Alpheidae, in the genus *Synalpheus*. Each shrimp has one enlarged claw as shown in Fig. 10.60, which produces a sharp transient acoustic signal with the socket on the propodius. As the claw closes, the plunger slides into the socket and causes a water jet to shoot out through a groove producing a jet stream of water in the forward direction. The closing of the claw produces a short transient acoustic signal that is extremely broadband with components up to 200 kHz (Au, 1981; Cato, 1993). Au and Banks (1998) performed detailed measurements of snapping shrimp sounds in a tank, measuring 10 snaps each from 40 different *Synalpheus parneomeris*. An example of a snap in both the time and

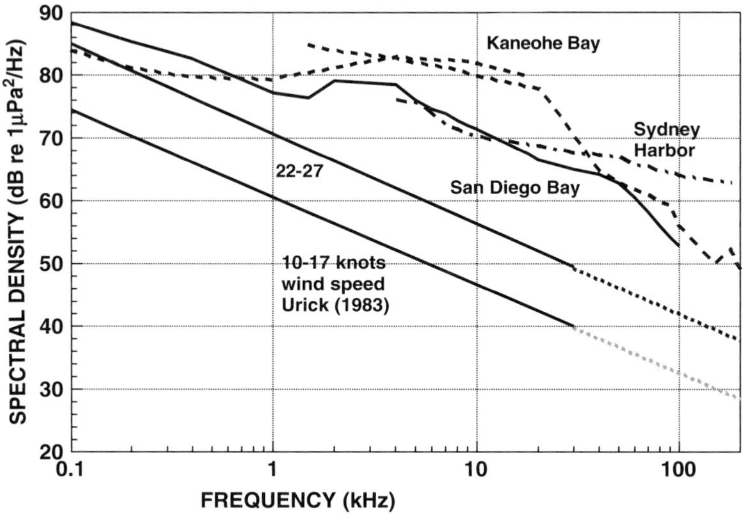

FIGURE 10.59. Snapping shrimp noise measured in Sydney Harbor, Australia (Cato and Bell, 1992), Kaneohe Bay, Hawaii (Albers, 1965; Au, 2003), and San Diego Bay, California (courtesy of J. Finneran). The shallow wind induced underwater noise is also shown (Urick, 1983).

frequency domains is shown in Fig. 10.61. A low-frequency, low-intensity precursor can be seen in the time wave form preceding the high-intensity oscillations by approximately 275 μs, which has never been observed prior to the work of Au and Banks (1998). The frequency spectrum indicates a very broad frequency signal, extending from zero to greater than 200 kHz. Only 20 dB separate the maximum and minimum spectral level over the 200 kHz frequency range, an extremely broadband signal, perhaps the broadest of any sounds produced by any organism.

The mechanism of sound production was determined recently by Versluis et al. (2000), who used a high-speed video system and an acoustic detection system simultaneously to examine the sound production mechanism of snapping shrimp. Their findings can best be discussed with the help of Fig. 10.62, showing points on the waveform of a snapping shrimp snap and the corresponding video image of the large claw. At point 1, the claw is closing and the plunger is entering the orifice on the opposite face of the claw. At point 2, the plunger has entered the orifice and the jet stream is in the process of being generated. The precursor signal first observed by Au and Banks (1998) seems to be associated with the process of the plunger entering the orifice and the jet stream being generated. A negative pressure region is created as water is pushed through the orifice creating a jet stream. The negative pressure allows a bubble to form and grow at point 3. Finally, the bubble cavitates to create the peak in the acoustic pressure at point 4. The cavitation process is so energetic that a photon of light is produced (Lohse et al., 2001). The species of snapping shrimp observed

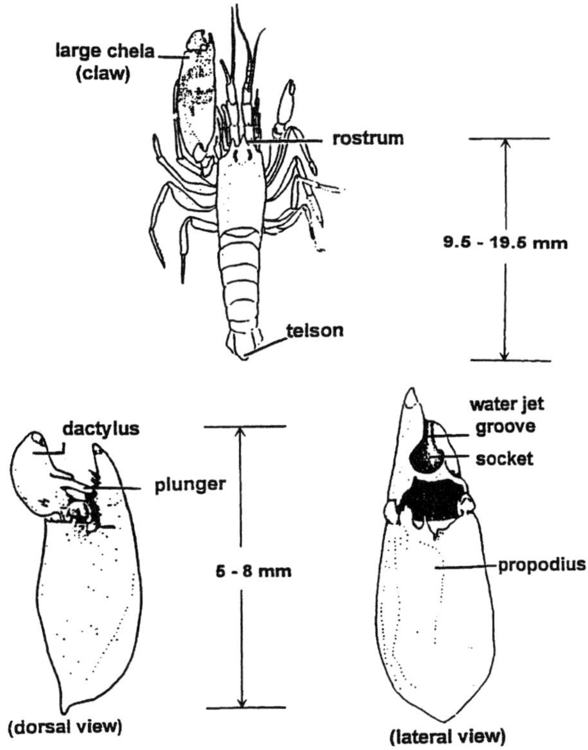

FIGURE 10.60. Drawing of a snapping shrimp showing the top view of the whole shrimp and the dorsal and lateral views of the large claw. The lateral view is with the dactylus removed to show the water jet groove (after Knowlton and Moulton, 1963).

by Versluis et al (2000) and Lohse et al. (2001) was *Alpheus heterochaelis*, which is about 2.6 times larger than *Synalpheus parneometris* used by Au and Banks (1998), and therefore produced a louder snap. Two other characteristics of snapping shrimp snaps that we will discuss are the source level/source energy and the peak and center frequencies. The mean and standard deviation of the source level/source energy flux density for ten snaps per *Synalpheus parneometris* and for 40 animals are shown in Fig. 10.63. The results suggest that the amplitude of the snaps do not vary very much from animal to animal. The mean source level varied at most by 7 dB from the maximum and minimum levels. Next, the peak frequency and center frequency variations between 40 animals are shown in Fig. 10.64. Both the peak and center frequency do not vary much between animals. Most of the peak frequency is between 3 and 5 kH, which can also be seen in the spectrum of Fig. 10.64. The precursor and slow rise time of the high-frequency portion of the snap are responsible for the peak frequency. The center frequency varied between 35 and 50 kHz, suggesting that most of the energy is in the ultrasonic range, although the peak frequency is rather low.

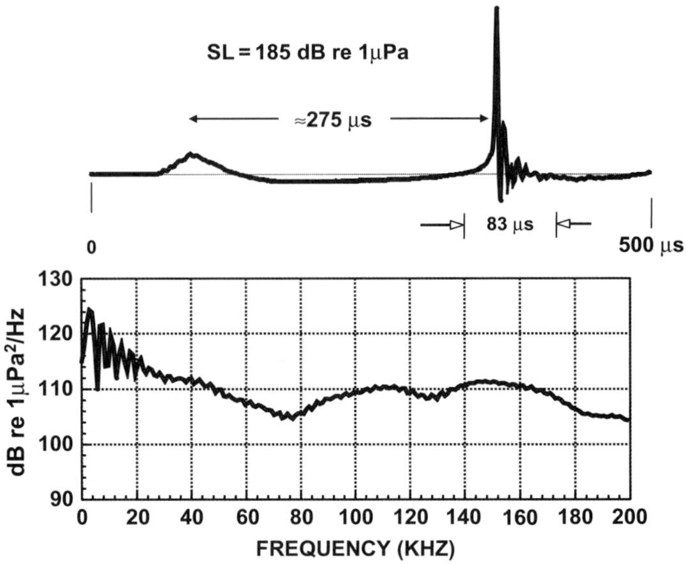

FIGURE 10.61. An example of the waveform and frequency spectrum of a typical snap from *Synalpheus parneometris* (adapted from Au and Banks, 1998).

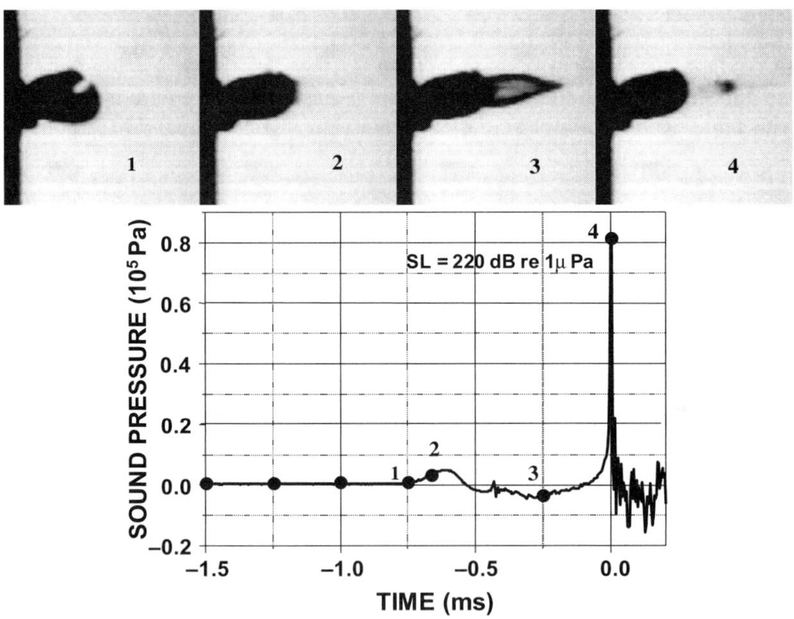

FIGURE 10.62. Corresponding positions of the large claw with the waveform of a snap (adapted from Verslius et al., 2000).

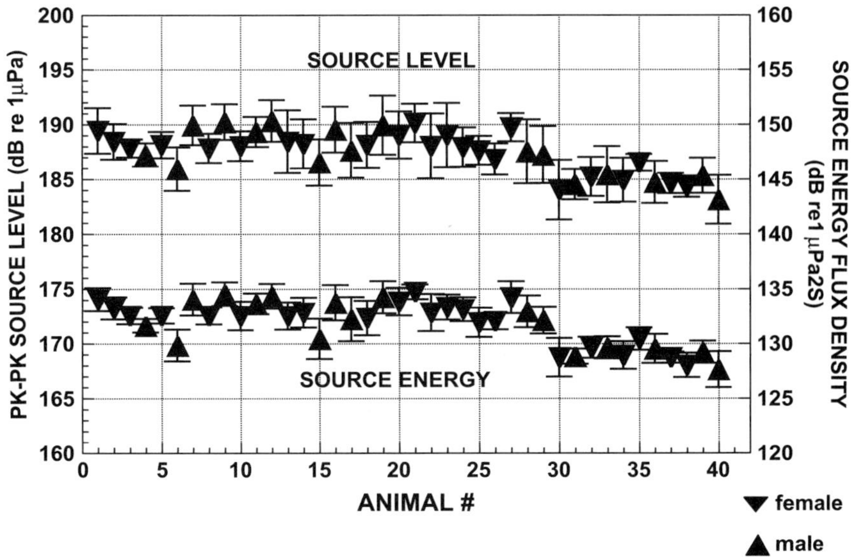

FIGURE 10.63. The mean and standard deviation of 10 snaps per animal and for 40 animals (adapted from Au and Banks, 1998).

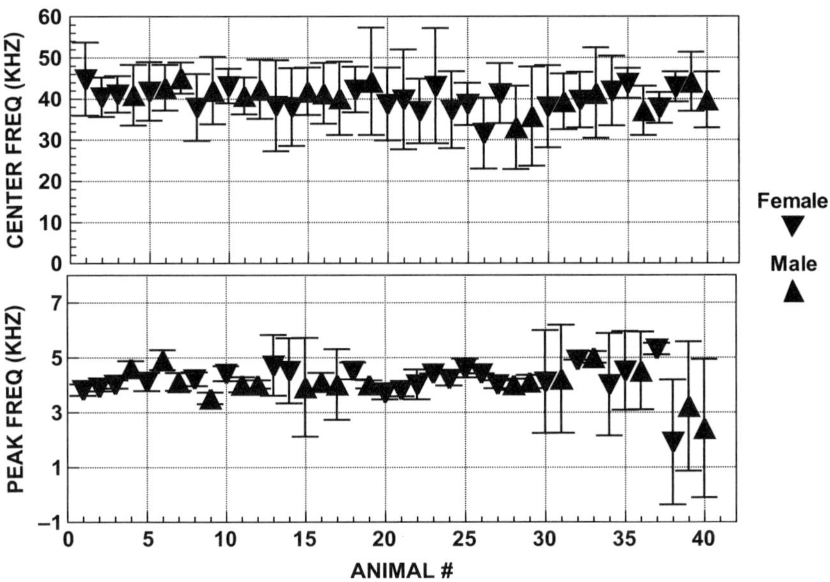

FIGURE 10.64. The mean and standard deviation of the peak and center frequencies from10 snaps and 40 animals (adapted from Au and Banks, 1998).

References

Albers, V. M. (1965). *Underwater Acoustics Handbook-II* (The Pennsylvania State University Press, University Park, PA).

Alling, A. K. and Payne, R. (1991). "Song of the Indian Ocean blue whale in Balaenoptera musculus" in *Cetaceans and Cetacean Research in the Indian Ocean Sanctuary*. S. Leatherwood and G. P. Donovan, eds. pp. 33–65. United Nations Environment Programme, Marine Mammal Technical Report No. 3. 1991.

Alling, A. E., Dorsey, E. M., and Gordon, J. C. (1991). "Blue Whales (*Balaenoptera musculus*) off the Northeast Coast of Sri Lanka: distribution, feeding, and individual identification," in *Cetaceans and Cetacean Research in the Indian OceanSanctuary*, S. Leaderwood and G. P. Donovan, eds. (UNEP Mar. Mammal Tech. Rep. 3), pp. 247–258.

Amundin, M. (1991). "Sound production in odontocetes with emphasis on the harbor porpoise *Phocoena phocoena*," Ph.D. Dissertation, University Stockholm, Stockholm, Sweden.

Amundin, M. and Andersen, S. H. (1983). "Bony nares air pressure and nasal plug muscle activity during click production in the harbour porpoise, *Phocoena*, and the bottlenosed dolphin, *Tursiops truncatus*," J. Exp. Biol. **105**, 275–282.

Aroyan, J. L., McDonald, M. A., Webb, S. C., Hildebrand, J. A., Clark, D., Laitman, J. T., and Reidenberg, J. S. (2000). "Acoustic model of sound prodution and progagation," in *Hearing by Whales and Dolphins*, W. W. L. Au, A. N. Popper, and R. R. Fay, eds. (Springer-Verlag, New York), pp. 409–417.

Asselin, S., Hammill, M. O., and Barrette, C. (1993). "Underwater vocalizations of ice breeding grey seals," Can. J. Zool. **71**, 2211–2219.

Au, W. W. L. (2003). "A comparison of the sonar capabilities of bats and dolphins," in *Echolocation in Bats and Dolphins*, J. Thomas, C. Moss, and M. Vater, (eds), (University of Chicago press, Chicago), pp. xiii–xxvii.

Au, W. W. L. and Banks, K. (1998). "The acoustics of the snapping shrimp *Synalpheus parneomeris*," J. Acoust. Soc. Am. **103**, 41–47.

Au, W. W. L., Lammers, M. O., and Aubauer, R. A. (1998). "A Portable Broadband Data Acquisition System for Use in Field Bioacoustics," Mar. Mammal Sci. **15**, 526–531.

Au, W. W. L. and Penner, R. H. (1981). "Target detection in noise by echolocating Atlantic bottlenose dolphins," J. Acoust. Soc. Am. **70**, 687–693.

Au, W. W. L., Penner, R. H., and Turl, C. A. (1987). "Propagation of beluga echolocation signals," J. Acoust. Soc. Am. **82**, 807–813.

Au, W. W. L., Pack, A. A., Lammers, M. O., Herman, L., Deakos, M., and Andrews, K. (2006). "Acoustic properties of humpback whale songs," J. Acoust. Soc. Am. **120**, 1103–1110.

Bazua-Duran, M. C. (1997). "Comparisons of whistles among different groups dolphins (*Tursiops truncates*) from the Gulf of Mexico," MS thesis, Universidad Nacional Autonoma de Mexico.

Beier, J. C. and Wartzok, D. (1979). "Mating behaviour of captive spotted seals (*Phoca larghal*)," Anim. Behav. **27**, 772–781.

Blevins, C. and Parkins, B. (1973). "Functional anatomy of the porpoise larynx," Am. J. Anat. **138**, 151

Bowles, A. E., Young, W. G., and Asper, E. D. (1988). "Ontogeny of stereotyped calling of a killer whale calf, *Orcinus orca*, During Her First Year," Rit. Fiskideildar **11**, 251–275.

Braham, H., Krongman, B., Johnson, J., Marquette, W., Rugh, D., Nerini, M., Sonntag, R., Bray, T., Brueggeman, J., Dalheim, M., Savage, S., and Geogel, C. (1980). "Population studies of the bowhead whale (*Balaena mysticetus*): Results of the 1979 Spring Research Season," Rep. Internat. Whaling Comm. **30**, 391–404.

Buck, J. R. and Tyack, P. L. (1993). "A quantitative measure of similarities for *Tursiops truncatus*signature whistles," J. Acoust. Soc. Am. **94**, 2497–2506.

Busnel, R. G. and Dziedzic, A. (1967).Resultats Metrologique Experimentaux de l' Echolocation Chez Le *Phocaena phocaena*et leur Compaison avex ceux de Certaines Chauves-Souris," in *Animal Sonar Systems: Biology and Bionics*, R. G. Busnel, ed. (Laboratoire de Physiologie Acoustique, Jouy-en-Josas,France), pp. 307–335.

Caldwell, M. C. and Caldwell, D. K. (1965). "Individualized whistle contours in bottlenosed dolphins (*Tursiops truncatus*)," Science, **207**, 434–435.

Caldwell, M. C. and Caldwell, D. K. (1979). "The whistle of the atlantic bottlenose dolphin (*Tursiops truncatus*)-Ontogeny," in *Behavior of Marine Animals* Vol 3, *Cetaceans*, H. E. Winn and B. L. Olla, eds. (Plenum Press, New York), pp. 369–401.

Caldwell, D. K., and Caldwell, M. C. (1970). "Etiology of the chirp sounds emitted by the Atlantic bottlenose dolphin: a controversial issue," Underwater Naturalist, **6**, 6–8.

Caldwell, M. C., Caldwell, D. K., and Tyack, P. (1990). "Review of the signature whistle hypothesis for the atlantic bottlenose dolphin, *Tursiops truncatus*," in *The Bottlenose Dolphin: Recent Progress in Research*, S. Leatherwood and R. Reeves, eds. (Academic Press, San Diego), pp. 199–234.

Cato, D. (1993). "The biological contribution to the ambient noise in waters near Australia," Acoust. Australia **20**, 76–80.

Cato, D. and Bell, M. J. (1992). "Ultrasonic ambient noise in Australian shallow waters at frequencies up to 200 kHz," Material Research Laboratory Technical ReportMRL-TR-91-23, Ascot Vale, Victoria, Australia.

Cato, D. and McCauley, R. D. (2002). "Australian research in ambient sea noise," Acoust. Australia, **30**, 13–20.

Clark, C. W. (1982). "The acoustic repertoire of the southern right whale: a quantitative analysis," Animal Beh. **30**, 1060–1071.

Clark, C. W. (1983). "Acoustic communication and behavior of the southern right whale," in *Behavior and Communication of Whales*, R. S. Payne, ed. (Westview Press, Boulder), pp. 643–646.

Clark, C. W. (1990). "Acoustic behavior of mysticete whales," in *Sensory Abilities of Cetaceans*, J. Thomas and R. Kastelein, eds. (Plenum, New York), pp 571–583.

Cleator, H. J., Stirling, I., and Smith, T. G. (1989). "Underwater vocalizations of the bearded seal (*Erignathus barbatus*)," Can. J. Zool. **67**, 1900–1910.

Crane, N. L. (1992). "Sound production of gray whales, *Eschrichtius tobustus* Along Their Migration Route," MS Thesis, San Francisco St. Univ., San Francisco, CA.

Cranford, T. W. (2000). "In search of impulse sound sources in odontocetes," in *Hearing by Whales and Dolphins*," W. W. L. Au, A. N. Popper, and R. R. Fay, eds. (Springer-Verlag, NY), pp. 155.

Cummings, W. C., Thompson, P. O., and Cook, R. (1968). "Underwater sounds of Migrating gray whales, *Eschrichtius glaucus* (Cope)," J. Acoust. Soc. Am. **44**, 278–1281.

Cummings, W. C. and Thompson, P. O. (1971). "Underwater sounds from the blue whale, *Balaenoptera musculus*," J. Acoust. Soc. Am. **50**, 1193–1198.

Cummings, W. C., Fish, J. F., and Thompson, P. O. (1972). "Sound production and other behavior of southern right whales, *Eubalena glacialis*," Trans. San Diego Soc. Nat. Hist. **17**, 1–13.

Cummings, W. C., Holliday, D. V., and Graham, B. J. (1983). "Measurement and localization of underwater sounds form the prudhoe region, Alaska, March, 1981, Tracor Appl. Sci. Report T-81-SD-013, San Diego, CA.

Cummings, W. C., Holliday, D. V., and Lee, B. J. (1986a). "Potential impact of Man-Made noise on ringed seals: vocalization and reactions," Outer Cont. Shelf Environ. Asses. Program, Final Rep. NOAA, Anchorage, AK **37**, 95–230.

Cummings, W. C., Thompson, P. O., and Ha, S. J. (1986b). "Sounds from Bryde's whale *Balaenoptera edeni*, and Finback, *B. Physalus*, whales in the gulf of California," Fish. Bull. **84**, 359–370.

Cummings, W. C. and Holliday, D. V. (1987). "Sounds and source levels from bow-head whales off Pt. Barrow, Alaska," J. Acoust. Soc. Am. **82**, 814–821.

Cummings, W. C. and Thompson, P. O. (1994). "Characteristics and seasons of blue and finback whale sounds along the U.S. West Coast as Recorded at SOSUS Stations," J. Acoust. Soc. Am. **95**, 2853.

Dalheim, M. E., Fisher, H., Dean, and Schempp, J. D. (1984). "Sound production by the gray whale and ambient noise levels in Laguna San Ignacio, Baja California Sur, Mexico," in *The Gray Whale*, M. Jones, S. Swartz, and S. Leatherwood, eds. (Academic Press, San Diego), pp. 511–541.

Demski, L. S., Gerald, J. W., and Popper, A. N. (1973). Central and peripheral mechanisms of teleost sound production," Am. Zool. **13**, (4), 1141–1167.

Diercks, K. J., Trochta, R. T., Greenlaw, C. F., Evans, W. E. (1971). "Recording and analysis of dolphin echolocation signals," J. Acoust. Soc. Am., **49**, 1729–17

Dixon, W. J., Brown, M. B., Engelman, L., and Jennrich, R. I. (1990). BMDP Statistical Software Manual Vol. 2. (University of California Press, Berkeley, CA), pp. 891–902.

Dormer, K. J. (1979). "Mechanism of sound production and air recycling in delphinids: Cineradiographic evidence," J. Acoust. Soc. Am. **65**, 229–239.

dos Santos, M. E., Ferreira, A. J., and Harzan, S. (1995). "Rhythmic sound sequences emitted by aroused bottlenose dolphins in the Sado Esuary, Portugal.," in *Sensory Systems of Aquatic Mammals*, R. A. Kastelein, J. A. Thomas and P. E. Nachtigall, eds. (De Spil Publ. Woerden, The Netherland), pp. 325–334.

Dreher, J. L. (1966). "Cetacean communication: small-group experiment," in *Whales, Dolphins and Porpoises*, K. S. Norris, ed. (University of California Press, Berkeley, CA), pp 529–543.

Dreher, J. L. and Evans, (1964). "Cetacean Communication," in *Marine Bio-Acoustics*,. N. Tavolga, ed. (Pergamon, Oxford, England), pp. 373–393.

Driscoll, A. D. (1995). "The Whistles of Hawaiian Spinner Dolphins," MS Thesis, Univ. of Calif. Santa Cruz, CA.

Edds, P. L. (1982). "Characteristics of the Blue Whale, *Balaenoptera musculus*, in St. Lawrence River" J. Mamm. **63**, 345–347.

Edds, P. L. (1988). "Characteristics of Finback, *Balaenoptera physalus*, Vocalization in the St. Lawrence Estuary," Bioacoustics **1**, 131–149.

Edds, P. L., Odell, D. K., and Tershy, B. R. (1993). "Vocalization of a Captive Juvenile and Free-Ranging Adult-Calf Pairs of Bryde's Whales, *Balaenoptera edeni*," Mar. Mamm. Sci. **9**, 269–284.

Evans, W. E. (1967). "Vocalization among marine animals," in *Marine Bioacoustics* Vol 2, W. N. Tavolga, ed. (Pergamon, New York), pp. 159–186.

Evans, W. E. and Powell, B. A. (1967). "Discrimination of different metallic plates by an echolocating delphinid," in *Animal Sonar Systems: Biology and Bionics*,

R. G. Busnel, ed. (Lab. De Physiologie Acoustique, Jouy-en-Josas, France), pp. 363–382.

Fish, M. P., and Mowbray, W. H. (1970). *Sounds of Western North Atlantic Fishes.* (John Hopkins Press, London).

Ford, J. K. B. (1984). "Call traditions and dialects of killer whales (*Orcinus orca*) in British Columbia," Ph.D. Dissertation, Univ. Of British Columbia, Vancouver.

Ford, J. K. B. (1985). "Group-specific dialects of killer whales (*Orcinus orca*) in British Columbia," in *Communication and Behavior of Whales*, R. Payne, ed. AAAS Sel. Symp., 129–161.

Ford, J. K. B. (1991). "Vocal traditions among resident killer whales (*Orcinus orca*) in coastal waters of British Columbia," Can. J. Zool. **69**, 1454–1483.

Ford, J. K. B. and Fisher, H. D. (1978). "Underwater acoustic signals of the narwhal (*Monodon monoceros*)," Can. J. Zool. **56**, 552–560.

Ford, J. K. B. and Fisher, H. D. (1983). "Group-specific dialects of killer whales (*Orcinus orca*) in British Columbia," in *Communication and Behavior of Whales*, AAAS Sel Symp. **76**, (Westview Press, Boulder, CO.), pp. 129–161.

Gedamke, J., Costa, D. P., and Dunstan, A. (2001). "Localization and visual verification of a complex minke whale vocalization." J. Acoust. Soc. Am. **109**, 3038–3047.

Gilmore, R. G. Jr. (2002). "Passive acoustic transects: Mating calls and spawning ecology in East Florida Sciaenids," Proc. Of An International Workshop on the Application of Passive Acoustics in Fisheries, April 8–10, 2002.

Grimes, L. G. (1974). "Dialects and geographical variation in the song on the splendid sunbird *Nectarinia coccinigaster*," Ibis **116**, 314–329.

Hanggi, E. B. and Schusterman, R. J. (1992). "Underwater acoustic displays my male harbor seals (*Phoca vitulina*): initial results," in *Marine Mammal Sensory Systems*, J. A. Thomas, R. A. Kastelein and A. Ya. Supin, eds. (Plenum, New York), pp. 449–457.

Hanggi, E. B. and Schusterman, R. J. (1994). "Underwater acoustic displays and individual variations in male harbour seals, *Phoca vitulina*," Anim. Behav. **48**, 1275–1283.

Hartigan, J. A. (1975). *Clustering Algorithms* (John Wiley & Sons, New York).

Hartman, D. S. (1979). "Ecology and behavior of the manatee (*Trichechus manatus*) in Florida," Am. Soc. Mammal, Spec. Publ. 5.

Helweg, D. A. (1989). "Diurnal and Seasonal Patterns of Behavior and Abundance of Humpback Whales (*Megaptera novaeangliae*) in Hawaiian Waters," M.A. Thesis, Univ. Hawaii.

Helweg, D. A., Frankel, A. S., Mobley Jr., J. R. And Herman, L. M. (1992). "Humpback whale song: our current understanding," in *Sensory Abilities of Aquatic Mammals*, J. A. Thomas, R. Kastelein and A. Supin, eds. (Plenum, New York), pp. 459–483.

Herman, L. M. and Tavolga, W. N. (1980). "The Communication Systems of Ceta-ceans," in *Cetacean Behavior: Mechanisms and Function*, L. H. Herman, ed. (Wiley-Interscience, New York), pp. 149–209.

Herzing, D. L. (1996). "Vocalizations and associated underwater behavior of free-ranging atlantic spotted dolphins, *Stenella frontalis* and bottlenose dolphins, *Tursiops truncatus*," Aquatic Mamml. **22.2**, 61–79.

Herzing, D. L. (1988). "A quantitative description and behavioral associations of a burst-pulsed sound, the Squawk, in captive bottlenose dolphins, *Tursiops truncatus*," MA Thesis, San Francisco State Univ., San Francisco, CA.

Hollien, H., Hollien, P., Caldwell, D. K. and Caldwell, M. C. (1976). "Sound production by the atlantic bottlenose dolphin, *Tursiops truncatus*," Cetology, **26**, 1–7.

Itakura, F. (1975). "Minimum prediction residual principle applied to speech recognition," IEEE Trans. Acoustic, Speech and Signal Proc. **ASSP-32**,67–72.

Kamminga, C. and Wiersma, H. (1981). "Investigations of cetacean sonar II. Acoustical similarities and differences in odontocete sonar signals," Aquat. Mam. **8**, 41–62.

Knowlton, A. R., Clark, W. W., and Kraus, S. D. (1991). "Sounds recorded in the presence of sei whales, *Balaenoptera borealis*," in Proc. 9th Bienn. Conf. Biol. Mar. Mamm., Chicago, IL, Dec. 1991, p. 40.

Knowlton, R. E. and Moulton, J. M. (1963). "Sound production in the snapping shrimp *Alpheus*(*Crangnon*) and *Synalpheus*," Biol. Bull. **125**, 311–331.

Knudsen, V. O., Alford, R. S., and Emling, J. W. (1948). "Underwater ambient noise," J. Mar. Res. **7**, 410–429.

Krebs, J. R. and Kroodsma, D. E. (1980). "Repertoires and geographical variation in bird song," Adv. Study Behav. **11**, 143–177.

Lammers, M. O. and Au, W. W. L. (1996). "Broadband recording of social acoustic signals of the Hawaiian Spinner and spotted dolphins," J. Acoust. Soc. Am. **100**, 2609.

Le Beouf, B. J. and Peterson, R. S. (1969). "Dialects in elephant seals," Science **166**, 1654–1656.

Le Beouf, B. J. and Perinovich, L. (1974). "Dialects of northern elephant seals, *Mirounga angustirostris*: origin and reliability," Anim. Behav. **22**, 656–663.

Lemon, (1966). "Geographical variation in the song of cardinals," Can. J. Zool. **44**, 413–428.

Lilly, J. C. and Miller, A. M. (1961). "Sounds emitted by the bottlenosed dolphin," Science, **133**, 1689–1693.

Ljungblad, D. K., Thompson, P. O., and Moore, S. E. (1982). "Underwater sounds recorded from migrating bowhead whales, *Balaena mysticetus* in 1979," J. Acoust. Soc. **71**, 477–482.

Lohse, D., Schmitz, B., and Versluis, M. (2001). "Snapping shrimp make flashing bubbles". Nature, **413**, 477–478.

Mackay, R. S. and Liaw, C. (1981). "Dolphin vocalization mechanism," Science **212**, 676–678.

Mann, D. A. and Jarvis, S. M. (2004). "Potential sound production by a deep-sea fish," J. Acoust. Soc. Am. **115**, 2331–2333.

McCauley, R. D. "Biological Sea Noise in Northern Australia: Patterns of Fish Calling," Ph.D. Dissertation, James Cook University, Queensland.

McCowan, B. (1995). "A new quantitative technique for categorizing whistles using simulated signals and whistles from captive bottlenose dolphins (Delphinidae, *Tursiops truncatus*)," Ethology **100**, 177–193.

McCowan, B. and Reiss, D. (1995). "Quantitative comparison of whistle repertoires from captive adult bottlenose dolphins (Delphinid, *Tursiops truncatus*): a re-evaluation of the signature whistle hypothesis," Ethology **100**, 194–209.

McDonald, M. A., Hilderbrand, J. A., and Webb, S. C. (1994). "Blue and fin whales observed on a seafloor array in the Northeast Pacific," J. Acoust. Soc. Am.

Mohl, B., Terhune, J. M., and Ronald, K. (1975). "Underwater calls of the harp seal, *Pagophilus groenlandicus*," in *Biology of the Seal*, K. Ronald and A. W. Mansfield eds., Rapp. P. V. Reun. Cos. Int. Explor. Mer. **169**. 533–543.

Myberg, A. A., Ha, S. J., and Shamblott, M. J. (1993). "The sounds of bicolor damselfish *(Pomacentrus partitus*): predictors of body size and a spectral basis for individual recognition and assessment," J. Acoust. Soc. Am. **94**, 3067–3070.

Myrberg, A. A. Jr., and Riggio, R. J. eds. (1985). "Acoustically mediated individual recognition by a coral reef fish (*Pomacentrus partitus*). Animal Behav. **33**, 411–416.

Myrberg, A. A. and Fuiman, L. A. (2002). "The sensory world of coral reef fishes," *Coral Reef Fishes. Dynamics and Diversity in a Complex System*. P. F. Sale. Academic Press, San Diego, pp. 123–148.

Nishiwaki, M. and Marsh, H. (1985). "Dugong *Dugong dugon* (Müller, 1776)," in *Handbook of Marine Mammals*,Vol 3, S. H. Ridgway and R. Harrison, eds. (Academic Press, London), pp. 1–31.

Norris, K. S. (1964). "Some problems of echolocation in cetaceans," in *Marine Bio-Acoustics*, W. Tavolga, ed. (Pergamon, New York), pp. 317–336.

Norris, K. S. and Watkins, W. A. (1971). "Underwater sounds of *Arctocephalus philippii*, the Juan Fernández Fur Seal," in *Antarctic Pinnipedia*, (Antarct. Res. Ser.18. Am. Geophys. Union, Washington D.C.), pp. 169–171.

Norris, K. S., Dormer, K. J., Pegg, J., and Liese, G. T. (1971). "The Mechanism of Sound Production and Air Recycling in Porpoises: a Preliminary Report," in Proc. VIII Conf. Biol. Sonar Diving Mammals, Menlo Park, CA.

Norris, K. S., Wursig, B., Wells, R. S., Wursig, M., Brownlees, S. M., Johnson, C., and Solow, J. (1985). "The Behavior of the Hawaiian Spinner dolphin, *Stenella longirostris*," SWFC/NMFS Ad. Rep. LJ-85-06C, March 1985.

Nottebohm, F. (1969). "The song of the chingolo, *Zonotrichia capensis*, in Agentina: description and evaluation of a system of dialect," Condor **71**, 299–315.

Parks, S. E. and Tyack, P. L. (2005). "Sound production by North Atlantic right whales (*Eubalaena glacialis*) in surface active groups". J. Acoust. Soc. Am. **117**, 3297–3306.

Payne, R. S. and McVay, S. (1971). "Songs of humpback whales," Science **173**, 585–597.

Payne, R. S. and Payne, K. (1971). "Underwater sounds of southern right whales," Zoologica, **58**, 159–165.

Payne, R. S. and Payne, K. (1985). "Large scale changes over 19 years in songs of humpback whales in Bermuda," Z. Tierpsychol. **68**, 89–114.

Poulter, T. C. (1968). "Underwater vocalization and behavior of pinnipeds," in *The Behavior and Physiology of Pinnipeds*, R. J. Harrison, R. C. Hubbard, R. Peterson, C. E. Rice, and R. J. Schusterman, eds. (Appleton-Century-Crofts, New York), pp. 84.

Purves, P. E. (1967). "Anatomical and experimental observations on the cetacean sonar system," in *Animal Sonar Systems, Biology and Bionics*, R. G. Busnel, ed. (Laboratoire de Physiologie Acoustique, Jouy-en-Jouy), pp. 197–270.

Purves, P. E. and Pilleri, G. (1986). *Echolocation in Whales and Dolphins* (Academic Press, London).

Rankin, S. and Barlow, J. (2005). "Source of the North Pacific 'boing' sound attributed to minke whales." J. Acoust. Soc. Am. **118**, 3346–3351.

Ray, G. C. and Watkins, W. A. (1975). "Social function of underwater sounds in the walrus *Odobenus rosmarus*," in *Biology of the Seal*, K. Ronald and A. W. Mansfield eds. Rapp. P. V. Reun. Cos. Int. Explor. Mer. **169**. 524–526.

Renouf, D., Galway, G., and Gaborko, L. (1980). "Evidence for echolocation in harbour seals," J. Mar. Biol. Assoc. U.K. **60**, 1039–1042.

Richardson, W. J., Green Jr., C. R., Malme, C. I., and Thomson, D. H. (1995). *Marine Mammals and Noise* (Academic Press, San Diego, CA).

Ridgway, S. H., Carder, D. A., Green, R. F., Gaunt, A. S., Gaunt, S. L. L., and Evans, W. E. (1980). "Electromyographic and pressure events in the nasolaryngeal system of dolphins during sound production," in *Animal Sonar Systems*, R. G. Busnel and J. F. Fish, eds. (Plenum, New York), pp. 239–249.

Ridgway, S. H. and Carder, D. A. (1988). "Nasal Pressure and sound production in an echolocating white whale, *Delphinapterus leucas*," in *Animal Sonar: Processes and Performances*, P. E. Nachtigall and P. W. B. Moore, eds. (Plenum, New York), pp. 53–60.

Rivers, J. A. (1995). "Blue Whale, *Balaenoptera musculus*, Vocalizations from the Waters off Central California," Mar. Mamm. Sci. **13**, 186–195.

Rogers, T., Cato, D. H., and Bryden, M. M. (1995). "Underwater vocal repertoire of the leopard seal (*Hydrurga leptonyx*) in Prydz Bay, Antarctica," in *Sensory Systems of Aquatic Mammals*, R. A. Kastelein, J. A. Thomas and P. E. Nachtigall, eds. (De Spil, Woerden, The Netherlands), pp. 223–236.

Sayigh, L. S., Tyack, P., L., Wells, R. S., and Scott, M. D. (1990). "Signature whistles of free-ranging bottlenose dolphins, *Tursiops truncatus*: Stability and Mother-offspring Comparisons," Behav. Ecol. Sociobiol., **26**, 247–260.

Sayigh, L. S. (1992). "Development and Functions of Signature Whistles of Free-Ranging Bottlenose Dolphins, *Tursiops truncatus*," PhD. Dissertation, MIT/WHOI Joint Program.

Schenkkan, E. J. (1973). "On the Comparative anatomy and function of the nasal tract in odontocetes (Mammalia, Cetaea)," Bijdragen tot de Dierkunde, **43**, 127–159.

Schevill, W. E. and Watkins, W. A. (1965a). "Underwater calls of *Leptonychotes* (Weddell seal)." Zoologica, **50**, 45–47.

Schevill, W. E. and Watkins, W. A. (1965b). "Underwater calls of *Trichechus* Manatee)." Nature, **205**, 373–374.

Schevill, W. E., Watkins, W. A., and Ray, C. G. (1966). "Analysis of underwater *Odobenus*calls with remarks on the development and function of the pharyngeal pouches," Zoologica, **51**, 103–106.

Schevill, W. E. and Watkins, W. A. (1972). "Intense low-frequency sounds form an antarctic minke whale, *Balaenoptera acutorostrata*, " Breviora:Mus. Comp. Zool. **388**, 1–8.

Schultz, K. W. and Corkeron, P. J. (1994). "Interspecific differences in whistles produced by inshore dolphins in Moreton Bay, Queensland, Australia," Can. J. Zool. **72**, 503–509.

Schusterman, R. J., Gentry, R., and Schmook, J. (1967). "Underwater sound production by captive California Sea Lions, *Zalophus californianus*," Zoolgica **52**, 21–24.

Schusterman, R. J. (1978). "Vocal communication in pinnipeds," in *Behavior of Captive Wild Animals*, H. Markowita and V. J. Stevens eds. (Nelson-Hall, Chicago), pp. 247–308.

Sjare, B. L. and Smith, T. G. (1986). "The vocal repertoire of white whales, *Delphinapterus leucas*, summering in Cunningham Inlet, Northwest Territories," Can. J. Zool. **64**, 407–415.

Small, A. M. and McClellan, M. E. (1963). "Pitch associated with time delays between two pulse trains," J. Acoust. Soc. Am. **48**, 1246–1255.

Smolker, R. A., Mann, J., and Smuts, B. B. (1993). "Use of signature whistles during separations and reunions by wild bottlenose dolphin mothers and infants," Behav. Ecol Socio. **33**, 393–402.

Steel, C. and Morris, J. G. (1982). "The west Indian Manatee: an acoustic analysis," Am. Zool. **22**, 925.

Steen, J.B. (1970). "The swimbladder as a hydrostatic organ," in: *Fish Physiology*, Vol IV. W. S. Hoar and D. J. Randal, eds. (Academic Press, London New York), pp. 413–443.

Steiner, W. W. (1981). "Species-specific differences in pure-tonal whistle vocalizations of five Western North Atlantic dolphin species," Behav. Ecol. Sociobiol. **9**, 241–246.

Stimpert, A. K., Wiley, D. N., Au, W. W. L., Johnson, M. P., Arsenault, R. (2007). "Megapclicks": acoustic click trains and buzzes produced during nighttime foraging of humpback whales (Megaptera novaeangliae)," Biol. Lett., **3**, 467–470.

Stirling, I. and Sniff, D. B. (1979). "Underwater vocalization of leopard seals (*Hydorurga leptonyx*) and crabeater seals (*Lobodon carcinophagus*) near the South Shetland Islands, Antarctica," Can. J. Zool. **57**, 1244–1248.

Stirling, I. (1973). "Vocalization in the ringed seal (*Phoca hispida*)." J. Fish. Res.Board Can. **30**, 1592–1595.

Stirling, I., Calvert, W., and Spencer, C. (1987). "Evidence of stereotyped underwater vocalization of Male Atlantic walruses (*Odobenus rosmarus*)," Can. J. Zool. **65**, 2311–2321.

Taruski, A, G. (1979). "The whistle repertoire of the North Atlantic pilot whale (*Globicephala melaena*) and its relationship to behavior and environment," in *Behavior of Marine Animal: Current Perspectives in Research*, H. E. Winn and B. L. Olla, eds. (Plenum Press, New York), pp. 345–368.

Tavolga, W. N. (1971). "Acoustic orientation in the sea catfish, *Galeichthys felis*." Ann. NY Acad. Sci. **188**, 80–97.

Tavolga, W. N. (1976). "Acoustic obstacle detection in the sea catfish (Arius felis)," in: *Sound Reception in Fish*. A. Schuijf and A. D. Hawkins eds. (Amsterdam, Elsevier), pp. 185–204.

Terhune, J. M. and Ronald, K. (1973). "Some hooded seal (*Cystophora cristata*) Sounds in March," Can. J. Zool. **51**, 319–321.

Terhune, J. M. and Ronald, K. (1986). "Distant and near-range functions of harp seal underwater vocalizations," Can. J. Zool. **64**, 1065–1070.

Terhune, J. M. (1994). "Geographical variation of harp seal underwater vocalizations," Can. J. Zool. **72**, 892–897.

Thomas, J. A. and Kuechle, V. (1982). "Quantitative analysis of the underwater repertoire of the weddell seal (*Leptonychotes weddelli*)," J. Acoust. Soc. Am. **72**, 1730–1738.

Thomas, J. A., Zinnel, K. C., and Ferm, L. M. (1983). "Analysis of weddell seal (*Leptonychotes weddelli*) vocalization using underwater playbacks," Can. J. Zool. **61**, 1448–1456.

Thomas, J. A., Puddicombe, R. A., George, M., and Lewis, D. (1988). "Variations in underwater vocalizations of weddell seals (*Leptonychotes weddelli*) at the vestfold hills as a measure of breeding population discreteness," Hydrobiologia **165**, 279–284.

Thomas, J. A. and Golladay, C. L. (1995). "Geographic variation in leopard seal (*Hydrurga leptonyx*) underwater vocalizations," in *Sensory Systems of Aquatic Mammals*, R. A. Kastelein, J. A. Thomas and P. E. Nachtigall, eds. (De Spil, Woerden, The Netherlands), pp. 201–221.

Thompson, P. O. and Friedl, W. A. (1982). "A long term study of low frequency sounds from several species of whales off Oahu, Hawaii," Cetology **45**, 1–19.

Thompson, P. O., Cummings, W. C., and Ha, S. J. (1986). "Sounds, source levels, and associated behavior of humpback whales, Southeast Alaska," J. Acoust. Soc. Am. **80**, 735–740.

Thompson, P. O., Findlay, L. T., and Vidal, O. (1987). "Underwater sounds of blue whales, *Balaenoptera musculus*, in the Gulf of California, Mexico," Mar. Mamm. Sci. **12**, 288–293.

Thompson, P. O., Findley, L. T., and Cummings, W. C. (1996). "Underwater sounds of blue whales, *Balenoptera musculus*, in the Gulf of California, Mexico," Mar. Mamm. Sci. **12**, 288–293.

Thompson, T. J., Winn, H. E., and Perkins, P. J. (1979). "Mysticete sounds," in *Behavior of Marine Animal: Current Perspectives in Research*, H. E. Winn and B. L. Olla, eds. (Plenum Press, New York), pp. 403–431.

Tyack, P. L. (1981). "Interactions between singing Hawaiian Humpback whales and conspecifics nearby," Beh. Ecol. Sociobiol. **8**, 105–116.

Tyack, P. L. (1997). "Development and social functions of signature whistles in bottlenose dolphin (*Tursiops truncatus*)." Bioacoustics, **8**, 21–46.

Urick, R. J. (1983). *Principles of Underwater Sound (3rd Ed)* (Mcgraw-Hill, NY).

Versluis, M., Schmitz, B., von der Heydt, A., and Lohse, D. (2000). "How snapping shrimp snap: through Cavitating bubbles," Science, **289**, 2114–2117.

Wang, D., Wursig, B., and Evans, W. (1995). "Comparisons of whistles among seven odontocete species," in *Sensory Systems of Aquatic Mammals*, R. A. Kastelein, J. A. Thomas and P. E. Nachtigall, eds. (De Spil Publ. Woerden, The Netherland), pp. 299–323.

Watkins, W. A. (1966). "The harmonic interval: fact or artifact in spectral analysis of pulse trains," in *Marine Bioacoustics* Vol 2, W. N. Tavolga, ed. (Pergamon, New York), pp. 15–43.

Watkins, W. A. and Schevill, W. E. (1972). "Sound source location by arrival-time on a non-rigid three dimensional hydrophone array." Deep Sea Res. **19**, 691–706.

Watkins, W. A. and Schevill, W. E. (1977). "Sperm whale codas," J. Acoust. Soc. Am. **62**, 1485–1490.

Watkins, W. A. and Ray, G. C. (1977). "Underwater sounds from ribbon seal, *Phoca (Histriophoca) fasciata*," Fish. Bull. **75**, 450–453.

Watkins, W. A. and Schevill, W. E. (1979). "Distinctive characteristics of underwater calls of the harp seal, *Phoca groenlandica*, during the breeding season," J. Acoust. Soc. Am. **66**, 983–988.

Watkins, W. A. (1981). "The activities and underwater sounds of fin whales," Sci. Rep. Whales Res. Inst. **33**, 83–117.

Watkins, W. A. and Ray, G. C. (1985). "Sounds of the ross seal," J. Acoust. Soc. Am. **77**, 1598–1600.

Watkins, W. A., Tyack, P., Moore, K. E., and Bird, J. E. (1987). "The 20-Hz signals of finback whales (*Balaenoptera physalus*)," J. Acoust. Soc. Am. **82**, 1901–1911.

Weilgart, L. S. and Whithead, H. (1993). "Coda vocalizations in sperm whales (*Physeter macrocphalus*). Off the Galapagos Islands," Can. J. Zool. **71**, 744–752.

Weilgart, L and Whitehead, H. (1997). "Group-specific dialects and geographical Variation in coda repertoire in South Pacific Sperm whale," Behav. Ecol. Sociobiol. **40**, 277–285.

Wells, R. S., Irvine, A. B., Scott, M. D. (1980). "The social ecology of inshore odontocetes," in *Cetacean Behavior: Mechanisms and Functions*, L. M. Herman, ed. (Wiley, New York), pp. 263–317.

Wells, R. S., Scott, M. D., and Irving, A. B. (1987). "The social structure of free-ranging bottlenose dolphin," in *Current Mammalogy* Vol 1, H. Genoways, ed. (Plenum, New York), pp. 247–305.

Wells, R. S. (1991). "The role of long-term study in understanding the social structure of a bottlenose dolphin community," in *Dolphin Societies*, K. S. Pryor and U. Norris, eds. (California Press, Berkeley, CA), pp. 199–235.

Wenz, G. M. (1964). "Curious noises and the sonic environment in the ocean," in *Marine Bio-Acoustics*, W. N. Tavolga, ed., (Pergamon, New York), pp. 101–119.

Whitehead, H., Waters, S., and Lyrholm, T. (1991). "Social organization in female sperm whales and teir offspring: Constant companions and casual acquaintances," Behav. Ecol. Sociobiol. **29**, 385–389.

Winn, H. E., Perkins, P. J., and Poulter, T. C. (1971). "Population estimate of the humpback whale (*Magaptera novaenglia*)," in *Proceedings of the Seventh Annual Conference on Biological SONAR and Diving Mammals*, (Stanford Research Institute, Menlo Park, CA), pp. 39–52.

Winn, H. E. and Perkins, P. J. (1976). "Distribution and sounds of the minke whale, with a review of mysticete sounds," Cetology **19**, 1–11.

Winn, H. E. and Reichley, N. E. (1985). "Humpback Whale *Megaptera novaeangliae* (Borowski, 1781)," in *Handbook of Marine Mammals: The Sirenians and Baleen Whales,* (Vol. 3) S. H. Ridgway and R. Harrison, eds. (Academic Press, San Diego), pp. 241–273.

Winn, H. E. and Winn, L. K. (1978). "The song of the humpback whale *Megaptera novaeangliae* in the West Indies," Marine Biol. **47**, 97–114.

Wursig, B. and Clark, C. (1993). "Behavior," in *The Bowhead Whale*, J. J. Burns, J. J. Montague, and C. J. Cowles, eds., Spec. Publ. 2. (Soc. Mar. Mammal, Lawrence, KS), pp. 157–199.

Zelick, R., Mann D. A. et al. (1999). "Acoustic communication in fishes and frogs,". *Comparative hearing: Fishes and Amphibians,* R. R. Fay and A. N. Popper, eds., (New York), pp. 363–412.

11

Echolocation in Marine Mammals

Echolocation is the process in which an organism projects acoustic signals and obtains a sense of its surrounding from the echoes it receives. In a general sense, any animal with a capability to hear sounds can echolocate by emitting sounds and listening to the echoes. A person in an empty room can gain an idea of the size and shape of the room by emitting sounds and listening to the echoes from the different walls. However, we are using echolocation in a more specific sense in which an animal has a very specialized capability to determine the presence of objects considerably smaller than itself, discriminate between various objects, recognize specific objects and localize objects in three-dimensional space (determine range and azimuth). Dolphins, bats, and perhaps sperm whales, have this specialized capability of echolocation.

A dolphin's ability to survive and thrive in an aquatic environment is maximized by its ability to echolocate. Acoustic energy propagates in water more efficiently than almost any other form of energy, making the use of acoustics ideal for the aquatic environment. Many species of dolphins can be found in shallow offshore waters, bays, and rivers where the water can be extremely turbid with very limited underwater visibility. Under these conditions, the ability to echolocate becomes critical for navigating, locating and catching prey, and to avoid predators.

The capability of a dolphin to detect objects in noise and in clutter, and to discriminate between various objects, depends to a large extent on the information-carrying capabilities of the emitted signals. Also important are the extent to which the dolphin's auditory system can extract pertinent information from the echoes and the animal's cognitive capabilities. In order to make optimal use of acoustical information, the dolphin should have an auditory system that is very sensitive over a wide frequency range. The dolphin should also be sensitive in both quiet and noisy environments and should be able to detect short- and long-duration sounds. A good spectral analysis capability is important in discriminating and recognizing predator, prey, and other objects in the environment. Other important characteristics of a good sonar receiver include the capability to spatially resolve and localize sounds, reject externally generated interferences, and recognize temporal and spectral patterns of sounds.

W.W.L. Au, M.C. Hastings, *Principles of Marine Bioacoustics*,
DOI: 10.1007/978-0-387-78365-9_11,

The discussion of echolocation in this chapter will not be a thorough or exhaustive treatment of the subject. An entire volume has been written on this subject by one of the present authors and those interested in more details should consult his book, *The Sonar of Dolphins*, by Au (1993). This chapter will be an overview of the subject area and the intent is to provide the reader with some basic and fundamental characteristics of the dolphin sonar system.

11.1 The Dolphin Transmission System

Dolphins most likely produce echolocation signals within their nasal system and the signals are projected out through the melon (Cranford, 1992, 2000). The longstanding controversy on whether sounds are produced in the larynx or in the nasal system of odontocetes was discussed in the introductory section of Chapter 11. All experimental data with live dolphins indicate that sounds are produced in the nasal system and not in the larynx (Au, 1993). Cranford (2000) have recently obtained simultaneous synchronized video and acoustic recordings of sound production by a dolphin, which are strongly suggestive and compelling in pinpointing sound generation to the nasal area of a dolphin's head. They observed with an endoscope-synchronized vibration of the two tissue complexes, one located on either side and just above the membranous nasal septum, with the production of acoustic signals. Sounds are apparently generated by pushing air across sets of internal "lips" (also referred to as the monkey lips by Cranford, (1999)). Acoustic pulses occurred in coincidence with one oscillatory cycle of the lips whereas no other structures were found to vibrate in synchrony with each acoustic pulse event. The melon directly in front of the nasal plug may play a role in channeling sounds into the water, a notion first introduced by Wood (1964). Norris and Harvey (1974) found a low-velocity core extending from just below the anterior surface toward the right nasal plug, and a grated outer shell of high-velocity tissue. Such a velocity gradient could channel signals originating in the nasal region in both the vertical and horizontal planes. Using both a two-dimensional (Aroyan et al., 1992) and three-dimensional model (Aroyan, 1996) to study sound propagation in a dolphin's head, Aroyan has shown that echolocation signals most likely are generated in the nasal system and are channeled into the water by the melon.

Echolocation signals seem to fall into two general categories. The first category is associated with the signals produced by dolphins that can also produce whistle or frequency-modulated signals of relatively long duration (1–3 s), with the possible exception riverine dolphins and the sperm whales. The properties of these signals will be discussed first, followed by a discussion of the second category of echolocation signals associated with dolphins that do not produce whistles. The differences between these two classes of echolocation signals will be obvious as we discuss them.

11.1.1 Echolocation Signals of Dolphins Capable of Whistling

Most of the dolphin species are able to produce whistle signals as we discussed in Chapter 11. Among some of the species in this category in which echolocation signals have been measured include the bottlenose dolphin (*Tursiops* sp.), beluga whale (*Delphinapterus leucas*), killer whale (*Orcinus orca*), false killer whale (*Pseudorca crassidens*), Pacific white-sided dolphin (*Lagenorhynchus obliguidens),* Risso's dolphin (*Grampus griseus*), Tucuxi (*Sotalia fluviatilis*), Atlantic spotted dolphin (*Stenella frontalis*), Pacific spotted dolphin (*Stenella attenuata*), spinner dolphin (*Stenella longirostris*), pilot whale (*Globicephala sp*), rough tooth dolphin (*Steno bredanensis*). Most of the available data have been obtained for three species: the bottlenose dolphin, beluga whale, and false killer whale.

Evans (1973) reviewed the state of knowledge of dolphin echolocation and reported echolocation signals with peak frequency (frequency of maximum energy) between 30 and 60 kHz, and peak-to-peak source levels (sound pressure level 1 m from the animal) of about 170–180 dB for *Tursiops truncatus*. Prior to 1973, most echolocation signals of *Tursiops* were measured in relatively small tanks, and it was generally believed that peak frequencies of the sonar signals were as reported by Evans (1973). It was not until the study of Au et al. (1974) that certain features of biosonar signals used by *Tursiops* and other dolphins in open waters were discovered. We discovered that the signals had peak frequencies between 120 and 130 kHz, which were over an octave higher than previously reported peak frequencies between 30 and 60 kHz (Evans, 1973). We also measured an average peak-to-peak click source level on the order of 220 dB re 1 μPa at 1 m, which represents a level 30–50 dB higher than previously measured for *Tursiops*. Examples of typical echolocation signals emitted by *Tursiops truncatus* are shown in Fig. 11.1 for two situations. The top signal is typical of the signals used in the open waters of Kaneohe Bay, Oahu, Hawaii, and the second signal represents typical signals for a *Tursiops* in a tank. Signals measured in Kaneohe Bay regularly have duration between 40 and 70 s, 4–10 positive excursion, peak frequencies between 110 and 130 kHz, and peak-to-peak source levels between 210 and 228 dB re 1 μPa. The signals in Fig. 11.1 are not drawn to scale; if they were, the tank signal would resemble a flat line. We attributed the use of high-frequency echolocation signals to the high ambient noise environment of Kaneohe Bay caused by snapping shrimp. An example of a typical echolocation click train for a *Tursiops* performing a target detection task in open waters is shown in Fig. 11.2. The frequency spectra plotted as a function of time are shown on the left and the individual click waveform is displayed on the right. The frequency spectra represent the relative magnitude of the fast Fourier transform (FFT) of each click. The shape of the signals in a click train can be very repetitive and stereotypical.

Au et al. (1985) postulated that high-frequency echolocation signals were a by-product of the animals producing high intensity clicks to overcome

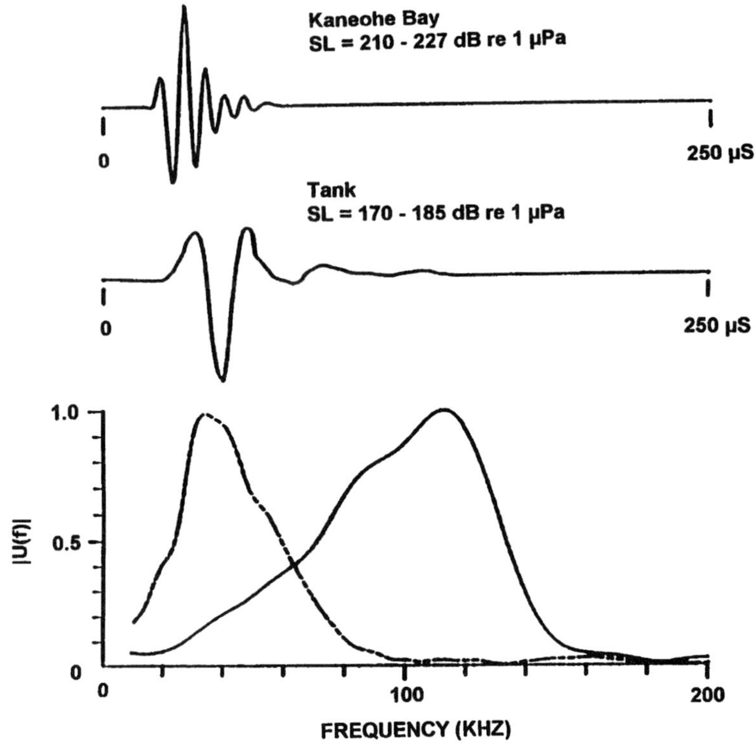

FIGURE 11.1. Example of echolocation signals used by *Tursiops truncatus* in Kaneohe Bay (adapted from Au, 1980), and in a tank.

snapping shrimp noise. In other words, dolphins can only emit high-level clicks (greater than 210 dB) if they use high frequencies. The effects of a noisy environment on the echolocation signals used by a beluga or white whale (*Delphinapterus leucas*) was vividly demonstrated by Au et al. (1985). The echolocation signal of beluga was measured in San Diego Bay, California, before the whale was moved to Kaneohe Bay. The ambient noise in Kaneohe Bay is between 15 and 20 dB greater than in San Diego Bay. The whale emitted echolocation signals with peak frequencies between 40 and 60 kHz and with a maximum averaged peak-to-peak source level of 202 dB re 1 μPa in San Diego Bay. In Kaneohe Bay, the whale shifted the peak frequency of its signals over an octave higher to 100 and 120 kHz. The source level also increased to over 210 dB re 1 μPa (Au et al., 1985). Examples of typical echolocation signals used by the whale in San Diego and in Kaneohe Bay are shown in Fig. 11.3a. Here, the signals are drawn to scale with respect to each other. Echolocation signals used by belugas in tanks also resemble the low-frequency signals shown in Fig. 11.3a (Gurevich and Evans, 1976; Kamminga and Wiersma, 1981). Turl et al. (1991) measured the sonar signals of a beluga

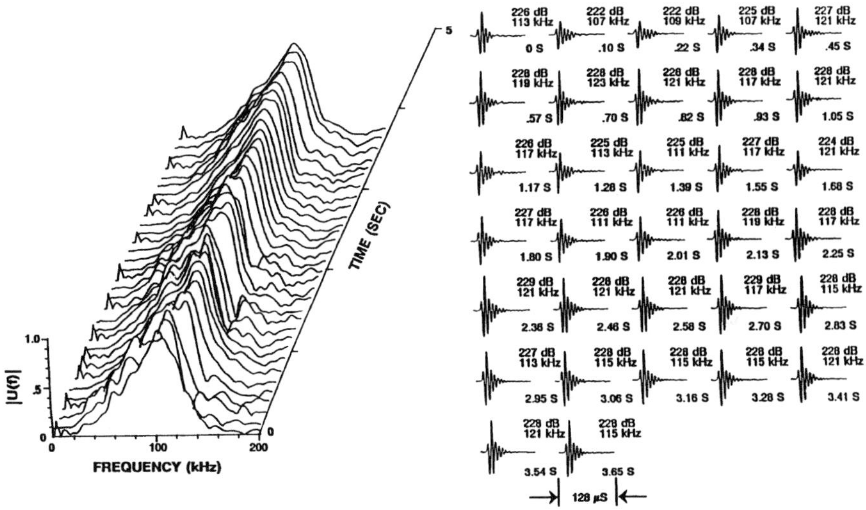

FIGURE 11.2. A typical sonar click train for a *Tursiops truncatus* performing a detection task in Kaneohe Bay. The peak-to-peak SPL in dB re 1 μPa at 1 m and the peak frequency in kHz is shown to the right of each signal. The time of occurrence of each click relative to the first click is shown below (adapted from Au, 1980).

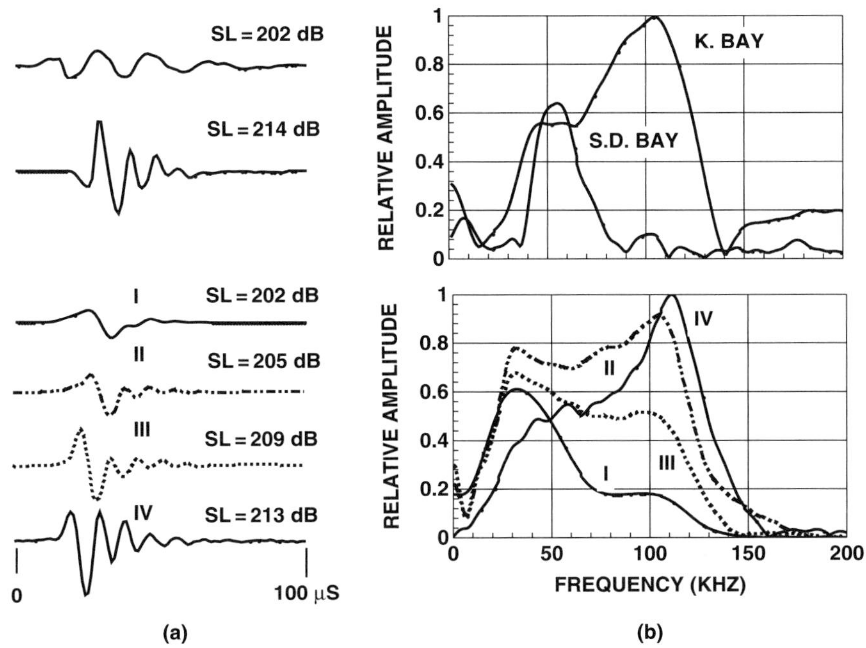

FIGURE 11.3. (a) Example of Beluga echolocation signals measured in San Diego Bay and in Kaneohe Bay (after Au et al., 1985). (b) Examples of *Pseudorca* echolocation signals, SL is the averaged peak-to-peak source level (after Au et al., 1995).

in a target in clutter detection task in San Diego Bay and found the animal using high frequency (peak frequency above 100 kHz) and high intensity (greater than 210 dB re 1 μPa) signals. Therefore, low-amplitude clicks of the beluga had low peak-frequencies and the high-amplitude clicks had high peak frequencies. The data of Moore and Pawloski (1990) for *Tursiops* also seem to support the notion that the shape of signal in the frequency domain and the amplitude of the signals are related.

Recent results with a false killer whale showed a clear relationship between the frequency content of echolocation signals and source level (Au et al., 1995). The *Pseudorca* emitted four basic types of signals, which are shown in Fig. 11.3b. The four signal types have spectra that are bimodal (having two peaks); the spectra in Fig. 11.3 are also bimodal. The type I signals were defined as those with the low-frequency peak (<70 kHz) being the primary peak and the high-frequency peak being the secondary peak with its amplitude at least 3-dB below that of the primary peak.. Type II signals were defined as those with a low-frequency primary peak and a high-frequency secondary peak, having an amplitude within 3 dB of the primary peak. Type III signals were those with a high-frequency primary peak (>70 kHz), and a low-frequency secondary peak having an amplitude within 3 dB of the primary peak. Finally, type IV signals were those with a high-frequency primary peak having an amplitude that was at least 3 dB higher than that of the secondary low-frequency peak.

The data of Thomas et al. (1988) and Thomas and Turl (1990) also indicated a similar relationship between the intensity and the spectrum of the signal. The echolocation signals of a *Pseudorca* measured in a tank had peak frequencies between 20 and 60 kHz and source levels of approximately 180 dB re 1 μPa (Thomas et al., 1988). Most of the sonar signals used by another *Pseudorca* performing a detection task in the open waters of Kaneohe Bay had peak frequencies between 100 and 110 kHz and source levels between 220 and 225 dB re 1 μPa (Thomas and Turl, 1990).

A bimodal spectrum is best described by its center frequency, which is defined as the centroid of the spectrum, and is the frequency that divides the spectrum into two parts of equal energy. From Fig. 11.3, we can see that as the source level of the signal increased, the frequency components at higher frequencies also increased in amplitude, suggesting a relationship between source level and center frequency. This relationship can be examined by considering the scatter diagram of Fig. 11.4, showing center frequency plotted against source level. The solid line in the in figure is a linear regression curve fit of the data and has a correlation coefficient of 0.80.

The bimodal property of the echolocation signals in Figs. 11.3 seems to suggest that the response of the sound generator may be determined by the intensity of the driving force that eventually causes an echolocation signal to be produced. When the intensity of the driving force is low, only signals with low amplitudes and low-frequency peaks are produced. Therefore, in small tanks, the signal resembles the tank signal of Fig. 11.1, and the bimodal

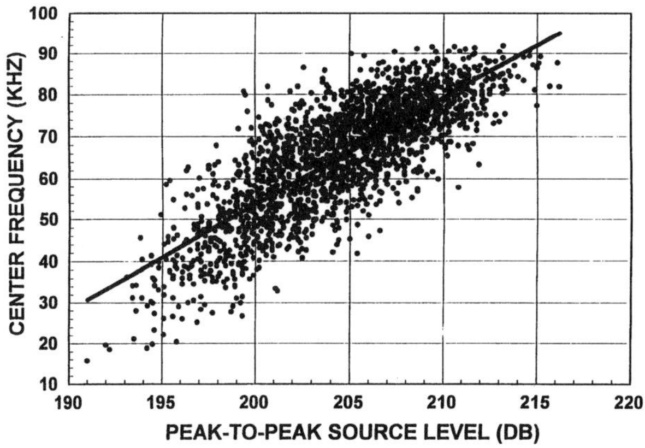

FIGURE 11.4. Center frequency of echolocation signals emitted by a *Pseudorca crassidens* as a function of the peak-to-peak source level (adapted from Au et al., 1995).

feature is suppressed since the high-frequency portion of the source cannot "kick in" for a low driving force. As the driving force increases to a moderate level, the low-frequency peak also increases in amplitude and the high-frequency portion of the signal begins to "kick" in. As the driving force increases further, the amplitude of the high-frequency peak becomes larger than that of the low-frequency peak, resulting in Type III signals. As the driving force continues to increase to a high level, the amplitude of the high-frequency peak becomes much greater than the amplitude of the low-frequency peak and completely dominates the low-frequency peak causing the bimodal feature to be suppressed.

11.1.2 Echolocation Signal of Dolphins That Do Not Whistle

The second class of echolocation signals are produced by dolphins and porpoises that do not emit any whistle signals. Their sound production is restricted to click signals that are high frequency, narrow bandwidth, and low intensity, as can be seen in the example of Fig. 11.5 showing echolocation signals used by five of these species. The duration of the signals is much longer with many oscillations, and the spectra much narrower (bandwidths are less than half as wide) than the echolocation signals used by whistle-producing dolphins. The signals of the harbor porpoise, *Phocoena phocoena* (Mϕhl and Andersen, 1973; Kamminga and Wiersma, 1981; Hatakeyama et al., 1988), and the finless porpoise, *Neophocaena phocaenoides*, (Kamminga, 1988) are similar in shape and time duration with peak frequencies between 120 and 140 kHz. Echolocation signals with peak frequencies between 90 and 115 kHz have been measured for a Dall's porpoise, *Phocoenoides dalli*, in a tank (Hatakeyama and Soeda, 1990) and with peak frequencies of

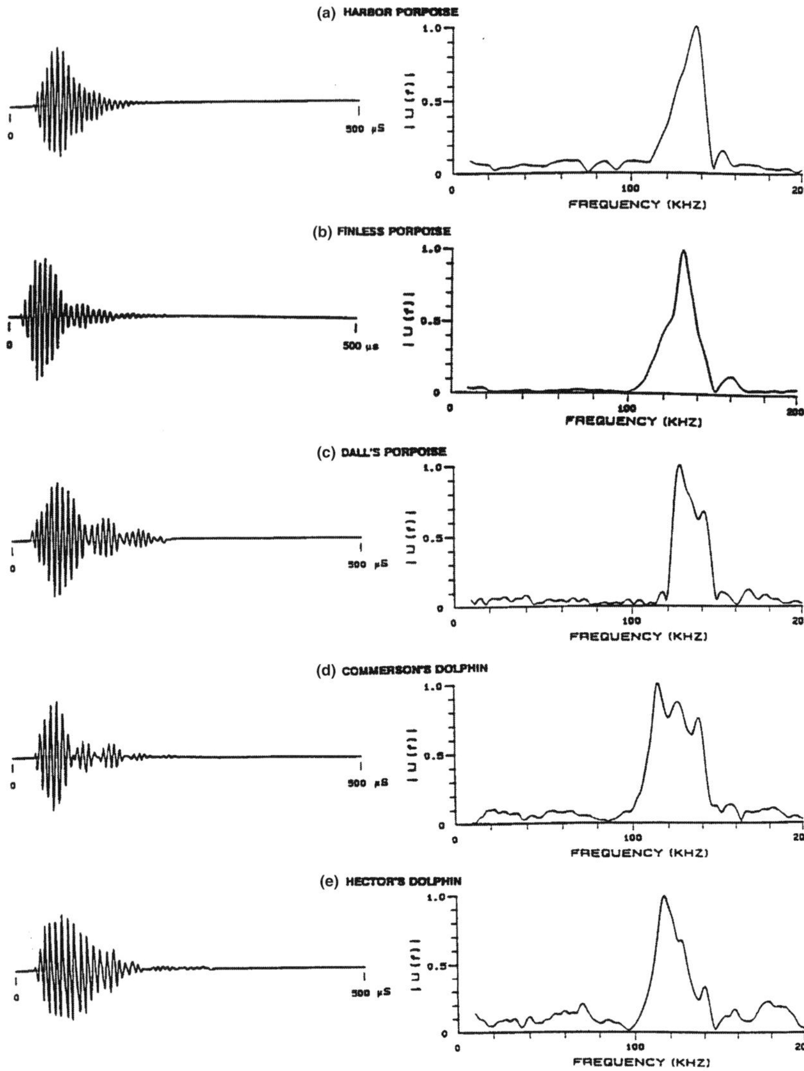

FIGURE 11.5. Examples of sonar signals of (**a**) harbor porpoise (from Kamminga and Wiersma, 1981), (**b**) finless porpoise (from Kamminga, 1988), (**c**) Dall's porpoise (from Hatakeyama and Soeda, 1990), (**d**) Commerson's dolphin (from Kamminga and Wiersma, 1981), and (**e**) Hector's dolphin (from Dawson, 1988).

120–160 kHz in the open sea (Awbrey et al., 1979; Hatakeyama and Soeda, 1990). Kamminga and Wiersma (1981) measured peak frequencies of about 124 kHz for a Commerson's dolphin, *Cephalorhynchus commersonii*, in a tank and Evans et al. (1988) measured peak frequencies of 130–140 kHz in the open sea. The pygmy sperm whale (*Kogia* sp.) also produces high-frequency (peak frequencies around 130 kHz), narrow bandwidth click signals similar to the other non-whistling odontoces (Carder et al., 1995). Kamminga and Wiersma (1981) also presented some *Phocoena* clicks that consisted of multiple pulses. Ripples in the spectrum of the Dall's porpoise and the Commerson's dolphins come from the multiple component nature of the signals. The signals of the species depicted in Fig. 11.5 often have multiple components as shown in (c) for the Dall's porpoise and (d) for the Commerson's dolphin (Evans et al., 1988) and Hector's dolphin (Dawson, 1988). However, these double pulses may be the result of a surface-reflected component, which is often difficult to discern. One reason for the use of long-duration, narrow bandwidth signals may be related to the relatively small size of the porpoises in the family *phocoenidae* and the dolphins in the genus *Cephalorhynchus*, since for a given amplitude the energy in a signal is directly proportional to duration.

11.1.3 Some Properties of Echolocation Signals

11.1.3.1 Transmission Beam Pattern

The outgoing echolocation signals of dolphins and other odontocetes are projected in a directional beam that have been measured in the vertical and horizontal planes for the bottlenose dolphin (Au et al., 1978, 1980, 1986), beluga whale (Au et al., 1987) and the false killer whale (Au et al., 1995). The composite beam pattern from the three measurements on *Tursiops* along with the averaged waveform from a single trial measured by 5 or the 7 hydrophones are shown in Fig. 11.6. The beam patterns for the beluga and false killer whale are shown in Fig. 11.7.

The 3-dB beamwidth for the bottlenose dolphin was approximately 10.2° in the vertical plane and 9.7° in the horizontal plane. The 3-dB beamwidth for the beluga whale was approximately 6.5° in both planes. Four beam patterns corresponding to the four signal types described in Fig. 11.3b are shown for the false killer whale. For the highest frequency (Type-IV signal), the 3-dB beamwidth in the vertical plane was approximately 10° and 7° in the horizontal plane. The beam axis in the vertical plane for the bottlenose dolphin and the beluga whale was +5° above the horizontal plane. For the false killer whale, 49% of the beam axis was at 0° and 32% at –5°.

The waveforms detected by the various hydrophones in Fig. 11.6 indicate that signals measured away from the beam axis will be distorted in comparison to the signals measured on the beam axis. The further away from the beam axis, the more distorted the signals will be. The frequencies shown with each waveform are the frequencies of local maxima in the spectrum. At some angles the averaged signal had multiple peaks in the spectrum, and these peaks are listed in an order of decreasing amplitude. The presence of multipath is apparent for the signals measured away from the beam axis. These

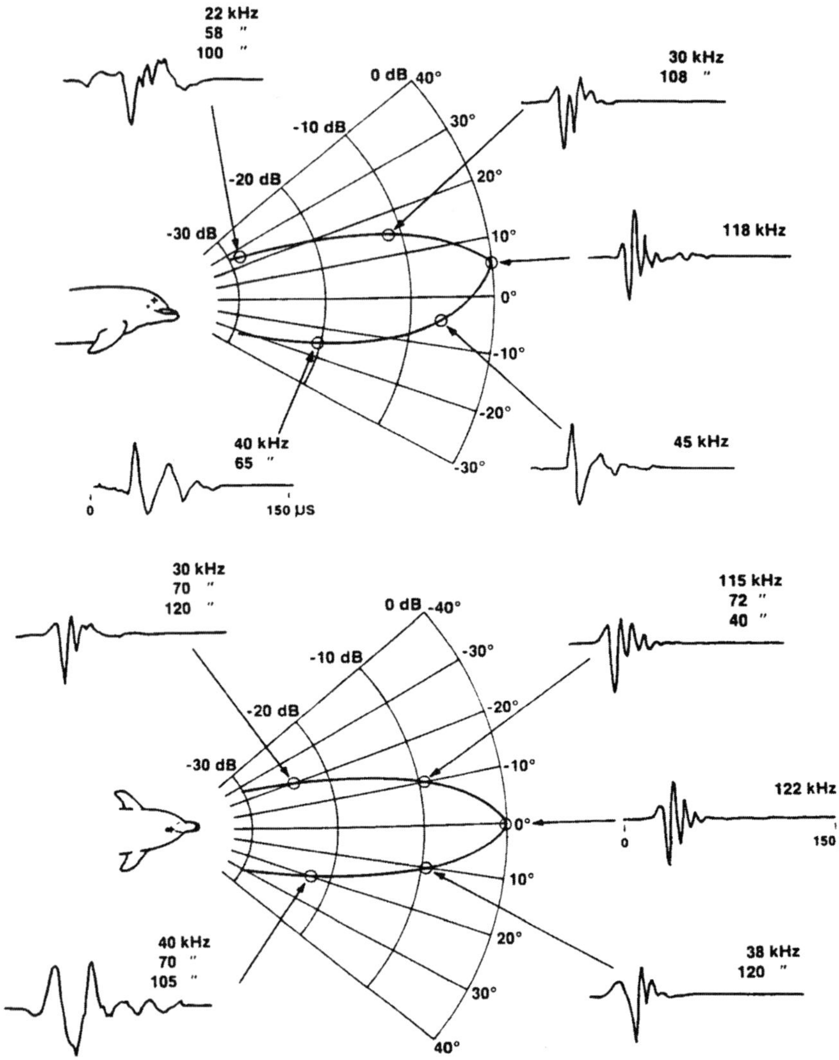

FIGURE 11.6. The transmission beam pattern of a *Tursiops truncatus* in the vertical and horizontal planes with the waveform of a click measured by 5 of 7 hydrophones (adapted from Au, 1993).

multi-paths may be due to internal reflection and refraction of the signals within the animal's head or may be due to signals being radiated from different portions of a finite source region, or to combinations of these two possibilities. Signals measured away from the beam axis for the beluga and false killer whales were also distorted in comparison to the on-axis signals.

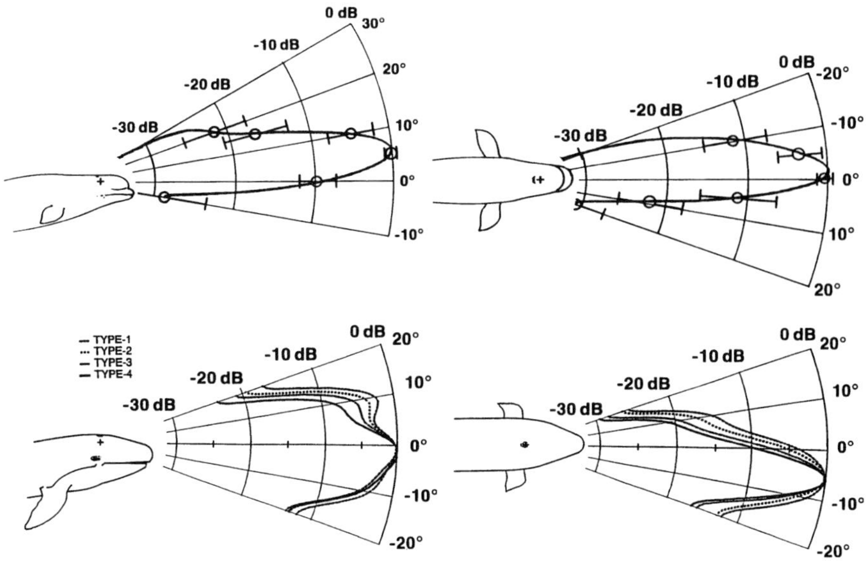

FIGURE 11.7. The transmission beam pattern of a *Delphinapterus leucas* (after Au et al., 1987) and a *Pseudorca crassidens* (after Au et al., 1995) in the vertical and horizontal planes.

11.1.3.2 Click Intervals

Dolphins typically echolocate in a pulse mode; they send out a click, receive, and process the echo, and after a specific lag time, emit the next click. Lag time is defined as the time difference between the reception of an echo and the emission of the next echolocating click. Evans and Powell (1967) and Johnson (1967) found an almost constant mean lag time of 15.4 ms for target ranges from 1.4 to 0.4 m, as the dolphin swam toward a target. When the dolphin's distance to a target closed from 0.4 to 0.03 m, the lag time decreased to a minimum of 2.5 ms. Morozov et al. (1972) reported a fairly constant lag time of 20 ms over a distance of 40 to 4 m as the dolphin swam toward a target.

The average click interval as a function of target range for four different experiments with *Tursiops* in Kaneohe Bay in which target range was fixed for each experimental session is shown in Fig. 11.8. Also included in the figure is the two-way transit time for a signal to travel from the dolphin to the target and back. Lag times associated with Fig. 11.8 and those measured for free - swimming dolphins tend to vary between 19 and 45 ms, except for very small target ranges (less than 0.4 m). When target range was randomly chosen from trial to trial between 40 and 120 m in 20-m increments, Penner (1988) found that *Tursiops* would start a trial using click intervals that were appropriate for the longest target range (120 m) and then adjust its click interval (if the target

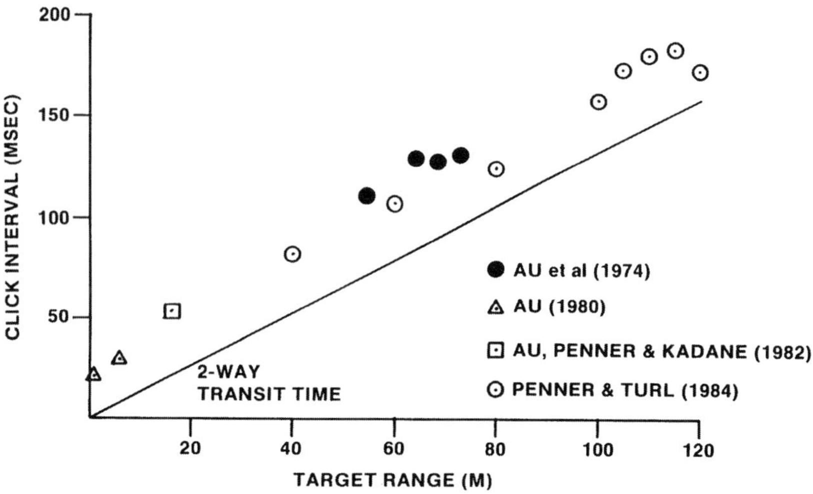

FIGURE 11.8. Click interval as a function of target range for the bottlenose dolphin. The value of the two-way transit time for any target range can be read off the click interval scale (adapted from Au, 1993).

range was less than 120 m) to match the target range plus a lag time varying between 19 and 30 ms.

Pseudorca crassidens also emit echolocation signals with similar click interval patterns as *Tursiops* (Thomas and Turl, 1990). However, *Delphinapterus leucas* seems to have a slightly different click interval pattern (Au et al., 1987). Three different patterns or mode of signaling were observed. Mode 1 signals had click intervals that were greater than the two-way transit time, mode 2 signals had click intervals that were less than the two-way transit time but greater than 5 ms, and mode 3 signals had click intervals between 1 and 2 ms. The mode 3 signals seemed suspiciously similar to burst pulse signals (Herzing, 1996) that are not typically used for echolocation; the amplitudes were more than 12 dB below the mode 1 signals. Turl and Penner (1989) found that when the target was greater than about 100 m, the beluga emitted packets of clicks in which the intervals between clicks in a packet were less than the two-way transit time but the intervals between packets were greater than the two-way transit time. An example of the beluga echolocation click intervals for a target range greater than 100 m is shown in Fig. 11.9. The packet results of Turl and Penner (1989) are for one specific beluga whale; it would be interesting to determine if the use of packet is common for beluga whales.

11.1.3.3 Click Source Levels

Dolphins can vary the amplitude of their echolocation emissions over a very large range, from as low as 150–160 μPa (peak-to-peak) to as high as 230 μPa

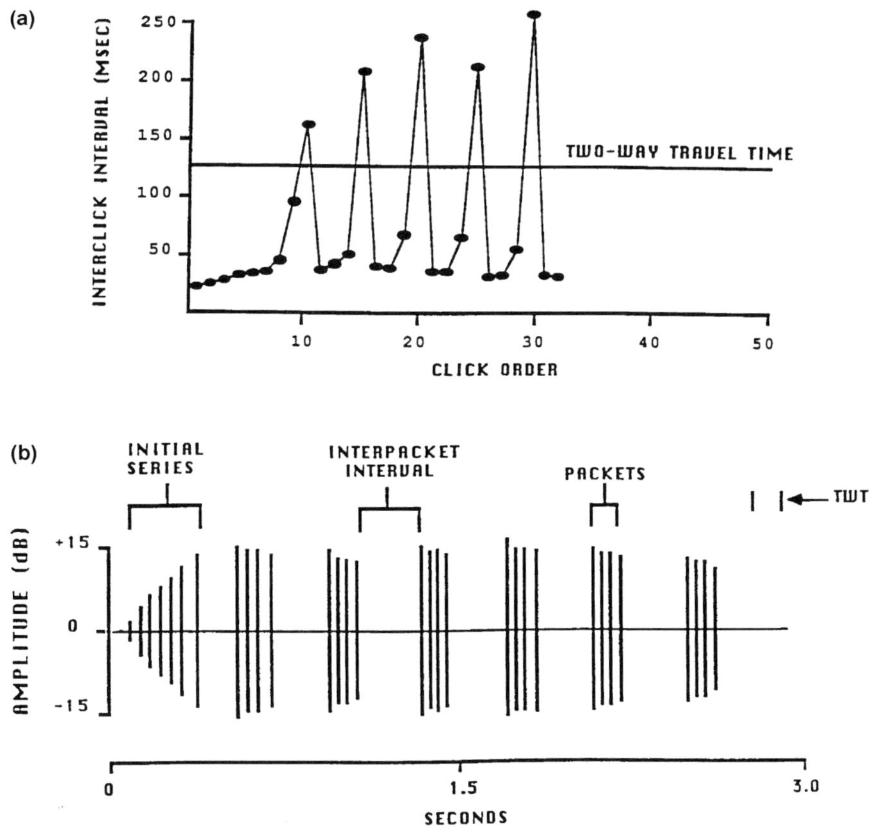

FIGURE 11.9. An example of a beluga click interval pattern for a target a 100 m. (a) Click interval vs. click interval no. (b) amplitude vs. time (adapted from Turl and Penner, 1989).

peak-to-peak (Au, 1980). We have found in open waters that two of the major influences on the source levels used by *Tursiops* are the amount of acoustic loss associated with a given task and the level of any masking noise. The amplitude of the echoes from a target (echo level) is dependent on the source level of the projected signal, the target range, and the reflectivity of the target (target strength). The peak-to-peak source level as a function of total loss due to transmission and target reflection losses for five different *Tursiops* under different circumstances is displayed in Fig. 11.10. The data shown in Fig. 11.10 fell into two distinct categories: one associated with total loss less than about 70 dB and the other with total loss greater than 70 dB. A least-square second-order polynomial curve is fitted to the data. For losses less than 70 dB, the source level was fairly constant.

As the total loss increased above 70 dB, the source level increased almost linearly with loss. However, the source level did not increase at the same rate

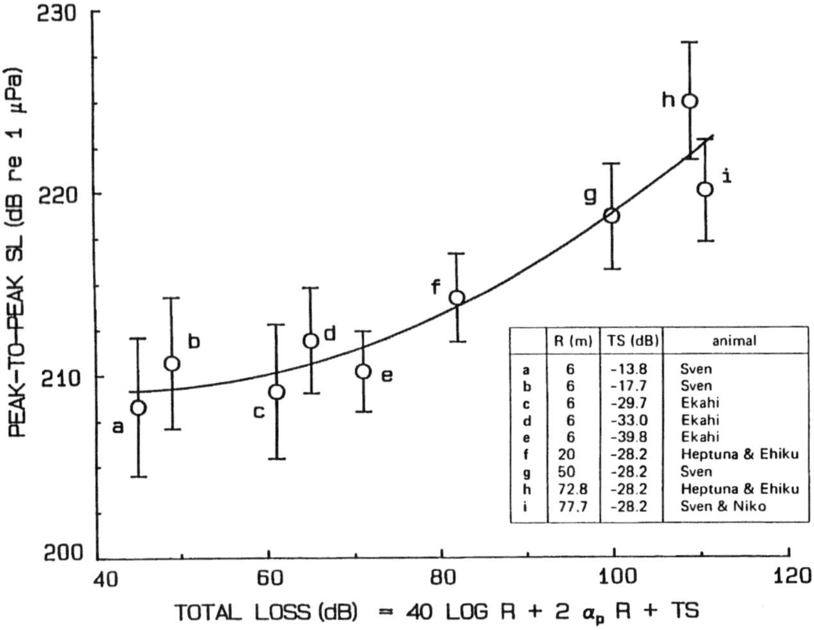

	R (m)	TS (dB)	animal
a	6	-13.8	Sven
b	6	-17.7	Sven
c	6	-29.7	Ekahi
d	6	-33.0	Ekahi
e	6	-39.8	Ekahi
f	20	-28.2	Heptuna & Ehiku
g	50	-28.2	Sven
h	72.8	-28.2	Heptuna & Ehiku
i	77.7	-28.2	Sven & Niko

FIGURE 11.10. Peak-to-peak source level as a function of total acoustic energy loss for five different *Tursiops* performing different echolocation tasks in Kaneohe Bay (adapted from Au, 1980).

as the total loss, indicating that the dolphins seem to prefer to operate at a high echo signal-to-noise ratio. The decrease in loss from case (i) to case (a) was over 62 dB, and the corresponding difference in source level was only 12 dB. The maximum average peak-to-peak source level was recorded at 227.6 dB (case h). The largest single click measured was 230 dB, emitted by Heptuna (case h). The largest single click measured for *Delphinapterus* was approximately 225 dB (Au et al., 1987) and for *Pseudorca*, 228 dB (Thomas and Turl, 1990). Møhl et al. (1990) measured high-intensity clicks from a Narwhal (*Monodon monoceros*) that had similar amplitude levels as those emitted by odontocetes in Kaneohe Bay.

11.2 Target Detection Capabilities

The capabilities of any echolocation system are usually divided into two general categories, target detection and target discrimination. The target detection capabilities of an echolocating dolphin will be limited by ambient or artificial noise, by reverberation, and by its own hearing sensitivity. Echolocation experiments to determine the target detection capabilities of dolphins in noise and reverberation have been performed by Au (1993).

However, target detection in a quiet environment in which a dolphin cannot hear the ambient noise in the frequency range of its echolocation signals has not been considered. Many different types of discrimination experiments have been performed (see Nachtigall, 1980; Au, 1993).

11.2.1 Target Detection in Noise

Three different types of target detection in noise experiments have been performed with dolphins. The first type is the simple one in which a specific target is moved progressively away from the position of an echolocating dolphin. Eventually, the target will be sufficiently far away that the echoes from the target will have amplitudes that are similar to the amplitudes of the ambient noise, making the target progressively harder to detect. The maximum detection range of two *Tursiops* was determined in Kaneohe Bay by Murchison (1980) using a 2.54-cm diameter solid steel sphere and by Au and Synder (1980) using a 7.62-cm diameter water-filled sphere. In both studies, an overhead suspension system with a movable trolley and pulleys was used to vary target range between two poles spaced 200 m apart. The results of both experiments are displayed in Fig. 11.11 with correct detection and false alarm rates plotted as a function of the target range. The detection threshold ranges (range at 50% correct detection) for the 2.54-cm and 7.62-cm diameter

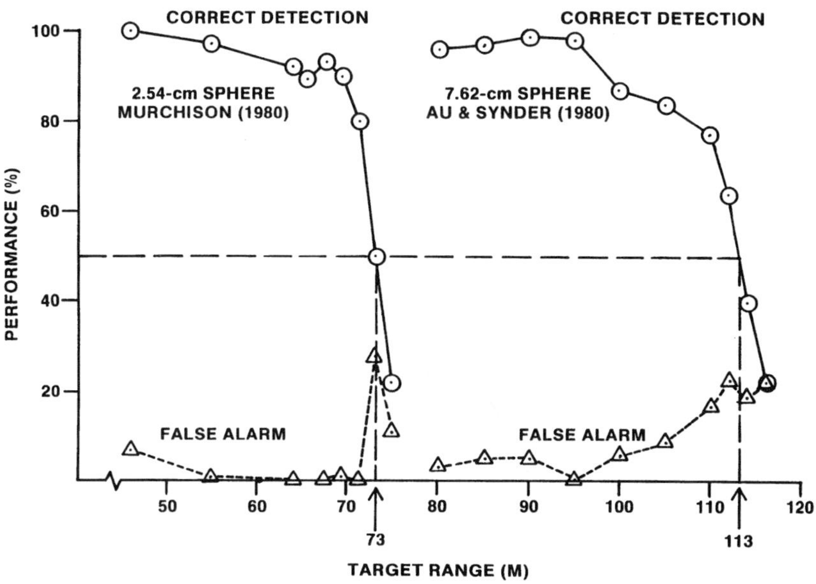

FIGURE 11.11. *Tursiops truncatus* target detection performance as a function of range. The 2.54 cm sphere results are from Murchison (1980) and the 7.62-cm sphere results are from Au and Snyder (1980).

spheres were 73 and 113 m, respectively. The animal's results for the two different targets are relatively consistent if target strength and transmission loss differences are considered. Thomas and Turl (1990) also used a 7.62-cm water-filled sphere to determine the detection threshold range of a *Pseudorca crassidens* in Kaneohe Bay. They obtained a threshold range of 119 m, which is relatively similar to our results with *Tursiops* (the difference in the transmission loss being only 1.4 dB). With a larger head, the receiving directivity index of the *Pseudorca* could easily be 1.4 dB larger than for *Tursiops*, so that the false killer whale may be receiving slightly less noise.

The target detection results shown in Fig. 11.11 are interesting but of limited value since it is dependent on the ambient noise in Kaneohe Bay, a body of water well known for its loud population of snapping shrimp (Albers, 1965). However, if the noise-limited form of the sonar equation is applied to the data in Fig. 11.11, then the echo energy to noise spectral density (E_e/N_0) can be determined for any specific performance level. The sonar equation for a dolphin projecting short-duration broadband signals was derived in Section 8.2, Eq. (8.12) of Au (1993), and can be expressed as

$$\mathrm{DT_E} = \mathrm{SE} - 2\,\mathrm{TL} + \mathrm{TS_E} - (\mathrm{NL} - \mathrm{DI}), \qquad (11.1)$$

where the detection threshold is written in terms of the energy-to-noise ratio commonly used in human psychophysics. SE is the source energy flux density and $\mathrm{TS_E}$ is the target strength based on energy. TL, NL, and DI have been previously defined as the transmission loss, ambient noise spectral density, and the receiving directivity index, respectively. For the type of signals used by *Tursiops* in Kaneohe Bay, the energy flux density can be approximated by the expression

$$E(\mathrm{dB}) \approx \mathrm{SPL_{pp}}(\mathrm{dB}) - 58. \qquad (11.2)$$

During an echolocation task, dolphins typically vary the amplitude of their echolocation signals by as much as 20 dB, making it difficult to estimate the detection threshold accurately. We initially approached this issue by using the maximum source energy flux density per trial (Au and Penner, 1981); however, later we were able to determine that a dolphin's detection threshold was approximately 2.9 dB below the maximum echo level produced by the maximum source level during a trial (Au et al., 1988). The echolocation signals of the dolphin involved with the range detection task of Fig. 11.11 were measured in the study of Au et al. (1974), where the maximum peak-to-peak source level averaged over 12 trials at the 77-m range was 225 dB re 1 μPa. The target strength of the 2.54-cm sphere and the 7.62-cm sphere was measured using simulated dolphin echolocation signals and the results are shown in Fig. 11.12. From Fig. 11.12, the target strength based on energy was –41.6 dB for the 2.54-cm solid sphere and –28.3 dB for the 7.62-cm water-filled sphere. The ambient background noise of Kaneohe Bay, at the location of the experiment and at a frequency of 120 kHz, was approximately 54 dB re $\mu\mathrm{Pa}^2/\mathrm{Hz}$. The receiving directivity index for *Tursiops truncatus* given by

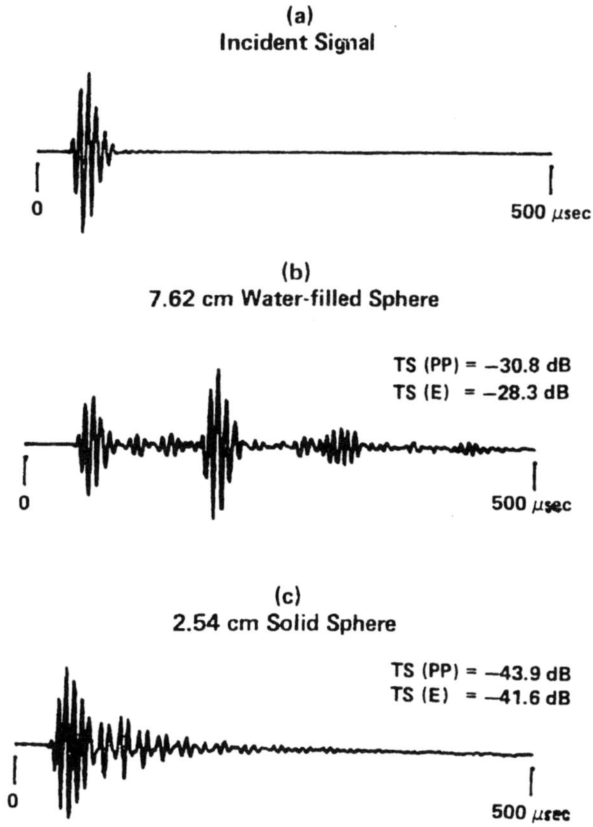

FIGURE 11.12. (a) Example of a simulated dolphin echolocation signal and the echo from the (b) 2.54-cm solid sphere and (c) 7.62-cm water-filled sphere (after Au and Snyder, 1980).

Eq. (10.10) at a peak frequency of 120 kHz is equal to 20.2 dB. Including all these parameters into Eq. (11.1) and (11.2), we arrive at the animal's performance as a function of the echo energy-to-noise spectral density in Fig. 11.13. The results indicated that the detection threshold occurred at $(E_e/N_0)_{COR}$ between 7.5 and 9.8 dB Au (1993). The target detection capability of *Tursiops* was measured by two other techniques. A target was positioned at a fixed range, and the dolphin's capability to detect it was measured as a function of the level of a wide band masking noise (Au and Penner, 1981; Au et al., 1988; Turl et al., 1987). In the third experiment, an electronic simulated echo generator was used to simulate a target at 20 m (a phantom target) and the level of the echo was progressively made smaller as the echolocating dolphin performed a detection task in a fixed noise field (Au et al., 1988). The experimental geometry for the phantom target detection experiment is shown in Fig. 11.14. The dolphin was trained to station in a hoop with

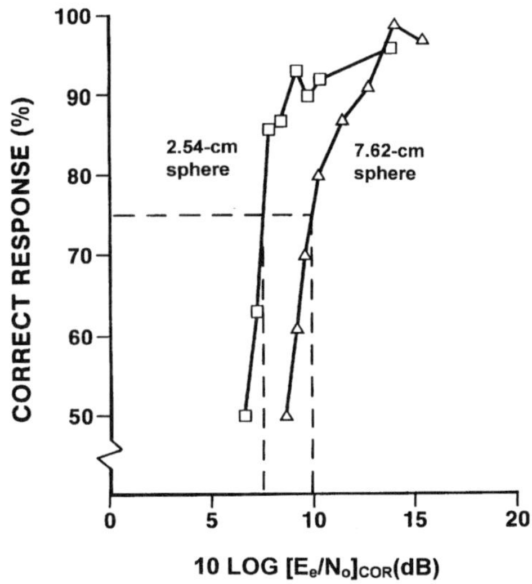

FIGURE 11.13. Target detection performance of a *Tursiops truncatus* as a function of the echo energy-to-noise ratio for the target detection data of Fig. 11.11 (after Au, 1993).

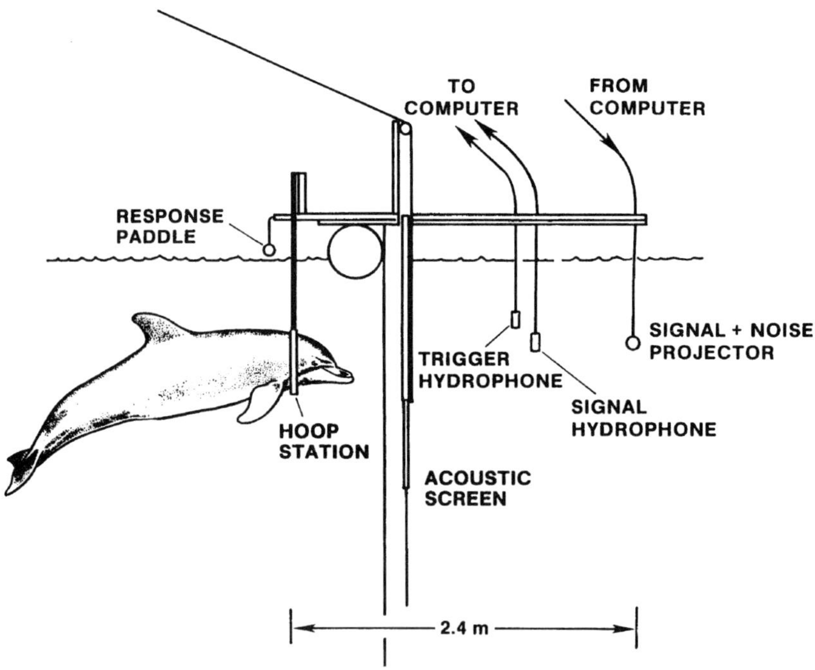

FIGURE 11.14. Experimental configuration with dolphin in hoop station and transducers used in electronic target simulation system (adapted from Au et al., 1988).

three hydrophones directly in front of the animal. When raised, it blocks the animal's sonar emissions from reaching the hydrophones. The dolphin's sonar emissions were detected by the first hydrophone, which triggered an A/D converter causing it to digitize the signal received by the second hydrophone located 1.9 m from the hoop. The A/D converter operated at a 1-MHz rate and digitized 128 points per signal with the digitized data An acoustic screen was located between the hoop and the hydrophones. When the screen was stored in a static random access memory (RAM). The output of the first hydrophone also triggered an external delay generator and flagged the computer that a signal had been emitted. After the appropriate delay corresponding to the simulated target range, the delay generator flagged the computer to project the signal stored in RAM. The number of echo highlights and the amplitude of each highlight were controlled by the computer. Masking noise was mixed with the RAM signal and projected from the third transducer located 2.4 m from the hoop. The projector was driven by an equalization circuit that flattened its transmit response, allowing it to transmit broadband noise and simulated dolphin signals with a minimum of distortion (Au et al., 1987).

The results of the three different methods of measuring a dolphin's target detection capability are shown in Fig. 11.15. The dolphin's 75% correct response threshold in Fig. 11.15 occurred at an $(E_e/N_0)_{COR}$ of 7.5. The signals in Fig. 11.16 can be used to visualize what an E_e/N_0 E_e/N_0 of 7.5 dB represents. The top waveform of Fig. 11.16 is the echo from a 7.62-cm sphere produced with a simulated dolphin echolocation signal. The second and third

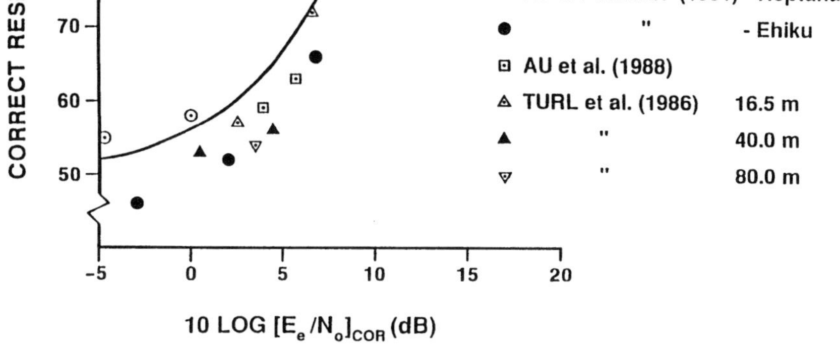

FIGURE 11.15. *Tursiops truncatus* performance results compared with the results of an energy detector, which is the solid line (adapted from Au, 1993).

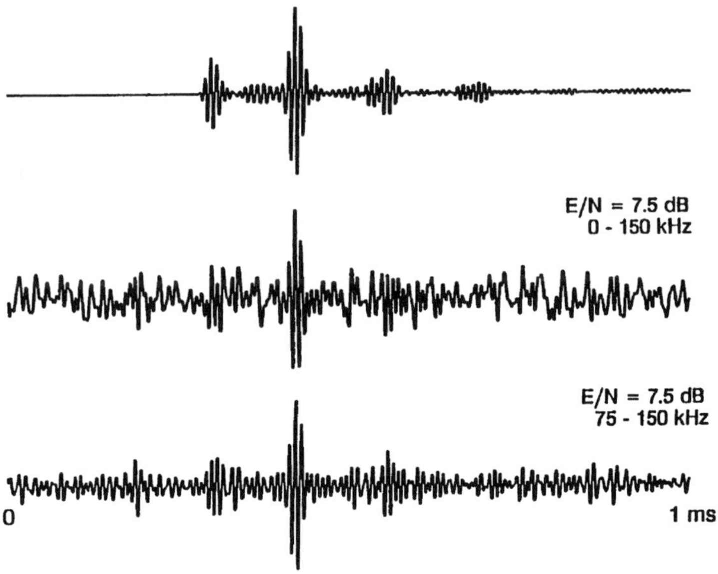

FIGURE 11.16. Target echo in noise at the dolphin's detection threshold.

waveforms depict the same echo for a 7.5 dB signal-to-noise ratio condition for two different bandpass filter settings. The largest highlight of the echo is observable in the noisy echo; however, the smaller highlights are masked by the noise and the acoustic quality of the echo was altered. The dolphins could probably hear the largest highlight, but the echo probably did not "sound" like the sphere they were trained to detect and consequently reported the target as not present. Therefore, it seems that a target detection experiment is probably not purely one of detecting a signal in noise but also involves discriminating the features of the echoes from a target.

Another important parameter in understanding the detection of targets in either noise or reverberation is the integration time of the dolphin's auditory system for broadband signals. We considered this issue by using the phantom target configuration of Fig. 11.14 and performed an experiment that would allow us to determine the integration time for an echolocating dolphin. The dolphin target detection threshold in noise was first measured by returning a single-click echo for each click emitted by the dolphin, and progressively lowering the amplitude of the phantom echo in a staircase manner. Then, double click echoes with seven different separation time between clicks varying from 50 to 600 μs were used and the shifts in threshold were observed as a function of the separation time between clicks. The specific separation time for each double click threshold measurement was randomized. The results of the auditory integration time experiment are displayed in Fig. 11.17. The ordinate is the E_e/N_0 at threshold and the abscissa is the separation time (ΔT) between the double clicks. The threshold shifted approximately 3 dB as the

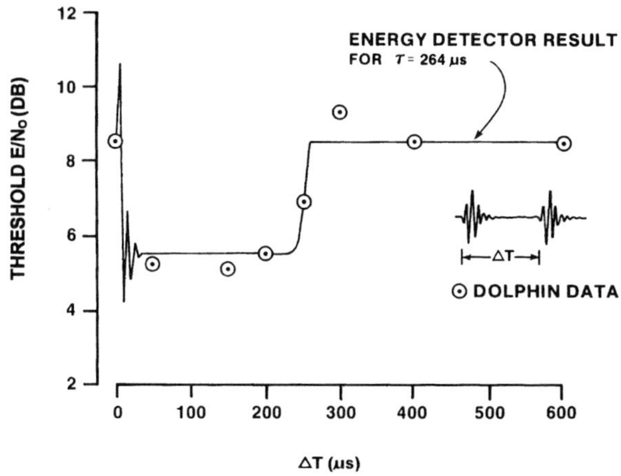

FIGURE 11.17. Results of integration time experiment showing the energy-to-noise ratio at threshold as a function of the separation time between the clicks of a double click echo. For a ΔT of 0 μs, the echo consisted of only a single click. The *solid curve* is the response of an ideal energy detector that best fits the dolphin's data (adapted from Au et al., 1988).

echo changed from a single click to a double click. This shift is exactly what would be expected from an energy detector, since there is 3 dB more energy in the double click echo than in a single click. As ΔT increased to 250 μs, the dolphin's double click threshold began to shift toward the single-click threshold, reaching it at a ΔT of about 300 μs and greater. Therefore, the presence of a second click separated by more than 300 μs from the first click did not help the dolphin to "detect" the phantom target. The dolphin's results were fitted with several curves derived for an ideal energy detector having different integration times, and the curve associated with 264 μs best fitted the data in Fig. 11.17. The oscillations of the left side of the curve were caused by the first and second clicks overlapping at small values of ΔT. A 264 μs integration time is considerably shorter than the several milliseconds indicated by Johnson (1968) results shown in Fig. 9.4. The integration time results suggest that dolphins may process broadband transient signals differently than narrow-band pure-tone signals.

11.2.2 Target Detection in Reverberation

A second way in which a sonar system can be limited is by the presence of reverberation. Reverberation differs from noise in several aspects. It is caused by the sonar itself, and is the total contribution of unwanted echoes scattered back from objects and inhomogeneities in the medium and on its boundaries. The spectral characteristics of reverberation are similar to those of the projected signal, and its intensity is directly proportional to the intensity of the projected signal. Therefore, in a reverberation-limited situation, target detection cannot be improved by increasing the intensity of the projected signal. Target detection becomes dependent on the ability of the system to discriminate between the target of interest and false targets and clutter that contribute to the rever-beration.

Titov (1972), as reported by Ayrapet'yants and Konstantinov (1974) and Bel'kovich and Dubrovskiy (1976), investigated the capability of *Tursiops* to detect targets in the presence of smooth rocks, varying in size from 5 to 30 mm in 5-mm increments forming two 40-cm circles on the bottom of a tank; each circle containing 300 rocks, 50 of each size. The dolphin could detect a 50-mm lead sphere lying on the bottom in the center of the clutter 75% of the time at a distance of 5 m. A 33-mm solid steel sphere had to be raised 1.7 cm above the largest rock before the animal could detect it 75% of the time at a distance of 5 m. The dolphin approached the target swimming close to the bottom rather than at the surface, presumably to minimize the reverberation. Murchison (1980) studied the effects of bottom reverberation on the target detection capabilities of two *Tursiops* in Kaneohe Bay. A 6.35-cm diameter solid steel sphere was used at depths varying from 1.2 to 6.3 m. At a depth of 6.3 m, the target was on the bottom. The animals' 50% correct detection threshold ranges for the different target depths are plotted in Fig. 11.18. As the target depth increased, the animals' detection ranges decreased, showing the effects of bottom reverberation.

The reverberation-limited form of the sonar equation was applied to the result of Murchison (1980) in which the target was laying on the bottom by Au (1992, 1993). The geometry of a dolphin ensonifying a target laying on the bottom is shown in Fig. 11.19. The generalized form of the sonar equation for such a reverberation-limited situation equates the detection

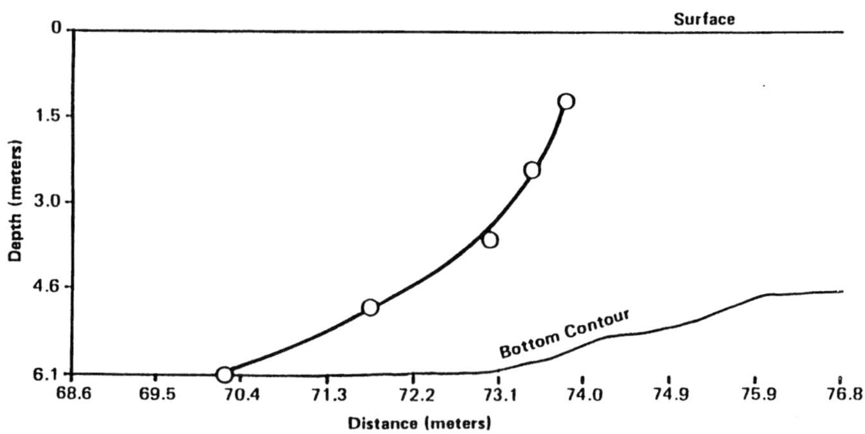

FIGURE 11.18. Fifty percent correct detection threshold range for two *Tursiops truncatus* as a function of the depth of the 6.35-cm solid spherical target (adapted from Murchison, 1980).

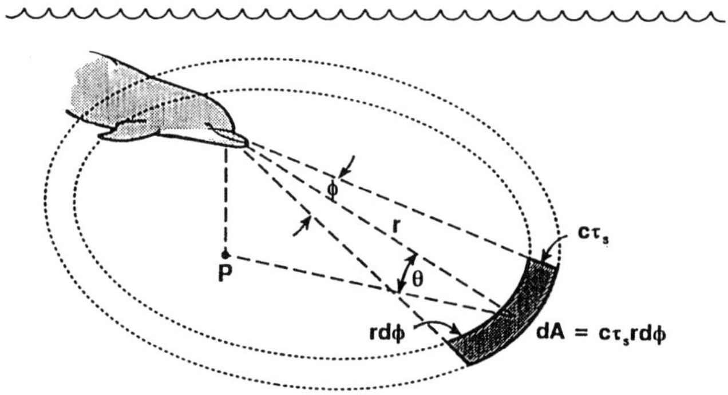

FIGURE 11.19. Geometry of a dolphin ensonifying an area dA of the bottom. θ is the angle between the beam axis and the horizon, τ_s is the duration of the bottom reverberation, which should be equal to τ_{inc}, the integration time of the dolphin, and c is the velocity of sound in water. For small values of θ, $r \approx r_h$, the horizontal range (adapted from Au, 1993).

threshold DT_R to the ratio of the echo and reverberation energy flux densities (Urick, 1983) and derived in Section 8.6.1, Eq. (8.46) of Au (1993), and can be expressed as

$$DT_{RS} = TS_E - S_{sE} - 10\log(\Phi r_h \tau_{inc} c) \tag{11.3}$$

where $S_{sE}$is the energy version of the scattering strength of the bottom reverberation, and Φ is an integral of the product of the transmit (b_t) and receive (b_r) beam patterns.

Au (1992) used a simulated dolphin sonar signals to measure the scattering strength of the bottom where Murchison's experiment was conducted. The target strength of the 6.35-cm diameter solid steel sphere was also measured with a simulated dolphin sonar signal and found to be approximately –33.4 dB. Inserting the appropriate values into the reverberation-limited form of the sonar equation, the echo-to-reverberation ratio at the dolphins' detection threshold was estimated to be approximately 4 dB.

The target echo and the bottom reverberation corresponding to an E_e/R_e of 4 dB are shown in Fig. 11.20. Also included in the figure are the sum of the target echo with the bottom reverberation and the spectra of the summed echo denoted by the solid line and the target echo denoted by the dashed line. The effects of the bottom reverberation in masking the signal in both the time and frequency domains can be seen in Fig. 11.20. The echo plus reverberation waveform does not contain the distinct highlight structure of the target echo measured in the free field. Only the two largest highlights of the target can be clearly seen in the echo. In the frequency domain, the distinctive peaks and nulls of the target echo are no longer distinguishable for frequencies below 100 and 120 kHz, in the spectrum of the summed echo. The dolphins

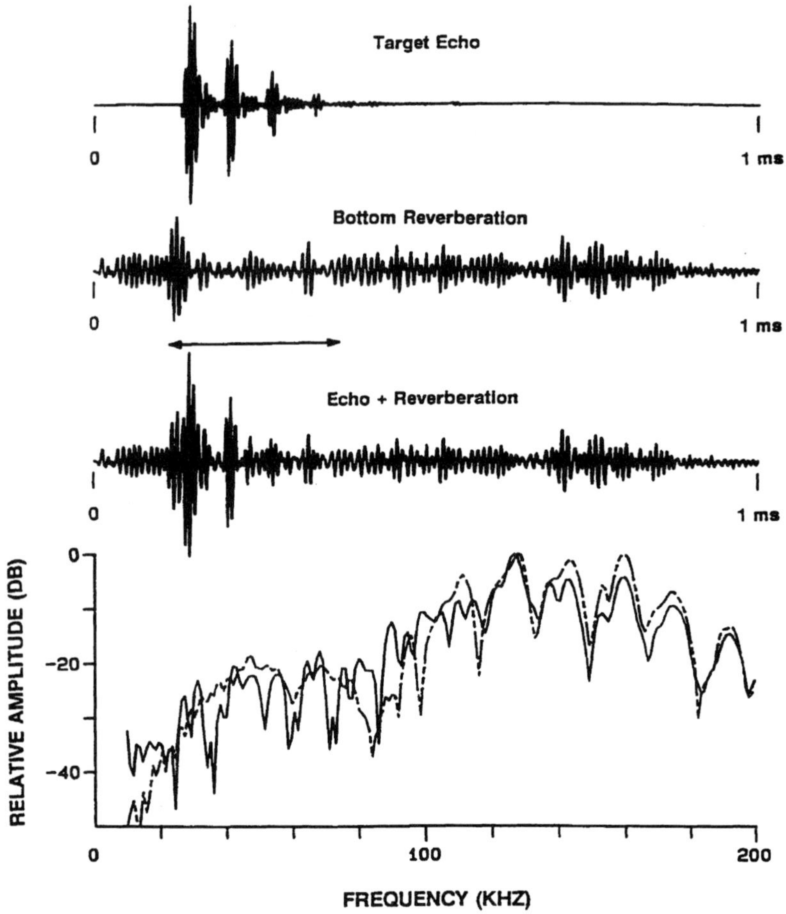

FIGURE 11.20. Echo from the 6.35-cm solid sphere target used by Murchison (1980) and the bottom reverberation measured by Au (1992). The relative amplitudes of the echoes were scaled for an E_e/R_e of 4 dB.

could probably hear the larger highlights from the target but the echo probably did not "sound" like the target they were trained to detect and consequently reported the target as not present. As in the detection in noise experiments, target detection in reverberation also involves recognition of the target echo.

Two studies investigated the capabilities of echolocating dolphins to detection targets in the presence of a clutter screen. Au and Turl (1983) used a clutter screen consisting of forty-eight 5.1-cm diameter cork balls suspended in a 6×8 rectangular array. The horizontal and vertical spacing between cork balls was 15.2 cm center-to-center. Turl et al. (1991) used a clutter screen consisting of 300 cork balls spaced the same distance apart as the screen used by Au and Turl (1983). In the study of Au and Turl (1983), the clutter screen was located at a distance of $6 + \Delta R$ m from a bottlenose dolphin and in the study of Turl et al. (1991) the clutter screen was $8 + \Delta R$ from an echolocating beluga whale, where ΔR is the distance of the target from the clutter screen. Au and Turl (1983) used hollow aluminum cylinders of 3.81-cm diameter and 0.32-cm wall thickness having lengths of 10, 14, and 17.8 cm, whereas stainless steel cylinders of the same diameter and wall thickness with lengths of 3, 5, 7, 10, and 14 cm were used by Turl et al. (1991). By using cylinders of the same diameter and wall thickness but of different lengths, the intensity of the echoes could be controlled while keeping the echo structure the same. Performance data were collected as a function of ΔR for each target. The animals' accuracy decreased both as the separation distance decreased and as the target got smaller.

The target strength of the targets and both clutter screens were measured with simulated dolphin echolocation signals. Au and Turl (1983) used an integration window of 1 ms to compute the reverberation level. This turned out not to be an accurate value since Au et al. (1988) measured an integration time of 264 μs for an echolocating *Tursiops truncatus*. Au (1993) later corrected this inaccuracy by reducing the reverberation strength measured by Au and Turl (1983) by 2 dB. The results of both studies for $\Delta R = 0$ are shown in Fig. 11.21. The 50% correction detection threshold for the bottlenose dolphin corresponded to an E_e/R_e of 2.3 dB and E_{pp}/R_{pp} of 2.5 dB. The value of E_e/R_e of 2.3 dB is in good agreement with an E_e/R_e of 4 dB determined by Au (1992) for the bottom reverberation situation with *Tursiops*. The 50% correct

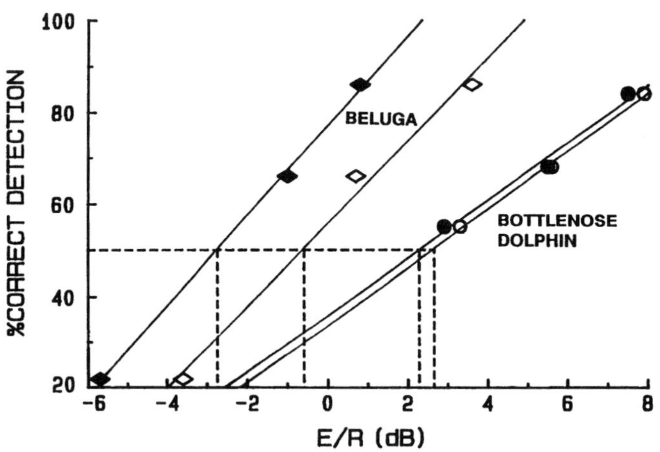

FIGURE 11.21. Bottlenose dolphin and beluga whale target detection performance as a function of the echo-to-reverberation ratio for $\Delta R = 0$ (targets were in the plane of the clutter screen). The E/R based on a 264 μs integration time are indicated by *closed symbols* and those based on peak-to-peak values are indicated by *open symbols*. The solid lines are linear curved fitted to the data (modified from the data presented by Turl et al., 1991).

detection threshold for the beluga whale corresponded to E_e/R_e of –2.7 dB and E_{pp}/R_{pp} of –0.6 dB. It is extremely interesting that the target echo can be smaller on both a peak-to-peak and energy basis than the clutter screen echo, and yet the beluga was able to detect the target. The results also indicated that the beluga was approximately 3.2–5.0 dB more sensitive than the bottlenose dolphin in detecting a target in reverberation.

Perhaps, one of the most amazing echolocation target detection and discrimination behavior exhibited by dolphins is the "crater-feeding" behavior of *Tursiops truncatus* in the waters of the Grand Bahama Island (Herzing, 1996; Rossbach and Herzing, 1997). These dolphins have been observed foraging for prey buried under the sand by positioning about 3 m above the bottom and rotating their heads alternately in a clockwise and counterclockwise motion while performing echolocation scanning, or by swimming about a 1 m above the bottom and scanning in a side-to-side motion with their bodies parallel to the bottom. A photograph of a bottlenose dolphin burrowing into the sandy sediment to capture a prey and the crater impressions from several such burrowing are shown Fig. 11.22. The buried prey is probably being detected by echolocation since clicks can be heard by swimmers observing the behavior. It is also possible that the dolphins are listening to prey movement, although if this was the case, the dolphins would probably forage in silence and not continuously emit clicks. If these dolphins are detecting their prey by echolocation, then these animals are performing an echolocation task in a highly reverberant environment. I am not aware of any man-made sonar that can detect and discriminate small buried objects as these bottlenose dolphins do in the Bahamas. Another interesting feature of this benthic feeding behavior is the observation that these animals rarely burrow without capturing a prey.

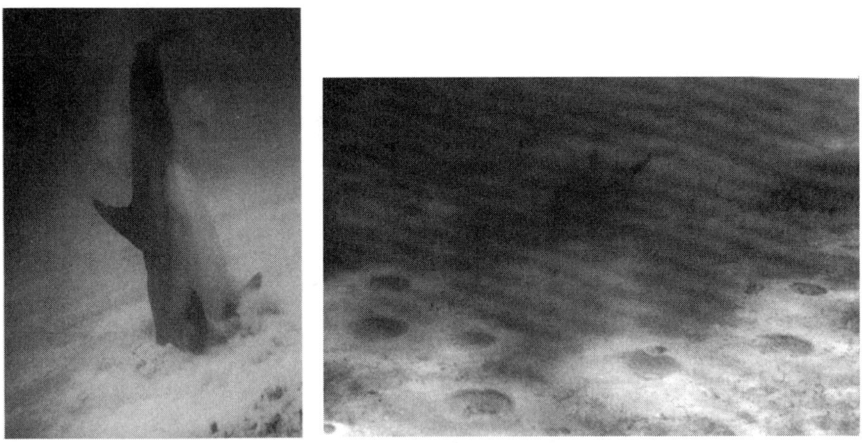

FIGURE 11.22. (*left panel*) A bottlenose dolphin burrowing into the sandy bottom to capture a prey, (*right panel*) crater left by several burrowing into the bottom by bottlenose dolphins (adapted from Rossbach and Herzing, 1997).

The Atlantic spotted dolphin (*Stenella frontalis*) has also been observed foraging for buried prey (Herzing, 1996) in the waters off the Grand Bahama Island. These animals typically align themselves almost vertically with their rostrum several cm off the bottom. Click signals can be heard while the dolphins forage off the bottom. However, the prey is usually only buried several enticetersm in the bottom so that the dolphins do not need to burrow more than about 6–10 cm into the sand.

11.3 Target Discrimination Capabilities

11.3.1 Range Resolution Capabilities

Murchison (1980) conducted a study to determine the capability of an echolocating dolphin to indicate which of two targets was at a closer range. The *Tursiops* was trained to wear rubber eyecups and station in a chin cup that could swivel from side to side and to echolocate two targets separated in azimuth by 40° directly in front of the chin cup. The targets were 7.62 cm polyurethane foam spheres with internal lead weights. A diagram of the dolphin–target geometry is contained in the insert of Fig. 11.23. The dolphin

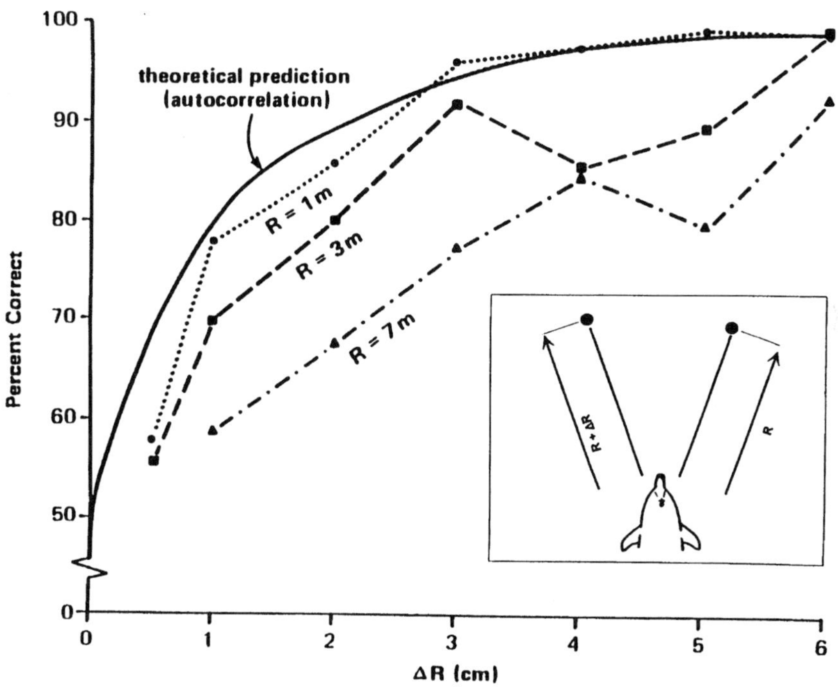

FIGURE 11.23. Target range difference discrimination results of an echolocating *Tursiops truncatus* (adapted from Murchison, 1980).

was trained to station in the chin cup and begin its sonar scan when an acoustic screen was lowered out of the way. Upon completing its sonar scan, the animal backed out of the chin up and responded by touching the paddle on the same side of the center line as the closer target. The dolphins relative range acuity was tested for absolute target ranges of 1, 3, and 7 m. The dolphin's performance results are shown in Fig. 11.23 with percentage correct plotted as a function of ΔR for the different absolute target ranges. Although the results clearly indicated that the ΔR threshold increased with absolute range, the $\Delta R/R$ became progressively smaller with range. The 75% correct response thresholds were at ΔRs of 0.9, 1.5, and 3 cm for absolute target ranges of 1, 3, and 7 m, indicating that the dolphin could resolve range differences of 0.9% at 1 m, 0.50% at 3 m, and 0.43% at 7 m. At the 1-m absolute range, the dolphin's performance approached the theoretical performance of a matched-filter (see Section 9.6 of Au, 1993). These results indicate that the temporal resolution capability of *Tursiops* is extremely fine. The dolphin is able to resolve time differences of 12 μs in 1333 μs, 20 μs in 4,000 μs, and 40 μs in 9,333 μs.

11.3.2 Target Structure Discrimination

11.3.2.1 Thickness and Material Composition of Plates

Evans and Powell (1967) demonstrated that a blindfolded, echolocating *Tursiops* could discriminate between metallic plates of different thickness and material composition. The dolphin was trained to recognize a 30-cm diameter circular copper disc of 0.22 cm thickness from comparison targets of the same diameter but different material or thickness. Both the standard copper disc and a comparison target were presented side by side separated by a center-to-center distance of 50 cm. The dolphin was required to swim toward the targets echolocating along the way and eventually touch the standard target with its rostrum. The performance of two *Tursiops* and a *Lagenorhynchus* reported later by Evans (1973) are shown in Fig. 11.24. All three dolphins performed about the same, being able to differentiate the standard from all the comparison targets at a performance level of 75% or better except for the 0.32-cm Cu and 0.32-cm brass targets. *Tursiops #1* discriminated the 0.32-cm Cu plate whereas *Tursiops #2* discriminated the 0.32-brass plate approximately 75% of the time.

Au and Martin (1988) examined the same metal plates of Evans and Powell (1967) with a simulated dolphin sonar signals. Backscatter results at normal incident and at 14° for the standard copper and three other plates of the same 0.32-cm wall thickness are shown in Fig. 11.25. The echo waveforms for normal incident were similar for all the plates; however, there were obvious differences in the echo waveform at 14°. Such obvious differences also existed between the standard copper and the other plates, including the other copper plates of different thicknesses. Although the intensity of the echoes at 14° was

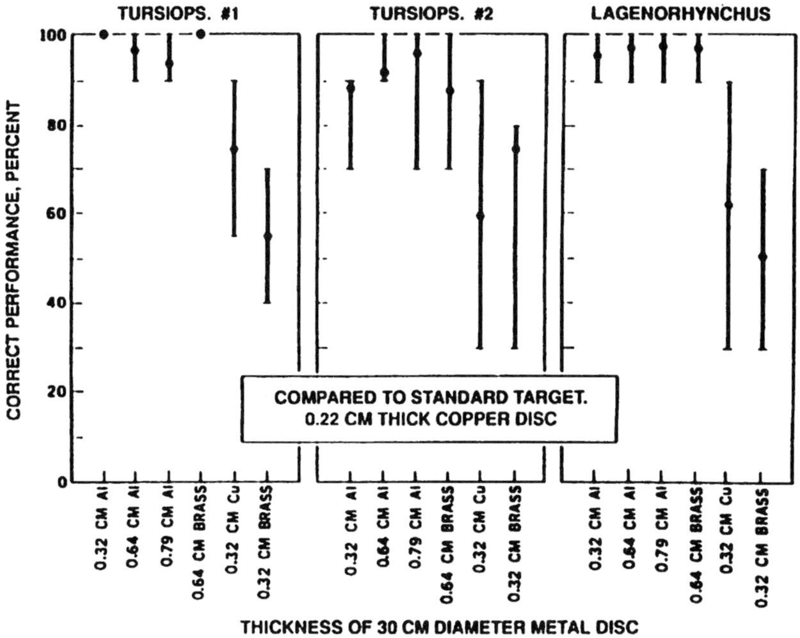

FIGURE 11.24. Comparative performance of two *Tursiops* and one *Lagenorhynchus* in the plate discrimination and recognition experiment, showing the mean and range of performance scores (adapted from Evans, 1973).

about 30 dB below that for normal incident, this reduction of echo strength is well within the large dynamic of a mammalian auditory system.

Two scattering processes were suspected of producing the multiple highlight echoes: "leaky" Lamb waves for the initial highlight and edge reflection of internally trapped waves for the secondary highlight. The time of arrival of the secondary echo components for the trapped waves is a function of the thickness and material composition (velocity of sound in the material) of the plates. This was verified by examining two aluminum plates of the same thickness but different diameters.

11.3.2.2 Structure and Material Composition of Hollow Cylinders

Hammer and Au (1980) performed three experiments to investigate the target recognition and discrimination capability of an echolocating *Tursiops*. The dolphin was trained to recognize the two hollow aluminum cylinders, 3.81 cm and 7.62 cm in diameter, and two coral rock cylinders of the same diameters as the aluminum cylinders. All of the cylinders were 17.8 cm in length. The coral rock targets were constructed of coral pebbles encapsulated in degassed epoxy. The targets were presented 6 m and 16 m from the animal's pen. The dolphin was required to echolocate the target and respond to paddle A if it

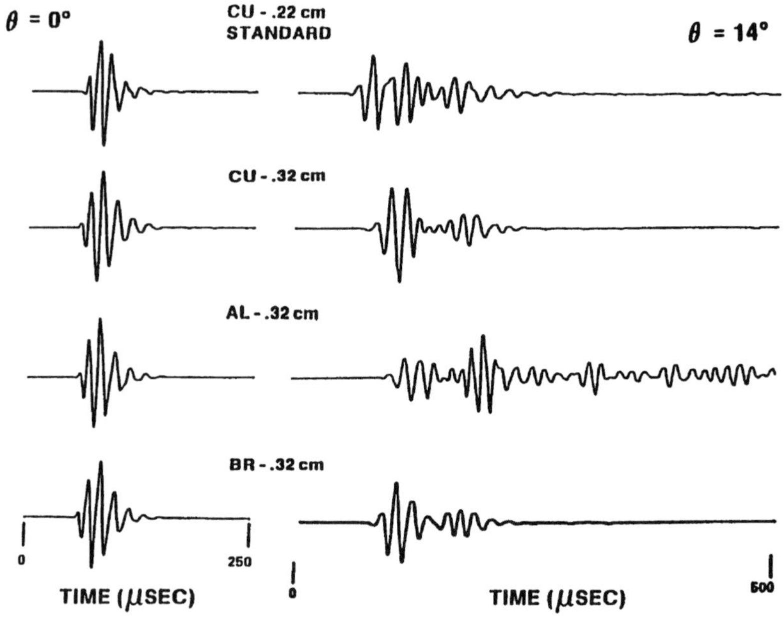

FIGURE 11.25. Echoes from four of the plates used by Evans and Powell (1967) at normal and 14° incident (after Au and Martin, 1988).

was one of the aluminum standards or paddle B if it was one of the coral rock standards. After baseline performance exceeded 95% correct with the standard targets, probe sessions were conducted to investigate the dolphin's ability to discriminate novel targets varying in structure and composition from the standards. All the probe targets were cylinders, 17.8 cm in length. Two probe targets were used in each probe session and only 8 of 64 trials of the session were used for probe trials, four for each probe target. The various probe targets used in all three experiments are described in Table 11.1.

The first experiment involved a general discrimination in which two solid aluminum cylinders with the same diameters as the two standard (hollow) aluminum cylinders, two hollow aluminum of different diameter and wall thickness than the aluminum standards, one coral rock with a larger diameter than either coral rock standards, one PVC tube, and two corprene (cork-neoprene) cylinders were used as probe targets. Initially, the dolphin classified three of the aluminum probes with the standard aluminum targets, but eventually classified all of the probes with all other targets as being "**not A**". In other words, the dolphin became a null detector and searched for the presence or absence of the aluminum standards.

In the second experiment, we investigated the dolphin's ability to discriminate wall thickness. Hollow aluminum probe targets with the same outer diameters but different wall thickness from the aluminum standards were used. The results

TABLE 11.1. Probe targets used in Hammer and Au experiments. Abbreviations for the material compositions are Al – aluminum, CPN – corprene, CRK – coral rock, PVC – poly vinyl chloride

1. General discrimination								
Probe No.	1	2	3	4	5	6	7	8
Composition	Al	CPN	AL	CPN	AL	CRK	AL	PVC
Wall (cm)	0.48	solid	solid	solid	0.64	solid	solid	0.79
O.D. (cm)	6.35	6.35	3.81	4.06	11.43	11.43	7.62	7.62
2. Wall thickness discrimination								
3.81-cm O.D. aluminum				7.62-cm O.D. aluminum				
Probe No.	1	2	3	4	5	6	7	8
Wall (cm)	0.32	0.48	0.79	0.95	0.32	0.69	1.29	7.62
3. Material composition discrimination								
3.81-cm O.D./0.32-cm wall				7.62-cm O.D./0.40-cm wall				
Probe No.	1	2	3	4	5	6		
Composition	bronze	glass	steel	bronze	glass	steel		

showed that the dolphin could reliably discriminate wall thickness differences of 0.16 cm for the 3.81-cm O.D. cylinders and 0.32 cm for the 7.62-cm O.D. cylinder. However, a thickness difference threshold was not determined.

In the third experiment, we investigated the dolphin's ability to discriminate material composition using bronze, glass, and stainless steel probe cylinders, which had the same dimensions as the aluminium standard. The dolphin mistook the small bronze and steel cylinders for an aluminium standard 25% and 13% of the time, respectively. The glass cylinders were mistaken for the aluminum standards 100% and 87.5% of the time for the small and large cylinders, respectively.

All of the targets used were examined acoustically in a test pool using simulated dolphin sonar signals. The echoes from the 3.81-cm O.D. cylinders are shown in Fig. 11.26. From a vis the aluminum standards. The dolphin mistook the small bronze and steel cylinders for an aluminal inspection of the echo structures; we can see that all of the targets have different arrival times for the secondary echo components, and therefore different echo structures. We suggested that the predominant cue used by the dolphin in discriminating the various cylinders was probably time- separation pitch (TSP) generated by the presence of the first and second echo components or highlights. Arrival time differences in the highlights may be perceived as TSP, especially if the echo components are highly correlated. Humans when presented with a correlated pair of sound pulses perceive a pitch that is equal to $1/T$, where T is the separation time between pulses (Small and McClellan, 1963; McClellan and Small, 1965). Signals with ripples in the frequency domain or highlights in the time domain can produce TSP in the human auditory system.

The glass and aluminum cylinders described in Table 11.1 were next used by Schusterman et al. (1980) in a two-alternative forced-choice discrimination experiment involving the same *Tursiops* used in the Hammer and Au (1980)

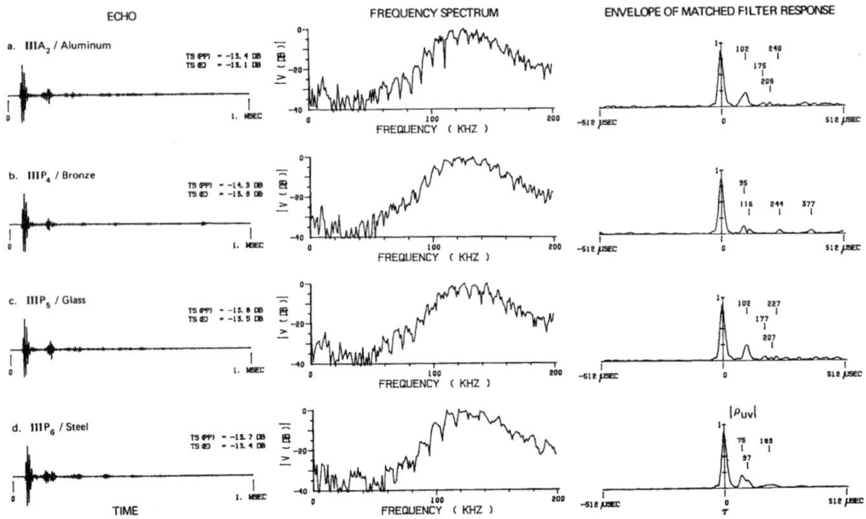

FIGURE 11.26. Echoes from the 3.81-cm O.D. cylinders used in the material composition discrimination experiment (adapted from Hammer and Au, 1980).

study. Targets were located at a range of 6 m from the front of the experimental pen and were submerged in a vertical orientation at a depth of 1 m. A trial consisted of having either a glass or aluminum cylinder presented and the animal was required to perform a sonar scan and respond to paddle A if it associated the target with either of the two aluminum cylinders or paddle B for the glass cylinders. After 30 sessions, the dolphin could perfectly discriminate between the small (3.61-cm O.D.) aluminum and glass cylinders. However, the animal was never able to discriminate between the large (7.62-cm O.D.) aluminum and glass cylinders.

Au and Turl (1991) performed a material composition experiment with cylinders to test if there existed aspect-independent cues that the dolphin could utilize to discriminate between cylinders. An aluminum cylinder with a 7.62-cm O.D., 0.40 cm wall thickness, and a 17.1 cm length was used as the standard target. Two comparison targets, a hollow stainless steel cylinder having the same dimensions as the standard and a coral rock encapsulated in degassed epoxy cylinder, were used. The standard and a comparison target were supported on the rotor bar with their longitudinal axis parallel to the bar and the horizon. The dolphin was required to swim into a hoop, begin echolocating when the trainer presented a tone, and respond to a specific paddle if the target was the aluminum standard and another paddle for the comparison targets.

The dolphin was trained to discriminate between two targets at the baseline aspect of 0 v, 45°, and 90°, where 0° was the broadside aspect (longitudinal axis perpendicular to the animal). After the dolphin performed the discrimination at a level close to 90% correct for the baseline aspects, probe trials at novel aspect angles were introduced and the animal's response was recorded. In the first discrimination task with aluminum and steel cylinders, the dolphin was able to achieve the 90% correct response criterion at the 0° aspect but was unable to reach the criterion for the 45° and 90° aspect. Therefore, probe sessions were not conducted with this set of targets, but additional sessions were conducted to obtain the animal's performance at 10° and 80° aspect angle. The dolphin's discrimination performance as a function of target aspect angle is shown in Fig. 11.27. The results indicate that the dolphin could easily discriminate between the aluminum and steel cylinders at different aspect angles. The worst performance occurred at an aspect angle of 45°, but even at this angle the dolphin's overall performance, which is the average between the performance for the individual targets, was above 80° correct.

An easier discrimination task between the aluminum and coral rock cylinders was next attempted. The 220 trials of baseline conditions per angle were obtained and the dolphin performance was at least 89% correct. Each novel aspect angles were used in a minimum of 25 probe trials except at 60°, where only 10 probe trials were conducted. The animal's performance was almost perfect with the probe trials. These results indicated that for a simple discrimination that there are aspect-independent cues that a dolphin can utilize in discriminating material composition.

Echoes from the three targets obtained with simulated dolphin sonar signals for three aspect angles, 0°, 45°, and 90° are shown in Fig. 11.28. Difference between the echoes can be clearly seen; however, at 45° and 90°, the echo structure was very complex. At the 0° aspect, the highlights occurred at different times for the aluminum and steel cylinders. The echoes for the aluminum cylinder also did not decay as rapidly as for the steel cylinder. At the 45° and 90° aspect angles, the time of occurrence of the different highlights were similar for the aluminum and steel cylinders. Differences in the

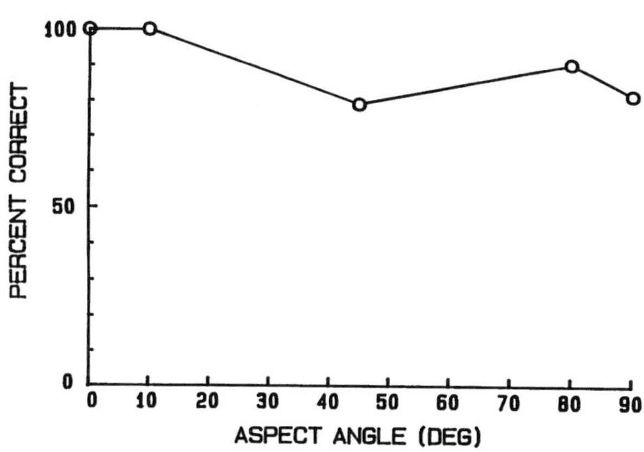

FIGURE 11.27. Dolphin's aluminum vs. steel discrimination performance as a function of the target aspect angle (adapted from Au and Turl, 1991).

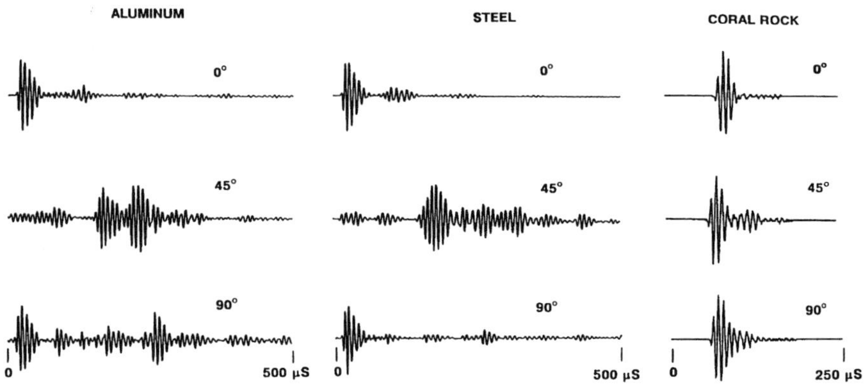

FIGURE 11.28. Echo waveform from the aluminum, steel, and coral rock cylinders at different aspect angles (adapted from Au and Turl, 1991).

amplitude and shape of the highlights can be seen. The difference between the echoes from the aluminum and coral rock cylinders are even more obvious. Secondary highlights tended to attenuate quickly for the coral rock target so that the echo structure from this target was not very complex.

11.3.2.3 Wall Thickness of Cylinders

The capability of an echolocating *Tursiops* to discriminate differences in the wall thickness of hollow steel cylinders was studied by Titov (1972). The outer diameter and length of the cylinders were 50 mm and wall thickness varied between 0.1 and 2 mm. The dolphin was trained to choose the thinner of two cylinders presented simultaneously at a range of 5 m. The dolphin was able to discriminate a wall thickness difference of 0.2 mm at the 75% correct response level.

The predominant cue for the dolphin in the wall thickness experiment of Titov (1972) was probably the difference in the arrival time of the echoes from the front and back walls of the cylinder. The time difference in the arrival of echoes from the front and back walls is

$$\tau = \frac{2\text{th}}{c_l} + \frac{2(\text{O.D.} - 2\text{th})}{c_0}\epsilon \tag{11.4}$$

where c_l is the longitudinal sound velocity of the cylinder material, c_o is the sound velocity in water, and th is the wall thickness of the cylinder. The difference in τ for two cylinders of different wall thickness can be expressed as

$$\Delta t = \left|\frac{2\Delta\text{th}}{c_l} - \frac{4\Delta\text{th}}{c_o}\right|, \tag{11.5}$$

where$\Delta' = \tau_1 - \tau_2$ and $\Delta\text{th} = \text{th}_1 - \text{th}_2$. Inserting the sound velocity for steel and sea water along with a discrimination threshold of 0.2 mm into Eqs. (11.4) and (11.5), we can estimate the dolphin's time difference discrimination threshold as being $)t \approx 5$ μs.

Au and Pawloski (1992) also examined the wall thickness discrimination capability of an echolocating *Tursiops*. Our experimental geometry consisted of a dolphin stationing in a hoop with two targets 8 m from the hoop, separated by 22° azimuth. The standard target was a 3.81-cm O.D. aluminum cylinder with a wall thickness of 6.35 mm. Comparison targets with wall thickness both thinner and thicker than the standard were used. The comparison targets had incremental differences in wall thickness of ± 0.2, ± 0.3, ± 0.4, and ± 0.8 mm from the standard target. The dolphin was required to echolocate and to respond to the paddle that was on the same side of the center line as the standard target. The dolphin's performance as a function of wall thickness difference is shown in Fig. 11.29. The 75% correct response threshold corresponded to a wall thickness difference of –0.23 mm for thinner targets and +0.27 mm for thicker targets. These results compared well with the results of Titov (1972) who used a different paradigm and targets of different sizes and material composition.

The echo waveform, envelope, and frequency spectrum for the standard target and the comparison targets having a wall thickness difference of –0.3 mm and –0.2 mm are displayed in Figs. 11.30 and 11.31, respectively. The dolphin was able to perform above threshold for the targets associated with Fig. 11.30, and was not able to perform above threshold for the targets associated with Fig. 11.31. The envelope curves suggest that if the dolphin used time-domain cues, it may be able to perceive incremental time

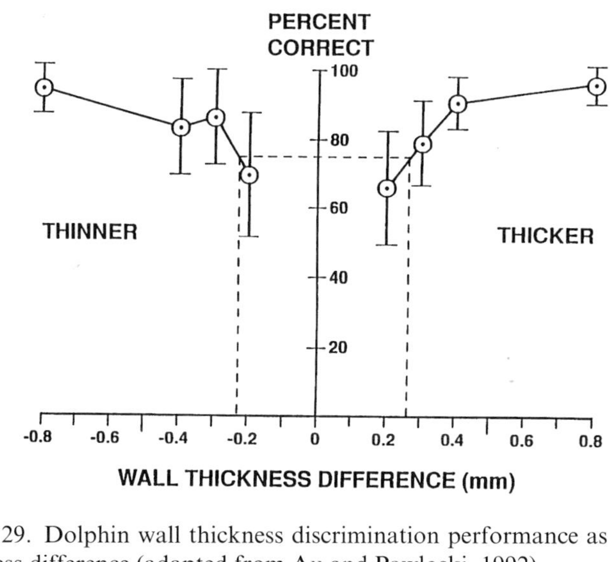

FIGURE 11.29. Dolphin wall thickness discrimination performance as a function of wall thickness difference (adapted from Au and Pawloski, 1992).

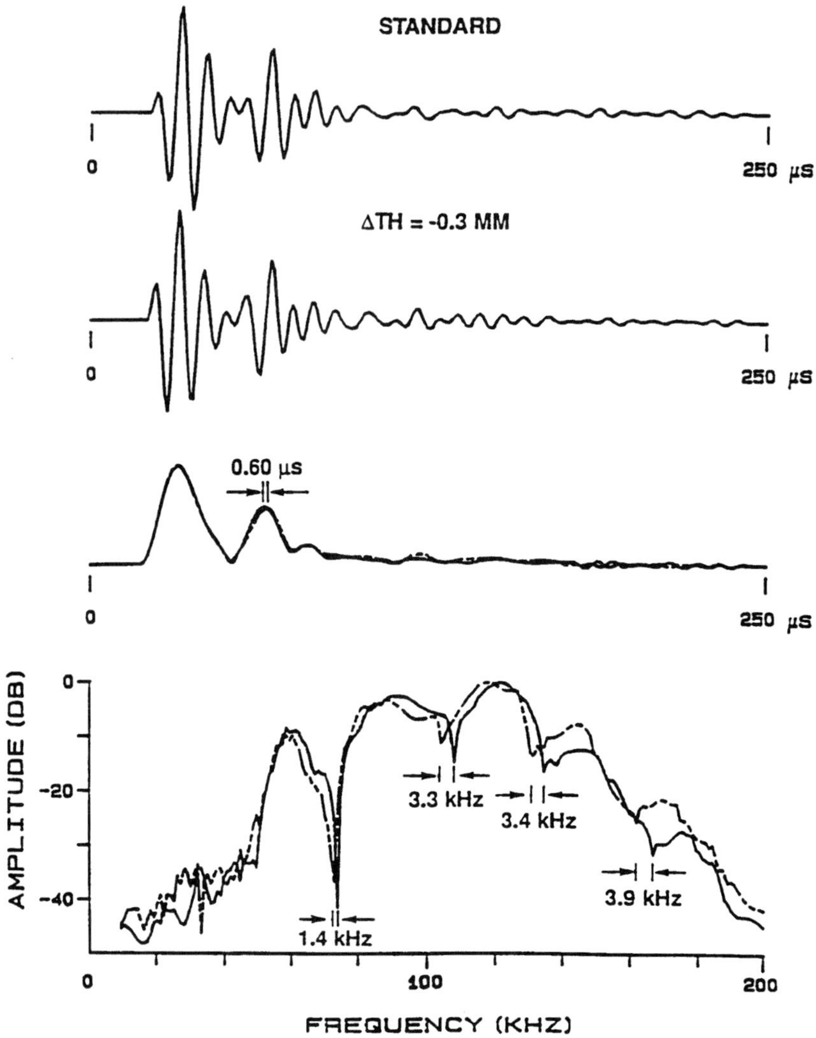

FIGURE 11.30. Echo waveform, waveform envelope, and frequency spectrum for the standard and the comparison target having a wall thickness difference of –0.3 mm. The *dashed envelop* and *spectrum curves* are for the comparison target (adapted from Au and Pawloski, 1992).

differences of approximately 0.5 μs between highlight intervals, agreeing with Δt of approximately 0.5 μs obtained with Titov's (1972) results.

Differences in the frequency spectra of the echoes from the standard and comparison targets can be seen in Fig. 11.31. The frequency spectra for the thinner comparison targets resembled the spectrum of the standard target, but were shifted slightly toward lower frequencies. The spectra for the thicker

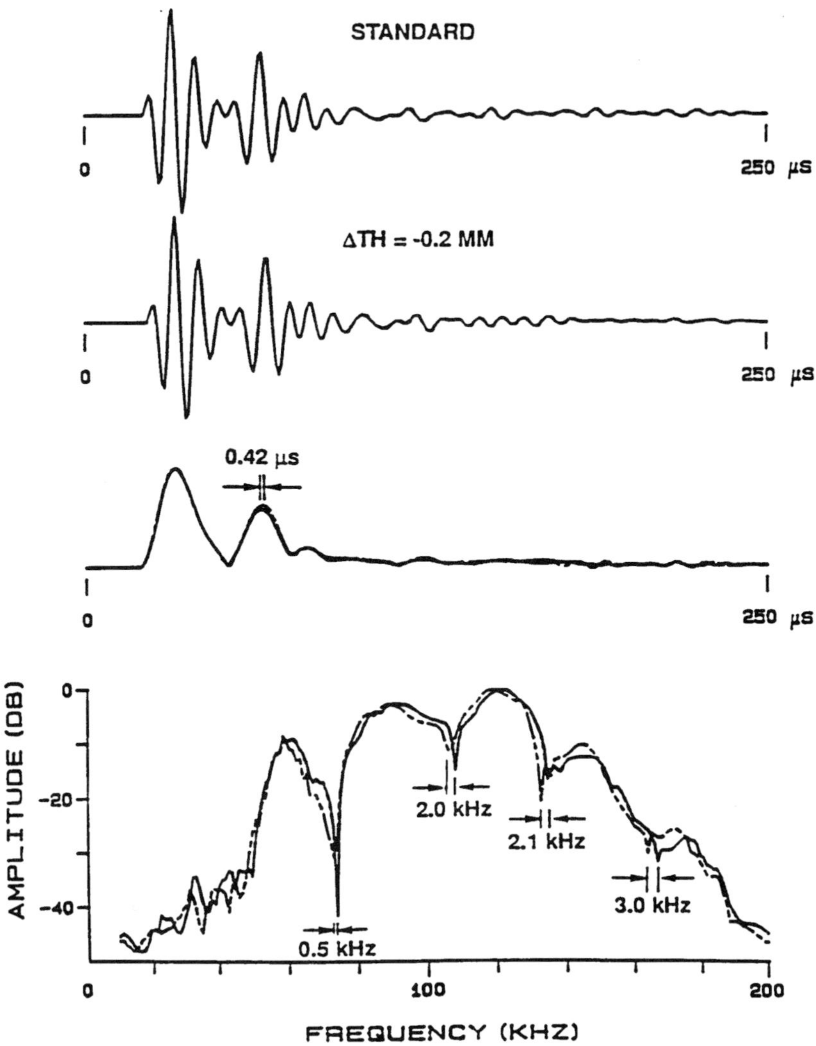

FIGURE 11.31. Echo waveform, waveform envelope, and frequency spectrum for the standard and the comparison target having a wall thickness difference of –0.2 mm. The *dashed envelop* and *spectrum curves* for the comparison target (adapted from Au and Pawloski, 1990).

comparison targets were shifted toward higher frequencies. The average frequency differences were 3.2 and 2.2 kHz for a wall thickness difference of –0.3 mm and –0.2 mm, respectively. If the dolphin used this shift in frequency spectra to discriminate wall thickness difference, then the spectral data suggest that the dolphin could perceive a shift of approximately 3.3 kHz, but not a shift of 2.1 kHz.

Another way that the dolphin may have discriminated the targets is by the use of TSP. The first two highlights in the echoes of Figs. 11.30 and 11.31 should generate a TSP of approximately 28.30 kHz for the standard, 27.94 kHz for the –0.2 mm, and 27.77 kHz for the –0.3 mm comparison targets. If the dolphin used differences in TSP, we can infer that the animal could discern a TSP difference of 530 Hz but not a difference of 360 Hz. Similar results would have been obtained by considering the thicker comparison targets.

11.3.3 Shape Discrimination by Echolocating Dolphins

11.3.3.1 Planar Targets

Barta (1969) trained a blindfolded *Tursiops* to choose between circular, squares, and triangular aluminum plates covered with neoprene. Using a two alternative forced-choice paradigm, the animal reliably discriminated circles from squares and triangles of the same cross-sectional area. Bagdonas et al. (1970) used targets made from ebonite (10 mm thick) and trained a *Delphinus delphis* to discriminate a 100 cm^2 square from a 50 cm^2 triangle.

The dolphins in the experiments of Barta (1969) and Bagdonas et al. (1970) probably received noticeable changes in echo amplitude as they scanned across different shaped targets. Such amplitude fluctuations could have provided the basis of the discrimination. Polar plots of the relative intensity of sound reflected from the different targets used by Barta (1969) are shown in Fig. 11.32. It is clear that the different shaped targets had different angular

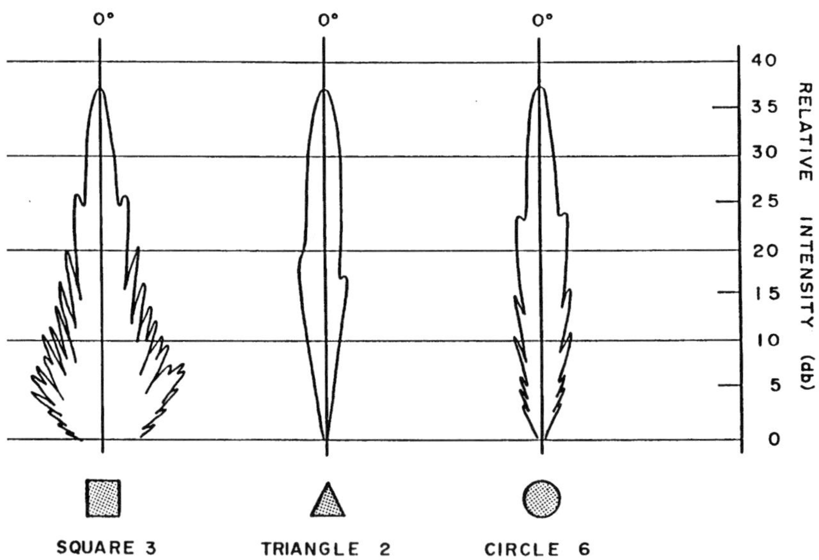

FIGURE 11.32. Polar plots of echo radiation patterns from the targets used by Barta (1969) (adapted from Fish et al., 1976).

variations in the echoes. The 6-dB difference in target strength could have also provided an additional cue in the targets used by Bagdonas et al. (1970).

11.3.3.2 Spheres and Cylinders

Au et al. (1980) conducted an experiment to determine if an echolocating dolphin could discriminate between foam spheres and cylinders located 6 m from a hoop station. Three spheres and five cylinders of varying size, but overlapping target strength, were used so that target strength differences would not be a cue. The dimensions of the spheres and cylinders, along with their corresponding target strengths, are listed in Table 11.2. Two spheres and two cylinders were used in each 64-trial session, with one target present per trial. The dolphin was required to swim into a hoop station and "echolocate on command" (Schusterman et al., 1980). Two response paddles, one associated with spheres and the other with cylinders, were used. The task was not difficult for the dolphin since it could perform the various discriminations at a correct response level of atleast 94%.

Examples of the echo from one of the spheres and cylinders are shown in Fig. 11. 33. A secondary reflection can be seen in the echo from the sphere, and is not present in the echo from the cylinder. A possible cue is the presence of circumferential waves following the specular reflection off the front surface of the targets. A circumferential wave component following the specular circumferential wave component of the echoes would arrive shortly after the specular reflection and would not be separable from the specular reflection. The cylinders had diameters that were much smaller than the diameter of the spheres and any circumferential wave component is not resolvable.

11.3.3.3 Cylinders and Cubes

Nachtigall et al. (1980) trained a blindfolded echolocating *Tursiops* to discriminate between foam cylinders and cubes. The animal was trained to

TABLE 11.2. Dimensions of the spheres and cylinders and their corresponding target strength used in the shape discrimination of Au et al. (1980)

Target	Diameter	Length	Target strength
Spheres			
S_1	10.2 cm	–	–32.1 dB
S_2	11.7 cm	–	–31.2 dB
S_3	15.2 cm	–	–28.7 dB
Cylinders			
C_1	1.9 cm	4.9 cm	–31.4 dB
C_2	2.5 cm	3.8 cm	–32.3 dB
C_3	2.5 cm	5.1 cm	–28.7 dB
C_4	3.8 cm	3.8 cm	–30.1 dB
C_5	3.8 cm	5.1 cm	–27.6 dB

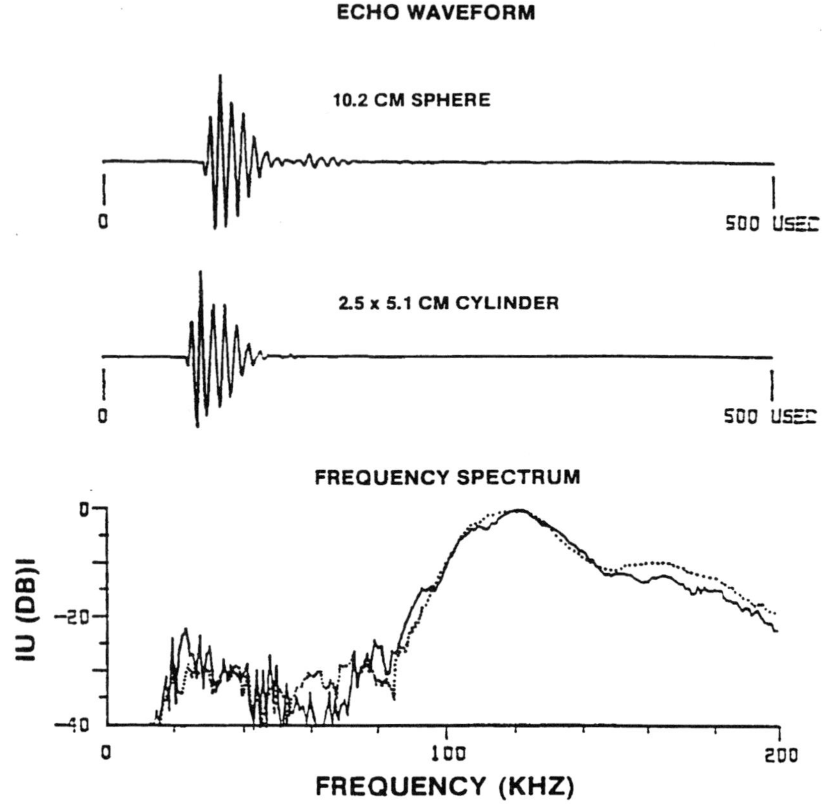

FIGURE 11.33. Results of backscatter measurements of a foam sphere and a foam cylinder used in the sphere–cylinder discrimination experiment of Au et al., (1980). The *dotted* frequency spectrum is for the cylinder echo.

station within a circular tank and echolocate down a water-filled trough. Two targets (a cylinder and a cube) were presented simultaneously 2 m from the entrance to the trough, and the animal was required to touch the response rod on the same side as the cylinder. Targets were placed between four 30 lb test monofilament lines attached, with stretchable elastic cord, to a wood frame. Three different sized cylinders were repeatedly paired with each of three different sized cubes. Once the ability of the animal to discriminate cylinders from cubes was well established with the targets in an upright position, a probe technique was used to examine the effects of changing the target aspect. Baseline performance trials were conducted on 56 of the 63 trials per session, but on the other seven trials one of the targets was presented either on its side or with the flat face toward the dolphin. The results indicated that the dolphin could correctly discriminate the upright cylinders and cubes in 91% of the trials, and that two of the probe orientations did not affect the animal's ability to discriminate the targets. However, the results also show the animal could not discriminate the targets when the probes were in the "flat-face forward" orientation.

Backscatter measurements indicated high variability in the echo amplitudes for the cubes and the cylinders in the flat-face forward orientation and low variability for the cylinder in the upright position. The standard deviations of 15 target strength measurements on the three cubes with the flat faces forward were 2.7, 5.9, and 6.2 dB compared with 1.2, 1.0, and 1.0 dB for the cubes standing upright. The targets were removed and placed back on the holder after each measurement to simulate the experimental situation. Measurements from cylinders with a flat-face forward yielded similar results as the cubes in a flat-face forward orientation. Standard deviations of 2.6, 4.9, and 5.9 dB were obtained. These differences in variability paralleled the performance of the dolphin. The animal most likely received echoes varying in amplitude when scanning across the flat surfaces of the cubes or the tops of the cylinders and received relatively uniform amplitude echoes when scanning across the curved portion of the cylinders.

11.3.4 Multidimensional Discrimination

Roitblat et al. (1990) performed a matching-to-sample experiment with an echolocating *Tursiops*, in which the dolphin was required to match targets from a sample array with targets from three comparison arrays, each array having similar targets. A target set consisted of a large hollow PVC tube (10-cm O.D., 25-cm length, 0.3-cm wall), a small hollow PVC tube (7.5-cm O.D., 15-cm length, 0.3-cm wall), a solid aluminum cone (10-cm base, 10-cm length), and a 5-cm water-filled sphere. The sample array was located 4.8 m directly ahead of the animal's station and the comparison arrays were suspended from a bar located 3.9 m from the observing aperture and spaced apart from each other. The blindfolded dolphin first echolocated a sample target while in a hoop station. Then after backing out of the station, one target from each comparison array was then lowered into the water; one of the targets matched the sample and two were different. The dolphin re-entered the station and echolocated the targets of the comparison arrays and responded by touching one of three response paddles to indicate the location of the matching target.

The dolphin's choice accuracy during the final 48 sessions of the experiment averaged 94.5% correct. Roitblat et al. (1990) were able to apply a sequential sample model using Bayesian decision rules and obtained results that corresponded well with the dolphin's performance in the matching-to-sample task. Differences in the target echoes using simulated dolphin signals were obvious. Easily discriminable targets were deliberately used to demonstrate that a dolphin could grasp the concepts of same–difference and matching-to-sample in a sonar modality (Nachtigall and Patterson, 1981). Another reason for the experiment was to study the dolphin's decision-making process.

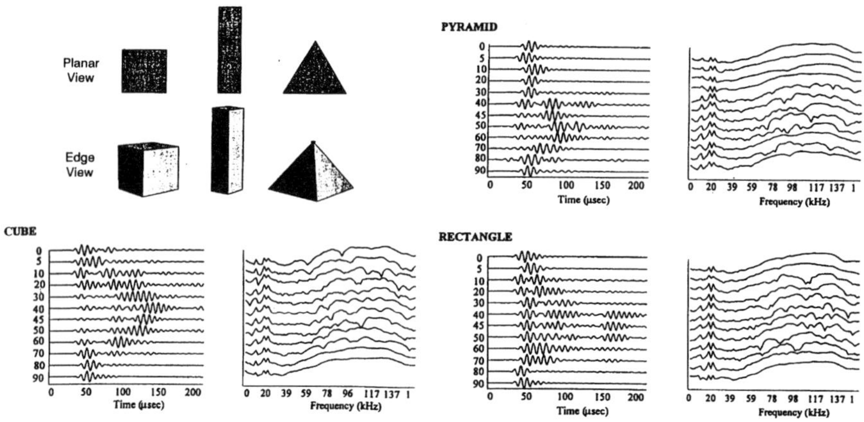

FIGURE 11.34. Targets and echoes of the targets used by Helweg et al. (1996).

Helweg et al. (1996) examined the capabilities of an echolocating dolphin to develop aspect-independent representation of acoustically nonsymmetrical targets from echolocation. Echoes from the foam targets measured in a test pool at aspects of 0–90° in 10° increments are shown in Fig. 11.34, where 0° corresponded to the planar view. The experimental setup of Roitblat et al. (1990) was used and a similar matching-to-sample paradigm was incorporated to test the dolphin. The orientation of the sample and the comparison targets were not controlled in the experiment so that the targets were free to revolve in a random fashion from trial to trial. The dolphin's correct response level for the different targets were 81% for the rectangle, 79% for the pyramid, and 17% for the cube. However, when the sample target was the cube, the animal had a strong bias toward responding rectangle and this occurred 63% of the time, suggesting that the dolphin could recognize the cube well above change but had a biased response. The results of study support the speculation that dolphins identify objects by averaging, or integrating, the information gleamed from successive echoes within an echolocation scan.

Harley et al. (1996) examined the dolphin's ability to integrate visual and echolocation information. A variety of different types of targets were used in a matching to sample paradigm to test the capability to recognize targets using echolocation only, vision only, echolocation with vision, and cross-modally between echolocation and vision. They found that the dolphin performed best when it was allowed to use both vision and echolocation. They also found that with familiar targets, the dolphin could perform cross-modal matching, suggesting that the dolphin has an object-based representation system. However, the dolphin performed considerably better when the task was to echolocate the sample and visually choose the matching target (echoic-vision), then vice-versa. For familiar targets, the dolphin's performance on one set of targets was 78% correct for the echoic-vision and 22% for the vision-echoic condition. For another set of familiar targets, correct performance was 99% for echoic-vision and 61% for vision-echoic, and for a third set of targets the correct performance was 67% for echoic-vision compared to 39% for vision-echoic. When novel targets were used, the dolphin performance was 39% correct for both echoic-vision and vision-echoic. Chance performance was 33%.

Investigation of cross-modal identification of targets was also performed by Pack and Herman (1995). They used target made out of short sections of gray PVC pipes having outer diameters of 1.27, 1.91, 2.54, and 3.81 cm. The short sections of pipes were filled with sand and glued together so that a variety of different target shapes were achieved. The targets were small enough to be handled rather easily by a person holding them with two hands. The dolphin was trained to do a matching to sample task visually, by echolocating, and cross-modally. In the experiment, the dolphin would be presented with a sample target, followed by two alternative targets, one of which was the same as the sample. Target presented underwater were dipped into a box 1.1-m wide by 0.6-m deep by 1.0-m high, with the front of the box facing the center of the pool consisting of 0.32-cm thick black plexiglass. After several years of training, the dolphin could do a visual-to-echoic and an echoic-to-visual matching to sample correctly on the first trial for many novel pairs of objects. Pack and Herman concluded that shape information registers directly in dolphin perception through vision or through echolocation. Harley et al. (1996) vision-echoic cross-modal results with familiar targets and both vision-echoic and echoic-vision results with novel targets were not consistent with those of Pack and Herman. Unfortunately, the study of Pack and Herman did not include any acoustic measurements. The dolphin's echolocation signals were not measured. The echoes from the various objects were not measured. The effects of the surrounding "target" boxes were not measured. Although they claimed that the box were anechoic, they provided no data to support that claim. Therefore, it is extremely difficult to evaluate the experiment of Pack and Herman. Target recognition may have been formed by some other cues than what the investigators have suggested. Certain features of the target echoes may have allowed the dolphin to make accurate cross modal choices that may not have anything to do with the shape of the targets. Therefore, until acoustic data of the targets are collected, this experiment cannot be properly evaluated.

11.4 Sperm Whale Echolocation

We will begin a discussion of sperm whale acoustics by first considering the signal generation mechanism. The geometry and physical properties of the forehead of a sperm whale (which makes up about 1/3 of the body length and weight) indicate that the head is a very unusual structure and it would be

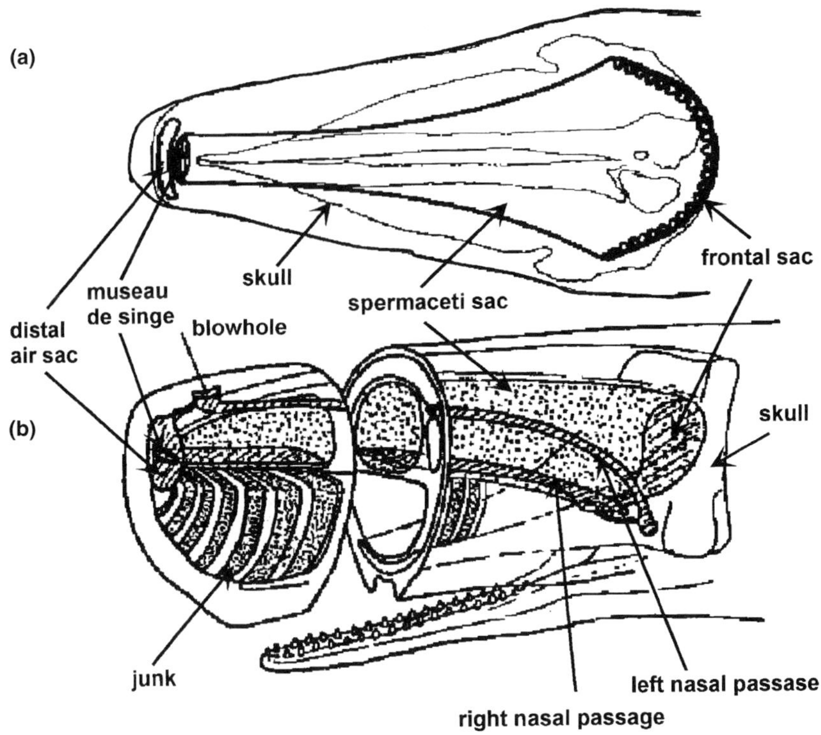

FIGURE 11.35. Anatomy of a sperm whale forehead: (a) a planar cross-sectional dorsal–ventral view (after Mohl and Amundin, 1991), (b) a three-dimension anterior-lateral cut-away view (after Clarke, 1979, copyrighted (c) 1979 by Scientific American Inc., redrawn by K. Benoit-Bird).

interesting to discuss a theory of how sounds are produced and propagate through the head of a sperm whale. A schematic of a planar cross-sectional view of the internal structure of a sperm whale forehead from a dorsal–ventral perspective, along with a three-dimensional anterior-lateral cut-away view are shown in Fig. 11.35. Norris and Harvey (1974), after a detailed anatomical study of the structure of a sperm whale head, proposed an elegant theory of sound production. The important structures involved with sound production according to Norris and Harvey (1974) are the distal and parabolic shaped frontal air sacs, the spermaceti organ located between the two air sacs, and the lip-shaped structure of connective tissue termed the "museau du singe" at the anterior end of the forehead. The museau de singe is a tough fibrous structure, which forms two tightly occluding lips at the anterior end of the right nasal passage. The Spermaceti organ is an elongated bag containing a complex mass of oil-filled connective tissue enclosed by a membrane case, which is surrounded by muscle and blubber. The liquid wax or oil within the spermaceti organ has a relatively high sound velocity of about 2.6 km/s.

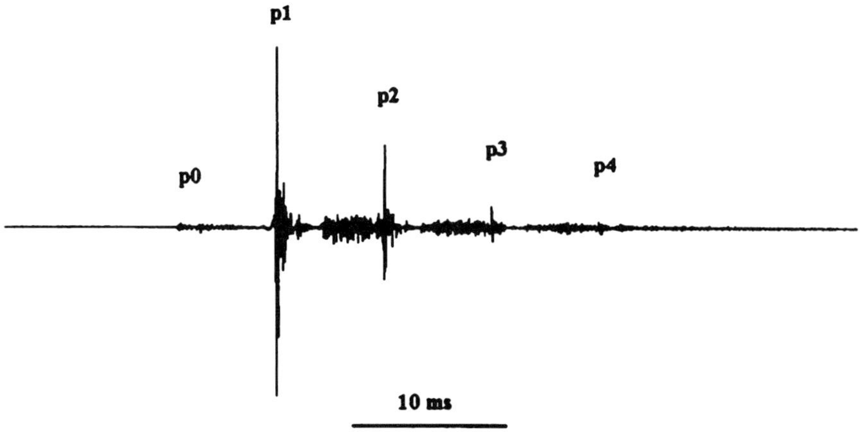

FIGURE 11.36. A classical multi-pulse structure of a sperm whale click with pulses labeled from p0 to p4. Any surface reflections were suppressed by editing (adapted from Mϕhl et al., 2003).

Norris and Harvey (1974) proposed that a single pulse is produced by the whale forcing a burst of air through tightly stretched lips of the museau du singe at the anterior of the forehead. This pulse is emitted into the water and also propagates through the spermaceti organ and is reflected and focused by the parabolic-shaped air sac at the posterior of the skull. Therefore, the acoustic signals of sperm whales consist of a number of clicks that are progressively attenuated as the original pulse reverberates between the two air sacs on both ends of the spermaceti organ, producing a signal as shown in Fig. 11.36. The theory of Norris and Harvey (1974) was no doubt inspired one of the studies to describe the structure of the sperm whale acoustic signals by Backus and Schevill (1966). They reported that sperm whale clicks actually consistes of a series of pulses that are separated by 2–4 ms. However, the geometry of the recording was never described and some of their data may actually contain surface-reflected components. The time interval between the successively decaying clicks is directly related to the distance between the two air sacs and can be used to estimate the size of a whale (Mϕhl et al., 1981; Gordon, 1991; Goold, 1996). Later recordings of sperm whale clicks by Mϕhl and Amundin (1991) and Goold and Jones (1995) also contained multiple pulses per click; however, the numbers of pulses and the amplitudes of the pulses decayed much faster with time than those of Backus and Schevill (1966). Mϕhl and Amundin (1991) also observed the p0 component shown in Fig. 11.36, a component never before observed. They argued that the museau du singe actually projected the initial click toward the back of the skull and p0 may be a part of the initial inward directed click after propagating around the air sac at the anterior end of the forehead.

Prior to the late 1990 s, the most prevalent understanding of sperm whale signals is that the clicks were broadband with peak frequencies between 4 and 8 kHz (Watkins, 1977; Levenson, 1974; Weilgart and Whitehead, 1988; Mϕhl and Amundin, 1991). There were only two reports on source levels. Dunn (1969) using sonobouys measured 148 sperm whale clicks from a solitary sperm whale and estimated an average peak-to-peak source level of 183 dB re 1 μPa. Levenson (1974) estimated peak-to-peak source level of 180 dB re 1 μPa. The clicks were also thought to be essentially nondirectional and projected in codas (Watkins, 1980). Therefore, the notion of sperm whales echolocating was somewhat questionable. Part of the problem was the lack of measurements of sperm whale signaling in conjunction with foraging and the fact that click signals can have more than one function, such as communication and echolocation. In a review paper, Watkins (1980) spelled out his rational for not supporting the notion of a sonar function for sperm whale clicks, "Other features of their sounds, however, do not so easily fit echolocation." Watkin's rational included his observations that clicks do not appear to be very directional, the interclick interval does not vary as if a prey or obstacle is being approached, solitary sperm whales are silent for long periods, the level of their clicks appears to be generally greater than that required for echolocating prey or obstacles, and the individual clicks are usually too long for good range resolution. It is important to state that most of Watkins measurements were conducted at low temperate zone latitudes where females and calves are found.

Sperm whale foraging for prey typically make nearly vertical dives swimming approximately 100 m/min and remain submerged up to 90 min (Goold and Jones, 1995). Watkins et al. (1993) used an acoustic transponder, which they were able to imbed into two sperm whales, and tracked the whales to depth of 400–600 m and more, including a dive to 1,185 m and one possibly up to 2,000 m. The whales studied by Goold and Jones (1995) off the Azores typically began clicking about 2–3 min after fluking up to initiate a dive. While in their dive, they produce clicks continuously with repetition rates between 0.5 and 2 clicks per second. At certain intervals during a dive, the repetition rate may increase up to 200 clicks per second and these rapid bursts are commonly referred to as "creaks." Goold and Jones (1995) found that the average click rate for bull males was about 1.2 clicks per second and 2.0 clicks per second for females. They suggested that this difference may be attributed to the larger males ranging at longer distances than the smaller female whales. The steady repetition rate of the clicks is a peculiar feature of sperm whale clicking. Except for the occasional creaks, the steady repetition rate, even as a whale closes in on a prey, seems to suggest either a completely different echolocation strategy than the smaller odontocetes or that the animals do not really echolocate as argued by Watkins (1980).

The clicks emitted by sperm whales in their foraging dives may also convey information to other whales. For example, a foraging mother would dive and remain submerged for many tens of minutes while her calves would remain on the surface but continuing to swim in the general direction of the mother. The mother may surface many kilometers from the fluke-up location, yet the calves will continue on and eventually meet up with her. There is a strong possibility that the calves are listening in on their mothers' clicks, being able to discriminate the appropriate click train from the click trains produced by other sperm whales and are able to rendezvous at the appropriate area.

The issue concerning sperm whale echolocation began to be more fully understood with the data obtained in ground breaking work by Bertel Mϕhl and his students from the University of Arhus, along with other Danish colleagues. They began to perform large array aperture measurements of large bull sperm whales foraging along the slope of the continental shelf off Andenes, Norway, beginning in the summer of 1977 (Mϕhl et al., 2000). Up to seven multi-platforms spaced on the order of 1-km apart were used in their study, with hydrophones placed at depths varying from 5 to 327 m (Mϕhl et al., 2003). They also came up with a unique but logistically simple scheme of obtaining GPS information to localize the position of each platform. Each platform continuously logged its position and time stamps on one track of a DAT recorder, the other track being used for measuring sperm whale clicks (Mϕhl et al., 2001). The GPS signals were converted to an analog signal by frequency shift keying (FSK) modulation. In this way, each platform can be operated essentially autonomously, and yet its location and time stamps could be related to all the other platforms.

An important finding of Mϕhl and his colleagues is the monopulsed nature of on-axis clicks emitted by the sperm whales rather than the classic multi-pulsed signal shown in Fig. 11.37. An example of an on-axis click is depicted in Fig. 11.38, showing mainly the p1 pulse. It is clear from the figure that the p1 dominates the other components. This pulse has a theoretical time resolution of about 71 μs corresponding to a range resolution of about 11 cm. The product of the time resolution and centralized rms bandwidth was calculated to be 0.29, numerically similar to what Au (1993) found for

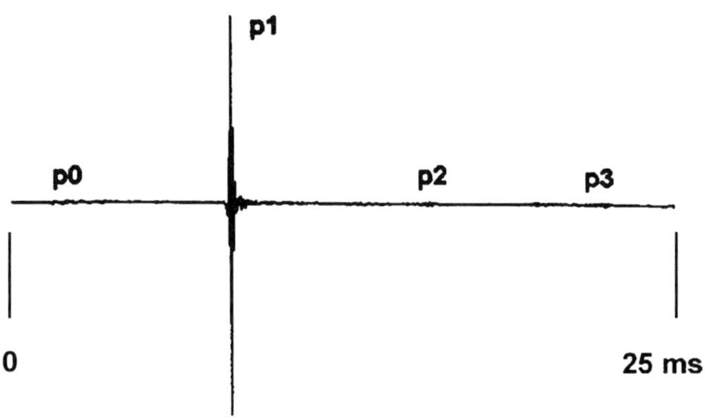

FIGURE 11.37. Time series of an on-axis click (adapted from Mϕhl et al., 2003).

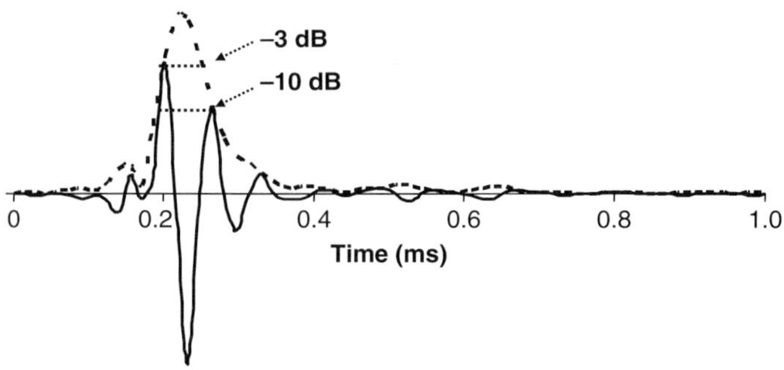

FIGURE 11.38. Waveform and envelop of an on-axis p1 pulse (courtesy of Bertel Mϕhl). The intervals between the –3 dB and –10 dB points of the envelope are also shown.

Tursiops. Ninety nine percent of the energy in an on-axis click is found in the p1 component.

Another important finding of Mϕhl et al. (2000) is the high source level that they recorded with their array, as high as 223 dB re 1 μPa per RMS. The per RMS measure is the level of a continuous sine wave having a peak-to-peak amplitude as the click. This measure is clearly an overestimate of the rms value of a sperm whale click. Nevertheless, the 223 dB reported by Mϕhl et al. (2000) can easily be given a peak-to-peak value of 232 dB (adding 9 dB to the per RMS value). In a follow-on study, Mϕhl et al. (2003) measured clicks with rms source levels as high as 236 dB re 1 μPa using the expression given in Eq. (1.44) that is repeated here

$$p_{\mathrm{rms}} = \sqrt{\frac{1}{T}\int_0^T p^2(t)\,\mathrm{d}t}. \tag{11.6}$$

For this measure, they used a time interval corresponding to the 3-dB down points of the signal waveform similar to the one shown in Fig. 11.38. For this waveform, an rms source level of 236 dB corresponds to a peak-to-peak source level of 243 dB re 1 μPa.

In our view, reporting an rms sound pressure level is problematic because the duration of the signal needs to be estimated. Using the interval between 3-dB point is an extremely conservative approach, since from Fig. 11.38 one can see that the signal is about four to five times longer than the time interval between the –3 dB points. In the waveform of Fig. 11.37, the actual duration of the signal is about four to five times longer than the duration corresponding to the –3 dB points. If the rms values is computed from time 0.11 to 0.39 ms, the rms value will be 228 dB re 1 μPa.

The third important finding has to do with the high directionality of the sperm whale emission beam. Mϕhl et al. (2003) used simultaneous data from

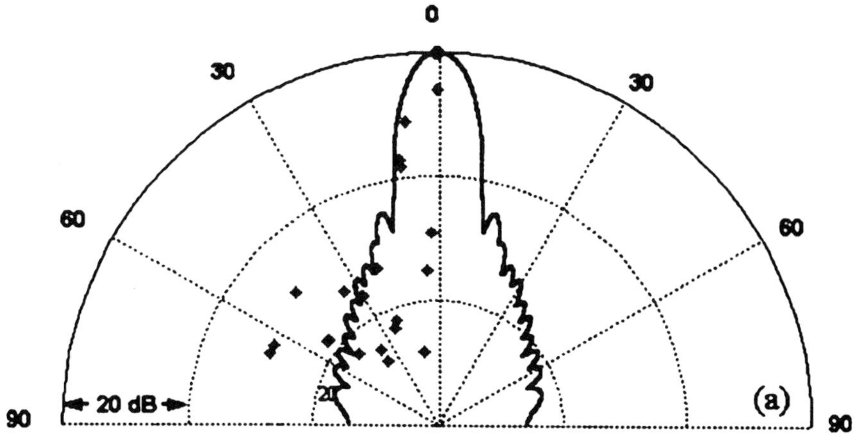

FIGURE 11.39. Composite directional beam pattern based on six clicks with on-axis properties. The *symbols* are plotted as the difference in level and angle from the on-axis recording to recording from other hydrophones for each click. The *thick line* is the theoretical radiation pattern of a circular disc having a diameter of 80 cm.

presumable seven hydrophones to come up with the composite directional pattern based on six clicks with on-axis properties shown in Fig. 11.39. The directivity pattern modeled as a circular disc with a diameter of 80 cm. Such a circular disk would have a 3-dB beamwidth of about 8°, well in line with the 10° for *Tursiops* (Au, 1993), and a directivity index of 27 dB.

Another important contribution from the Arhus team came from the acoustic tag results of Madsen et al. (2002). Acoustic recording tags were placed on four sperm whales from which data collected that previously did not exist. Reporting on the whale that had a tag attached during its whole dive, the following information were found: coda clicks were emitted only during the initial portion of the dive and ceased after a depth of 265 m, after which only usual clicks were used the center frequency of the usual clicks varied almost linearly with the strength of the signal and center frequencies as high as 26 kHz were detected. Madsen et al. (2002) proposed that a bimodal generation of clicks, one mode for coda clicks and another for usual clicks. They further suggested that the mechanism for generating coda clicks would lead to a less directional beam than for usual clicks.

The results obtained by Mϕhl et al. (2000, 2003) and Madsen et al. (2002) seem to answer almost all the objections enumerated by Watkins (1980) to sperm whale echolocation. Whether or not sperm whales echolocate is a question that cannot be answered definitely without a tightly controlled behavioral experiment. One can certainly infer echolocation by the properties of the signals used by sperm whales while they are foraging. Furthermore, we can utilize the sonar equation and hope to find some supportive evidence that would bring light to this issue, as was done by Goold and Jones (1995).

However, unlike Goold and Jones (1995), we will use the transient form of the sonar equation since the clicks produced by sperm whales are relatively brief and also broadband. Considering the noise-limited form of the sonar equation, we have

$$DT = SE - 2\ TL + TS_E - (NL - DI).$$

The unknown parameters are the detection threshold, DT, the source energy flux density, SE, the target strength of a prey, and the receiving directivity index, DI. Sperm whales primarily feed on deep sea squid.

11.5 Pinniped Echolocation

The question whether pinnipeds echolocate or not was a very controversial issue in the 1960 s. Thomas Poulter at the Stanford Research Institute published two art cles (Poulter, 1963a,b) providing evidence that the California seal lion, *Zalophus Californianus*, not only has the ability to echolocate but can do so extremely well. In experiments conducted in the dark, *Zalophus* was able to retrieve dead fish but would turn away from horse meat of similar sizes. All the while, the animals emitted clicks that were assumed to be sonar clicks. In the previous chapter, we discussed some of the properties of clicks produced by *Zalophus*. Clicks are produced in trains typically lasting about 2 s, although many trains had durations in the tens of seconds. Click repetition rate varies from less than 5 pps to 70–80 pps. When retrieving fish in the dark, Poulter (1963a,b) reported that the sea lion emitted clicks that resembled damped sinusoidal pulses, with the first cycle having the greatest amplitude and the subsequent cycles having a progressively lower amplitude. The average click interval was about 34 ms with a minimum pulse rate of 15 pps and a maximum of 80 pps. The duration of the clicks were typically between 3 and 5 ms, with a peak frequency between 3 and 13 kHz. Poulter also reported that both the click intervals and the amplitude decreased as the sea lion approached a fish, an acoustic behavior that resembles dolphins echolocating and swimming toward an object. Poulter and Jennings (1969) later used a tank lined with wooden wedges to provide an anechoic environment to study echolocation in sea lions. In an experiment performed in the dark, dead fish were lowered into the tank using a piece of fishing line threaded through the tail. The sea lion was required to remain at a specific spot in the tank and begin its search for the fish upon receiving a command. The fish line was held by hand so that the experimenter could detect any contact and know when the fish was taken. When a very small fish was suspended into the water, the clicks appeared to have two distinct peak frequencies, a lower peak frequency between 0.5 and 1 kHz and an upper frequency between 2.5 and 3.5 kHz. When no fish were being introduced into the tank, the sea lion produced much louder clicks at a nearly uniform rate. These clicks had a broader frequency range. The sea lion did not have any problems finding the fish and often swam directly toward it.

The fish was randomly placed at different locations in the tank, even extremely close to the walls. In the final sentence of their article, Poulter and Jennings (1969) stated, "the studies have demonstrated that the sonar discrimination ability of the California sea lion is nothing short of phenomenal."

Although Poulter's arguments for sea lion echolocation seemed very persuasive, other investigators performing other types of controlled behavioral experiments could not demonstrate echolocation by seal lions. Evans and Haugen (1963) conducted an experiment in which a sea lion was trained to retrieve 16.5-cm diameter rings made of 1-cm air-filled plastic tubing with a weight of lead shots inside the tubing. The weight located at one point of the ring would cause it to sink, but remain in an upright orientation on the bottom. The sea lion was trained to retrieve the rings while her retrieval time and search pattern were observed and any vocalization monitored acoustically. A total of 25 trials were run in dark conditions (0.03–0.95 foot candle) in which 18 trials were successful retrieval trials. The retrieval time during the "lights on" condition had an average of 8 s and during the dark condition the average time increased to 15 s. During many of the darkened trials in which the animal successfully retrieved the rings, the animal did not appear to emit any sounds. The search pattern during all test conditions consisted of a series of continuous loops, however the pattern was more complex during the completely dark condition. Finally, the sea lion was trained to wear a blindfold. Evans and Haugen (1963) found that when blindfolded the sea lion's ability to find fish introduced into the water was seriously impeded and the animal had difficulties in navigating and collided into the tank wall during the first blindfolded trial. Evans and Haugen (1963) concluded that their results did not support the theory that the California sea lion has a well-developed echolocation ability.

Schusterman (1967) performed three experiments to test the echolocation capabilities of *Zalophus*. In the first experiment, two sea lions were trained to visually discriminate two circular discs, one having an area of 736 cm^2 and the other 16 cm^2 that were presented side by side with a barrier between the disc as shown in Fig. 11.40. In a given trial, the sea lion task was to swim toward, and touch, one of the specific disc. One sea lion was trained to touch the larger disc while the other was trained to touch the smaller disc. The barrier could be extended from 1.5 to 2 cm from the plane of the two discs, forcing the sea lion to make its decision at a distance no closer than the extent in which the barrier protruded from the plane of the two discs. The right–left position of the discs was randomized from trial to trial. After the sea lion could perform this task perfectly in clear water, one of the sea lion was tested in water made turbid by the introduction of a harmless vegetable dye, attenuating the visibility to a distance of approximately 2.8 m (the maximum distance at which a diver could perceive the largest target). The sea lions performance dropped to 60% in the turbid water, clearly demonstrating an inability to perform the task and a lack of any sonar capability.

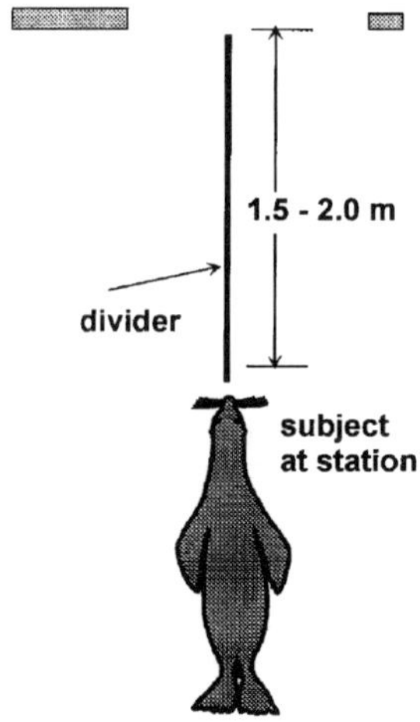

FIGURE 11.40. A schematic of Schusterman (1967) experiment to test the echolocation capability of a sea lion.

Schusterman (1967) considered the possibility that the sea lion could not perform the previous task because of having to make its decision at a distance between 1.5 and 2 m from the targets, and that the sea lion sonar may be operational only at very close distances from a target. Therefore, in the second experiment, two plexiglass disks of the same diameter and thickness were used in a discrimination experiment. One of the disc had an air space 0.32-cm thick sandwich between two other plexiglass discs of 0.32-cm thickness to give a total thickness of 0.96 cm. The other disc had the same radius but was constructed of a solid piece of plexiglass, 0.96-cm thick. The motivation for using such targets was based on the fact that the plexiglass with the air space would have a stronger acoustic reflection than the solid plexiglass disc. Both discs were painted black and appeared to be visually identical. The sea lion was first trained with only the air-filled disc present in the water and the sea lion was also trained to emit clicks and swim toward and touch the air-filled disc. Eventually, test trials were performed in which both discs were presented to the sea lion and its task was to swim and touch the air-filled disc. The animal's performance remained at a chance level throughout the course of 2,050 test trials. This performance occurred even with the animal emitting clicks on every trial. Clicking would continue as the animal approached to within 15–61 cm of a target. Again, in this second experiment, the sea lion failed to demonstrate an echolocation capability.

In the third test, a sea lion was allowed to view a dead herring being thrown into the test tank filled with clear water and was immediately permitted to enter the pool. Three different animals were used and, in ten trials per animal, each animal swam directly toward the herring and ate it. Then the visibility of the water was reduced to a distance of approximately

10–20 cm and each sea lion was tested for 10 additional trials. At the end of the 10 additional trials, the tank was drained and 6–9 untouched fish were recovered. Underwater acoustic monitoring during both clear and turbid water testing indicated a complete absence of clicking sounds. Schusterman (1967) then used live fish. Thirty live fish about 13–15 cm in length were thrown in a tank of clear water while a sea lion watched the process. It was then allowed to enter the pool and it caught and ate 18 fish during the first 15 s, 9 during the next 15 s, and 3 during the last 6 s, thus consuming 30 live fish in 36 s. The procedure was repeated in turbid water and the sea lion was reintroduced and remained in the pool for a period of 135 s. At the end of this time period, the pool was drained and 17 fish were found, indicating that the sea lion caught and ate only 13 of the 30 live fish during a 135-s period. To test if the animal was perhaps satiated in the turbid water test, the pool was refilled with clear water and 9 live fish were thrown into the pool. The sea lion caught and ate all 9 fish within 10 s of reentering the pool. The same procedure with live fish was repeated for two other sea lions with similar results. These results indicated that the three *Zalophus* were considerably more efficient in catching live fish in clear water than in extremely turbid waters. Two of the three animals were able to find almost 50% of the available fish without emitting a single click.

Schusterman (1967) concluded that *Zalophus* failed to demonstrate active sonar under conditions that would logically motivate its use. He also contended that clicking is part of a single system of calling, which is elicited by stimulus conditions arousing intense or prolonged attention. Therefore, the production of underwater clicks is related to a general arousal or activation of the sea lion as well as being one aspect of the orientation reflex.

The issue of echolocation by pinnipeds became passé in the 1970 s with a general consensus that there was no compelling evidence to demonstrate echolocation by any species of pinnipeds. During this time period, Oliver (1978) performed an interesting maze navigation experiment with the gray seal (*Halichoerus grypus*) to examine the functional utilization of high-frequency sounds produced. Five parallel pipes 1-m apart were suspended 60 cm above a 6.1 m diameter pool. PVC pipes painted black with both ends corked and sealed could be suspended from the overhead system of the five parallel pipes. By varying the position of 18 rods, Oliver (1978) was able to construct four different obstacle courses. In 100 trials in which the gray seal had to navigate in the tank, 24 poles were displaced: 7 in the light and 17 in the dark. A total of 196 vocalizations were recorded, 9 in the light and 187 in the dark. The majority of the signals were isolated clicks with the remainder being several clusters of clicks and 3 hisses. Oliver stated that avoidance of the obstacles by the seal was impressive since only 24 errors were committed out of a possible 3,600 errors. He concluded that the performance by the seal in trials in which sounds were recorded provided no evidence for an active echolocation system. Seal sounds were recorded only in 39% of the dark trials, yet in these trials 17% of all errors occurred.

Scronce and Ridgway (1980) performed two experiments to study the possibility of echolocation by a gray seal. In the first experiment, the seal wore a blindfold and was required to retrieve an air-filled plastic ring 20 cm in diameter placed at random locations in a 5 × 1 m section of a 10-m diameter red wood tank. The blindfolded seal performed this task at a 99% correct response level in 427 trials. However, its latency was 6.5 s compared with 3.8 s when not wearing a blindfold. Head scanning movements were observed on about half of the blindfolded trials and click trains were detected on only about 10% of the trials. The strongest indicator of possible echolocation was the head scanning movements. However, the animal was able to retrieve the ring just as well when no clicks were emitted and no head scans were used.

In the second experiment, Scronce and Ridgway (1980) used two 25-cm diameter foam discs in which one disc had its flat face oriented toward the seal and the other disc had its edge facing forward. The discs were placed on either side of a divider (similar to the geometry shown in Fig. 11.37). The blindfolded seal was required to approach the divider (which placed the seal at least 1.5 m from each disc) and required to indicate which side of the divider the disc with the flat-face forward was located. Without blindfold, the seal's correct response was almost perfect, but with blindfold the seal's performance was only 46% correct after 617 trials.

Scronce and Ridgway (1980) concluded that there was no evidence to suggest echolocation by the gray seal. They also concluded that the gray seal had a good passive acoustic capability and good tactile senses via its vibrissae. Although the seal made sounds underwater, it did not use the sounds for echolocation.

The possibility of pinnipeds echolocating was brought up again in 1982 when Renouf and Davis (1982) published an article that claimed to have provided evidence of echolocation by the harbor seal (*Phoca vitulina*). They performed their experiment in an outdoor 7.5-m diameter tank and used two rings for targets, each ring had an outer diameter of 17 cm and an inner diameter of 9 cm, and were constructed out of 4-cm diameter hollow black plastic tubing. One ring was internally ballasted with 600 g of lead and was air-filled and the other was ballasted with 101 g of lead and was water-filled. The negatively buoyant rings were supported by strings in such a way that the lead weights lay at the bottom and the rings were presented side by side approximately 0.5 m apart. The rings were slipped into the water while the seal was at station on the opposite side of the tank with its head out of the water. The seal was trained to swim across the tank and retrieve the air-filled ring. Testing was done at night with a tarpaulin placed over the tank to eliminate ambient light. Eventually, all the experimental sessions were conducted in daylight since both rings were visibly identical. Vocalization was monitored during all the experimental sessions.

The seal's averaged correct discrimination rate toward the end of the 26-session experiment was 75%. After these sessions, both rings were filled with water and the seal performance fell to chance. From these results, Renouf and Davis concluded that their results suggested that the seal was using echolocation, although they admitted that they could not state unequivocally the possibility of some unaware cues being present.

Wartzok et al. (1984) in a letter to the editor responded by stating that the evidence reported by Renouf and Davis for echolocation was not convincing enough to require a reassessment of earlier studies of pinniped echolocation. They pointed to a field test, which demonstrated that a blindfolded spotted seal (*Phoca vitulina largha*) could find and surface through a 50-cm diameter hole in the ice from a starting point over 30 m away without producing any vocalization. They also referenced Oliver's (1978) experiment in which a negative correlation was found between sound emission and improved navigational ability of a gray seal swimming through four mazes, and the study of Scronce and Ridgway (1980) in which no evidence of echolocation was obtained. They also pointed out to the possibility that although both rings used by Renouf and Davis were visually identical, the dynamics of motion in the water of the water-filled and air-filled rings could be very different, providing important cues to the seal. Wartzok et al. (1984) were also critical of the small number of sessions (5) conducted with both rings filled with water, since the seal originally performed the discrimination between water-filled and air-filled rings close to chance and eventually improved its performance. The same type of learning may have been possible for the case in which both rings were water-filled.

11.6 Baleen Whale Echolocation

The sounds of some baleen whales can be heard over a thousand miles, especially if the sounds enter into the SOFAR channel (Gagnon and Clark, 1995). Cummings and Thompson (1971) indicated that they could receive fin whale 20-Hz sounds over a 100 miles without any special signal-processing technique other than normal filtering. Therefore, it would not be inconceivable that certain whales may be able to hear echoes of their sound emissions from large objects. It this case, we are not considering echolocation as with bats and dolphins, but in a more general case in which large obstacles such as the ocean bottom, sea mounts, islands, and continental shelves may be detected. We said in the beginning of this chapter that any organism that produces sounds will, in certain circumstances, be able to hear echoes of their sounds, as many of us have experienced while shouting into a canyon or hollow structure. Since we do not know how sensitive baleen whales are to sounds at different frequencies, we will not be able to come to any definitive conclusions regarding this issue. However, by considering this problem we may arrive at an understanding of the critical issues and perhaps pave the way to eventually obtaining data that will be relevant to this question.

Let us consider the blue whale, the largest mammal on this planet, which probably produces the most powerful acoustic signals of any mammals (Cummings and Thompson, 1971). An example of typical blue whale sounds are shown in Fig. 10.42, showing part A and part B signal separated by about 24.5 s, which represents a two-way travel distance of 18 km. Echoes from the part A emission can return before the part B sound is emitted and echoes associated with the part B sound can return before the sequence is repeated. If echoes from the part A signal are used for sounding purposes, the maximum sounding range would be no more than 18 km, since a sonar function would not be too meaningful if the whale could not get a sense of the range to an obstacle. However, since the part A and B sounds have duration of about 19 s, this indicates that a whale would not be able to hear an echo that arrives within 19 s of the sound emission, since the echo would arrive while a sound is still being transmitted. The two-way travel time for a 19 s long signal is approximately 14 km. Therefore, if the part A sounds are used for a sonar function, it would be restricted to obstacles that have a minimum range of 14 km and a maximum range of 18 km, a highly unlikely scenario. If the part B sounds are used for echoranging, then the obstacle must have a minimum range of 14 km for the echo to arrive after the cessation of the sound emission.

We can now examine what kind of echo levels would be associated with a range of 14 km and compare the echo level with the ambient noise level. The noise-limited form of the sonar equation can be expressed as

$$\mathrm{DT} = \mathrm{SL} - 2\,\mathrm{TL} + \mathrm{TS} - (\mathrm{NL} - \mathrm{DI}). \tag{11.7}$$

If we consider a blue whale emitting a pulse sound with a center frequency of 20 Hz and a source level of 188 dB and a large obstacle with a target strength of 0 dB, then the echo level at a range of 14 km, assuming spherical spreading propagation (neglecting absorption losses which will be very low at 20 Hz), will be approximately 22 dB re 1 μPa. At low frequencies between 10 and 150 Hz, ambient noise is believed to be caused mainly by ocean crossing ships. In an area of light shipping, the noise level is will about 68 dB re 1 $\mu\mathrm{Pa}^2/\mathrm{Hz}$. The directivity index can be set to zero since the wavelength of 75 m is much larger than the typical head size of a whale. The noise received by a whale is equal to NL + Δf, where Δf is the filter width of the whale at 20 Hz. Assuming that the filter width is equal to 1/3 octave (which is generally accepted as a good filter bandwidth for the mammalian auditory system), the echo level would be considerably below the ambient noise level. A 1/3 octave band filter with a center frequency of 20 Hz will have a bandwidth of about 4.6 Hz or 6.7 dB. If the whale is in an area where there is no shipping noise, the noise at sea state 0 will be about 22 dB re 1 $\mu\mathrm{Pa}^2/\mathrm{Hz}$. If we add the 1/3 octave bandwidth to the ambient noise, the whale would receive an ambient noise level of about 29 dB. Therefore, even for a sea state 0 ambient noise condition with no shipping noise, the received echo level will still be below the received noise level, making echoranging implausable. However, if the whale was in relatively shallow water with depth less than about 5 wavelengths, we can use the cylindrical spreading loss model of $10 \log R + 10 \log H$, where H is the water depth. Let us assume a water depth of 300 m, the echo level for a 0 dB target strength and a SL of 188 dB will be about 55 dB. In this cause, the whale would hear an echo only if there was an absence of shipping noise.

Our preliminary conclusion from this simple analysis is that blue whale probably do not use any form of sonar. This is based mainly on the fact that the minimum range must be about 14 km, and at that range, the receive echo from a large target would only be heard if no shipping noise was present. The long duration of the blue whale pulse works against the notion of echo sounding by these animals.

Other whales produce pulses that are much shorter than the blue whale and these whales may have an echoranging capability. The fin whale, for example, produces a 20 Hz moan that is about 1 s in duration and has a source level up to about 186 dB (Watkins et al., 1987). With such a pulse, the minimum obstacle range would be about 750 m so that, with spherical spreading, the echo level for a 0 dB target will be about 71 dB. This level will still be below the noise in areas where shipping noise contributes to the ambient noise. However, in area that has no shipping noise the echo level will be above the received ambient noise level. Then the issue becomes whether a fin whale can hear a 71 dB level signal at 20 Hz. If the whale was in shallow waters (300 m depth) and we can consider cylindrical spreading loss, then an obstacle at 750 m will result in an echo of about 79 dB. This level is above the ambient noise level even with shipping noise present. The maximum range would depend again on the noise level, the propagation conditions and the animal's hearing sensitivity. For the sake of argument, let us make the same assumptions as in the blue whale case and consider a light shipping condition so that the ambient noise level will be about 68 dB. With a 1/3 octave bandwidth, the received noise level will be about 75 dB. Assuming a cylindrical spreading transmission model, the echo level for a 0 dB target, the echo level at a range of 1.2 km will be about 75 dB, the same as the received noise level. Therefore, if the fin whale has a hearing sensitivity of at least 75 dB at 20 Hz, it should hear the echo.

From this simple and general analysis, we should keep open the possibility that some whales may be able to echorange. The length of its signal will dictate the minimum range and the ambient noise will probably dictate the maximum range. The notion of echolocation by whales is certainly not a new one and it would be impossible to determine who first originated this notion. However, Cummings and Thompson (1971) did use the term echolocation in regards to sounds produced by the blue whale, suggesting that some scientists were already considering this possibility.

Baleen whales are not usually known to produce broadband click sounds. Some transient click sounds have been recorded in the presence of minke whales (4–7 kHz, 7 clicks/s; Beamish and Mitchell, 1973), but without any documentation of concurrent behavior. One broadband pulse train has been reported from humpbacks (Thompson et al., 1986; 1 ms pulses and irregular interpulse interval), which was attributed to baleen rattle. The use of broadband clicks in association with foraging has never been documented for any baleen whale or pinniped, until Stimpert et al. (2007) reported on humpback whale emitted broadband clicks while foraging in the Gulf of Maine. Acoustic D-tags (this tag will be discussed in the last chapter) were placed on two foraging humpback whales, designated as mn177a and mn178a, using a suction cup attachment technique.

An example of the waveform and spectrum of "Megapclicks" is shown in Fig. 11.41. The clicks were grouped together into bouts and a total of 101 bouts (17 from one whale and 84 from the other), containing 34,026 clicks, were identified from two whales on different days. The sounds had peak frequencies of approximately 800 and 1,700 Hz, and received levels at the DTAG of 143 +/− 5dB re 1 μPa pp one whale and 154 +/− 5 dB re 1 μPa pp for the other whale.

Approximately, 98% of interclick intervals during click bouts were between 19 and 200 ms. The shortest ICIs generally occurred at the end of bouts as part of an acceleration in click rate. This pattern was aurally similar to an odontocete or bat "terminal buzz" (e.g., Griffin et al., 1960; Johnson et al., 2004). Buzzes contained megapclicks with ICIs of 25 ms or less and had durations of at least 0.5 s. Four click bouts with one or more buzzes were noted from one whale and 35 from the other whale. All megapclick bouts occurred during nighttime hours (between sunset at 20:26 and sunrise at 05:05). Megapclick bouts generally occurred near the bottom of dives, and average whale depth at time of bout production was 38 m +/− 12 m, which was near the sea floor in this area.

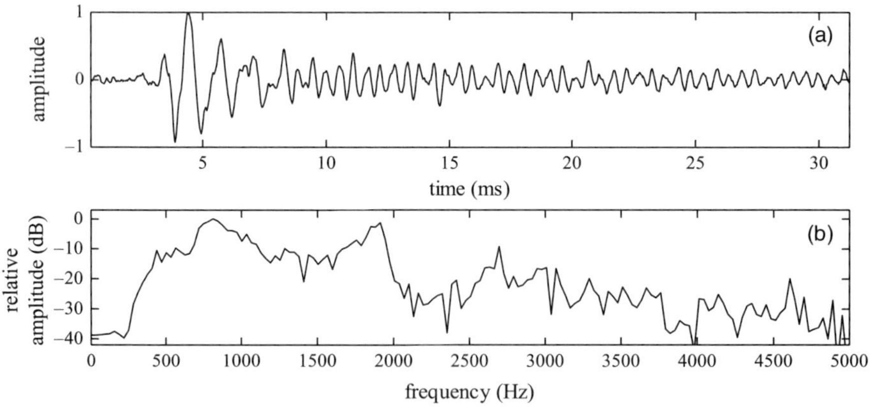

FIGURE 11.41. An example of the waveform and spectrum of a click produced by a humpback whale while foraging in the Gulf of Maine (adapted from Stimpert et al., 2007).

Several possibilities for the role of the click signals were given by Stimpert et al. (2007). The superficial similarity of the interclick intervals of megapclicks to odontocete echolocation sounds suggest a possible echoranging function. The temporal pattern of many of the bouts was reminiscent of odontocete clicks in echolocation-mediated foraging, where terminal buzzes after click bouts appear to be associated with prey capture attempts (e.g., sperm whale "creaks" in Mill er et al., 2004). Depending on hearing sensitivities and directionality of the humpback sound production system, sounds with these characteristics could be useful for some form of rough acoustic detection such as identifying the sea-floor or other large target. Since these clicks were recorded at night, with the whales approaching the bottom, it would not be far fetched for the whales to receive echoes from the bottom in order to avoid "crashing" into the bottom. Another echoranging possibility may be associated with ranging in on schools of herring.

References

Albers, V. M. (1965). *Underwater Acoustics Handbook II* (Penn St. U. Press, Univ. Park, PA).

Aroyan, J. L., Cranford, T. W., Kent, J., and Norris, K. S. (1992). "Computer Modeling of Acoustic Beam Formation in (*Delphinus delphis*)," J. Acoust. Soc. Am. **95**, 2539–2545.

Aroyan, J. L. (1996). "Three-Dimensional Numerical Simulation of Biosonar Signal Emission and Reception in the Common Dolphin," Ph.D. Dissertation, U.C. Santa Cruz.

Au, W. W. L., Floyd, R. W., Penner, R. H., and Murchison, A. E. (1974). "Measurement of Echolocation Signals of the Atlantic Bottlenose Dolphin (*Tursiops truncates*) Montagu, in Open Waters," J. Acoust. Soc. Am. **56**, 1280–1290.

Au, W. W. L., Floyd, R. W., and Haun, J. E. (1978). "Propagation of Atlantic Bottlenose Dolphin Echolocation Signals," J. Acoust. Soc. Am. **64**, 411–422.

Au, W. W. L. (1980). "Echolocation Signals of the Atlantic Bottlenose Dolphin (*Tursiops truncatus*), in Open Waters," in *Animal Sonar Systems*, R. G. Busnel and J. F. Fish, eds. (Plenum, New York), pp. 251–282.

Au, W. W. L., Schusterman, R., and Kersting, D. A. (1980). "Sphere-Cylinder Discrimination via Echolocation by (*Tursiops truncates*)," in *Animal Sonar Systems*, R. G. Busnel and J. F. Fish, eds. (Plenum Press, New York), pp. 859–862.

Au, W. W. L. and Snyder, K. J. (1980). "Long-range Target Detection in Open Waters by an Echolocating Atlantic Bottlenose Dolphin (*Tursiops truncatus*)," J. Acoust. Soc. Am. **68**, 1077–1084.

Au, W. W. L. and Penner, R. H. (1981). "Target Detection in Noise by Echolocating Atlantic Bottlenose Dolphins," J. Acoustic Soc. Am. **70**, 687–693.

Au, W. W. L. and Turl, C. W. (1983). "Target Detection in Reverberation by an Echolocating Atlantic Bottlenose Dolphin (*Tursiops truncatus*)," J. Acoust. Soc. Am. **73**, 1676–1681.

Au, W. W. L., Carder, D. A., Penner, R. H., and Scronce, B. L. (1985). "Demonstration of Adaptation in Beluga Whale Echolocation Signals," J. Acoust. Soc. Am. **77**, 726–730.

Au, W. W. L., Moore, P. W. B., and Pawloski, D. (1986). "Echolocation Transmitting Beam of the Atlantic Bottlenose Dolphin," J. Acoust. Soc. Am. 688–691.

Au, W. W. L., Penner, R. H., and Turl, C. W. (1987). "Propagation of Beluga Echolocation Signals," J. Acoust. Soc. Am., **83**, 807–812.

Au, W. W. L. and Martin, D. (1988). "Sonar Discrimination of Metallic Plates by Dolphins and Humans," in *Animal Sonar: Processes and Performance*, P. E. Nachtigall and P. W. B. Moore, eds. (Plenum, New York), pp. 809–812.

Au, W. W. L., Moore, P. W. B., and Pawloski, D. A. (1988). "Detection of Complex Echoes in Noise by an Echolocating Dolphin," J. Acoust. Soc. Am. **83**, 662–668.

Au, W. W. L. and Turl, C. W. (1991). "Material Composition Discrimination of Cylinders at Different Aspect Angles by an Echolocating Dolphin," J. Acoust. Soc. Am. **89**, 2448–2451.

Au, W. W. L. (1992). "Application of the Reverberation-Limited Form of the Sonar Equation to Dolphin Echolocation," J. Acoust. Soc. Am. **92**, 1822–1826.

Au, W. W. L. and Pawloski, D. (1992). "Cylinder wall thickness difference discrimination by an echolocating atlantic bottlenose dolphin," J. Comp. Physiol. A. (172), 41–47.

Au, W. W. L. (1993). *The Sonar of Dolphins* (Springer-Verlag, New York).

Au, W. W. L., Pawloski, J., Nachtigall, P. E., Blonz, M., and Gisiner, R. (1995). "Echolocation Signal and Transmission Beam Pattern of a False Killer Whale (*Pseudorca crassidens*)," J. Acoust. Soc. of Am. **98**, 51–59.

Awbrey, F. T., Norris, J. C., Hubbard, A. B., and Evans, W. E. (1979). "The Bioacoustics of the Dall's Porpoise-Salmon Drift Net Interaction," H/SWRI Technical Report, pp. 79–120.

Ayrapet'yants, E. S. and Konstantinov, A. I. (1974). *Echolocation in Nature* (Nauka, Leningrad).

Backus, R. H. and Schevill, W. E. (1966). "*Physeter* Clicks," in *Whales, Dolphins and Porpoises*, K. S. Norris, ed. (Univ. of California Press, Berkeley, CA), pp. 510–528.

Bagdonas, A. P., Bel'kovich, V. M., and Krushinskaya, N. L. (1970). "Interaction Between Delphinid Analyzers in Discrimination," J. Higher Neural Act. **20**, 1070–1074.

Barta, R. E. (1969). "Acoustical Pattern Discrimination by an Atlantic Bottlenosed Dolphin," unpublished manuscript (Naval Undersea Center, San Diego, CA).

Beamish, P. and Mitchell, E. (1973). "Short pulse length audio frequency sounds recorded in the presence of a minke whale (*Balaenoptera acutorostrata*)," Deep Sea Res. **20**, 375-386.

Bel'kovich, V. M. and Dubrovskiy, N. A. (1976). *Sensory Basis of Cetacean Orientation* (Nauka, Leningrad).

Carder, D., Ridgway, S., Whitaker, B., and Geraci, J. (1995). "Hearing and Echolocation in a Pygmy Sperm Whale *Kogia*," *Eleventh Biennial Conf. Biol. Mar. Mamm.*, Dec. 14–18, (Orlando, FL. (A)).

Clarke, M. R. (1979). "The Head of the Sperm Whale," Sci. Am. **240**, 106–117.

Cranford, T. W. (1992). "Functional morphology of the odontocete forehead: implications for sound generation," Ph.D. Dissertation, Univ. Calif, Santa Cruz, CA.

Cranford, T. W. (1999). "Evidence for multiple sonar signal generators in odontocetes," 13th Biennial Conference of the Biol. of Mar. Mamml. Wailea, Maui, HI.

Cranford, T. W. (2000). "In search of impulse sound sources in odontocetes," in *Hearing in Dolphins and Whales*, W. W. L. Au, A. N. Popper, and R. R. Fay, eds., (Springer-Verlag, NY), pp. 109–155.

Cummings, W. C. and Thompson, P. O. (1971). "Underwater Sounds from the Blue Whale, *Balaenoptera musculus*," J. Acoust. Soc. Am. **50**, 1193–1198.

Dawson, S. M. (1988). "The High Frequency Sounds of Free-Ranging Hector's Dolphin, *Cephalorhynchus hectori*," Rep. Int. Whal. Commn (Special Issue 9), 339–341.

Dunn, J. L. (1969). "Airborne Measurements of the Acoustic Characteristics of a Sperm Whale," J. Acoust. Soc. Am. **46**, 1052–1054.

Evans, W. W. and Haugen, R. M. (1963). "An Experimental Study of the Echolocation Ability of a California Sea Lion, *Zalophus californianus* (Lesson)," Bull. Southern Calif. Acad. Sci. **62**, 165–175.

Evans, W. W. and Powell, B. A. (1967). "Discrimination of Different Metallic Plates by an Echolocating Delphinid," in *Animal Sonar Systems: Biology and Bionics*, R. G. Busnel, ed. (Labaoratoire de Physiologie Acoustique, Jouy-en-Josas, France), pp. 363–382.

Evans, W. E. (1973). "Echolocation by Marine Delphinids and One Species of Fresh-Water Dolphin," J. Acoust. Soc. Am., **54**, 191–199.

Evans, W. E., Awbrey, F. T., and Hackbarth, H. (1988). "High Frequency Pulse Produced by Free Ranging Commerson's Dolphin (*Cephalorhynchus commersonii*) Compared to those of Phocoenids," Rep. Int. Whal. Commn. (Special Issue 9), 173–181.

Fish, J. F., Johnson, C. S., and Ljungblad, D. K. (1976). "Sonar Target Discrimination by Instrumented Human Divers," J. Acoust. Soc. Am. **59**, 602–606.

Gagnon, G. I. and Clark, C. W. (1995). "The Use of U.S. Navy IUSS Passive Sonar to Monitor the Movement of Blue Whales," *Tenth Bien. Conf. Biol. Mar. Mamm.*, Nov. 11–15, 1993 (Galveston, Texas).

Goold, J. C. and Jones, S. E. (1995). "Time and Frequency Domain Characteristics of Sperm Whale Clicks," J. Acoust. Soc. Am. **93**, 1279–1291.

Goold, J. C. (1996). "Signal Processing Techniques for Acoustic Measurement of Sperm Whale Body Lengths," J. Acoust. Soc. Am. **100**, 3431–3441.

Gordon, J. C. D. (1991). "Evaluation of A Method for Determining the Length of Sperm Whales, *Physeter catodon*, from Their Vocalizations," J. Sool. London **224**, 301–314.

Griffin, D. R., Webster, F. A., and Michael, C. R. (1960). "The Echolocation of flying Insects by Bats," Animal Behav. **8**, 141–154.

Gurevich, B. S. and Evans, W. E. (1976). "Echolocation Discrimination of Complex Planar Targets by the Beluga Whale (*Delphinapterus leucas*)," J. Acoust. Soc. Am. **60**, S5.

Harley, H. E., Roitblat, H. L., and Nachtigall, P. E. (1996). "Object Representation in the Bottlenoise Dolphin (*Tursiops truncatus*): Integration of Visual and Echoic Information," J. Exp. Psychol. Anim. Behav. Proc. **22**, 164–174.

Hatakeyama, Y., Ishii, K., Soeda, H., Shimamura, T., and Tobayame, T. (1988). "Observation of Harbor Porpoise's Behavior to Salmon Gillnet," (Document submitted to the International North Pacific Fisheries Commission.) Fisheries Agency of Japan, Tokyo, Japan, 17 pp.

Hatakeyama, Y. and Soeda, H. (1990). "Studies on Echolocation of Porpoises Taken in Salmon Gillnet Fisheries," in *Sensory Abilities of Cetaceans*, J. A. Thomas and R. Kastelein, eds. (Plenum, New York), pp. 269–281.

Helweg, D. A., Au, W. W. L., Roitblat, H., and Nachtigall, P. E. (1996). "Acoustic Basis for Recognition of Aspect-dependent Targets by an Echolocating Atlantic Bottlenose Dolphin," J. Acoust. Soc. Am. **99**, 2409–2420.

Herzing, D. L. (1996). "Vocalizations and Associated Underwater Behavior of Free-ranging Atlantic Spotted Dolphins (*Stenella frontalis*) and Bottlenose Dolphins (*Tursiops truncatus*)," Aquatic Mamm. **22**(2), 61–79.

Johnson, C. S. (1967). "Discussion," in *Animal Sonar Systems: Biology and Bionics*, R. G. Busnel, eds. (Labaoratoire de Physiologie Acoustique, Jouy-en-Josas, France), pp. 384–398.

Johnson, S. C. (1968). "Relation between absolute threshold and duration of tone-pulse in the bottlenosed porpoise,? J. Acoust. Soc. Am. **43**, 757–763.

Johnson, M., Madsen, P. T., Zimmer, W. M. X., Aguilar de Soto, N., Tyack, P. L. (2004). "Beaked whales echolocate on prey," Proc. R. Soc. Lond. **B 271**, pp. S383–386.

Kamminga, C. (1988). "Echolocation Signal Types of Odontocetes," in *Animal Sonar: Processes and Performance*, P. E. Nachtigall and P. W. B. Moore, eds. (Plenum, New York), pp. 9–22.

Kamminga, C. and Wiersma, H. (1981). "Investigations on Cetacean Sonar V: The True Nature of the Sonar Sound of *Cephaloryncus Commersonii*," Aquatic Mamml. **9**, 95–104.

Levenson, C. (1974). "Source level and bistatic target strength of the sperm whale (*Physeter catodon*) measured from an oceanographic aircraft," J. Acoust. Soc. Am. **55**, 1100–1103.

Madsen, P. T., Payne, R., Kristensen, N. U., Wahlber, M., Kerr, I., and Mϕhl, B. (2000). "Sperm Whale Sound Production Studied with Ultrasonic Time/Depth Recording Tags," J. Exp. Biol. **205**, 1899–1906.

Madsen, P. T., Payne, R., Kristiansen, N. U., Wahlberg, M., Kerr, I., and Mϕhl, B. (2002). "Sperm Whale Sound Production Studied with Ultrasound Time/Depth-recording Tags," J. Exp. Biol. **205**, 1899–1906.

McClellan, M. E. and Small, A. M. (1965). "Time-Separation Pitch Associated with Correlated Noise Burst," J. Acoust. Soc. Am. **38**, 142–143.

Miller, P. J. O., Johnson, M., Tyack, P. L., and Terray, E. A. (2004). "Swimming gaits, passive drag, and buoyancy of diving sperm whales (*Physeter macrocephalus*)," J. Exp. Biol., **207**, 1953–1967.

Mϕhl, B. and Andersen, S. (1973). "Echolocation: High-frequency Component in the Click of the Harbor Porpoise (*Phocoena ph. L.)*," J. Acoust. Soc. Am. **54**, 1368–1372.

Mϕhl, B., Larsen, E., and Amundin, M. (1981). "Sperm Whale Size Determination: Outlines of an Acoustic Approach," in No. 3, *Mammals of the Seas, Vol III*, FAO Publications, Rome, pp. 327–331.

Mϕhl, B., Surlykke, A., and Miller, L. A. (1990). "High Intensity Narwhal Clicks," in *Sensory Abilities of Cetaceans*, J. A. Thomas and R. A. Kastelein, eds. (Plenum, New York), pp. 295–303.

Mϕhl, B. and Amundin, M. (1991). "Sperm Whale Clicks: Pulse Interval in Clicks from a 21 m Specimen," in *Sound Production in Odontocetes with Emphasis on the Harbour Porpoise, Phocoena phocoena*, Amundin PhD Dissertation, Stockholm Universtiy.

Mϕhl, B., Wahlberg, M., Madsen, P. T., Miller, L. A., and Surlykke, A. (2000). "Sperm Whale Clicks: Directionality and Source Level Revisited," J. Acoust. Soc. Am. **107**, 638–648.

Mϕhl, B., Wahlberg, M., and Heerfordt, A. (2001). "A GPS-linked Array of Independent Receiver for Bioacoustics," J. Acoust. Am. **109**, 434–437.

Mϕhl, B., Wahlberg, M., Madsen, P. T., Heerfordt, A., and Lund, A. (2003). "The Monpulsed Natue of Sperm Whale Clicks," J. Acoust. Soc. Am. **114**, 1143–1154.

Moore, P. W. B. and Pawloski, D. A. (1990). "Investigations on the Control of Echolocation Pulses in the Dolphin (*Tursiops truncatus*)," in *Sensory Abilities of Cetaceans Laboratory and Field Evidence*, J. A. Thomas and R. A. Kastelein, eds. (Plenum, New York), pp. 305–316.

Morozov, B. P., Akapiam, A. E., Burdin, V. I., Zaitseva, K. A., and Solovykh. (1972). "Tracking Frequency of the Location Signals of Dolphins as a Function of Distance to the Target," Biofiika **17**, 139–145.

Murchison, A. E. (1980). "Maximum Detection Range and Range Resolution in Echolocating Bottlenose Porpoise (*Tursiops truncatus*), in *Animal Sonar Systems*, R. G. Busnel and J. F. Fish, eds. (Plenum, New York), pp. 43–70.

Nachtigall, P. E. (1980). "Odontocete Echolocation Performance on Object Size, Shape and Material," in *Animal Sonar Systems*, R. G. Busnel and J. F. Fish, eds. (Plenum, New York), pp. 71–95.

Nachtigall, P. E., Murchison, A. E., and Au, W. W. L. (1980). "Cylinder and Cube Discrimination by an Echolocating Blindfolded Bottlenose Dolphin," in *Animal Sonar Systems*, R. G. Busnel and J. F. Fish, eds. (Plenum, New York), pp. 945–947.

Nachtigall, P. W. and Patterson, S. A. (1981). "Echolocation and Concept Formation by an Atlantic Bottlenosed Dolphin: Sameness-difference and Matching-to-Sample," (Abstract), *Fourth Biennial Conf. Biol. Mar. Mamm.*, (San Francisco, Ca).

Norris, K. S. and Harvey, G. W. (1974). "Sound Transmission in the Porpoise Head," J. Acoust. Soc. Am. **56**, 659–664.

Oliver, G. W. (1978). "Navigation in mazes by a grey seal, *Halichoerus grypus* (Fabricius)." Behaviour **67**, 97–114.

Pack, A. A. and Herman, L. M. (1995). "Sensory Integration the Bottlenosed Dolphin: Immediate Recognition of Complex Shapes Across the Senses of Echolocation and Vision," J. Acoust. Soc. Am. **98**, 722–733.

Penner, R. H. (1988). "Attention and Detection in Dolphin Echolocation," in *Animal Sonar: Processes and Performance*, P. E. Nachtigall and P. W. B. Moore, eds. (Plenum, New York), pp. 707–712.

Poulter, T. C. (1963a). "Sonar Signals of the Sea Lion," Science, **139,** 753–755.

Poulter, T. C. (1963b). "The Sonar of the Sea Lion," IEEE Trans. Ultrasonics Eng. **10**, 109–111.

Poulter, T. C. and Jennings, R. A. (1969). "Sonar Discrimination Ability of the California Sea Lion (*Zalophus californianus*)," Proc. Calif. Acad. Sci. **XXXVI**, 381–389.

Renouf, D. and Davis, M. B. (1982). "Evidence that seals may use echolocation," Nature (London) **300**, 635–637.

Roitblat, H. L., Penner, R. H., and Nachtigall, P. E. (1990). "Matching-to-Sample by an Echolocating Dolphin," J. Exp. Psychol. Anim. Behav. Proc. **16**, 85–95.

Rossbach, K. A. and Herzing, D. L. (1997). "Underwater Observations of Benthic-feeding Bottlenose Dolphins (*Tursiops truncatus*) near Grand Bahama Island, Bahamas," Mar. Mamm. Sci. **13**, 498–504.

Schusterman, R. J. (1967). "Perception and Determinants of Underwater Vocalization in the California Sea Lion," in *Animal Sonar Systems:Biology and Bionics*, R.-G. Busnel, ed. (Labaoratoire de Physiologie Acoustique, Jouy-en-Josas, France), pp. 535–617.

Schusterman, R. J., Kersting, D. A., and Au, W. W. L. (1980). "Stimulus Control of Echolocation Pulses in (*Tursiops truncates*)," in *Animal Sonar Systems*, R. G. Busnel and F. Fish, eds. (Plenum, New York), pp. 981–982.

Scronce, B. L. and Ridgway, S. H. (1980). "Grey Seal (*Halichoerus*): Echolocation not Demonstrated," in *Animal Sonar Systems*, R. G. Busnel and F. Fish, eds. (Plenum, New York), pp. 991–993.

Small, A. M. and McClellan, M. E. (1963). "Pitch Associated with Time Delay Between Two Pulse Trains," J. Acoust. Soc. Am. **35**, 1246–1255.

Stimpert, A. K., Wiley, D. N., Au, W. W. L., Johnson, M. P., and Arsenault, R. (2007). "Megapclicks: Acoustic Click Trains and Buzzes Produced During Nighttime Foraging of Humpback Whales (*Megaptera novaeangliae*)," Biol. Lett. **3**, 467–470.

Thomas, J. A., Stoermer, M., Bowers, C., Anderson, L., and Garver, A. (1988). "Detection Abilities and Signal Characteristics of Echolocating False Killer Whales (*Pseudorca crassidens*)," in *Animal Sonar Processing and Performance*, P. E. Nachtigall and P. W. B. Moore, eds. (Plenum, New York), pp. 323–328.

Thomas, J. A. and Turl, C. W. (1990). "Echolocation Characteristics and Range Detection by a False Killer Whale (*Pseudorca crassidens*)," in *Sensory Abilities of Cetaceans Laboratory and Field Evidence*, J. A. Thomas and R. A. Kastelein, eds. (Plenum, New York), pp. 321–334.

Thompson, P. O., Cummings, W. C., and Ha, S. J. (1986). "Sounds, source levels, and associated behavior of humpback whales, Southeast Alaska," J. Acoust. Soc. Am. **80**, 735–740.

Titov, A. A. (1972). "Investigation of Sonic Activity and Phenomenological Characteristics of the Echolocation Analyzer of Black Sea Delphinids," Candidatorial dissertation, Karadag.

Turl, C. W., Penner, R. H., and Au, W. W. L. (1987). "Comparison of Target Detection Capabilities of the Beluga and Bottlenose Dolphin," J. Acoust. Soc. Am. **82**, 1487–1491.

Turl, C. W. and Penner, R. H. (1989). "Differences in Echolocation Click Patterns of the Beluga (*Delphinapterus leucas*) and the Bottlenose Dolphin (*Tursiops truncatus*)," J. Acoust. Soc. Am. **68**, 497–502.

Turl, C. W., Skaar, D. J., and Au, W. W. L. (1991). "The Echolocation Ability of the Beluga (*Delphinapterus leucas*) to detect Targets in Clutter," J. Acoust. Soc. Am. **89**, 896–901.

Urick, R. J. (1983). *Principles of Underwater Sound* (McGraw-Hill, NY).

Wartzok, D., Schusterman, R. J., and Gailey-Phipps, J. (1984). "Seal echolocation?" Nature (London). **308**, 753.

Watkins, W. A. (1980). "Acoustics and the Behavior of Sperm Whales," in *Animal Sonar Systems*, R.-G. Busnel and J. F. Fish, eds. (Plenum, New York), pp. 283–289.

Watkins, W. A., Tyack, P., Moore, K. E., and Bird, J. E. (1987). "The 20-Hz Signals of Finback Whales (*Balaenoptera physalus*)," J. Acoust. Soc. Am. **82**, 1901–1912.

Watkins, W. A., Daher, M. A., Fristrup, K. M., and Howald, T. J. (1993). "Sperm Whales Tagged with Transponders and Tracked Underwater by Sonar," Mar. Mammal Sci. **9**, 55–67.

Wood, F. G. (1964). "Discussion" in *Marine Bio-Acoustics Vol II*, W. Tavolga, ed. (Pergamon, Oxford, England), pp. 395–396.

12

Some Signal Processing Techniques

By this time we should be very aware of the fact that marine animals produce a wide varieties of different type of sounds. These sounds include clicks, tone pulses, AM pulses, FM chirps, and combination signals containing AM, FM, and constant frequency components. Up to this point, signals have been analyzed and characterized mainly by using the FFT to determine their spectral characteristics. In this chapter, we will introduce some additional signal processing techniques that may be helpful in further characterizing and analyzing these signals. Some concepts from radar signal processing will be used to analyze the properties of dolphin sonar signals. We will assume that echoes are the result of reflection from point targets so that their waveforms will be the same as the transmitted signals but with lower amplitudes. We will also assume that echoes are received and processed by a matched filter receiver. Although a matched-filter point of view will be taken in much of this section, it has already been shown (Chapter 8, Au, 1993) that dolphins do not process echoes like a matched filter; however, the procedure taken in this section can be helpful in understanding some of the ideal capabilities of the dolphin sonar signals. The concept of matched filtering will also be discussed in this chapter. For a deeper study of radar signal processing, the readers are referred to two books (Burdic, 1968; Rihaczek, 1969). Although some of the topics are geared toward dolphin echolocation, in reality the technique discussed in this chapter can be applied to any type of signals.

12.1 Some Useful Signal Processing Concepts

12.1.1 The Envelope Function

It is often useful to determine the envelope of the signal being analyzed to study how the amplitude varies with time. By analyzing the envelope of the signal, we essentially ignore the higher frequency oscillations. The envelope of a signal can be computed by taking the "real" time domain signal and making it complex or "analytic" using the Hilbert transform. Any actual time domain signal is real (does not have an imaginary part). Its frequency domain representation obtained through the FFT is complex. A complex representation of

W.W.L. Au, M.C. Hastings, *Principles of Marine Bioacoustics*,
DOI: 10.1007/978-0-387-78365-9_12, © Springer Science+Business Media, LLC 2008

a real signal (having a real and imaginary part) can be created by adding an imaginary part (having special properties) to the real signal, creating a new signal that is referred to as an analytic signal. The analytic signal is created in such a way that the negative frequency portion of the real signal spectrum is suppressed and the positive frequency components are doubled. Let $s(t)$ be the real signal and $\hat{s}(t)$ be the imaginary part so that

$$s_{\mathrm{a}}(t) = s(t) + j\hat{s}(t). \tag{12.1}$$

The requirements that the analytic signal have zero negative frequencies and that its positive frequency components are doubled that of the real signal imply that

$$\Im[\hat{s}(t)] = \begin{cases} -jS(f) & \text{for } f > 0 \\ jS(f) & \text{for } f < 0 \end{cases}, \tag{12.2}$$

where $S(f)$ is the Fourier transform of $s(t)$. Equation 12.1 can also be expressed as

$$s_{\mathrm{a}}(t) = |s_{\mathrm{a}}(t)|e^{j\phi(t)}, \tag{12.3}$$

where

$$|s_a(t)| = \sqrt{s^2(t) + \hat{s}^2(t)}. \tag{12.4}$$

$$\phi(t) = \arctan\frac{\hat{s}(t)}{s(t)} \tag{12.5}$$

The magnitude of the analytic signal in Eq. (12.4) is the envelope of the real signal $s(t)$. Therefore, the envelope of a signal can be obtained by simply adding a complex portion having the properties expressed in Eq. (12.2) to the real signal and then determining the absolute value of the analytic signal using Eq. (12.4). The result of performing the mathematical operation of Eqs. (12.2)–(12.4) is depicted in Fig. 12.1, where the envelope of the echo from the 7.62-cm water-filled sphere shown in Fig. 13.12 was calculated. The functions $s(t)$ and $\hat{s}(t)$ are also related through the Hilbert transform, where $\hat{s}(t)$ is the Hilbert transform of $s(t)$ and $s(t)$ is the inverse Hilbert transform of

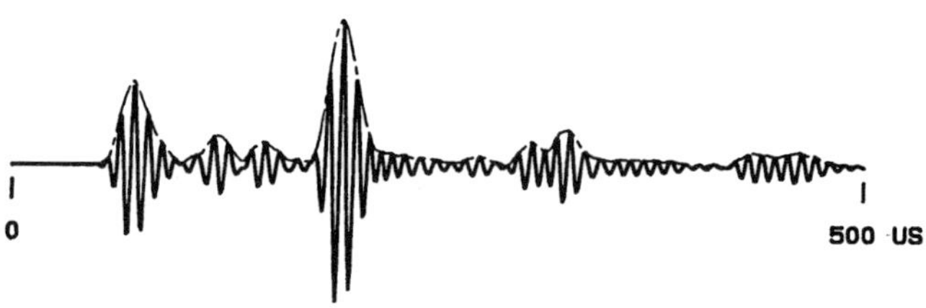

FIGURE 12.1. Echo signal from a 7.62-cm water-filled sphere and its envelope (*dashed curve*).

$\hat{s}\,(t)$. For a more comprehensive and thorough discussion of the analytic signal, the reader should consult Rihaczek (1969) and Burdic (1968).

12.1.2 Matched Filtering

Signals contaminated with noise can be analyzed by first processing the signal with a filter having a response that is matched to the expected signal. For dolphin echolocation signal and typical target reflections (see Fig. 11.12), the echo highlights tend to resemble the transmitted signal so that a matched filter would be a good way to reduce the effects of white Gaussian noise. A matched filter has a response that is matched to the time-reversed form of the expected signal. If $x(t)$ is the input to a filter with a response $h(t)$, the output at time t is the convolution integral

$$y(t) = \int_0^t x(\tau)h(t-\tau)d\tau. \tag{12.6}$$

The response of a matched filter will be

$$t(t) = s(T-t) \qquad \text{for } 0 \le t \le T. \tag{12.7}$$

From Eq. (12.6) the output of the matched filter for an input $x(t)$ will be

$$y(t) = \int_0^t x(\tau)s(T-t+\tau)\,dt. \tag{12.8}$$

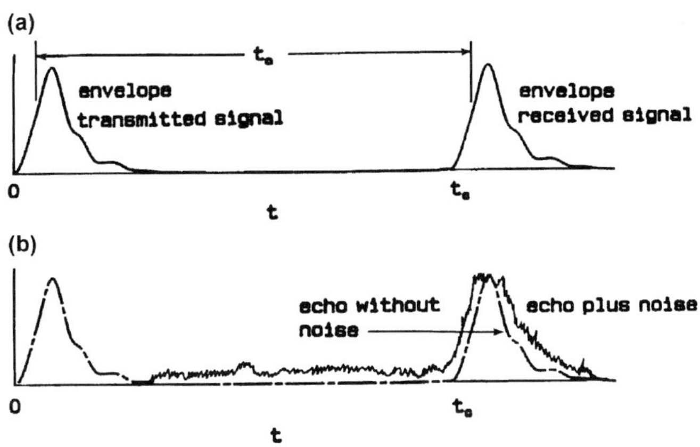

FIGURE 12.2. (**a**) Relationship between envelopes of the transmitted signal and received echoes in the absence of noise, (**b**) envelope of received echo in the presence of noise.

The output of the matched filter will build up with time until a maximum value is reached at time $t = T$, and then begin to drop. At time $t = T$, the output is

$$y(T) = \int_0^T x(t)s(t)\mathrm{d}t, \tag{12.9}$$

which is equal to the correlation integral of the product $x(\mathrm{t})s(t)$. Therefore, if $x(t)$ is the outgoing signal, then the response of the matched filter represented by Eq. (12.7) will be $s(T-t)$. Let $X(f)$ and $S(f)$ be the Fourier transform of the echo $x(t)$ and the transmitted signal $s(t)$, respectively, then according to the Correlation Theorem in Fourier Analysis (Brigham, 1988) the Fourier transform of the output can be written as

$$\Im[y(t)] = Y(f) = X(f)S^*(f), \tag{12.10}$$

where $S^*(f)$ is the complex conjugate of $S(f)$. The output of the matched filter in the time domain is the inverse Fourier Transform of Eq. (12.10),

$$y(f) = \Im^{-1}[X(f)s^*(f)]. \tag{12.11}$$

The envelope of $y(t)$ can then be calculated to determine the spacing and relative amplitude of the highlights in the echo. The matched filter is the most optimal receiver for detecting a specific signal in Gaussian noise.

12.1.3 Center Frequency and RMS Bandwidth

The frequency characteristics of signal are usually described in terms of peak frequency and 3-dB bandwidth. Both parameters are relatively easy to determine. The peak frequency is defined as the frequency at which the spectrum has its maximum value. The 3-dB bandwidth is defined as the frequency band between the lower and upper half-power (3 dB down from the maximum value) points in the frequency spectrum. In signal analysis, a broadband signal is often described by its center or centroid frequency instead of its peak frequency. Let $s(t)$ be the signal waveform and $S(f)$ its Fourier transform. The center frequency of the signal is defined as

$$f_0 =< f >= \frac{\int_{-\infty}^{\infty} f|s(f)|^2 df}{\int_{-\infty}^{\infty} |s(f)|^2 df} \tag{12.12}$$

The center frequency f_0 divides the energy in the spectrum into two equal parts so that the energy in upper frequency portion of the spectrum is equal to the energy in lower frequency portion of the spectrum. Equation 12.12 can also be expressed in terms of energy density, since by Rayleigh or Parseval Theorem in Fourier Analysis (Bracewell, 1978)

$$E = \int_{-\infty}^{\infty} |s(t)|^2 \, dt = \int_{-\infty}^{\infty} |S(f)|^2 \, df. \quad (12.13)$$

However, any actual physical waveform $s(t)$ is real and its energy spectrum is an even function of frequency and is fully determined by its values for $f \geq 0$ (Bracewell, 1978). Therefore, we can rewrite Eq. (12.12) as

$$f_0 = \frac{\int_0^{\infty} f|S(f)|^2 df}{\int_0^{\infty} |S(f)|^2 df} = \frac{1}{E/2}\int_0^{\infty} f|S(f)|^2 df \quad (12.14)$$

The bandwidth of a signal can be described by first defining the mean-square value of the function $S(f)$ as (Bracewell, 1978)

$$\langle f^2 \rangle = \frac{\int_{-\infty}^{\infty} f^2 |S(f)|^2 df}{\int_{-\infty}^{\infty} |S(f)|^2 df}. \quad (12.15)$$

The mean square bandwidth β^2 about the centroid of $|S(f)|^2$ can now be expressed as

$$\beta^2 = \frac{\int_{-\infty}^{\infty} (f - f_0)^2 |s(f)|^2 df}{\int_{-\infty}^{\infty} |s(f)|^2 df} \quad (12.16)$$

which reduces to

$$\beta^2 = \frac{\int_0^{\infty} f^2 |S(f)|^2 df}{\int_0^{\infty} |S(f)|^2 df} - f_0^2 = \langle f_0^2 \rangle - f_0^2. \quad (12.17)$$

Echolocation signals with a bimodal distribution in the frequency domain will typically have center frequencies that are lower by several kHz than the peak frequencies. The narrow band echolocation signals used by nonwhistling dolphins tend to have the peak and center frequencies that are relatively similar in value. The rms bandwidth of the broadband echolocation signals of whistling dolphins will tend to be smaller than the 3-dB bandwidth, but the reverse is true for the signals of the nonwhistling dolphins and porpoises.

12.1.4 Accuracy in Target Range Determination

An important piece of information a sonar should obtain from the received signal is the target range. One approach to determining the target range is to detect the envelope of the received echo and measure the time difference, t_0, between two corresponding points on the transmitted and received envelope, as depicted in Fig. 12.2a. This process can be performed very accurately in the absence of any noise. However, the presence of noise will distort the received waveform, as shown in Fig. 12.2b, so that corresponding points on the envelopes of the transmitted signal and received echoes cannot be accurately identified. The time of arrival of the noisy echo can be determined by shifting a replica of the transmitted signal along the time axis and calculate the position for which the integrated squared difference between this shifted waveform and the received noisy echo is a minimum. This procedure is equivalent to finding the best-fit solution in comparing the two waveforms. Let the transmitted signal and the received noise be analytic (see Section 12.1), so that they can be expressed as $s_a(t)$ and $n_a(t)$, respectively. The received echo plus noise is then

$$r(t) = A\,s_a(t) + n_a(t), \tag{12.18}$$

where $A << 1$ because of transmission and reflection losses. The time of arrival of the echo can be determined by finding the minimum of the integrated squared difference between the received echo and the shifted transmitted waveform. The upper limit of integration T represents the maximum

$$\varepsilon^2 = \int_0^T |s_a(t+\tau) - r(t)|^2 dt \tag{12.19}$$

range of interest and is assumed to be large compared with the signal duration. Expanding Eq. (12.19) we obtain

$$\varepsilon^2 = K - 2\mathrm{Re}[\int_0^T s_a^*(t+\tau)r(t)\mathrm{d}t], \tag{12.20}$$

where $s_a^*(t+\tau)$ is the complex conjugate of $s_a(t+\tau)$ and K is a constant based on the energy in the transmitted and received signal. ε will be minimum at the value of τ for which the integral has a maximum positive value. If t_0 is the delayed time between the transmit and received signal and τ_1 is the estimate of the delay time based on minimizing, the standard deviation of the error in estimating t_0 can be expressed as (Woodward, 1953)

$$\sigma_\tau = (\tau_1 - t_0)_{rms} = \frac{1}{2\pi\,\beta_0\sqrt{2\frac{E}{N}}} \tag{12.21}$$

where β_0 is the rms bandwidth of Eq. (12.16), E is the energy flux density, and $N_0/2$ is the noise spectral density for real noise waveform. Since $t_0 = 2R/c$, the standard deviation in the range estimation error can be expressed as

$$\sigma_R = \frac{1}{2\pi\beta_0\sqrt{2\frac{E}{N_0}}}\frac{c}{2} \tag{12.22}$$

The uncertainty in estimating range is inversely proportional to the product of the rms bandwidth and the square root of the signal-to-noise ratio.

In Section 11.3.1 we discussed the range determination experiment of Murchison (1980) using a *Tursiops truncatus* named Heptuna. The dolphin's signals typically had an rms bandwidth of 21.5 kHz (Au, 1993), and processing the echo from a point target with a matched filter receiver, the uncertainty in estimating range from Eq. (12.12) is plotted in Fig. 12.3 as a function of the signal-to-noise ratio. The results indicate that the dolphin sonar signal has a good range determination capability.

For E/N_0 of 10 dB, a matched filter receiver would have an uncertainty of only 0.18 cm in determining range. Let us now examine the relative range accuracy experiment of Murchison (1980) discussed in Section 13.1.1. The echo signal-to-noise ratio in dB can be calculated by using Eq. (13.1), which is rewritten here as

$$10\log(E_e/N_0) = \mathrm{SE} - 2\mathrm{TL} + \mathrm{TS} - (\mathrm{NL} - \mathrm{DI}). \tag{12.23}$$

Measurements of Heptuna's signal when the dolphin was performing the range accuracy experiment at 1 m indicated an average SL of 181 dB, which

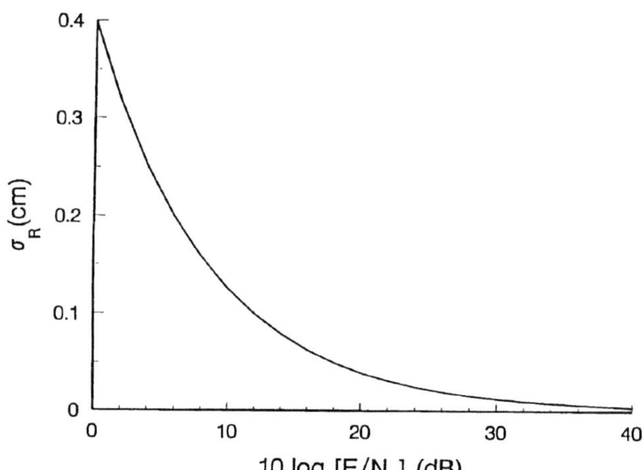

FIGURE 12.3. Range error estimate for a matched filter using a *Tursiops* sonar signals.

from Eq. (13.2) gives an SE of 123 dB. The target strength of a soft sphere can be calculated from the expression (Urick, 1983)

$$\mathrm{TS} = 20\,\mathrm{Log}(d/4), \tag{12.24}$$

where d is the diameter in meters. This expression results in a target strength of –34 dB for a 7.62-cm diameter sphere. From Eq. (10.10), the receiving directivity index for a peak frequency of 117 kHz is approximately 21 dB. The ambient noise level in Kaneohe Bay at a peak frequency of 117 kHz is equal to approximately 55 dB. Inserting the appropriate values into Eq. (12.23), we obtain $E_e/N_0 \approx 54$ dB. With such a high signal-to-noise ratio, a matched-filter receiver would be extremely accurate in determining target range, much more accurate than the dolphin.

12.1.5 Range Resolution

The range resolution capability of a sonar is another important parameter in evaluating its effectiveness for target identification and recognition. Range resolution involves resolving the echoes from two targets (or two highlights from an extended target) that are along the same azimuth with respect to the sonar. The range resolution capability of a signal is the minimum distance two point targets can be separated before it is impossible to determine whether one or two targets (or highlights) are being detected by the receiver. Let $s_a(t)$ be the transmitted target and the echo from two stationary point targets be $A{\cdot}s_a(t + t_0)$ and $A{\cdot}s_a(t + t_0 + \tau)$, where $t = 0$ is the time origin for transmission, t_0 is the round-trip delay for the first target, and τ is the difference in delay between the two echoes. We can use the integrated square magnitude of the difference between the two echoes as a measure of the distinguishability of the two returns:

$$\varepsilon^2 = A^2 \int_{-\infty}^{\infty} |s_a(t + t_0) - s_a(t + t_0 + \tau)|^2 dt \tag{12.25}$$

Expanding Eq. (12.25), we have

$$\begin{aligned}\varepsilon^2 = A^2 \int_{-\infty}^{\infty} (|s_a(t + t_0)|^2 - |s_a(t + t_0 + \tau)|^2) dt \\ -A^2 \int_{-\infty}^{\infty} [s_a(t + t_0) s_a^*(t + t_0 + \tau) - s_a^*(t + t_0)\, s_a(t + t_0 + \tau)] dt\end{aligned} \tag{12.26}$$

The first integral is the sum of the energy in each echo and is equal to $2A^2$ times the energy in the transmitted signal. Therefore, Eq. (12.26) can be reduced to

$$\varepsilon^2 = 2A^2 \int_{-\infty}^{\infty} |s_a(t)|^2 \mathrm{d}t - 2A^2 \mathrm{Re}\left[\int_{-\infty}^{\infty} s_a^*(t+\tau)\mathrm{d}t\right] \tag{12.27}$$

The first integral is a constant and the second integral is the autocorrelations function of $s_a(t)$. The difference between the echoes will be minimum when the autocorrelation function is maximum. The second integral is defined as the "range ambiguity function," $c(\tau)$:

$$c(\tau) = \int_{-\infty}^{\infty} s_a(t) s_a^*(t+\tau)\mathrm{d}t. \tag{12.28}$$

The degree of target resolution is measured by $|c(\tau)|^2$. The maximum value of $|c(\tau)|$ occurs at $\tau = 0$, and if $|c(\tau)| = c(0)$ for any value of $\tau \neq 0$, then the two target cannot be resolved. If $|c(\tau)|$ is close to $c(0)$ for some τ, then the two targets can be resolved with difficulty. Equation 12.28 can be expressed in a form more convenient for computation by using the Correlation Theorem in Fourier Analysis (Burdic, 1968),

$$c(\tau) = \Im^{-1}[S_a(f)\, S_a^*(f)]. \tag{12.29}$$

An example of a dolphin sonar click with its autocorrelation function and corresponding envelope are shown in Fig. 12.4. The time resolution capability of a sonar signal can be defined by the "time resolution constant" of Woodward (1953), which is related to the range ambiguity function by the expression

$$\Delta\tau = \frac{\int_{-\infty}^{\infty} |c(\tau)|^2 \mathrm{d}\tau}{c^2(0)}. \tag{12.30}$$

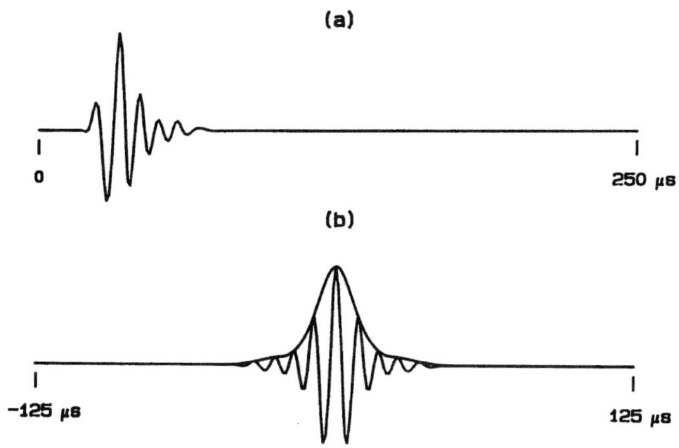

FIGURE 12.4. (**a**) *Tursiops truncatus* sonar signal, (**b**) autocorrelation function and its envelope.

The relative resolution capabilities of different waveforms can be compared by using Eq. (12.30) to calculate the time resolution constant. The waveform having the smallest value for $\Delta\tau$ can be assumed to have the greatest potential for resolving between two signals in time. From the correlation theorem in Fourier analysis (Bracewell, 1978),

$$\Im[c(\tau)] = S_a^*(f)S_a(f) = 4|S(f)|^2. \tag{12.31}$$

From Parseval Theorem, the denominator of Eq. (12.30) can be expressed as

$$c(0) = \int_{-\infty}^{\infty} | s_a(t) |^2 \, \mathrm{d}t = 2\int_0^{\infty} |S(f)|^2 \, \mathrm{d}f. \tag{12.32}$$

Substituting Eqs. (12.31) and (12.32) into Eq. (12.30), we obtain for Woodward time resolution constant

$$\Delta\tau = \frac{\int_0^{\infty} |S(f)|^4 \, \mathrm{d}f}{[\int_0^{\infty} |S(f)| \mathrm{d}f]^2}. \tag{12.33}$$

The time resolution constant can also be used to define another type of bandwidth referred to as the effective bandwidth,

$$\beta_e = 1/\Delta\tau. \tag{12.34}$$

Typical values for *Tursiops* signals are between 12 and 15 μs, which translate to a resolvable range of approximately 1.0–1.1 cm between point targets. The time resolution problem can be visualized with the aid of Fig. 12.5, which depicts the envelope of the autocorrelation function of echoes from two point targets separated by various time τ. Heptuna's signal shown in Fig. 12.3 was used to calculate the autocorrelation function. The normalized autocorrelation envelopes are not distinguishable for $\tau = 10$ μs, and barely distinguishable for τ between 14 and 15 μs, which corresponds well with a $\Delta\tau = 14.5$ μs. The relatively small time resolution constants for *Tursiops* probably play an important role in the animal's ability to make fine target discrimination.

12.1.6 Wideband Ambiguity Function

Up to this point, we have considered only stationary targets. Targets moving radially away or toward a sonar will affect the time or range resolution characteristics of the signal because of Doppler effects. A narrow band signal will experience a Doppler shift toward a higher frequency if the target is approaching, and toward a lower frequency if the target is moving away from the sonar source. Broadband sonar signals, typical of dolphin sonar, will not experience Doppler frequency shifts but will rather experience Doppler stretching or compression

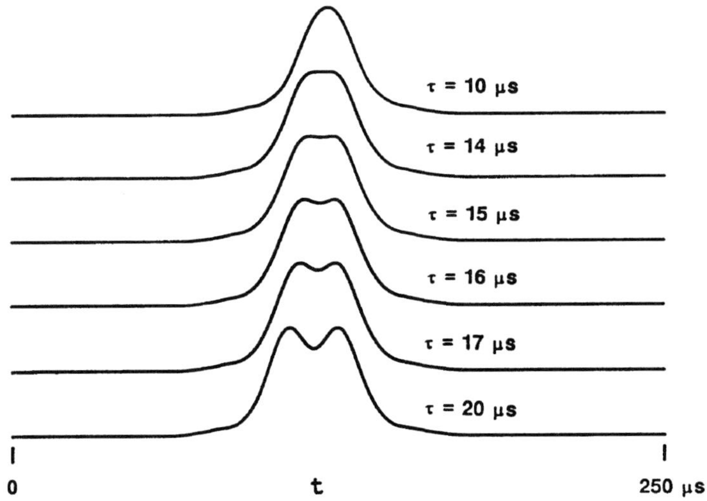

FIGURE 12.5. The normalized autocorrelation function envelopes of echoes from two point targets separated in time by the parameter τ. Heptuna's signal shown in Fig. 12.3 and having a $\Delta\tau$ of 14.5 μs was used to calculate the autocorrelation function (adapted from Au, 1993).

(Kelly and Wishner, 1965). If the radial velocity of a point target is v, and the velocity of sound in the water is c, the Doppler scale factor η is defined as

$$\eta = \frac{1 + v/c}{1 - v/c}. \tag{12.35}$$

If we let $s_a(t)$ be the transmitted signal, the echo from a radially moving point target can then be expressed as $As_a[\eta(t + \tau)]$, where τ is the time delay between the transmitted and received signals. Once again we can use the integrated square magnitude of the difference between the transmitted and received echo as a measure of the distinguishability between the two signals:

$$\begin{aligned} \varepsilon^2 &= \int_{-\infty}^{\infty} |\, s_a(t) - As_a[\eta(t+\tau)] \,|^2 \, \mathrm{d}t \\ &= \int_{-\infty}^{\infty} |\, s_a(t) \,|^2 \, \mathrm{d}t + A^2 \int_{-\infty}^{\infty} |\, s_a[\eta(t+\tau)] \,|^2 \, \mathrm{d}t \\ &\quad - 2\,A^2\, Re \int_{-\infty}^{\infty} s_a(t) s_a^*[(\eta(t+\tau)] \mathrm{d}t \end{aligned} \tag{12.36}$$

The first two terms are the energy in the transmitted and reflected signals, which are constants for a given signal, target, and range. The difference ε^2 will be minimum when the last integral is maximum. This integral represents the cross-correlation between the emitted signal and its delayed reflected echo from a target moving at a velocity v. The third integral, when normalized, is the broadband ambiguity function derived by Kelly and Wishner (1965)

$$\chi(\eta,\tau) = \sqrt{\eta}\int_{-\infty}^{\infty} s_a(t)s_a^*[\eta(t-\tau)]\mathrm{d}t. \tag{12.37}$$

The ambiguity function can be expressed in a more convenient form for computational purposes by using the Shift Theorem and Parseval Theorem in Fourier Analysis (Bracewell, 1978)

$$\begin{aligned}\chi(\eta,\tau) &= \frac{1}{\sqrt{\eta}}\int_{-\infty}^{\infty} S_a(f)S_a^*(f)\mathrm{e}^{j2\pi f\tau}\mathrm{d}f \\ &= \frac{1}{\sqrt{\eta}}\Im^{-1}[S_a(f)S_a^*(f)].\end{aligned} \tag{12.38}$$

Some of the properties of the wideband ambiguity function have been described by Altes (1973). A convenient technique to evaluate Eq. (12.39) using a digital computer has been described by Lin (1988). The envelope of the wideband ambiguity function for the *Tursiops truncatus* signal shown in Fig. 12.3 is plotted in Fig. 12.6 for different values of the Doppler scale factor η. The target velocity for $\eta = 1.03$ is approximately 22 m/s. The wideband ambiguity function plotted in Fig. 12.5 indicates that the range resolution capability of a typical dolphin sonar signal is not affected by the velocity of the target so that the signal can be considered to be Doppler tolerant. The ambiguity diagram is also given by Dziedzic et al. (1969, 1979). These signals were also Doppler-tolerant with no velocity resolution capability.

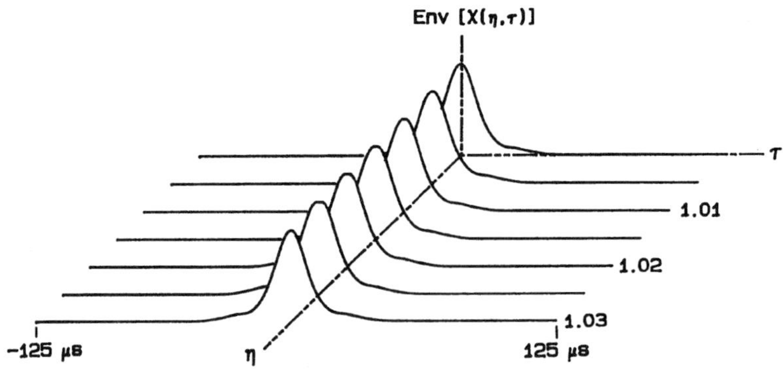

FIGURE 12.6. The wideband ambiguity diagram of the *Tursiops truncatus* signal found in Fig. 12.3 (adapted from Au, 1993).

The sonar signals of some bat species that contain a constant frequency (CF) portion have ambiguity plots that suggest very good velocity resolution capabilities. The sonar signals of *Rhinolophus ferrumequinum* were found by Beuter (1980) to have a velocity resolution capability as fine as indicates that the dolphin signal cannot be used to resolve the velocity of the target. Ambiguity function plots for the sonar signals of *Phocoena phocoena* and *Delphinus delphis* have been presented by 10 cm/s. Lin (1984, 1988) also showed that the signal of *Myotis mystacinus* has good velocity resolution. Therefore, from a single echo, these bats may be able to estimate the radial velocity of their prey. However, such a capability may not be necessary for dolphins, since the speed of sound in water is approximately five times as much as in air. Therefore, echoes from prey at comparable ranges return to the dolphin much sooner than for a bat allowing the dolphin to track the trajectory of the prey by receiving many consecutive echoes.

12.1.7 Time–Bandwidth Product

The *time–bandwidth* product of a signal is another parameter that is often important to calculate. The mean-square bandwidth is defined in Eq. (12.16), and in a similar manner, the *mean-square* duration of the signal waveform can be defined as (Bracewell, 1978)

$$\tau_d^2 = \frac{\int_{-\infty}^{\infty} (t - t_0)^2 |s(t)|^2 \mathrm{d}t}{\int_{-\infty}^{\infty} \frac{t|s(t)|^2 \mathrm{d}t}{|s(t)|^2 \mathrm{d}t}}, \tag{12.39}$$

where

$$t_0 = \frac{\int_{-\infty}^{\infty} t|s(t)|^2 \mathrm{d}t}{\int_{-\infty}^{\infty} |s(t)|^2 \mathrm{d}t} \tag{12.40}$$

is the centroid of the time waveform. Since we are concerned only with the width of the time waveform and frequency spectrum, the analysis can be greatly simplified by relocating the centroid of the waveform and spectrum to the origin of their respective coordinate system (essentially letting t_0 and $f_0 = 0$). The product of both mean-square function can now be expressed as

$$\tau_d^2\beta^2 = \frac{\int_{-\infty}^{\infty} t^2 |s(t)|^2 \mathrm{d}t \int_{-\infty}^{\infty} |S(f)|^2 \mathrm{d}f}{\int_{-\infty}^{\infty} |s(t)|^2 \mathrm{d}t \int_{-\infty}^{\infty} |S(f)|^2 \mathrm{d}f}. \tag{12.41}$$

Equation 12.41 can be further simplified in a straightforward but involved process using the derivative theorem, Rayleigh's theorem, and Schwartz's inequality found in Fourier Analysis. The derivation will not be covered here; those interested in the details should refer to Bracewell (1978). After the appropriate manipulation of Eq. (12.41), the time–bandwidth product can be expressed by the following inequality

$$\tau_d\beta \geq \frac{1}{4\pi}. \tag{12.42}$$

Equation 12.42 states that the time–bandwidth product must be greater than or equal to a minimum value of 1/4π or 0.08. The time–bandwidth product is often expressed as being ≥1, with the 4π factor cancelled out by defining the mean-square duration and bandwidth with a 4π constant appended to the right-hand side of Eqs. (12.16) and (12.39).

Equation 12.42 is often expressed as the radar or sonar uncertainty principle deriving its name from Heisenberg's Uncertainty Principle of quantum mechanics. The inequality in Eq. (12.42) places a constraint on the rms duration and bandwidth of a signal. For example, the range estimation accuracy of a signal expressed in Eq. (12.22) is inversely proportional to its rms bandwidth; however, the bandwidth cannot be made arbitrary large without regard to the rms duration. Gabor (1947) has shown that the only function that satisfies the equality in Eq. (12.42) is a constant frequency sinusoidal pulse having a Gaussian envelope. Gabor called this function the *elementary signal* and its equation is given below as Eq. (12.43). Gabor suggested that it could be used as the basis for signal

$$s(t) = A \begin{bmatrix} \cos(2\pi f_0 t + \phi) \\ or \\ \sin(2\pi f_0 t + \phi) \end{bmatrix} e^{-\pi^2\frac{(t-t_0)^2}{\Delta t^2}}, \tag{12.43}$$

where

A = relative amplitude
f_0 = peak frequency
t_0 = centroid of the signal
$\Delta\tau$ = rms duration of the signal
ϕ = phase shift

decomposition instead of the sinusoidal function in Fourier analysis.

The time–bandwidth product of dolphin capable of whistling are closer to the theoretical lower limit than the signals of the smaller nonwhistling dolphins and porpoises. However, even for the bottlenose dolphin signal, $\tau_d\beta$ is greater than the lower limit by 38–75%. These values do not support the notion of Kamminga and Beitma (1990) that dolphin echolocation signals have $\tau_d\beta$ values that approach the lower theoretical limit to with 20%. As $\tau_d\beta$ approaches the lower limit, the number of possible signal shapes will progressively become smaller until $\tau_d\beta$ equals the lower limit, in which case, only the Gabor elementary signal is possible. However, this may be a moot point since considering the time–bandwidth product of the transmitted signal is not realistic or practical. We have seen in

Section 13.2.1 that echoes arriving back to a dolphin will typically contain many highlights and will have a much longer duration than the transmitted signal. Therefore, the $\tau_d\beta$ product of echoes should be more important than the $\tau_d\beta$ product of the transmitted signal. If an echo has a duration equal to the broadband integration time for *Tursiops* (264 μs), the time–bandwidth product can be as large as 7, a value almost 90 times larger than the lower limit of 0.08.

12.2 Mammalian Auditory System Modeled as an Energy Detector

Research with humans and other mammals suggest that the mammalian auditory system behaves like an energy detector. Green and Swet (1966) have shown that an energy detector is a good analog of the human auditory detection process. Like wise, the auditory threshold versus duration and the critical ratio experiments of Johnson (1968a,b) with *Tursiops truncatus* indicate that the dolphin's inner ear functions like the human inner ear and that the animal integrates acoustic energy in the same way as humans. Therefore, it seems reasonable to approach the auditory system of dolphins as well as other marine mammals as an energy detector. The study of Au et al. (1988) discussed in the Section 13.2 indicates that for short-duration, broadband, high-frequency sonar signals, the auditory system of *Tursiops* behaves like an energy detector with an integration time of 264 μs.

12.2.1 Urkowitz Energy Detection Model

Urkowitz (1967) examined the detection of a deterministic signal in white Gaussian noise, using an energy detector and derived expressions for the *P*(Y/n) and *P*(Y/sn). In this section, the derivation of Urkowitz will be discussed and his results applied to dolphin target detection in masking noise results discussed in the previous section. In Urkowitz's model, the signal $y(t)$ is squared and then integrated over a time T. The noise is a zero mean Gaussian random variable with a flat band-limited spectral density N_o (rms noise level per Hz), and a bandwidth of W Hz. The Gaussian probability density function of the noise can be expressed as

$$f_N(y_i) = \frac{1}{\sqrt{2\pi\,\sigma^2}} e^{\frac{-y_i^2}{2\sigma^2}}, \tag{12.44}$$

where the noise variance σ^2 is equal to the rms noise level, so that

$$\sigma^2 = N_0 W = 2N_{02} W. \tag{12.45}$$

The parameter N_{02} is the two-sided noise spectral density and was used by Urkowitz (1967). The signal plus noise has a probability density function of

$$f_{SN}(y_i) = \frac{1}{\sqrt{2\pi\sigma^2}} e^{-\frac{(y_i - s)^2}{2\sigma^2}}. \tag{12.46}$$

The detection is a test of the following two hypotheses.

1) H_0: The input $y(t)$ is noise alone:

   a) $y(t) = n(t)$

2) H_1: The input $y(t)$ is signal plus noise:

   a) $y(t) = n(t) + s(t)$
   b) $E[n(t) + s(t)] = s(t)$

$E[y]$, the expected value of the random variation y, is zero for the noise since its mean is zero, and is $s(t)$ for the signal plus noise. The output of the energy detector taken over an interval T is denoted as V and is taken as the test statistic, with

$$V = \frac{1}{N_{02}} \int_0^T y^2(t)\,\mathrm{d}t \tag{12.47}$$

The incoming signal is sampled at an interval of Δt, and from Shannon's sampling theorem, a band-limited signal containing no frequency components above W Hz is uniquely determined by samples taken at intervals of $\Delta t = 1/(2W)$. According to Shannon's sampling theorem (Schwartz, 1980), the sampled noise can be expressed as

$$n(t) = \sum_{i=1}^{2TW} \frac{\sin[\pi(2WT - i)]}{\pi\ (2WT - i)} \tag{12.48}$$

where n_i is the i-th sample of the noise at time $i/(2W)$. The energy in the noise can be written as

$$\int_0^T n^2(t)\,\mathrm{d}t = \int_0^T \sum_{i=1}^{2TW} n_i \frac{\sin[\pi\,(2WT - i)]}{\pi\,(2WT - i)} \sum_{j=1}^{2tw} \frac{\sin[\pi\,(2WT - j)]}{\pi(2WT - j)} \mathrm{d}t. \tag{12.49}$$

However, the integral of Eq. (12.49) can be simplified since

$$\int_0^T \frac{sin[\pi(2WT - i)]}{\pi(2WT - i)} \frac{sin[\pi(2WT - j)]}{\pi(2WT - j)} \mathrm{d}t = \begin{cases} \frac{1}{2W} \text{when } i = j \\ 0 \text{when } i \neq j. \end{cases} \tag{12.50}$$

Therefore, when the input to the energy detector is noise only, using Eqs. (10.33) and (1034), the test statistic of Eq. (10.31) can be expressed as

$$V = \sum_{i=1}^{2TW} b_i^2, \tag{12.51}$$

where

$$b_i = \frac{n_i}{\sqrt{2\,WN_{02}}}. \tag{12.52}$$

Thus, V is the sum of the square of 2TW Gaussian random variables, each with a zero mean and unity variance and such a random variable has a χ_i^2 or chi-square distribution (Meyer, 1965). V has a chi-square distribution with 2TW degrees of freedom.

Now consider the situation when input to the energy detector is only the signal $s(t)$. Using the sampling theorem the signal can be expressed as

$$s(t) = \sum_{i=1}^{2TW} \frac{\sin[\pi(2WT - i)]}{\pi(2WR - i)}. \tag{12.53}$$

Following the same line of reasoning concerning Eqs. (12.49) and (12.50), we can write

$$\frac{1}{N_{02}}\int_0^T s^2(t)\,\mathrm{d}t = \frac{1}{2W}\sum_{i=1}^{2TW}\beta_i^2, \tag{12.54}$$

where

$$\beta_i = \frac{s_i}{\sqrt{2\,WN_{02}}} \tag{12.55}$$

Using Eqs. (12.51) and (12.54), the total input $y(t)$ with the signal plus noise present can be written as

$$y(t) = \sum_{i=1}^{2TW}(n_i + s_i)\frac{sin[\pi(2WT - i)]}{\pi(2WT - i)} \tag{12.56}$$

Under the hypothesis H_1, the test statistic V can now be written as

$$V = \frac{1}{N_{02}}\int_0^T y^2(t)\mathrm{d}t = \sum_{i=1}^{2TW}(b_i^2 + \beta_i^2)^2. \tag{12.57}$$

The sum in Eq. (10.57) has a noncentral chi-square distribution with 2TW degrees of freedom and a noncentrality parameter λ given by

$$\lambda = \sum_{i=1}^{2TW}\beta_i^2 = \frac{1}{N_{02}}\int_0^T s^2(t)\mathrm{d}t = \frac{E_s}{N_{02}}. \tag{12.58}$$

λ is merely the energy signal-to-noise ratio.

The probability of false alarm will be related to the area under the chi-square distribution curve. The probability of false alarm for a given threshold V_T derived by Urkowitz (1967) is given by

$$P(Y/n) = Pr\{V > V_T|H_0\} = Pr\{\chi^2_{2TW} > V_T\} \tag{12.59}$$

For the same threshold V_T, the probability of detection is given by

$$P(Y/sn) = Pr\{V > V_T|H_1\} = Pr\{\chi^2_{2TW}(\lambda) > V_T\} \tag{12.60}$$

The symbol $\chi^2_{2TW}(\lambda)$ indicates a noncentral chi-square variable with 2TW degree of freedom and noncentrality parameter λ. The expression for the probability of false alarm can be rewritten as

$$P(Y/n) = 1 - Pr\{V_T \leq \chi^2_{2TW}\}, \tag{12.61}$$

where Pr is the area under the chi-square distribution curve with 2TW degrees of freedom. Similarly, the expression for the probability of detection can be rewritten as

$$P(Y/sn) = 1 - Pr\{V_T/G \leq \chi^2_D\}, \tag{12.62}$$

where

$$D = (2TW + E/N_{02})/(2TW + 2E/N_{02}). \tag{12.63}$$

$$G = (2TW + 2E/N_{02})/(2TW + E/N_{02}) \tag{12.64}$$

Pr is now the area under the noncentral chi-square distribution with a modified number of degrees of freedom *D* and a threshold divisor *G*.

12.2.2 Application of Urkowitz Model

These expressions derived by Urkowitz (1967) was applied by Au (1988) to the dolphin target detection in noise data discussed in the Section 11.2.1. In order to apply Urkowitz's model, we need to assumed that the dolphin behaved like an unbiased detector, which means that

$$P(Y/sn) = 1 - P(Y/n). \tag{12.65}$$

The first step in applying Urkowitz's model is to choose a desired value of 2TW and calculate, by an iterative method, the V_T necessary to obtain different specific values of *P*(Y/n) using Eqs. (12.59) or (12.61). Since from Eq. (12.65), once *P*(Y/sn) is specified, *P*(Y/sn) is also specified, so that the necessary E/N_{02} to obtain *P*(Y/sn) for a given 2TW and V_T can be determined from Eq. (12.60) or (12.62). The values E/N_{02} necessary in order to achieve different specific *P*(Y/sn) can be compared with the dolphin performance data. The process is continued for different 2TW values until the values of *P*(Y/sn) as a function of E/N_{02} which best fitted the dolphin data is found.

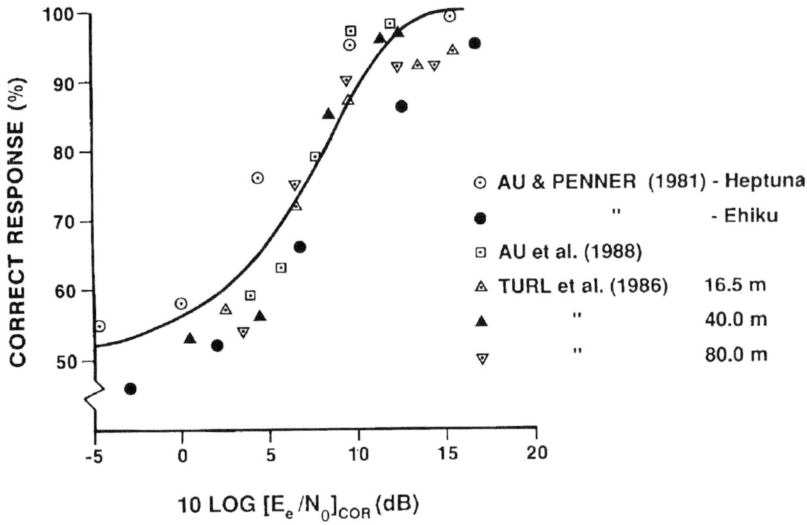

FIGURE 12.7. Dolphin performance results compared with results from Urkowitz energy detector model for 2TW = 20 (adapted from Au, 1993).

The result of such an iterative process using Urkowitz model is plotted in Fig. 12.7, along with the performance results of the three dolphin target detection in noise experiments discussed in the previous section. Urkowitz's energy detector model agrees well with the dolphin's performance results, further supporting the notion that the dolphin behaves like an energy detector. If we insert the dolphin's integration time of 264 μs into 2TW = 20, we get a bandwidth of 37.9 kHz or 45.8 dB for the energy detector. Au and Moore (1990) measured a critical ratio of approximately 18 kHz or 42.6 dB at 120 kHz for *Tursiops*, which is in relatively good agreement with the bandwidth for the energy detector model. The unbiased detector assumption used to derive $P(Y/sn)$ is good for signal-to-noise ratio conditions that correspond to performance at or above the 75% correct response threshold as can be seen in Fig. 11.15 of Au (1993).

12.3 Signal Processing Models for Signal Recognition

In this section we will consider methods by which broadband sonar signals like those used by dolphins can be processed in order to discriminate and recognize targets. One again we would like to stress that although the examples considered in this section pertain to dolphin sonar, the methods are valid for any type of signals. Three different models will be considered, although there are no doubt other models that can be used to process broadband sonar signals. All three signal processing models possess a common approach with their major difference being the manner in which features of targets are described and stored. The three major steps involved in developing and using a signal processing model for target recognition are as follows:

1. Use a mathematical model to identify and extract pertinent features of targets from the echoes.
2. Collect echoes from known targets and store features in a library.
3. Process echoes from an unknown target using the mathematical model developed in (1) and identify the target features. The features from the unknown targets are then compared with features of known targets stored in the library. The library search and comparison process should lead to the identification or recognition of the unknown target if its features have been previously collected and stored in the library, or should indicate targets that are closest to the unknown target.

If we are only interested in performing a discrimination task to determine if targets A and B are different, steps 2 and 3 can be skipped. The features for both targets can be determined in step 1 and used directly to compare the echoes from both targets.

12.3.1 Energy Detection in a Filter Bank

The first signal processing model consist of characterizing a target by passing either its echo or its transfer function through a bank of contiguous filters and calculating the energy within each filter (Chestnut et al., 1979). If we let the frequency boundaries of the filters be $f_0, f_1, f_2, \ldots f_N$, $s(t)$ = the transmitted signal, $y(t)$ = the echo, then the energy contained in the i-th filter can be expressed as

$$E_i = \int_{f_{i-1}}^{f_i} \left| Y(f) \right|^2 \mathrm{d}f \quad \text{(Echoes)}. \tag{12.66}$$

$$E_i = \int_{f_{i-1}}^{f_i} \left| \frac{Y(f)}{S(f)} \right|^2 \mathrm{d}f \quad \text{(TransferFunction)}, \tag{12.67}$$

where $S(f)$ and $Y(f)$ are the Fourier transform of the transmitted signal and received echo, respectively. The collection of N energy terms, $E_1, E_2, \ldots, E_N$, will be the feature set for a target. The function in the integral is usually normalized to unity so that only features associated with the frequency spectrum will be computed and amplitude will not be a factor. The advantage of transfer function form comes from the fact that the same signal does not have to be used to characterize the target for library storage and for examining unknown targets. Any well-characterized broadband signal can be used to measure the features of a target, and another known signal can be used in the field. In order to use Eq. (12.66), the same signal needs to be used to both measure the target features and ping off targets in the field. The advantage of using only the echo comes from not having to perform a complex division of the received echo spectrum by the incident signal spectrum.

Two basic type of filters can be used, constant interval and constant-Q filters. For the constant interval filters, the parameters that need to be specified are the lower and upper frequencies of interest, the bandwidth of each filter, and the number of filters. Since echo data will no doubt be processed with a computer and an FFT algorithm will be used to estimate the Fourier transform in Eq. (12.66), the parameters should be chosen to coincide with the FFT bin size and the boundaries of the FFT bins. If we let SR = analog-to-digital sample rate, N = the number of samples used to characterize the time domain echo, then the FFT bin size from Eq. (1.26) will be

$$\Delta f = \frac{\mathrm{SR}}{N} = \frac{1}{N\Delta t} \tag{12.68}$$

where Δt = time interval between samples. Therefore, the boundaries of the FFT bins will be at frequencies corresponding to

$$f_i = (i-1)\Delta f, \tag{12.69}$$

where $i = 1, 2, \ldots N/2$. Therefore, it would be convenient to choose the lower and upper frequencies of the filter bank to coincide with Eq. (12.66), and to require that the filter interval be a integer multiple of Δf.

The second type of filter, the constant-Q filter, is closely related to the dolphin's auditory system since the critical ratio of *Tursiops truncatus* (Fig. 10.6) measured by Johnson (1968b) can be approximated by a bank of constant-Q filters. From Eq. (3.4), $Q = f_o/\Delta f$, where f_o is the center frequency of the filter and Δf is its bandwidth. Let f_{i-1} be the lower frequency limit, and f_i be the upper frequency limit of the *i*-th filter, then the center frequency will be the geometric mean between the frequency boundaries, or $f_o = (f_i - f_{i-1})/2$, and the bandwidth $= f_i - f_{i-1}$. Substituting these expressions into the definition of Q, the frequency boundaries of the *i*-th filter can be expressed as

$$f_i = \frac{2Q+1}{2Q-1} f_{i-1}. \tag{12.70}$$

Let f_l = the lower frequency limit and f_u = the upper frequency limit in the bank of constant filters bank, we can then expressed the ratio of the lower to upper frequency limit of each filter as

$$\frac{f_{i+1}}{f_i} = \frac{2Q+1}{2Q-1} \quad 1^{\text{st}} \text{ filter}$$
$$\frac{f_{i+2}}{f_{i+1}} = \frac{2Q+1}{2Q-1} \quad 2^{\text{nd}} \text{ filter}$$
$$\ldots\ldots\ldots\ldots\ldots \tag{12.71}$$
$$\ldots\ldots\ldots\ldots\ldots$$
$$\frac{f_u}{f_{u-1}} = \frac{2Q+1}{2Q-1} \quad \text{N}^{\text{th}} \text{ filter}$$

Multiplying each frequency ratio with each other and canceling common terms, we get

$$\frac{f_u}{f_l} = \left[\frac{2Q+1}{2Q-1}\right]^N. \tag{12.72}$$

Solving for Q in Eq. (10.55), we arrive at

$$Q = \frac{1}{2}\frac{\left(\frac{f_u}{f_l}\right)^{\frac{1}{N}}+1}{\left(\frac{f_u}{f_l}\right)^{\frac{1}{N}}-1}. \tag{12.73}$$

There are three parameters which can be varied, f_l, f_u, and the number of filters, N. According to Chestnut et al. (1979), there is redundancy in the vectors of energy values used in the constant-Q filter bank, implying that the method is not as efficient in terms of the amount of information stored.

12.3.2 Measure of Feature Recognition

The features used to characterize targets, such as the energy in a bank of filters, can be consider a multielement vector, having some pattern of distribution. In order to discriminate, recognize or classify echoes, some sort of pattern recognition scheme must be used in which features of received echoes can be compared against archetypal features vectors of echoes from known targets. The problem becomes one of recognizing the patterns of feature distribution and determining a quantitative measure of the similarity or dissimilarity of a feature vector from a host of other feature vectors. There are many pattern recognition schemes that can be used; however, only one, the Euclidean distance measure, will be considered here. Let E_i be the features of an echo just received by a sonar and $E_i(k)$ be the features of target-k stored in an archetype library, where $i = 1, 2, \ldots, N$, then the similarity between the feature vectors E_i and $E_i(k)$ can be expressed by the Euclidean distance

$$d_k = \sqrt{\sum_{i=1}^{N} [E_i - E_i(k)]^2}. \tag{12.74}$$

Features of the received echo are compared with corresponding features of all targets stored in the archetype library, and the pair of targets having the smallest Euclidean distance will be the most similar. Chestnut et al. (1979) also discussed some other pattern recognition methods.

Chestnut et al. (1979) performed a sonar target recognition experiment using the energy in a bank of filter and the Euclidean distance measure as one of the signal processing scheme. They also used an all-pole filter model to characterize echoes; however, only the results of the filter bank experiment will be discussed. Echoes from the 16 simple geometrical targets listed in Table 12.1 were collected in a test pool using the low-frequency broadband signal having a peak frequency of 23 kHz as shown in Fig. 12.8. A total of 240 echoes, 10 or 20 from each target, were collected. Four echoes from each target were used in the design set and the rest were used for the test set. The overall results in comparing the echoes in the design set with the echoes in the test set as a function of the signal-to-noise ratio are shown in Fig. 12.9. The probability of rejection (solid curves) and probability of misclassification (dashed curves) as a function of signal-to-noise -ratio are plotted in Fig. 12.9. The results indicate that as the signal-to-noise ratio increases the probability of making an error decreases. The results also indicate that less errors were progressively made as the number of filter increases to 30. The best results were obtained using 30 filters on the interval from 15 to 45 kHz. There were only slight differences between the uniformly spaced intervals and the constant-Q intervals.

12.3.3 Time Domain Highlight Features

Martin and Au (1988) described a time domain feature extraction model, which uses cues previously identified as salient for humans in Section 9.6. This model consists of extracting highlight separation and highlight amplitude

TABLE 12.1 Targets used in the sonar experiment of Chestnut et al. (1979)

Target	Diameter	Length	Wall
Hollow aluminum cylinder	3.81	17.78	0.48
Hollow aluminum cylinder	3.81	17.78	0.79
Hollow aluminum cylinder	3.81	17.78	0.95
Hollow aluminum cylinder	7.62	17.78	0.32
Hollow aluminum cylinder	7.62	17.78	0.64
Hollow aluminum cylinder	7.62	17.78	0.95
Hollow aluminum cylinder	7.62	17.78	1.58
Hollow aluminum cylinder	10.16	17.78	0.64
Hollow aluminum cylinder	6.36	17.78	0.64
Solid rock cylinder	3.81	17.78	–
Solid rock cylinder	6.35	17.78	–
Hollow PVC cylinder	7.62	17.78	1.27
Hollow aluminum sphere	7.62	–	–
Hollow aluminum sphere	15.14	–	–
Aluminum biconic	–	12.27	–

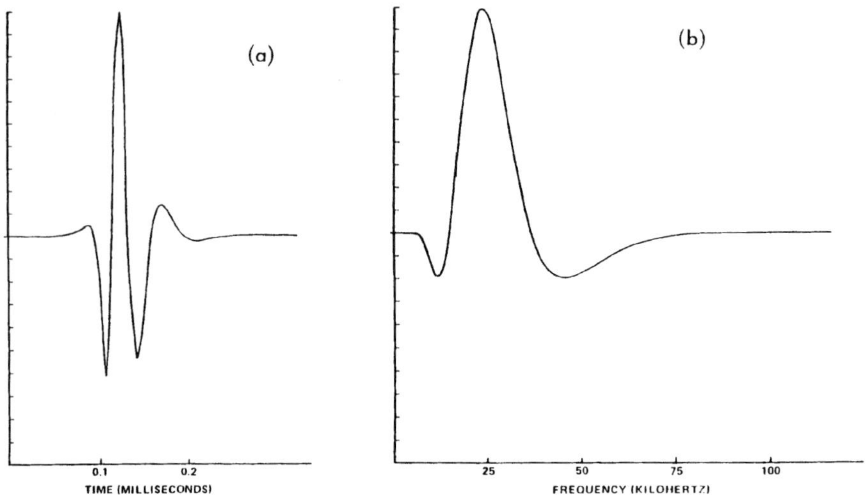

FIGURE 12.8. (**a**) The transmitted pulse and (**b**) the real part of the Fourier Transform of the transmitted pulse (adapted from Chestnut et al., 1979).

ratios from echo envelopes to form the target feature sets. The first stage consists of passing a target echo through a matched-filter with the transmitted signal as the reference. Let $s(t)$ be the transmitted signal and $y(t)$ be the received target echo, the output of the matched-filter will be

$$x(t) = \Im^{-1}[Y(f)\, S^*(f)], \tag{12.75}$$

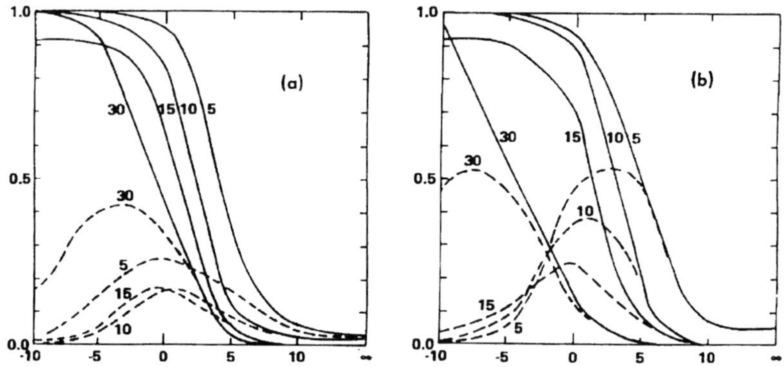

FIGURE 12.9. Probability of rejection (*solid curves*) and probability of misclassification (*dashed curves*) as a function of the signal-to-noise ratio for (**a**) constant filter width and (**b**) constant-Q model. The parameters are the number of filters (adapted from Chestnut et al., 1979).

where $S(f)$ and $Y(f)$ are the Fourier Transform of the time domain signals. From Eq. (9.4), the envelope of the matched-filter output can be expressed as

$$Env[x(t)] = |x_a(t)| = \sqrt{x^2(t) + \hat{x}^2(t)}, \tag{12.76}$$

where from Eq. (12.2), $\hat{x} = \Im^{-1}[-jX(f)]$ for $f > 0$. The resultant signal envelope is processed by a peak detector that determines the location of highlights by finding points where the slope of the envelope changes from positive to negative. Small-amplitude extrema in the immediate neighborhood of a larger maximum are rejected. After obtaining a list of highlights for a given echo, the absolute maximum is assigned a time separation of zero and an amplitude of one. The other highlights are assigned negative or positive time separations according to the position before or after the largest highlight. Amplitude ratios for each highlight are calculated with respect to the maximum. In the second stage, reference vectors are created from a set of echoes by aligning the features across the signals. Absolute maxima are aligned, so that every group of signals has a highlight occurring with amplitude ratio of one and time separation zero. Other highlights are also aligned, and amplitude ratios and time separation become statistical quantities represented as means and standard deviations for the group of signals.

Martin and Au (1988) performed a sonar target recognition experiment using their time-domain feature model. Target echoes were collected in a test pool using a simulated dolphin sonar signal having a peak frequency of 120 kHz similar to the incident signal shown in Fig. 13.12. The echoes were digitized at a sample rate of 1 MHz. Reference and test data were represented by vectors of 25 features; each feature represented the time separation and relative amplitude of an echo highlight. Reference data were means of the feature vectors from 10 echoes. The time axis was partitioned into bins of 20 points or 20 μs each, and each bin was assigned an amplitude value determined by highlights in that portion of the echo. When any of the 25 time windows contained an echo highlight, that element was assigned the value of the highlight amplitude ratio relative to the maximum. If the partition did not contain a highlight, a value of 0 was assigned. When more than one highlight was present in a partition, the largest amplitude was used. The Euclidean distance measure of Eq. (12.74) was used to compare echoes from the reference set with echoes from the test set.

Echoes from the cylinder material composition discrimination experiment of Hammer and Au (1980), and a solid aluminum cylinder, were tested by Martin and Au (1988). Performance of the feature extraction and pattern recognition algorithms are shown in Tables 12.2 and 12.3 as confusion matrix. Scores along the diagonal of the matrix represent correct responses; off-diagonal elements represent confusions. The signal processing scheme was 90% correct for material composition discrimination in Table 10.2 and 100% correct for internal structure discrimination in Table 12.3. Chance performance was 12.3% correct

TABLE 12.2 Confusion matrix for cylinder material composition discrimination using time domain features. Cycl-1 had outer diameters of 3.61-cm, Cycl-2 and the solid aluminum cylinder had diameters of 7.62-cm. All of the cylinders had lengths of 17.78 cm (adapted from Martin and Au, 1988)

TEST TARGET ECHOES \ REFERENCE TARGET ECHOES	ALUM CYL-1	STEEL CYL-1	BRONZE CYL-1	ALUM CYL-2	STEEL CYL-2	BRONZE CYL-2	SOLID ALUM CYL
ALUM CYL-1	97%					3%	
STEEL CYL-1		100%					
BRONZE CYL-1			100%				
ALUM CYL-2				93%		7%	
STEEL CYL-2				37%	63%		
BRONZE CYL-2						100%	
SOLID ALUM CYL							89%

(1 in 7) for material composition and 25% correct (1 in 4) for the internal structure discrimination. When echoes from two different glass cylinders were added to the data of Table 12.2, to make a 9 by 9 confusion matrix, performance dropped to 62% correct; chance dropped to 11% correct (1 in 9). Many glass test echoes were incorrectly identified, and some echoes from other cylinders were wrongly identified as glass. Identification of echoes from the glass cylinders was also poor for both humans (Au and Martin, 1989) and dolphins (Hammer and Au, 1980; Schusterman et al., 1980). The study of Martin and Au (1988) was performed in a noise-less environment.

12.3.4 *Spectrogram Correlation Model*

In Chapter 10 we described a spectrogram as a time–frequency representation of a signal in which frequency spectra corresponding to different portions of the signal in the time domain are displayed. Therefore, a spectrogram represents the short-time spectral history of a signal (Altes, 1980a) and is plotted in a three-dimensional form like the ambiguity function shown in Fig. 12.6,

TABLE 12.3 Confusion matrix for internal structure discrimination. All of the cylinders had out diameters of 7.62 cm and lengths of 17.78 cm (adapted from Martin and Au, 1988)

TEST TARGET ECHOES \ REFERENCE TEST TARGET	ALUM CYL-2 AIRFILLED	CORAL CYL-2 SOLID	ALUM CYL-2 WATERFILLED	ALUM CYL-2 SOLID
ALUM CYL-2 AIRFILLED	100%			
CORAL CYL-2 SOLID		100%		
ALUM CYL-2 WATERFILLED			100%	
ALUM CYL-2 SOLID				100%

but with the velocity axis being replaced by a frequency axis. The features of a target can be determined by using Eq. (12.67), for each spectrum of the spectrogram, thereby creating a time–frequency description of the target with transfer function $H(f)$ and impulse response $h(t)$. Therefore, the time–frequency density function of the target can be expressed as (Rihaczek, 1969)

$$e_{\mathrm{hh}}(t,f) = h(t)H^{*}(f)e^{(-j2Bft)}. \tag{12.77}$$

According to Altes, the time energy density function provides a useful description of a target because it contains information about both highlight and energy spectrum. Highlight ranges can be obtained from $e_{\mathrm{hh}}(t,f)$ from the integral

$$|h(t)|^2 = \int_{-\infty}^{\infty} e_{hh}(t,f)\,\mathrm{d}t. \tag{12.78}$$

Similarly, the frequency spectrum of the target impulse response is

$$|H(f)|^2 = \int_{-\infty}^{\infty} e_{hh}(t,f)\,\mathrm{d}t. \tag{12.79}$$

The features of a spectrogram obtained from a sonar return can be denoted as E_{ij}, where i refers to the energy in the i-th frequency bin within the j-th time bin. The features from the k-th target in a library of archetypes can be expressed as $E_{ij}(k)$. The locally optimum detector is a spectrogram correlator (Altes, 1980a), which computes the quantity

$$l(\underline{e}) = \sum_{i=1}^{N}\sum_{j=1}^{M} E_{ij}\, E_{ij}(k). \tag{12.80}$$

Altes and Faust (1978) applied the spectrogram model to echoes from all of the targets listed in Table 12.1 along with echoes from a 6-cm foam cube (0°, 45°), a 6 × 6 cm foam cylinder (0°, 90°), and a 6.35 cm O.D. × 17.8 cm long cylinder at incident angles 0–90° in 5° increments. They used 16 frequency bins and 27 time bins to characterize the spectrogram of each target echo. Their results comparing the 6.35 × 17.8 cm cylinder versus all other targets, cylinders versus noncylinders, and metal versus nonmetal are shown in Fig. 12.10. The probability of misclassification versus the signal-to-noise ratio is shown in Fig. 12.10. The results indicate that the target classification scheme of Altes (1980a) using spectrogram correlation can indeed be used to classify sonar echoes. However, this signal processing technique is extremely time consuming to implement since N × M features are required to characterize a target compared with N features for the filter bank model and M features for the time domain model.

12.3.5 Comparative Evaluation of Target Recognition Models

The three signal processing models considered for target recognition all seem to have a capability to discriminate, recognize, and classify target echoes. The relative effectiveness of these three models as well as other mathematical models has not been adequately assessed under rigorous test conditions. The relative effectiveness of these three models has not been compared with the dolphin sonar. All of the discriminations considered with the mathematical models could also be performed relatively readily by the dolphin, with the exception of the hollow glass versus aluminum discrimination performed by Martin and Au (1988). Dolphin sonar target recognition experiments need to be conducted in which the animal's performance varies from very good to

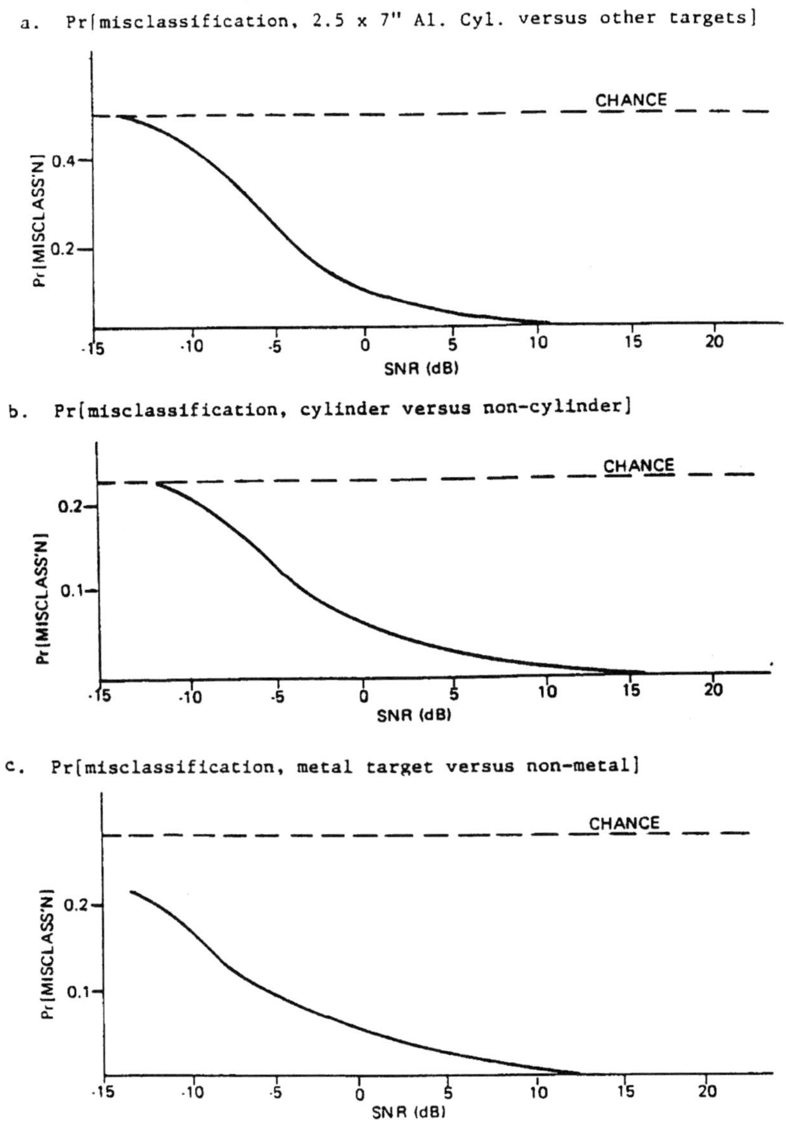

FIGURE 12.10. Classification performance of the spectrogram model in white noise (adapted from Altes, 1980a).

chance as a target parameter is varied. These targets can then be used to compare the performance of different signal processing models with the dolphin's performance. Initially, such tests should probably be performed with echoes collected in a stable, quiet environment of a test pool. Models that compare well with the dolphin's performance should then be further tested in a more realistic environment, and in real time. In a real environment, factors such as target motion, transducer motion, ambient noise, and reverberation can have adverse effects on any sonar function.

12.4 Artificial Neural Network and Signal Recognition

We will now consider an area of research that has attracted considerable attention in the past several years, and has shown considerable promise in solving pattern recognition problems. Although research into artificial neural networks (ANN) began in late 1940 s, interest in this area only began to mushroom in the mid-1980s, as investigators began to achieve success in solving certain problems that have been essentially unsolvable or, have not been solved successfully or reliably. Only a brief overview of ANNs will be presented in this book. Those interested in a deeper presentation should refer to the many excellent books of the subject such as those written by Wasserman (1989), Hecht-Nielsen (1989), and Dayhoff (1990).

The human brain and nervous system regularly perform massive amounts of parallel processing using an estimated 10^{11} neurons in perhaps 10^{15} interconnections over a relatively small volume (size of the brain). Humans can solve complex pattern recognition problems on which computational methods and models have only achieved moderate and uninspiring success. Artificial neural networks are computational models that attempt to emulate the massive parallel processing nature of the nervous system and brain. The networks are built around mathematical analogs of interconnected neurons, with different weights (emphasis) attached to each connection.

The neuron is the most basic cellular unit of the brain and its mathematical model represents the most basic unit in an artificial neural network. A schematic of an artificial neuron called a processing element (PE) is presented in Fig. 12.11. In discussing the mathematical model of a neuron, the biological equivalent parts or function for a real neuron will be given in parenthesis. A simulated neuron, referred to as a processing element (PE), has input

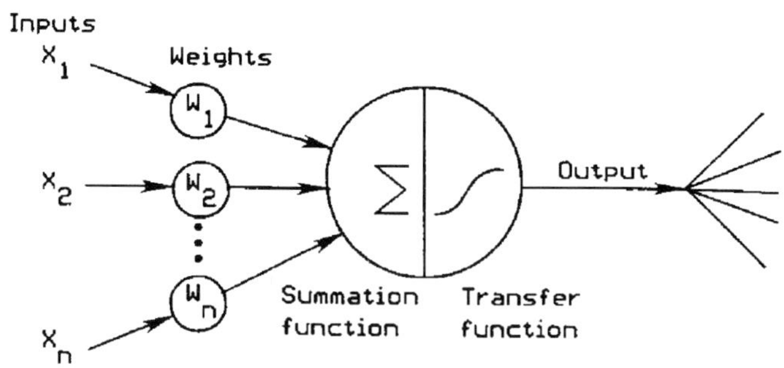

FIGURE 12.11. A schematic of a processing element in an artificial neural network.

connections (dendrites) through which it receives activations from other units, a summation function that combines the various inputs into a single activation, a transfer function that converts this summation into output activation, and an output connection (axon) by which a unit's output activation arrives as input to other units. Each input has a weight (synaptic strength) that modifies each the input signal so that the summation by the j-th processing element can be thought of as a weighted sum

$$I_j = \sum_{i=1}^{N} W_{ij} X_i . \tag{12.81}$$

The weighted sum is processed by a transfer function $F(x)$, often referred to as a "squashing function" to produce an output. A variety of squashing functions exist to limit the output level of the PE. One popular squashing function is the "sigmoid" (meaning S-shaped) function, $F(x) = 1/(1 + e^{-x})$. The sigmoid function is plotted in Fig. 12.12. The output of the j-th PE with a sigmoid squashing function will be

$$O_j = F(I_j) = 1/(1 + e^{-I_j}). \tag{12.82}$$

The squashing function serves to limit the output of the PE to within certain bounds. For large positive values of I, $F(I)$ in Eq. (12.82) will approach 1. For large negative values of I, $F(I)$ will approach 0. Therefore, the sigmoid squashing function will limit the output of a PE to values between 0 and 1. Another popular squashing function is the hyperbolic tangent or tan $h(x)$ function, which will limit the output of a PE to be between –1 and 1.

An artificial neural network consist of many interconnected PEs that operate in parallel, simulating the process in which the human brain may encode information. The processing elements are organized into a sequence of layers with full connections between successive layers. There is an input

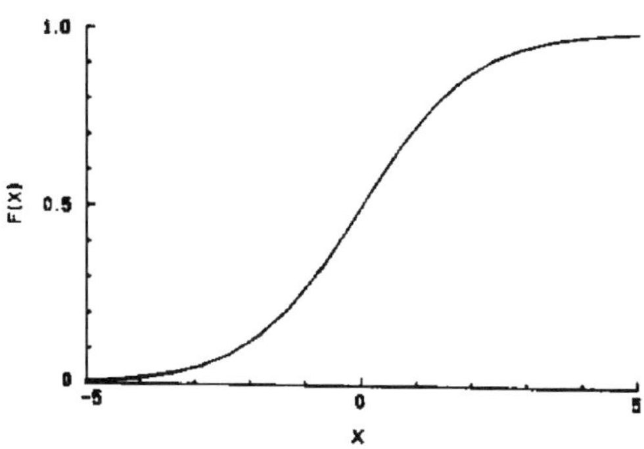

FIGURE 12.12. The sigmoid squashing or transfer function.

layer where data are presented, an output layer that holds the response of the network to a given input, and "hidden layers" that are neither input nor output layers where the bulk of the processing is performed. An artificial neural network is not programmed to solve specific problems but is "taught" or trained to give acceptable answers to specific inputs. Training can be supervised and unsupervised. In supervised training, each input is paired with a desired output, forming a training pair. Input data are given to the network and an output is calculated and compared with the desired output. The difference or error is fed back through the network and the weights of each connection adjusted according to a specific learning rule. The process is repeated continuously until the error is minimal. The resultant weights corresponding to a satisfactory solution constitute a learned response. which the network can remember by storing values of the weights.

In unsupervised training no desired outputs are required and the training set consists solely of input data. Training consists of modifying the network weights to produce outputs that are consistent. This process leads to internal data clustering.

12.4.1 Backpropagation Network

There are various ways in which processing elements can be configured to form a specific network. In this book, only two network architectures will be considered, the backpropagation and the counterpropagation networks. The backpropagation network is probably the most widely used neural network paradigms and has been applied successfully to solve a variety of problems. Werbos (1974) was the first to present a conceptual basis of backpropagation. Parker (1982) unknowingly reinvented the backpropagation concept. Rumelhart and McClelland (1986), unaware of the previous works of Werbos (1974) and Parker (1982), popularized the backpropagation network in their book, *Parallel Distributed Processing*. A simplified diagram of a three-layer backpropagation network consisting of an input layer, an output layer, and one hidden layer is shown in Fig. 10.12. Each PE would be similar to the PE shown in Fig. 12.13. Variations to this circuit would include addition hidden layers and different amount of PE in each layer. Initially, all the weights are assigned small random values and training begins by input of a training vector set (Xs) and computing the output using the initial random weights. The calculated output vector is compared with the desired target vector and the error between both vectors is calculated and appropriate correction factors are determined that would minimize the amount of error. The weights of the network are adjusted by "backpropagating" the error adjustment terms progressively from the output toward the input as shown schematically in Fig. 12.13. The process is repeated by applying another set of input vector resulting in a new set of output and a new set of error terms, which will be backpropagated through the network again. This iterative process is continued indefinitely until the amount of error falls within a desired limit. When

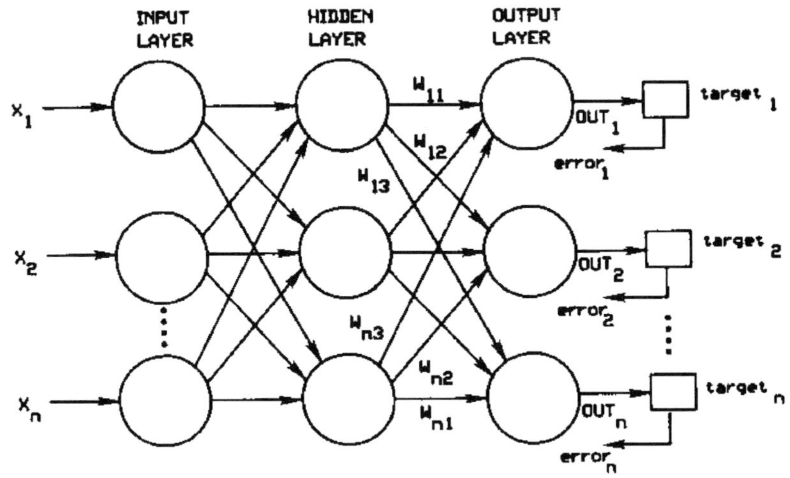

FIGURE 12.13. Diagram of a simplified three-layer backpropagation artificial neural network.

training is terminated, the weights associated with each PE are saved and can be used later to classify data in a test set.

There are various learning rules by which error and associated weight adjustment can be calculated; however, the delta learning rules is probably the most used with the backpropagation network. The delta rule is a simple and elegant learning heuristic that gives a network an ability to form and modify its own connections in ways that often approach optimum performance. In the backpropagation network, the error correction that will be backpropagated through the network must be done differently for the output and hidden layers. The difference between the target output T_j and the actual output O_j for the j-th PE in the output layer of Fig. 12.11 is the output error and the error value, δ_j, that is backpropagated from the output layer is given by

$$\delta_j = (T_j - O_j)F'(I_j), \tag{12.83}$$

where $F'(I_j)$ is the derivative of the squashing function and I_j is the weighted sum of the j-th PE in the output layer. The quantity $(T_j - O_j)$ is the amount of error and $F'(I_j)$ scales the error to force a stronger correction when the sum I_j is near the rapid rise in the sigmoid curve of Fig. 12.12. The derivative of the Sigmoid function $F'(I)$ is simply

$$F'(I) = F(I)(1 - F(I)) = O(1 - O), \tag{12.84}$$

where O is the output of the PE. The error for a PE in the hidden layer is more complicated and can be explained with the diagram in Fig. 12.14, showing a simplified situation involving the j-th PE in the hidden layer. The error will be

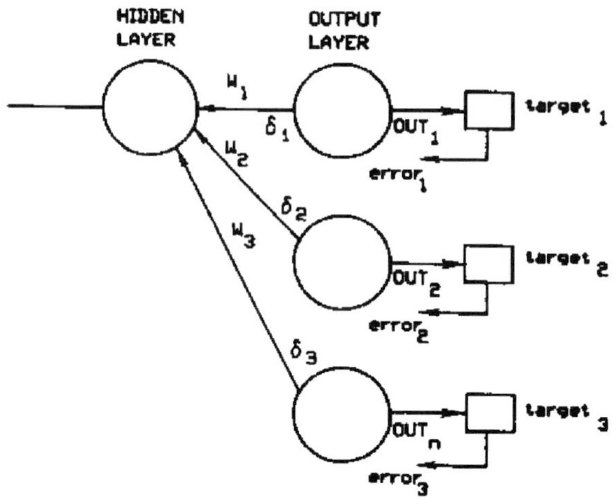

FIGURE 12.14. Simplified diagram involving the calculation of the error for a PE in the hidden layer.

the weighted sum of error associated with each PE in the output layer multiplied by $F'(I)$

$$\delta_j = F(I_j)(\delta_1 W_1 + \delta_2 W_2 + \delta_3 W_3). \tag{12.85}$$

Let η be a "learning rate" coefficient, then the adjustment Δ to the weights of the PE will be

$$\Delta_j = \eta \delta O_j. \tag{12.86}$$

If $w_j(n)$ is the weight before and $w_j(n+1)$ is the weight after adjustment, the relationship between the weights before and after adjustment is

$$W_j(n+1) = W_j(n) + \Delta_j. \tag{12.87}$$

The delta rule can adjust weights for targets and desired outputs of either polarity and for both continuous and binary data.

12.4.2 *Counterpropagation Network*

A network that is useful in separating and learning categories is the counterpropagation network devised by Hecht-Nielsen (1987). For pattern classification problems, the counterpropagation network is a viable alternative to the backpropagation network with potentially faster learning and is nearly as optimal in solving problems. The network is a combination of two well-known algorithms: the self-organizing map of Kohonen (1988) and the Grossberg (1969) outstar. A diagram of a three-layer counterpropagation network consisting of an input layer, a Kohonen layer, and a Grossberg layer is shown in Fig. 12.15. The PEs of the input layer perform no computation but serve a fan-out function connecting the input vector to each Kohonen neuron in the Kohonen layer. A set of weights is associated with each input to each Kohonen neuron (as shown in Fig. 12.15) so that the output of a Kohonen neuron will be the weighted sum of its input as expressed by Eq. (12.81). The Kohonen neuron with the largest output is set to 1 and the outputs of the other Kohonen neurons are set to 0. Therefore, the Kohonen layer serves as a "Winner-take-all" function and is often referred to as a competitive layer.

The Kohonen layer classifies the input vectors into similar groupings by adjusting the weights so that similar input vectors will activate the same Kohonen neuron. The training is a self-organizing algorithm that operates in an unsupervised mode. The input vectors are usually normalized by dividing each component by the vector's length so that

$$x_i = \frac{x_i}{\sqrt{x_1^2 + x_2^2 + \ldots + x_n^2}}. \tag{12.88}$$

This operation converts an input vector into a unit vector pointing in the same direction. Each normalized input vector can be thought of as being on the surface of a hypersphere of unit radius. The winning PE in the Kohonen layer will be the unit with weight vectors that most closely matches the input vector and its weights are adjusted by the equation

$$W_{\text{new}} = W_{\text{old}} + \alpha(x - W_{\text{old}}), \tag{12.89}$$

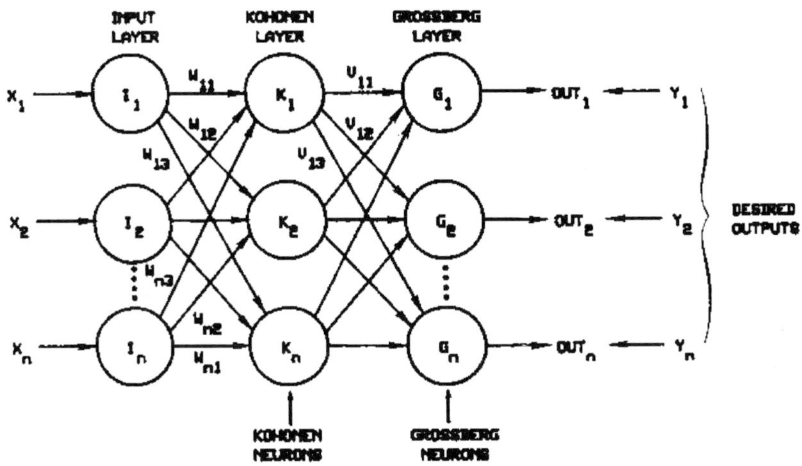

FIGURE 12.15. Diagram of a three-layer counterpropagation network.

where α is a training rate coefficient. Initially, the weights are randomized and normalized. As training progresses, each weight associated with the winning Kohonen neuron is changed by an amount proportional to the difference between its output and input values to minimize the difference between the weight and its input. The winner then becomes more likely to win the competition when the same or similar input vector is presented to the network.

The output of the Kohonen layer, $k_1, k_2, \ldots, k_2$ are applied to the Grossberg layer through a series of weights depicted as $V_{11}, V_{21}, \ldots, V_{nl}$. The output of the j-th Grossberg neuron can be expressed as

$$\text{out}_j = \sum_{i=1}^{n} k_i V_{ij}, \tag{12.90}$$

where the vector **V** is the Grossberg layer weight matrix. Since only one PE in the Kohonen layer will have an output of 1 and the rest 0, the calculation for the Grossberg layer is rather simple. Training for the Grossberg layer is also relatively simple. After the output of the Grossberg neurons are calculated by Eq. (12.90), each weight is adjusted only if it is connected to a Kohonen neuron having a nonzero output. The new weight for the j-th PE is related to the old weight by the equation

$$v_{\text{ij new}} = V_{\text{ij old}} + \beta(y_j - V_{\text{ij old}})^{k_i}, \tag{12.91}$$

where k_i is the output of Kohonen neuron i and y_j is the component j of the vector of desired outputs. Initially, β (training rate) should be set to approximately 0.1 and gradually reduced as training progresses. Training in the Grossberg layer is supervised since the algorithm has a desired output to which it trains. The weights of the Grossberg layer will converge to the average values of the desired outputs whereas the weights of the Kohonen layer are trained to the average values of the inputs.

12.4.3 Application to Cetacean Social Signals

There have been at least three studies in which some type of neural network was used to process the social sounds of cetacean. Potter et al. (1994) used an ANN approach to undetermine the presence of bowhead whale songs in a data set of bowhead whale recordings. Murray (1997) used an ANN approach to categorize the sounds of the false killer whale, and Erbe (1997) used an ANN to detect the presence of beluga phonation in the presence of masking noise produced by an icebreaker ship.

12.4.3.1 Discrimination of Bowhead Whale Sounds

Potter et al. (1994) applied a backpropagation neural network with a single hidden layer to detect bowhead whale (*Balaena mysticetus*) song notes. The inputs to the neural network consisted of time–frequency values obtained from spectrograms. A spectrogram of a bowhead whale endnote consisting of 184 frequency points and 128 time points (23,552 total points) is shown in Fig. 12.16a. In order to limit the number of input neurons, they had to constrain both the time–frequency range and resolution. In order to decide the frequency and time range, spectrograms were calculated for all the data and those for endnotes were geometrically averaged (arithmetic mean in dB space) to make a mean endnote spectrogram. The lowest frequency in the initial inflection was used to time align each spectrogram (see time alignment point in Fig. 12.16a). Potter et al. (1994) also chose a frequency resolution of 63.5 Hz, which corresponded to a Q of 6 at the center frequency. The time resolution was chosen to be 128 ms so that the average absolute gradient of an endnote $<|df/dt|> \approx 1$ so that the endnote center frequency would change, on average, one frequency bin over the period of one time bin. These choices

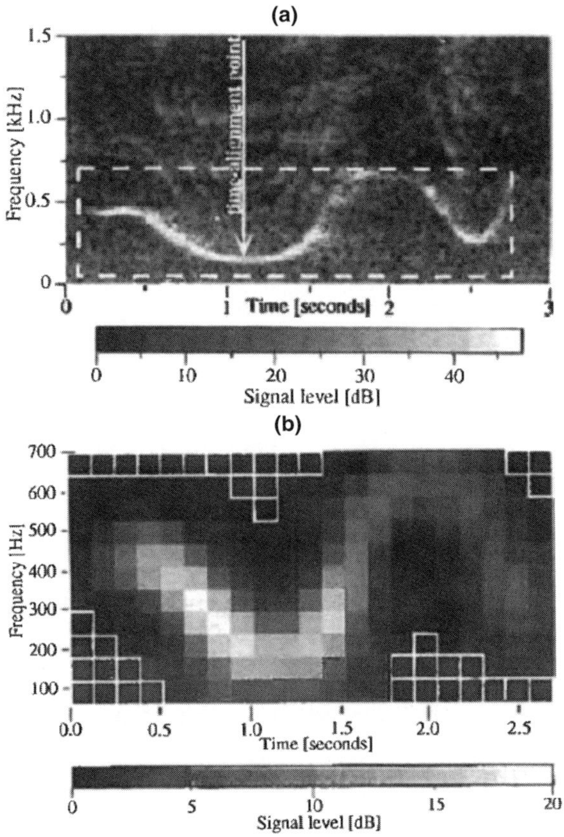

FIGURE 12.16. (**a**) Spectrogram of a typical bowhead whale endnote for the 1988 song. The *dashed white box* indicates the range used as input *t* the ANN, (**b**) the dB-averaged call spectrogram (adapted from Potter et al., 1994).

limited the feature space to 11 × 21 time-frequency bins. Some peripheral bins were discarded, reducing the input dimension to 192, as can be seen in Fig. 12.16b. A training set of 395 endnotes and 395 interference segments of noise, comprising 54% of the 1,475 sounds in the data set, was used. All input values were scaled to the range of values in the entire data set for each pixel. This procedure "equalized the importance of each pixel, so that numerically small but important differences in pixel whose values are nearly constant have potentially equal impact as big difference in pixels with large variations." The network was tested on the remaining 46% of the total sounds. The results of the testing are shown in Fig. 12.17 along with the results from a spectrogram correlation model similar to what was discussed in Section 12.3.4. The neural network performed at about one half the error rate of the SCF.

12.4.3.2 Categorizing False Killer Whale Sounds

Murray (1997) used two types of *unsupervised neural networks* to categorize sounds of a false killer whale (*Pseudorca crassidens*) into whistles, burst pulses, and clicks. The first type used a self-organizing *competitive learning* where the neurons in a competitive layer compete for the opportunity to

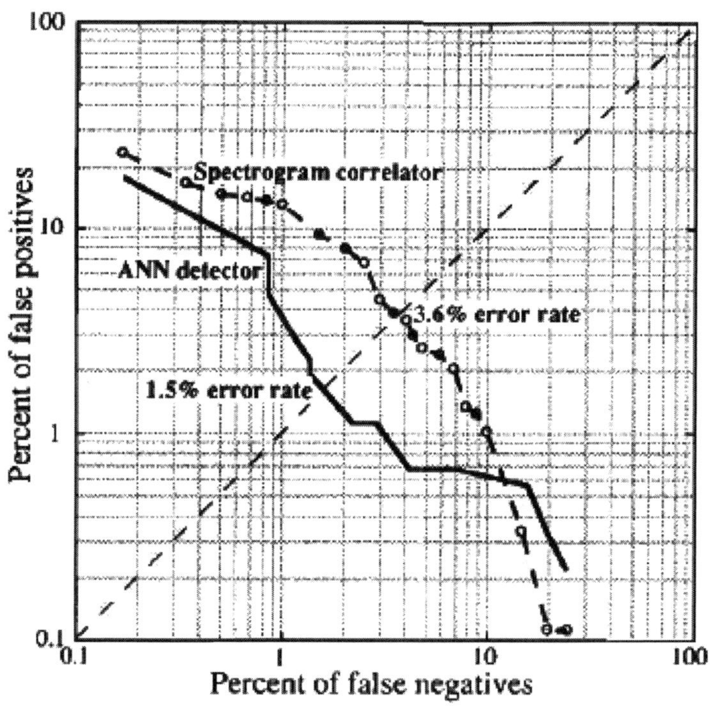

FIGURE 12.17. Performance of the trained neural network along with the performance of a spectrogram correlation filter (SCF). The *diagonal dashed line* indicates equal false positive and negative error rates (adapted from Potter et al., 1994).

respond to an input pattern. The winning unit adjusts its values to more closely match the value of the input pattern and, consequently, it is more likely to win the next time the same or a similar pattern is presented to the network. The second type used was a two-dimensional, self-organizing feature map, which is similar to the competitive network with the neurons ordered topographically in a two-dimensional square grid. The feature maps used by Murray was a 5 × 5 grid. Both networks were trained with three different types of inputs: (1) duty cycle measure, (2) peak frequency values, and (3) combination of duty cycle and peak frequencies. He partitioned the sounds into 512 point blocks (representing 11.6 ms at a sample rate of 44.1 kHz). Each 512 point blocks were further subdivided into 30 segments. A duty cycle measure and the peak frequency of each of the 30 segments per 512 point block were used and the input vectors. The duty cycle measure took into account the degree to which the amplitude of a signal deviated from a non-amplitude-modulated sine wave. A value of 1.0 represented a sine wave and a 0.0 indicated no signal. Values between 1.0 and 0.0 reflected the degree of pulsing. A typical click train might have a value of 0.05, a burst pulse between 0.3 and 0.6, and a whistle may have a value of 0.8 or greater. With this technique, a sine wave with varying amplitude could be assigned a value less than 1. Each point of the signal was expressed as a percentage of the maximum of the signal. When considering duty cycle and peak frequency, the 30 duty cycle values were followed by the 30 peak frequency values for each sound. All of the input vectors were scaled to z-scores using the global mean and standard deviation.

A total of 500 individual sounds were used with the neural network. Two hundred fifty randomly selected sounds were used for training and the other 250 were used for testing. Initially, forty neurons were used in the competitive learning network. However, with the three different types of input vectors, only five of the neurons learned which was indicative of their weights changing values during the training process. Thereafter, only five neurons were used in the network. The performance of the competitive learning network was evaluated by calculating the cosine of the angle between each input vector and its respective weight vector. This calculation can be interpreted as the correlation between vectors with the cosine approaching 1 as the angular distance between the two vectors approached 0. Overall, the competitive network performed well and the most stable categories were formed when the duty cycle values were used for the input vectors. The network trained with the duty cycle values had the highest performance of 0.79 with the test data. Peak frequency inputs had a slightly lower performance level of 0.67. The combined duty cycle and peak frequency values were used, the network performed at a level of 0.72, placing it approximately intermediate of the other two input types. The main categories seem to be ascending whistles, low-frequency pulse trains, and high-frequency pulse trains. The network was also able to recognize transition of whistles evolving into pulse trains.

The self-organizing feature map also performed well with a 5 × 5 array of neurons. The distribution of weights in the two-dimensional feature map was determined and the correlation between the test set and training set distribution for the duty cycle only inputs was 0.97. With the peak-frequency input, the correlation between the distribution of the test and training set was 0.89. The same correlation was obtained with the combination duty cycle and peak frequency inputs. Both type of network demonstrated that the self-organizing approach is an effective way of categorizing dolphin phonations.

12.4.3.3 Detection of Beluga Phonation in Noise

Erbe (1997) obtained behavioral performance measure for a beluga whale (*Delphinapterus leucas*) detecting beluga phonations that were collected in the wild, in which the calls were masked by two different types of icebreaker ship noise as well as natural ice cracking noises. The first type of noise was the bubbler noise, which consists of high-pressure air striking the water to prevent broken ice from jamming into the side of the ship. The second noise is ramming noise, which is actually a propeller cavitation noise. She then used a back-propagation neural network to determine how a network would perform in detecting the phonations and how the performance compared with that of the beluga whale. A training matrix was created by adding noise to the beluga calls in the averaged spectrogram domain. The training matrix was the assembly of 850 noisy call spectrograms and 850 pure noise. Each spectrogram being divided into 20 time bins and 20 frequency bins, resulting in a 20 × 20 matrix per spectrogram. The desired output as 1 for the noisy calls and 0 for the pure noise.

The neural network results compared very well with the beluga results, although the neural network responses had much steeper slopes than the animal's behavioral data (Erbe, 2000). She also considered several different signal analysis procedure including spectrogram correlation, matched filtering, visual discrimination of spectrograms, and an adaptive filter technique. Only the neural network results followed along the same line as the animal result; the performance of the network matched the whale's performance as the order of the noises from strongest to weakest masking and also the degree of masking was very similar between the neural network performance and the whale's performance. The signal-to-noise (SNR) ratio was determined by calculating the rms value of the signal versus the rms value of the noise signal. The animal and the neural network could operate below well with negative SNR be the noise spectra were not like white noise but had specific shapes that did not coincide with the spectra of the beluga calls.

12.4.4 Application to Dolphin Sonar Discrimination

The principal use of neural networks in sonar signal processing involves recognition of the patterns distribution of features used to characterize a target. The features obtained with the three signal processing models discussed in the previous section can be used as input to a neural network. The performance of the network will depend on the features presented to it. In a sense, a neural network is a substitute for the Euclidean distance measure used in the previous section, or any other pattern recognition scheme or algorithm. Table 12.4 lists the various neural network modeling of dolphin echolocation.

Roitblat et al. (1989) were the first to apply an artificial neural network to emulate a dolphin performing a sonar recognition task. Echoes from the target used in the delayed matching-to-sample experiment discussed in Section 13.3.4 were used in a counterpropagation network. The frequency spectra of echoes were divided into twenty equally spaced frequency bins between 63 and 162 kHz.

Examples of echoes from the four targets used by Roitblat et al. (1989) are shown in Fig. 12.19. Since the echoes were sampled at a 1 μs time interval and 1,024 points were collected per echo, the frequency spacing of the FFT calculation was 977 Hz. Each frequency bin for the neural network data was the averaged amplitude (on a linear scale) of five contiguous FFT bins, so the frequency spacing of the neural network data was 4.9 kHz. Ten echoes per target were used for the learning set and another ten echoes per target were used for the test set. The input layer of the counterpropagation network consisted of 20 input units corresponding to the 20 frequency bins of the data. The second layer contained 21 units and was used to normalized the inputs. The third layer was the Kohonen layer consisting of eight units. The fourth layer was the output layer consisting of four units, each unit representing a specific target. All of the echoes from the training set (40 echoes, ten echoes

TABLE 12.4 Studies of neural network modeling of dolphin echolocation

Reference	Type of networks	Target features used	Reference to dolphin experiment being modeled
Roitblat et al. (1989)	counterpropagation	spectral/constant bandwidth	Matching-to-Sample Roitblat et al. (1990)
Moore et al. (1991)	integrated gateway and backpropagation	spectral/constant bandwidth	Matching-to-Sample Roitblat et al. (1990)
Roitblad et al. (1992)	integrated gateway and backpropagation	spectral/frequency difference limen	Matching-to-Sample
Au (1994)	counterpropagation	spectral/constant-Q	Wall Thickness Diff Au and Pawloski (1990)
Au et al. (1995)	backpropagation	time–frequency/ constant-Q	Wall Thickness Diff Au and Pawloski (1990)
Roitblat et al. (1996)	integrated gateway and backpropagation	time–frequency/ constant time bin and constant bandwidth	Buried Targets Unpublished

per target) were randomly fed into the network and a total of 5,000 iterations of the data were used to train the network. When the 40 echoes from the test set were presented to the trained network, the network was able to classify the echoes with a 100% accuracy. Recall from Section 13.3.4 that the dolphin's accuracy in performing the matching-to-sample task was 94.5%.

In the next phase of the study, Roitblat et al. (1989) used echoes from the target collected while the dolphin echolocated on the sample targets. A directional hydrophone was placed adjacent to the observation aperture at the same depth as the center of the aperture. Echoes from the sample targets resulting from the dolphin's sonar signals were recorded on a instrumentation tape recorder operated at 152 cm/s (60 ips). Unlike the test pool echoes, these echoes exhibited fluctuations in amplitude and shape that caused the signal-to-noise ratio to vary from relatively high to very poor. These fluctuations probably resulted from variations of the animal–target–hydrophone alignment caused by wave- and wind-induced motions of the target and pen, and movement by the dolphin within the observation aperture. Ten echoes per target with high signal-to-noise ratios were digitized at 1 MHz sample rate, and used as the training set. Ten additional echoes per targets with high signal-to-noise ratios were also digitized and used as the test set. Only three of the original targets (large tube, sphere, and cone) were used. The natural echoes were used in the same counterpropagation network used with the test pool echoes, except that the frequency range was changed to include frequencies between 40 and 138 kHz. The network, after training for 5,000 iterations, was able to correctly classify 29 or 30 echoes (96.7%). Another set of echoes was collected 3 months after the original set. After digitizing selected echoes that had high signal-to-noise ratios, and using these echoes in the neural network, the results were similar to the results from the first set of natural echoes. The network was able to correctly classify 97% of the test echoes. A backpropagation network consisting of 20 input units, an 8-unit hidden layer, and a 3-unit output layer was also trained for 5,000 iterations with the second set of natural echoes. This network was also able to correctly classify the echoes in the test set at an accuracy of 97%.

Moore et al. (1991) took a slightly different approach in emulating the dolphin with a neural network in the delayed matching-to-sample experiment. Instead of using only selected natural echoes having high signal-to-noise ratios, they used sets of consecutive echoes. Therefore, the data included signals having a variety of signal-to-noise ratios. They used a backpropagation network that was modified to include an integration layer that preprocessed the data before being introduced to a more standard backpropagation network. The frequency spectrum was divided into 30 bins of 1.95 kHz per bin from 31.3 to 146.5 kHz. The information in each frequency bin for each target was averaged from echo to echo before being applied to the backpropagation network. The network was trained with six sets of ten successive echoes selected from the ends of randomly chosen echo trains. Two sets of echoes were chosen for each of three targets (large tube, sphere, and cone). The network was trained with 12,300 iterations before the criterion RMS output error of

0.05 was achieved. The complete set of 1,335 sequential echoes was presented to the trained network and the network was allowed to classify each echo train. The results were determined in term of "confidence ratio" defined as the ratio of the activation level of the correct classification unit versus the total activation level of the three classification units in the output layer. The results of Moore et al. (1991) study are shown in Fig. 12.18, with the confidence ratio plotted as a function of the number of clicks used. Overall, the dolphin's performance was better than the network, which achieved correct classification performance between 90% and 93%. However, the network tended to make correct classifications with fewer echoes than the dolphin.

The data from Moore et al. (1991) were also applied to a standard back-propagation network without the preprocessing integration layer. The same number of input units, hidden layer units, and output units were used. In this situation, the network correctly classified the targets with only a 63% accuracy. Therefore, preprocessing of the data by summing or integrating the signal in each frequency bin across the different echoes in the training set was very important for accurate target classification.

Roitblat et al. (1992) modified the approach of the two previous studies by using the spectral information from the target echoes but weighted the effective gain of the filters or frequency bins according to the audiogram of a *Tursiops truncatus* obtained by Johnson (1967). They also departed from the use of constant bandwidth filter, and modeled the filters to have

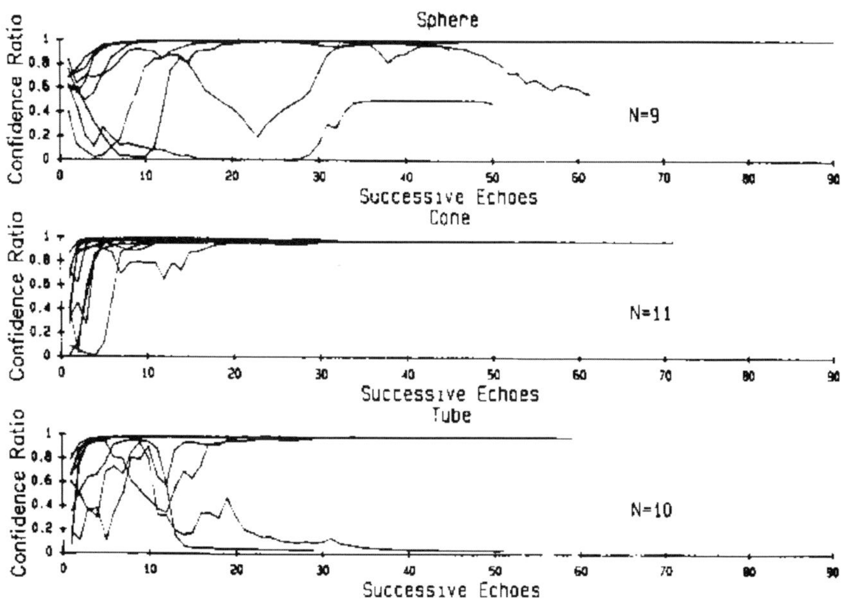

FIGURE 12.18. Results of generalization testing in the form of the confidence of the network in assigning the echo train to the proper category (adapted from Moore et al., 1991).

varying bandwidths corresponding to the frequency discrimination capability of *Tursiops truncatus* determined by Herman and Arbeit (1972) and Thompson and Herman (1975). The filter bandwidth increased linearly from 0.22 to 2.7 kHz as the center frequency increased from 20 to 140 kHz. A total of 30 filters were used to process echo data before introducing the information to the integrated gateway network of Moore et al. (1991). However, this model using a *Tursiops* hearing sensitivity and comparing with results of an echolocation experiment in Kaneohe Bay, Oahu, was not valid since dolphins in Kaneohe Bay are masked by the ambient background noise out to frequencies beyond 100 kHz, making the actual hearing sensitivity of the dolphin irrelevant (Au and Nachtigall, 1995).

Au (1994) modeled the dolphin wall thickness difference discrimination experiment of Au and Pawloski (1992), which was discussed in Section 13.3.2. This was the first neural network modeling of the dolphin sonar system using dolphin performance data that varied from high performance level (greater than 95% correct) to threshold. In the effort to provide information to neural networks that are biologically relevant, Au (1994) deviated from the three previous studies in some significant ways. First, the target echoes were collected with a transducer mounted on the dolphin's pen so that echoes could be collected in the same environment and under similar conditions as the dolphin. This technique along with the projection of simulated dolphin echolocation signals resulted in echoes that had over 20 dB higher signal-to-noise ratio than achievable with natural echoes. Fifty echoes were collected for each target. Second, the auditory system of the dolphin was modeled by a bank of constant-Q or proportional bandwidth filters. Since the pioneering work of Fletcher (1940) on critical ratio and critical bandwidth, a commonly accepted model of the mammalian auditory filter system has been a bank of constant-Q filters. The critical ratio and critical bandwidth are two different measures that estimate the filter bandwidth of a mammalian auditory system. Third, the gain of each constant-Q filter was kept at unity instead varying in a manner corresponding to the audiogram of a dolphin. The major reason for using a constant gain across the filter bank stems from the characteristic of the snapping shrimp noise in Kaneohe Bay. Therefore, the shape of the audiogram is not important when considering echolocation in Kaneohe Bay, as well other bodies of water inhabited by snapping shrimp.

The energy in each constant-Q filter was used as the input to a counter-propagation neural network. In the previous studies, the amplitude of the spectrum was used rather than energy. The use of energy has a subtle effect on the frequency spectrum of a signal. Features (energy from each constant-Q filter) from 20 echoes for each target were used for the training set and features from the remaining 30 echoes of each target were used for the test set. The targets were considered in pairs, with the standard target paired off with different comparison targets. Training was performed in two stages, first in the Kohonen layer for a predetermined number of iterations, and then in the Grossberg outstar layer. The number of iterations required for training depended on the learning rate parameter and the similarity between the echoes of the standard target and the specific comparison target. In this study, the default number of iterations and learning rate parameters provided by Neural Ware for training the Kohonen layer was used. Three thousand iterations divided in three increments of 1,000 iterations each were used in training the Kohonen layer, with the learning rate coefficients made progressively smaller in the second and third increments. Training of the Grossberg outstar layer started immediately upon completion of the 3,000 iterations used to train the Kohonen layer. Training continued until the rms output error fell below 0.001. Typical training required between 3,100 and 5,000 iterations. The network's capability to discriminate the standard from each comparison target was determined for different values of Q and N.

The performance of the counterpropagation network for the free-field echoes and Q values of 4 and 5 is shown in Fig. 12.19, along with the dolphin's performance. The number of filters, N, was equal to $Q - 1$ because of the specific upper and lower frequency limits chosen for the filter bank. The network's performance for a Q of 4 was not as good as the dolphin when comparing the standard with the –0.3 and –0.2 mm comparison targets. However, for a Q of 5 the network's performance was better than the dolphin for almost all of the thinner comparison targets. For the cases involving the thicker comparison targets, the results were mixed. For a Q of 5, the network performed better than the dolphin when comparing the standard with the 0.3

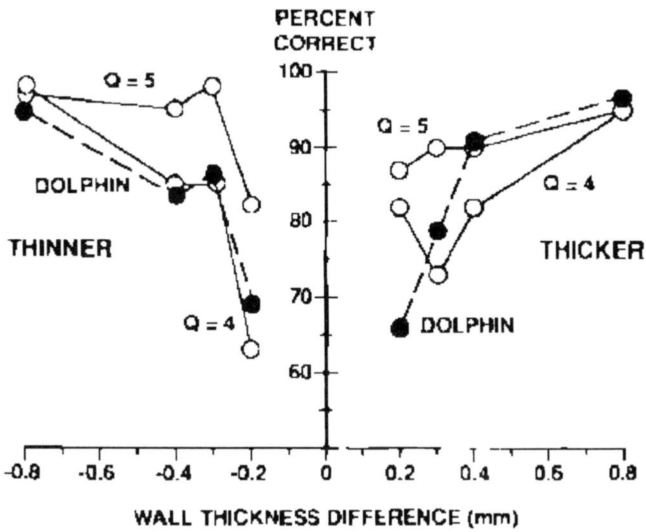

FIGURE 12.19. Target discrimination results in the free-field for the neural network (*open circles*) using different constant-Q values and for the dolphin (*closed circles*) (adapted from Au, 1994).

and 0.2 mm thicker targets, and slightly worse for the 0.4 and 0.8 mm thicker targets. However, for a Q of 4, the network did poorer than the dolphin when comparing the standard target with the 0.3, 0.4, and 0.8 mm thicker comparison targets. These results indicate that a constant-Q filter bank consisting of 4 filters, each having a Q of 5, produced target features that allowed the network to perform as well as the dolphin in discriminating between the standard and the comparison targets. The network results in Fig. 12.20 departed slightly from a monotonic decrease as wall thickness differences decreased. These departures may be attributed to ping-to-ping variations in the echoes caused motion of the pen and targets produced by wind and wave action. The statistical nature of the counterpropagation algorithm may have also contributed to some performance fluctuations. Environmental noise was not a factor since the signal-to-noise ratio of the echoes was very high.

The performance of the network with noise added to the echoes was also examined. Our experience has shown that many discrimination schemes can be used to discriminate echoes with high signal-to-noise ratios; however, only a few will continue to be robust in the presence of noise. The performance of the network with noisy echoes is shown in Fig. 12.20 for signal-to-noise ratios of 19 and 15 dB. The dolphins emitted signals that vary considerably in amplitude during a trial; however, we assumed that only the higher level clicks were used to make the target discrimination. Therefore, the echo-to-noise ratio in Fig. 12.21 was the averaged of the maximum amplitude per trial for a number of trials in the dolphin experiment. Only the thinner targets were used here and in the dolphin experiment (Au and Pawloski, 1992). For a Q of

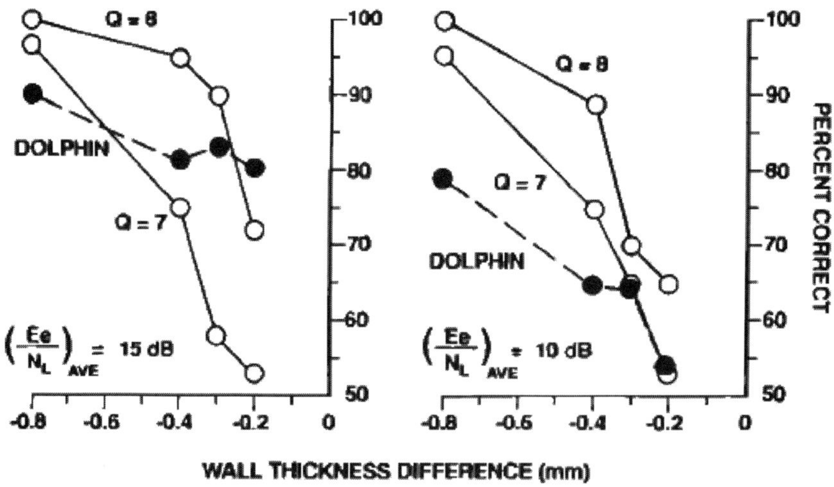

FIGURE 12.20. Target discrimination results in noise for the neural network (*open circles*) and for the dolphin (*closed circles*). The echo energy flux density to noise spectral density ratio of 19 and 15 dB corresponds closely to the dolphin's situation in which $\overline{(E_e/N_L)}_{\max}$ = 19 and 15 dB, respectively (adapted from Au, 1994).

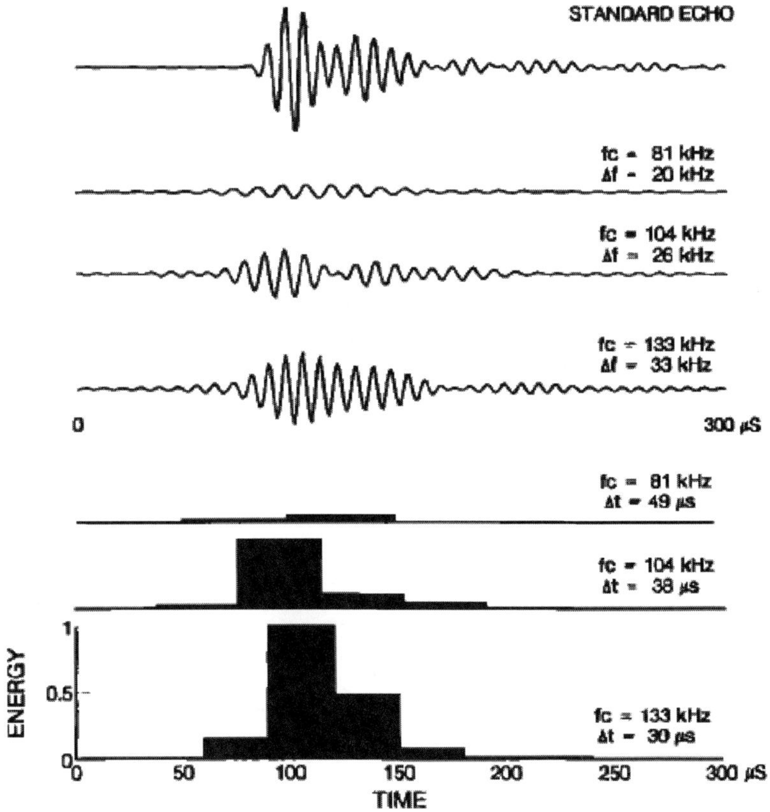

FIGURE 12.21. The waveform of the echo from the target before and after passing through contiguous constant-Q filters with a Q value of 4, and the relative energy of the filtered signals. The parameters on the right are the filter center frequency (f_c), the filter 3-dB bandwidth (Δf), and the time increment (Δt) for the energy calculation (adapted from Au et al., 1995).

8 and a signal-to-noise ratio of 10 dB, the network performed better than the dolphin in discriminating the standard from all comparison targets. When the signal-to-noise ratio increased to 19 dB, the network performed better than the dolphin except for the –0.2 mm comparison target. For a Q of 7 and a signal-to-noise ratio of 15 dB, the network generally performed better than the dolphin. When the signal-to-noise ratio increased to 15 dB, the network was generally worse than the dolphin. The performance of the network did not improve much with the increased signal-to-noise; however, the dolphin's performance improved significantly when the signal-to-noise ratio increased from 19 to 15 dB.

Insight into the cues used by the neural network to discriminate the cylinders was obtained by examining the target echoes. Examples of echoes from the standard and the ±0.2 mm comparison targets are shown in Figs. 12.21 and 12.22. Small differences in the spectra for the comparison target (dashed line) compared to the standard target can be seen. The spectrum of the –0.2 mm comparison target was shifted slightly toward lower frequencies with respect to the standard target. The shift in the spectrum of the comparison targets can be best seen by observing the nulls in the spectra for frequencies between 100 and 140 kHz. The spectrum of the +0.2 mm comparison target shifted slightly toward higher frequencies with respect to the standard target. The shift in the spectra of the comparison targets will cause the energy detected by some or all of the constant-Q filters to be different for the comparison and standard targets. Au (1994) showed mathematically the reason for the spectral shifts.

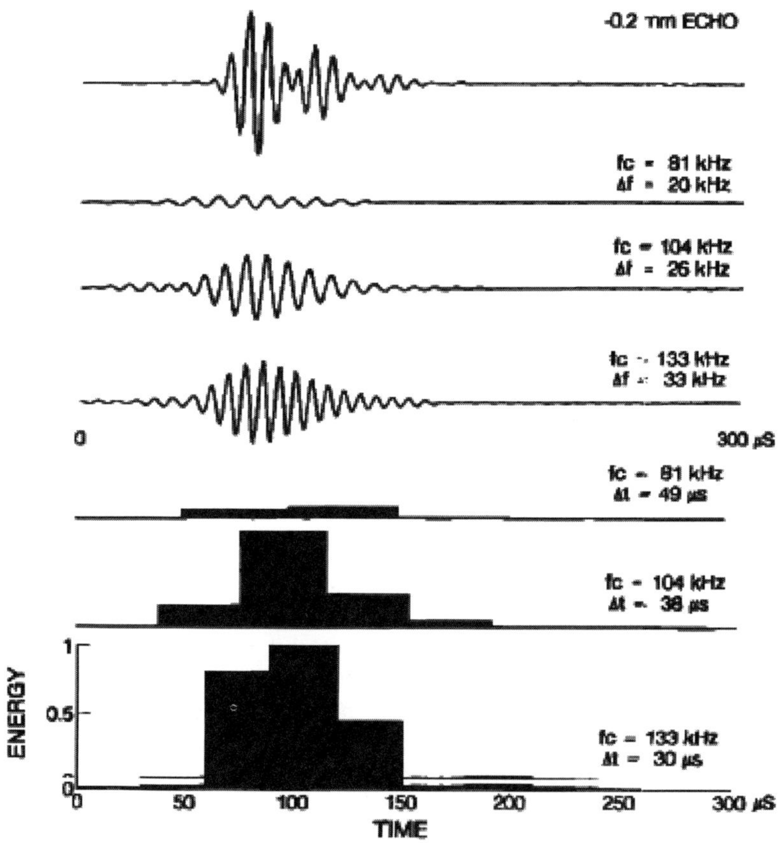

FIGURE 12.22. The waveform of the –0.20-mm comparison echo before and after passing through contiguous constant-Q fiters with a Q value of 4, and the relative energy of the filtered signals. The parameters on the right are the filter center frequency (f_c), the filter 3-dB bandwidth (Δf), and the time increment (Δt) for the energy calculation (adapted from Au et al., 1995).

Au et al. (1995) continued their modeling of the wall thickness experiment by using time–frequency information from a bank of contiguous filters, each filter having the same constant-Q. Digital finite impulse response filters having a cutoff skirt of 163 dB per octave were used to filter the echoes and the energy at the output of each filter was used to characterize each target. The upper frequency limit of the filter bank was chosen to be 150 kHz and the lower frequency limit was 62 kHz. With these constraints, the number of filters used was equal to $Q - 1$. An example of the output of three filters used with a Q of 4 is shown in Fig. 12.21 for the standard and Fig. 12.22 for the comparison target having a wall thickness difference of –0.2 mm. Both counterpropagation and backpropagation networks were used along with different Q values. The results for a Q of 4 are shown in Fig. 12.24. The Euclidean distance measure of Eq. (12.74) was used to compare how it would function with respect to the neural networks. The dolphin's performance is also included in the figure. The results indicated that even with a low Q of 4, the neural networks were able to do better than the dolphin. The results in Fig. 12.23 also indicated that the back propagation network did better than the counterpropagation network and both networks did better than the Euclidean distance measure.

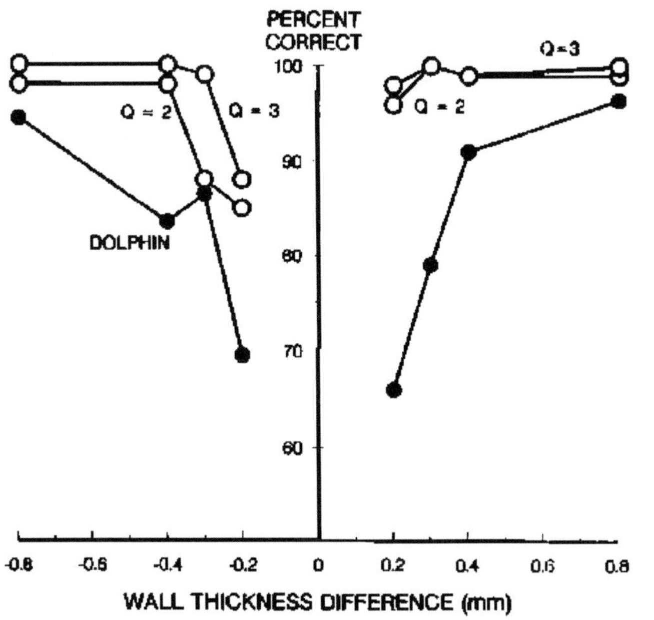

FIGURE 12.23. Wall thickness target discrimination results in the free-field for various models (*open circles*) using temporal and spectral information from a bank of constant-Q filters with a Q of 4, and the dolphin's performance results (adapted from Au et al., 1995).

The performance of the network with noise added to the echoes are shown in Fig. 12.24. Similar to the first study of Au (1994), the neural network had considerably more difficulties making the wall thickness discrimination when noise was added to the echoes. Comparing the results of both neural network modeling of the wall thickness experiment, the results indicated that a network can perform better if both time and frequency information are provided rather than just frequency information.

The final neural network modeling study we will discuss is that of Roitblat et al. (1996) in which a study of how well an echolocating dolphin can discriminate different targets buried in bottom sediment. The data analysis and neural network modeling were performed using RIPPEN^{TM}, a graphical real-time interactive programming and processing system produced by Orincon Corp. In Fig. 13.22, we showed a bottlenose dolphin burrowing into the ocean bottom to capture a prey. The dolphin behavorial results have not been published yet, but the neural network simulation of the dolphin experiment has. Targets were four cylinders, each 10.16 cm in height and 3.81 cm in diameter of different material composition. The three metallic cylinders were constructed of

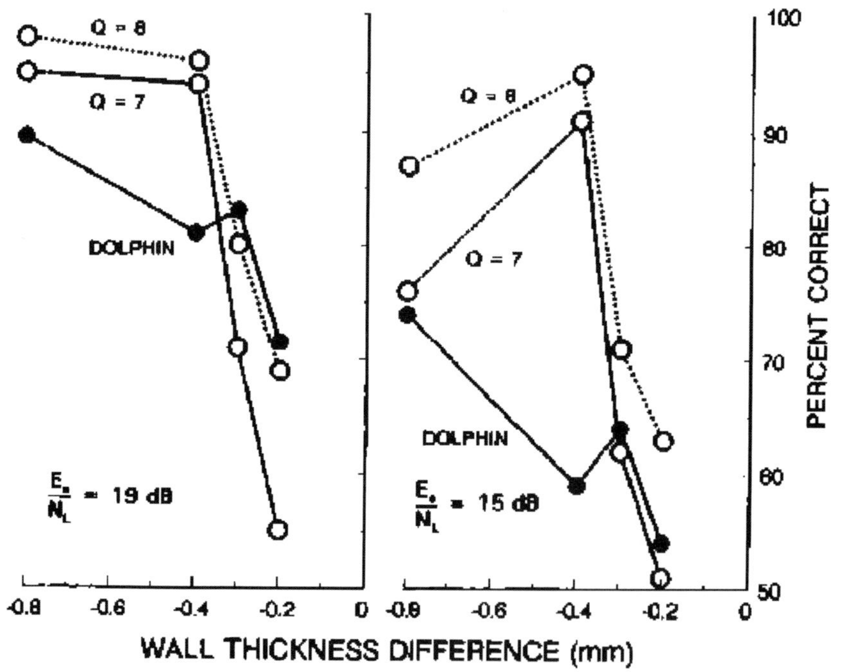

FIGURE 12.24. Target discrimination results in noise for the neural network (*open circles*) and for the dolphin (*closed circles*). The echo energy flux density to noise spectral density ratio of 19 and 15 dB corresponds closely to the dolphin's situation in which $\overline{(E_e/N_L)}_{\max} =$ 19 and 15 dB, respectively (adapted from Au et al., 1995).

aluminum, one hollow and one air-filled, and a stainless steel, each having a wall thickness of 0.64 cm. The fourth cylinder was made of coral pebbles embedded in degassed epoxy. These objects were placed in a sediment chamber consisting of a redwood box with a front face of 91.5 cm × 91.5 cm. The front and back faces were originally covered with a rubber sheet. Eventually, the rubber sheet was removed when it was discovered the bottom sediment contained enough clay to keep the water-sediment interface free standing.

Acoustic reflections of each target buried in the sediment was obtained by placing a transducer 1.2 m from the face of the box, transmitting simulated dolphin sonar signals, similar to the signal shown in Fig. 13.12. The targets were placed about 50 cm into the mud and their longitudinal axis was oriented to be perpendicular to the incident signal. Examples of the echoes from the targets in the mud chamber are shown in Fig. 12.25. The spectrum of each echo was estimated using both an FFT and autoregressive (AR) spectral estimator (Marple, 1987), in 256 point segments. After processing the first segment, the window was shifted by 16 points and another 256-point segment was processed. This procedure was repeated 64 times for each 1,024 point echo, producing a spectrogram having 93.7% overlapped in the frequency domain with each spectrum separated by 16 μs. The AR processes was done in a similar fashion to the FFT process in 256-point segments.

The output of the FFT and AR processing procedures went to separate backpropagation neural network, as can be seen in Fig. 12.26 showing the processing chain. For the FFT processing chain, the input to the neural network consisted of 600 units representing 20 time scans by 30 frequency bins covering a range of 58.6–175.8 kHz. The AR inputs to the neural network consisted of 960 units representing 20 time scan by 48 frequency bins. From Fig. 12.26, we can see that the output of each backpropagation network were sent to a single modified probabilistic neural network (PNN; Specht, 1990), which estimated

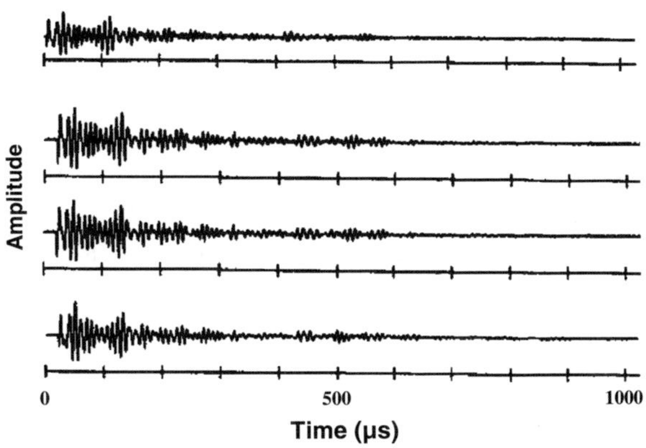

FIGURE 12.25. Echoes of the targets submerged in the mud chamber (adapted from Roitblat et al., 1996).

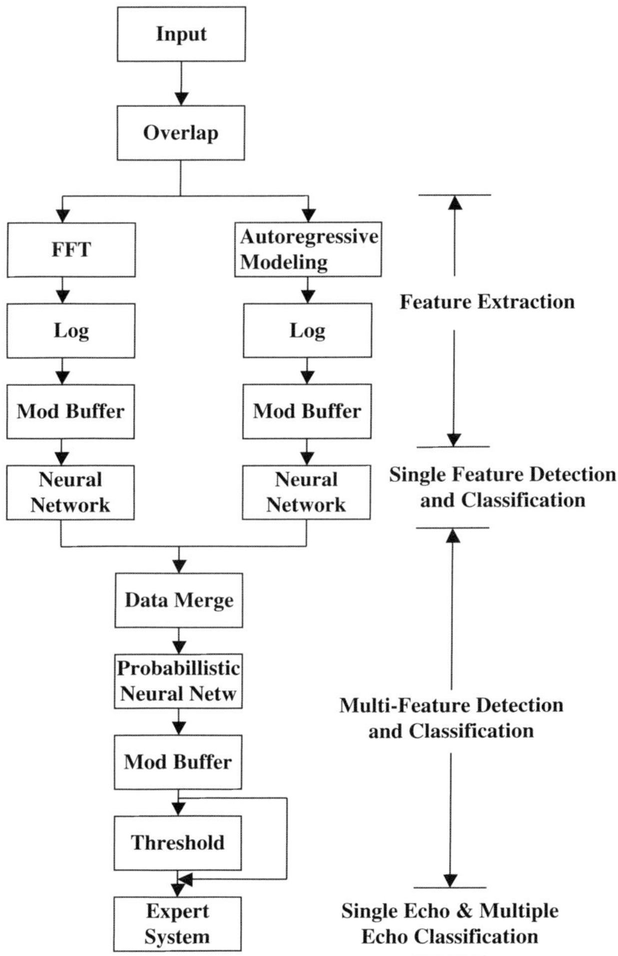

FIGURE 12.26. The RIPPEN processing chain used to discriminate buried targets (adapted from Roitblat et al., 1996).

the probability density functions from training samples and provided a Bayes-optimal mapping of input patterns to output classes. A detailed discussion of the PNN network is and the expert system is beyond the scope of this book, and interested readers should consult Specht (1990) and Roitblat et al. (1996).

Two hundred echoes were collected for each target and an additional 200 echoes were collected without any target present (clutter). Half of the echoes were used for the training set, and half were used for the generalization test set. Performance of the network was tested with 100 echoes from each target and 100 echoes without any target present. Target detections normally occurred contiguously over several time slices, but because of noise, sometime slices would be missing. If the number of skipped scans exceed the maximum allowed, then subsequent detections were counted as part of a separate detection. Multiple echoes indicating a target were considered to be returns from the same object if their estimated ranges was within the range specified by the range tolerance in scans parameter. The network was able to discriminate the different targets at a performance level that exceeded 90% correct.

12.5 Concluding Remarks

The material presented in this chapter is but a small part of a much wider field of study concerning signal processing. There are many other methods and techniques available to process any type of signals that are beyond the scope of this book. Wavelet transforms have become very popular in the analysis of music and speech. Wavelets have a interesting property of having constant-Q "filters" that could be a good analog to a mammalian auditory system. There are also a variety of time–frequency distribution techniques that we have not discussed. The Wigner-Ville, Choi-Williams, Zhao-Atlas-Mark, and Born-Jordon distributions are some of the many that one can investigate. Those interested in time–frequency analysis should consult Cohen (1994) for a good start. The two-dimensional Gabor transform also has some interesting properties that could be applicable to signals produced by marine animals. Many of these signal processing techniques are becoming available as software packages for PC. However, we would like to insert a word of caution in using these or any signal processing technique. The user should attempt to understand the mathematical underpinning of any signal analysis process before applying that process to one's data. Each technique has its own limitations and traps that an uninformed user can easily be unaware of and subsequently use the techniques or software very inappropriately.

References

Altes, R. A. (1973). "Some invariance properties of the wideband ambiguity function," J. Acoust. Soc. Am. **53**, 1154–1160.

Altes, R. A. and Faust, W. J. (1978). *Further Development and New Concepts for Bionic Sonar. Vol. 1 – Software Processors*, Naval Ocean Systems Center Technical Report 404, San Diego, CA 92152.

Altes, R. A. (1980a). "Models for echolocation," in *Animal Sonar Systems*, R. G. Busnel and J. F. Fish, eds. (Plenum, New York), pp. 625–671.

Au, W. W. L. (1988). "Detection and recognition models of dolphin sonar systems," in *Animal Sonar: Processes and Performance*, P. E. Nachtigall and P. W. B. Moore, eds. (Plenum Press, New York), pp. 753–768.

Au, W. W. L. (1993). *The Sonar of Dolphins* (Springer-Verlag, NY).

Au, W. W. L. (1994). "Comparison of Sonar Discrimination: Dolphin and an Artificial Neural Network," (W. Au), J. Acoust. Soc. Am. **95**, 2728–2735.

Au, W. W. L., Moore, P. W. B., and Pawloski, D. A. (1988). "Detection of complex echoes in noise by an echolocating dolphin," J. Acoust. Soc. Am. **83**, 662–668.

Au, W. W. L. and Martin, D. W. (1989). "Insights into dolphin sonar discrimination capabilities from human listening experiment," J. Acoust. Soc. Am. **86**, 1662–1670.

Au, W. W. L. and Moore, P. W. B. (1990). "Critical ratio and critical bandwidth for the Atlantic bottlenose dolphin," J. Acoust. Soc. Am. **87**, 1635–1638.

Au, W. W. L. and Pawloski, D. (**1992**). "Cylinder wall thickness difference discrimination by an echolocating atlantic bottlenose dolphin," J. Comp. Physiol. A. **172**, 41–47.

Au, W. W. L. and Nachtigall, P. E. (1995). "Artificial neural network modeling of dolphin echo location," in *Sensory Systems of Aquatic Mammals*, R. A. Kastelein, J. A. Thomas, and P. E. Nachtigall, eds. (De Spil Publishing, Woerden, The Netherlands), pp. 183–199.

Au, W. W. L., Andersen, L., Rasmussen, A., Roitblat, H., and Nachtigall, P. E. (1995). "neural network modeling of a dolphin's sonar discrimination capabilities," J. Acoust. Soc. Am. **98**, 43–50.

Burdic, W. S. (1968). *Radar Signal Analysis* (Prentice-Hall, Englewoord Cliffs, NJ).

Beuter, K. J. (1980). "Echo evaluation in auditory system of bats," in *Animal Sonar Systems*, R. G. Busnel and J. F. Fish, eds. (Plenum Press, New York), pp. 747–761.

Bracewell, R. N. (1978). *The Fourier Transform and Its Applications*, (Second Edition) (McGraw-Hill, New York).

Brigham, E. O. (1988). *The Fast Fourier Transform and Its Applications* (Prentice Hall, Englewood Cliffs, NJ).

Chestnut, P., Landsman, H., and Floyd, R. W. (1979). "A sonar target recognition experiment," J. Acoust. Soc. Am. **66**, 140–147.

Cohen, L. (1994). *Time Frequency Analysis: Theory and Applications*. (Prentice-Hall, NY).

Dayhoff, J. (1990). *Neural Network Architectures An Introduction* (Van Nostrand Reinhold, New York).

Dziedzic, A., Escudie, B., Guillard, P., and Hellion, A. (1969). "Mise en evidence de la tolerace a l'effet Doppler de l'emission sonar de (*Delphinus delphis*)," Cetace odontocete. C.R. Acad. Sci. **279**, 313–1316.

Dziedzic, A., Chiollaz, M., Escudie, B., and Hellion, A. (1979). "Sur queslques Proprietes des Signaux onar Frequence du Dauphin (*Phocoena phocoena*)," Acustica **37**, 258–266.

Erbe, C. (1997). "Zones of masking around icebreakers affecting beluga whales," J. Acoust. Soc. Am. **102**, 3102.

Erbe, C. (2000). "Detection of whale calls in noise: performance comparison between a beluga whale, human listeners, and a neural network," J. Acoust. Soc. Am. **108**, 297–303.

Fletcher, H. (1940). "Auditory patterns," Rev. Mod. Phys. **12**, 47–65.

Gabor, D. (1947). "Acoustic quanta and the theory of hearing," Nature **159**, 591–594.

Green, D. and Swet, J. (1966). *Signal Detection Theory and Psychophysics* (Krieger Pub. Co., Huntington, New York).

Grossberg, S. (1969). "Some networks that can learn, remember and reproduce any number of complicated space-time patterns," J. Math. Mech. **19**, 53–91.

Hammer, Jr. C. E. and Au, W. W. L. (1980). "Porpoise echo-recognition: an analysis of controlling target characteristics," J. Acoust. Soc. Am. **68**, 1285–1293.

Hecht-Nielsen, R. (1987). "counterpropagation networks," in *Proceedings of the IEEE First International Conference on Neural Networks*, M. Caudill and C. Butler, eds. Vol 2, (SOS Printing, S.D.), pp. 19–32.

Hecht-Nielsen, R. (1989). *Neurocomputing* (Addison-Wesley, Reading Ma).

Herman, L. M. and Arbeit, W. R. (1972). "Frequency difference limens in the bottlenose dolphin: 1-70 kHz," J. Audit. Res. **12**, 109–120.

Johnson, C. S. (1967). "Sound detection thresholds in marine mammals," in *Marine Bioacoustics*, (Pergamon Press, NY), pp. 247–260.

Johnson, S. C. (1968a). "Relation between absolute threshold and duration of tone pulse in the Bottlenosed Porpoise," J. Acoust. Soc. Am. **43**, 757–763.

Johnson, S. C. (1968b). "Masked tonal thresholds in the Bottlenosed Porpoise," J. Acoust. Soc. Am. **44**, 965–967.

Kamminga, C. and Beitsma, G. R. (1990). "Investigations on cetacean soanr IX remarks on dominant sonar frequencies from *Tursiops truncates*," Aquatic Mamml. **16**, 14–20.

Kelly, E. J. and Wishner, R. P. (1965). "Matched-filter theory for high-velocity, accelerating targets," IEEE Trans. Microwave Theory Tech. **9**, 56–69.

Kohonen, T. (1988). *Self-Organizing and Associative Memory* (Springer-Verlag, New York).

Martin, D. W., and Au, W. W. L. (1986). *Investigation of Broadband Sonar Classification Cues*, Naval Ocean Systems Center Technical Report 1123, San Diego, Ca. 92152.

Martin, D. W. and Au, W. W. L. (1988). "An automatic target recognition algorithm using time-domain features," in *Animal Sonar: Processes and Performance*, P. E. Nachtigall and P. W. B. Moore, eds. (Plenum, New York), pp. 829–833.

Lin, Z.-B. (1984). "A method for computation of wideband ambiguity function and the numerical analysis of the bat's sonar signal," IEEE ICASSP Proc. **3**, 47.11.1–47.11.4.

Lin, Z.-B. (1988). "Wideband ambiguity function of broadband signals," J. Acoust. Am. **83**, 2108–2116.

Rumelhart, D. E. and McClelland, J. L. (1986). *Parallel Distributed Processing*, Vol 1 (MIT Press, Cambridge, MA).

Marple, Jr. S. L. (1987). *Digital Spectral Analysis with Applications* (Prentice-Hall, Englewood Cliff, NJ).

Meyer, P. L. (1965). *Introductory Probability and Statistical Applications* (Addison Wesley, Reading, Mass).

Moore, P. W. B., Roitblat, H. L., Penner, R. H., and Nachtigall, P. E. (1991). "Recognizing successive dolphin echoes with an integrator gateway network," Neural Networks, **4**, 701–709.

Murchison, A. E. (1980). "Detection Range and Range Resolution of Porpoise," in *Animal Sonar Systems*, R. G. Busnel and J. F. Fish, eds. (Plenum, New York), pp. 43–70.

Murray, S. O. (1997). "The graded structure and neural network classification of false killer whale (*Pseudorca crassidens*) vocalization," M.S. Thesis, Univ. of Hawaii.

Parker, D. B. (1982). "Learning logic," Invention Report S81-64, File 1, Office of Technology Licensing, Stanford University, Stanford, CA.

Potter, J. R., Mellinger, D. K., and Clark, C. W. (1994). "Marine mammal call discrimination using artificial neural networks," J. Acoust. Soc. Am. **96**, 1256–1262.

Rihaczek, A. W. (1969). *Principles of High-Resolution Radar*, (McGraw-Hill, New York).

Roitblat, H. L., Moore, P. W. B., Nachtigall, P. E., Penner, R. H., and Au, W. W. L. (1989). "Natural echolocation with an artificial neural network," Inter. J. Neural Net. **1**, 239–248.

Roitblat, H. L., Moore, P. W. B., Helweg, D. A., and Nachtigall, P. E. (1992). "Representation and processing of acoustic information in a biomimetic neural network," *Proceedings of the Second International Conference on Simulation of*

Adaptive Behavior, J. A. Meyer, H. L. Roitblat, and S. W. Wilson, eds. (MIT Press, Cambridge, MA), pp. 90–99.

Roitblat, H., Au, W. W. L., Nachtigall, P. E., Shizumura, R., and Moons, G. (1996). "Sonar recognition of targets embedded in sediment," Neural Networks, **8**, 1263–1273.

Schusterman, R. J., Kersting, D. A., and Au, W. W. L. (1980). "Stimulus control of echolocation pulses in *Tursiops truncatus*," in *Animal Sonar Systems*, R. G. Busnel and J. F. Fish, eds. (Plenum Press, New York), pp. 981–982.

Schwartz, M. (1980). *Information Transmission, Modulation, and Noise*, 3rd edition (McGraw Hill, New York).

Specht, D. F. (1990). "Probalistic neural networks," Neural Networks, **3**, 109–118.

Thompson, R. K. R. and Herman, L. M. (1975). "Underwater frequency discrimination in the bottlenose dolphin (1-140 kHz) and human (1-8 kHz)," J. Acoust. Soc. Am. **57**, 943–947.

Urick, R. J. (1983). *Principles of Underwater Sound*, 3rd edition (McGraw Hill, New York).

Urkowitz, H. (1967). "Energy detection of unknown deterministic signals," Proc. IEEE. **55**, 523–531.

Wasserman, P. D. (1989). *Neural Computing Theory and Practice* (Van Nostrand Reinhold, New York).

Werbos, P. J. (1974). "Beyond regression: new tools for prediction and analysis in the behavioral sciences," Masters thesis, Harvard Univ., Cambridge, MA

Woodward, P. M. (1953). *Probability and Information Theory with Applications to Radar* (Pergamon Press, New York).

13

Some Instrumentation for Marine Bioacoustics Research

In this final chapter, we will discuss some unique instrumentations used for bioacoustics research. The focus in this chapter will be on instrumentation and not on the data and information that various instruments have collected. In some cases, experiments are in progress as this chapter is being written or have just recently been completed and the results have not been published yet. Many bioacoustics measuring and monitoring instruments tend to be of the one of a kind variety designed from scratch or assembled with different types of modules and equipment not originally designed for the specific purpose that a researcher is interested in. Therefore, researchers must be very innovative, imaginative, and have a willingness to continually look for new technological developments in a variety of fields. In writing this chapter, we take the risk of being dated because of continual and rapid technological development occurring in many fields such as communications, electronics, and computer science. Techniques and devices that we could only dream of in the recent past are fast becoming reality, expanding our capabilities to perform better and more in-depth experiments and studies. Laptop and transportable lunch-box PCs continue to increase in speed, with ever increasing memory and disc space. The introduction of compact, low-power microcontroller systems containing all the necessary drivers and modules to be a stand-alone computer minus a keyboard and a monitor have made it possible to create long-term monitoring systems that can gather interesting data associated with the natural history of marine mammals. Closely related to microcontrollers are digital signal processing (DSP) chips that can be programmed to perform data acquisition functions as well as perform high-speed computations. The introduction of GPS (global positioning satellite) navigational system along with continual improvement in monitoring devices has provided another valuable tool for researchers and others interested in studying the natural behavior of animals. Compact single board computers are also coming on the scene and will enable researchers to have considerably more capabilities than a typical microcontroller system, but perhaps at the expense of battery power.

W.W.L. Au, M.C. Hastings, *Principles of Marine Bioacoustics*,
DOI: 10.1007/978-0-387-78365-9_13,

13.1 Some Instrumentation for Marine Mammal Acoustics

13.1.1 The Bioacoustic Probe: Acoustic Recording Tag

Many marine animals rely on acoustics to capture prey, avoid predators, reproduce, and navigate, yet we know very little of the type of acoustic signals marine mammals encounter in the open ocean. The ocean is a very noisy environment, especially at low frequencies. In order to measure and record the noise field that marine mammals swim in, Burgess et al. (1998) developed an acoustic recording tag that could be attached to northern elephant seals. Northern elephant seals regularly haul out on land, allowing easy access for attachment and recovery of instrumentation packages. These animals migrate annually, swimming thousands of kilometers north and west from California, and during this migration they experience a wide variety of acoustic environments (Le Bouef et al., 1993). The earliest version of the Bprobe was used by Fletcher et al. (1996) to monitor the low-frequency sounds from the ATOC (acoustic thermography of ocean climate) source as tagged elephant seals swam in the vicinity of the source.

Since its first implementation, the Bprobe has undergone several modifications resulting in the latest version shown in Fig. 13.1. It combines a hydrophone, pressure (depth), temperature and acceleration sensors, a data acquisition unit, data storage, and a field replaceable battery in a single self-contained package. The heart of the Bprobe is a programmable microcontroller chip. A 16-bit analog-to-digital converter that can sample the hydrophone output at rates up to20 kHz and store the results in a 1 Gbyte flash memory is used. The user can select a hydrophone amplifier gain of 0, 10, and 20 dB. The Bprobe can be programmed to sample at specified intervals for a specified duration, and between sampling intervals the probe can be put to sleep to conserve battery power. A total of approximately 41 hours of operation time can be achieved. The probe is small, light weight, and is encapsulated in polyurethane epoxy. Communication between the probe and the "outside" world for setting

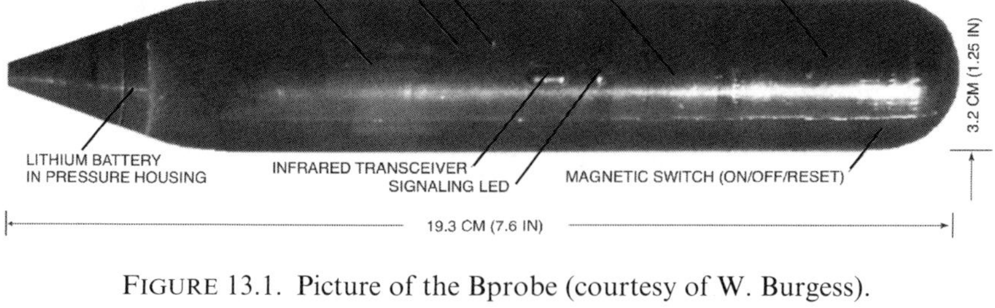

FIGURE 13.1. Picture of the Bprobe (courtesy of W. Burgess).

the A/D sampling rate and the hydrophone gain is accomplished via an infrared serial link that operates at a speed of 5.3 kbytes/s.

With this instrumentation package, researchers can determine whether diving marine mammals make active sounds, measure the frequencies and levels of sounds diving seals encounter in their environment, and have the acoustic data related to diving behavior of elephant seals (Burgess et al., 1998). A picture of an elephant seal carrying a Bprobe on its back is shown in Fig. 13.2. An example of the acoustic signal received by a Bprobe is shown in Fig. 13.3, with the dive depth shown above the color sonogram. Most of the received signals had frequencies in the range of 20–200 Hz. Snapping shrimp, cetacean sounds, boat noise, and seal swim strokes and heartbeats are clearly audible in some of the data. Flow noise is correlated with swim speed, suggesting that the optimal time for acoustic sampling would be when the seals are swimming slowly. Results of several deployments have indicated that it is also feasible to obtain long-term, reliable, quantitative, and noninvasive cardiac monitoring of elephant seals and other marine mammals (Burgess et al., 1996). This capability has been an important bonus to the project.

In 1997, three early versions of the Bprobe were mounted on northern elephant seals just prior to their annual migration from California to Alaska. Two of the packages were recovered after over 4 months at sea (Burgess et al., 1997). The hard discs contained measurements of pressure, temperature, ambient noise, and acoustic signatures of swim speed, swim stroke rate, respiration, and cardiac function. One subject swam across the northeastern Pacific, averaging 58 dives per day with a maximum dive depth of 780 m

FIGURE 13.2. An elephant seal carrying an instrumentation package preparing for its winter migration (courtesy Burney LeBeouf).

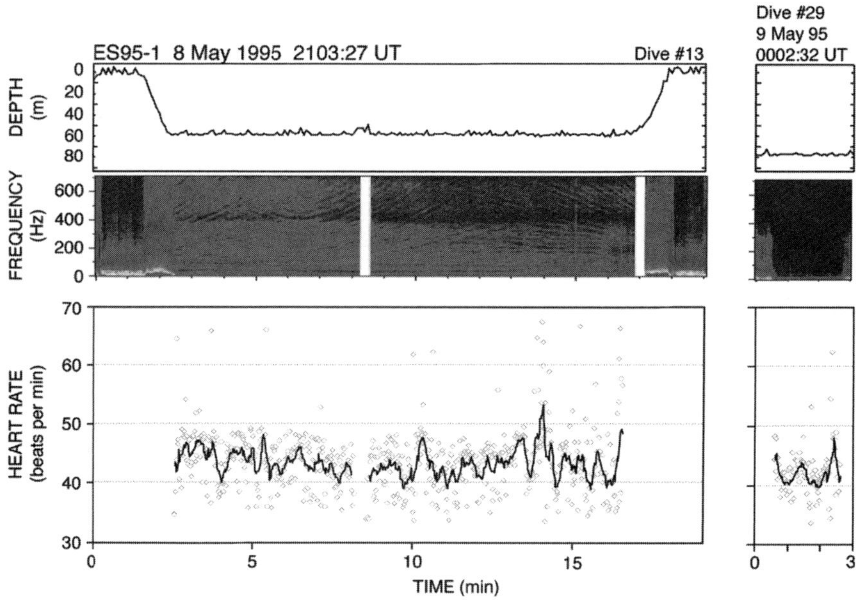

FIGURE 13.3. Example of the acoustic signal received by the instrumentation package on a diving elephant seal (courtesy of W. Burgess).

during the 26 days that the logger batteries supported data acquisition. The other subject swam along the West Coast, diving 81 times per day with a maximum dive depth of 770 m. The results suggest that electroacoustic packages offer a comprehensive and reliable means of sampling acoustic stimuli and associated behavior for free-ranging marine animals over long periods at sea.

A unique application of the Bprobe was recently devised by Thode et al. (2004), in which a number of them were used as the elements of a vertical line array to measure the songs emitted by humpback whales in Australian waters. The use of Bprobes in an array configuration allows for a tremendous amount of flexibility, since the sensor spacing can be readily changed and the requirement of a multi-conductor power and signal-carrying cable is eliminated. In order to utilize this "insta-array," Thode et al. (2004) had to develop a procedure to time-synchronize the recorded data to within a millisecond or less. The raw acoustic data may be offset in time by several seconds because the Bprobes cannot be precisely activated at the same time. Thode et al. (2004) first made use of an external broadband signal that would allow synchronization of probes spaced 3 m apart to within 10 ms by calculating the cross-correlation function of the signal measured by a pair of probes. They next utilized a global inversion algorithm to maximize the fit between measured acoustic data and the output of a propagation model, a process referred to as "geoacoustic inversion" or "focalization". Finally, they were able to exploit the spatial coherence of ocean ambient noise. Provided that the Bprobes are not spaced too far apart, there should be a high correlation of the ambient noise recorded by each probe. The relative difference in timing of each probe can be determined by cross-correlating the signals from each probe with the other probes in the array. The tilt in the line array caused by current was also monitored so that correction for tilt could be made.

13.1.2 Digital Acoustic Recording Tag: D-Tag

Another successful acoustic tag or probe, developed by Mark Johnson at Woods Hole Oceanographic Institute (Johnson and Tyack, 2003), is called the D-tag. It has a complementary function to the tag developed by Burgess. It too uses a DSP module to control acoustic data acquisition and storage as well as the measurement of various parameters such as acceleration, depth, temperature, orientation, and magnetic field strength. The principle differences between the Bprobe and the D-tag are imbedded in the design objectives of both tags. The Bprobe was designed to be deployed over a long period of time, on the order of months, and be used with animals that emit low-frequency sounds and encounter low-frequency noise. The D-tag was designed to measure high-frequency sound emissions on a continuous basis for a short period of time on the order of several hours. Sampling rates as high as 98 kHz for a 12-bit A/D have been achieved with the D-tag and still higher sampling rates are being considered (Tyack, personal communications). The D-tag was designed to be flexible in terms of modifications and therefore not necessarily "user"-friendly except to a small cadre of well-trained users. The Bprobe sacrificed flexibility for user-friendliness and simplicity in operation. The D-tag is packaged in a bag of oil so that modifications can be done as needed. A picture of the D-tag electronics is shown in Fig. 13.4. A complete tag with suction cup mounts is shown in Fig. 13.5. A burn-wire attachment between the housing and the suction cup is used to release the vacuum seal so that the tag can be released off the animal.

The D-tag has been used with northern right whales, sperm whales (Johnson and Tyack, 2003), Blainville's beaked whales, *Mesoplodon densirostris*, and Cuvier's beaked whales, *Ziphius cavirostris* (Johnson et al. 2004; Madsen et al., 2005). The deployment of the D-tags on beaked whales resulted in some very interesting data, providing extremely important insights into the echolocation process of beaked whales. One Cuvier's beaked whale performed one foraging dive of 50 min to 824 m. One of the Blainville's beaked whales made six foraging dives to between 655 and 975 m in the 15.4 hours over which the tag was on the animal. The second Blainville's beaked whale made two deep dives to 730 and 815 m in the 3 hours that the tag was attached to the animal. Echolocation signals were not detected

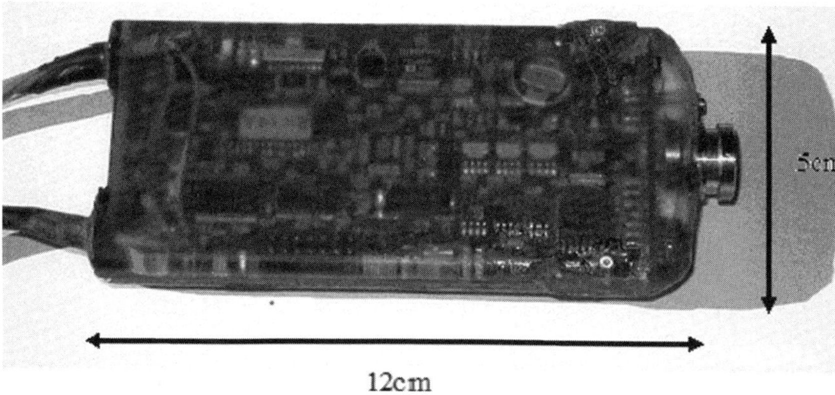

FIGURE 13.4. An encapsulated electronic package of the D-tag (adapted from Johnson and Tyack, 2003).

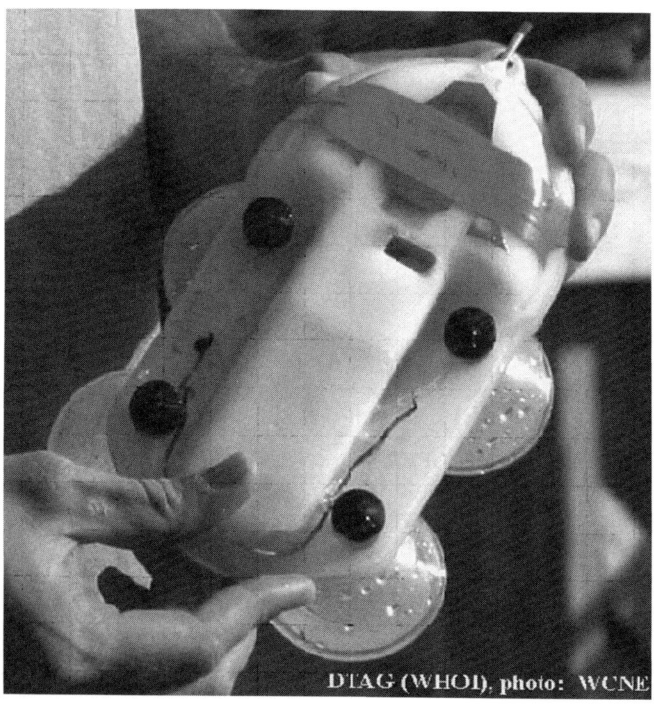

FIGURE 13.5. Complete tag including plastic fairing floatation and two suction cups (adapted from Johnson and Tyack, 2003).

until the whales were at least 200 m deep after which they clicked continuously. The *Ziphius* started clicking at an average depth of 475 m and stopped clicking when they started their ascent at an average depth of 400 m. The *Mesoplodon* began clicking at an average depth of 400 m and stopped clicking when they started their ascent at an average depth of 720 m. Click intervals during much of a dive varied between 0.2 and 0.4 s. As the whales apparently closed in on their prey, the click rate increased to about 250 clicks/s.

Johnson and Tyack were also able to record signals that may have been emitted by conspecifics. Two of these signals are shown in Fig. 13.6. The spectra of the two clicks shown in Fig. 13.6 suggest that these beaked whales emit echolocation clicks with peak frequencies between 30 and 40 kHz, and that the spectra of the clicks can extend beyond 45 kHz (the Nyquist frequency of the data acquisition system). These two clicks are the widest band clicks recorded for beaked whales. Besides measuring clicks from conspecifics, the D-tag has also been able to detect the echoes from prey and other organisms

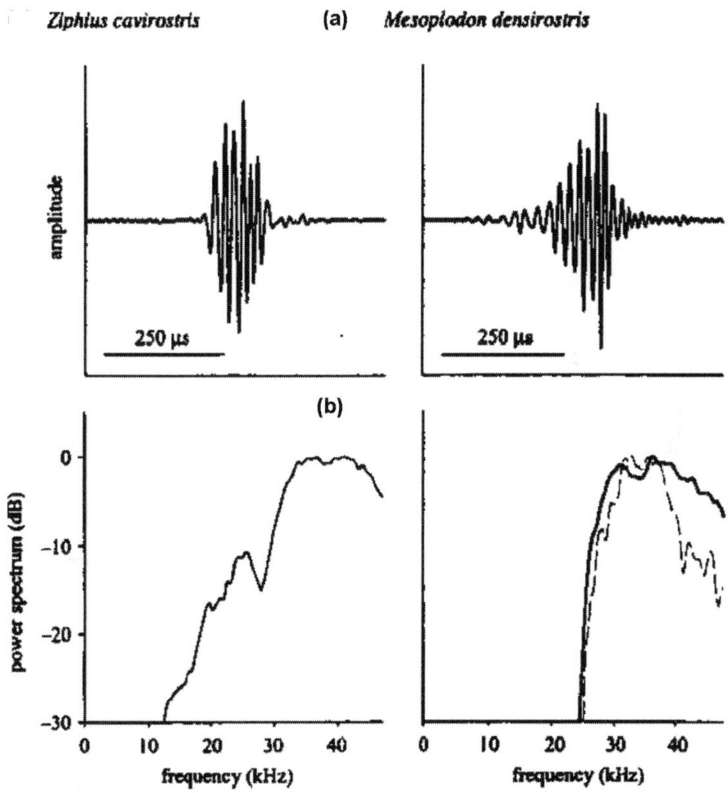

FIGURE 13.6. Waveforms and spectra of echolocation clicks seemingly emitted by conspecifics (**a**) *Ziphius cavirostris* and (**b**) *Mesoplodon densirostris* (adapted from Johnson and Tyack, 2004).

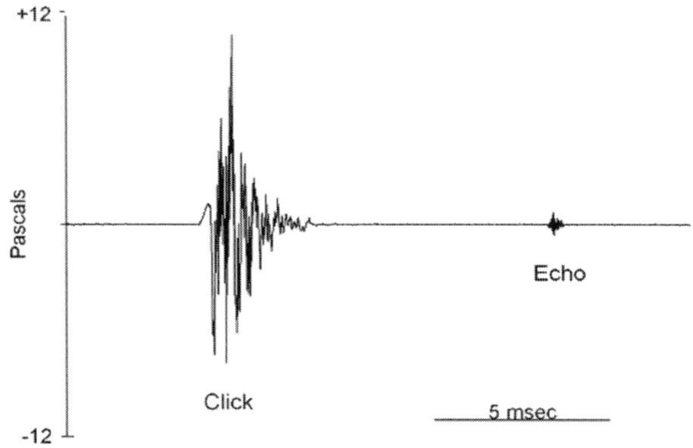

FIGURE 13.7. An echolocation signal measured by the D-tag located behind the blow hole of a Blainville's beaked whale (*Mesoplodon densirostris*) and the corresponding echo from a prey (adapted from Madsen et al., 2005).

(Madsen et al., 2005). The outgoing signal (measured in the back of the sound source) and the echo from a prey are shown in Fig. 13.7.

13.1.3 Other Odontocete Tags

Other customized special purpose acoustic tags have been fabricated by researchers to obtain data of interest. Madsen et al. (2002) developed an acoustic tag to study sperm whale echolocation. This tag had a 12 bit-A/D converter that operated at a sample rate of 62.5 kHz with the data stored on a 192 MByte compact flash card. The data acquisition process was controlled by a MAX AD7870 μcontroller from Maxim. A pressure sensor was also incorporated into the tag so that simultaneous depth and acoustic data could be collected. A 10-cm diameter aluminum housing with a syntactic foam tail was used to house the electronics. This tag was attached to four sperm whales and provided valuable data on how the whales use their echolocation signals for foraging. The results of Madsen et al. (2002) were previously discussed in Chapter 11, Section 11.4.

Akamatsu et al. (2000) used a simple tag to study the echolocation behavior of the finless porpoise (*Neophocaena phocaenoides*) and the Chinese river dolphin, baiji (*Lipotes vexillifer*). A peak-hold circuit was used to capture the peak output of the echolocation signal and that peak was recorded by a Sony ICD-80 integrated circuit recorder. With this simple device, the time of occurrence and peak amplitude of echolocation signals could be recorded. During nonecholocation periods greater than 1 s, the recorder was turned off to conserve battery power. The data logger was used by capturing the subject, attaching the tag with a suction cup, and then releasing the animal. A second tag or data logger that measured behavioral information such as depth, swim speed, and the tilt angle of the subjects was also attached to subjects (Akamatsu et al., 2000).

13.2 Special Techniques to Localize and Track Vocalizing Marine Mammals

13.2.1 Radio Synchronization of Hydrophone Stations

Tracking of vocalizing marine mammals can be done on a large scale involving several ten, sometimes hundreds of miles with relatively gross location resolution using permanently installed bottom-mounted hydrophones, towed arrays, or sonobouys. Tracking can also be done on a more localized scale, covering only several miles but with finer location resolution than long-range tracking. Whales often enter and reside for short periods of time in small bodies of water typical of a bay, inlet, harbor, or some specific body of water. Hayes et al. (1997) have developed a relatively inexpensive multi-hydrophone array system to track the small-scale movement of vocalizing animals over a confined area. The system consists of multiple sparbuoys, each buoy containing an inexpensive monitoring system. The sparbuoy is made of inexpensive PVC tubing with a 15-cm outside diameter as shown in Fig. 13.8. Each buoy contains a hydrophone that is connected to one channel of a small portable dual-channel DAT recorder. The second channel of the DAT recorder is used to record the signal from a small portable VHF radio. The VHF radio link is

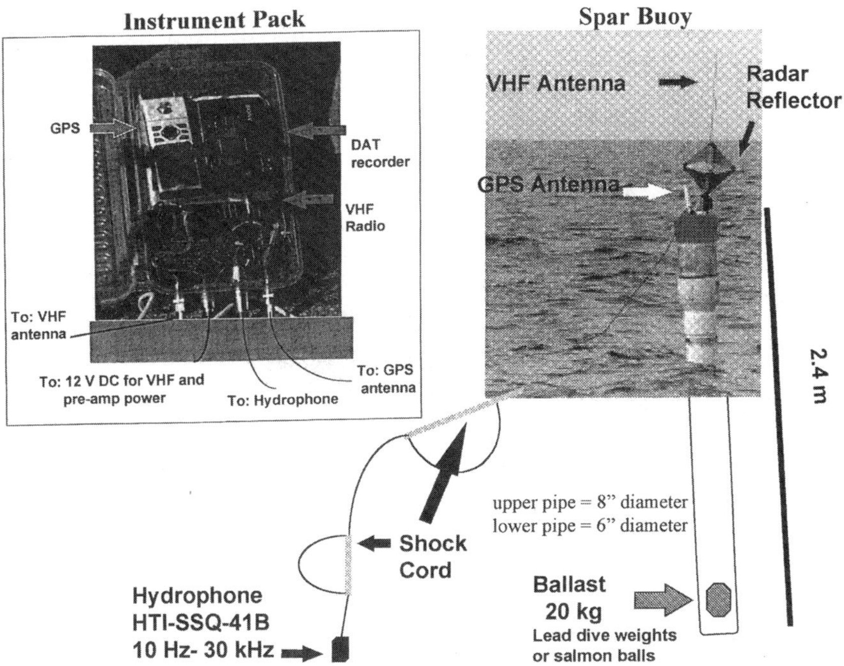

FIGURE 13.8. Inexpensive spar-buoy sound monitoring system (adapted from Hayes et al., 1997).

used for time-synchronization purposes so that accurate time-of-arrival differences between the hydrophones in the array can be determined. During a tracking session, an operator will periodically key the transmitter and send a signal that all the VHF receivers on different buoys will receive essentially simultaneously so that an accurate time fix can be obtained. Finally, a GPS with a built-in logger is also housed in each buoy and records the position of each bouy at any given time.

When a marine mammal of interest is sighted, the instrumented sparbuoys are deployed from a small boat and spaced several kilometers apart. The deployment procedure usually is completed within an hour, the duration of deployment being dependent on how long it takes to boat from one buoy location to the other locations. With the DAT running on its slowest speed (sampling at 32 kHz), about 6 hours of recording can be achieved, with about 5 hours reserved for the recording of vocalizations by all the hydrophones in the array. In calibration tests with three buoys deployed in a triangle having a spacing of about 1.8 km, Hayes et al. (1997) could achieve range accuracies to within 0.2 km. They used weighted light bulbs that would explode upon sinking to their crush depths for the sound source in their calibration. So far, Hayes et al. (1997) have tracked eight marine mammals with this system.

13.2.2 GPS Technique of Hydrophone Position

Mohl et al. (2000) devised an extremely clever method to obtain accurate time synchronization and location information on a measuring system that consisted of a high number of hydrophones on different boats. They were able to modify a GPS receiver to obtain the "raw" GPS signals, which contained a good time synchronization signal and the location or position of the receiver. The hydrophone signals were fed into one channel of a two-channel DAT recorder, and the GPS information was fed into the other channel. Mohl et al. (2001) outfitted a large number of boats with a hydrophone and the GPS-DAT recording system, and these boats were free to position themselves anywhere within a specified area. When sperm whales swam into the area emitting echolocation signals, the signals were recorded by a number of hydrophones along with the GPS synchronization signals and location information. Therefore, the precise location of each boat could be determined (to within GPS accuracies) and the relative time difference of the click signals could be determined in post-data analysis. Mohl et al. (2001) enlisted a number of "friends" with boats to form a flotilla of measuring stations, each essentially operating independent of each other but all synchronized in time.

13.2.3 Broadband Measurement of Dolphin Social Sounds

Animals that have the capability to echolocate, such as dolphins and bats, can typically hear high-frequency sounds up to 100–150 kHz depending on the specific species (Au, 1993; Moss and Schnitzler, 1995). These animals also emit signals with energy components that can extend to their maximum frequency of hearing and beyond. The large bandwidth of the acoustic signals emitted by these animals imposes a severe constraint on any signal measurement system used for field studies. Bioacousticians are constantly on the lookout for acoustic instrumentation that is broadband, portable (lightweight and can be operated with batteries), and inexpensive.

Digital audio tape (DAT) recorders have been used extensively in the field because of their small size and inexpensive cost. However, the fastest sampling rate for standard DAT recorders is 48 kHz, limiting the analog bandwidth to approximately 22 kHz. Some manufacturers have produced DAT recorders that can operate at two times the standard speed and sample rate so that the analog bandwidth is increased to 44 kHz, but the recording time is reduced by 1/2 of the standard recording time. One manufacturer has produced a DAT recorder operating at eight times the standard speed and sample rate to achieve an analog bandwidth of 176 kHz, but with 1/8th the recording time and an almost 30-fold increase in price.

Au et al. (1999) developed an acoustic recording system that satisfies the stringent requirements for field studies by being portable, broadband, and relatively inexpensive. This portable broadband data acquisition system (PBDAS) was developed to record the social signals of wild dolphins but can also be used to record echolocation signals of bats and other animals that emit ultrasonic signals. The PBDAS consists of two major subsystems: (a) an analog signal processing (ASP) unit which includes an amplifier, filter, and an external trigger unit and (b) a laptop computer with a PCMCIA data acquisition card that can operate at a sample rate up to 500 kHz. A block diagram of the PBDAS system, not including the PCMCIA card, is shown in Fig. 13.9. The modules inside the dashed rectangle are part of the ASP unit, which is contained in an aluminum chassis ($9 \times 15 \times 25$ cm) along with two 12 v (800 mAh) rechargeable batteries. The ASP consists of a differential amplifier input to reduce common-mode signals that may be induced into the hydrophone cable, an adjustable analog high-pass filter that is used to reduce unwanted engine and flow-noises, followed by an adjustable gain (42–84 dB) signal amplifier. The output of the gain amplifier is fed to two-switch capacitor low-pass filters in series. The switch-capacitor filters are used for anti-aliasing. The output of the second low-pass filter is divided into three paths, one going to the analog-to-digital (A/D) converter of the PCMCIA card, the second going to the peak-hold and level detector module that drives a light-emitting-diode (LED) bar graph display, and the third going to an adjustable gain audio amplifier that drives a set of head phones. The voltage level detector and LED bar graph are incremented into 3 dB steps. The final component is a push-button switch connected to an electronic circuit that removes any switch bounces and provides an external trigger for the A/D converter circuitry in the PCMCIA card.

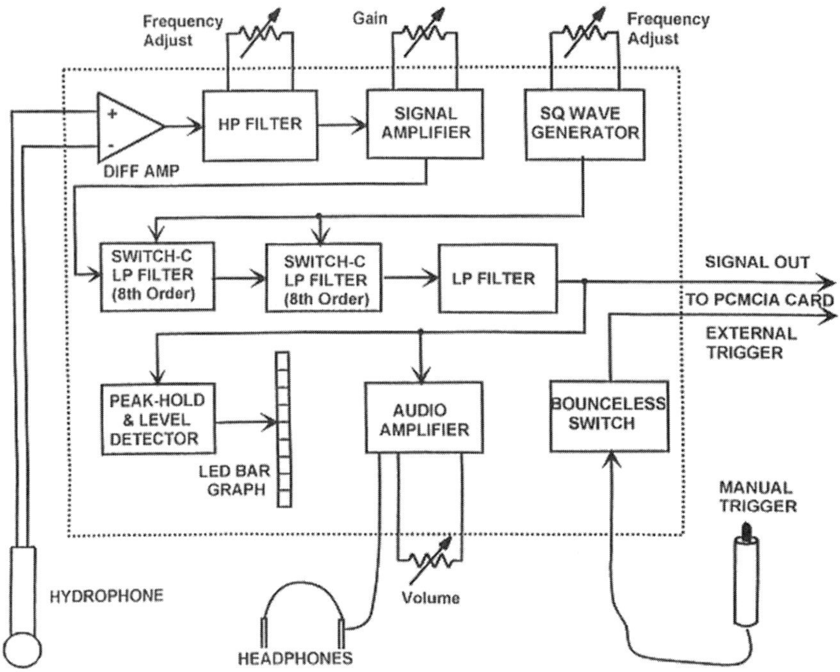

FIGURE 13.9. Block diagram of the PBDAS excluding the laptop computer and PCMCIA card. The modules within the dashed rectangle are housed in an aluminum chassis (adapted from Au et al., 1999).

The functioning of the PCMCIA card is controlled by software via the laptop computer. The PBDAS operates by creating a data array that can function as a circular buffer within the laptop computer. At the beginning of data acquisition, the A/D converter digitizes continuously with its digital output read into the laptop and temporarily stored in the circular buffer array of N_{total} points as shown in Fig. 13.10. Once an acoustic signal is heard and/or detected by the LED bar graph indicator, the operator presses the manual trigger, which causes the PCMCIA card to acquire only N_{post} more points and then cease any further data acquisition. Therefore, the number of pretriggered points (N_{pre}) in the circular buffer will be $N_{total} - N_{post}$ and the number of post trigger points will be N_{post}, where $N_{pre} + N_{post} = N_{total}$. The N_{total} digitized values in the circular buffer are then stored onto a disc file starting with the data at the start address, which corresponds to trigger address + N_{post} +1, and ending with the data at the end address which corresponds to the trigger address + N_{post}, as depicted in Fig. 13.10.

The data acquisition technique using a circular buffer and a manual trigger ensures that only N_{total} points will be acquired per signal. Digitizing signals at a high sample rate requires the availability of large amounts of memory, and

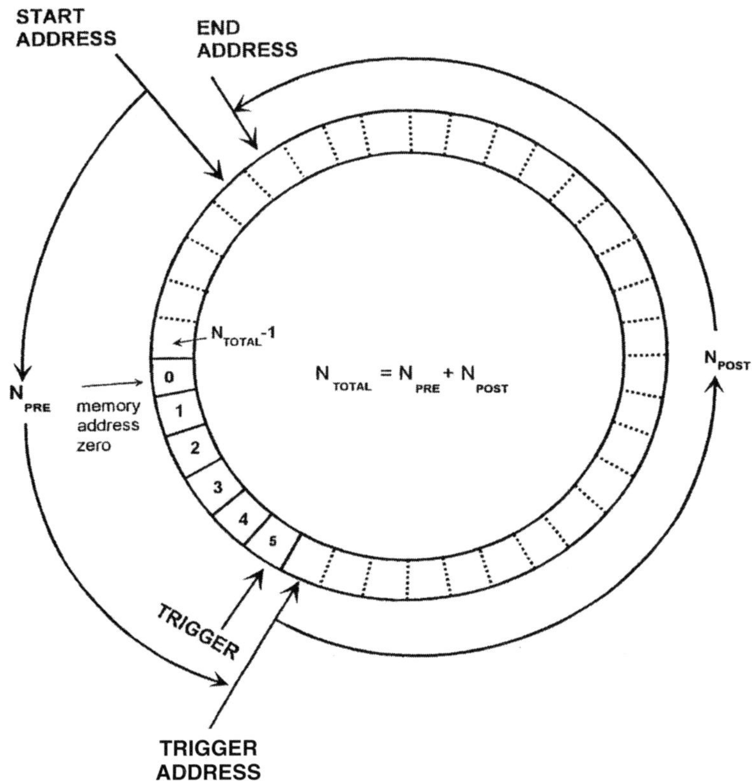

FIGURE 13.10. Schematic of a circular buffer depicting the data acquisition and storage process.

special precautions must be taken so that digital data are stored only when an acoustic signal is present. For example, if a unit is digitizing continuously for 2 min at a 500 kHz sample rate, 120 Mbytes of memory would be required. Therefore, trigger considerations that affect the data acquisition process are extremely important to optimize the amount of data that is typically captured and stored.

Under an ideal situation, the whole process can be automated by programming the PCMCIA card to initiate its own "manual trigger" when the input voltage to the A/D converter exceeds a specific threshold level. However, we have found that this technique does not perform well in an environment where snapping shrimp and other extraneous noise sources are present. With manual triggering, the operator can both listen to incoming signals and observe the LED display to detect signals of interest. The LED display holds the maximum signal deflection for a minimum time of 250 ms, which is sufficient to detect a dolphin's whistles, echolocation clicks, and burst pulses. A transient signal from a snapping shrimp will be on the LED display for only a brief time and will likely be heard, so the operator can ignore the signal. Burst pulse signals with frequency components in the audible range are easily recognizable. Some burst pulse signals having only ultrasonic frequency components will not be heard, but their presence will appear on the LED display if only for a short time. If the operator observes a brief flash on the LED and does not hear any signals, then there is a high likelihood that the system has acquired burst pulses. Echolocation click trains tend to last between 1 and 2 s and are often audible. Whistles also typically last about 1–2 s. Therefore, using the manual triggering technique, integration of the acoustic information can be achieved by using both the audio signals and the LED display before a decision is made on whether or not to collect and store the data.

In our application, we use a PCMCIA data acquisition card that can theoretically sample up to 500 kHz with a 12-bit resolution. However, the maximum sample rate can only be achieved for short signals that fit into the internal 1,024 points first-in-first-out (FIFO) buffer memory of the PCMCIA card. We use a Pentium-based laptop computer operating at 133 MHz, which allows a continuous sampling rate of 260 kHz. Our sampling rate is limited by the processor's ability to read data out of the PCMCIA memory buffer. The software is written in National Instrument's Labview environment using the Labview object-oriented programming technique. The number of points that can be collected per signal depends on the available memory of the computer, which in our case translates to 60 s worth of data. To capture an entire signal and to allow for reasonable human processing and reaction times, we typically collect 3 s worth of data, which includes between 1 and 1.5 s of pretrigger data. Therefore, each signal stored to file requires a storage space of 1.56 Mbytes. The maximum sample rate achievable will depend on the specific PCMCIA card, the laptop computer, and the language in which the controlling software is written.

We have used the PBDAS with a hydrophone that has a 1.3-cm diameter spherical PZT crystal, to record the social signals of the Hawaiian spinner dolphin (*Stenella longirostris*), and pan tropical spotted dolphin (*Stenella attenuata*) in waters off the leeward coast of Oahu, Hawaii. A 17′ ft Boston whaler propelled by an outboard engine was used to approach and record groups of animals socializing and traveling near-shore. Measurements were made with the whaler underway at relatively slow speed while flanking a group of animals traveling at less than 3 knots or while stationary with the engine off. Examples of measurements made with the PBDAS can be found in Fig. 10.18 in Chapter 10.

13.2.4 Measurement of Echolocation Signals: Wild Dolphins

Making accurate measurements of the echolocation signals of wild dolphins can be a difficult task because of how echolocation signals propagate from a dolphin's head. Unless a hydrophone is on the axis of the transmit beam, the signals measured will likely be distorted as is illustrated. Therefore, the relative orientation of the dolphin with respect to the measuring hydrophone is very important. Another issue that needs to be considered is the source level of the signal being measured. From Figs. 11.3 and 11.4, we know that the spectrum of an echolocation signal is directly related to the source level of the signal. Therefore, a relatively accurate estimate of the dolphin's range to the hydrophone must be made in order to have an understanding of the relative source level of a received signal with respect to the maximum source level for a specific species. Finally, measurement equipment most likely must be portable, light in weight, and relatively small, but must also have a bandwidth on the order of 170 kHz or more in order to accurately measure echolocation signals.

Au and Herzing (1997) used a symmetrical star array of four hydrophones to measure the echolocation signals of wild spotted dolphins (*Stenella frontalis*) in the waters of the Grand Bahama Island. PVC tubings attached to a nylon rectangular piece were used as the arms of the symmetrical star with the center hydrophone attached to the nylon center piece as shown in Fig. 13.11. Each hydrophone was spherical in shape with a diameter of 2.2 cm. A video camera was attached to the nylon center piece and was used to ascertain the orientation of an echolocating dolphin with respect to the array. The array electronics housed in a water-tight container attached to the nylon center piece consisted of a 20 dB gain amplifier and a line driver for each hydrophone. A multi-conductor cable consisting of 5 RG-174 coaxial cable along with two plain wires extended from the underwater electronic housing and a

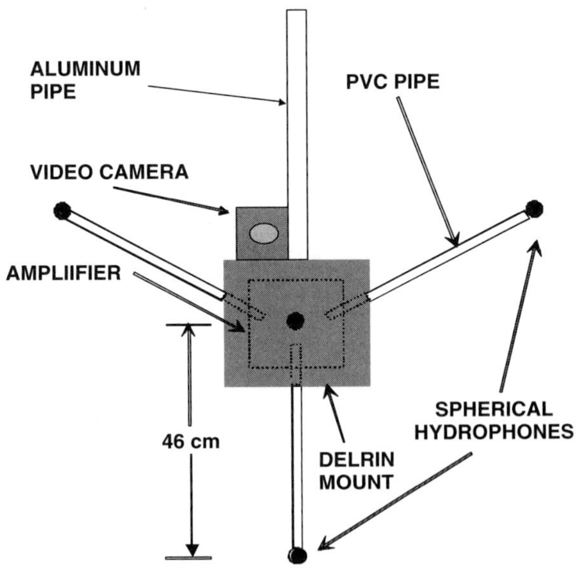

FIGURE 13.11. A schematic diagram of an array of four hydrophones arranged as a symmetrical star used by Au and Herzing (1997).

boat. Two different lengths of cable were used, one that was 76 m in length was used when the array was carried by a snorkeled swimmer and a short 3-m length was used when the array was extended into the water from a boat.

The multi-conductor cable was connected to two data acquisition boards housed in a lunch-box personal computer. The two data acquisition boards had two channels a piece and were linked together in a master–slave relationship, allowing four channels of simultaneous analog-to-digital conversion at sampling rates up to 20 MHz. We used a sample rate of 500 kHz in order to conserve disc space since the output of four hydrophones were being stored. The data acquisition board could be programmed to save a specified number of pretrigger and post-trigger points. The clocks of the PC and a portable high 8-mm video recorder were manually synchronized and the time at which each click was detected was also stored.

The range to an echolocating dolphin can be determined by measuring the difference in the time of arrival at the center hydrophone with the hydrophones on the tip of the star. From Eqs. (5.54) and (5.55) along with Table 5.1, the range of a sound source from the center hydrophone of the array can be expressed as

$$R = \frac{3d^2 - c^2(t_{21}^2 + t_{31}^2 + t_{41}^2}{2c(t_{21} + t_{31} + t_{41})}. \tag{13.1}$$

We tested the accuracy of the symmetrical star array by playing a simulated dolphin echolocation signal through a projector and progressively increased the distance between the projector and the array. The results of these measurements for an array geometry in which the separation distance between the hydrophones at the tips of the star and the center hydrophone was 61 cm are shown in Fig. 13.12. The calibration results indicate that for a sound source located along a projection from the center hydrophone that is perpendicular to the plane of the array, the symmetrical star can accurately estimate the range of the source. We have used this system to measure the echolocation signals of the Atlantic spotted dolphin, *Stenella frontalis,* (Au and Herzing, 1997), and of the Pacific spotted dolphin, *Stenella attenuata* and spinner dolphins, *Stenella longirostris*.

The U.S. Navy has developed dolphin systems that are used to echolocate and detect ordinances laying on the sea floor. These dolphin systems are extremely good at detecting and discriminating certain types of mines and exceed the capabilities of modern sonars. In order to study the echolocation process of dolphins as they perform their search and discrimination tasks, Sigurdson (1996, 1997) has developed a microprocessor controlled portable acoustic measurement system that is attached to a bite-plate that a dolphin can be trained to carry during its sonar search. The instrumentation package is shown in Fig. 13.13. The dolphin is presented with the instrumentation just prior to a sonar search, and the magnetic azimuth of the animal's head is measured at a rate of 10 times per second, the roll, pitch, and depth of the

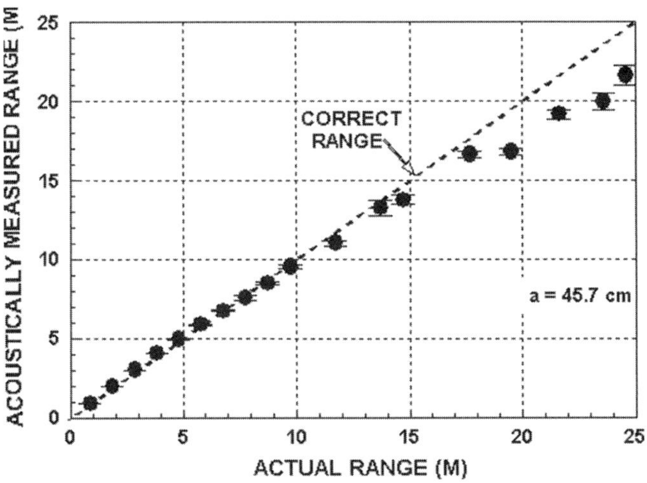

FIGURE 13.12. Results of measuring the range of a source with a symmetrical star array.

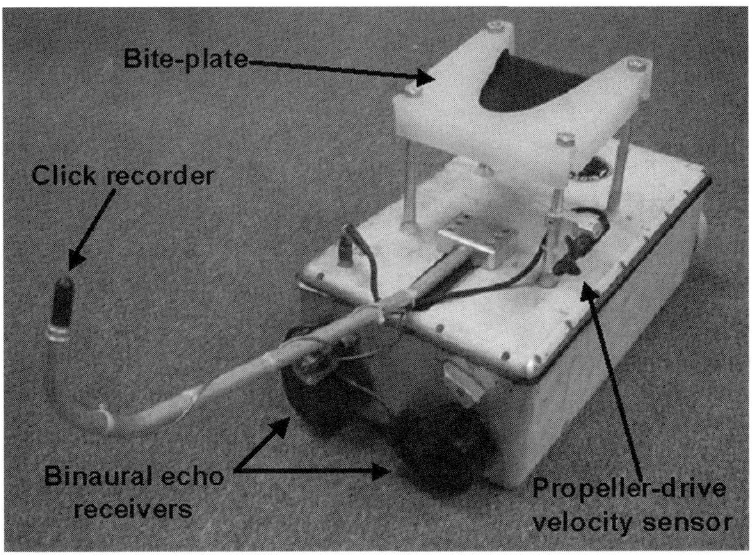

FIGURE 13.13. A photograph and schematic of the echolocation instrumentation package developed to monitor the echolocation process of a trained dolphin in open waters (adapted from Sigurdson, 1997).

head are measured at a rate of 100 times per second, the interclick intervals are measured with an accuracy of ±0.5 ms and the echolcoation signal is digitized with a 12-bit accuracy at a sampling rate of 0.9 MHz.

A B&K-8103 hydrophone is shown extending about 0.55 m ahead of the tip of the upper rostrum and about 4 deg down from the axis of the transmit beam. Three accelerometers with adjustable sensitivity are mounted to the frame of the package on mutually orthogonal axes. A pressure gage on the back of the unit is used to obtain depth information. A fluxgate compass mounted on a custom gimbals that allows for ±80° rotation in each axis is housed in the front portion of the package, as can be seen in Fig. 13.9. All of the different measurements and data storage functions are controlled by a single board computer with a 10 MHz RTX-2000 processor. The processor is programmed in the Forth language stored in read-only-memory. With this instrumentation package, Sigurdson (1976, 1977) has been able to measure the dynamics of motion of an echolocating dolphin searching for bottom ordinance in a manner that was previously not possible. An example of the type of data that the instrumentation package can record are shown in Fig. 13.14. In this figure, the outgoing signal is recorded along with binary echoes collected by the two hydrophones.

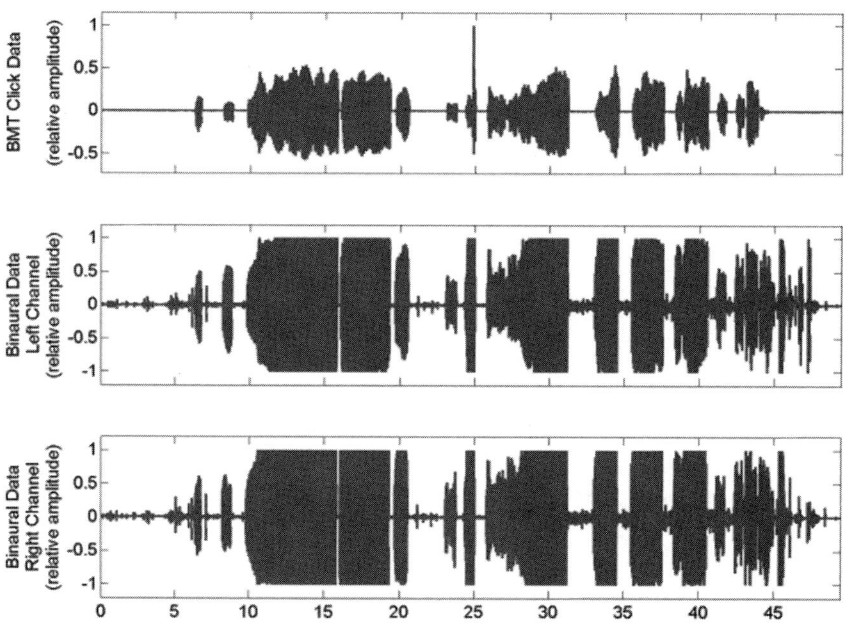

FIGURE 13.14. Examples of the outgoing echolocation signal and the corresponding echoes measure by the instrumentation package developed by Sigurdson (1976, 1977).

13.2.5 Dolphin Phantom Echo Sonar Experiment

In behavioral experiments where real targets are used to investigate dolphin echolocation, it is often very difficult to extract the relevant echo parameters that the animals use to discriminate or classify targets. The complex relationship between the physical dimensions and the reflection characteristics of real targets prevents separate control of various echo parameters of the stimuli presented in an echolocation experiment. Aubauer and Au (1998) devised a new echo simulation method using a phantom electronic simulated target. Dolphin echolocation sounds are transformed with the target impulse response in the time domain or transfer function in the frequency domain into artificial echoes, which are played back to the animal. The configuration for such an experiment is similar to that of Fig. 11.14, except only one hydrophone is used and the projector is further away from the animal in the hoop. The dolphin emits a signal that is captured and digitized and the phantom echo is played back after an appropriate time delay. The procedure is similar to what was described in conjunction with Fig. 11.14 with the major exception being the phantom echo in this case will represent specific targets, whereas in Fig. 11.14 the animals click was played back as if it reflected off a planar surface.

In order to measure a target's transfer function, we performed a backscatter measurement of the target in a test tank using an impluse-like signal, $s_i(t)$ having a Fourier transform of $S_i(f)$. If we let $e_t(t)$ in the time domain and $E_t(f)$ in the frequency domain be the echo from the target, then the transfer function of the target will be

$$H_t(f) = \frac{E_t(f)}{S_i(f)}, \tag{13.2}$$

and the impulse response of the target will be

$$h_t(t) = \Im^{-1}\left(\frac{E_t(f)}{S_i(f)}\right). \tag{13.3}$$

In the experiment with the dolphin, every time the dolphin emits an echolocation click, the signal is digitized and Fourier transformed and multiplied with the target transfer function to produce a target echo in the frequency domain. If we let $s_d(t)$ be the dolphin echolocation click having a Fourier transform of $S_d(f)$, then the target echo that is desired can be expressed as

$$e_i(t) = s_d(t) * h_t(t) = \Im^{-1}\{S_d(f)H_t(f)\}. \tag{13.4}$$

However, in most cases the projector that is used to transmit the "echo" will not have a flat transmit frequency response and would need to be equalized. In order to equalize the projector, its transfer function will need to be measured. Let $s_i(t)$ be input to the power amplifier–projector combination and let the projected signal be $s_p(t)$ in the time domain and $S_p(f)$ in the frequency domain, where the shape of $s_p(t)$ will not be similar to the shape of $s_i(t)$, then the transfer function of the projector will be

$$H_p(f) = \frac{S_p(f)}{S_i(f)}. \tag{13.5}$$

In order to project an undistorted version of $s_i(t)$ into the water to simulate an echo from a target, $s_i(t)$ must be convolved with the reciprocal of the impulse response of the projector, which is the inverse Fourier transform of Eq. (13.5) so that the input to the power amplifier–projector combination should be

$$s_e(t) = \Im^{-1}\left(\frac{E_i(f)S_d(f)H_t(f)}{H_p(f)}\right). \tag{13.6}$$

The signal processing required in the phantom echo system is implemented on a digital signal processing board using a Texas Instrument TMS320C40 DSP chip operating at 40 MHz and interacts with its host PC. The experimenter has full programmable control over the echo generating process and the echo structure itself. Echoes of several underwater targets were simulated to evaluate the quality of the method, and the results are shown in Fig. 13.15. A comparison of simulated echoes with the original echoes demonstrated a very good agreement independent of the incident signal.

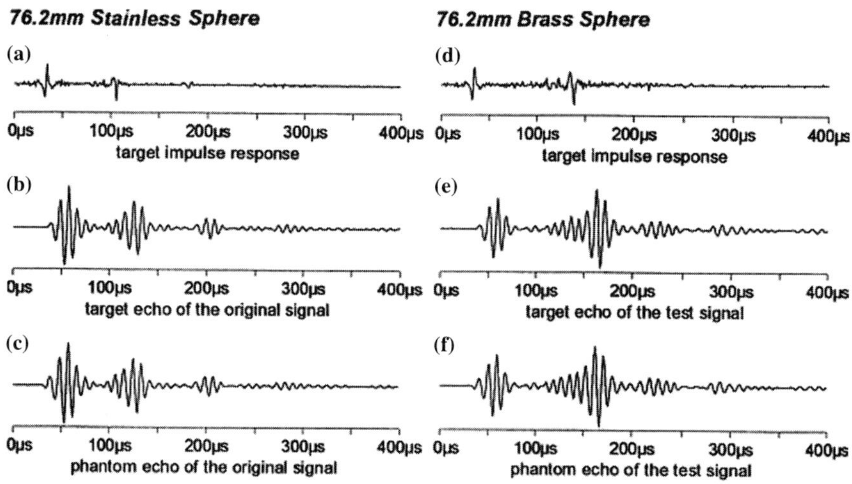

FIGURE 13.15. Example of the phantom echoes generation to simulate echoes from a solid stainless-steel sphere and a brass sphere. (**a**) The impulse response of the target, (**b**) echo from the real target, and (**c**) using the phantom echo after it is projected into the water (adapted from Aubauer and Au, 1998).

13.3 Some Instrumentation for Fish Acoustic Research

13.3.1 Acoustic Tags Satellite Tag, Chat Tag, Pop-up Satellite Tag

Acoustic tags are essentially miniature ultrasonic pingers that are attached to marine animals for tracking purposes. Tags have been used with fish, pinnipeds, turtles, sea snakes, and alligators. With these tags, many parameters associated with a species can be learned. Important information such as the size and dimension of the home range, site fidelity, diurnal changes in distribution and behavior, foraging patterns, the effects of topography, and physical oceanographic variables (Holland et al., 1992) can be obtained by using acoustic tags. Although acoustic tags have been used in fish research as early as 1970 (Yuen, 1970), the earlier versions were bulky and heavy and could be used only with large subjects. It was not until the 1980 s that miniaturization reached the level which made it practical for a variety of species of different sizes. Acoustic tags have been used with both pelagic species (Carey, 1983; Holland et al., 1985) and coastal species (Holland et al., 1992) of fish. Modern tags are generally very small and lightweight. Holland et al. (1985) have used a cylindrical tag that was 8.0 cm long, 1.6 cm in diameter, and weighed 27.7 g in air and 11.7 g in seawater with pelagic fish. An example of a ultrasonic tag attached to the dorsal surface of a tuna is shown in Fig. 13.16. Holland et al. (1992) have used smaller tags (4-cm long, 0.8 cm diameter) with hammerhead sharks. The transmitters are inserted into the fish gut by sticking a hollow tube into the fish's mouth and having it extended into the gut region. The transmitters are then dropped into the tube and kept in place with a ramrod inserted down the tube. The range of a tag is

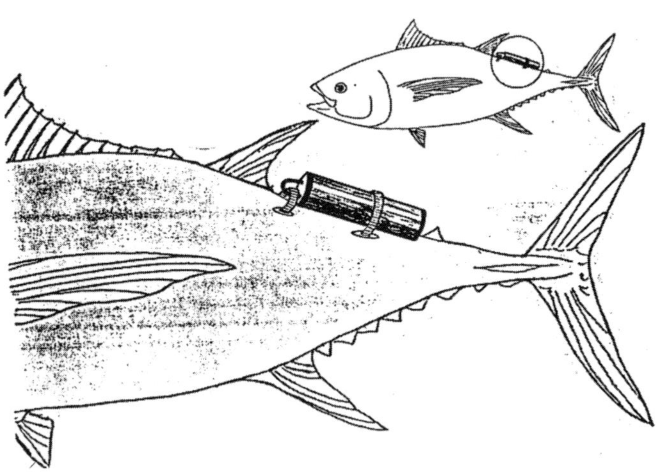

FIGURE 13.16. A schematic of an ultrasonic transmitter attached to the dorsal surface of a tuna (adapted from Holland et al., 1985).

dependent on the transmitted power, which in turn depends on the amount of battery power and battery life desired. For example, a 3-day tag may be detectable out to about 0.8 miles whereas a 22-day tag may be detectable to only 0.5 miles (Holland et al., 1985). However, battery technology is increasing at a steady pace with ever increasing energy density so that the life and range of acoustic tags will also most likely continue to increase.

Besides obtaining tracking data, additional information can also be obtained by clever design of an acoustic tag. Tags have been used in conjunction with a pressure sensor to control the rate at which tone pulses are produced, so that as the fish swims deeper the rate of tone pulse emission also increases and vice versa. The frequency of the tone pulses are generally controlled with a crystal so that tones with precise frequencies can be produced, making it possible to have a very narrow but precisely located (in the frequency domain) bandpass filter in the receiver, which would be helpful in reducing the effects of ambient noise on receiving the tag's signal. Lowe et al. (2008) have recently developed an acoustic tag that can be used to both locate a fish and to measure the fish tailbeat frequency. The tailbeat sensor consists of a pivoting vane with a small magnet that activates a reed switch, as shown in Fig. 13.17, mounted on a juvenile hammerhead shark (*Sphyrna lewini*). The vane tends to remain relatively stationary by the water flow around the moving fish, and as the tail sweeps side to side, the reed switch passes under the vane and magnet causing the switch to close. Each time the switch closes, a pulse tone burst of 18 ms duration with a carrier frequency of 76.8 kHz is generated, and there are 2 pulses per tailbeat cycle. Another tone burst of 9 ms is also produced at a regular rate of 1 tone burst per 5 s, for tracking purposes and especially in situations when the shark stops swimming for a brief period. The receiver contains an onboard microprocessor that separates the location tone and tailbeat tone bursts and also controls the logging of GPS receiver data and the tailbeat and location data to a laptop computer.

Acoustic tags continue to grow in sophistication as new types of tags continue to appear in the market place. There are two basic types of coding used with pingers, continuous and coded. Continuous pingers emit signals continuously until the battery is depleted. Information such as depth, temperature, velocity, heart rate, and acceleration can be encoded by changing the repetition rate or pulse interval of the emitted signal. Multiple amounts of information can be transmitted by having a sync interval of a given length, which will indicate the beginning of a data transmission sequence. Therefore, if one variable is encoded by the time interval between successive tone pulses and another variable is encoded by the interval between the next tone pulse, etc., the data can be sorted out by knowing when the data sequence started. An example of this process is shown in Fig. 13.18.

Coded pingers use digital coding to identify individual pingers operating at the same frequency. Coding is achieved by using N-tone pulses, providing $N-1$ time intervals followed by either a semi-random or fixed time delay before the same sequence of N-pulses are transmitted again. The identification of a specific

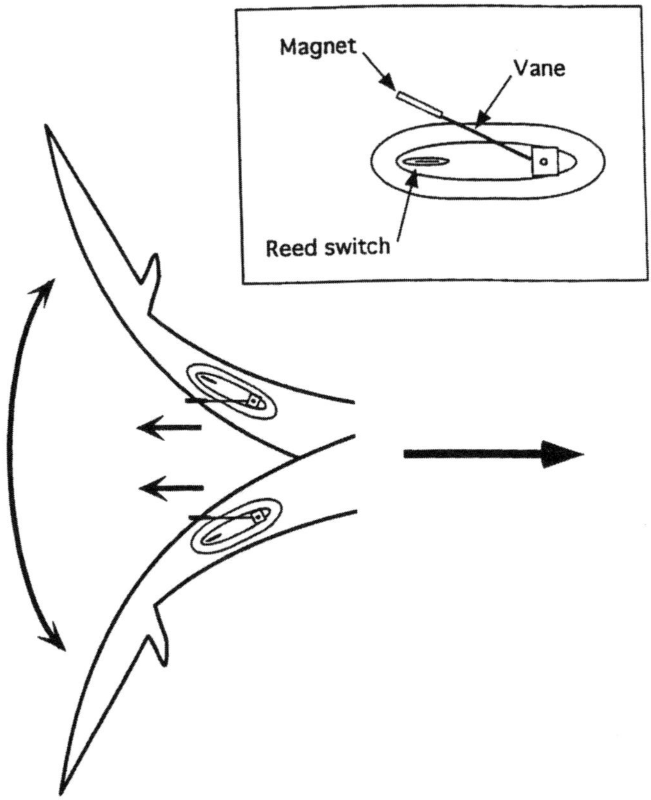

FIGURE 13.17. Schematic of the dorsal view of the tailbeat sensor (pivoting vane) on a shark's tail (adapted from Lowe et al., 2008).

pinger is contained in the pattern of time intervals between the successive tone pulses. Coded pingers can be used with a number of fishes in the same general location of a body of water. However, the receiver must be able to receive several sequences of coded pulses from one pinger before receiving coded pulses from other pingers. If two or more coded signals are overlapped at the receiver, neither code will be detected. The delay between coded sequences in a pinger generally varies between tens of seconds to over 100 seconds.

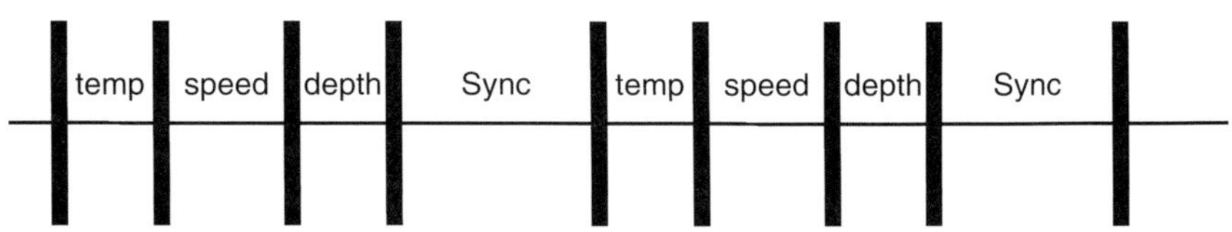

FIGURE 13.18. Example of coding temperature, speed, and depth information in a continuous signal.

This delay between coded sequences will also serve to prolong battery life in comparison to the continuous transmission technique. We have covered some of the basic principles involved with acoustic tags used in fishery science. It would be beyond the scope of this book to go into further detail. Instead, interested readers can check the web site of some of the leading manufacturers of fish acoustic tags to examine newer products.

13.3.2 Active Acoustic Mooring

Benoit-Bird and Au (2003) developed a unique remote echosounder system to study the horizontal migration dynamics of the mesopelgaic boundary community (MBC) organisms that spinner dolphins forage on in waters of the Hawaiian and other Pacific islands. They used a relatively inexpensive Computrol Tournament Master echosounder that projected signals at 200 kHz. The MBC organisms begin the shoreward migrate at about dusk and reach as close as 1 km from shore at about midnight and then head back toward deeper water. Therefore, the echosounder is needed only between dusk and dawn. A CF-1 single board computer from Persistor Instrument Incorporated controlled the time at which the echosounder was turned on and also was used to digitize the envelope of the echoes and store the data in compact flash memory.

A block diagram of the measurement system is shown in Fig. 13.19. When the system first turns on, the CF-1 will be in the suspend mode, drawing only about 0.25 ma and the rest of the electronics with the exception of a CMOS flip-flop will be off. A consumer model digital alarm clock with its own battery is used to wake the CF-1 up at dusk by connecting the alarm output to the CF-1 (this simple wake-up system is extremely user-friendly and easy to use). When the CF-1 wakes up, it triggers a flip-flop that closes a solid-state relay to apply power to the electronic circuit and the echosounder. Immediately after power is applied to the echosounder, it will begin to ping and receive echoes. The output trigger pulse of the echosounder triggers the analog-to-digital converter on the CF-1 and the envelope of the echoes is digitized at a 10 kHz sample rate and the results sent to the CF-1 for storage. The CF-1 was programmed to receive echoes from ten pings, after which the system is turned off and the CF-1 enters into a sleep period, waking up every 15 min to repeat the cycle of pinging and collecting echoes. During the sleep mode, the CF-1 draws about 2 ma of current. In the active mode, it draws between 5 and 50 ma of current. Finally, the system is turned completely off and the CF-1 enters into the suspend mode at dawn. The number of pings per wake up period, the length of the sleep period, the time to wake-up from the suspend mode, and the time to enter the suspend mode are all variables that can be input through a serial connection from a PC. A picture of the echosounder, the CF-1 Persistor, and compact flash memory is shown in Fig. 13.20. The electronics were stored in the underwater housing as shown in Fig. 13.20.

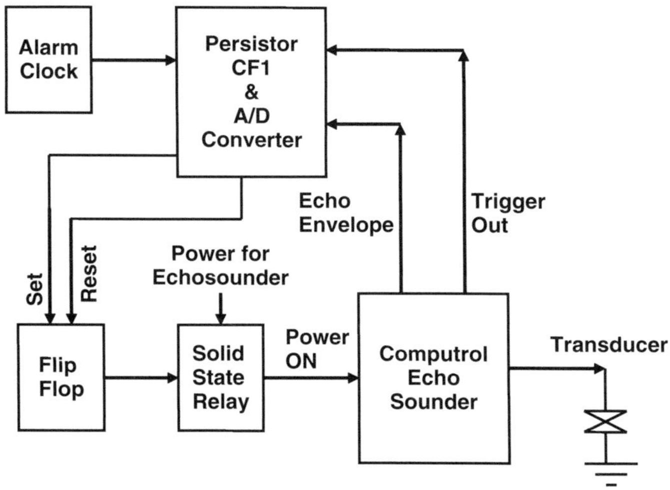

FIGURE 13.19. Block diagram of the echoranging system of Benoit-Bird and Au (2003).

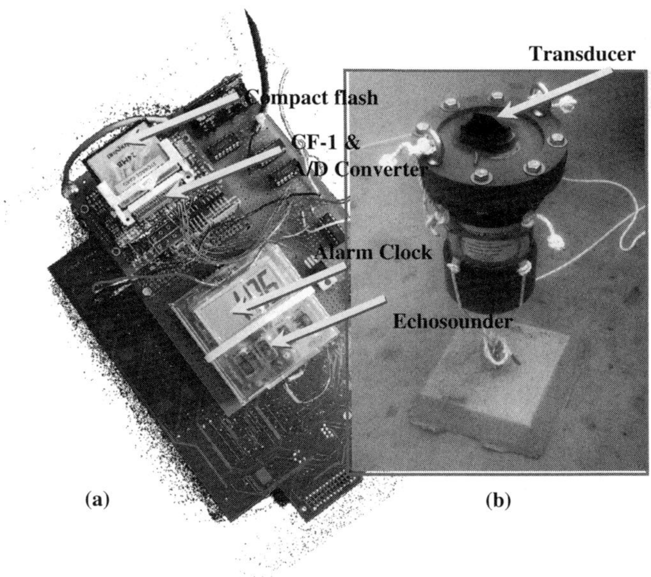

FIGURE 13.20. (**a**) Echosounder board, CF-1 Persistor single board computer with an analog-to-digital converter, and the compact flash memory card, (**b**) transducer on top of the Mod-1 underwater housing.

A single active acoustic mooring would be of limited use. However, a number of active acoustic moorings all operating in synchrony can provide important information on the dynamics of the horizontal migration of the MBC. Benoit-Bird and Au (2003) distributed five active acoustic moorings perpendicular to shore with the first mooring about 1 km from shore and the others spaced out at increments of 0.5 km. A summary of the results of four nights of observations are shown in Fig. 13.21. From the results, one can see that maximum density occurred at about 1 km from the shore at about midnight. The shape, density, and extent of the MBC layer perpendicular from shore can also be seen. The rate of the horizontal component of the migration could also be calculated by determining when the front of the layer is first detected for the shoreward portion of the migration and the when the tail of the layer is last detected for the outward portion of the migration by the different sensors.

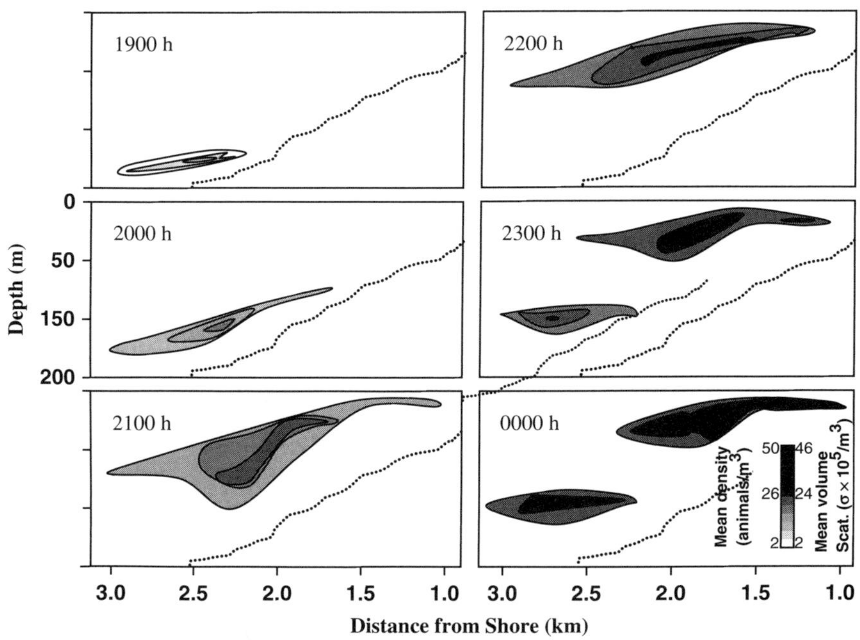

FIGURE 13.21. The mean distribution of the boundary layer at all five moorings every hour until the approximate midpoint of the migration pattern with linear interpolations between moorings. The edges of the layer are outlined in black. Areas of significantly different (as determined using the Webster method) mean mesopelagic animal density (determined through echo-energy integration assuming the layer is primarily composed of myctophids) are shown in *gray scale*. The *gray scale* could also represent areas of significantly different volume backscattering. The ocean bottom is denoted by the *dash line* (adapted from Benoit-Bird and Au, 2003).

13.3.3 Active Impedance Control

The effects of low-frequency sounds on marine animals has recently been a topic of high interest for many people concerned about the increased level of artificial noise being introduced into the ocean environment. Many species of fish are sensitive to both acoustic pressure waves and particle velocity (in the near field of a sound source) so that it is important to perform experiments with fish using plane traveling acoustic waves. For plane traveling wave propagation, the relationship between acoustic pressure and intensity will simply be $I = p^2/\rho c = pu$, instead of the more complex expression requiring a knowledge of the pressure gradient, where u is the particle velocity. However, because of the large wavelengths involved with low-frequency sounds below 1 kHz, it is very difficult to have plane traveling wave propagation in tanks and other enclosures. In a highly reverberant enclosure, standing waves will be established and the acoustic situation will no longer be simple.

Finneran and Hastings (1997, 1999) have developed a cylindrical waveguide that can produce low-frequency traveling plane waves by a using an adaptive active impedance control to eliminate reflections from one end of the waveguide. The waveguide consists of 14-m acrylic tubing with an 18-cm inner diameter and a 6.35-mm wall thickness. A schematic of the waveguide with a Naval Research Laboratory J-13 projector as the primary acoustic source on one end and another J-13 projector as the active termination or secondary source at the other end is shown in Fig. 13.22. The two hydrophones, B&K 8103, are shown in the test area of the waveguide. Because the acrylic is relatively flexible compared to water, there is larger attenuation of the acoustic wave along with a slower sound speed in the waveguide as compared to open water. These conditions help to achieve anechoic end conditions by reducing both the amplitude of a wave and the reflection coefficient at the waveguide termination. However, at low frequencies there can still be a considerable amount of reflection from the end of the tubing, requiring the aid of an active termination to obtain traveling wave conditions at the test site.

Let the position of the hydrophone in the test section closest to the primary source be defined as x_1 and the acoustic pressure it measures as $p_1(t)$, and the position of the other hydrophone as x_2 and its pressure as $p_2(t)$. We define the difference in the hydrophone distances as $s = x_2 - x_1$ and the midpoint between the two hydrophones is $barx = (x_1 + x_2)/2$. Then the transfer function within the test section can be expressed as

$$H(f) = \frac{P_2(f)\,P_1^*(f)}{P_1(f)P_1^*(f)}. \tag{13.7}$$

Because the waveguide is absorptive, the wavenumber $\mathbf{k}$ will be complex,

$$\mathbf{k} = k - j\alpha = \frac{2\pi f}{c} - j\alpha. \tag{13.8}$$

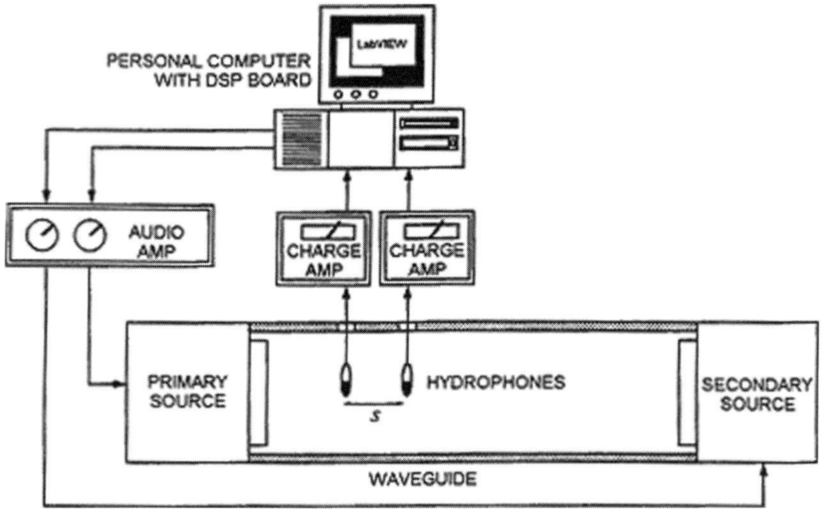

FIGURE 13.22. Schematic of the acrylic acoustic waveguide showing a primary and secondary acoustic source along with two measuring hydrophones in the test section, and accompanying electronics and computer instrumentation (adapted from Finneran and Hastings, 1999).

Therefore, the transfer function will also be complex (Finneran and Hastings, 1999). If there were no reflections at the termination of the waveguide, the transfer function for a traveling wave would merely be

$$H(f) = e^{-jks}. \tag{13.9}$$

The active impedance control operates by measuring the acoustic pressure received by each hydrophone, determining the transfer function, and then computes the error between the measured transfer function and the ideal transfer function. The controller then makes a gain and phase adjustment to the secondary source and the process is repeated continuously until the secondary source behaves in such a way as to reduce any reflective components in the test section so that only a traveling wave component will be present. A pattern search algorithm (Elliott et al., 1987) is used to reduce the error in the measured transfer function. The pattern search takes incremental steps, called pattern moves, after suitable directions have been found by local exploration. If the search progresses well, the step size is increased, otherwise the step size is reduced. When the step size is reduced below a set value, the search is ended (Arby and Dempster, 1974). An example of the effectiveness of the active control system showing the reflection coefficient in the test section is depicted in Fig. 13.23. The reflection coefficient has been reduced to below 0.05 over a frequency range of 12.5–400 Hz so that a traveling wave will exist in the test section.

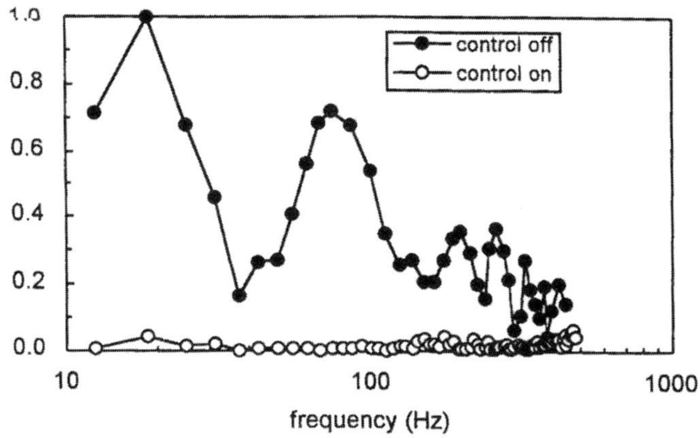

FIGURE 13.23. Reflection coefficient measured with and without active control. The hydrophone spacing was 1.8 m for $f < 38$ Hz, 0.4 m for $38 < f < 112$ Hz and 0.1 m for $f > 112$ Hz (adapted from Finneran and Hastings, 1999).

The active impedance waveguide has been used to study the effects of low-frequency sound on the peripheral auditory system of gold fish and oscar (Finneran, et al., 1996; Derenburger et al., 1997). These studies were done in conjunction with the non-invasive ultrasonic measurement system that will be discussed in the next section.

13.3.4 Non-Invasive Ultrasonic Measurement System

Ultrasonic techniques are routinely used to examine the internal structures of living organisms. These techniques rely upon acoustic impedance mismatches within the organism to scatter and reflect sounds. The swim bladder, Weberian ossicles, and otoliths of fishes including the goldfish provide an acoustic impedance mismatch that is significant enough to scatter incident ultrasonic waves. Finneran et al. (1996) and Finneran and Hastings (2004) have refined an ultrasonic technique developed by Rogers and Hastings (1989) to measure the amplitude and phase of acoustically induced vibrations. The basic approach of the technique can be explained with the aid of Fig. 13.24, showing a low-frequency acoustic source, an ultrasonic acoustic source, an ultrasonic receiver, and a target. If a fish is placed within a continuous low-frequency sound field of frequency f_L, the organ being investigated will vibrate with a displacement $x_L(t)$ that can be expressed as

$$x_L(t) = X_L \cos(2\pi f_L t - \phi_L), \tag{13.10}$$

where X_L is the displacement amplitude and ϕ_L is the displacement phase. The ultrasonic receiver output in volts will be

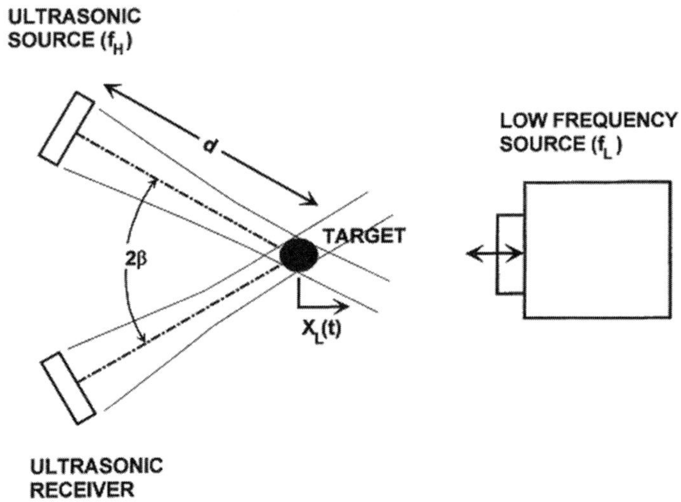

FIGURE 13.24. Non-invasive ultrasonic measurement system (adapted from Finneran and Hastings, 2004).

$$e_r(t) = K_e R A_i \cos[2\pi f_H t - \phi_r], \tag{13.11}$$

where K_e is a constant, R is the magnitude of the pressure reflection coefficient, and A_i is the amplitude of the ultrasonic signal incident on the target, and ϕ_r is the phase of the received ultrasound (relative to the phase of the transmitted ultrasonic signal). For a stationary target, the phase is the product of the wave number and the source–target–receiver length plus the phase changes due to the reflection process and can be expressed as

$$\phi_r = 2k_H d + \phi_R, \tag{13.12}$$

where k_H is the ultrasonic wave number, d is the distance from the source and receiver to the target, and ϕ_R is the phase of the pressure reflection coefficient. The improvement to the non-invasive ultrasonic measurement system of Rogers and Hastings (1989) by the addition of a phase measurement allows for determining the relative motion of different parts of a specimen.

The effect of the low-frequency vibration is to introduce sidebands into the received echo spectrum. The sidebands are located at frequencies of $(f_H + f_L)$ and $(f_H - f_L)$ and will have an amplitude of $k_H X_L \cos\beta$ relative to the high-frequency carrier amplitude. By measuring the ratio of the sideband amplitude, the amplitude X_L of the low-frequency vibration may be determined from

$$\frac{E_{\text{sideband}}}{E_{\text{carrier}}} = k_H X_L \cos\beta. \tag{13.13}$$

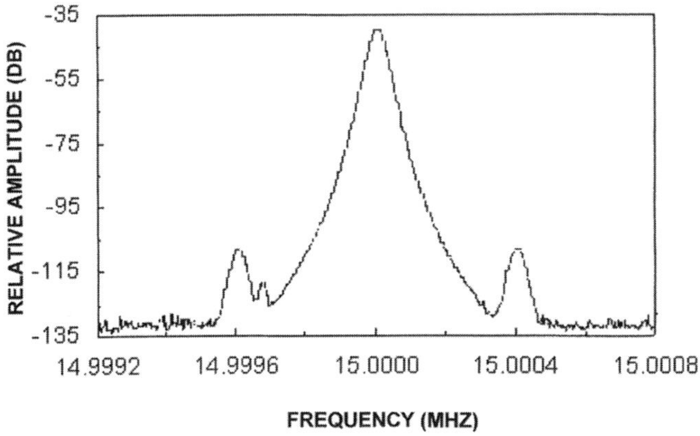

FIGURE 13.25. Received ultrasonic spectrum from a goldfish anterior swim bladder showing the relationship between the carrier and the sidebands. For this case, the low-frequency source was 400 Hz and the ultrasonic source frequency was 15 MHz (adapted from Finneran and Hastings, 2004).

With a 15-MHz transmitter and a spectrum analyzer having a dynamic range of 80 dB, the smallest measurable displacement is about 1.5 nm. An example of the signal received from a goldfish anterior swim bladder vibrating at 400 Hz using a 15-MHz ultrasonic source is shown in Fig. 13.25. For this example, the amplitude ratio $E_{\text{sideband}}/E_{\text{carrier}}$ is approximately –68 dB and $\beta \cong 30°$, so that $X_L \cong 7.3$ nm.

The low-frequency source operates at only a small fraction of the ultrasonic frequency so that the receiver output must be demodulated in order to measure the phase shift. Demodulating the received ultrasonic signal will shift the carrier and sideband to f_L where it may be examined with conventional instrumentation. Phase demodulation is achieved using a phase detector consisting of a double-balanced mixer and low-pass filter, as depicted in Fig. 13.26. The phase detector output is a time-varying dc signal proportional to the cosine of the instantaneous phase difference between two inputs and can be expressed as

$$e_\phi(t) = K_\phi \cos(\phi_1 - \phi_2), \tag{13.14}$$

where ϕ_1 and ϕ_2 are the phases of the applied inputs and K_ϕ is the phase detector maximum dc output. If the ultrasonic carrier and receiver signals are the phase detector inputs, the output is

$$e_\phi(t) = -K_\phi \cos(\phi_\phi) \tag{13.15}$$

or

$$e_\phi(t) = -K_\phi \cos[2k_H d + 2k_H X_L \cos\beta \cos(2\pi f_L t - \phi_L) + \phi_R]. \tag{13.16}$$

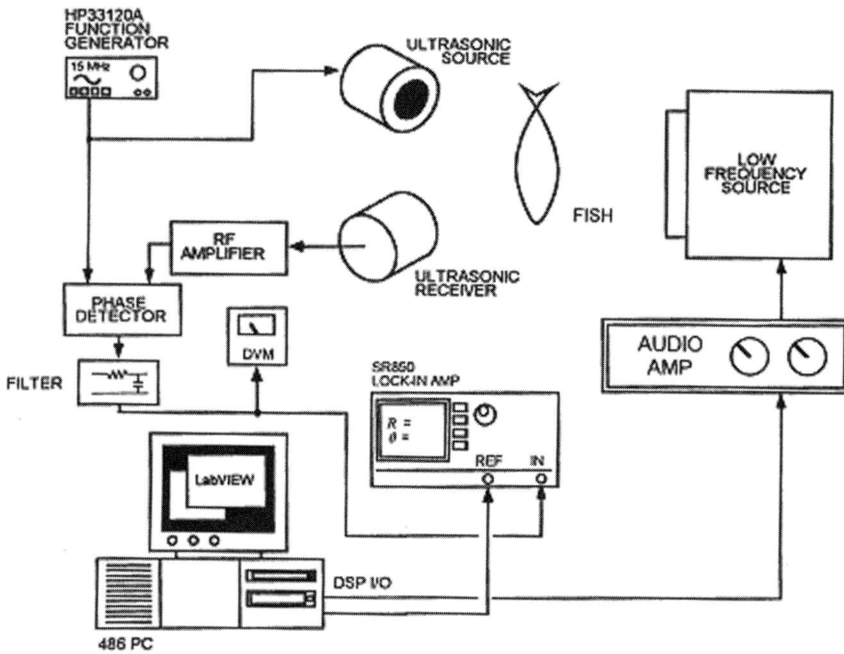

FIGURE 13.26. Experimental setup used to measure the relative phase of the received ultrasonic signal (after Finneran and Hastings, 2004).

The phase ϕ_L is therefore obtained by measuring the phase of $e_\phi(t)$. The lock-in amplifier in Fig. 13.26 is necessary because $e_\phi(t)$ will contain components at frequencies other than f_L.

13.4 General Instrumentation

13.4.1 Autonomous Acoustic Recorder – Bottom Resting

There are times when there is a need to monitor the acoustic signals of whales for an extended period at many different locations, and the use of cables from the sensor to a recorder on land is not economically feasible. In such a situation, Greene (1997) developed an autonomous acoustic recorder for a shallow water environment in the arctic to measure the sounds of bowhead whales as well as other ambient noise sounds. The device used by Greene (1997) is depicted in Fig. 13.27, showing a tubular housing supporting a hydrophone, signal conditioner and recorder and a bank of batteries. The autonomous remote device has been deployed on the bottom to depths of 44 m. A weighted line attached to the rear of the housing and extending away from the housing is used for recovery of the device by grappling hooks. The exact location of each device is determined by differential GPS measurements at the time of deployment so that recovery by grappling hooks is feasible.

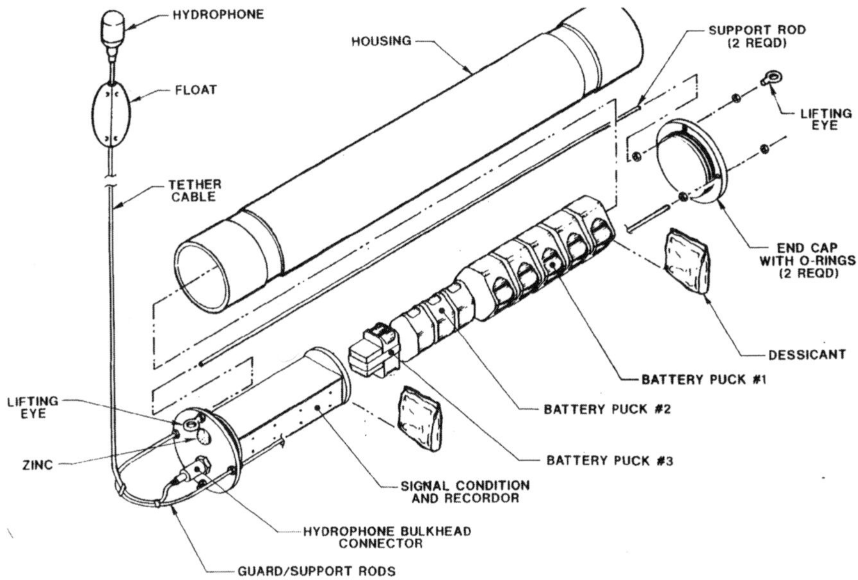

FIGURE 13.27. An autonomous remote acoustic recording system (adapted from Greene, 1997).

Over 100 units have been deployed and recovered after 15 days in the Alaskan Beaufort Sea.

The signal electronics consists of amplifiers, filter, A/D converter and a 4-Gbyte hard disc, all controlled by a microcontroller card. Acoustic signals are sampled continuously at a sampling rate of 1 kHz. However, for 1 min out of every 14 min and 24 s, the sampling rate is doubled. The storage capacity of the hard disc for the sampling rates used is 24 days. The higher sampling rate is used to monitor the ambient noise over a larger frequency range. The lower sampling rate is used to monitor the sounds of air gun pulses used in seismic surveys and bowhead whale calls. With a large number of remote units deployed over a wide area, Greene and his colleagues can obtain a fairly accurate assessment of the effects of seismic exploration on the behavior and well-being of bowhead whales.

13.4.2 Cornell Pop-Ups

The Cornell Bioacoustics Laboratory developed an autonomous remote acoustic recorder that can be deployed to a depth of 6,000 m and later retrieved by sending a special acoustic signal from the surface to detach it from its mooring allowing it to pop up to the surface, and hence was given the name "pop up". The electronics consist of a Tattletale 8 microcontroller from Onset Computer Corp. that has an onboard 8 channel analog-to-digital converter with a throughput of 100 kHz to acquire acoustic data from the hydrophone which is

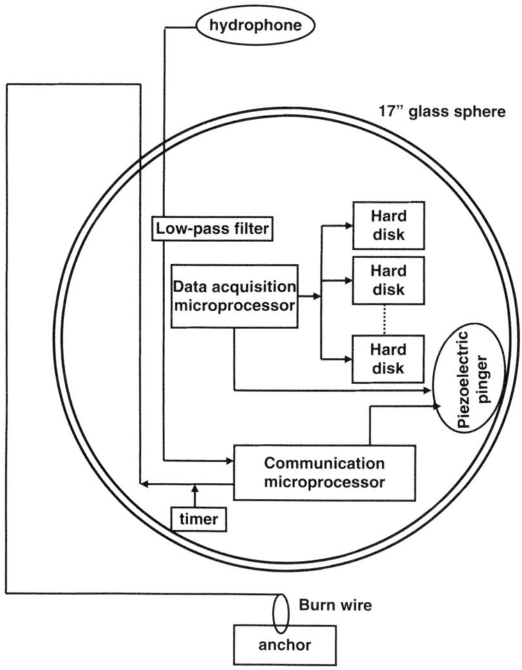

(a)

(b)

FIGURE 13.28. (**a**) Schematic of the "pop-up" and (**b**) a pop-up in a shipping container.

connected to it, with the data being stored on an 80 Gbyte hard disk. A schematic of the pop-up subsystems are shown in Fig. 13.28, with the electronics housed in a 17-in diameter glass sphere. The microcontroller can control the turn-on and record phase and the turn-off and sleep phase under software control. Therefore, the battery power can be minimized and the unit deployed for an extended period until either the capacity of the hard drive is reached or the batteries are drained.

A deployed pop-up is connected to an anchor with a stainless steel wire, which can be "burned" to release the pop-up form the anchor. Acoustic communications from the surface to the pop-up occur with the use of a surface controller unit and a hydrophone. When the pop-up receives the appropriate signal from the surface, it acknowledges by emitting its own acoustic response signal. Depending on what signal is sent from the surface vessel, the pop-up responds either with its acoustic response alone or by triggering the burn wire to release the anchor. A VHF radio beacon is housed with the pop-up unit, which will begin transmitting as soon as the unit reaches the surface and the antenna is out of the water. A high-intensity strobe light is also automatically turned on when the device reaches the surface so that the unit can be easily spotted and retrieved. Once the pop-up is retrieved, the unit can be refurbished by removing the hard disc and downloading the acoustic data to a computer. The information on the disk is then erased, the disk replaced, new batteries installed, and the unit is ready for redeployment.

13.4.3 HIMB/PIFSC Ecological Acoustic Recorder (EAR)

A remote autonomous acoustic recorder has been developed jointly between the Hawaii Institute of Marine Biology (HIMB) and the Pacific Islands Fisheries Science Center (PIFSC) and has been used in the field since 2006. It was designed to be a bottom-moored passive acoustic logger with a capability for long-term monitoring of the underwater ambient sound field (Fig. 13.29). The EAR is a digital recorder based on a Persistor™ CF2 microprocessor. It is a low-power system that records on a programmable duty cycle, but is also capable of responding to and recording acoustic events that meet specific criteria (Lammers et al., 2008). The EAR has an analog recording bandwidth of 30 kHz. Its deployment life is determined by the programmable recording duty cycle and the number of battery packs included, but typical deployment durations range between 3 and 12 months.

To date, the EAR has been used with success to document the occurrence of spinner dolphins in and around a daytime resting bay on the island of Oahu.

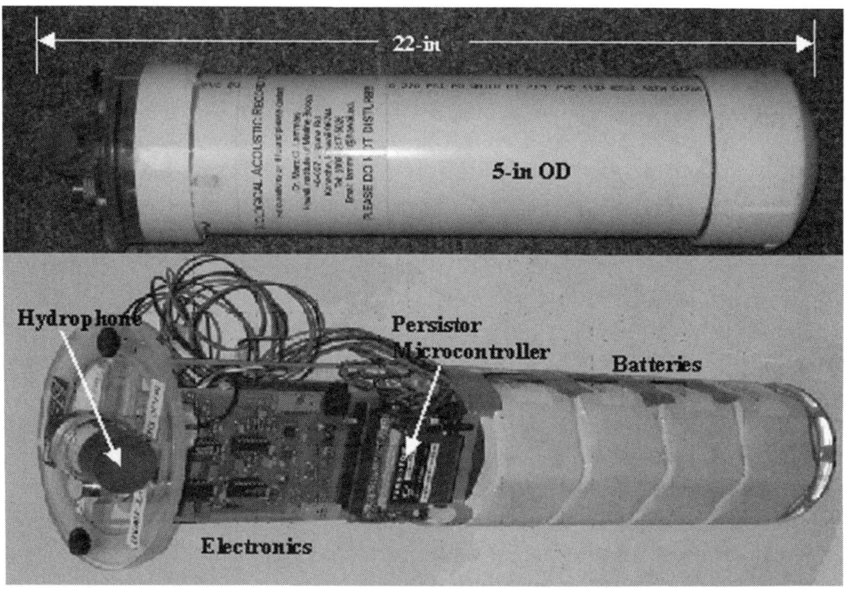

FIGURE 13.29. *Top* – EAR packaged for mounting on the bottom of the ocean, *bottom* – Internal electronics showing a mixed signal preprocessing board with analog amplification-filtering and analog-to-digital conversion controlled by a Persistor CF2 microcontroller with data stored on a hard drive.

A six-week deployment of the EAR in February/March 2006 yielded 203 separate detections of dolphin whistles and clicks. The data reveal that the occurrence of dolphins at the site was not uniform throughout the deployment period. Rather, detections were few and sporadic during the initial part of the deployment, relatively consistent for a period of approximately 2 weeks during the second half of the deployment, and nearly absent during the last several days. This suggests that spinner dolphins use the area episodically rather than continuously, perhaps choosing other daytime resting locations based on environmental conditions or the availability of nearby prey.

Although the EAR was designed for long-term bottom mounted deployments, in the simplest scenario it can be easily incorporated into an ocean glider almost as is, or modified to use the glider power source. Multiple CF2 Persistor microcontrollers along with associated electronics can be ganged together with one master CF2 providing the data acquisition clock timing for synchronous, multi-channel data acquisition. The EAR would be programmed for quasi-continuous recording of acoustic signals, with time-out intervals to transfer data from the flash card on the Persistor card to the hard drive.

13.4.4 Scripps HARP

A high-frequency acoustic recording package (HARP) has been developed by Scripps Institution of Oceanographic to perform continuous long-term monitoring in remote locations under various weather conditions and independent of daylight, (Wiggins and Hildebrand, 2007). Development of the HARP was motivated by the need for a broader-band, higher-data capacity system capable of autonomously recording toothed whales and other marine mammals for long periods. A picture of the HARP system deployed on the bottom is shown in Fig. 13.30a, and the HARP module acoustic package is shown in Fig. 13.30b. The acoustic recorder is controlled by a 32-bit 20 MHz Motorola microcontroller with an Analog Devices 16-bit A/D converter used to digitize acoustic signals detected by the hydrophones. Eight sample rates ranging from 2 to 200 kHz are available with corresponding anti-aliasing filters for each sample rate. The sampled data are stored temporarily into a data buffer consisting of 16 2 Mbyte SRAM chips until about 30 Mbytes of data are collected and then the data are sent to one of 16 laptop type hard drives for permanent storage via an eEthernet 10BaseT link. A total of 1.92 Tbyte of data storage capacity is available so that 55 days of continuous sampling at a sample rate of 200 kHz can be achieved. Lower sampling rates will allow for longer total recording time and so would schedule sampling where the recorder is turned on for a period of time between off or sleep periods.

The HARP comes with two hydrophones, one for low frequencies from 10 Hz to 2 kHz and a high frequency one from 1 to 100 kHz. An International Transducer ITC-1042, spherical omni-directional transducer is used for the high-frequency hydrophone. The low-frequency hydrophone consists of six cylindrical Benthos AQ-1 transducers connected in series for increased

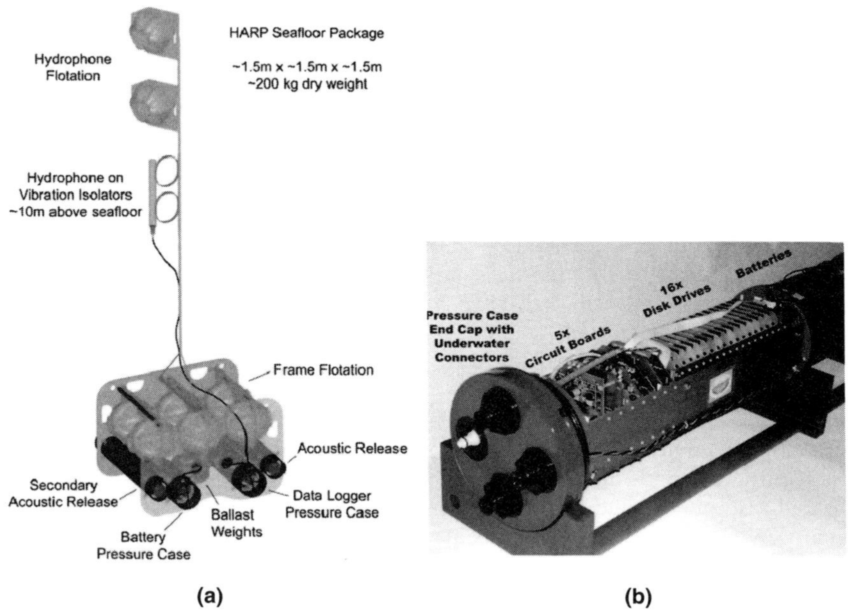

FIGURE 13.30. (**a**) A schematic of the HARP system deployed on the ocean bottom, (**b**) the internal configuration of the HARP recording package.

sensitivity. A 40-dB gain preamp is for the low-frequency recordings and an 80-dB gain preamp is used for the high-frequency recordings. Both signals are pre-whitened for the frequency variation of typical ocean ambient noise.

13.5 Concluding Remarks

The purpose of this chapter is to provide examples of the various types of instrumentation that have been used in marine bioacoustics research. Technology associated with embedded microcontroller systems what are basically miniatures or microcomputers continue to advance at a rapid pace, making this chapter essentially obsolete as it is being written. Microcontrollers, DSP chips, FPGA (field programmable gate arrays) are rapidly becoming smaller, have greater capabilities and greater speed. Flash memory chips and various forms of memory devices and chips are also becoming smaller or can hold ever increasing amount of memory and new devices continue to come into the market place. The key to better measuring devices is strongly coupled to the size, speed and capability of electronic chips that have "computer-like" capabilities and their associated memory. The internet along with cell phone developments will also contribute to new ways of obtaining data from remote locations and will increase our abilities to examine questions that just recently were too complicated to even imagine. The slang "you ain't seen nothing yet" is completely apropos in regards to instrumentation for marine bioacoustics research.

References

Arby, P. R. and Dempster, M. A. H. (1974). *Introduction to Optimization Methods* (Chapman and Hall, Longdon).

Akamatsu, T., Wang, D., Wang, K., and Yahuhiko, N. (2000). "A method for individual identification of echolocation signals in free-ranging finless porpoises carrying data logger," J. Acoust. Soc. Am. **108**, 1353–1356.

Au, W. W. L. (1993). *The Sonar of Dolphins* (Springer-Verlag, New York).

Au, W. W. L. and Herzing, D. L. (1997). "Measurement of the echolocation signals of the Atlantic Spotted Dolphin *Stenella frontalis* in the waters off the Grand Bahamas," J. Acoust. Soc. Am, **101**, 3137–3138.

Au, W. W. L., Lammers, M. O., and Aubauer, R. (1999). "A portable broadband data acquisition system for field studies in bioacoustics," Marine Mammal Sci. **15**, 526–531.

Aubauer, R. and Au, W. W. L. (1998). "Phantom echo generation: a new technique for investigating dolphin echolocation," J. Acoust. Soc. Am., **104**, 1165–1170.

Benoit-Bird, K. J. and Au, W. W. L. (2003). "Fine-scale diel migration dynamics of an island-associated sound-scattering layer," Deep Sea Research Part I, **51,** 707–719.

Burgess, W. C., Tyack, P. L., Le Boeuf, B. J., and Costa, D. P. (1996). "Acoustic measurement of cardiac function on northern elephant seals, " J. Acoust. Soc. Am. **100**, 2709.

Burgess, W. C., Tyack, P. L., LeBoeuf, B. J., Costa, D. P., and Crocker, D. E. (1997). "A microprocessor-based tag to record acoustics experienced and generated by migrating northern seals," J. Acoust. Soc. Am. **102**, 3197.

Burgess, W. C., Tyack, P. L., Le Boeuf, B. J., and Costa, D. P. (1998). "A programmable acoustic tag and first results from free-ranging northern elephant seals," Deep-Sea Res. **45**, 1327–1351.

Carey, F. G. (1983). "Experiments with free swimming fish," in *Oceanography, the Present and the Future*, P. G. Brewer, ed. (Spinger-Verlag, New York), pp. 57–68

Elliott, S. J., Curtis, A. R. D., Bullmore, A. J., and Nelson, P. A. (1987). "The active minimization of harmonic enclosed sound fields, Part III: Experimental verification," J. Sound Vib. **117**, 35–38.

Finneran, J. J., Hastings, M. C., and Derenburger, C. (1996). "Low-frequency response of the peripheral auditory organs in the goldfish (*Carassius auratus*)," J. Acoust. Soc. Am. **100**, 2709–2710.

Finneran, J. J. and Hastings, M. C. (1997). "Active impedance control in a water-filled waveguide for low-frequency bioacoustic testing," J. Acoust. Soc. Am. **102**, 3196–3197.

Finneran, J. J. and Hastings, M. C. (1999). "Active impedance control within a cylindrical waveguide for generation of low-frequency underwater plane traveling waves," J. Acoust. Soc. Am. **105**, 3035–3043.

Finneran, J. J. and Hastings, M. C. (2004). "A continuous-wave ultrasound system for displacement amplitude and phase measurement," J. Acoust. Soc. Am., **115**, 3202–3209.

Derenburger, C., Finneran, J. J., and Hastings, M. C. (1997). "Comparison of the low-frequency response of the peripheral auditory organs in the goldfish and oscar," J. Acoust. Soc. Am. **101**, 3163.

Greene, Jr. C. R. (1997). "An autonomous acoustic recorder for shallow artic water," J. Acoust. Soc. Am. **102**, 3197.

Fletcher, S., Le Boeuf, B. J., Costa, D. P., Tyack, P. L., and Blackwell, S. B. (1996). "Onboard acoustic recording from diving northern elephant seals," J. Acoust. Soc. Am. **100**, 2531–2539.

Hayes, S., Costa, D. P., Croll, D., Mellinger, D. K., and Borsani, J. F. (1997). "Inexpensive passive acoustic animal tracking system," J. Acoust. Soc. Am. **102**, 3212.

Holland, K., Brill, R., Ferguson, S., Chang, R., and Yost, R. (1985). "A small vessel technique for tracking pelagic fish," Mar. Fish. Review **47**, 26–32.

Holland, K. N., Lowe, C. G., Peterson, J. D., and Gill, A. (1992). "Tracking coastal sharks with small boats: hammerhead shark pups as a case study," Aust. J. Mar. Freshwater Res. **43**, 61–66.

Johnson, M. and Tyack, P. (2003). "A digital acoustic recording tag for measuring the response of wild marine mammals to sound," J. Oceanic Eng. **28**, 3–12.

Johnson, M., Madsen, P. T., Zimmer, W. M. X., Aguilar de Soto, N., and Tyack, P. L. (2004). "Beaked whales echolocate on prey," Proc. R. Soc. Lond. B **271**, 383–386.

Lammers, M. O., Brainard, R. E., Au, W. W. L., and Wong, K. B. (2008). "An ecological acoustic recorder (EAR) for long-term monitoring of biological and anthropogenic sounds on coral reefs and other marine habitats," J. Acoust. Soc. Am, **123**, 1720–1728.

Le Bouef, B. J., Crocker, D. E., Blackwell, S. B., Morris, P. A., and Thorson, P. H. (1993). "Sex differences in diving and foraging behavior of northern elephant seals," Symp. Zool. Soc. London, **66**, 149–178.

Lowe, C. G., Holland, K. N., and Wolcott, T. (2008). "A new acoustic tailbeat transmitter for fishes," submitted for publication.

Moss, C. F. and Schnitzler, H.-U. (1995). "Behavioral studies of auditory information processing," *Hearing by Bats*, A. N. Popper and R. R. Fay, eds. (Springer-Verlag, New York), 87–145.

Madsen, P. T., Johnson, M., Aguilar, DeSto M., Zimmer, W. M. X., and Tyack, P. (2005). "Biosonar performance of foraging beaked whales (*Mesoplodon densirostris*)," J. Exp. Biol. **208**, 181–194.

Møhl, B., Wahlberg, M., Madsen, P. T., Miller, L. A., and Surlykke, A. (2000). "Sperm whale clicks: directionality and Source level revisited," J. Acoust. Soc. Am. **107**, 638-648.

Rogers, P. H. and Hastings, M. C. (1989). "Noninvasive vibration measurement system and method for measuring amplitude of vibration of tissue in an object being investigated," U. S. Patent No. 4,819,643.

Sigurdson, J. E. (1996). "Open-water echolocation of bottom objects by dolphins (*Tursiops truncatus*)," J. Acoust. Soc. Am. **100**, 2610.

Sigurdson, J. E. (1997). "Biosonar dynamics of the bottlenose dolphin in VSW search and detection tasks," J. Acoust. Soc. Am. **102**, p. 3123.

Thode, A. M., Gerstoft, P., Gerra, M., Stokes, D. M., Burgess, W. C., Noad, M. J. and Cato, D. H. (2004). "Range-depth tracking of humpback whales using autonomous acoustic recorders," J. Acoust. Soc. Am. **116**, 2809.

Wiggins, S. M. and Hildebrand, J. A. (2007). "High-frequency acoustic recording package (HARP) for broad-band, long-term marine mammal monitoring." *Int. Symp. Underwater Tech.* Japan, Institute of Electrical and Electronics Engineers, pp. 551–557.

Yuen, H. S. H. (1970). "Behavior of Skipjack Tuna, *Katsuwanus pelamis*, as determined by tracking with Ultrasonic Devices," J. Fish. Res. Board Can. **27**, 2071–2079.

Index

Printed in the United States of America